Artificial Intelligence in Design '96

Artificial Intelligence in Design '96

Edited by

John S. Gero

and

Fay Sudweeks

Department of Architectural and Design Science,
University of Sydney,
Australia

KLUWER ACADEMIC PUBLISHERS
DORDRECHT / BOSTON / LONDON

A C.I.P. Catalogue record for this book is available from the Library of Congress

ISBN-13: 978-94-010-6610-5   e-ISBN-13: 978-94-009-0279-4
DOI: 10.1007/ 978-94-009-0279-4

Published by Kluwer Academic Publishers,
P.O. Box 17, 3300 AA Dordrecht, The Netherlands.

Kluwer Academic Publishers incorporates
the publishing programmes of
D. Reidel, Martinus Nijhoff, Dr W. Junk and MTP Press.

Sold and distributed in the U.S.A. and Canada
by Kluwer Academic Publishers,
101 Philip Drive, Norwell, MA 02061, U.S.A.

In all other countries, sold and distributed
by Kluwer Academic Publishers Group,
P.O. Box 322, 3300 AH Dordrecht, The Netherlands.

Printed on acid-free paper

contents

preface

Change is one of the most significant parameters in our society. Designers are amongst the primary change agents for any society. As a consequence design is an important research topic in engineering and architecture and related disciplines, since design is not only a means of change but is also one of the keystones to economic competitiveness and the fundamental precursor to manufacturing. The development of computational models founded on the artificial intelligence paradigm has provided an impetus for much of current design research - both computational and cognitive.

These forms of design research have only been carried out in the last decade or so and in the temporal sense they are still immature. Notwithstanding this immaturity, noticeable advances have been made both in extending our understanding of design and in developing tools based on that understanding. Whilst many researchers in the field of artificial intelligence in design utilise ideas about how humans design as one source of concepts there is normally no attempt to model human designers. Rather the results of the research presented in this volume demonstrate approaches to increasing our understanding of design as a process. The goal in most of this research is to make the computer more useful in design since it is clear when looking at designs produced by unaided humans that they often fail to perform satisfactorily. The expectation is that computer-aided human designers will produce better designs. The research methods employed are closely linked to the scientific method but that does not imply that the activity of designing is scientific.

The papers in this volume are from the Fourth International Conference on Artificial Intelligence in Design held in June 1996 in Stanford, California. They represent the state-of-the-art and the cutting edge of research and development in this field. They are of particular interest to researchers, developers and users of computer systems in design. This volume demonstrates both the breadth and depth of artificial intelligence in design and points the way forward for our understanding of design as a process and for the development of computer-based tools to aid designers. The papers describe advances in both theory and application.

The forty papers are grouped under the following headings:

Case-Based Design
Conceptual Design
Creativity and Innovation in Design
Design Objects
Design Spaces
Distributed Design
Genetic Algorithms/Genetic Programming in Design
Grammars in Design
Learning in Design
Representations in Design
Reuse of Designs
Rules, Models and Theories in Design
Spatial and Layout Planning in Design

All papers were extensively reviewed by three referees drawn from a large international panel. Thanks go to them, for the quality of these papers depends on their efforts. They are listed below. After the papers were reviewed, a small panel considered the reviews prior to making a final recommendation.

John S. Gero
University of Sydney
March 1996

International Panel of Referees

1

representation in design

Multi-level molecular representation
Patrick Olivier, Keiichi Nakata, Malcolm Landon
Text analysis for constructing design representations
Andy Dong, Alice M. Agogino
Learning genetic representations as alternative to hand-coded shape
grammars
Thorsten Schnier, John Gero

J. S. Gero and F. Sudweeks (eds), Artificial Intelligence in Design '96, 3-20.
© 1996 *Kluwer Academic Publishers.*

MULTI-LEVEL MOLECULAR REPRESENTATION

Kinematic synthesis using an object-centred spatial decomposition

PATRICK OLIVIER, KEIICHI NAKATA AND MALCOLM LANDON
Centre for Intelligent Systems
University of Wales
Aberystwyth Dyfed SY23 3DB
United Kingdom

Abstract. In the initial molecular representation proposal for kinematic reasoning (Gupta and Jakiela, 1994), object contours were represented using a list of touching circles each of equal diameter. Gupta and Jakiela characterise procedures by which: (1) kinematic analysis can be performed by advancing the driving contour through a small displacement and moving the driven object in such a manner as to minimise divergence in the spatial relationship between the two contours; and (2) kinematic synthesis can be effected by deforming a *blank* component with respect to the known half of a kinematic pair. We have further developed this approach and employed a multi-level molecular representation, and have considerably improved both the synthesis and analysis procedures. In this we paper we describe our kinematic synthesis and analysis algorithms, and give an account of how the multi-level representation can be maintained with minimum effort during the synthesis process.

1. Approaches to Kinematic Analysis and Synthesis

Kinematic pairs are pairs of objects whose motion is dependent on each other by virtue of their position, shape and the contact that results from their relative motion. Lower pairs maintain a constant contact over the whole mating surface, for example, bearings and prismatic joints are typical lower pairs. Higher pairs, however, are characterized by the absence of full contact over their mating surfaces. That is, the points of contact between components of a higher pair change in the course of their relative motion, as is the case in meshing gears and cam-follower mechanisms.

Established approaches in engineering include *special case* analytical techniques, which are highly tuned for the class of higher pair to be either analyzed or synthesized; and graphical techniques based on the the interference constraint, that is, that neither component of the pair can occupy the same space. The latter constraint is the motivation for our approach, as it also is for all approaches originating in computer graphics[1]. Kinematic reasoning in artificial intelligence has been primarily motivated by two subfields: robot-motion planning and qualitative reasoning. In classical robot motion planning chains of lower kinematic pairs (or one or more mobile robots and a fixed environment (Erdmann and Lozano-Perez, 1986)) are analyzed with the aim of computing obstacle-free paths in the robot workspace. Whereas in qualitative reasoning the aim is to obtain a high-level description of mechanism behavior (eg. (Faltings, 1990; Faltings, 1992; Forbus *et al.*, 1987; Joskowicz, 1988; Joskowicz and Sacks, 1991)). Both problems are addressed using configuration space calculations (Lozano-Perez, 1983).

Most approaches to kinematic analysis that originate in computer graphics are based upon checks for interpenetration between the object models (see (Hahn, 1988; Baraff, 1989) and for an example of a relevant application (Garcia-Alonso *et al.*, 1994)). Objects themselves are typically represented by piece-wise continuous segments, and when interpenetration occurs rules for determining the resulting motion are invoked. Most of these approaches simplify the interpenetration computation using the assumptions that vertex-to-vertex, vertex-to-edge and edge-to-vertex penetrations occur with a very low probability, an assumption that has been shown to be overly simplistic in the domain of real mechanisms (Krishnasamy and Jakiela, 1993). Quadtree and octree representations have been used before to speed up interpenetration determination for polygon models, but not, as the actual representation with which the analysis and synthesis is performed.

2. Multi-level Molecular Approach: Interference Detection and Analysis

In the initial molecular approach (Gupta and Jakiela, 1994), objects contours are represented using points of notional diameters (which resemble a chain of molecules, hence the naming). Gupta and Jakiela characterise procedures both by which kinematic synthesis and analysis can be performed. In their analysis procedure the driving contour is advanced through some through a small *virtual* displacement and the driven object is displaced in such a manner as to minimise divergence in the spatial relationship between the two contours. We have developed our own analysis and synthesis procedures based on a multi-level representation, and show in the following sections the considerable benefits to be gained from such an approach.

2.1. INTERFERENCE DETERMINATION

Contact is detected within a kinematic pair when one of the molecules of one object contour overlaps (*interferes*) with one of the molecule of the other. That is, molecules interfere when the distance between their centres is less than one dia-

---

[1]However, in section 3 we diverge markedly from established techniques in outlining how this very constraint can be utilized to use an multi-level molecular for component synthesis.

meter in length. The maximum number of checks to identify contact is the product of the number of molecules comprising each object. Interference determination will prove to be the crucial component of both kinematic analysis and synthesis, and in this this section we discuss improvements on the *worst case* brute force comparison.

2.2. INTERFERENCE: SINGLE-LEVEL MOLECULAR REPRESENTATION

Gupta and Jakiela (1994) describe a method of reducing the number of checks needed to identify interfering molecules. On checking the distance between two molecules (*distance*, in units of molecule diameters), if it is less than one diameter then the molecules interfere. If the molecules do not interfere, then the argument can be used that even if the second object's contour ran in a straight line directly towards the molecule of the first object, it could not interfere with it for at least $(distance - 1)$ molecules.

This is shown in Figure 1. The two black molecules are checked for overlap and clearly do not interfere. The second object's contour runs in a straight line towards the molecule in the first object. Knowing that the objects are a continuous chains of touching molecules we therefore know that the next $(distance - 2)$ molecules of the second object can not interfere with the black molecule on the first object. We can therefore skip along the second object contour by $(distance - 1)$ molecules, greatly reducing the number of interference checks required in general.

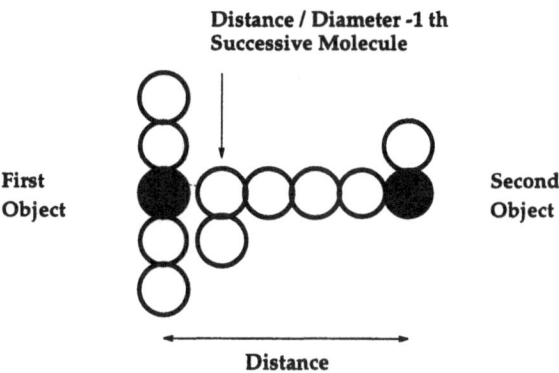

Figure 1. Molecule skipping during interference checking.

2.3. MULTI-LEVEL REPRESENTATION

In the multi-level approach the contour of each object is represented at the *base level* by a continuous chain of circles, not overlapping, but just touching. Figure 2(a) shows a small square represented in this manner using thirty six molecules. These base level molecules are then contained inside a level of larger molecules. Each larger molecule has a diameter three times that of its immediate children. Its centre is the centre of the middle molecule contained within it.[2] Figure 2(b) shows the square with the addition of larger molecules. This process of creating larger molecules to contain three smaller ones is continued until the whole object is contained within one or two molecules (Figures 2(c) and 2(d) illustrate this).

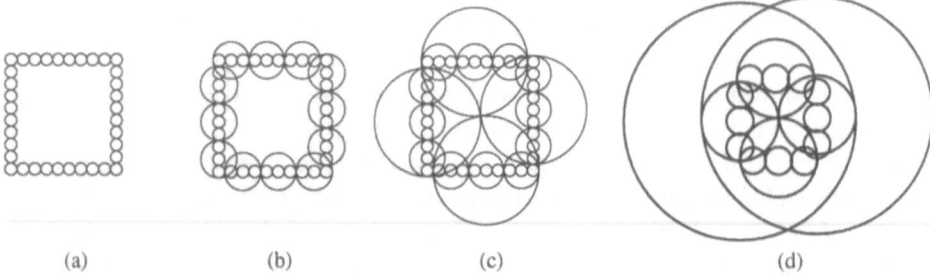

(a)   (b)   (c)   (d)

Figure 2.   An example of multi-level molecular decomposition.

2.4. INTERFERENCE: MULTI-LEVEL MOLECULAR REPRESENTATION

Using a multi-level molecular representation leads to a great reduction in the number of molecules which have to be checked to establish interference between components. In this approach the largest molecules of two objects are first checked for overlap. Interference between these molecules does not mean that contact has been detected, but that the smaller molecules contained within them should be compared. Only the smaller molecules contained within the interfering larger molecules need to be checked. Smaller molecules contained within larger molecules that do not interfere can not themselves interfere with each other.

Figure 3 illustrates this; only two of the initial six molecules in Figure 3(a) interfere; consequently only the molecules contained within them are checked against each other in Figure 3(b). This procedure is applied recursively (see Figure 3(c)) until either interference at the base level has been established, or no interference is detected.

---

[2]The choice of the number of molecules to be contained in a larger molecule, in this case three, is rather arbitrary. However, by choosing an odd number, we do not have to recompute the centre of the larger circle, and three is the smallest reasonable odd number.

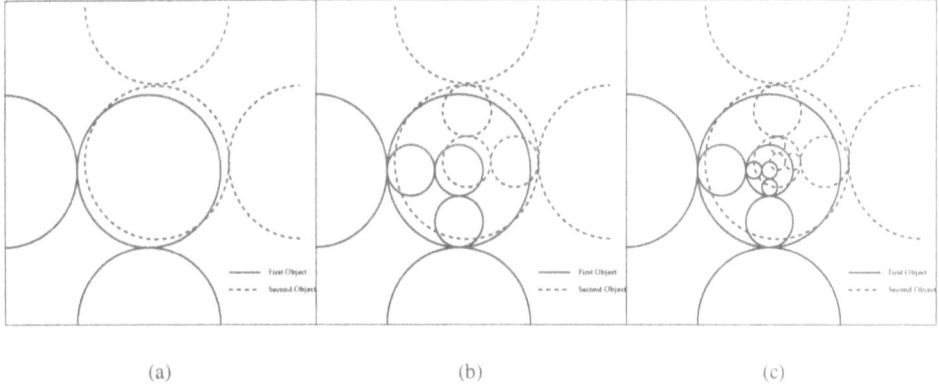

(a)   (b)   (c)

Figure 3. Exploiting molecular decomposition in interference determination.

To reduce the number of checks needed even further, the single- and multi-level approaches can be combined. When checking the three smaller molecules contained in a larger molecule against the three smaller molecules contained in an interfering larger molecule, nine checks for interference are needed. But within each of the three molecules, there is a linear ordering and Gupta and Jakiela's skipping algorithm can be applied. At best the number of checks within the pair of three molecules is reduced from nine to three (see Figure 4).

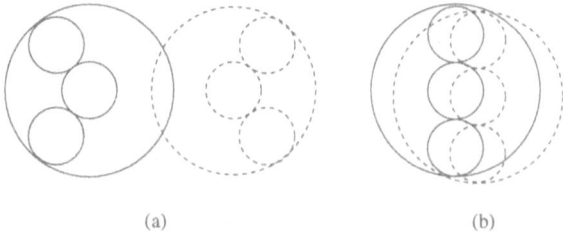

(a)   (b)

Figure 4. Exploiting linear ordering in the multi-level interference checking. Figure (a) illustrates the best case performance (3 checks) and Figure (b) worst case (9 checks).

Whilst precise comparison of performance is very much dependent on the geometry of the interference problem at hand, Table 1 contrasts the number of interference checks required using both the single-level and the multi-level molecular representations on the problem depicted in Figure 5, at varying granularities.

2.5. MOTION INFERENCE

In analysis the driving object is advanced in each simulation increment by an angle that will displace the molecule farthest from the the centre of rotation (for

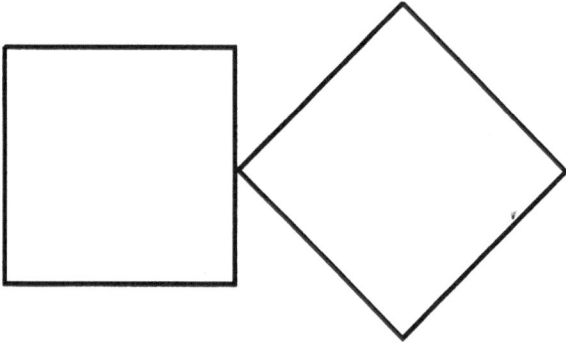

Figure 5. Test case for comparison of the single- and multi-level molecular representations.

TABLE 1. Number of interference checks for the single- and multi-level molecular representations.

No. molecules per object	324	972	2916	8748
Single-level checks	2349	7044	21132	63418
Multi-level checks	414	507	688	849

rotational degree of freedom objects) by no more than one molecule diameter. Thus the granularity and component geometry of the model place a constraint on the number of angular increments to perform a complete analysis. When interference is detected at the base level in the molecular representation, the nature of the resulting motion must be inferred.

Unlike the *virtual motion* mentioned in the previous section, we preprocess each object and encode on each molecule the sense of the motion that contact with it gives rise to. For example, for a molecular representation of a gear, the motion resulting from contacting any molecule of the contour is independent of the orientation or position of the gear in a mechanism, it is dependent on the relationship between the normal at the point on the contour and its vector position relative to the degree of freedom. We therefore precompile this qualitative motion (anti- or clockwise) into each molecule. In the case where multiple overlaps occur at the base level and the driven object molecules have different qualitative motions compiled into them, it can be inferred that the motion of the driver is blocked.

3. Multi-level Approach: Synthesis

3.1. OVERVIEW

We propose a synthesis algorithm that differs in many respects from that proposed by Gupta and Jakiela (1994). The class of synthesis problem addressed assumes full knowledge of one component in the pair, the nature and location of the degree of freedom of the component to be synthesised, and the input/output function for the pair. The multi-level properties of the representation are once again exploited in the interference detection problem, but in this approach we avoid the need to compress and expand the contour of the component being synthesised and adopt a simpler procedure. In our procedure the unknown contour (initially a feature-less circle) and the known contour are overlaid in some relative position and orientation satisfying the required input/output function. Molecules in the unknown contour that overlap with the known object contour are deleted and replaced with molecules that precisely trace the known contour between consecutive overlaps (see Figure 6). This trace is the external contour of the known object and is pre-compiled prior to synthesis.

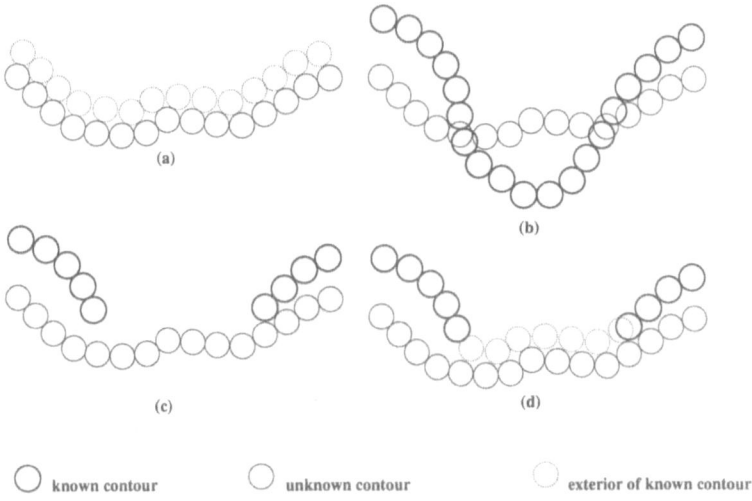

Figure 6. Molecular synthesis. In (a) the known contour is depicted with its associated exterior contour; (b) shows the overlap of an unknown contour, of the component being synthesized and the known contour; in (c) the overlapping molecules have been deleted; in (d) the new contour is formed using the molecules in the exterior contour of the known component.

3.2. EXAMPLE OF SYNTHESIS PROCEDURE

Synthesis starts by placing the known shape and the unknown shape into their initial positions and cutting the unknown to the known shape as shown in Figure 8.

```
Foreach angular displacement of known and unknown component

   Locate intersections and overlaps
   Create a blank list to contain 'cut' sections

   For each pair of intersections on known contour
        If corresponding unknown contour molecules
          are on unknown contours list then
             Remove section from between corresponding molecules
             on the unknown contour.
             Place the 'cut' section on the cut list
             Generate new section, following the known contour
             between the two intersections and put it into the
             unknown contour to replace the 'cut'
        Elseif corresponding unknown contour molecules
          are on unknown contours list then
             Remove section from between corresponding molecules
             on the unknown contour in the cut list - the remaining
             sections become two lists on the cut list
             Generate new section, following the known contour
             between the two intersections, making the section
             just 'cut' a complete contour
             Add the new contour to the unknown contour list
        Endif
   Endfor

   For each overlapping section
        Remove overlapping section of the unknown contour
        Generate a new section, following the known contour
        for the length of the overlap, and place it into the
        unknown contour to replace section removed
   Endfor

Endforeach
```

Figure 7. Synthesis algorithm.

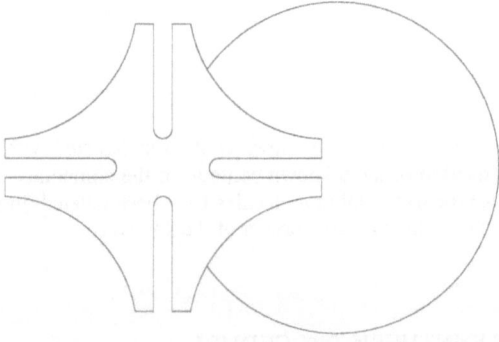

Figure 8. Initial positions of unknown and known shapes.

Both shapes are then rotated, resulting in a number of intersections between them. The intersections are labelled, first by following the known shape as shown in Figure 9 and then by following the unknown contour as shown in Figure 10. Table 2 shows how these labels correspond.

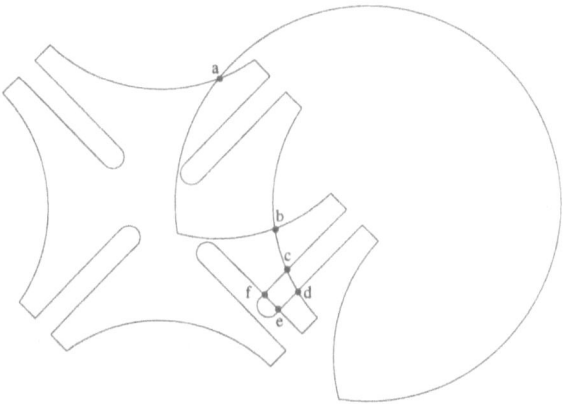

Figure 9.   Intersections numbered by following known contour.

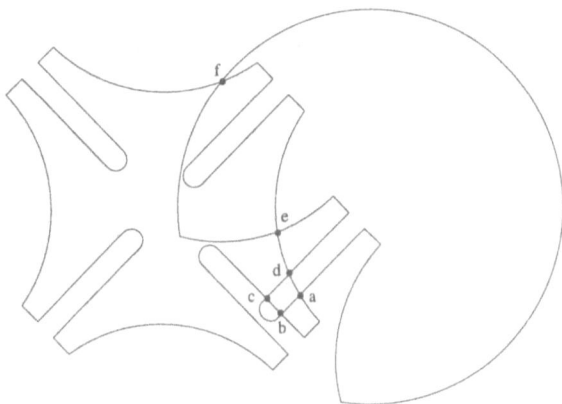

Figure 10.   Intersections labelled by following unknown contour.

TABLE 2.   Corresponding labels between known and unknown contours.

Known Shape	a	b	c	d	e	f
Unknown Shape	f	e	d	a	b	c

The intersections can be thought of as nodes and the contours between them as arcs. The following explains the notation used to refer to the nodes and arcs in this example.

κ_a Known shape molecule at node A. It is the last interfering molecule when the known contour moves from the outside to the inside of the unknown shape. It is the first interfering molecule when the known contour moves from the inside to the outside of the unknown shape.

$\kappa_{a+1}...\kappa_{b-1}$ The molecules making the arc between molecules κ_a and κ_b.

v_c Unknown shape molecule at node C. It is the first interfering molecule when the unknown contour moves from the outside to the inside of the known shape. It is the last interfering molecule when the unknown contour moves from the inside to the outside of the known shape.

$v_{c+1}...v_{d-1}$ The molecules making the arc between molecules v_c and v_d.

v_e^* The new node molecule after cutting the unknown contour. It is the outside molecule related to κ_i where i is the corresponding intersection on the known contour.

$v_{e+1}^*...v_{h-1}^*$ The molecules making the new arc between molecules v_e^* and v_h^*.

At the start of the example the contours are as in Table 3. It then follows the algorithm outlined in Figure 7.

TABLE 3. Contours at the initial position.

Known Contours:	$\kappa_a, \kappa_{a+1}...\kappa_{b-1}, \kappa_b, \kappa_{b+1}...\kappa_{c-1}, \kappa_c, \kappa_{c+1}...\kappa_{d-1},$
	$\kappa_d, \kappa_{d+1}...\kappa_{e-1}, \kappa_e, \kappa_{e+1}...\kappa_{f-1}, \kappa_f, \kappa_{f+1}...\kappa_{a-1}$
Unknown Contours:	$v_a, v_{a+1}...v_{b-1}, v_b, v_{b+1}...v_{c-1}, v_c, v_{c+1}...v_{d-1},$
	$v_d, v_{d+1}...v_{e-1}, v_e, v_{e+1}...v_{f-1}, v_f, v_{f+1}...v_{a-1}$
Cut Contours:	Empty.

STEP 1

Known contour nodes κ_a and κ_b match with unknown contour nodes v_f and v_e. Unknown contour nodes v_f and v_e are on a contour in the unknown contour list. Remove the nodes and the contour between them placing them on the cut list. Generate new nodes v_f^* and v_e^* and a new arc $v_{e+1}^*...v_{f-1}^*$ to follow arc $\kappa_{a+1}..\kappa_{b-1}$ and place them into the unknown contour.

Figure 11 shows this in detail. At the intersection between κ_a and v_f the known contour moves from outside to inside the unknown shape. κ_a is therefore the last interfering molecule on the known contour. The unknown contour moves from outside to inside the known shape. v_f is therefore the first interfering molecule on the unknown contour. At the intersection between κ_b and v_e the known

TABLE 4. Contours after Step 1.

Unknown Contours:	$v_a, v_{a+1}...v_{b-1}, v_b, v_{b+1}...v_{c-1}, v_c, v_{c+1}...v_{d-1},$
	$v_d, v_{d+1}...v_{e-1}, v_e^*, v_{e+1}^*...v_{f-1}^*, v_f^*, v_{f+1}...v_{a-1}$
Cut Contours:	$v_e, v_{e+1}...v_{f-1}, v_f$

contour moves from inside to outside the unknown shape. κ_b is therefore the first interfering molecule on the known contour. The unknown contour moves from inside to outside the known shape. v_e is therefore the last interfering molecule on the unknown contour. The molecules v_f and v_e together with all the molecules between them are removed from the unknown contour and placed on the cut list. v_f^* is the outside molecule associated with κ_a and v_e^* is the outside molecule associated with κ_b. These molecules, together with all the molecules on the outside contour between them, replace the cut section in the unknown contour. Figure 12 shows the changes to the two contours during Step 1. The result is summarised in Table 4.

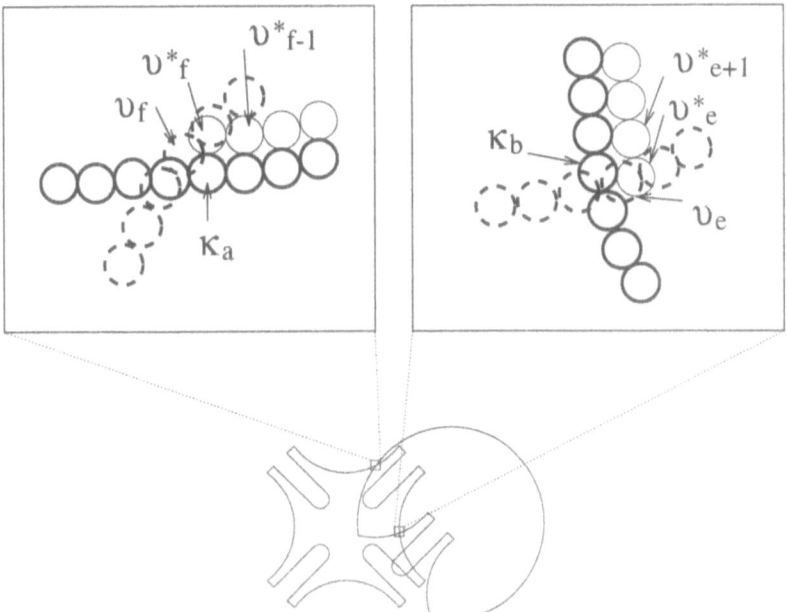

Figure 11. A close up of the intersections for Step 1.

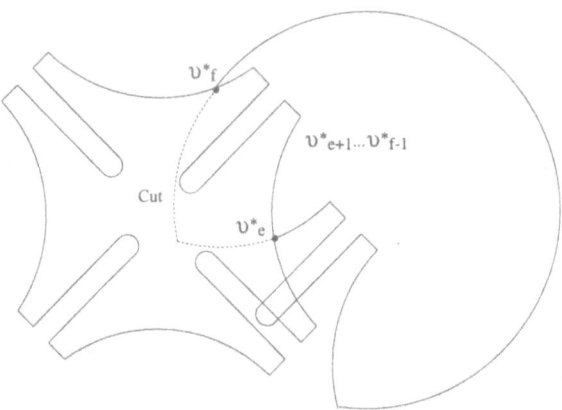

Figure 12. After Step 1.

STEP 2

Known contour nodes κ_c and κ_d match with unknown contour nodes v_d and v_a. Unknown contour nodes v_d and v_a are on an unknown contour. Remove the nodes and the contour between them placing them on the cut list. Generate new nodes v_d^* and v_a^* and a new arc $v_{a+1}^*...v_{d-1}^*$ to follow arc $\kappa_{c+1}..\kappa_{d-1}$ and place them into the unknown contour. Figure 13 shows this in detail. At the intersection between κ_c and v_d the known contour moves from outside to inside the unknown shape. κ_c is therefore the last interfering molecule on the known contour. The unknown contour moves from outside to inside the known shape. v_d is therefore the first interfering molecule on the unknown contour. At the intersection between κ_d and v_a the known contour moves from inside to outside the unknown shape. κ_d is therefore the first interfering molecule on the known contour. The unknown contour moves from inside to outside the known shape. v_a is therefore the last interfering molecule on the unknown contour.

The molecules v_d and v_a together with all the molecules between them are removed from the unknown contour and placed on the cut list. v_d^* is the outside molecule associated with κ_c and v_a^* is the outside molecule associated with κ_d. These molecules, together with all the molecules on the outside contour between them, replace the cut section in the unknown contour. Figure 3.2 shows the changes to the two contours during Step 2, and Table 5 summarises the result.

STEP 3

Known contour nodes κ_e and κ_f match with unknown contour nodes v_b and v_c. Unknown contour nodes v_b and v_c are not on an unknown contour; they are on a cut contour. The contour between the nodes is removed from the cut contour, leaving behind the two end molecules. What is removed is placed into the blank contour list as a separate contour. New nodes v_b^* and v_c^* are generated and a new

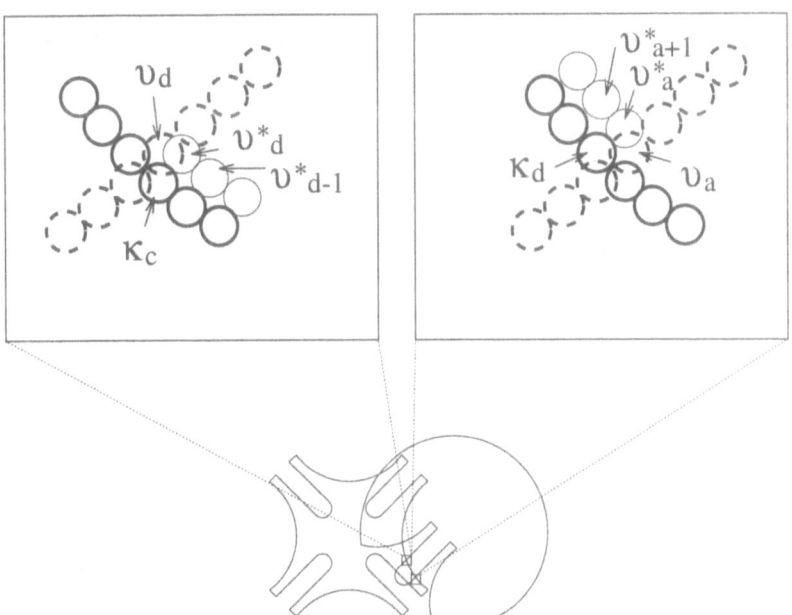

Figure 13. A close up of the intersections for Step 2.

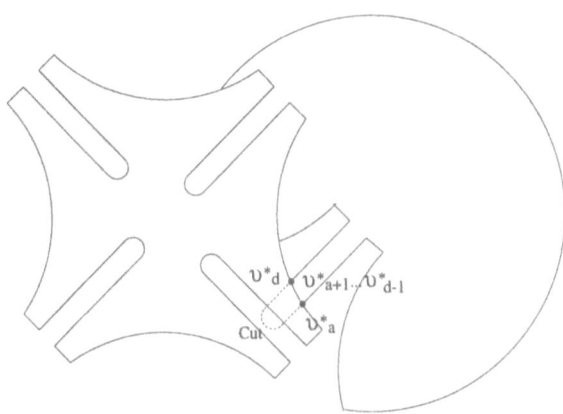

TABLE 5. Contours after Step 2.

Unknown Contours:	$v_a^*, v_{a+1}^* \ldots v_{d-1}^*, v_d^*, v_{d+1} \ldots v_{e-1}, v_e^*, v_{e+1}^* \ldots v_{f-1}^*, v_f^*, v_{f+1} \ldots v_{a-1}$
Cut Contours:	$v_e, v_{e+1} \ldots v_{f-1}, v_f,$
	$v_a, v_{a+1} \ldots v_{b-1}, v_b, v_{b+1} \ldots v_{c-1}, v_c, v_{c+1} \ldots v_{d-1}, v_d$

arc $v^*_{c+1}...v^*_{b-1}$ to follow arc $\kappa_{e+1}..\kappa_{f-1}$ is added to them to complete the new contour. Figure 14 shows this in detail. At the intersection between κ_e and v_b the known contour moves from outside to inside the unknown shape. κ_e is therefore the last interfering molecule on the known contour. The unknown contour moves from outside to inside the known shape. v_b is therefore the first interfering molecule on the unknown contour.

At the intersection between κ_f and v_c the known contour moves from inside to outside the unknown shape. κ_f is therefore the first interfering molecule on the known contour. The unknown contour moves from inside to outside the known shape. v_c is therefore the last interfering molecule on the unknown contour. The molecules v_b and v_c are on a cut contour. The molecules between them are restored as a separate contour. v^*_b is the outside molecule associated with κ_e and v^*_c is the outside molecule associated with κ_f. These molecules, together with all the molecules on the outside contour between them, are added to the new contour. There are now two separate contours making up the unknown shape. Figure 15 shows the changes to the two contours during Step 3, and the result is summarised in Table 6.

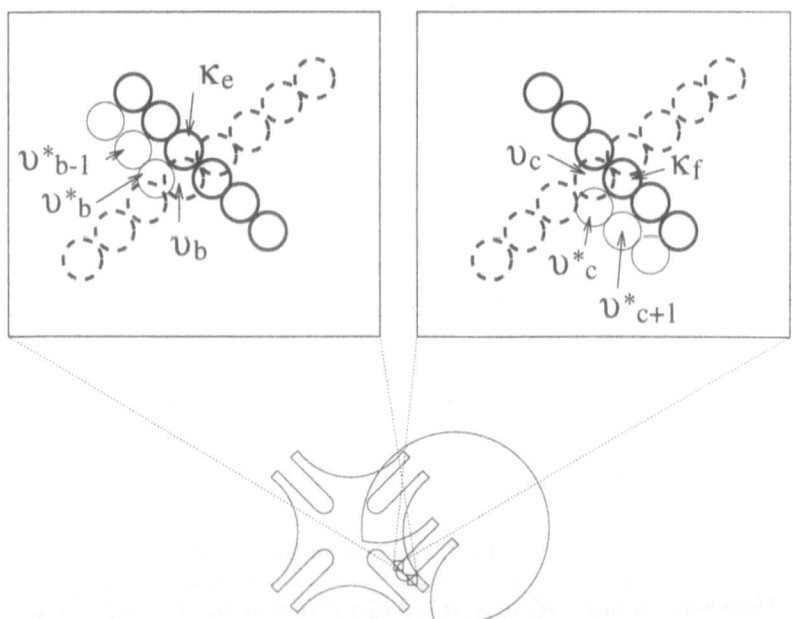

Figure 14. A close up of the intersections for Step 3.

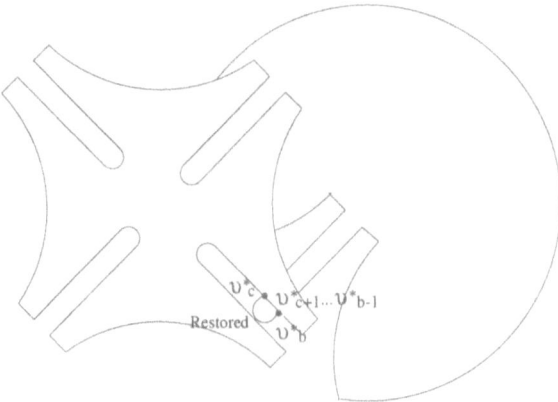

Figure 15. After Step 3.

TABLE 6. Contours after Step 3.

Unknown Contours:	$v_a^*, v_{a+1}^* \ldots v_{d-1}^*, v_d^*, v_{d+1} \ldots v_{e-1}, v_e^*, v_{e+1}^* \ldots v_{f-1}^*, v_f^*, v_{f+1} \ldots v_{a-1}$, $v_b^*, v_{b+1} \ldots v_{c-1}, v_c^*, v_{c+1}^* \ldots v_{b-1}^*$
Cut Contours:	$v_e, v_{e+1} \ldots v_{f-1}, v_f$, $v_a, v_{a+1} \ldots v_{b-1}, v_b$, $v_c, v_{c+1} \ldots v_{d-1}, v_d$

4. Reconstructing the Multiple Levels During Synthesis

A crucial factor in the efficient performance of the synthesis procedure is that interferences must be rapidly detected using the hierarchical procedure detailed earlier. However, during synthesis, base level molecules of the unknown contour are constantly being added to and replaced. This section shows how a data structure, based on a multi-linked B^+ tree (i.e., B^+ tree with bidirectional links), containing the multiple levels of representation of an object can be maintained efficiently in the course of the contour deletions and insertions during synthesis.

An example object contains twenty seven molecules, but molecules fourteen to seventeen are to be removed and replaced with twelve new molecules. The initial state of the data structure is shown in Figure 16.

The four molecules being replaced are overwritten by the first four new molecules. Molecule 14 is overwritten by the first of the new molecules, 1', 15 by 2', 16 by 3' and 17 by 4'. The intermediate state of the data structure is shown in Figure 17.

To speed up the process we wish to insert larger molecules containing three of the smaller ones. A new larger molecule can not be placed inside another larger

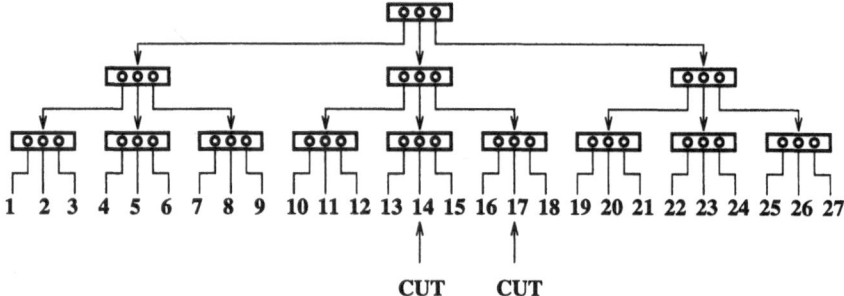

Figure 16. Initial data structure.

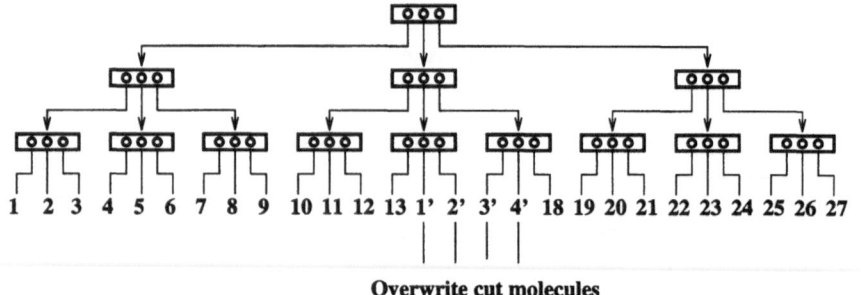

Overwrite cut molecules

Figure 17. Removed molecules are overwritten.

molecule, so the insertion point must be moved to the start of a larger molecule. This is done by creating a new larger molecule and placing in it old smaller molecules after the insertion point. In this example, a new larger molecule is created and 18 is placed in it. The two larger molecules are then filled, the original with the next smaller molecules to be inserted, 5', and the new with the last molecules to be inserted, 11' and 12'. This is shown in Figure 18. Now the remaining molecules are built into new larger molecules, in Figure 19.

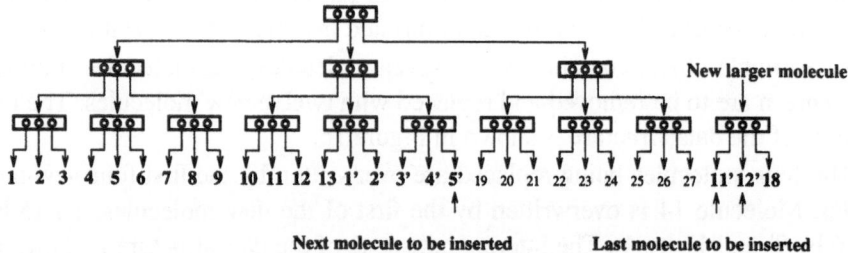

Next molecule to be inserted   **Last molecule to be inserted**

Figure 18. Ready to insert larger molecules.

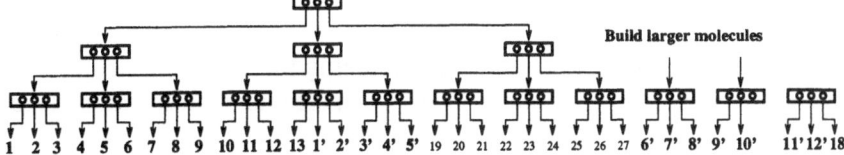

Figure 19. New larger molecules.

The new larger molecules are then inserted into the list of larger molecules in the same way as the smaller molecules were inserted into them.

There are no molecules to be removed, so none can be overwritten. The insertion point is already at the start of a larger molecule, so the three molecules to be inserted are built into a new larger molecule to be inserted into the larger molecules, as shown in Figure 20.

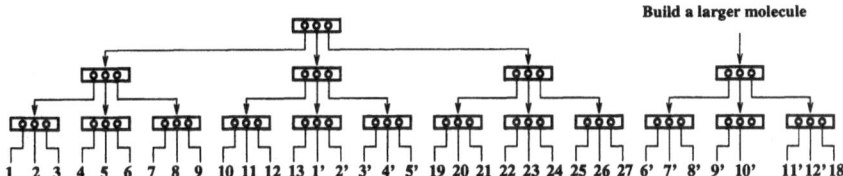

Figure 20. New larger molecule.

No molecules can be overwritten, but the insertion point is not at the start of a larger molecule. A new larger molecule is created, and the molecule after the insertion point is placed in it. The molecule to be inserted is placed in the original larger molecule to fill it, but there are no further molecules to insert which can be used to fill the newly created larger molecule. The new molecule is inserted into the top level. As the top molecule does not have a parent a new level is created containing it and the molecule being inserted. Figure 21 shows the completed data structure.

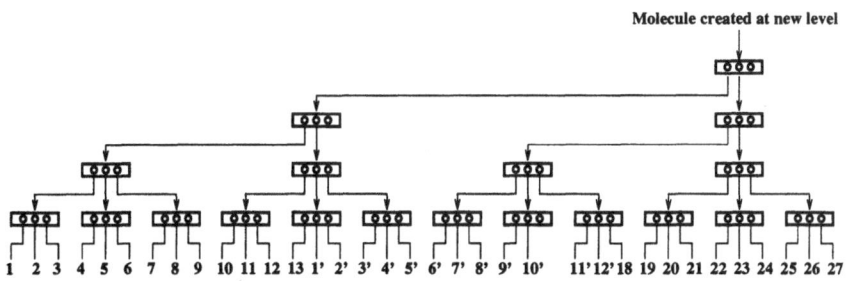

Figure 21. The completed data structure.

5. Closing Comments

Our current efforts are directed towards developing the representation for three-dimensional problems. The original single-level molecular representation depended very heavily on the linear ordering of molecules to facilitate the *skipping* component of the interference detection algorithm. In the multi-level case there is no such dependence. It is likely that points distributed on the nodes of a regular triangular mesh may be aggregated into spheres enclosing hexagonal collection of points as in the two-dimensional case. Synthesis in three-dimensions will need a further extension in the form of a data structure by which areas of such spheres may be added and deleted the hierarchical decomposition efficiently maintained.

References

Baraff, D.: 1989, Analytical methods for dynamic simulation of non-penetrating rigid bodies, *Computer Graphics*, **23**(3), 223–232.

Erdmann, M. and Lozano-Perez, T.: 1986, On multiple moving objects, *IEEE Robotics and Automation Conference*, pp 1419–1424,

Faltings, B.: 1990, Qualitative kinematics in mechanisms, *Artificial Intelligence*, **44**, 89–120.

Faltings, B.: 1992, A symbolic approach to qualitative kinematics, *Artificial Intelligence*, **56**, 139–170.

Forbus, K. D., Neilsen, P. and Faltings, B.: 1987, Qualitative kinematics: A framework, *Proceedings Tenth International Joint Conference on Artificial Intelligence*, Milan, Italy, pp. 430–436.

Garcia-Alonso, A., Serrano, N. and Flaquer, J.: 1994, Solving the collision detection problem, *IEEE Computer Graphics and Applications*, **14**(3), 36–43.

Gupta, R. and Jakiela, M.: 1994, Simulation and shape synthesis of kinematic pairs via small-scale interference detection. *Research in Engineering Design*, **6**, 103–123.

Hahn, J. K.: 1988, Realistic animation of rigid bodies, *Computer Graphics*, **22**(4), 299–308.

Joskowicz, L. and Sacks, E.: 1991, Computational kinematics, *Artificial Intelligence*, **51**, 381–416.

Joskowicz, L.: 1988, *Reasoning about Shape and Kinematic Function in Mechanical Devices*, PhD Thesis, New York University.

Krishnasamy, J. and Jakiela, M.: 1993, Computer simulation of vibratory parts feeding and assembly, *Proceedings of the 2nd International Conference on Discrete Element Methods*, Cambridge, Massachusetts, pp. 403–411.

Lozano-Perez, T.: 1983, Spatial planning: A configuration space approach, *IEEE Transaction on Computers*, **C-32**(2), 289–120.

J. S. Gero and F. Sudweeks (eds), Artificial Intelligence in Design '96, 21-38.

TEXT ANALYSIS FOR CONSTRUCTING DESIGN REPRESENTATIONS

ANDY DONG AND ALICE M AGOGINO
University of California at Berkeley
Department of Mechanical Engineering
5136 Etcheverry Hall
Berkeley CA 94720-1740 USA

Abstract. An emerging model in concurrent product design and manufacturing is the federation of workgroups across traditional functional "silos." Along with the benefits of this concurrency comes the complexity of sharing and accessing design information. The primary challenge in sharing design information across functional workgroups lies in reducing the complex expressions of associations between design elements. Collaborative design systems have addressed this problem from the perspective of formalizing a shared ontology or product model. We share the perspective that the design model and ontology are an expression of the "meaning" of the design and provide a means by which information sharing in design may be achieved. However, in many design cases, formalizing an ontology before the design begins, establishing the knowledge sharing agreements or mapping out the design hierarchy is potentially more expensive than the design itself. This paper introduces a technique for inducing a representation of the design based upon the syntactic patterns contained in the corpus of design documents. The association between the design and the representation for the design is captured by basing the representation on terminological patterns in the design text. In the first stage, we create a "dictionary" of noun-phrases found in the text corpus based upon a measurement of the content carrying power of the phrase. In the second stage, we cluster the words to discover inter-term dependencies and build a Bayesian belief network which describes a conceptual hierarchy specific to the domain of the design. We integrate the design document learning system with an agent-based collaborative design system for fetching design information based on the "smart drawings" paradigm.

1. Motivation

The design of complex mechanical systems requires an intimate understanding of the interactions among the different disciplines and subsystems so that cross-disciplinary tradeoffs can be made. Any change that might have been precipitated explicitly by modifying a requirement or implicitly by observing a failed simulation will propagate a chain of interaction between designers, manufacturing engineers, process planning engineers, and sales and marketing professionals. Knowing the role of individual functional and physical design elements and their association to other elements in the overall design helps the product design team "understand" the design from the perspective of other members.

In reality, to "know" the interaction between design elements, designers expend a considerable amount of effort in accessing and absorbing design information. One can characterize this scenario roughly as a three-step process. First, the designer looks for possible related elements such as interdependent design functions or physical components. Next, the designer analyzes and interprets the relations between them, relations that might be explicitly stated in mathematical equations, rules, or implied by design standards and "best-practices." Finally, the designer decides which of the associations is plausible in some sense. If there is no reason to reject or defer, then the association is accepted (Baya et al., 1993). Unfortunately, few CAD applications have begun to address the problem of reducing the time designers spend understanding the design, including absorbing design information, keeping up with design changes and reconciling problems or sharing information (Toye et al., 1993). According to Akman (1994), only systems which embed advanced reasoning capabilities will be able to deal with the complexity arising from the management of large quantities of design data.

Since this assessment is typically achieved by reading natural language texts such as memos and design specifications associated with the design model (Ullman, 1988), we would like to build a program to automate this process. This research introduces an automated technique to acquire a representation of the design based upon contextual clues in the design documents. By allowing the current context of the design to influence the representation, we eliminate the *a priori* determination of a structured hierarchy or design language and permit dynamic updating of the design vocabulary.

The research was motivated by a desire to take advantage of existing design information to assist in collaborative design. Current CAD tools adequately capture the final design details such as specifications and analysis results. Still, we need to develop tools that learn the interconnections between well-documented design elements so that federated workgroups can have access to relevant information without necessarily having to be an expert in each area of the design. The underlying aim of the research then is to discover the terminological patterns in design text as a basis for constructing a meaningful engineering model of the design.

2. Prior Research

The kernel of design information systems is the ontology which describes the product model. The ontology is a repository of information and provides a means by which concurrency in design may be achieved. The evolving STEP standard (ISO CD 10303-1) highlights the thrust towards product modeling and a common ontology in product models. Product modeling-based systems have been quite successful at setting up complex rules which describe in detail the possible underlying structures of a design (Wong and Sriram, 1993; Szykman and Cagan, 1992); at the same time ontology-based systems are trying to define semantic relations and to model the functional and behavioral structures underlying the synthesis of a design for representing stereotypical information (Olsen et al., 1995; Shah, 1993). A similar design-document learning system to the one proposed here is being pursued by Reich (1993) except that the relationships between words are not learned but rather negotiated by the designers. How the words fit together into a structure communicated an idea.

We agree that an ontology provides a means for sharing information. However, the approach presented in this research differs from that taken by other researchers in the design community who developed specialized grammars and shared ontologies (Gruber, 1992) or product-models in that it derives from the design documents. Information models should capture and represent product information to give the reader an "understanding" of the design the model represents. But they must also be dynamic to reflect the evolutionary nature of design. Even though one could argue that the addition of new ontology and negotiated agreements makes the ontology-based or product modeling-based systems dynamic with the design, since the "meaning" of design elements changes with an evolving design, modifying the model or adding new ontology to reflect the changes in real-time might be difficult. In fact, the evolutionary and uncertain nature of design require representations that operate on meaning, not expression (Wood and Agogino, 1995).

Part of the problem of these systems is that they assume that the "meaning" of a design could be computed as a function of the constituents. To "understand" a design, designers must take advantage of a variety of mechanisms that use all sorts of knowledge to fill in any necessary information. In making a computer model of design knowledge, this presents a serious problem. On the one hand, it is impossible to isolate all aspects of domain-dependent knowledge from the others. On the other hand, it is clearly undesirable to give the program all the knowledge related to the design. In this research, the dilemma is resolved by inferring plausible conclusions by relating the various elements of the design using the design documents themselves as a complete and accurate representation of the current state of the design.

We propose the architecture of an intelligent agent in a collaborative design environment which dynamically learns the current status of the design. One application of the agent is the retrieval of relevant information to the current needs of the designer. The system achieves the learning and understanding of the design using the design documents as the "model of the world." We present a theory of design discourse as a theoretical premise for generating a model of the design based on the design documents, and illustrate how to integrate learning the design within a collaborative design framework for bringing relevant design information to the decision-maker based on the "smart drawings" system presented in a prior paper (Dong et al., 1995).

3. Methodology

3.1. GENERAL THEORY

In discourse, people take advantage of a variety of mechanisms that depend upon the existence of an intelligent hearer who will use all sorts of knowledge to fill in any necessary information (Wilensky, 1983). To make an intelligent agent understand the design as communicated by the designers through design documents, then, we must construct a framework within which the agent has a sufficient search space to formulate an adequate understanding of the design (Dong et al., 1995). In order for the agent to fill in necessary information regarding the design, though, it must learn the connections between the functions or components of the design. Currently, the solution strategy is to have experts construct both the ontology and describe the decomposition of the design to the agent. However, we argue that this information is in fact available and contained in the design documents themselves. Research in full-text retrieval systems (Lewis, Croft and Bhandaru, 1989) verify how certain syntactic patterns in documents refer to meaningful concepts, and how language-oriented techniques for information retrieval can build the relationships between categories, category instances and relations of those concepts. These categories define a model of the design. By reading the design documents periodically, the agent excerpts the current connections between the different design elements.

In building this agent, we assume to a first-approximation that the linguistic content (words) of the design documents provide a useful index to the composition and structure of key design concepts at the current state of the design. Second, it is assumed that every statistical association derives from causal interaction; therefore logical coherence is based on statistical coherence. Based upon these assumptions, we propose the following theory of design discourse as the theoretical foundation for the learning algorithm:

> *A theory of design discourse*—The content of design documents is related to a conceptual structure of the design, whose communication comprises the goal of the designer.

The claim is the agent can induce a model of the design, including the functions and components of the design and their relations by learning over the design text associated with the product. Eastman (1991) identifies several criteria for describing engineering product models: (1) the semantics, which describe the functions, components and attributes of the design; and (2) class structures, which describe both the generalizations (properties relevant to any design element), and the decomposition (how functions and components are inter-related) of the design.

While the product model derived by this method is not the same as that proposed by Eastman, these criteria serve as a guideline for learning the design. In essence, the sequence of operations in the program are (as shown in Figure 1) to: (1) extract the natural language text annotations to CAD drawings to excerpt the semantics of the design; (2) generate the class structures describing which properties are relevant to any function of a design using clustering; and (3) build a decomposition of the design which in this method is accomplished with belief networks.

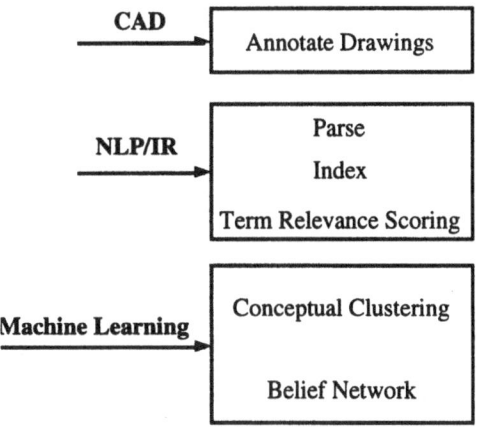

Figure 1. Process Flow Chart—The figure outlines the sequence of operations of the program in learning the content and structure of the design. The research proposes a methodology for annotating CAD documents to create "smart drawings," techniques for extracting the design vocabulary from the design text using natural language processing and information retrieval, and model learning and inference by applying machine learning.

3.2. STAGE 1: TEXT ANALYSIS

The general method to discover terminological patterns in design documents, which act as a basis for constructing the design model, is to parse the document text, cluster inter-term dependencies and build a conceptual hierarchy.

First, the text was passed through a parser and indexer, freeWAIS-sf[1] (Pfeifer and Huynh, 1994) *waisindex*, which returns a dictionary of every word in the text except for common "stop words."[2] We then filter this set of terms to develop a set of content-carrying terms. The filtering process is based upon a word score metric similar to that described by the CLARIT method (Evans et al., 1991). The scoring equation is based on the freeWAIS-sf term relevance score (TRS) metric shown in Equation 1. The primary statistics include (1) a frequency count of the number of times the word was encountered in individual documents in the corpus; and (2) an inverted weighted distribution measurement for the number of documents containing a particular term. The idea is that the frequency measurement correlates with the text semantics. Words that occur often in a text are better indicators of what the text is about. More terms can always describe the document concepts better, but too many terms dilute the importance of any individual concept. Thus the distribution (or inverted document frequency) of the terms in the documents captures the intuition that words which have high frequency across documents are "general" in the domain and do not serve as good discriminators of concepts.

$$TRS = \frac{(\log(tf) + 10) x idf}{number\_of\_terms\_in\_a\_document}$$

$$tf = 0.5 + \frac{0.5 \times \sum_{doc} word}{\max \sum_{doc} word}$$

$$idf = \frac{1}{\sum_{doc} word}$$

Equation 1. The freeWAIS-sf TRS Metric—The TRS metric is based upon the term frequency *tf* which counts the number of times the word appears over all documents, the inverted document frequency *idf* which counts the number of documents containing the word (a measurement of distribution) and normalized by the number of terms in a document, to account for the rarity of a word.

---

[1] One advantage to using a WAIS (Wide Area Information Server) program such as freeWAIS-sf for full-text document parsing, indexing and retrieval is that documents can be queried and retrieved over the Internet using the Z39.50 V2 protocol.

[2] Stop words include conjunctions and articles such as "a", "the," "since" and other words frequently used in natural language to connect terms but not necessarily to distinguish topics or provide contextual cues for topics.

The score does not account for variations in author style or the presentation of the text. For example, one might score words which are typed in bold face or italicized or words from more recent design documents higher than others. Other factors such as the person who wrote the document, paragraph headers or document titles could be used as additional word weights; however, the efficacy and numerical value of these weights is difficult to quantify. Further, this complicates the clustering. For example, "recent" terms might be associated by time rather than meaning which violates the purpose of the algorithm. Thus the algorithm has limited sensitivity to the organization and presentation of the text.

Then, the program computes the average score and standard deviation. Words whose score fall above the mean become the inventory of index terms for the corpus, the certified terminology. The system filters out words which are relatively frequent, have less value in forming good topic discriminators than relatively rare words, and words which are seldom used since they are probably not conditionally dependent upon the concepts described or vice versa. We will explain later why this conditional relevance is important in building a dependency matrix of concepts which forms the basis of the representation.

Finally, based on the set of certified, content-carrying terms, the system determines their contextual similarity by measuring the frequency of occurrence of any two of the certified terms in the documents. That is, the program generates a nxn matrix, where n is the number of certified words, which scores how "often" the certified words co-occur. This matrix is created by executing a *waisquery* consisting of the query string "[word-A] AND [word-B]". The query sums up the score for similarity between the query string and the document base. The conjecture is that if the query string appears frequently over the entire document base then the words have a shared contextual dependency. In freeWAIS-sf, document similarity is measured as a vector product formula. The similarity between the query string Q and the document D is given by

$$similarity(Q, D) = \sum_k (w_{qk} \times w_{dk})$$

Equation 2. The freeWAIS-sf Similarity Metric

where w_{qk} is the weight assigned to term k in the query and w_{dk} is the weight assigned to term k in the document D.

3.3. STAGE 2: CLUSTERING AND INDUCING A BAYESIAN NETWORK

Once the system has developed a prescribed vocabulary, the program maps the terms into context descriptors. The words themselves have no "meaning" outside the context in which they appear. In fact, research in full-text information retrieval has shown that words which appear in the same context tend to have a shared dependency (Gardiner, Riedl and Slagle, 1994). Thus, we need to map the relevance between the assigned terms and the context in which they appear.

For this process, we apply two machine learning techniques. In the first portion, we classify related terms into "conceptual cells" using unsupervised learning. These cells represent terms which are self-similar in the documents. This determination is based on the observation that terms which appear together (in the same context) in documents typically connote similar meaning (Gardiner, Riedl and Slagle, 1994; Lewis, Croft and Bhandaru, 1989). Since the matrix measures closeness based on the spread of data or distance between words, a convenient distance-based clustering technique is the K-means algorithm in Table 1 (Duda, 1973). The variable x_i is the score in the matrix for the pairwise occurrence of two words in the document collection.

TABLE 1. K Means Algorithm.

procedure K_MEANS
    (Initialize the cluster centers w_j, j=1, 2, . . . , N_1)
    (repeat
           ; Group the patterns with the closest cluster center
           (for all x_i do

$$\text{(Assign } x_i \text{ to } \Theta_{j*}, \text{ where } w_{j*} = \min_j \left\| x_i - w_j \right\|$$

        endloop)
      ; Compute the sample means
      (for all w_j do

$$w_j = \frac{1}{m_j} \sum_{x_i \in j} x_i$$

    endloop)
    until there is no change in cluster assignments from one iteration to the next
)
end ; { K_MEANS }

Next, the goal is to obtain a decomposition that explicitly reveals as much information regarding the conditional independence of design elements as possible. The key feature of belief networks is their explicit representation of the conditional independence among events (Pearl, 1988). That is, they can explicitly and compactly represent the dependency of design elements. Topological transformations (through arc reversals and node absorption for example) can answer questions concerning possible causal relations or dependencies between design elements. Since the Bayesian network conveys an intuitive understanding of how the reasoning process works, the designer can also follow the reasoning process of the design based upon the dependencies/independencies of the events to determine how the change in any one element might affect any other element.

The general method for constructing belief networks is to draw arcs from causal nodes to effect nodes and then attach a probability to that arc (Russell and Norvig, 1995). While techniques exist for constructing the most probable belief network B_S given a database D of instances (often called the *maximum a posteriori* structure) based on assumptions of a uniform distribution of belief network structures (Cooper and Herskovits, 1992), the Bayesian Dirichlet likelihood equivalent metric (Heckerman and Geiger, 1995) and minimum description length (Lam and Bacchus, 1993), we generate an initial network using a heuristic approach. We plan to apply one of the metrics to optimize the network locally about a network structure which correctly represents the design.

The heuristic used to construct the Bayesian network is based upon the conjecture that seeing a lower TRS word with respect to a word that it shares contextual similarity causes the system to update the belief that the higher TRS word will appear (Evans et al., 1991). This causal influence and contextual similarity is found by pairing words with the highest TRS in the co-occurrence matrix. The strategy for building the network is to link the highest associated words in their own clusters first then to link the words between clusters. The algorithm is outlined in the Table 2.

TABLE 2. Network Algorithm—In the first box, the table illustrates the general method for creating belief networks based on expert knowledge. In the bottom box, the table outlines the heuristic algorithm employed by the program.

General Procedure

1. Choose the set of relevant variables X_i that describe the domain
2. Choose an ordering on the variables
3. While there are variables left:
   (a) Pick a variable X_i and add a node to the network for it
   (b) Set Parents (X_i) to some minimal set of nodes already in the net such that the conditional independence property is satisfied (direct causal influence)
   (c) Define the conditional probability table for X_i

Network Algorithm

1. Define a variable X_i for each word
2. Order the variables X_i in their respective clusters by ascending TRS
3. While there are variables left in the cluster
   (a) Select the variable X_i with the lowest TRS and add as node in the network
   (b) Set X_j as Parent Of(X_i) where X_i and X_j have the highest similarity in the co-occurrence matrix and $TRS(X_j) > TRS(X_i)$
   (c) Select next node in ordering as X_{i+1} and continue; repeat for each cluster

4. Order the clusters by ascending cumulative TRS
5. While there are variables left in the cluster
   (a) Select a variable X_i from the lowest TRS cluster
   (b) Set X_j as ParentOf(X_i) as the node from the next cluster with the highest similarity in the co-occurrence matrix
   (c) Select next node in ordering as X_{i+1} and continue; repeat for each node and cluster
6. Define the conditional probability table for X_i

3.4. AGENT ARCHITECTURE FOR DESIGN INFORMATION RETRIEVAL

Figure 2 depicts the agent architecture for learning the design based on the documents. The architecture augments the "smart drawings" system presented in a previous paper (Dong, Agogino, Moore and Woods, 1995).

The agent environment consists of the database of design documents, including the CAD drawings, design specifications, design notes and memos and e-mails written between designers. The agent reads the text periodically to generate the list of content-carrying words. By manipulating the list and using the document database for additional data, the agent constructs the inter-term clusters and belief networks to build a model of the design. The model helps the agent to understand the design by finding out what properties are relevant to a function in the design and the decomposition of the design. In response to requests from the user, the agent can retrieve relevant design information.

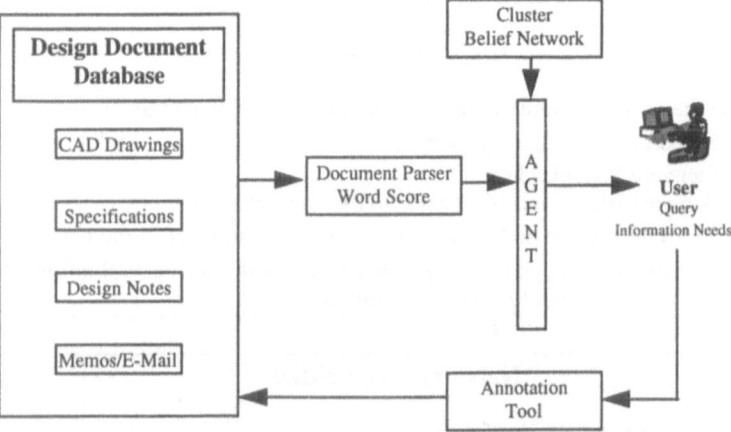

Figure 2. Agent Architecture—The user annotates and adds design documents to the document database. The agent interacts with the document database by parsing and scoring the words in the document. The agent uses the data to create the clustering and belief network to learn the connections between the design elements. The user can then ask for relevant information with respect to current information needs by having the agent search for related design components.

4. Experimentation and Results

For this project, we created a machine-readable form of *The Mechanical Engineers' Handbook* (Kutz, 1986) which was scanned and run through an optical character recognition (OCR) software ("dirty")[3] to output the final text. The program was then run on the chapters on controller design to derive a model of controllers.

The cluster results are shown in Figure 3. These clusters indicate which properties are relevant to any particular function or element of the design, giving the agent knowledge of relevant issues in the design. The clusters indicate, for example, that the main content of the documents is the *design* of a *controller* or the *control* of a *system*. The third cluster reveals that the performance of the system is influenced by the *gain* and *order* of the control as well as any *damping* in the system while the sixth cluster indicates that the *position* seems to be the *variable* to be controlled in the system as it is tightly related to the *feedback, input* and *output*. One critique of the clusters is that *zero* appears with *performance* and *root* appears with *stable*, whereas it is known that both the *zero* and *root* of the system affect the stability. However, in the document collection, *zero* statistically appears more often with *performance* and *root* with *stable* since, for example, the documents discuss more often that a zero affects steady-state error (a measurement of controller performance) whereas the closed-loop roots determine the stability of the system. The cluster results agree with known knowledge of the relevant properties of the functions and attributes of controllers.

```
((system design controller control)
(transfer function time error state signal response plant)
(zero integral gain order damping performance steady action)
(stable root frequency process model loop)
(valve pressure power pneumatic motor displacement)
(variable value position feedback input output)
(disturbance diagram constant) ... )
```

Figure 3. Cluster Results—The cluster results for the chapters on controller design.

Finally, the system generates the belief network shown in Figure 4 and the conditional probability table associated with the network. The states for each of the event nodes (words) are 0, when the word (or design element) is not present in the document, and 1 otherwise. The conditional probability table for the network is based on frequency counts. For example, the probability of the word *controller* co-occurring with the word *design* is given by:

---

[3] "Dirty" OCR refers to documents un-modified after the OCR process, i.e. no spell check.

$$P(controller \mid design) = \frac{\#\ \text{occurrences controller and design}}{\#\ \text{occurrences design}}.$$

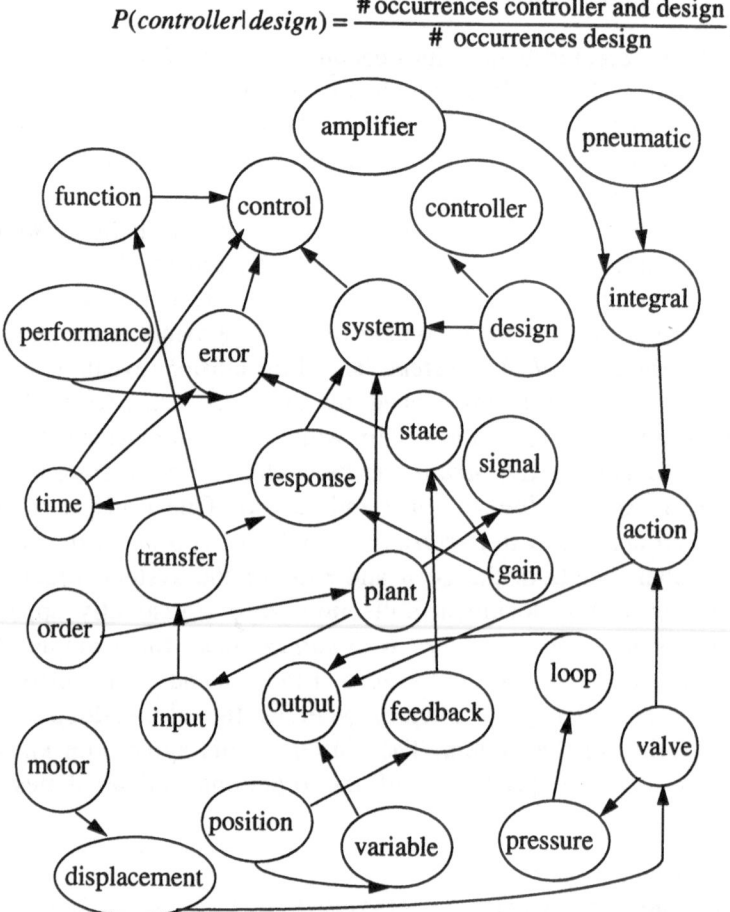

Figure 4. Initial Belief Network (partial)—This figure illustrates parts of the belief network generated using the heuristic algorithm. For purposes of clarity, not all arcs and nodes are shown.

For clarity, not all nodes and their associated arcs are shown in Figure 4. For the nodes shown, the arcs are complete. One can read some interesting inferences off of the network. The first inference expresses the dependency of the design elements. The expression of dependency describes the decomposition of the design.

1. The *system* to be controlled is characterized by the desired *response* and the *controller design*. The *control* law is conditioned on the *transfer function*, the *error* and the desired *response* of the *system*.

The second inference illustrates the degree of dependency between design elements. These types of inferences relate both information and the degree of relevance based on the amount of evidence available.

2. The concept of *system response* is more dependent upon *gain* in this controller design than the specific *input* criteria.

The third is perhaps the most interesting since it shows how the system could infer the interaction of several elements in the design which produce a certain function. Therefore, if the designer were interested in increasing the pressure in the controller, one of the design elements to modify is the motor and followed by checking if the valve could handle the increased stress. More notable is that without explicitly telling the system these design element connections or the design topic, the system correctly extracted from the text that these chapters discussed controller design using pneumatic devices.

3. The *motor* changes the *displacement* of the *valve* which affects the *pressure*.

While the arc directions could change through topological transformations, the above network and associated inferences illustrate two important ideas. First, inspection of the network indicates that the heuristic generates a network with arcs between elements in the direction of physical causality, as illustrated by the third example in that the *motor* causes *displacement* rather than vice versa. Second, the network illustrates the more important problem of capturing the dependencies between design elements. By capturing these dependencies, the system is more efficient in searching for meaningful and relevant design information. The combination of the cluster information and the belief network augments the search by finding closely associated design elements (cluster information) which may not actually appear in the designer's query while removing less relevant information if less evidence supports the association between the design elements (belief network).

The program was then integrated with a "smart drawing" (Dong et al., 1995) system as shown in Figure 5. Some preliminary tests were conducted to test how well the system learned the design data. One of the tests asked the system to retrieve relevant information to the "Lyapunov stability of the controller."[4] Based on the clustering results, the program knows that the *roots* of the system affect *stability*, so that documents which discuss *roots* frequently should also be returned and scored high in relevance. By expanding the query to include closely related terms which in this example indicate closely related attributes to the stability of the system, the program can find documents related to *stability* that may not mention the word

---

[4] One aesthetic limitation of the current implementation is that the user is given only the path to the document rather than the document title, for example. By selecting one of the documents, though, the system automatically brings up a viewer for the document type, such as a text file or AutoCAD drawing.

stability in the document. Those elements which have shared dependencies in the belief network are scored higher. Without expanding the query based on the learned data (i.e. a standard freeWAIS query), only documents which frequently discuss both *stability* and *controller* would have scored high. That is, the dynamically learned design structure augmented retrieval to include information not explicitly cast in the query but which should be reported together by virtue of design dependency. In this case, design documents which discuss any property shown relevant based on the clusters to *stability* or *controller* score high.

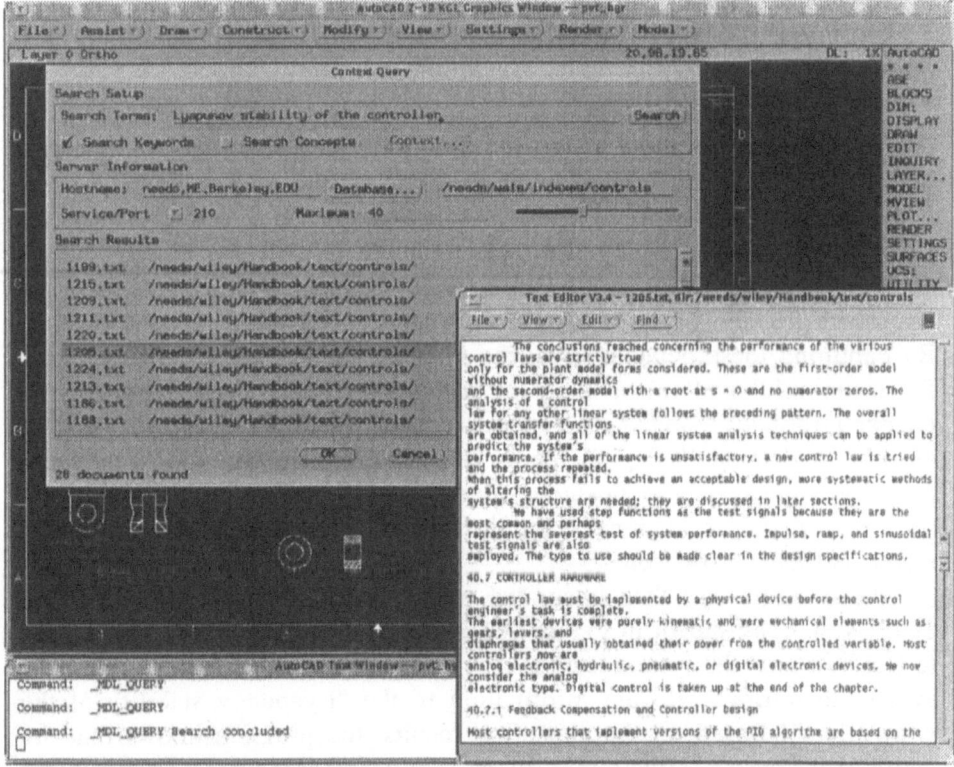

Figure 5. Smart Drawings Desktop—The agent learns the content of the design data based on the design documents using the learning methodology outlined above. Then, when prompted, the agent can retrieve relevant design information based on the current information needs of the designer using the information content of the active document as the query.

The role of the clusters and belief network for design information retrieval is similar to the purpose of the decision dependency network presented by Garcia, Howard and Stefik (1994) in the Explanation interface to their ADD system. The Explanation interface displays related information by retrieving documents that are generally reported together. The key differentiation is that the dependency network is based on a pre-processed parametric design model for the design domain which seems to violate their thesis that the evolution of the design description via documents relates to the evolution of the design. For example, to capture design rationale, ADD prompts the designer for decisions which deviate from the preferred norm. This strategy for design rationale capture suggests that changes in the design affect how the design should have been modeled or parametrized, that, in fact, the design model dynamically evolves with the design.

Systems such as ADD and the one proposed which address the problem of accessing design information by employing a structured design model to augment the retrieval of unstructured design documents can improve recall over those which have only an unstructured model (such as freeWAIS-sf) or only structure (ontology-based systems). However, the important metrics for evaluating these systems should include both the overhead for creating the structured model to account for the dynamic nature of the design as well as the performance in retrieving relevant information compared to baseline systems which employ no structure. The design learning methodology proposed illustrates a preliminary system which addresses both metrics.

While this is only a preliminary test of how well the system learns the design data, what these tests suggest is the ability to augment design information search by finding related information based upon meaning, not just how the search request is expressed in the query. Second, the clustering and belief network open the possibility of organizing the retrieved data in a manner which is more meaningful to the designer than just straight frequency metrics, such as ordering by related concepts. We are currently investigating how to integrate the utility of the information to the designer based on the preferences of the designer and the structure of the design model in the belief network to improve the relevance ranking of the returned information beyond simple frequency count measures. In particular, we are implementing a "concept query" mechanism which more closely analyzes what concepts would be interesting to the designer and the cost of obtaining that information.

5. Summary and Future Directions

This research develops a computable learning method to extract the content of the design model to facilitate information sharing among designers. The premise of the methodology is that the design specifications and solutions as communicated through design documents are related to a model of the design. Certain combinations of the chosen properties of the design give rise to the corresponding combinations of design descriptions in the design text. Therefore, by learning these descriptions (words) through text analysis, the system induces a model of the design. The learning algorithm is based upon natural language processing text analysis to extract content-carrying terms, and then applying techniques from machine learning to cluster inter-dependent terms and decompose the design into dependent elements using belief networks. The model derived for the controller design example was plausible and correct based upon knowledge of controllers.

What this research emphasizes is that CAD systems cannot ignore the communication of design information with respect to the current and relevant information needs of the design based on the annotation of the drawings (Ullman, 1990). That is, the effect of techniques which implement inductive learning techniques such as the one proposed to generate new knowledge structures about the design rather than techniques that improve the efficiency of problem-solving (explanation-based learning techniques) is tantamount to improving CAD systems. By putting the knowledge of design components in a form in which we can explicitly express the connections between the different parts of the system's knowledge, we enrich the possibility of interaction for collaborative design.

The methodology explored in this paper only begins to explore the possibilities of full-text analysis for deriving a model of the design and its application. For example, one could augment the learned design structure with formally derived ontologies or use the learned structure as the basis for a formal ontology (Gruber, 1993). In particular, enhancing the parsing ability of the program and augmenting the co-occurrence measurement strategy to consider the number of words between two contextually similar words (Grefenstette, 1992) promise to improve the efficiency of the algorithm and achieve finer granularity in representing the design data. We are currently investigating these issues as well as testing the relevance of the learned knowledge in design documents from mechanical engineering design courses.

6. Acknowledgements

The authors would like to acknowledge William H Wood III for his valuable comments and converting the scanned document images into ASCII text, and John Wiley and Sons, Inc. for their permission to scan and OCR the text used for research and testing. We would like to thank in particular our industrial partners, Sun Microsystems, Inc., and Autodesk, Inc., not only for financial and equipment support but for valuable collaboration. This research was sponsored by the NSF Concept Database grant #DDM-9300025.

References

Akman, V., ten Hagen, P. J. W., and Tomiyama, T.: 1994, Desirable functionalities of intelligent CAD systems, in C. H. Dagli and A. Kusiak, (eds), *Intelligent Systems in Design and Manufacturing*, New York: ASME Press, pp. 119-138.

Baya, V., Gevins, J., Baudin, C., Mabogunje, A., Toye, G., and Leifer, L.: 1992, An experimental study of design information reuse, *Proceedings of the ASME Conference on Design Theory and Methodology*, DE-Vol. 42, pp. 141-147.

Cooper, G. R., and Herskovits, E.: 1992, A Bayesian method for the induction of probabilistic networks from data, *Machine Learning*, **9**, 309-347.

Dong, A., Agogino, A. M, Moore, F. Woods, C.: 1995, Managing design knowledge in enterprise-wide CAD, *in* J. S. Gero and F. Sudweeks, (eds), *Preprints Advances in Formal Design Methods for CAD*, Key Centre of Design Computing, University of Sydney, Sydney, Australia, pp. 330-347.

Duda, R. O. and Hart, P. E.: 1973, *Pattern Classification and Scene Analysis*, New York: Wiley.

Eastman, C. M., Bond, A. H., and Chase, S. C.: 1991, A formal approach for product model information, *Research in Engineering Design*, **2**, 65-80.

David A. E., Ginther-Webster, K., Hart, M., Lefferts, R. G., and Monarch, I. A.: 1991, Automatic indexing using selective NLP and First-Order Thesauri, in *Intelligent Text and Image Handling* , *in* A. Lichnerowicz, (ed.), *Proceedings of a Conference on Intelligent Text and Image Handling 'RIAO 91*, Barcelona, Spain, pp. 624-643.

Garcia, A. C. B., Howard, H. C., and Stefik, M. J.: 1994, Improving design and documentation by using partially automated synthesis, *Artificial Intelligence in Engineering Design and Manufacturing*, **6**(1), 335-354.

Gardiner, D., Riedl, J., and Slagle, J.: 1994, TREC-3: Experience with conceptual relations in information retrieval, *Proceedings of the Third Text Retrieval Conference (TREC-3)*, Gaithersburg, MD.

Gruber, T. R.: 1993, Toward principles for the design of ontologies used for knowledge sharing, *in* Guarino and Poli, (eds), *Formal Ontology in Conceptual Analysis and Knowledge Representation*, Kluwer Academic Publishers, Dordrecht; *Technical Report KSL 93-04*, Knowledge Systems Laboratory, Stanford University.

Gruber, T. R., Tenenbaum, J. M., and Weber, J. C.: 1992, Toward a knowledge medium for collaborative product development, *in* J. S. Gero (ed.), *Artificial Intelligence in Design '92*, Kluwer Academic Publishers, Dordrecht.

Grefenstette, G.: 1992, Use of syntactic context to produce term association lists for text retrieval, *in* N. Belkin, P. Ingwersen, and A. M. Pejtersen, (eds), *Proceedings of the Fifteenth Annual International ACM SIGIR Conference on Research and Development in Information Retrieval*, pp. 89-97.

Heckerman, D., and Geiger, D.: 1995, Learning Bayesian Networks, *Microsoft Corporation Technical Report MSR-TR-95-02*.

Mechanical engineers' Handbook, Myer Kutz, (ed), John Wiley, New York.

Lam, W., and Bacchus, F.: 1993, Using causal information and local measures to learn Bayesian Networks, *in* D. Heckerman and A. Mamdani, (eds), *Proceedings of the Ninth Conference Uncertainty in Artificial Intelligence*, Morgan Kauffman, San Mateo, pp. 243-250.

Lewis, D. D., Croft, W. B., and Bhandaru, N.: 1989, Language-oriented information retrieval, *International Journal of Intelligent Systems*, **4**, 285-318.

Olsen, G. R., Cutkosky, M., Tenenbaum, J. M., and Gruber, T. R.: 1995, Collaborative engineering based on knowledge sharing agreements, *Concurrent Engineering: Research and Applications*, **2**(3), 145-159.

Pearl, J.: 1988, *Probalistic reasoning in intelligent systems: networks of plausible inference*, Morgan Kaufmann, San Mateo.

Reich, Y., Konda, S. L., Levy, S. N., Monarch, I. A., and Subrahmanian, E.: 1993, New roles for machine learning in design, *Artificial Intelligence in Engineering*, **8**, 165-181.

Russell, S. J. and Norvig, P.: 1995, *Artificial Intelligence: A Modern Approach*, Prentice Hall, Englewood Cliffs, New Jersey.

Shah, J. J., Bliznakov, P. and Urban, S. D.: 1993, Development of a machine understandable language for design process representation, *Proceedings of the ASME Conference on Design Theory and Methodology 1993*, DE-Vol. 53, pp. 15-24.

Szykman, S., and Cagan, J.: 1992, A computational framework to support design abstraction, *Proceedings of the ASME Conference on Design Theory and Methodology*, DE-Vol. 42, pp. 27-39.

Pfeifer, U., and Huynh, T.: FreeWAIS-sf, ftp://ls6-www.informatik.uni-dortmund.de/pub/wais/freeWAIS-sf.1.0.tgz.

Ullman, D. G., Wood, S., and Craig, D.: 1990, The importance of drawing in the mechanical design process, *Computers and Graphics*, **14**(2), 263-274.

Ullman, D. G., Dietterich, T. G., and Stauffer, L. A.: 1988, A model of the mechanical design process based on empirical data, *Artificial Intelligence in Engineering Design and Manufacturing*, **2**(1), 33-52.

Wilensky, R.: 1983, *Planning and Understanding: A Computational Approach to Human Reasoning*, Addison-Wesley, Reading, Massachusetts.

Wood, W. H. and Agogino, A. M.: 1995, A case-based conceptual design information server, *Journal of Computer Aided Design*.

Wong, A., and Sriram, D.: 1993, SHARED: An information model for cooperative product development, *Research in Engineering Design*, **5**, 21-39.

J. S. Gero and F. Sudweeks (eds), Artificial Intelligence in Design '96, 39-57.

LEARNING GENETIC REPRESENTATIONS AS ALTERNATIVE TO HAND-CODED SHAPE GRAMMARS

THORSTEN SCHNIER AND JOHN S. GERO
Key Centre of Design Computing
Department of Architectural and Design Science
University of Sydney NSW 2006
Australia

Abstract. Shape grammars have been used to analyze and describe designs, and to create new designs that are similar in style to the designs the grammar is based on. The grammars are created by hand, involving a large amount of research about the designs and the design process. This paper proposes a different approach, where a system is given design examples, and in a bottom-up process learns stylistic features of the examples. This is achieved by using an evolutionary system that is able to change the representation it is using. With the creation of a more and more complex evolved representation, the search space of the evolutionary process is transformed so that the search for new designs is biased towards designs similar to the design examples.

1. Introduction

Shape grammars (Stiny, 1980) have been introduced as a method for formal descriptions of designs. Shape grammars consist of an alphabet of shapes, a starting shape, and shape rules that define spatial relations between shapes.

The power of shape grammars to analyze and describe designs has been shown in a variety of design areas, from architectural design (examples include Palladian Villas, Frank Lloyd Wright Houses, Wren's City Church designs and Japanese tearooms (Stiny and Mitchell, 1978; Knight, 1981; Koning and Eizenberg, 1981; Buelinckx, 1993), over garden landscaping (Stiny and Mitchell, 1980; Knight, 1990) to de Stijl style paintings (Knight, 1989).

In all these examples, however, the translation from a set of designs into a shape grammar (and the reverse) has been done by hand. Very few attempts have been made to automatize the process. Chase (1989) showed how the automatic generation of shapes from (given) shape grammars can be realized. Mackenzie (1989) described a system that is able to produce grammars from example designs, if the designs are described in terms of their basic components and their topology. The transformations used in that paper are not unique, many different grammars are possible for any given set of designs. The system uses a 'utility' function to distinguish between good and bad representations. This demonstrates a general problem: to create a 'sensible' shape grammar, a large amount of high level knowledge is required. Intentions of the designer, logical units in the design, logical stages in the design process are all represented in the grammars.

The purpose of this paper is to describe an alternative, more computationally oriented view. In the spirit of artificial life research, it uses a bottom-up approach, where complex shapes are created by assembling smaller sub-parts.

2. Evolving coding and Frank Lloyd Wright houses

In Gero and Schnier (1995), we described an evolutionary system which produces problem solutions that are based on example designs. In evolutionary systems, the results of a search process are very much influenced by the representation of the problem space in the coding. In usual implementations, this can pose a serious problem, because the representation might bias the results too much into certain directions. The system described in Gero and Schnier (1995), on the other hand, makes use of this by intentionally biasing the solution space towards a set of potentially interesting solutions. It does this in a two stage process. In the first step, the system is given a set of example designs. The goal of the evolutionary process in this step is to create individuals that resemble the example designs as closely as possible. To do this, the fitness function measures what and how much of the example designs are described by the individuals. The coding of the individuals is chosen to be very low-level, using very simple 'basic' genes.

While the individuals are evolved, they are at the same time analyzed, and successful combinations of low level genes are identified. For every such gene combination, a new gene is created (an 'evolved' gene) and introduced into the population. In the course of the evolution, the evolved genes which are produced aggregate more and more basic and lower-level evolved genes, encoding more and more complex aspects of the example designs. The coding itself, therefore, contains information about the example applications. Any evolutionary system using this coding is biased towards solutions similar to the design examples. This is used in the second step, where a conventional evolutionary system creates solutions to a design problem, using both original basic genes and the evolved genes. The system therefore produces solutions that incorporate aspects of the example designs, but are adapted to the new design requirements. Figure 1 illustrates the idea: beginning with a basic representation, the system creates an evolved coding based on a set of design examples. This evolved representation is then used to create new designs that show design features from the examples.

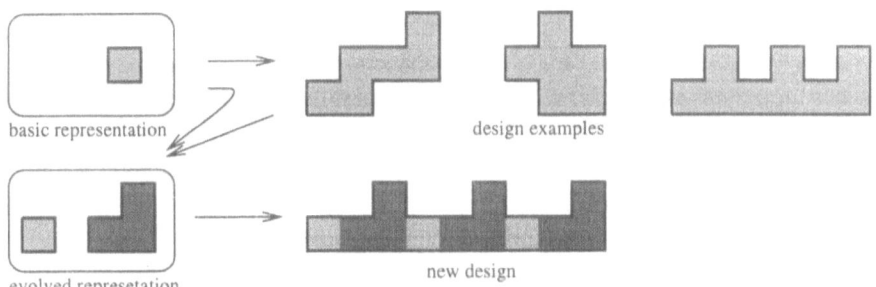

Figure 1. Use of evolved representation to capture and use typical features of a set of example designs.

2.1. INDIVIDUAL STYLE AND THE STYLE OF FRANK LLOYD WRIGHT'S PRAIRIE HOUSES

In Chan (1995), the author explained that the distinct features of a design are produced by both common features and common procedures used by the architect. Describing a style as a function of how it is generated therefore requires a deep understanding of the design process, usually supported by comments from the designer. Shape grammars usually take this approach, they represent both the common procedures (in the rules and the sequences of rules that are possible), and the common features (in the shapes manipulated by the rules). However, Chan also noted that "common features present in an architects work are indeed used by viewers to categorize the architect's style. . . . a style is said to be the function of common features". This means that, even without knowledge about the design process, it is possible to infer important aspects of a style common to a set of designs.

Chan (1992) has analyzed the style of Frank Lloyd Wright's prairie houses more closely. Some of the aspects that are of interest for the work presented here are:

1. Floorplans are always based on a grid, the grid size depends on the project.
2. The fireplace is at the center of the composition, all spaces extend from there.
3. One major shape in the floorplan is long and narrow, much of the house is only one room in depth.
4. The prairie houses have similar topological arrangements.
5. The first design step after developing an abstract of the space, is to create a geometric pattern (based on the grid).
6. The next design step integrates the functional requirements.
7. The elevation follows directly from application of an 'elevation grammar' to the plan.

Item 1 allows us to use a basic coding that is based on unit length horizontal and vertical lines. The next two items are aspects that we wish our system to pick up from the example designs, together with some of the topological arrangements from item 4. Items 5 and 6 describe the first steps in the design process, the plans used here represent in their level of detail approximately the ones produced after these steps.

The elevations are not considered in this work. As item 7 states, they are a result of the development of a floorplan. Learning features in elevations without having a given floorplan does therefore not seem sensible.

2.1.1. *Shape rules for Frank Lloyd Wright houses*

The work in this paper is based on the analysis of Frank Lloyd Wright's Prairie Houses by Koning and Eizenberg (1981). Based on the layouts of 11 prairie houses, Koning and Eizenberg develop a shape grammar that can be used to construct 10 of these houses, as well as many others that show a similar style. Their work is a typical example of the top-down analysis described above. The design using the rules is separated into 24 different steps. Roughly, the following phases can be distinguished: starting with the fireplace, a basic composition is created (18 rules). This composition is further elaborated by adding corners and porches, and detailing the interior layout (16 rules). More exterior details are added (22 rules), and the design is extended into the third dimension (12 rules). The roof is established (19 rules), together with some more details (4 rules). With the 8 rules to manipulate labels, 99 rules are necessary to create the ten different layouts.

The focus of this paper is the designs that are created by the first 34 rules: 2-dimensional layouts, with a developed basic layout, organized into function zones, and some detailing.

2.2. SEMANTICS

An important aspect of the designs we are looking at is the distinction between different function zones. The layouts have zones representing living space, service space and porches. Of central importance is the location of the fireplace. In the shape grammar used in Koning and Eizenberg (1981), the zones for service and living space are established around the fireplace with the first rules, and detailed at the end of the first 38 rules. At the same time, porches are added.

2.2.1. *Semantics in basic coding*

As described, both shapes and functional organization can be important aspects of a style of a set of designs. A system like the one described in Gero and Schnier (1995) that uses only four primitives to describe outlines (line, step, right turn, left turn) would therefore not be sufficient. To capture information about the functional organization, the evolving coding has to be able to integrate information about the semantics of the shapes. To do this, the basic coding has to be changed, so that semantic information can be attached to the outlines.

This is done by changing the set of primitives coded by the basic genes. The elements to change direction (left turn, right turn) remain unchanged. But instead of having only either a drawn line or a step ahead, the changed basic coding now includes a set of lines of different types, with the step ahead represented as a line of type 'invisible'. The number of types is not restricted, the number of types used depends on the application.

The line types in the the basic coding are interpreted to represent the different semantics or functions of the rooms. The fitness functions used in both steps of the evolution reflect this. In the first step, to score for a certain part of a design, any individual produced has to fit the design not only in line types, but also in shape. This also means that the number of line types in the basic coding can be higher than the number used in any specific example. Individuals exhibiting unused line types don't score any fitness, the evolving genes therefore do not incorporate any combinations of basic genes that produce these line types. In the second phase, the way the results are interpreted depends on the way the line types are used. If the line types specify the function of an enclosed room, the function of the room is defined by the line type that has the majority. If the line type encodes a detail in the outline, e.g. a certain wall type or a window, then the result can be used directly without further interpretation. Both functions can be mixed, as seen in the example used here.

The line types can be represented by different colours, the coding then has some similarities to colour grammars (see e.g. Knight, 1994). In colour grammars, however, the colours don't have any semantics attached, and colours can be mixed.

3. Learning evolved genes

3.1. GENETIC ENGINEERING

One foundation for evolving representations is genetic engineering. It is derived from genetic engineering notions related to human intervention in the genetics of natural organisms. If a group of similar organisms can be seperated into two sets distinguished by a difference in one particular attribute, then a comparison of the genetic codes of the organisms in the two groups can reveal what genes or gene groups are responsible for the difference. This knowledge can be used to modify that attribute, and introduce it into or eliminate it from organisms by manipulating its genetic material (see Gero and Schnier (1995) for a more detailled discussion).

A useful notion related to genetic engineering is the definition of 'genotype' and 'phenotype'. The genotype is the set of genetic instructions that make up the genetic code, while the phenotype is the structure that is produced as the result of the interpretation of the genotype (Langton, 1988).

3.2. EVOLVING REPRESENTATION

The starting point in the development of a system that creates and makes use of an evolving coding is a standard evolutionary system. The first step is to create a population of randomly created individuals. The coding of these individuals, the 'basic' genes, is chosen to be very low-level, putting as little domain knowledge into the coding as possible, and making sure not to exclude any interesting part of the search space. The individuals are then subjected to the standard evolutionary cycle of replication with errors and survival of the fittest. But at the same time, an additional operation screens the population, identifying particularly successful combinations of genes. For every such gene combination, a new, 'evolved' gene is created that represents this combination, and is introduced into the population. Figure 2 shows pseudo-code describing the algorithm. More detailed explanations to some of the steps can be found in the following sections.

```
begin
    create-population
    while not end-criterion
            select 2 parents
            create offspring
            if offspring not already in population
               then add offspring to population
                      register offspring for shared fitness fi
            if n new individuals produced
               then select best gene combination
                      create evolved gene
                      add evolved gene
                      replace all occurences of gene combination with evolved gene fi
            if m new individuals produced
               then recalculate weights
                      recalculate all fitnesses fi
    end
end
```

Figure 2. Pseudocode for an evolutionary system used to produce an evolved representation.

Since the goal during the development of a representation is variety and not optimization, all offspring created in the variation function are kept if they:

− are not empty, i.e. they draw at least one segment.
− match (as described below) the design case at least at one position.
− no other individual already in the population has a genotype that codes for the same phenotype as the new individual. If such an individual exists, then the individual with the shorter genotype is kept. The use of evolved genes

is hereby encouraged, since evolved genes usually lead to shorter genotypes. This is the only instance where an individual in the population can be replaced.

As a result, there is no step in the pseudo-code where individuals are deleted.

In the first few cycles, the evolved genes will be composed from basic genes, but in later cycles most evolved genes will represent combinations of other, lower-level evolved genes, or combinations of those with basic genes. This growing hierarchy of representations gives rise to a more and more complex and abstract coding, which is increasingly adapted to the application. In other words, the process gradually collects application-specific knowledge and codes it into the representation, rather than being coded into it by the user in the first place.

What does this mean in terms of search space? The length of the genotype is only restricted by the size of the computer memory. The search space is extremely large with respect to the number of states that can be evaluated in a limited computation time, and can therefore be seen as having infinite size. However, since the alphabet used for genotypes is finite at any state during the evolution, the set of possible designs that can be defined by a genotype of a certain length is limited. The search space can therefore be illustrated by a number of concentric circles, each defining the space of designs that can be defined by a genotype of a certain length. The inner circle contains the genotypes of length one, i.e. the basic building blocks. The further away a design (or part of design) is from the centre, the more difficult it is to find by means of generate and search. Every time an evolved gene is created, the structure of the search space is changed. The state of the new gene in the search space is moved into the centre, all design states in the next circle that can be derived from that state are moved into the second circle, and so on. Figure 3 illustrates this: the original search space is illustrated in Figure 3(a), with the four basic building blocks in the centre. The building blocks code for vectors of one unit length with the directions up, down, left and right. The arrow points to the starting point of the next element drawn (if any). The second circle shows all designs that can be derived from genes of length two (i.e., using two building blocks). The other circles give some examples of designs using genotypes of length three, four and five. If now the two closed shapes in the fourth circle are identified as particularly successful and an evolved gene is introduced for them, then the search space changes as shown in Figure 3(b). The squares now become basic building blocks, and the shapes on the fifth circle that are derived from the squares, can now be found in the second circle. The more evolved genes a design state involves, the more it is moved towards the centre. For example, the shape with the four squares that is now on the fifth circle (i.e. can be constructed from genotypes of length five) would have been on the fourteenth circle before[1].

---

[1] fourteen lines, the shape cannot be drawn without drawing two lines twice

Since the introduction of a new gene increases the size of the alphabet of the coding, the size of the circles also grows.

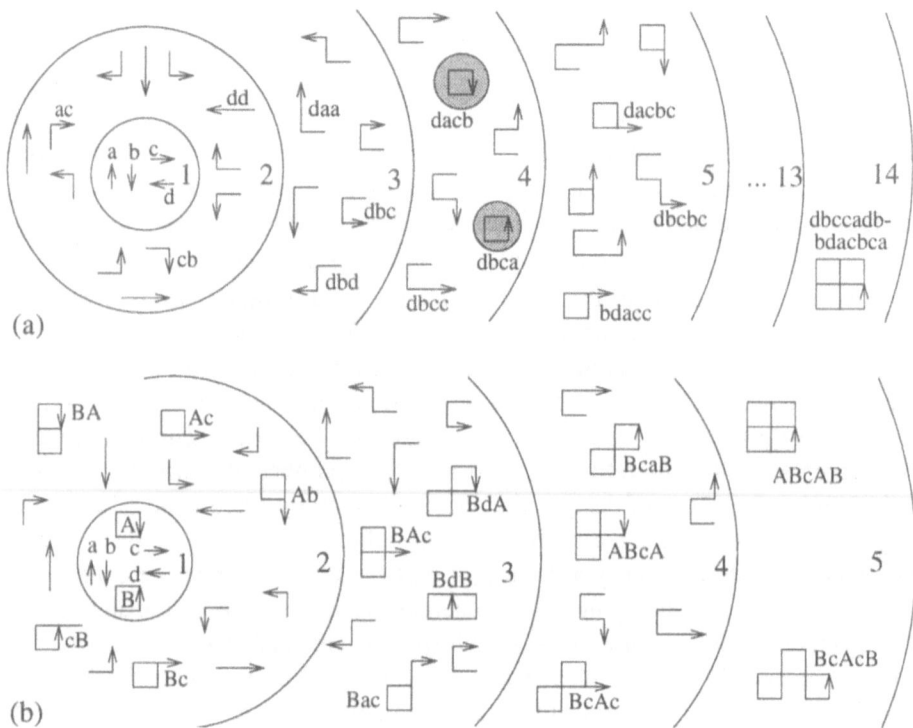

Figure 3. Example of an evolving representation: (a) original representation and (b) representation with evolved genes. Some of the corresponding genotypes are give, capital letters denote evolved genes. The transformation from phenotype to genotype is not always unique, e.g. the genotypes 'ABc' and 'BAc' produce the same phenotype. Arc segments indicate that only part of the space is shown.

The introduction of evolved genes obviously changes the probability that a gene sequence maps onto a useful feature. The number of different genes that can be used in a genotype expands, but at the same time the length of the genotype that is necessary to describe a feature shrinks, effectively reducing the size of the search space. For example, the floorplan shown in Figure 10 (a) was produced by a genotype of length eight, using two basic and 326 evolved genes. Expressed in basic genes only, the genotype has a length of 445. The space of designs that can be coded by genes of length 445 using two basic genes is about 110 orders of magnitude larger than the space of genes of length eight using 328 genes.

4. Learning genetic representations of floorplans

4.1. EXAMPLE FLOORPLANS

In the examples of the Frank Lloyd Wright prairie houses, lines that enclose living spaces use a different line type from lines that enclose service spaces or porches, and the fire place has a type of its own. Since only the main floor is considered here, no bedroom zones occur in the designs.

From the eleven Prairie Houses analyzed by Koning and Eizenberg (1981), four have been selected as examples for the evolving coding: the Henderson house, the Thomas house, the Martin house and the Baker house (Koning and Eizenberg, 1981). Since the basic coding only allows horizontal or vertical lines, the diagonal lines at the wings of the Henderson house have been changed into a stepped shape. Figure 4 shows the floorplans used. As a comparison, Figure 5 shows the plans of the Thomas house as given in Koning and Eizenberg (1981).

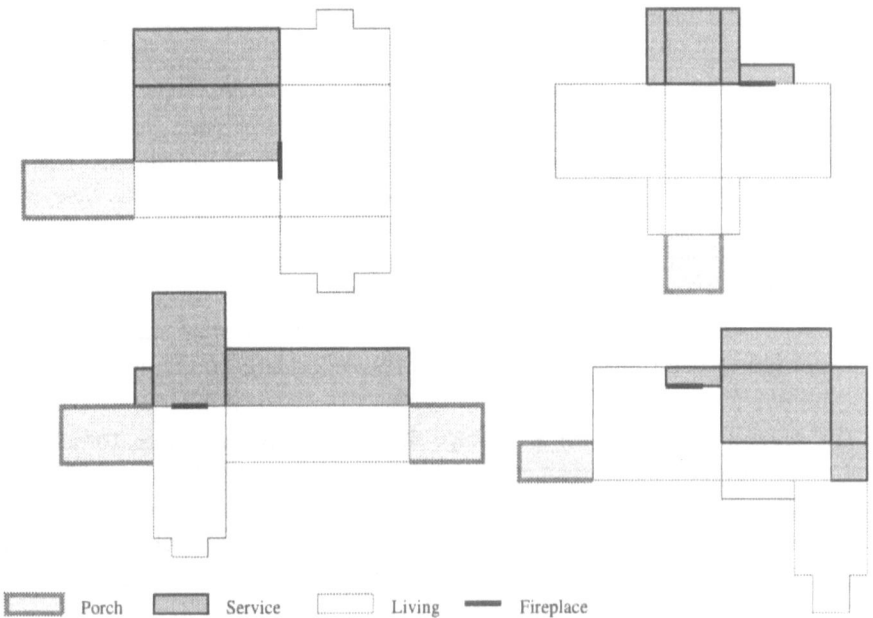

Figure 4. Frank Lloyd Wright Houses used to create the evolved coding: Henderson house (top left), Martin house (top right), Baker house (bottom left), Thomas house (bottom right).

4.2. BASIC CODING

The basic coding has to allow for lines with a variable number of line types. The coding presented in Gero (1994) used four basic genes, each coding for a different basic element: a left turn, a right turn, a line ahead, and a step ahead. One possibility for including line types is to increase the number of different basic genes,

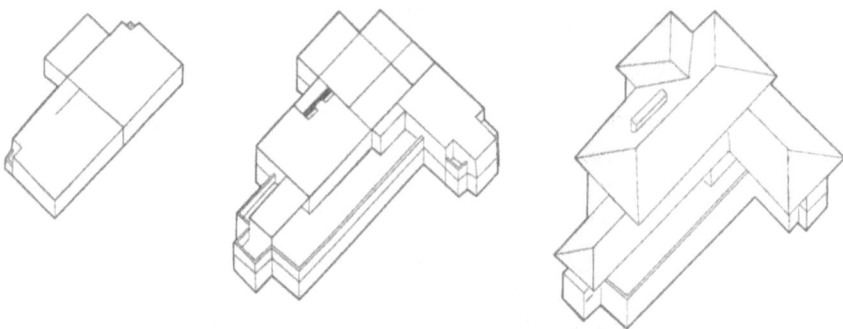

Figure 5. Thomas house defocused and reduced to four function zones, bedroom level, main floor level, and external form (Koning & Eizenberg 1981).

so that one basic gene is used for every additional type. If the number of types increases, or if additional basic elements like diagonal lines are introduced, the number of different basic genes used grows.

Another possibility is to keep the number of different basic genes constant, and use two or more successive basic genes to code for the elements. The first basic gene on a genotype would select the type of primitve used, in this case either turn or line. The following one or more basic genes then code for an attribute value. If the first gene coded for a turn, the attribute represents either of the two values 'left' or 'right'. For lines, the attribute defines the type of the line, steps ahead are treated as a line of type 'invisible'. The number of successive basic genes needed to code for the line types varies depending on the number of line types and on the number of basic genes used.

The second coding has the advantage that it is easily extensible, for example to introduce curved lines, only a new type of primitives would be added. This type could have one or more additional attributes. Similarly, for diagonal lines, one could change the number of values the turn can represent to eight. At the same time the coding remains very simple, this is one of the goals in the design of the basic coding.

A major difference from the first coding is that the meaning of basic genes is not totally position independent anymore.

In the example presented here, the second approach was chosen. Two basic genes (values 0 and 1) are used, the basic coding (without evolved genes) is therefore a binary coding. The first basic gene selects the type of the primitve:

0 A line. The attribute can take five value (five line types, including the 'step ahead', requiring three basic genes. The eight possible values are taken modulo five, three line types are therefore produced by two different values.

1 A turn. The following basic genes distinguishes between left and rigth turn.

The basic coding is shown in Figure 6.

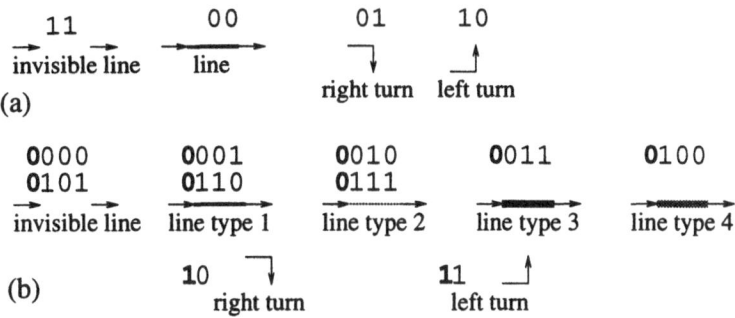

Figure 6. Basic coding, (a): original coding, (b) coding expanded to allow for different line types

4.3. CREATING EVOLVED GENES

To create evolved genes, the algorithm shown in Figure 2 is used. A special fitness function is used that measures what percentage of the examples is represented by an individual, while at the same time preventing convergence.

4.3.1. *Fitness Calculation*

The fitness calculation for a new individual during the evolution of the represent-ation stage involves the following steps.

1. Transform the genotype of the individual into a phenotype, i.e. line-drawing.
2. Find all positions where the phenotype 'matches' the design case. A match is declared if and only if for all line segments in the phenotype there is a corresponding line segment in the design case (but not necessarily the other way round).
3. At all matching positions, draw the phenotype as a partial drawing.
4. Create the sum of the weights associated with all line segments in the design case that are represented in the partial drawing (see below for a description how the weights are calculated).
5. This sum is the current fitness value for the individual. Whenever the weights for the segments in the design case are recalculated, the fitness values of all individuals in the population have to be updated.

As an example, Figure 7(a) shows a design case with associated weights, and the shapes produced by two different individuals. Both individuals can be applied at two different positions, resulting in fitnesses of 24 and 18.

This fitness alone would lead to a convergence of individuals that describe only some aspects of the design case. To prevent this, another analogy from evol-ution in nature is used. The different aspects of the design case are seen as a re-source (for example food) that has to be shared between all individuals using it.

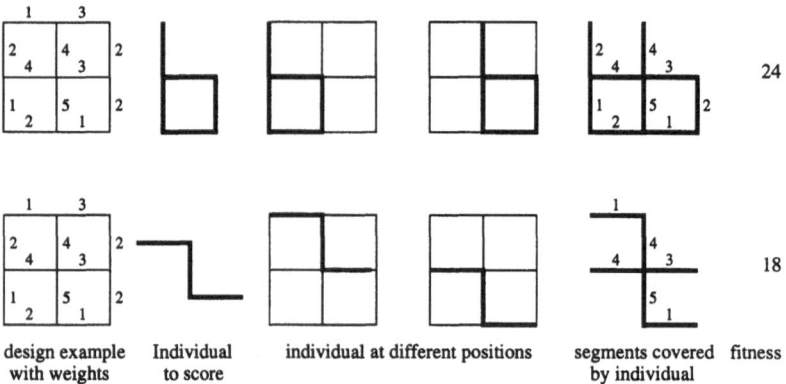

Figure 7. Example of fitness calculation for two different individuals.

Individuals therefore get high rewards if they describe aspects of the case that are covered only by few other individuals (evolutionary niches), and only little additional fitness for aspects that are described by many individuals.

To create the 'niching' effect, every line segment in the design case has a weight associated with it (the values in Figure 7). This weight is calculated regularly as a fixed value divided by the number of individuals in the population that can be used to 'stamp' that segment. If for example only two individuals code for a 'stamp' that can be used for a certain part in the design case, both get 50% of the constant value as fitness for that part. If 20 individuals do so, each one gets only a fitness of 5%. This effectively prevents convergence towards only some features in the design case (e.g. only horizontal lines). The effect can be seen in Figure 8: without sharing, the evolving genes develop mainly in a very small region, fitness sharing leads to a much better distribution of evolved individuals.

4.3.2. *Results*

Figure 9 shows examples of evolved genes created from the four example designs. Shown are some of the last evolved genes created from the examples. Clearly visible are the shapes of rooms, and the different line types, associated with the different functions. Two of the evolved genes shown (310 and 318) have the fireplace as part of the line-drawing they code for (in this case from the Henderson house).

4.4. CREATING NEW FLOORPLANS USING THE EVOLVED REPRESENTATION

In the second phase, the representation evolved from the example cases is used to create new designs that show similarities in style to the example cases. For this, a standard evolutionary system is used, with the set of basic and evolved genes used in the coding.

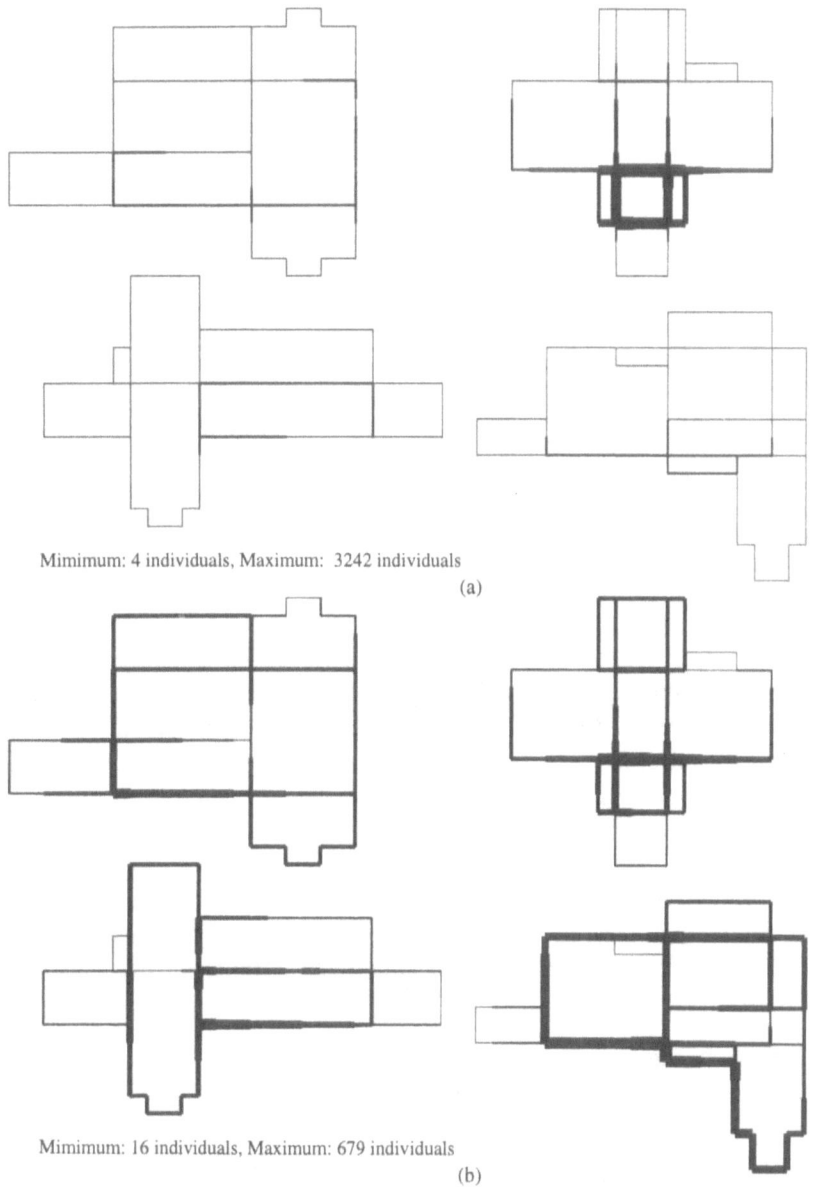

Mimimum: 4 individuals, Maximum: 3242 individuals

(a)

Mimimum: 16 individuals, Maximum: 679 individuals

(b)

Figure 8. Distribution of individuals in population, (a) without fitness sharing, (b) with fitness sharing. Thicker lines represent more individuals covering that segment.

4.4.1. *Fitnesses*

The evolved coding, as exemplified in Figure 9, captures information about shape and function of parts of the example designs. However, the way these parts are assembled to create new designs is only influenced by the fitness function that

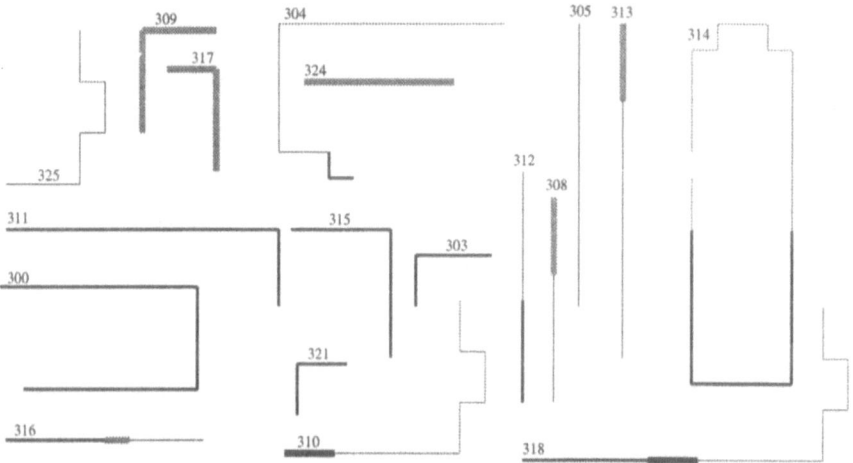

Figure 9. Examples of evolved genes created from the example designs shown in Figure 4. The labels refer to the numerical sequence in which the genes evolved.

evaluates new designs. This means that any gene that codes for a room of a certain type, for example, has no influence on what other room is next to it; and there is nothing preventing a design from using two evolved genes that each include a fireplace. As a result, topological constraints are not automatically satisfied by using the evolved coding. If designs created by the evolutionary system are to fulfil topological constraints, they have to be included in the fitness function (see Section 5 for how we plan to integrate more topology information into the evolved coding).

Frank Lloyd Wright's prairie houses follow a number of topological constraints, and they all have to be made part of the fitness. For the results presented here, fourteen different aspects influence the fitness. The following list shows the fitnesses used:

- One porch, size between 9 and 12 units
- porch connected to living area, and not connected to service area
- two to four rooms in the service area, total size between 45 and 60 units
- two to four rooms in the living area, total size between 55 and 70 units
- only one service and one living area, i.e. all rooms of that type are connected
- one fireplace, two units length, between living and service area
- no 'dead ends', i.e. lines that do not enclose any room.

4.4.2. *Pareto optimization*

For a human designer, it is relatively easy to find designs that fulfill all the fitness conditions. For a standard evolutionary system, the fitnesses have to be integrated into one fitness. This could for example be done by calculating a value between

0 and 1 for every individual fitness, and adding or multiplying them into a single fitness value. Unfortunately, by integrating all fitnesses into one value, the information about what fitness conditions are fulfilled and what conditions are not is lost to the system. As a result, the system ends up converging towards a population that is good in some respects while individuals good in different aspects are lost. Even after a very high number of individuals have been produced, the system is not able to find satisfying solutions.

A better way to handle a high number of individuals is therefore to utilize 'Pareto optimization' (see for example Radford and Gero, 1988). In a Pareto optimization process, only a partial ranking between two individuals can be established. If two individuals are compared, one individual is better than the other (dominates it) only if it is better or equal in all fitness criteria and better in at least one criterion. The comparison therefore often ends up in a draw. To select individuals that are used to produce offspring in the genetic operations, two individuals are picked randomly from the population and compared to a randomly picked reference subset (10% of the population). If one of the individuals is dominated by one of the reference individuals while the other is not, the second individual is selected as the parent. Otherwise, neither of the individuals is preferred.

This selection alone is not sufficient to prevent all individuals clustering as a small subset of possible, good solutions. As an additional measure to prevent convergence, 'niching' is used (Horn and Nafpliotis, 1993). Here, candidate individuals are compared with a number of other individuals in the population. For every individual, the distance between the fitness values is calculated. The number of individuals with a distance smaller than a threshold value is called the niche-count. In niching Pareto optimization, in order to select between two individuals that either both dominate the reference set or are both dominated by at least one individual in the reference set, the individual is chosen that has a smaller niche-count.

If a newly created individual dominates another individual in the population, it replaces it. If not, and the new individual is dominated by at least one other individual in the population, it is rejected. The third possibility is that the new individual populates a new part of the Pareto optimal front, and is therefore neither dominated nor dominates another individual. In this case, the individual has to be added to the population without replacing another individual, leading to a growing population. As an example, in one of the runs presented below, the population grew from 500 to 1581 individuals.

4.4.3. *Results*

Two runs where done using a set of 326 individuals created from the floorplans in Figure 4.

Run 1 ran for 60.677 loops, each loop creating two offspring individuals. The initial population was 1.000 individuals, the final population consisted of 1.652 individuals. From some 120.000 produced and tested individuals, 14.359 individuals where good enough to be introduced into the population.

Run 2 ran for 99.207 loops, the population grew from 500 to 1.581 individuals. Of the nearly 200.000 individuals produced, 12.502 made it into the population. Again, all 326 evolved genes where used.

The first result from Run 1 (Figure 10(a)) has a perfect fitness. The fitness function does not check if the fireplaces are straight, therefore a corner-fireplace could result. Since none of the floorplans in the example drawing have a corner fireplace, this feature cannot have been part of the evolved coding. It therefore must be coded in basic genes. The second-best result from Run 1 (Figure 10(b)) has a penalty due to one segment of 'dead end' close to the fireplace, but fulfills all other fitness criteria.

Both results of Run 2 (Figure 10(c) and (d)) have perfect fitnesses. Again, the system has taken advantage of a small weakness in the fitness function, that allowed it to put the porch inside the living space.

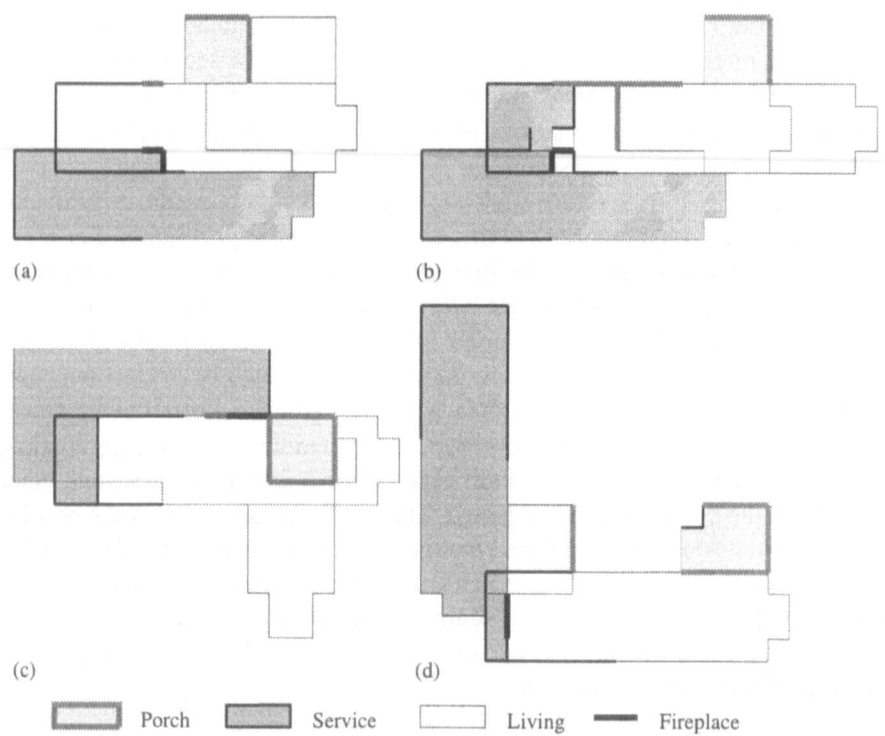

Figure 10. Floorplans created using the evolved genes from the example designs shown in Figure 4, (a) and (b) initial population 1000 individuals, (c) and (d) initial population 500 individuals.

Figure 11 shows how one of the results (the second of Run 1, see Figure 10(b)) can be extended into three dimensions by a graphic artist. The resulting house is obviously quite similar to the Thomas house (see Figure 5).

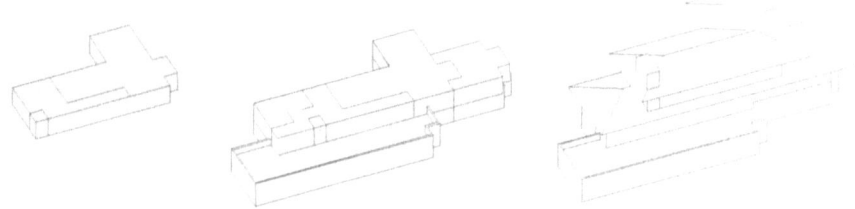

Figure 11.   Floorplan from Figure 11 (b) manually extended into three dimensions; shown are bedroom level, main floor level, and roof view.

5. Discussion

5.1. FRANK LLOYD WRIGHT PRAIRIE HOUSE PLANS

As described above, no aspects of the topology are coded in the evolved coding. This results in the fact that some aspects of the example design that the system could have learned have to be added as fitness. It also shows in the results: shapes that have been outer walls in the original drawings are used in the inside (e.g. the stepped line separating the right two parts of the living room in Figure 10(b), or the outer walls of the porch in Figure 10(a)).

5.1.1. *Possible improvement: more line types*
One way to improve the 'knowledge-content' of the evolved coding is to use a higher number of line types. Different line types could be used depending on the functions of both of the rooms a wall separates, and again different line types for outer walls. This way, knowledge that for example three out of four sides of the porch are outer walls, and the fourth wall is a wall to a living space, could be integrated into the evolved coding.

An example of the Thomas house drawn with this enhanced coding is shown in Figure 12.

To realize this, no changes other than changing the parameter for the number of line types used and modifying the example drawings are required in the first step. In the second step, the fitness function would have to be added that checks if lines are used in a correct context.

A system using this coding would avoid problems like the two problems with the design results in Figure 10. It would also reduce the number of fitness criteria required.

5.2. LEARNING REPRESENTATIONS

What has been successfully presented has been both the concepts and a demonstration example of the evolutionary learning of a genetic representation of a set

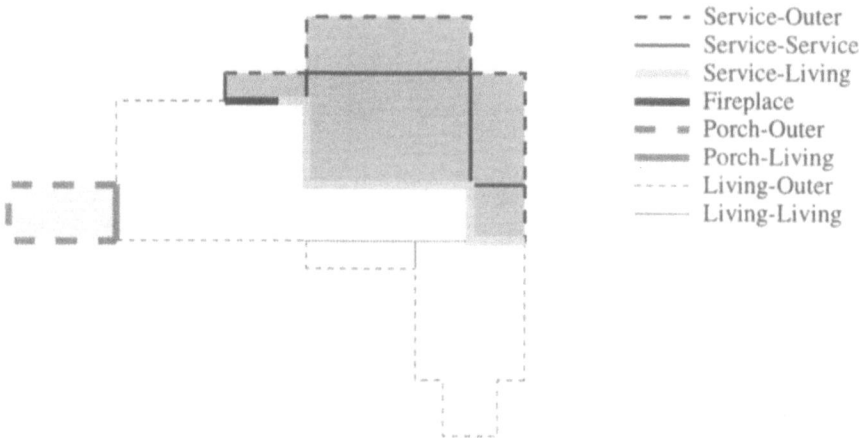

Figure 12. Thomas house drawn as design example with increased number of line types.

of building layouts. This representation can then be used to generate layouts with a similar style. The approach is based on the use of genetic engineering concepts to evolve not just the solution to a problem but also to allow the genes which are used to represent it also to evolve. This results in the evolution of complex genes, genes which are capable of forming increasingly large and complex parts of the phenotype or design. Evolutionary systems commence with a knowledge-lean representation that often contains little or no domain knowledge. What is happening here is that the evolved genes increasingly contain knowledge about the domain under consideration. They turn a knowledge-lean representation into a knowledge-rich representation. From an evolutionary viewpoint there are two distinct activities. In the first the genes are allowed to evolve with a fitness associated with the designs which act as exemplars. Then these evolved genes are used to generate designs with a completely different set of fitnesses. The resulting designs embody the knowledge which has been encoded in the evolved representation. This opens up possibilities in case-based design as well as an alternative approach to the generation of design grammars.

Acknowledgements

This work is supported by a grant from the Australian Research Council and by a University of Sydney Postgraduate Research Award.

References

Buelinckx, H.: 1993, Wren's language of city search designs: a formal generative classification, *Environment and Planning B* **20**, 645–676.

Chan, C.-S.: 1992, Exploring individual style through Wright's designs, *Journal of Architectural and Planning Research* **9**(3), 207–238.

Chan, C.-S.: 1995, A cognitive theory of style, *Environment and Planning B: Planning and Design* **22**, 461–47.

Chase, S. C.: 1989, Shapes and shape grammars: from mathematical model to computer implementation, *Environment and Planning B* **16**, 215–242.

Gero, J. S.: 1994, Towards a model of exploration in computer-aided design, *in* J. S. Gero and E.Tyugu (eds), *Formal Design Methods for CAD*, North-Holland, Amsterdam, pp. 315–336.

Gero, J. S. and Schnier, T.: 1995, Evolving representations of design cases and their use in creative design, *in* J. S. Gero, M. L. Maher and F. Sudweeks (eds), *Preprints Computational Models of Creative Design*, Key Centre of Design Computing, University of Sydney, Sydney, Australia, pp. 343–368.

Horn, J. and Nafpliotis, N.: 1993, Multiobjective optimization using the niched pareto genetic algorithm, *Technical Report 93005*, Illinois Genetic Algorithms Laboratory (IlliGAL), University of Illinois at Urbana-Champaign.

Knight, T.: 1989, Transformations of De Stijl art: the paintings of Georges Vantongerloo and Fritz Glarner, *Environment and Planning B* **16**, 51–98.

Knight, T.: 1990, Mughul gardens revisited, *Environment and Planning B* **17**, 73–84.

Knight, T.: 1994, Shape grammars and color grammars in design, *Environment and Planning B: Planning and Design* **21**, 705–735.

Knight, T. W.: 1981, The forty-one steps, *Environment and Planning B* **8**, 97–114.

Koning, H. and Eizenberg, J.: 1981, The languages of the prairie: Frank Loyd Wright's prairie houses, *Environment and Planning B* **8**, 295–323.

Langton, C. G.: 1988, Artificial life, *in* C. G. Langton (ed.), *Artificial Life*, Vol. VI of *SFI Studies in the Sciences of Complexity*, Addison-Wesley, Reading, pp. 1–47.

Mackenzie, C. A.: 1989, Inferring relational design grammars, *Environment and Planning B* **16**, 253–287.

Radford, A. D. and Gero, J. S.: 1988, *Design by Optimization in Architecture and Building*, Van Nostrand Reinhold, New York.

Stiny, G.: 1980, Introduction to shape and shape grammars, *Environment and Planning B: Planning and Design* **7**, 343–351.

Stiny, G. and Mitchell, W. J.: 1978, The Palladian grammar, *Envirnoment and Planning B: Planning and Design* **5**, 5–18.

Stiny, G. and Mitchell, W. J.: 1980, The grammar of paradise: on the generation of Mughul gardens, *Envirnoment and Planning B: Planning and Design* **7**, 209–226.

2

design objects

J. S. Gero and F. Sudweeks (eds), Artificial Intelligence in Design '96, 61-75.
© 1996 *Kluwer Academic Publishers.*

REACTIVE DESIGN AGENTS IN SOLID MODELLING

BRUNO FEIJÓ[1,2], NICK LEHTOLA[1]
[1]Intelligent CAD Laboratory, Dept of Computing, PUC-Rio
Rua Marquês de São Vicente, 225, 22453-900, Rio de Janeiro,
RJ, Brasil
[2]Concurrent Engineering Laboratory, UERJ
Rua São Francisco Xavier, 524 - Rio de Janeiro, RJ, Brasil

JOAO BENTO
Department of Civil Engineering, Instituto Superior Técnico
Av. Rovisco Pais, 1096 Lisboa Codex, Portugal

AND

SERGIO SCHEER
Center of Studies in Civil Engineering, UFPR
Curitiba, Paraná, Brasil

Abstract. This paper proposes a reactive agent architecture for the integration of solid modelling processes into more general design processes. The basic idea is to focus on reactivity rather than on symbolic representations of design knowledge. Also an Application Programming Interface is proposed to help developers writing intelligent CAD systems with links to any open architecture geometric modeller. In the proposed approach, solid modelling processes are formally immersed in the design process with the concept of modified CSG trees. Furthermore, solids are considered to be reactive design agents. A working system is also presented.

1. Introduction

In the 1980s, the research in solid modelling focused on geometric and topological problems of isolated objects. Nowadays, the focus is on complete engineering models and a number of concepts has been emerged to support a new generation of CAD systems such as *product modelling, feature-based modelling, tolerance modelling, constraint modelling, variational geometry, geometric reasoning* and *parametric methods* (Wozny et al., 1990). In the CAD market, developers have been concentrated on feature-based parametric modelling and most of them have been using the paradigm of object-oriented architectures and the techniques of knowledge representation

(Haase, 1992). In the design research arena, there are promising proposals for constraint modelling based on grammar formalisms (Brown et al., 1994). However, despite all those advances in the area of CAD systems there is still a gap between design research and solid modelling. This paper explores some possibilities to narrow this lacuna with a broader view of objects and modelling processes. Firstly solid modelling processes are formally immersed in the design problem space. Secondly, solids are considered reactive agents whose intelligence emerges mainly from the interactions with other agents. Thirdly, an Application Programming Interface is proposed to link the agent environment with generic solid modelling systems.

2. Design Problem Space

Design was first identified with problem solving in Simon (1969). According to his approach, a state space represents all possible states of the problem (i.e.: all possible problem descriptions) that need to be considered when a solution is attempted. Besides, he claims that it is practically computable to cover all the space.

Design as problem solving is a search process within a state space. In the context of traditional search, design knowledge is to be expressed in terms of goals and operators. Such a plain concept of searching does not directly address the characteristics of the design problem (Maher, 1990). The difficulties in this case are related to the variations of goals during the problem solving process and the problem of predetermining the relevant operators. Therefore, the notion of design as problem solving needs to be presented in a broader sense. For instance, the proposal of *design as exploration* by Smithers and Troxell (1990) can be understood as a meta-search process within the design problem space. This approach opens a promising research area to restore the concept of meta-planning for design perceived by Simon (1969) more than two decades ago. Generally speaking, the implementation methods for design as problem solving do not need to use the classical binomial goal-operator or even traditional planning techniques. In this case, the minimum requirement is to conform to the general principles of improving upon blind trial-and-error search, that is: (1) the progress principle (i.e. the ability to detect when progress is being made); (2) goals and subgoals (i.e. the decomposition procedure that reduces the problem space); and (3) the use of knowledge (i.e. if one knows how to solve a problem, then one can avoid search entirely) (Minsky, 1988). Most of the implementation methods in intelligent CAD systems are strongly based on a symbolic representation of the design world. In the sake of efficiency, the authors do not entirely support this approach and believe that a trade-off between procedural methods and symbolic representation can be achieved. One possible way of accomplishing this trade-off is through the use of hybrid agent architectures.

3. Agents

Agent technology has being applied in distributed AI (Bond and Gasser, 1988), in groupware (Baecker, 1993), in virtual environments (Bates et al, 1992) and in robotics (Brooks, 1990). Also agent-oriented programming has been proposed as a post-object paradigm (Shoham, 1993). Agent theory is not mature yet and leads to several definitions of agents and their properties. A complete survey on agent theories, architectures and languages can be found elsewhere (Wooldridge and Jennings, 1994).

In this work, the authors adopt the definition of agents as active objects described by the intentional stance. Indeed the notion of agency is bound to that of action. Therefore, agents are active objects, because they originate actions that affect their environment. To ascribe the intentional stance to agents means that they possess beliefs and desires.

Intention can be formally defined in terms of non-classical logic, such as the multi-modal logic proposed by Cohen and Levesque (1990) for their rational agents. However, this is not the scope of the present work.

There are three approaches to build agent-based computer systems: *deliberative, reactive* and *hybrid* architectures. The deliberative architecture is based on the classical symbolic AI paradigm. Examples of this approach can be found in Wood (1993) and Vere and Bickmore (1990). In this case, the symbolic model of the world is explicitly represented and the agents act via explicit logical reasoning. Usually, in this approach, an AI Planning system is the central component of the agent. This architecture, however, has several drawbacks: (1) the frame problem renders the knowledge difficult to represent in practice; (2) it is computationally inefficient; (3) it cannot cope with unpredictable events such as the actions of other agents; (4) it always requires that plans be too detailed, although one generally acknowledges that no system could produce completely detailed plans in domains of realistic complexity (Agre and Chapman, 1989). This scenario gets even worst if one thinks about meta-planning for design as problem solving. Therefore, alternative approaches to agent architecture have been proposed.

The reactive architecture is an alternative approach that breaks with the traditional symbolic AI paradigm. This sort of architecture is strongly advocated by Rodney Brooks who claims that intelligence can emerge without having explicit manipulable internal representation or explicit reasoning systems (Brooks, 1991). This architecture is based on reactive agents that must respond dynamically to changes in their environment.

The hybrid architecture attempts to harmonize the classical architecture with the reactive approach (Arkin, 1990; Georgeff, Lansky and Schoppers,

1987). The authors support the idea of a hybrid agent architecture for solid modelling environments, although only the reactive side is presented in this paper. Also the authors believe that a special agent language for solid modelling should be developed. However, this is an issue not yet fully investigated by the authors.

4. The Agency Principles

The principle underlying the reactive agent architecture proposed in this work are after Brooks (1991), that is: emergence and situatedness. The principle of emergence states that the intelligence of the agent system emerges from the interaction of agents among themselves and with their environment (Steels, 1990). As pointed by Brooks (1991, p.16): "It is hard to identify the seat of intelligence within any system, as intelligence is produced by the interactions of many components. Intelligence can only be determined by the total behavior of the system and how that behavior appears in relation to the environment. The key idea of emergence is: *Intelligence is in the eye of the observer*". This principle can also be identified in the work by Marvin Minsky (1988) where he proposes that intelligence emerges from a society of mindless agents.

Situatedness is also an idea proposed by Brooks (1991) who claims that the agent's intelligence is situated in the world and not in any formal model of the world built in the agent. Therefore, an agent uses its perception of the world rather than deductions based upon a symbolic representation of this world (such as those found in theorem provers or expert systems). This is a dramatic change from traditional AI paradigm and it is not fully investigated in the present paper. However, the authors believe that maintaining the traditional AI means that design agents will always have access to direct and perfect perceptions/actions and, consequently, no external world will really exist with its surprises, creative moments and ongoing design history.

5. Design Agents, States and Design History

From the design point of view, goals can be decomposed in terms of **form** (mainly physical attributes) or **function** (functional specifications describing the functions to be performed by the form). The question of decomposition (form *vs* function) may pose a dilemma for goal decomposition. However, as discussed by Maher (1990), in a problem solving approach to design this situation does not occur because representations of goals may capture both the notions of form and function. In this paper, the authors propose to represent goals by means of reactive design agents as a consequence of the previous investigation in design process models found in Feijó and Bento, (1991), Bento (1992), Scheer (1993), Prates (1993). The following definitions hold:

Def. **design agent** is an object *ai(I,F,f)* where *I* stands for identification attributes, *F* stands for a set of *form attributes* and *f* stands for a set of *function attributes*.

Def. A **state** *Tj* is a set of design agents *ai*.

Def. A **design history** H is a sequence of states *T0, T1, ..., Tn* where *T0* is the input specification and *Tn* is the artifact description.

Def. The **design problem space** *DPS* is the set of all possible design histories.

In this context, design is an evolutionary process that starts with a set of input specifications, *T0*, generates a kernel idea in the early stages of the process and refines it (by decomposition, generation or transformation) towards the artifact description *Tn*. Relationships between design agents and constraints within or across design agents may be defined both in terms of *F* and *f*.

6. Design Views

The authors of this paper claim that the solid modelling process can be viewed as a semantic tree called **modified CSG tree**. A classical CSG tree can be viewed in the example of Figure 1a. The proposed modifications consider the following changes: (1) the representation is hybrid (CSG/Brep); (2) one-place nodes are allowed for local operations (e.g. face extrusion); (3) both local and global operations are implemented based on the same set of operators (e.g. Euler operators); (4) primitive solids are instances of Brep (Boundary Representation); (5) the binary tree is exhibited in reverse order; (6) each modelling state has one of the following relationships with other solids: *part-of*, *is-a*, *trans-of* and *term-of*; (7) general nodes are allowed (that is, other than topological ones). Figure 1b shows a modified CSG tree.

The relation *trans-of* is associated to local operations found in Brep and the relation *term-of* is associated to terms of a global operations (such as the boolean difference between two solids).

The idea of the modified CSG tree is to permit the integration of solid modelling process into more general design process models. The nodes of the tree can be viewed as design agents with specific intentions. Furthermore, the authors propose the concept of *design views* where at least two orthogonal views co-exist when one is carrying out a solid modelling process. The first one is the **geometric view** in which lies the modified CSG tree and the second one is the **construction view** from where more general design agents drive the modelling view. Figure 2 illustrates those two views. Other possible views represent specific relationships such as *part-of* and *is-a*.

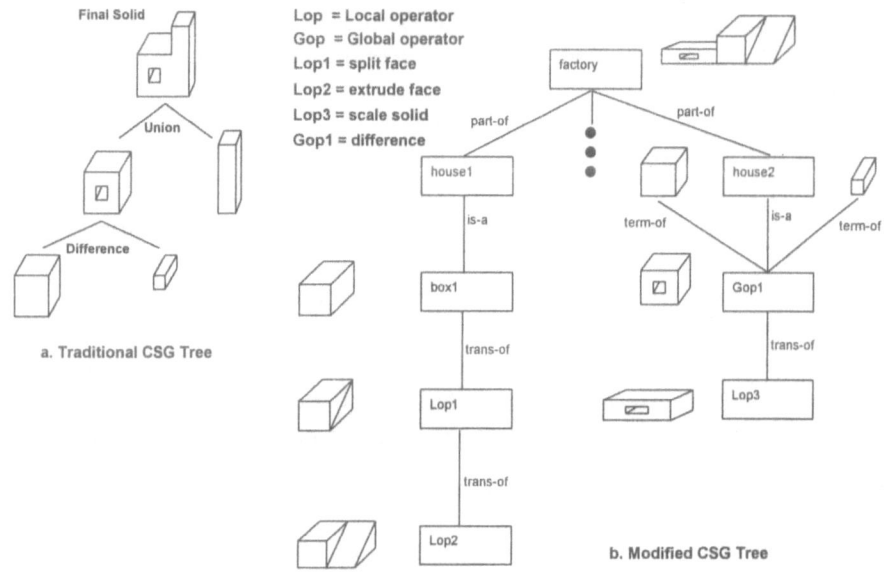

Figure 1. Traditional CSG tree and modified CSG tree.

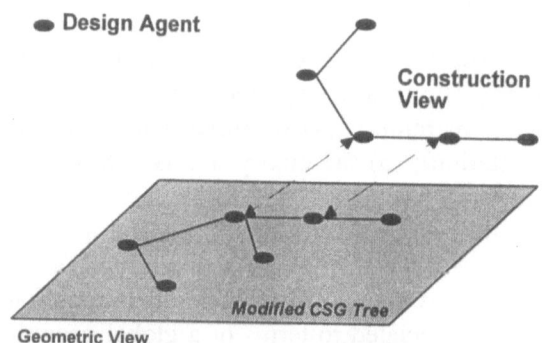

Figure 2. Design views.

In a system with those two views, a design agent can inherit properties from both the geometric view and the construction view. This integration allows a designer to specify a change in the radius of a cylinder either from the construction view environment or from a mouse movement in the 3D environment of the solid modeller. In any case, if the system is reactive, there will be a chain of dynamic modifications according to pre-established conditions or intentions.

7. The Object Paradigm

Design agents are based on the object paradigm. However, they do not commit themselves to any specific object-oriented language. For instance, inheritance of agent properties is automatically established when an agent is made child of another agent. Also inheritance is overridden by the explicit local inclusion of a specific method or attribute, with a null or non-null value. In this case, searching for a value or a method in the ancestry tree is prevented by the simple local occurrence of the attribute(s)/method(s) in question. Furthermore, inheritance is totally dynamic during execution time. Also design agents are mutable objects in the sense that their properties may change with the passing of time. Another characteristic of the proposed agents is that an object can be defined as an aggregation of parts which are themselves other objects. Objects formed in this manner are called, in the scope of this work, composite agents and the type of relationship amongst them are called *part-of*. The method for implementing composite agents is by defining agent names as the values of composite agents' attributes.

The ability of supporting mutable objects implies the propagation, or at least the communication, of changes occurring in the mutant object to those other objects that have references to it. This propagation of changes is implemented by procedures attached to attributes called *attached predicates*. Attached predicates are used for the purpose of triggering procedures on variable access, drawing on the style of *active values* and *access-oriented* programming techniques (Stefik, Bobrow and Khan, 1986) (Inference Corporation, 1985). This idea is also motivated by the new concept of active database systems (Abiteboul et al., 1995) (Widom and Ceri, 1995) (Picouet and Vianu, 1995). The understanding of the active objects context used by DBMS workers might help one situating the present work.

Active database systems provide "trigger systems" that execute actions in response to specified events according to rules in ECA form, that is: on <Event> if <Condition> then <Action>. These rules have three methods of firing: immediate (i.e. a rule is fired as soon as its event and condition becomes true); deferred (i.e. rule application is delayed until a specific state is reached); concurrent (i.e. a separate process is spawned for the rule action and executed concurrently with other processes). In relational active database systems, the action involves a sequence of insertions, deletions and modifications, and in object-oriented active systems it involves one or more method calls. From these definitions, one can notice that DBMS workers and AI workers share some fundamental concepts, although they use different sets of terminology. However, the context, problems and goals of those areas are substantially distinct. Moreover, the lack of extensive theoretical works in those areas make difficult to look for a common formal framework (e.g., see

Aiken et al. (1992), Beeri and Milo (1991), Hull and Jacobs (1991) and Picouet (1995) for the case of DBMS).

Only one kind of attached predicate is implemented in this work, called *if-changed*. This kind of predicate is a procedure that must be executed in the case of an attribute value being tentatively changed, before the assertion of the attribute value takes place. This predicate behaves like "watchdogs" of the attributes to which they are attached.

The attached predicate *if-changed* are used to implement internal reactivity within an agent. For instance, the attempt of changing the radius of a cylinder may cause a change in the value of its height if some relation is previously imposed on these attributes (say, height = 5 * radius), as illustrated in Fig. 3.

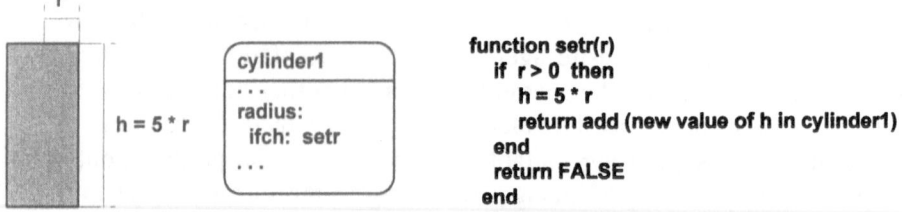

Figure 3. Internal reactivity.

However, more interesting cases of reactivity involves several agents. The rest of the paper considers this kind of reactivity.

8. A Reactive Design Agent Architecture

The authors propose the following taxonomy for the attributes of design agents:

identification attributes (*I*): [label] [description] [status]
form attributes (*F*)
    relationship attributes:[is-a] [children] [part-of] [link-to] [alternative] [version] [trans-of] [term-of]
        structure attributes: [physical] [geometric] [behavioural]
    function attributes (*f*): [intent] [functional specification] [performance specification]

In this taxonomy, *description* is a short note in text format or even in audio format; *status* is the current situation of an agent (alive, alternative or version); *trans-of* and *term-of* are used by the modified CSG tree mentioned above; structure attributes may be *physical* (e.g. color), *geometric* (e.g. radius), *behavioural* (e.g. temperature = 35 C, obtained from a thermal analysis); *intent* describes the designer's intention. In the evolutionary design process *functional specifications* (e.g. pleasant temperature) tend to be transformed into *performance specifications* (e.g. 18 temperature 25).

The data structure for design agents used in this work follows the taxonomy above, that is:

agent { label; description; parents; children; part_of_list; link_list; attribute_list };
attribute { label; datatype; attribute_type; ifch; putfn; value }.

Figure 4 illustrates the proposed agent architecture. In this architecture, **GeoObj** is part of the solid modeller application and **Hagent** is a hybrid agent representing a class of solids. **Hagent1** is, for example, an instance of **Hagent**. Each hybrid agent in the Geometric view has a counterpart in the Construction view and the integration between these views is made by binding them. This binding is established by associating addresses of objects in the Geometric view (void *) with labels in the Construction view (char *). In this case, a design agent **Hagenti** can inherit properties from both the geometric view and the construction view.

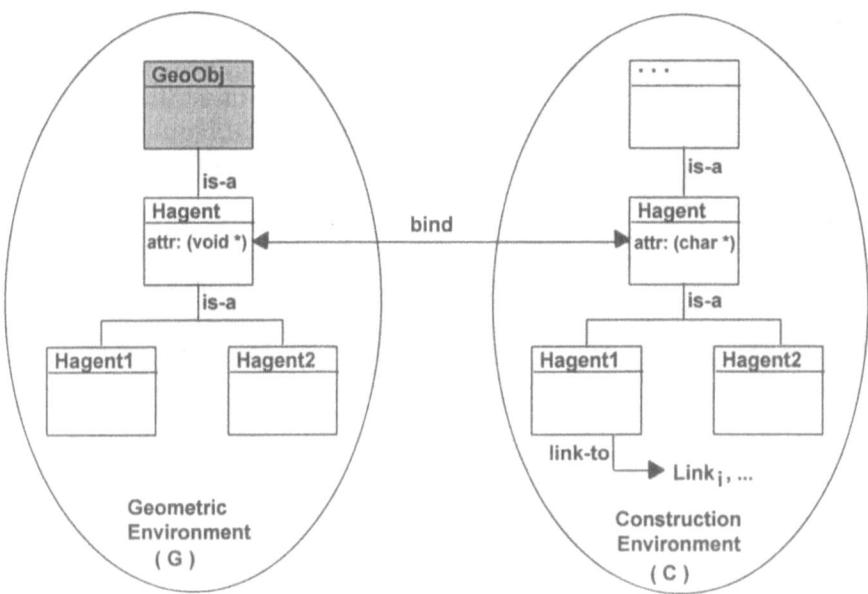

Figure 4. Proposed agent architecture.

Design agents are reactive in solid modelling through the relation *link-to*. The members *linki* of the list of links *link_list* of a design agent are twofold, that is:

$$linki = \{agentk; reactionk\}, \quad i = 1, n$$

where *n* is the number of linked agents and *reactionk* is the reaction of
agentk. For instance, if the agent *cylinder1* would be linked to the agents
box1 and *box2*, its *link_list* could be: {{*box1,reactbox1*}, {*box2,reactbox2*}}.
In this case, *reactionk* is defined as follows:

> *Def.* Given the attributes *attr* and *attrk* of the agents *agent* and *agentk*
> and a set of other attributes of these agents {*t1,...,tn*}, *reactionk*(*agent*,
> *agentk*) establishes that if the *condition attr* = **f**(*attrk, t1, ...*) is not
> satisfied, then the *intention* of *agent* should be imposed over *agentk*
> by the *inverse attrk* = **f̄o**(*attr, t1, ...*) followed by an *action* and the
> addition of new attributes values through the predicate *add_value*.
> *action* is optional and can be any procedure returning or not a new
> attribute value.

Figure 5 presents the pseudo-code version of a generic predicate *reactionk*
and an example for the case where the *radius* of *cylinder1* is intended to be
equal to 1/3 of the *height* of *box2* and an *action* is imposed in order to move
box2 along the axis X.

Before adding a new value to an attribute of an agent, the predicate
add_value executes every predicate *reactionj* found in the *link_list* of the
agent. This mechanism guarantees the full propagation of the changes.

In order to have reciprocity between two agents, one should define the
predicate *reaction* for both agents. In the example of Figure 5, one should
define another *reaction* with the condition *height* = 3.0 * *radius* and an
inverse *radius* = *height*/3.0.

```
function reactionk (agent, agentj)              function reactbox1 (cylinder1, box2)
    get attribute values                           get radius of cylinder1
    ...                                             get height of box2
    ...                                             get positionx of box2
    if condition is not satisfied then             if radius < > (height / 3.0)  then
        find the inverse of condition and get          height = 3.0 * radius
           new attribute value                         newpos = radius + positionx
        perform action                                 return add_value (height to box2) and
        return add_value (the new attribute values)           add_value (newpos to box2)
    ...                                            end
    end                                            return TRUE
    return TRUE                                 end
end
```

(a) General Template   (b) An Example

Figure 5. The function *reaction*

9. A Development Tool for Design Agents

In this paper, an abstraction layer is proposed for the development of CAD
systems based on an architecture of design agents. The integration of the
construction environment into the geometric one (and vice-versa) is done by

registering functions and manipulating them through their labels. This abstraction layer is valid for any solid modeller or geometric modeller that use an open architecture. Two in-house solid modellers were tested and a commercial one (ACIS, from Spatial Technology) is under investigation. This abstraction layer uses the Application Programming Interface (API) presented in Table 1.

TABLE I. Proposed API.

- Control functions
add_agent	(label)
add_attr	(label, attr, datatype)
add_if_ch	(label, attr, ifch)
add	(label, pred, def)
del_agent	(label)
del_attr	(label, attr)

- Information and modification functions
list_agent	( )
list_pred	( )
list_parent	(label)
list_children	(label)
list_parts	(label)
list_links	(label)
list_attr	(label)
has_pred	(pred)
has_attr	(label, attr)
has_attr_val	(label, attr)
has_if_nd	(label, attr)
has_if_ch	(label, attr)
has_parents	(label, prt)
has_part	(label, prt)
has_link	(label, lnk)
has_link_pred	(label, lnk)

- Relationship functions
is_a	(chd, prt)
part_of	(subpart, part)
link_to	(chd, prt, rprt, rchd)
del_is_a	(chd, prt)
del_part_of	(chd, prt)
del_link_to	(chd, prt, rprt, rchd)

- Assignment and retrieval functions
add_value	(label, attr, value)
put_value	(label, attr, value)
get_value	(label, attr)
jask_value	(label, attr)
ask_datatype	(label, attr)
ask_number	(value)
ask_agent	(label)

- Auxiliary functions
next_label	(list, pos)
next_agent	(pos)
next_attr	(obj, pos)

- Graphics interface functions
hist_forward	( )
hist_backward	( )
display_view	(view, clc)
display_curr_state	(view, clc)
find_mM	(view, xm, xM, ym, yM)
pick	(view, x, y)
get_node_level	(view, label)
get_node_depth	(view, label)

- Solid modeller interface functions
bind_g_modeller	(char *appl, void (*allconst)(void), void(*allgeo)(void))
bind_g_agent	(char *Hcagent, void *Hgagent)
bind_g_attr	(char *Hcagent, char *attr, char *datatype, char *putfn)
unbind_g_modeller	(char *appl)
register_g_action	(char *name, void *act)
register_c_action	(char *name, void *act)
do_action	( )
set_action	(char *appl, void *act)

The hybrid nature of the agents (Figure 4) is established by the *solid modeller interface functions*. For instance, *bind_g_modeller* binds a geometric modeller to the agent environment by specifying the following parameters: the application name (e.g. ACIS); a pointer to a function that registers all predicates in the construction view; and a pointer to a function that registers all predicates in the geometric view. *bind_g_agent* binds the geometric agent *Hgagent* to the construction agent *Hcagent*. *bind_g_attr* binds a geometric attribute to its counterpart in the construction view through the function *putfn*. *putfn* is a procedure to update an attribute in the geometric view when a change is made in the construction view. *register_g_action* registers a function that should be executed by the geometric modeller to manipulate agents in the construction view. This function is determined by its name and a pointer to it. *register_c_action* registers a function to be executed by the construction environment to manipulate the geometric modeller.

The authors developed a construction environment called DObEd (Design Object Editor) integrated into a solid modeller written in C++ and using the proposed API. Fig. 6a shows the 3D environment of the solid modeller with 3 solids, **HBoxLeft**, **HCyl** and **HBoxRight** (H standing for Hybrid), which are instances of the classes **HCylinder** and **HBox**. Figure 6b illustrates the environment to work with the construction view. In this example, the geometric attributes of the solids are as follows: *radius* and *height* of the cylinder; *length*, *height* and *width* of the boxes; coordinates *posx*, *posy* and *posz* of the center of mass of the solids. The following agent intentions are defined:

> *cyl_bxr* of HCyl over HBoxRight:
>   height of HBoxRight = 3.0 * radius of Hcyl
>   push HBoxRight
> *bxr_cyl* of HBoxRight over Hcyl:
>   radius of Hcyl = height of HBoxRight / 3.0
> *cyl_bxl* of Hcyl over HBoxLeft:
>   height of HBoxLeft = radius of Hcyl / 2.0
> *bxl_cyl* of HBoxLeft over HCyl:
>   radius of HCyl = 2.0 * height of HBoxLeft

The lists of agents, predicates, attributes and predicates can be browsed through the Design Object Interface developed with the proposed API, as illustrated in Figure 7.

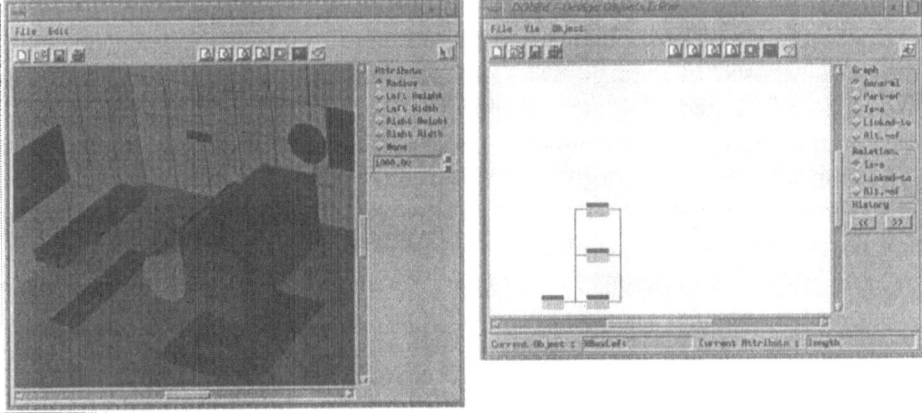

Figure 6. Geometric and construction environments.

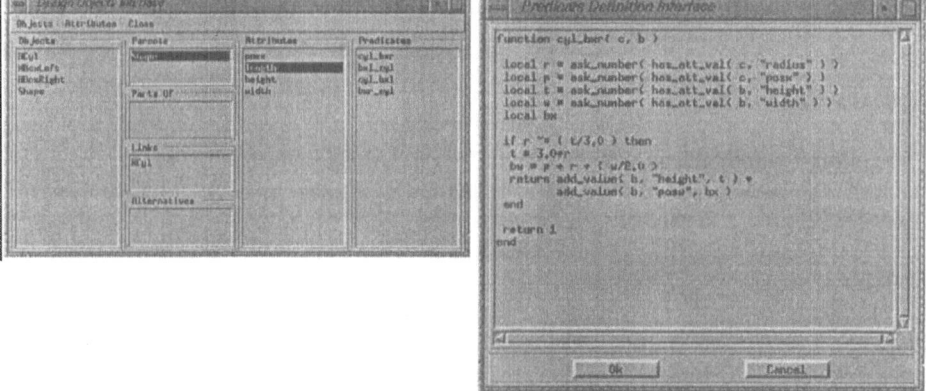

Figure 7. Design object interface and predicate definition.

10. Conclusions

The authors of this work believe that design automation systems should have their focus on reactivity rather than on symbolic representations of design knowledge. The complexity of the design process, its surprises, creative insights and large number of variables seem to be more adequately manipulated within an environment that can be adjusted without interruptions during design state evolution.

The authors started investigating this sort of reactive environment in the domain of solid modelling. Traditional constraint solid modelling (Feng and Kusiak, 1995; Solano and Brunet, 1994) cannot cope gracefully with several types of events, such as the creation of a new solid as a consequence of changing the height of another one or the call of an entire design code for conformance checking. The approach proposed by the authors does not exclude the use of mathematical programming or any other formalism, once they can be used in specific tasks. In fact, the authors are investigating the use of an API based on logic and objects in order to have logical deductions within the reactive environment. Anyway, the authors do not have the intention of pursuing optimal solutions from a set of constraints or preserving completeness properties. However, impossible sets of intention can be easily identified within few steps of execution. Again the idea here is to leave tasks and decisions for the reactive nature of the environment.

Future work also include the development of other design views that could manipulate geometric agents in a more autonomous way. This would be the case where an agent defines the intention of another agent.

The authors have not carried out extensive tests to compare the agent-based approach with other symbolic modelling approaches. However, the prototype reveals high degrees of functionality, speed and very low space requirements, specially because it uses straightforward procedural programming. Moreover, in reactive agent-based approaches there are no time consuming inference processes.

Acknowledgments

The authors would like to acknowledge CNPq, CAPES and JNICT for the financial support. The authors would also like to thank Dr. Sebastião A. L. de Andrade and Dr. Pedro Vellasco for the valuable discussions.

References

Agre, P. E. and Chapmen, D.: 1989, What are plans for?, *A.I. Memo* 1050a, Artificial Intelligence Laboratory, MIT.
Aiken, A., Widom, J. and Hellerstein, J. M.: 1992, Behavior of database production rules: Termination, confluence, and observable determinism, *Proceedings ACM-SIGMOD International Conference on Management of Data*, pp. 59-68.
Arkin, R. C.: 1990, Integrating behavioral, perceptual and world knowledge in reactive navigation, *in* P. Maes (ed.), *Designing Autonomous Agents: Theory and Practice from Biology to Engineering and Back*, MIT Press, Cambridge, MA, pp.105-122.
Baecker, R. M. (ed.): 1993, *Readings in Groupware and Computer-Supported Cooperative Work*, Morgan Kaufmann.
Bates, J., Loyall, A. B. and Reilly, W. S.: 1992, Integrating reactivity, goals and emotion in a broad agent, *Technical Report CMU-CS-92-142*, School of Computer Science, Carnegie Mellon University, Pittsburgh, PA.
Beeri, C. and Milo, T.: 1991, A model for active object-oriented databases, *Proceedings of International Conference on Very Large Data Bases*, pp. 337-349.
Bento, J.: 1992, *Intelligent CAD in Structural Steel: a Cognitive Approach*, PhD Thesis, Expert Systems Laboratory, Imperial College, London, UK.
Bond, A. H. and Gasser, L. (eds.): 1988, *Readings in Distributed Artificial Intelligence*, Morgan Kaufmann.
Brooks, R. A.: 1990, Elephants don't play chess, *in* P. Maes (ed.), *Designing Autonomous Agents: Theory and Practice from Biology to Engineering and Back*, MIT Press, Cambridge, MA.

Brooks, R. A.: 1991, Intelligence without reason, *A.I. Memo 1293*, Artificial Intelligence Laboratory, MIT.

Brown, K. N., McMahon, C. A. and Williams, J. H.: 1994, Contraint unification grammars: specifying languages of parametric designs, *in* J.S. Gero and F. Sudweeks (eds.), *Artificial Intelligence in Design '94*, Kluwer, Dordrecht, pp. 239-256.

Cohen, P. R. and Levesque, H. J.: 1994, Intention is choice with commitment, *Artificial Intelligence*, **42**, pp. 213-261.

Feijó, B. and Bento, J.: 1991, A cognitive approach to design, *CMEST Report* AI 5/91, IST, Lisbon.

Feng, C. and Kusiak, A.: 1994, Constraint-based design of parts, *Computer-Aided Design*, **27**(5), pp. 343-352.

Georgeff, M. P., Lansky, A. L. and Schoppers, M. J.: 1987, Reasoning and planning in dynamic domains: an experiment with a mobile robot, *Technical Report 380*, Artificial Intelligence Centre, SRI International, Menlo Park, CA.

Haase, B.: 1992, Who and what is smart: Intelligent CAD capabilities range from associativity to "KBE", *Design Net*, **1**(6), pp. 19-25.

Hull, R. and Jacobs, D.: 1991, Language constructs for programming active databases, *Proceedomgs of International Conference on Very Large Data Bases*, pp. 455-468.

Inference Corporation: 1985, *ART Programming Manual*, Inference Corporation, Los Angeles.

Maher, M. L.: 1990, Process models of design synthesis, *AI Magazine*, **Winter**, pp.49-58.

Minsky, M. L.: 1988, *The Society of Mind*, Pan Books, London, UK.

Picouet,P.: 1995, *Puissance d'expression et Consistance Sémantique de Systèmes de Triggers*, PhD Thesis, Ecole Nationale Supérieure de Télécommunications, Paris.

Picouet,P and Vianu,V.: 1995, Semantics and expressiveness issues in active databases, *Proceedings of the 14th ACM Symposium on Principles of Database Systems (PODS)*, San Jose, California, USA, pp. 126-138.

Prates, A. J.: 1993, *Fundamentos e Especificaÿão de um Ambiente de Design baseado em Lógica e Objetos*, Tese de Mestrado, Laboratório de CAD Inteligente, Dept. de Eng. Civil, PUC-Rio.

Scheer, S.: 1993, *Uma Análise Crítica sobre o Tratamento Cognitivo de Design em Sistemas de CAD*, Tese de Doutorado, Laboratório de CAD Inteligente, Dept. de Informática, PUC-Rio.

Shoham, Y.: 1993, Agent-oriented programming, *Artificial Intelligence*, **60**(1), pp. 51-92.

Simon, H. A.: 1969, *The Sciences of the Artificial*, MIT Press, Massachusets.

Simithers, T. and Troxell,W.: 1990, Design is intelligent behaviour, but what's the formalism?, *AI EDAM*, **4**(2), pp. 889-98.

Solano, L. and Brunet, P.: 1995, Constructive constraint-based model for parametric CAD systems, *Computer-Aided Design*, **26**(8), pp. 614-621.

Steels, L.: 1990, Towards a theory of emergence functionality, *Proceedings First International Conference on Simulation of Adaptive Behavior*, MIT Press, Cambridge, MA, pp. 451-461.

Stefik, M., Bobrow, D. and Khan, K.: 1986, Integrating access-oriented programming in a multi-paradigm environment, *IEEE Software*, IEEE Inc, New Jersey.

Vere, S. and Bickmore, T.: 1990, A basic agent, *Computational Intelligence*, **6**, pp. 41-60.

Widom, J. and Ceri, S.: 1995, *Active Database Systems: Triggers and Rules for Advanced Database Processing*, Morgan Kaufmann, San Francisco, California.

Wood, S.: 1993, *Planning and Decision Making in Dynamic Domains*, Ellis Horwood Ltd.

Wooldridge, M. J. and Jennings, N. R.: 1994, Agent theories, architectures, and languages: a survey, *Proceedings ECAI94 Workshop on Agent Theories, Architectures and Languages*, Amsterdam, The Netherlands, pp. 1-32.

Wozny, M. J., Turner, J. U. and Preiss, K. (eds.): 1990, *Geometric Modelling for Product Engineering*, Elsevier Science, The Netherlands.

J. S. Gero and F. Sudweeks (eds), Artificial Intelligence in Design '96, 77-96.
© 1996 *Kluwer Academic Publishers.*

A FRAMEWORK FOR DESIGN OBJECT EVOLUTION

Building and cataloging artefact prototypes

NIGEL R. BALL, TIM N. S. MURDOCH AND KEN M. WALLACE
*Engineering Design Centre, Department of Engineering,
University of Cambridge, Trumpington Street, Cambridge,
CB2 1PZ, UK.*

Abstract. This paper presents current work on Product Data Modelling in the Cambridge Engineering Design Centre (EDC) that offers a novel approach to circumventing some of the known problems with the Object Oriented paradigm in the design domain. A data driven approach to object based design is described that allows the designer to build class prototypes during the design process and capture these prototypes onto a catalogue. Catalogue class entries can be reused in an evolving product configuration through a process of selection and specialization with new characteristics. New classes generated during the design can be instantiated as part of the developing product design object and also written back onto the catalog as new prototypes. Catalogues implicitly cluster design objects into abstraction hierarchies that are maintained and developed by the designer rather than a computer programmer. The paper illustrates the technique with an industrial case study and discusses how the approach is being used to develop applications within and without the EDC.

1. Introduction

Design problems are multidimensional and highly interdependent. It is rare for any part of a design to serve only one purpose and it is frequently necessary to devise a solution which satisfies (not necessarily optimally) a whole range of requirements. Any attempt to balance design decisions across an entire product configuration to obtain total functionality and design optimization involves a complicated process of data processing. For example modification to an element in one sub-assembly may result in unpredictable consequences and unresolved conflicts among various others. The task of synthesizing, analyzing and evaluating a self-contained design system is difficult as it needs a vast amount of knowledge and information from diverse sources. Further complications arise when various sub-systems within a design environment have been implemented using different design philosophies, computer languages and system platforms. As a result, the

design, implementation and maintenance of such complex design
environments is a costly, lengthy process that has yet to be fully achieved
(Wallace, 1992).

2. Design Data Representations

A design system must be able to represent and supply useful amounts of well
understood and well structured objects for use in design. For example
knowledge base objects can be categorised by their structural, functional and
causal relationships. A structural relationship states how two (geometric)
objects are physically connected. A functional relationship determines how
two or more (not necessarily directly connected) objects contribute to the
behaviour of an overall system in responding to a particular set of initial
states. A causal relationship identifies the dynamic nature of two objects in
qualitative terms eg. what is the behaviour of object B if object A behaves in
certain way. It is instructive to consider the types of design knowledge
supported in the CAD domain with respect to required input, output and
constraints. Table 1 presents four distinct generations of CAD tool
functionality with current CAD tools belonging to the second generation
(Burgess and Wallace, 1995).

TABLE 1. Generations of CAD functionality.

GENER-ATION	INPUT REQUESTS	INPUT CONSTRAINTS	OUTPUTS	CAD TOOL TYPE
Fourth	Transmit power	Functional 3D Geometric Engineering	Working principle	Functional Synthesiser
Third	Shaft, keyway	3D Geometric Engineering	Dimensions Materials	Functional Modeller
Second (present)	Cylinders, slots	3D Geometric	Dimensions	Solid Modeller
First	Lines,circles	2D Geometric	Dimensions	2D Draughting

The types of design objects representing the input to each generation are
characterized by an increasing level of abstraction against a geometric
understanding of product breakdown. Current generation CAD systems
focus on variant design ie. manipulation of object parameters rather than
redefinition of the object itself. However variant design is at the lowest level
of the three levels of design identified by Pahl and Beitz (1984) as
• original design which involves elaborating an original solution principle;

- adaptive design which involves elaborating a known system;
- variant design which involves varying the size or arrangement of certain aspects of the chosen system;

and object manipulations in the design context (Ahmed et al, 1991) extend beyond simple parametric change to encompass -

- addition / removal / renaming of a instance or class variable;
- changes to the type of a class variable;
- changes to the default value or range of class variable;
- addition of super / sub classes in a class hierarchy;
- re-ordering of a class hierarchy;
- addition / removal of classes from a class library.

If we consider the types of object manipulation that are needed to support different design activity (Table 2) then variant design can be seen as requiring simple parametric change.

TABLE 2. Class manipulation as a function of design.

	ORIGINAL DESIGN	ADAPTIVE DESIGN	VARIANT DESIGN
change instance variable	√	√	√
change class variable	√	√	x
change class method	√	√	x
change class hierarchy	√?	x	x
change class library	√?	x	x

To support designers beyond variant design new CAD tools require a flexible class representation that permits manipulation of generic container classes to support at least the first three types of manipulation without recourse to library re-compilation. Essentially this library should allow the designer to construct "design objects" during the process of design in an evolutionary fashion. Note that original design does not necessarily imply completely new class representations - much creative design is feasible without radical changes to the design object representation.

3. Object Oriented Approach

Design can be considered to be object-oriented, constructive and incremental in that designers use basic components and simple mechanisms to construct larger and more complicated systems. A thorough understanding of basic components, their function, behaviour, and relationships in a dynamic situation forms a good basis for creating new designs. Computer-based design support systems need sophisticated knowledge representation schemes, powerful inferencing systems and efficient control methods in order to cope with the complexity of real world designs.

3.1 OBJECT ORIENTED TECHNOLOGY IN DESIGN

Object Orientation (OO) is a paradigm which attempts to overcome the limitations of conventional computational models by bridging the gap between a piece of data and its operations (Khoshafian and Abnous, 1990). In an object-oriented system, objects represent dynamic entities in computer memory that define data state. An object can typically serve to group data that pertains to one real world entity and encapsulate both state and behaviour by having a set of procedures that specify permissible operations. Sets of similar objects are grouped together under classes. This simplifies association of knowledge within objects by keeping the implementation details private within each class, thus allowing interactions between objects of different classes to be easily controlled and manipulated. The information about how an object behaves is hidden from the behaviours of other objects, only their interactions and relationships in different circumstances are described globally. An Object-Oriented approach is applicable in the design domain because of features such as abstract data typing and polymorphism but has significant weaknesses (Nguyen and Rieu, 1991) such as

- generic relationships are fixed at the class level;
- semantic relationships are difficult to represent in composite objects;
- object variants can only be modelled by multiple instantiation;
- object evolution / reclassification requires class library recompilation.

These weaknesses have led to a number of extensions to the paradigm (Nguyen and Rieu, 1992; Demaid and Zucker, 1992; MacKellar and Peckham, 1992; Donaldson and MacCallum, 1994) such as

- semantic relationships - extending object relationships beyond IS-A and HAS-A;
- multiple object perspectives - allowing an object to belong simultaneously to several points of view;
- dynamic reclassification of objects - support for object migration through a class hierarchy;
- dynamic evolution of class definitions - changes to class data members and function members;

The latter is an essential pre-requisite in applications supporting design synthesis and is the subject of this paper.

3.2 RESEARCH AT THE CAMBRIDGE ENGINEERING DESIGN CENTRE

The research aim of the Cambridge Engineering Design Centre (EDC) is to support designers and design teams throughout the design process by providing them with knowledge-based tools. Complex mechanical engineering systems have been targetted because of their large geometric content and requirements for team-based design that supports evolving design knowledge at different stages of the design process.

The major research efforts applying Artificial Intelligence-based design techniques have focused on:

• resolving the tensions between formalising design data and ensuring its usefulness by developing the Cambridge Product Data Model (CPDM) (Murdoch and Ball, 1994; Murdoch, 1995) to provide a distinct representation of generic and domain specific knowledge (using multi-layer and multi-perspective knowledge structures of design process) and achieve a high degree of reusability;

• developing a theoretically consistent process model (PROSUS) (Blessing, 1993) that can provide a good basis for systematizing design methodologies in the mechanical engineering domain;

• validating computational design techniques via case studies involving actual design data capture and design result evaluation;

• understanding design activities and capturing design experience and expertise though collaborative projects with industrial partners.

Object-oriented techniques and tools offer a useful way of coping with the complexity inherent in CAD/CAE projects. Clear strategies are being adopted at the EDC in the design of object-oriented systems by

• identification of objects (class elicitation);
• identification of class hierarchies (class structuring);
• establishment of message protocols (interactions between objects);
• mappings of methods - functions (reasoning about objects).

3.3 EXAMPLE CASE STUDY

The mature and complex nature of aero engine design has lead to the development of a wide range of specialist analysis techniques and tools. A good example of the nature of the tasks undertaken in aero engine design can be found in the high pressure turbine cooling air system (HPT/CAS), a portion of which is shown in Figure 1. The disc rim and blade root from the high pressure turbine are shown shaded. The combustion chamber lies to the top left of the figure, the centre line of the engine below it and the lower pressure turbines to the top right. The turbine blade requires extensive cooling in order not to melt in the main air stream. The high pressures at this point mean that cooling air must be drawn from the high pressure compressor situated up-stream of the combustion chamber. The arrows indicate the main direction of cooling flows for this cooling air configuration.

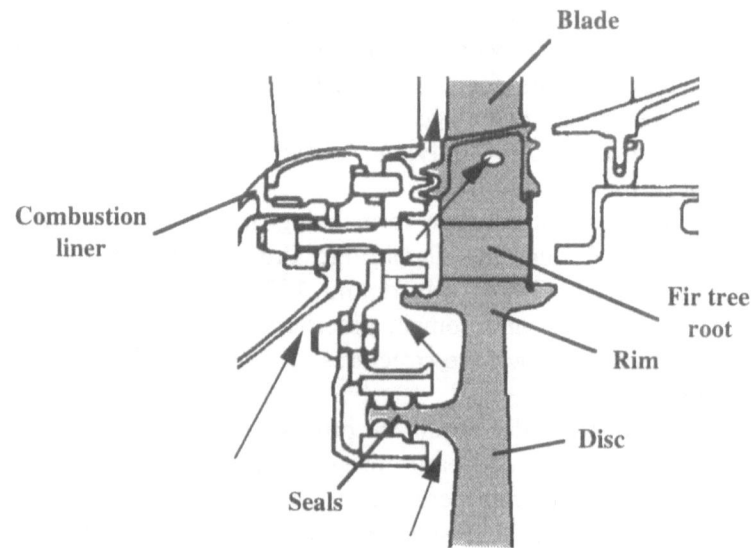

Figure 1. Section through a 1970's high pressure turbine cooling air system.

The traditional mechanism for grouping product data using a Bill of Materials breakdown was found to be inadequate to the task of representing this system. Consequently the EDC initiated a Product Data Modelling project to address the issues of capturing and indexing complex technical systems. Figure 2 shows a schematic for part of the HPT/CAS developed using the CPDM class libraries. The figure shows two groupings of design information : firstly the geometry of the HPT/CAS components and parts (represented by a simple 2D view) and secondly the flow of cooling air through the system (represented by flow lines between key points on each component). Both groupings represent functional interactions between physical components, one force transfer and one air flow, which must be supported by the CPDM representation.

Another important issue in grouping design data presented itself during the HPT/CAS case study. The HPT/CAS bleeds high pressure cool air from the last stage of the HP compressor, feeds it inside the combustion annulus and through to the HP turbine where it is used to cool the shaft, disc and blade. The design of the system takes advantage of parts from the HP compressor, combustor and HP turbine assemblies. This implies that these parts are being designed by two design teams with different functional requirements and perspectives of the design problem. This is a typical scenario in the design of complex and mature products where secondary systems are required to support the overall functionality without increasing cost or reducing performance. References within the paper to the HPT/CAS are shown in a **bold typeface**.

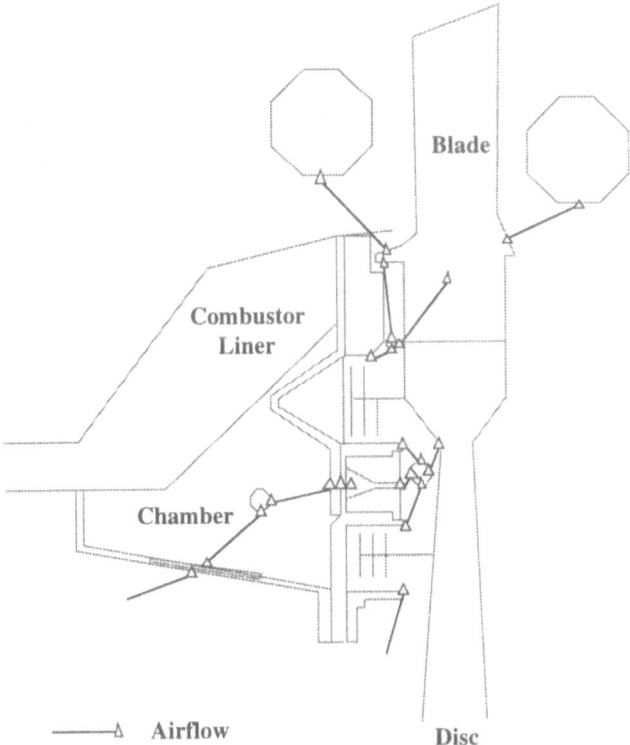

Figure 2. Schematic of HPT/CAS configuration.

4. The Product Data Model Architecture

The Cambridge Product Data Model (CPDM) forms the kernel of the Integrated Design Framework (Ball and Bauert, 1992) that supports the integration of heterogeneous systems capturing the creation and evolution of common design information. The following requirements have guided the design and development of the system architecture :

- support for the simple product breakdown tree;
- support for part decomposition in terms of components and features;
- modelling both physical and functional interactions between assemblies, parts and components;
- support for multiple perspectives on the product data;
- support for independent but interlinked product breakdowns;
- support the attachment of simple static data;
- capture of geometric data and CAD / CAM files;
- support the attachment of existing methods of representing technical data;
- adapt to new schema data as the product and the design process develops.

Attempts to satisfy the above list of requirements have produced an object-oriented system implemented in C++ that represents engineering design entities such as parts, components and assemblies. Each design entity has links to resource objects that can be instantiated to capture data for a specific design application. A neutral 'data entry' application (discussed in section 5) is built into the CPDM as a design knowledge acquisition tool for the designers to extend the object classes in the hierarchy or create instances of design object classes to be used in the new design. The definitions of objects and the way in which they can be structured in the CPDM also provides basic guidelines for defining engineering design objects. The CPDM is intended to form the basis for a number of engineering design tools being developed to share data through a common object-based database.

The development of CPDM represents the EDC's formulation of engineering design knowledge in an object-oriented way so that it can be shared by domain-specific design application systems. The key issues addressed in the development of this system are generality and reusability of engineering design objects. The underlying construct behind this approach is the prototype (Gero, 1990) - a generalized artefact that can be manipulated during the design process by the designer allowing co-development of the design object representation and the actual product. The design prototype plays two roles in this approach - a representation schema for collecting and integrating information relating to a design concept and an operationalization mechanism for the concept. To fulfill these roles a design prototype must capture descriptions of function, behaviour and structure as well as embedding knowledge which supports the reasoning behind design synthesis, analysis and refinement.

The relationship of this work to other research in the field is compared in Table 3 by considering the types of extension (described in 3.1) to the OO paradigm offered by three other systems - SHOOD, SORAC and CFS.

TABLE 3. Generations of CAD functionality.

	CPDM	SHOOD	SORAC	CFS
semantic relationships	√	√	√	√
multiple perspectives	√	√	√	√
dynamic reclassification	x	√	x	x
class schema evolution	√	√	x	√

SHOOD (Nguyen and Rieu, 1992) is an object-oriented data model designed to support highly dynamic applications. It implements support for object schema evolution (at both class and instance level), user defined semantic relationships between objects, and support for multiple object representations.

SORAC (MacKellar and Peckham, 1992) is a semantic modelling tool that supports active semantic relationships modelling parts and connections and allows a designer to specify behaviours associated with relationships within an OO representation.

CFS (Donaldson and MacCallum, 1994) is an OO frame system that supports the evolutionary development of a design concept model and provides a testbed for prototype-based representation. Each concept in CFS is an instance of a composite class whose 'slots' point to 'feature' objects.

All of these systems support the development of semantic relationships and multiple perspectives albeit using different mechanisms. Dynamic schema evolution is not supported in SORAC because the emphasis is on rich semantic modelling rather than novel class structures. The issue of dynamic object reclassification is comprehensively addressed in SHOOD but not in the other systems. The remainder of this section describes the architecture of the CPDM and highlights how this architecture supports

- <u>semantic relationships</u>, by specification of Equation objects;
- <u>multiple perspectives</u> by specification of System objects;
- <u>class schema evolution</u> by specification of Resource objects.

The issue of dynamic reclassification is discussed in section 5.

4.1. DESIGN OBJECT HIERARCHY

The framework for indexing a single layer of artefact data is shown in Figure 3. In common with many product data models the *product* (**three shaft gas turbine**) is broken down into *assemblies* (**HP module**) and *assemblies* into either *parts* (**turbine blade**) or further *assemblies* (**HP turbine**) (Murdoch and Ball, 1994). A *part* is broken down in terms of *components* (**aerofoil section**) where the actual geometry is defined. The geometry of a *component* may be enhanced by the addition of *features* (**drilled holes**). *Interfaces* (**fir tree root - disk connection**) are shown between nodes on the product breakdown and describe the connections between artefact elements.

The component network supports two different types of perspective. One is a traditional product hierarchy where parts are collected into assemblies, and assemblies into further assemblies until the product is complete. This conforms to a traditional Bill of Materials view of modelling product data in clustering elements according to physical relationships. The breakdown is strictly hierarchical with each node being referred to only once in the tree structure (**eg. the core of a three-shaft gas turbine is described in terms of three compressor assemblies, a combustor assembly and three turbine assemblies**).

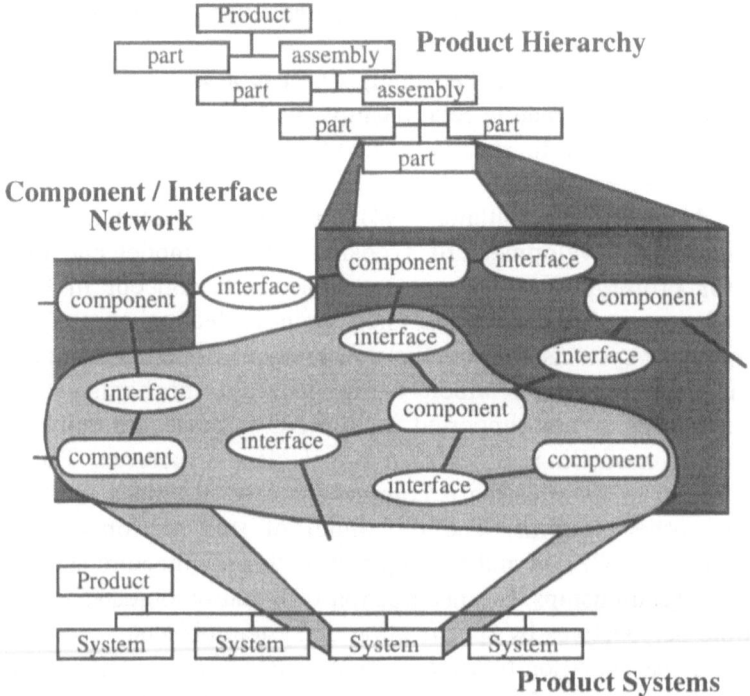

Figure 3. CPDM Framework: A hierarchical network of product elements.

The second type of perspective is product systems, where components are clustered according to functional relationships (see section 4.3).

4.2. LAYERS

The product schema shown in Figure 3 has been described using an example from a mechanical breakdown of the product. This combined tree and network breakdown may also be used to support other types of information. Figure 4 demonstrates the requirement for multiple layers of artefact data during the life of a product. The first layer is shown supporting the functional description of the product and others are shown supporting embodiment, detail and life cycle information. Each layer uses a entity structure to support different but compatible breakdowns of a single product similar to Andreasen's chromosome model (Mortensen and Andreasen, 1993). Links between layers map causal relationships between entities. Compatibility is defined in terms of a consistent mapping between layers **(eg. the mapping between the function 'compress air' and product assembly 'HP compressor' is shown by linking the compress air nodes in the function layer to the HP compressor node in the detail layer).**

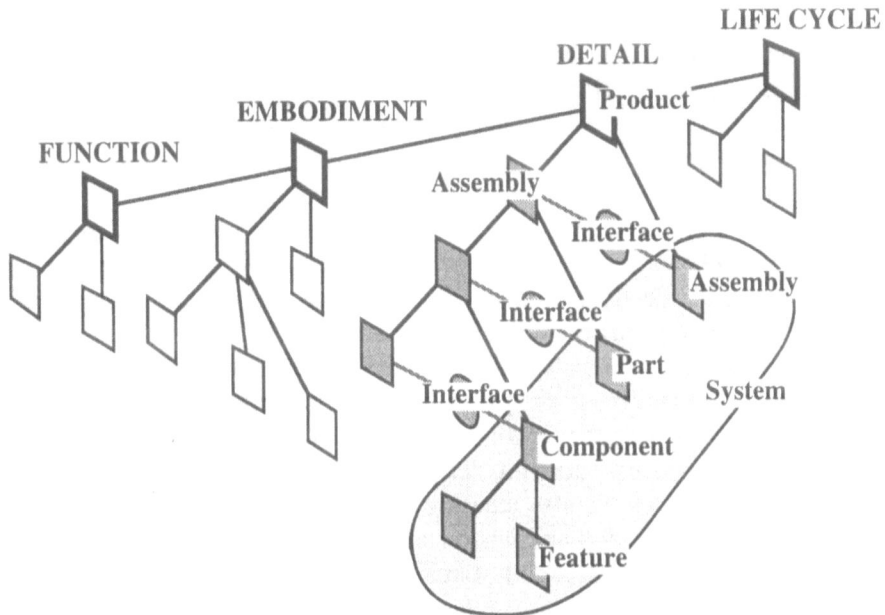

Figure 4. Multiple Layers of cross linked artefact data.

4.3. MULTIPLE PERSPECTIVES

The product definition (Figure 3) also supports the concept of systems that provide new groupings of existing nodes in the product breakdown. *Systems* **(cooling air system)** may be broken down into their constituents of *assemblies* **(compressor, combustor, turbine),** *parts* and *components* or further *systems* **(HPT cooling air system)**. Systems in the CPDM support "tightly coupled views" in that all processing occurs in the base objects rather than in the System object itself. This constrasts with more sophisticated approaches supporting "loosely coupled views" where processing occurs in the view directly (MacKellar and Peckham, 1994).

The use of multiple layers of data (Figure 4) and multiple perspectives within each layer (Figure 3) enables management of data ownership. The design of complex technical systems requires multi-disciplinary design teams. Each team requires its own perspective on the design data. By declaring these perspectives in the form of system networks the ownership of a specific item of product data can be managed effectively.

4.4. RESOURCES

There are at least five types of core product data, found in several of the stages of the design process :

• **Specification**: a description of requirements for other product data;

- **Function**: a description of the behavioural properties defined by the interaction among and between product elements and users;
- **Geometry**: a description of shape and material properties;
- **Attribute**: a description of other physical and abstract properties;
- **Production and Use Processes**: a description of how, when and where to make, transport, use, maintain and retire the product.

Specification information is relevant to all stages of the design process and is applicable to all types of design data. Whether the specification is used to direct the creation of a product concept or to support the evaluation of competing layout designs, its key element is the list of requirements.

The description of functionality and behaviour are central to both the conceptual and embodiment stages of design. Function structures and state-transition diagrams are frequently used during conceptual design and whilst working in the process domain. The information held and manipulated within these methods is also central to embodiment design and whilst working in the artefact domain, where recognisable, though abstract, physical shapes are used. Thus product data concerning functionality must be capable of being shared between tasks and either added to or abstracted from other types of product information.

Geometry, combined with the structure of a technical system, defines many physical aspects of the artefact. The definition of geometry can be divided into shape and dimension. While parameters can be used to define the specific sizes and material properties, shape requires more sophisticated methods. Attributes capture the remaining internal properties such as ergonomics and aesthetics. Recognising the separation of parameter and attribute properties is important in understanding the difference between design properties (those which the design team can manipulate) and internal and external properties (those which the design team can effect). Other whole-life properties are captured under the production and use processes.

Specification, Function, Geometry and Attribute data has been captured through the implementation of a number of Resource sub-classes which can be attached to any Artefact via list structures. These subclasses provide the basic container objects that designers can use to build specific "design objects" from the CPDM class library. This permits class schema evolution using a language defined in terms of Resource sub-classes without requiring reconstruction of the CPDM libraries. Two Resource subclasses are described in the following sections: Criteria and Characteristic.

4.4.1. Criteria

In the research literature, specification information is defined in terms of either a free text design brief or lists of specific design requirements (Pahl & Beitz, 1984, Hauser & Clausing, 1988). Methodical strategies for developing a specification result in structured text or requirements lists. Free text can be supported by simply referencing document filenames and directories. Requirements, however, must be specifically defined. The majority of specification data is modelled using lists of requirement and exchange rate objects in the Resource sub-class Criteria. The data definition of the class Requirement, shown in Figure 5, was developed from that used in the undergraduate teaching tool SpecBuilder (Thomas & Wallace, 1990) and the configuration optimisation tool KATE (Murdoch, 1993).

```
class Requirement: {

    character string            keyword;
    character string            requirement;
    character string            who;
    character string            when;

    Property                    property;

    List of real                target values;
    List of real                importance weightings;

    Descriptor                  type of requirement;
};
```

Figure 5. Class requirement embedded in the Resource sub-class Criteria

The first four entries are similar to those found in the SpecBuilder program which also captures the importance of the requirement in terms of either a demand or weighted wish. This and further numeric information is captured here in a list of several importance weightings associated against target values. The type of requirement captures whether the requirement is attainment of a specific goal or optimization of a property value. The pointer to a property is used to reference data stored elsewhere and by methods within the Requirement class to determine how well the current property value meets the specified requirement.

4.4.2. Characteristic

The functional, geometric and attribute properties of an artefact are stored in the class Characteristic shown in Figure 6. The class contains lists of three properties (function, parameter and attribute) and several items capturing other geometric information.

Functions capture the potential input and output energy, signal and material flows to an artefact (**labyrinth seal - air-output**). Parameters capture the sizes of certain geometric and material features (**radial position**).

```
class Characteristic: public Resource {

    List of Functions              functions;
    List of Parameters             parameters;
    List of Attributes             attributes;

    List of Points Interest        points interest;

    Origin                         origin;
    Orientation                    orientation;

};
```

Figure 6. Characteristic sub-class of Resource.

Attributes capture information which may be derived or measured from other artefact properties (**seal leakage**). The Points Interest capture physical points in space in and around the artefact and are used to assemble artefact objects in 3D space. They also define the position of the functional input and output points. The origin and orientation objects store the position and orientation of the artefact in local coordinates.

The Parameter class is shown in Figure 7. This class overlays four different types of information (real, integer, text or object reference) onto a single data member. Thus a parameter may store wall thickness, number of holes, name of surface colour or simply point to a class containing further and more extensive information. This latter option has been provided to support inter-change of standard shapes and materials. Other entries support the definition of ranges for variation and tolerances on the current parameter value and a state descriptor to capture variability and inter-linkage to other product data. The final item is a list of parametric Equation objects. These equations may be either equalities or inequalities. To maintain a determinate shape, only one equality equation is allowed to support parametric geometry. Any inequalities listed model constraints in the product parameters.

Equation objects may be used to construct links between any Characteristic property of an Artefact thus enabling the designer to model semantic relationships between design objects at any level in the Product hierarchy (**labyrinth seal: maximum-rotation-speed <= maximum-rubbing-speed/2*π*radius**). Capturing the relationships between parameter and attribute characteristics constitutes a major design activity and is supported in the CPDM through mappings based on *Equation* objects, *Analytical* (tabular) *Data* objects, C++ code subroutines and external standalone tools.

The list of Parameters in Characteristic define geometric variables which need to be mapped to an actual physical shape in order to define a specific

3D body. The mapping of parameters to geometry within the CPDM has been implemented by building up a body from intersecting primitives and swept laminas using the ACIS solid modelling kernel (Spatial Technology, 1995).

```
class Parameter: {

    character string              name;

    union {
    real                          Rvalue;
    integer                       Ivalue;
    character string              text;
    Pointer                       pointer;
    };
    descriptor                    code;

    real value                    min, max;
    real value                    tolerance;

    descriptor                    state;

    List of equation              parametric equations;
};
```

Figure 7. Parameter class.

5. Building Design Objects - The Design Object Catalogues

One of the motivations behind the CPDM research is to enable the evolution of design representations during the process of design rather than as a pre / post design activity. Specification of the representation before the process is often premature since a designer's grasp of the problem will be incomplete and probably biased towards past solutions. Documentation of the representation after the design is complete may be too late if there is pressure to move onto the next problem. Hence the underlying philosophy is one that **"doing the design is capturing the data"**.

The basic approach taken in the EDC to capture design data has been to enable direct modelling of design object through

- browsing of existing catalogues (by domain type);
- selecting a catalogue entry on the basis of similarity in Resource space;
- adapting the entry by adding new Resource subclass member definitions;
- saving the new node back onto the catalogue as a new class prototype ;
- adding the entry onto the design product tree as a new leaf node;
- specializing the node by entering variable data (slot filling).

The first four activities are performed using an in-house CPDM application called *Compdef*. This application supports the building of design prototypes (as described in section 4) within catalogues classified by product domain such as aerospace or civil engineering. The last two activities are performed using a design application that is targetted at instantiating specific Resource sub-class objects.

Each catalogue entry is a design object prototype that represents an intermediate state between class and instance. It is a generic object that can be instantiated within the context of a specific design. The implicit relationship between different prototypes in a catalogue is equivalent to an abstraction hierarchy where child members have been constructed by a designer through specification of new Resource objects rather than predefined as part of a fixed library. As described in 3.1 dynamic reclassification of abstraction hierarchies may have a role to play within the CPDM as a clustering mechanism for catalogue prototypes based of Resource object configuration. Such hierarchies will be one of a group of clustering perspectives available to the designer. The efficacy of each mechanism in the group will be dependent on design context. As yet no research has been conducted into identifying the membership of this group.

Methods mapping data members within a class prototype are modelled using the Equation and Analytical Methods subclasses. If these are inappropiate then a Tool class is available to link in external methods.

6. Integration of design applications using the Cambridge PDM

The CPDM class libraries support the storage of project information on a series of databases. This information can be accessed directly by workbench applications developed within the EDC using predefined database query and access routines. Applications developed outside this environment require a wrapper to access the database and translate the project information to and from the CPDM protocol. The wrapper also enables certain workbench tools to communicate with the stand-alone tools directly as part of the Integrated Design Framework. The supporting knowledge-base uses the same CPDM schema and data resources, combined with embedded design knowledge to provide catalogues of re-usable design objects. These objects are defined using workbench tools designed specifically for knowledge capture.

6.1 CURRENTLY AVAILABLE APPLICATIONS

Applications possess various levels of design integration. Those developed in the EDC using the CPDM schema and database query routines can be said to be fully integrated. Stand-alone applications, however, require other methods of communication and data transfer, aspects which form part of a bespoke wrapper. The following comprise the operational EDC application set:

Name	Description	Type
Compdef	Design prototype builder	CPDM
BuildSite	Generic Configuration Builder	CPDM
Compgeom	Solid Model Visualization	CPDM
KATE	Configuration Optimization	CPDM
CMS	Cambridge Materials Selector	Stand-alone (PC)
PROSUS	Design Event capture	CPDM
Review	Design Guidelines database	Stand-alone (PC)
FUNCSION	Functional Synthesiser	Stand-alone (Unix)

+ commercial CAD and FE systems using IGES interfaces.

6.2 INTEGRATION STRATEGY

This combination of existing stand-alone and newly developed design applications has been integrated into the IDF design workbench shown in Figure 8. The project database and supporting knowledge database are shown as parallel 'object buses' carrying information to and from various design tools. These databases currently comprise a Lisp environment supporting a number of functional modelling design tools and a C++ environment supporting the main CPDM data definition and configuration optimisation and process integration design tools.

The EDC design tools stand between these two databases showing various levels of integration and data sharing capabilities. Examples of these tools are listed in the key. Below the project databases are a number of translation modules. One specifically translates the Lisp based functional information into the PDM data definition. Others support the transfer of information to and from the commercial design tools shown at the base of the figure.

No mechanism has yet been developed to maintain the consistency of the CPDM across the workbench application set. The FUNCSION application uses an Assumption-Based Truth Maintenance System (Tang, 1995) to monitor data integrity and this approach may be applied to the CPDM.

6.3. IMPLEMENTATION DETAILS

The system has been implemented on a Local Area Network of Sun SPARCs running under SunOS and Solaris. All CPDM class libraries and PDM applications are written in C++4.0.1 with GUIs built using Sun's DevGuide tool. Two additional third party libraries are used within the PDM classes - Sun's XGL (a 2D geometry library) and Spatial Technoloy's ACIS (a 3D Geometry library). The SPARCWORKS 3.0.1 debugging environment is

used for all C++ development. PC-based applications are currently supported
under the Solaris Windows emulator WABI.

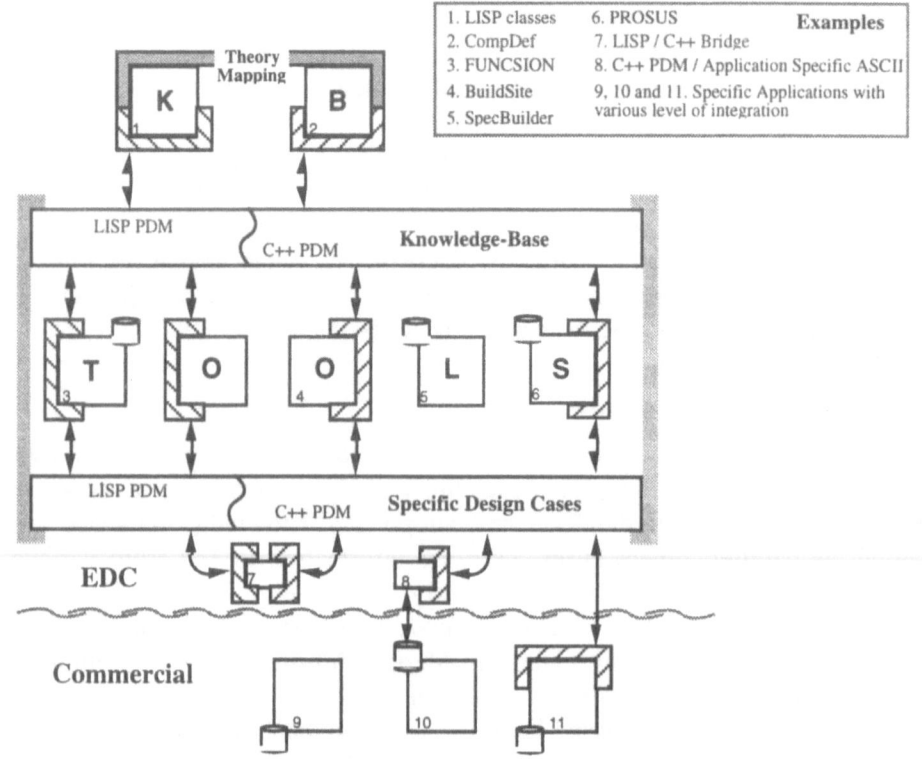

Figure 8. EDC Design environment.

The database supporting this work is Quillion Systems Limited's Object-
Based system called QuikTrieve (Quillion, 1992). Object persistence is
supported on the CPDM through the development of a C++ interface (QT-
IDF) that maps objects in an EDC application's virtual memory space onto
Quiktrieve data objects. This interface does not require bespoke coding
within each application and enables rapid linkage of an application to the
database via a few simple function calls - open, close, get, put, delete.
The schema and data definition described in this report results in very fine
granularity of data that allows participants to select only the information
specifically required for their activity.

6.4. PROJECT STATUS

Projects involving the CPDM are both industrially and academically driven.
A number of Design for X projects are being pursued in conjunction with

industrial clients in aerospace and civil engineering. Most of this work is categorised under the theme of 'Configuration Optimization' and the application of CPDM libraries is providing a focussed approach to a number of diverse projects such as 'design of reliability' and 'design for technical merit' (Stephenson and Wallace, 1995; Murdoch and Wallace, 1995).

A case study from an undergraduate design course, the Integrated Design Project, is being used to demonstrate some of the key requirements of product data modelling. Teams of six students design, build and test autonomous vehicles that are able to navigate a course marked out by a white lines and perform various pallet handling tasks. The three distinct systems of these vehicles - mechanical, electronic and software - are being analysed and modelled using the CPDM with *Compdef* and *Buildsite* tools (Murdoch and Ball, 1995).

7. Conclusions

The CPDM class libraries are a research laboratory for experimentation into the evolution of design objects during the design process. The underlying aim of the research is to empower the designer by allowing flexible class definition as well as instantiation without recourse to rebuilding the systems environment. The output of the research is to extend the level of design activity beyond that currently offered by CAD systems from variant into adaptive design.

Since being made persistent through the application of an OO Database Management System, the CPDM libraries have also become an integration medium that supports interfaces between EDC and commercial applications. Rapid prototyping of new applications (particularly in the DFX domain) is becoming possible through the addition of new Resource sub-classes to the base CPDM libraries. This is giving significant leverage to the implementation effort within the EDC and providing an software environment for research students to construct experimental systems.

Acknowledgements

This work is supported by funding from the Engineering and Physical Sciences Research Council.

References

Ahmed, S., Wong, A., Sririam, D. and Logcher, R.: 1991, A comparison of object-oriented database management systems for engineering applications, *Research Report R91-92*, Intelligent Engineering Systems Laboratory, MIT.

Ball, N. and Bauert, F.: 1992, The integrated design framework: supporting the design process using a blackboard system, *in* J. S. Gero (ed.), *Artificial Intelligence in Design '92*, Kluwer, Dordrecht, pp. 327-348.

Blessing, L. T. M.: 1993, *A Process-Based Approach to Computer Supported Engineering Design*, PhD Thesis, University of Twente, The Netherlands.

Burgess, S. C. and Wallace, K. M.: 1995, An overview of the functionality needed in CAD tools, *Proceedings of the International Conference on Engineering Design ICED95*, Heurista, Zurich, pp. 1296-1301.

Demaid, A. and Zucker, J.: 1992, Prototype-oriented representation of engineering design knowledge, *Artificial Intelligence in Engineering, 7*, 47-61.

Donaldson, I. and MacCallum, K.: 1994, The role of computational prototypes in conceptual models for engineering design, *in* J. S. Gero, and F. Sudweeks (eds), *Artificial Intelligence in Design '94*, Kluwer, Dordrecht, pp. 3-20.

Gero, J. S.: 1990, Design prototypes: A knowledge representation schema for design, *AI Magazine*, 11(4), 26-36.

Hauser, J. R. and Clausing, D.: 1988, The house of quality, *Harvard Business Review*, May-June, 63-73.

Khoshafian, S. and Abnous, R.: 1990, *Object-Orientation - Concepts, Languages, Databases, User Interfaces*, Wiley, New York.

MacKellar, B. K. and Peckham, J.: 1992, Representing design objects in SORAC: a data model with semantic objects, relationships and constraints, *in* J. S. Gero (ed.), *Artificial Intelligence in Design '92*, Kluwer, Dordrecht, pp. 201-220.

MacKellar, B. K. and Peckham, J.: 1994, Specifying multiple representations of design objects in SORAC, *in* J. S. Gero and F. Sudweeks (eds), *Artificial Intelligence in Design '94*, Kluwer, Dordrecht, pp. 555-572.

Mortensen, N. and Andreasen, M.: 1993, Structuring product data based on the chromosome model, *Technical Report*, Technical University of Denmark.

Murdoch T.: 1993, *Configuration Evaluation and Optimisation of Technical Systems*, PhD Thesis, University of Cambridge.

Murdoch T.:1995, Sharing design data, *Technical Report CUED/C-EDC/TR28*, Cambridge University.

Murdoch, T. and Ball, N.: 1994, Developing an EDC product data model, *Technical Report CUED/C-EDC/TR21*, Cambridge University.

Murdoch, T. and Ball, N.: 1995, A layered framework for sharing design data, *Proceedings of the International Conference on Engineering Design ICED95*, Heurista, Zurich, pp. 1471-1476.

Murdoch, T. and Wallace, K.: 1995, Design for technical merit, *in* G. Q. Huang (ed.), *Design for X: Concurrent Engineering Imperatives*, Chapman Hall (in press).

Nguyen, G. T. and Rieu, D.: 1991, SHOOD: a design object model, *in* J. S. Gero (ed.), *Artificial Intelligence in Design '91*, Butterworth-Heinemann, pp. 367-386.

Nguyen, G. T. and Rieu D.: 1992, Representing design objects, *in* J. S. Gero (ed.), *Artificial Intelligence in Design '92*, Kluwer, Dordrecht, pp. 221-240.

Pahl, G. and Beitz, W.: 1984, *Engineering Design*, The Design Council.

QuikTrieve Reference Manual: 1992, Quillion Systems Limited.

ACIS Geometric Modeller Application Guide: 1995, Spatial Technology Inc.

Stephenson, J. and Wallace, K.: 1995, Design for reliability, *in* G. Q. Huang (ed.), *Design for X : Concurrent Engineering Imperatives*, Chapman Hall (in press).

Tang, M. X.: 1995, Development of an integrated ai system for conceptual design support, *in* J. Sharpe (ed.), *AI System Support for Concept Design*, Springer-Verlag, pp. 153-169.

Thomas, R. and Wallace, K.: 1990, *Specbuilder*, Cambridge University Engineering Department, Teaching Software.

Wallace, K.: 1992, Some observations on design thinking, *in* N. Cross, K. Dorst and N. Roozenberg (eds), *Research in Design Thinking*, Delft University Press.

J. S. Gero and F. Sudweeks (eds), Artificial Intelligence in Design '96, 97-116.
© 1996 *Kluwer Academic Publishers.*

CREATING DESIGN OBJECTS FROM CASES FOR INTERACTIVE SPATIAL COMPOSITION

IAN SMITH, RUTH STALKER AND CLAUDIO LOTTAZ
AI Lab (LIA)
Computer Science Department (DI - Ecublens)
Federal Institute of Technology (EPFL)
1015 Lausanne, Switzerland

Abstract. This paper describes IDIOM, a system for composing layouts using cases. Layouts are interactively composed by users rather than automatically generated as has been proposed by previous research. The design is incrementally parameterized as cases are added and therefore, case adaptation, user interpretation and model activation can occur at any stage. IDIOM supports designers through reducing constraint complexity and through managing design preferences, thereby restraining proposed solutions and further adaptation within *globally* feasible design spaces. Improvements to the algorithm over previous implementations have increased reliability. In general, designers, who currently carry out spatial composition tasks using standard drawing tools, have reacted favourably to the system, providing useful feedback for further work.

1. Introduction

Design systems that support case-based design (CBD) have the potential to help designers reuse previous designs in new contexts. This approach is one that they have always employed for creative and routine design activities. Therefore, CBD has been studied extensively and applied to a range of fields. For example, CBD systems are proposed for mechanical engineering, civil engineering and architecture (Bahktari and Bartsch-Spörl, 1993; Flemming, 1994; Goel and Chandrasekaran, 1989; Goel and Kolodner, 1991; Maher and Zhang, 1991; Navinchandra, 1988; Sycara and Navinchandra, 1991). Although early adaptation work did not concentrate upon support for geometrical aspects, recent research, particularly studies associated with building design, have included geometrical aspects and much progress has been made, e.g. (Adani, 1995; Coulon, 1995; Dave et al., 1994; Gero and Schnier, 1995; Giretti et al., 1994; Hua, 1994; Zhang and Maher, 1993). We provide a further contribution by concentrating on interactivity, use of preferences and sound computational algorithms for continuous variables.

Designers usually employ information from several designs in order to complete tasks. Therefore, research into case adaptation has evolved into studies of case combination (Dave et al., 1994; Purvis and Pu, 1995; Sycara and Navinchandra, 1991; Zhao and Maher, 1992) which involves three processes (Smith et al., 1995) *analysis* of cases for applicable information, *interpretation* of this information in the new context and *resolution* of conflicts between the case and the new context for a feasible solution. Since the efficiency of the analysis depends upon the way cases are interpreted and how conflicts are resolved, our work focuses upon the last two of these processes, *interpretation* and *resolution*.

In many fields, the spatial configuration of design components determines design cost and in-service functionality. In multi-story apartment building design, once the floor layouts have been determined, it is estimated that 90% of the final cost is fixed for "standard" construction.[1] Computer support for layout configuration has been studied for more than twenty years. Studies include techniques such as mathematical programming (Mitchell et al., 1976), optimization (Mitchell et al., 1976), space discretizations (Voss, 1994), genetic evolution (Gero and Schnier, 1995), graphs (Coulon, 1995; Choi and Flemming, 1995), hierarchical generate and test (Flemming et al., 1988), natural language declarations (Fujii, 1995) and constraint satisfaction (Baykan and Fox, 1992; Medjdoub and Yannou, 1996; Tommelein, 1989). Rather than automate the configuration task, we have developed a system which supports designers as they compose designs *themselves* from parts of previous designs. As discussed later, practicing designers who were interviewed within the scope of this study emphatically did *not* wish to have computer systems perform automatic layout generation.

Some design requirements are expressed as preferences. Preferences reflect requirements that cannot be modelled more precisely, such as social and political considerations, as well as control knowledge that helps designers explore design spaces. Preferences differ from default information because if deactivated, they may be *reinstated* as opportunities arise. Models have been proposed which use assumption-based truth maintenance (Logan et al., 1991) for discrete variables. Borning (Borning et al., 1992; Wilson and Borning, 1993) used hierarchies of constraint sets in order to resolve contradictions in an interactive drawing system. Preferences have also been employed for complex Pareto optimality problems (D'Ambrosio and Birmingham, 1995). In WRIGHT, Baykan and Fox (1992) allow for constraint weakening in over-constrained situations. Deactivating requirements in order to explore design spaces was first proposed by Navinchandra (1991). Although design exploration has been investigated by several other researchers (Gero and Kazakov, 1996; Logan and Smithers, 1993; Maher and Poon, 1995), it is agreed that more work is needed; few validated and tested implement-

---

[1] "Standard" construction is intended to refer to construction that is most commonly found in a given socio-economic region.

ations are available that support practical design tasks. In our research, preferences are used to support exploration of alternatives for adaptation of spatial configurations.

We combine new ideas with successful parts of previous work (Hua, 1994; Hua et al., 1992) to support interactive spatial composition using existing designs and explicitly defined domain models. Preferences are included in these models and are combined with reliable and fast algorithms for constraint solving in order to produce an interactive system. Various aspects of this system are illustrated using apartment layouts. The next section contains a general description of the system and describes the cases employed. Section 3 describes how case combination is carried out using algorithms that have been improved over previous implementations. Section 4 discusses how the system has been conceived to interact with the designer and the last three sections discuss implementation details, testing with users, and related work.

2. IDIOM

We have developed a system called Interactive Design using Intelligent Objects and Models (IDIOM) in order to study design interactivity, the use of preferences, and explicit domain modelling for case adaptation. Model-based adaptation was first proposed by Goel (1989) for discrete variables. The term, IDIOM, was chosen because its meaning reflects a goal of this research. A dictionary definition (from Longman) for the word "Idiom" is

A phrase which means something different from the meaning of the separate words

This definition provides a useful analogy. We aim to support incremental composition of design cases while employing user interaction and domain models to include holistic considerations of groups of objects. Models are applied to designs several ways. They are activated when certain groups of objects are present in the design, they are used to interpret designs in certain contexts and they are incrementally introduced by the designer as the design is composed.

Our current research into case-based building design is motivated by two factors. The first factor is the observation that although building designers frequently reuse designs, they rarely wish to adapt whole building cases. Often, the cases which are most useful are spaces and collections of spaces (Schmitt, 1993).

The second factor is that most design domains cannot be modelled completely due to a complex consideration of social, political and economic factors. As a result, it can be frustrating to designers when a system performs automatic design and proposes just one solution. A much better role for computer systems is to provide support for defining *allowable spaces of acceptable designs*. When exploring these spaces, designers are able to introduce *their interpretation* of what is not modelled through user interaction.

Figure 1. An example of a case in IDIOM.

These two factors lead to the definition of an **intelligent object** that is used in this paper: an intelligent object is a *part of a successful design* which has been interpreted by designers for each new design task through constraint posting, declaration of neighbourhood relationships, adaptation and model activation. Therefore, an object becomes intelligent at run-time. This interpretation is used to accommodate additional objects during subsequent design stages. The notion of an intelligent object is not new, for example see Rigopoulos and Oppenheim (1992). An example of an intelligent object is a living room taken from a design of a previously built apartment building. This living room becomes an intelligent object when i) the user interprets it in a new context by imposing conditions such as neighbourhood relationships and ii) when the user activates domain models to add additional constraints, such as the size of the living room needed for the number of inhabitants in the apartment. More detail of the models employed is provided in Section 2.2.

2.1. CASES IN IDIOM

Cases in IDIOM are parts of designs of constructed apartment buildings. Cases have been carefully selected by an architect for flexibility, compatibility and success as designs of parts of existing buildings. They are grouped into types such as living rooms, kitchens, bathrooms and bedrooms. They contain windows, furniture and doors. An example of a case is shown in Figure 1.

Grey rectangles within spaces represent furniture elements. The size of these rectangles include the size of the element plus additional space necessary for adequate use. For example, the size of a rectangle representing a dining room table includes an allowance for chairs as well as adequate room for sitting in them. Other elements shown in Figure 1 are the window in the right wall and the door on the left wall. The outer dimensions of the case as well as the positions of elements such as windows, doors and furniture are treated as variables. Sizes of elements within cases are fixed. All variables start with default values that correspond to their values in the original design. The origins of the case are described by the location of the building and the name of the architect.

2.2. MODELS IN IDIOM

Models in IDIOM are causal mappings from structural parameters to behaviour related to individual objects (interpreted cases) and object groups. Behaviour is interpreted for a given context to correspond to a desired function. Therefore, model formulation follows the no function-in-structure principle (de Kleer and Brown, 1984; Gero, 1990). The definition of function, behaviour and structure follows (Gero, 1990).

We employ models to provide domain knowledge as configurations are composed. Models are abductively implemented through causal inversion (desired behavior to required structure). Since abduction is unreliable when a closed-world assumption is inaccurate, models in IDIOM are interactively activated, thereby providing one of several ways for the designer to introduce a problem-specific interpretation of the context.

In order to illustrate these mappings and their interpretations, four examples of models used are given below. These models reflect the scope of domain knowledge that can be included in the system. Models may cover strict rules which are simplified from physical principles (adequate natural lighting), guidelines (subsidized housing), technological considerations (economical facades) and personal designer preferences (luxury construction).

- *Subsidized housing* Government authorities publish specifications for buildings to qualify for registration as subsidized housing. Since designers know that the value of a building is reduced if these specifications are not met, they often consider them to be minimum requirements. For example, minimum room sizes are specified for the number of people living in an apartment.
- *Economical facades* When facades are continuous along one face, that is, no discontinuities or intermediate corners, the building envelope behaves better (reduced risk of leaking, deterioration, etc.) than if intermediate corners are present.
- *Adequate natural lighting* Local authorities specify a minimum ratio of window area to floor area in order to ensure that there is adequate natural lighting in rooms.
- *Luxury construction* Most building designers can provide specifications related to what they believe to correspond to above average construction standards. Parameters such as sizes of rooms and widths of hallways are linked to a behaviour which provides above average comfort.

Figure 2 shows the same design with and without activation of a model for luxury construction. Examples of constraints included when this model is activated are : minimum area of single room $= 16m^2$ (top object in Figure 2) and min-

Figure 2. The same design without (left) and with (right) activation of a model for luxury construction. (Text annotations have been disabled.)

imum area of kitchen $= 12m^2$ (bottom right object in Figure 2). Since this model reflects personal preferences, it should be elaborated upon and modified for each user. The use of preferences is discussed further in the next two sections.

3. Case Combination through Constraint Solving

Spatial composition of intelligent objects requires consideration of many interacting relationships between variables. Case combination is supported through incrementally solving relevant constraints, thus taking advantage of inter-relationships to reduce complexity.

Arrangements of intelligent objects and their elements such as doors, windows and pieces of furniture are defined by sets of constraints. Constraint sets have to be solved rapidly in order to allow interactive use, therefore we restrict these to linear and simple non-linear relationships. Relationships can be equalities or inequalities.

One of the most important aspects of the solver in IDIOM is its compatibility with interactive adaptation. When another case is added, IDIOM finds a solution whilst maintaining positions and sizes in the current design wherever possible. Many algorithms in linear programming cannot do this. For example, those which employ pre-defined objective functions cannot dynamically add parametric values to the optimization criteria.

3.1. SOURCES OF CONSTRAINTS

There are three sources of constraints: the library of cases, the interpretation of the design by the user and domain models. When a case is introduced into a design, all its associated constraints are added to the current set of constraints. The user can then add further constraints in order to interpret the case in its new environ-

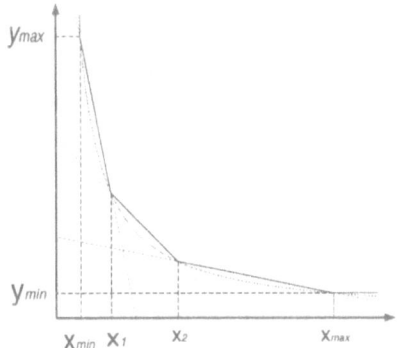

Figure 3. Linear approximation of $y = \frac{A_{min}}{x}$

ment. The most important constraint is the specification of the topology of the design, done by defining neighbourhood relations between objects (as described in Section 4). In addition, the user can specify constraints on the sizes, distances and alignments between objects and their elements. Before a new solution is calculated for the layout, constraints from active domain models are added to the current set of constraints. All constraints restrict values of continuous variables.

When all the constraints present in the system are linear, calculations can be completed in a reasonable time (less than five seconds). Certain non-linear constraints, such as minimum areas can be approximated by linear relationships. Consider the constraint, $xy \geq A_{min}$, where x and y are the length and width of an object and A_{min} is the minimum area imposed by the constraint (illustrated in Figure 3). Using the minimum sizes, all objects must have, together with the above constraint, a maximum value for x for consideration of the constraint as follows: $x_{max} = A_{min}/y_{min}$. If values of x are larger than x_{max}, then the constraint defining a minimum on y implies that there are always acceptable values for xy. Thus, it is sufficient to approximate $xy \geq A_{min}$ in the interval $[x_{min}, x_{max}]$. IDIOM employs a logarithmic relationship, to determine the points, x_1 and x_2 for linear approximations as shown in Figure 3. Typical constraints in IDIOM which have a form similar to $xy \geq A_{min}$ can be approximated with an error of less than 5% using only three linear constraints.

3.2. DIMENSIONALITY REDUCTION

Equalities in the constraint set reduce the degrees of freedom of design spaces. This approach has been used in statistics (Krishnaiah and Kanal, 1982) and image recognition (Saund, 1989) and was proposed for case-based design (Faltings, 1991). Subsequent development established that equalities can be used to reduce the number of variables occurring in inequalities (Hua, 1994).

IDIOM uses Gauss-Jordan elimination to perform dimensionality reduction and to identify dependent and independent variables. In the inequalities, dependent variables are substituted by independent ones, thereby finding the matrix of coefficients of the equalities and inequalities.

$$
\begin{bmatrix}
1 & 0 & \cdots & 0 & 0 & a_{n_e+1,1} & \cdots & a_{n_v,1} \\
0 & 1 & \cdots & 0 & 0 & a_{n_e+1,2} & \cdots & a_{n_v,2} \\
\vdots & \vdots & \ddots & \vdots & \vdots & \vdots & & \vdots \\
0 & 0 & \cdots & 1 & 0 & a_{n_e+1,n_e-1} & \cdots & a_{n_v,n_e-1} \\
0 & 0 & \cdots & 0 & 1 & a_{n_e+1,n_e} & \cdots & a_{n_v,n_e} \\
0 & 0 & \cdots & 0 & 0 & a_{n_e+1,n_e+1} & \cdots & a_{n_v,n_e+1} \\
\vdots & \vdots & & \vdots & \vdots & \vdots & & \vdots \\
0 & 0 & \cdots & 0 & 0 & a_{n_e+1,n_e+n_i} & \cdots & a_{n_v,n_e+n_i}
\end{bmatrix}
$$

where n_v is the number of variables, n_e is the number of linear independent equalities and n_i is the number of inequalities. After the elimination, inequalities contain a reduced number of variables; thus increasing system performance.

Gauss-Jordan elimination has been proved to be a polynomial time method for exact calculus (Schrijver, 1986), while for floating-point arithmetic its complexity is $O(n^3)$. In IDIOM, the algorithm is implemented using sparse matrices, thus improving efficiency (more than 95% of the coefficients in typical problems are zero).

3.3. TREATMENT OF INEQUALITIES

Design spaces are defined by inequalities that have been simplified using dimensionality reduction. CADRE employs recursive transformation (RT) of all violated inequalities into equalities in order to define the parameterization for adaptation (Hua, 1994). This method may omit correct solutions.

For example, consider two objects in a design having minimum vertical dimensions, y_{1min} and y_{2min}, as shown on the left-hand side of Figure 4. The neighbourhood relationship in the middle of the figure requires that Object 2 has at least the same size as Object 1. The right hand side of Figure 4 gives an obvious solution. Among the inequalities describing this example the following three cause RT to report a conflict although there isn't one.

$$y_1 \geq y_{1min}$$
$$y_2 \geq y_{2min}$$
$$y_2 \geq y_1$$

y_{1min} and y_{2min} are constants, y_1 and y_2 are the vertical sizes of Object 1 and Object 2. When RT is used, it detects that the current values shown on the left of Figure 4 violate the second and third inequality and will transform these into equalities. This forces both objects to take a size of y_{2min}, resulting in a conflict with the minimum size of Object 1. No solution is reported. However in reality, one exists as shown on the right-hand side of Figure 4.

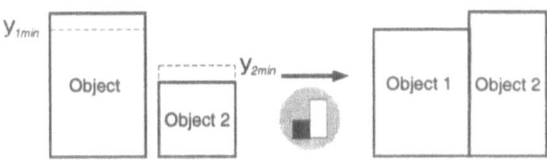

Figure 4. Example of two violated inequalities.

IDIOM avoids this by using the Fourier-Motzkin elimination method which is an algorithm for solving inequality-systems (Motzkin, 1936). The procedure involves eliminating all variables one by one until a simple inequality-system with only one variable is found (Schrijver, 1986). For each variable, the Fourier-Motzkin elimination calculates the following inequalities:

$$x_i \leq \sum_{j=i+1}^{n_v} c_{kj}x_j + b_k \quad ; \quad k = 1 \ldots l_i$$

$$x_i \geq \sum_{j=i+1}^{n_v} c_{kj}x_j + b_k \quad ; \quad k = l_i + 1 \ldots m_i$$

where c, b, l and m are constants determined by the Fourier-Motzkin elimination and n_v is the number of variables. These inequalities allow the solver to calculate an interval of possible values for variable x_i the bounds of which depend only on $x_{i+1} \ldots x_{n_v}$, where the interval for x_{n_v} is given by constants. To find a solution for the inequalities, the solver starts by choosing a value for x_{n_v}. If this value is chosen within the interval for x_{n_v} the Fourier-Motzkin elimination guarantees that, for x_{n_v-1}, an interval of possible values can also be found. Therefore the solver can recursively determine values for all variables.

Using intervals of possible values, it is easy to find a solution which is as near to the current solution as possible. The solver chooses a value for a variable by checking its interval of possible values. If the current value of the variable is within the interval the solver will use this value. If the value is outside it will be set to the nearest interval boundary.

In general, this algorithm generates an exponential number of inequalities. However, Nelson (Schrijver, 1986) showed that if each inequality involves only two variables, the Fourier-Motzkin elimination method has a complexity of $O(mn^{(2 \log n + 3)} \log n)$. Unfortunately the form of inequalities in IDIOM cannot be restricted in this way. Nevertheless, the use of sparse matrices and the reduction of redundant constraints (Lassez et al., 1993), have improved performance for problems treated by IDIOM. Table 1 shows the effect of redundancy reduction on a small example. It involves 3 rooms with few elements; 20 variables, 7 equalities and 41 inequalities.

TABLE 1. Constraints generated by Fourier-Motzkin elimination, with and without redundancy reduction (rr).

	w/o redundancy reduction (rr)	with rr
Generated constraints	1.6e06	132
Stored constraints	2784	58

3.4. PREFERENCE ACTIVATION

Constraints in IDIOM may be fixed or preferred, hereafter referred to respectively as fixed constraints and preferences. Fixed constraints must be fulfilled while preferences may be deactivated if they are in conflict with other preferences or fixed constraints. Preferences are reactivated when possible. The priority of a preference can be defined and preferences may have equal priority. IDIOM fulfils all fixed constraints and as many preferences as possible using the following heuristics :

- A preference that conflicts with fixed constraints is deactivated
- If two preferences with different priorities conflict, the higher priority preference is activated
- If two preferences with the same priority conflict, IDIOM activates the preference which conflicts with fewer lower priority preferences
- IDIOM re-activates preferences whenever possible

Preferences are divided into groups of equal priority and activated in order of importance. For example, six preferences are divided into three groups according to priority. The most important group g_1 contains p_1, p_2 and p_3, the second group g_2 contains p_4 and p_5 and the least important group g_3 contains p_6.

The activation of preferences starts with none activated; as many preferences as possible are activated in the first group through checking feasibility with all fixed constraints. This is performed incrementally for each preference. Several feasible combinations of preferences may have the maximum number of preferences activated and therefore these are stored into a list of solutions. In this example, two preferences out of g_1 can be activated and the following combinations are possible: $\{p_1, p_3\}$ and $\{p_2, p_3\}$. Then L, the list of solutions after treatment of g_1, is :

$$L = (\{p_1, p_3\}, \{p_2, p_3\})$$

The activation of preferences then sequentially considers all entries in the list with additions from g_2, and stores all combinations which have the maximum number of preferences activated. Thus the combination $\{p_1, p_3\}$ is considered first and IDIOM finds that only p_4 can be added. Then preference activation treats the combination $\{p_2, p_3\}$ and finds for instance, that only p_5 can be activated together with this second combination. Thus two solutions are found and a new list is created :

$$L = (\{p_1, p_3, p_4\}, \{p_2, p_3, p_5\})$$

After treating all preference-groups in this manner, preference activation terminates with a list of feasible combinations which contain as many important preferences as possible. One of these is then used to recalculate the new values of the design's parameters and for subsequent adaptation. For example the preference in g_3 can be added with the second combination in L, but not with the first combination. The final list contains one combination of feasible preferences which is used in further calculations.

$$L = (\{p_2, p_3, p_5, p_6\})$$

4. Designer Interaction

Since it is impossible to model everything which influences complex design tasks, interactive design systems are essential. Interactivity must not be understood to indicate an absence of reliable computational methods for automating certain tasks. Users wish to interpret designs and their contexts at intermediate stages and such input is essential for successful designs. Moreover, this interpretation is what designers enjoy doing best and because of this, they will never use a system which does not allow for such interaction.

When constructing intelligent design systems, the following three decisions must be taken:

- How much of the design task will be completely automated ?
- What tasks will be supported through interaction between the system and the designer ?
- When will users be required to perform tasks independent of computer support ?

No two systems propose the same answers to these questions. Our goal is to develop conditions where designers feel *encouraged* to explore the space of feasible design solutions. Support for design exploration is an essential element of intelligent design support (Gero and Kazakov, 1996; Logan and Smithers, 1993; Maher and Poon, 1995; Navinchandra, 1991). In our work, constraints are viewed as useful representations of the *boundaries* of possible design spaces. Since it is impossible to model all design knowledge, these constraints cannot sufficiently

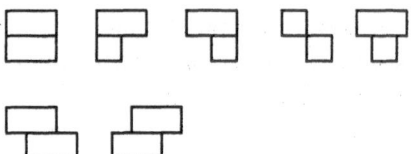

Figure 5. Examples of primitive topologies. These topologies form the basis for neighbourhood relationships.

define what is feasible – instead they are a *partial* description of what is *not* feasible. User interpretation is employed to refine the definition of design spaces for particular contexts.

IDIOM supports interaction with designers in the following ways :

- User interpretation
- Active design support
- Design critiquing

The next three subsections describe these aspects in more detail.

4.1. USER INTERPRETATION

A designer can interpret a given design, or a group of objects by activating models. This is done by choosing the desired model from a pull-down menu. For example, if the designer wishes to have a luxury apartment by introducing the relevant model, the minimum amount of space required for certain objects in the apartment would increase.

The designer may define neighbourhood relationships between two rooms and cause the design to change. A neighbourhood relationship can be specified for each pair of adjoining objects and is done so according to primitive topologies, as in Figure 5. These relationships are declared, changed and removed by choosing two objects consecutively with the mouse. The choice of neighbourhood relationship may affect the size and shape of both of the two adjoining objects. The most direct method of user interpretation is constraint posting. This is done by double clicking on an object which produces a dialog box containing the current values of the object. The designer posts constraints into the box. The constraints that can be fixed for each object are specifications such as minimum width and length, fixed and therefore absolute, width and length and fixed minimum area. Preferences are posted similarly and are given a priority. In this way, the designer can specify that the minimum area of a dining room is more important than the preferred size of a single bedroom. The designer can weaken and strengthen priorities as required by reorganizing their order in a dialog box. For example, if the user decides that amongst the preferences given in Section 3.4, group g_2 is more important than g_1,

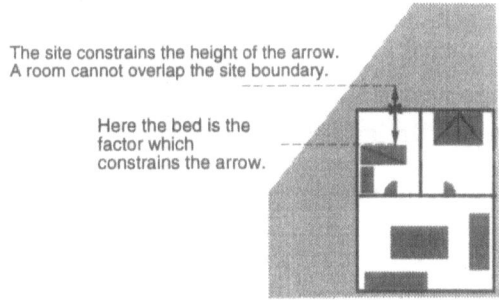

The site constrains the height of the arrow.
A room cannot overlap the site boundary.

Here the bed is the
factor which
constrains the arrow.

Figure 6.   The double arrow indicates the range of permitted adaptation

the list of feasible preference sets becomes :

$$L = (\{p_1, p_4, p_5\}, \{p_2, p_4, p_5\})$$

The list, L is determined because p_4 and p_5 are activated together (being the most important and not in mutual conflict), p_3 and p_6 are in conflict with p_4 and p_5 and only one of p_1 and p_2 is compatible at a time. Since L is now different from the solution given in Section 3.4, a new design space is available for exploration.

4.2. ACTIVE DESIGN SUPPORT

Active design support is most apparent in the displaying of arrows to indicate how far a given wall, window or piece of furniture can be moved. The element is clicked on with the mouse, an arrow appears indicating the permitted range and, the element is moved using the mouse. Figure 6 shows an arrow whose length was calculated through consideration of the site boundary and the position of the bed in the room.

Arrows were also used in Dave et al. (1994). IDIOM reuses the idea of arrows but extends them to deal with elements as well as walls in a design.

Active support is also provided in the form of hints. These hints give advice mainly on topology. For example, it may be suggested that the user puts the living room on the south side of the building, or placing the bathroom in the night zone of the apartment. These hints can be toggled on and off as requested. If hints have been provided and then not followed, then the system will notify the user in terms of critiquing.

4.3. DESIGN CRITIQUING

Design critiquing is a well established form of user interaction (Stolz, 1994). Users are advised of non-critical inconsistencies in their design. Critiques are provided immediately after the user has declared a neighbourhood relationship that precludes compliance with a hint provided prior to this declaration. Currently, hints and critiques are available in a dialog box that can be turned on and off by the user.

5. Implementation and Design Scenario

IDIOM is implemented in C and C++ with OpenGL and Motif as the user interface platform. The following is an example of a possible design scenario using the system (each step performed by the user):

1. Define the dimensions of the site where the layout must be placed
2. Choose a case from the case browser and place it into the site. At this point, constraints contained in the case and those activated by models are added to the constraint set
3. Define neighbourhood relationships with adjacent objects. This action automatically adds more constraints to the constraint set
4. Where needed, post additional constraints
5. Request solution. Here the system calculates the feasible solution space through conflict resolution with preferences and dimensionality reduction and selects a solution that involves minimal changes to the case and to the current design
6. Interactively adapt positions of walls, furniture, windows and doors to obtain configuration required
7. Return to step two

The screens shown in Figure 7 refer to step 2 on the left and step 5 on the right. On the left, a double bedroom is being added to the design. After user interpretation, in this case specifying that the hall should share the length of the right wall through declaration of a neighbourhood relationship, the solution proposed is shown on the right. Note that the vertical dimensions of both the hall and the bedroom have changed.

Once a solution is proposed, the user may wish to change positions of walls and elements within objects. This is carried out through clicking on a wall or element. The results of the dimensionality reduction are used to calculate the range of adaptation possible, as described in Section 3. The screens in Figure 8 show that moving a wall may change other dimensions that are linked in the parameterization. Here, the living room has been constrained to have the same length to width ratio, the bathroom dimensions have been fixed and the kitchen wall is required to share the whole right wall of the living room. Therefore, moving the living room wall results in a reduction in size of the single room at the top.

6. Testing and Validation

Testing with architects has produced mixed reactions. More traditional architects who are used to working within well defined schemas and grids find that IDIOM does not reflect what they do and therefore, the system provides little support. In

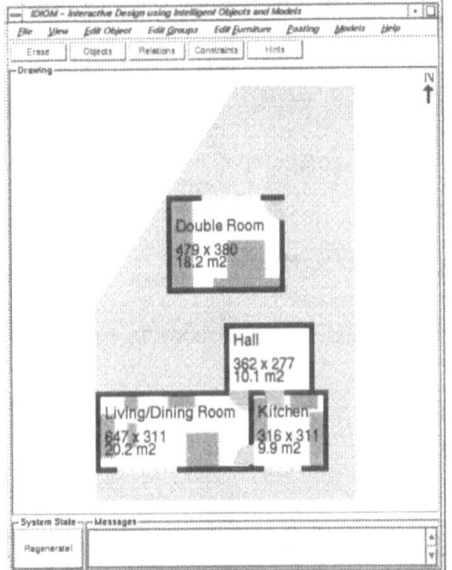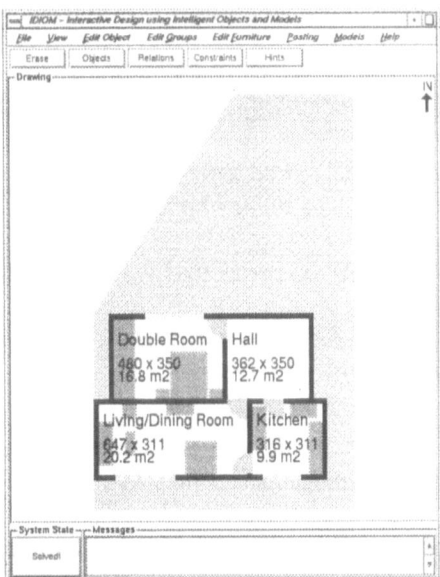

Figure 7. Adding a case to a design (left) and solution proposed after user interpretation and resolution with relevant constraints (right).

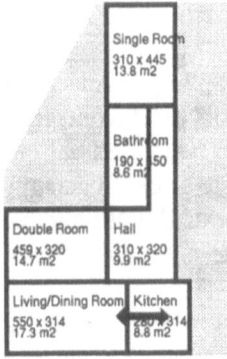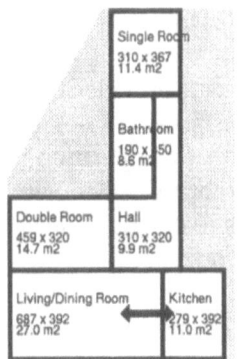

Figure 8. The figure illustrates Step 6, interactive adaptation. The user clicks on the right wall of the living/dining room (left) and drags it to the desired position (right). Note the changes to dimensions of the single room at the top.

contrast, other architects have found that IDIOM provides them with an opportunity to get away from traditional methods, thus allowing them to explore new architectural approaches.

The following comments are a sample of those which have been made by designers after becoming acquainted with the system :

"I prefer to compose building parts for my particular problem rather than adapt existing configurations such as complete layout designs."

"Automatic layout generation systems are not very attractive. I want to introduce constraints as the layout is composed."

"Grids are too restrictive when initial layouts are being examined."

"Use of models are interesting provided that non-essential models can be turned on and off."

"Preferences play an important role in our decision making. Support for preference management is helpful"

"Rational management of dimensions seems to be the biggest advantage of IDIOM."

"Neighbourhood relationships between objects need to be made more visual. Adaptation was occasionally blocked because we had overlooked a restrictive condition on topology."

"Sometimes we got stuck when there was no feasible solution."

The first two comments provide support for an important starting assumption we have made: that designers do not want automatic layout generation. The third comment suggests that computation needs to be carried out in terms of continuous variables rather than working with discrete grids. In general, these comments have encouraged us to continue development of IDIOM and have helped fix priorities for further work.

Limitations   The dimensional parameterization described is currently limited to rectangular spaces and elements. Only values for continuous variables are manipulated in IDIOM. Complex non-linear constraints slow the system down to the point where interactive design becomes difficult. For interactive use, constraints are formulated to be as close as possible to linear relationships. The current implementation of IDIOM allows for linear and simple non-linear constraints, such as those applied to areas of objects. When no solution can be found, the system currently provides little help identifying constraints that when modified, would lead to solution. This task is far from trivial and is doubtful that a general approach will be found. Nevertheless, some support should be possible under certain conditions. Our current work is focused on addressing these issues and on improving user interaction.

7. Related Work

The system most closely related to IDIOM is CADRE (Dave et al., 1994; Hua, 1994; Hua et al., 1992). Similarities include i) the use of dimensionality reduction and run-time parameterization to simplify adaptation and ii) certain aspects of user interaction, such as the use of arrows for defining feasible modifications. IDIOM differs from CADRE in the following ways: i) IDIOM employs intelligent objects to compose topological configurations where the CADRE implementation combines predefined configurations, ii) IDIOM accommodates preference constraints whereas in CADRE, all constraints are fixed, iii) in IDIOM, elements within spaces, such as furniture, doors and windows are included in the parameterization whereas in CADRE, only spaces and structural elements are included, iv) as explained in Section 3, IDIOM employs a more reliable algorithm for accommodation of inequalities during case combination, v) IDIOM employs explicitly defined domain models that are activated by the user whereas in CADRE domain knowledge was loaded into the system at the beginning and finally, vi) the opportunities for interactivity in IDIOM correspond more closely to the needs of building designers who were interviewed than in CADRE. Perhaps the most important difference between IDIOM and CADRE is that in IDIOM, the topology is determined interactively by the user, thereby avoiding difficulties of complexity experienced with CADRE when generating topologies.

The FABEL project, coordinated by GMD, St. Augustin (Bakhtar, 1993) focuses upon the application of case based design to heating and ventilating configuration of buildings. Although most of the effort in this project has been concerned with case indexing, recent work includes a study of three adaptation methods (Börner, 1995). FABEL uses fixed grids to model spatial information and does not perform case-combination. Other work includes the SEED project (Flemming, 1994) where large numbers of cases are stored and indexed for retrieval using functional units. Although a case editor is a available for adaptation, no other computational support is reported. Our approach is different to these two systems due to our capabilities to combine complex objects through run-time parameterizations that include design preferences, constraints from user-activated models and other opportunities for user interpretation.

An extension to CADSYN (Zhang and Maher, 1993) employs constraint satisfaction techniques for verification and repair of adapted designs. CADSYN ensures local consistency between constraints, thereby limiting its effectiveness to constraint networks where risks of divergence, looping and empty solution spaces are low. Our experience with geometric design has revealed that relevant constraint networks are highly interdependent and therefore, local consistency approaches are unreliable.

WRIGHT is another constraint based system created for layout synthesis (Baykan and Fox, 1992). Layouts are automatically generated and local consistency is achieved through use of the Waltz algorithm, thus risking cycles and divergence. In a comparison with methods based on hierarchical generate and test (Flemming et al., 1992), it was concluded that constraint propagation techniques are more efficient under certain circumstances. Since WRIGHT performs monotonic search, soft constraints are never reactivated if weakened. IDIOM differs from WRIGHT in the following ways : i) layouts are not generated in IDIOM – the user defines topology incrementally, ii) IDIOM does not propagate constraints but *solves* them, and together with the Fourier-Motzkin algorithm, identifies globally feasible solutions without running the risk of propagation cycles and iii) IDIOM may demonstrate non-monotonic behaviour as cases are added since preferences may be reactivated.

Work on layouts is being performed by Giretti et al. (1994). They report on a CBD system for architecture that supports graphical interaction. Their "theories" are similar to the models in IDIOM and "scenes" are analogous to groups of intelligent objects. However, run-time parameterization and subsequent dimensionality reduction is not performed. Therefore, performance problems would be expected for designs of realistic size. In addition, it is not clear whether local or global consistency is achieved during constraint solving. Finally, grids are used to adapt previous designs and topological adaptation is performed automatically. These functionalities were avoided during development of IDIOM because designers who were interviewed thought they would hinder rather than help layout design.

8. Conclusions

IDIOM provides a useful framework for supporting interactive spatial composition. Through solving constraints contained in cases, those generated during user interaction and those obtained from model activation, the user is able to explore a range of design solutions within globally consistent design spaces. This exploration is further enhanced by the accommodation of preferences in constraint sets and the opportunity to alter their priorities interactively. Algorithms are sufficiently fast and reliable to support exploration in an interactive manner. Reactions from designers indicate that IDIOM provides a good mix of computation and user support for spatial configuration.

Acknowledgements

The funding for this project was provided by the Swiss Priority Programme in Computer Science (SPP-IF). This project was performed in collaboration with CAAD, Federal Institute of Technology, Zurich. Beginning ideas related to intelligent objects arose during discussions with Boi Faltings and Gerhard Schmitt and Boi Faltings initially proposed an investigation of model-based case adaptation for this work. The authors would like to thank Nathanea Elte and David Kurmann for their collaboration as well as our practising architects: Geninasca - Delfortrie Architects, Neuchatel; Atelier d' Architecture, Lausanne; and Archilab, Lausanne for providing comments and for helping with testing and validation.

References

Adani, P.: 1995, Adaptation by active autonomous objects, Modules for design support, *FABLE Report No 35*, GMD.

Bahktari S. and Bartsch-Spörl B.: 1993, Our perspective on using CBR in design problem solving, *1st European Workshop on CBR*, Kaiserslauten.

Baykan, C. A. and Fox, M. S.: 1992, WRIGHT: A constraint based spatial layout system, *AI in Engineering Design*, **1**, 395–432.

Bakhtari S. et al.: 1993, EWCBR93: Contributions of FABEL, *Fabel Report No. 17*, GMD, Sankt Augustin

Börner, K.: 1995, Modules for design support, *FABLE Report No. 35*, GMD.

Borning, A., Freeman-Benson, B. and Wilson, M. Constraint hierarchies, *Lisp and Symbolic Computation*, **5**, 223–270.

Coulon, C-H.: 1995, Automatic indexing, retrieval and reuse of topologies in architectural layouts, *CAAD Futures'95*, Singapore.

Choi, B. and Flemming, U.: 1995, Adaptation of a layout design system to a new domain, *CAAD Futures'95*, Singapore.

D'Ambrosio, J. G. and Birmingham, W. P.: 1995, Preference directed design, *AIEDAM*, **9**, 219–230.

Dave, B., Schmitt, G., Faltings, B. and Smith, I.: 1994, Case-based design in architecture, *in* J. S. Gero and F. Sudweeks (eds), *Artificial Intelligence in Design '94*, Kluwer, Dordrecht, pp. 145–162.

de Kleer, J. and Brown, J. S.: 1984, A qualitative physics based on confluences, *Artificial Intelligence*, **24**.

Faltings, B.: 1991, Case based representation of architectural design knowledge, *Computational Intelligence*, **2**.

Flemming U.: 1994, Case-based design in the SEED System, *1st Computing Congress*, American Society of Civil Engineers, Washington.

Flemming, U., Coyne, R., Glavin, T. and Rychener, M.: 1988, A generative expert system for the design of building layouts, *AI in Engineering*, 445–464.

Flemming, U., Baykan, C. A., Coyne, R. F. and Fox, M. S.: 1992, Hierarchical generate and test vs constraint-directed search, *in* J. S. Gero (ed.), *Artificial Intelligence in Design '92*, Kluwer, Dordrecht, pp. 817–838.

Fujii, H.: 1995, Incorporation of natural language processing and a generative system, *CAAD Futures '95*, Singapore.

Gero, J. S.: 1990, Design prototypes: a knowledge representation schema for design, *AI Magazine*, **11**(4), 26–36

Gero, J. S. and Kazakov, V.: 1996, An exploration-based evolutionary model of a generative design process, *Microcomputers in Civil Engineering* (to appear), available at http://www.arch.su.edu.au/ john/publications.html.

Gero, J. S. and Schnier, T.: 1995, Evolving representations of design cases and their use in creative design, *in* J. S. Gero, M. L. Maher and F. Sudweeks (eds), *Preprints Computational Models of Creative Design*, Key Centre of Design Computing, University of Sydney, pp. 343–368.

Giretti, A., Spalazzi, L. and Lemma, M.: 1994, A.S.A.: An interactive assistant to architectural design, *in* J. S. Gero and F. Sudweeks (eds), *Artificial Intelligence in Design '94*, Kluwer, Dordrecht, pp. 93–108.

Goel, A. K. and Chandrasekaran, B.: 1989, Use of device models in adaptation of design cases, *DARPA CBR Workshop*, pp. 100–109.

Goel, A. K. and Kolodner, J. L.: 1991, Towards a case-based tool for aiding conceptual design problem solving, *DARPA CBR Workshop*, pp. 109–120.

Hinrichs, T. R. and Kolodner, J. L.: 1991, The roles of adaptation in case-based design, *DARPA Case-based Reasoning Workshop*, pp. 121–132.

Hinrichs, T. R.: 1992, *Problem Solving in Open Worlds*, Lawrence Erlbaum, Hillsdale, NJ.

Hua, K.: 1994, *Case-Based Design of Geometric Structures*, Thesis No. 1270, Swiss Federal Institute of Technology, Lausanne.

Hua, K., Smith, I., Faltings, B., Shih, S., and Schmitt, G.: 1992, Adaptation of Spatial Design Cases, *in* J. S. Gero (ed.), *Artificial Intelligence in Design '92*, Kluwer, Dordrecht.

Krishnaiah, L. and Kanal, P.: 1982, *Handbook on Statistics, Volume 2*, North-Holland, Amsterdam.

Lassez, J-L., Huynh, T. and McAloon, K.: 1993, Simplification and elimination of redundant linear arithmetic constraints, *Constraint Logic Programming*, MIT Press.

Logan, B. S., Corne, D. W. and Smithers, T.; 1991, Enduring support: On defeasible reasoning in design support systems, *in* J. S. Gero (ed.), *Artificial Intelligence in Design '91*, Butterworth-Heinemann, Oxford, pp. 433–454.

Logan, B. and Smithers, T.: 1993, Creativity and design as exploration, *in* J. S. Gero and M. L. Maher (eds), *Modelling Creativity and Knowledge Based Design*, Lawrence Erlbaum, pp. 139–175.

Maher, M. L. and Poon, J.: 1995, Modelling design exploration as co-evolution, accepted for the *Special Issue of Microcomputers in Civil Engineering on Evolutionary System in Design*, available from http://www.arch.su.edu.au/ josiah/CoGA.html.

Maher, M. L. and Zhang, D. M.: 1991, Case-based reasoning in design, *in* J. S. Gero (ed.), *Artificial Intelligence in Design '91*, Butterworth-Heinemann, Oxford, pp. 137-150.

Medjdoub, B. and Yannou, B.: 1996, Towards a new generation of architectural CAD softwares, accepted for *ITCSED-96*, Glasgow, Scotland.

Mitchell, W. J. Steadman, J. P. and Ligget, R. S.: 1976, Synthesis and optimisation of small rectangular floor plans, *Environment and Planning B*, **3**, 37–70.

Motzkin, T. S.: 1936, *Beiträge zur Theorie der linearen Ungleichungen*, PhD Thesis, University of Basel, Germany.

Navinchandra, D.: 1988, Case based reasoning in CYCLOPS, *DARPA Case-Based Reasoning Workshop*, pp. 286–291.

Navinchandra, D.: 1991, *Exploration and Innovation in Design: Towards a Computational Model*, Springer-Verlag.

Purvis, L. and Pu, P.: 1995, Adaptation using constraint satisfaction techniques, *CBR Research and Development, Lecture Notes in AI 1010*, Springer-Verlag, pp. 289–300

Rigopoulos, D. R. and Oppenheim, I. J.: 1992, Intelligent objects for synthesis of structural systems, *Journal of Computing in Civil Engineering*, **6**, 266-281.

Saund, E.: 1989, Dimensionality reduction using connectionist networks, *IEEE Trans. PAMI*, **11**, 304-331.

Schmitt, G.: 1993, Design reasoning with cases and intelligent objects, *International Association of Bridge and Structural Engineering, Report 68*, pp. 77-87.

Schrijver, A.: 1986, *Theory of Linear and Integer Programming*, Wiley, Chichester.

Smith, I., Lottaz, C. and Faltings, B.: 1995, Spatial composition using cases: IDIOM, *CBR Research and Development, Lecture Notes in AI 1010*, Springer-Verlag, pp. 88–97.

Stolz, M.: 1994, Visual critiquing in domain oriented design environments: showing the right thing at the right place, *in* J. S. Gero and F. Sudweeks (eds) *Artificial Intelligence in Design '94*, Kluwer, Dordrecht, pp. 467–482.

Sycara, K. P. and Navinchandra, D.: 1991, Influences: A thematic abstraction for creative use of multiple cases, *DARPA CBR Workshop*, pp. 133-144.

Tommelein, I. D.: 1989, *SightPlan - An Expert System for Designing Construction Site Layouts*, PhD Thesis, Stanford University.

Voss, A.: 1994, The need for knowledge acquisition in case-based reasoning—some experiences from an architectural domain, *11th ECAI*, Wiley, pp. 463–467.

Wilson, M. and Borning, A.: 1993, Hierarchical constraint logic programming, *Journal of Logic Programming*, **16**, pp. 277–31.

Zhang, D. M. and Maher, M. L.: 1993, Using CBR for the synthesis of structural systems, *Inter. Assoc. for Bridge and Structural Engineering, Report 68*, pp. 143-152.

Zhao, F. and Maher, M. L.: 1992, Using network-based prototypes to support creative design by analogy and mutation, *in* J. S. Gero (ed.), *Artificial Intelligence in Design '92*, Kluwer, Dordrecht, pp. 773-793.

3

genetic algorithms/genetic programming in design

Integrating a genetic algorithm into a knowledge-based system
for ordering complex design processes
James L. Rogers, Collin M. McCulley, Christina L. Bloebaum
AI in control system design using a new paradigm for design
representation
Sourav Kundu, Seiichi Kawata
Automated design of both the topology and sizing of analog
electrical circuits using genetic programming
John Koza, Forrest H. Bennett, David Andre, Martin A. Keane

J. S. Gero and F. Sudweeks (eds), Artificial Intelligence in Design '96, 119-133
© 1996 Kluwer Academic Publishers.

INTEGRATING A GENETIC ALGORITHM INTO A KNOWLEDGE-BASED SYSTEM FOR ORDERING COMPLEX DESIGN PROCESSES

JAMES L. ROGERS
NASA Langley Research Center
Mail Stop 159
Hampton, VA 23681 USA

AND

COLLIN M. MCCULLEY AND CHRISTINA L. BLOEBAUM
State University of New York at Buffalo
1009 Furnas Hall
Department of Mechanical & Aerospace Engineering
Buffalo, NY 14260 USA

Abstract. The design cycle associated with large engineering systems requires an initial decomposition of the complex system into design processes which are coupled through the transference of output data. Some of these design processes may be grouped into iterative subcycles. In analyzing or optimizing such a coupled system, it is essential to be able to determine the best ordering of the processes within these subcycles to reduce design cycle time and cost. Many decomposition approaches assume the capability is available to determine what design processes and couplings exist and what order of execution will be imposed during the design cycle. Unfortunately, this is often a complex problem and beyond the capabilities of a human design manager. A new feature, a genetic algorithm, has been added to DeMAID (Design Manager's Aid for Intelligent Decomposition) to allow the design manager to rapidly examine many different combinations of ordering processes in an iterative subcycle and to optimize the ordering based on cost, time, and iteration requirements. Two sample test cases are presented to show the effects of optimizing the ordering with a genetic algorithm.

1. Introduction

Many engineering systems are large and multidisciplinary and require a complex design cycle. Before a design cycle begins, the possible couplings among the design processes must be determined. After these possible couplings have been defined, a design cycle can be decomposed to identify its multilevel structure. The Design Manager's Aid for Intelligent Decomposition (DeMAID) is a knowledge-based software tool for ordering the sequence of design processes and for identifying a possible multilevel structure for a design cycle (Rogers 1989). The DeMAID software displays the processes in a design structure matrix format (DSM) in which an element on the diagonal is any process that requires input and generates an output (Steward 1981). Off-diagonal elements indicate a coupling between two processes. The primary advantage of the DSM over display tools such as Program Evaluation and Review Technique (PERT) or process flowcharts is the ability to group and display the iterative subcycles that are commonly found in the design cycle. After the iterative subcycles have been determined, their processes must be ordered in a manner that will produce a design in the least time and at minimum cost. The original DeMAID software employs a knowledge base to handle this task; however, the knowledge-based approach only examines a limited number of orderings, which provides the user a starting point from which to interactively search for the optimum sequence. This paper introduces a genetic algorithm (GA) capability that has been added to DeMAID. This GA examines a large number of orderings of processes in each iterative subcycle and optimizes the orderings based on cost, time, and iteration requirements.

2. Design Structure Matrix

The DSM is used to display the sequence of processes (Steward 1981). A sample DSM is shown in Figure 1. In the DSM, the processes are shown as numbered boxes on the diagonal. Output from a process is shown as a horizontal line that exits a process box, and input is shown as a vertical line that enters a process box. The off-diagonal squares that connect the horizontal and vertical lines represent couplings between two processes. Couplings in the upper triangle portion of the DSM represent feedforward data; couplings in the lower triangle part of the matrix represent feedback data. A feedback implies an iterative process in which an initial guess must be made. The knowledge base within DeMAID which is written with the C Language Integrated Production System (CLIPS, Giarratano and Riley 1989) orders the processes to eliminate as many feedbacks as possible. However, in many cases, not all of the feedbacks can be eliminated. If any feedbacks remain, DeMAID groups the processes into iterative subcycles called circuits. In Figure 1, processes 1-3, 5-19, 21-25, and 26-29 are grouped into circuits.

The DeMAID software also identifies crossovers. Crossover, in this context, occurs when feedback from one process crosses that of another process without an exchange of data through the intersection (no off-diagonal square). Crossovers are only defined in terms of feedbacks. For example, in Figure 1 a crossover occurs when the feedback from process 14 to process 7 crosses the feedback from process 17 to process 12. Crossovers should be avoided if possible because they obscure when to end one iterative loop and begin another. The DSM shown in Figure 1 contains 20 feedbacks and 3 crossovers.

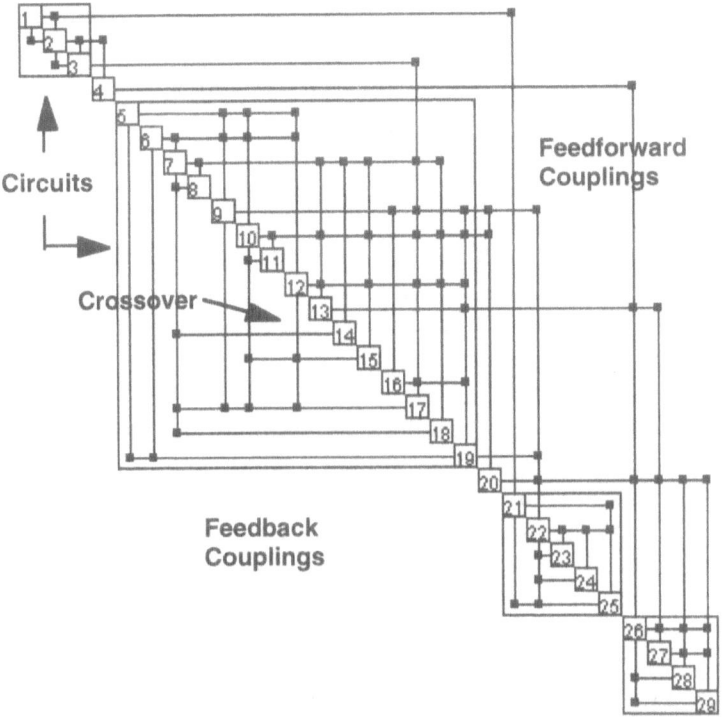

Figure 1. A design structure matrix.

In the original version of DeMAID, a knowledge base was used to minimize feedbacks and group processes into circuits. Crossovers were identified but were not minimized. No time factors, cost factors, or iteration factors (i.e. the number of iterations required for convergence) were applied. After the circuits were identified, DeMAID attempted to minimize the feedbacks within a circuit. In most cases, although more than one ordering could produce the minimum amount of feedbacks, only one ordering was identified.

A large circuit such as the one shown in Figure 1 that contains processes 5-19 can be very expensive to converge because the iterative loops defined by the feedbacks are nested, which require numerous executions of potentially expensive processes. Thus, a new technique is needed that rapidly examines many different orderings of processes within a circuit and selects the best ordering based on cost, time and iteration requirements. The GA capability that has been added to DeMAID meets this need.

3. Coupling Strengths

In the original version of DeMAID, a coupling either existed or not. The strength of the coupling could not be quantified. In the latest version of DeMAID, seven levels are used to quantify coupling strengths They are: extremely weak, very weak, weak, nominal, strong, very strong, and extremely strong. These strengths can be supplied by the user or they can be determined through sensitivity analysis (Bloebaum 1992; Rogers and Bloebaum 1994) and quantified according to rules in the knowledge base. The rules for quantifying are based on a statistical analysis of the normalized sensitivities. Recommendations are made as to which processes and couplings might be removed (or temporarily suspended) from the problem without a loss of solution accuracy.

The rules for removing or retaining processes are listed here. All processes with at least one coupling of nominal strength or greater are retained. Processes with only extremely weak coupling strengths are recommended for removal. Other recommendations depend on the relationships among the processes. For example, in figure 1, if the maximum coupling strength of process 19 is very weak, then in order to be retained, one of the processes to which it is coupled (process 5, 6, or 22) must have an extremely strong coupling strength. Otherwise, process 19 is recommended for removal. Similar rules exist for removing or retaining couplings.

The DeMAID software also has the capability to display the DSM with color codings for coupling strengths. To eliminate the use of black boxes to represent couplings in the off-diagonal elements, a color scheme can be used (i.e. extremely weak, red; very weak, pink; weak, yellow; nominal, green; strong, light blue; very strong, blue; and extremely strong; black). The user can interactively move processes along the diagonal to place the weaker couplings which require fewer iterations for convergence into the feedback positions.

After the complexity of the problem has been reduced by removing processes and/or couplings, another examination can be made of the remaining circuits. An iteration factor is identified that relates the coupling strengths to the number of iterations required for convergence. The default values are shown in Table 1. The user can override these default values if necessary. If coupling strengths are not available, the assumed number of iterations for computational purposes is 1.

TABLE 1. Relation of coupling strengths to iterations required for convergence.

Coupling strength	Default Iterations
Extremely weak	2
Very weak	3
Weak	4
Nominal	5
Strong	6
Very strong	7
Extremely strong	8

4. Cost and Time Requirement Calculation

Rules were added to the DeMAID knowledge base to determine the total cost and time required for a given design process. The DSM in Figure 2 is a circuit taken from a larger design project. Each process has been assigned a cost and a time (units depend on the user). The numbers in the left-hand column correspond to the original process numbers assigned by the user. The sequence of processes has been reordered by DeMAID. This circuit contains eight feedbacks and no crossovers. Coupling strengths were not used to estimate the required number of iterations for convergence for this problem; thus each iteration factor is 1.

###	Time	Cost
11	30	10
18	40	20
21	10	20
22	20	30
20	20	10
19	30	10
1	50	10
23	30	40
17	50	30
7	30	40
8	40	30
2	40	20
6	20	50
14	20	40
13	10	30
12	20	20
3	30	30
15	30	50
16	40	40
5	10	50
4	20	40
10	40	10
9	50	20

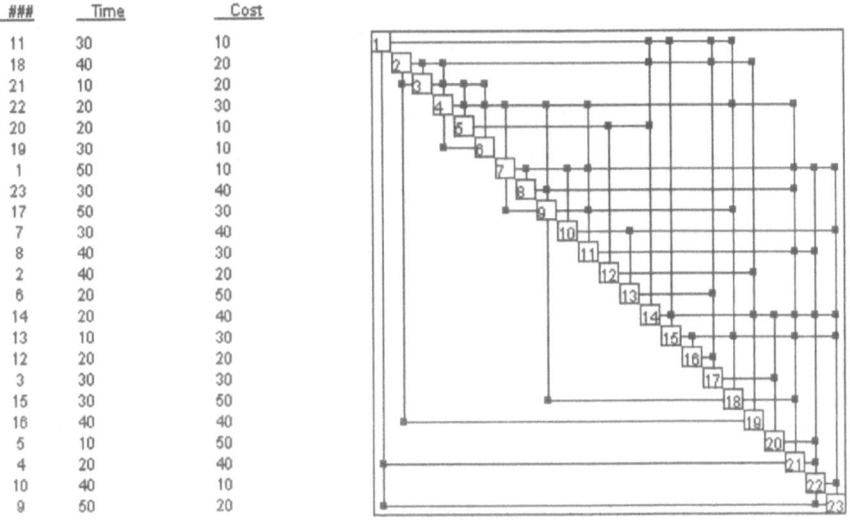

Figure 2. A design structure matrix minimized for feedbacks and crossovers.

Numerous nested iterative processes are evident within this circuit. The DeMAID software sums the time and cost of each process contained in a feedback loop and multiplies those sums by the iteration factor for the feedback. For example, the costs and times for processes 9-18 would be summed and multiplied by the iteration factor (1 in this case) for the feedback coupling from process 18 to process 9. The same would be accomplished for processes 2-19 using the iteration factor (again 1) for the feedback from process 19 to process 2. This computation continues until the contributions from all eight feedbacks have been summed. The drawback to this capability is that it only examines one ordering and makes no attempt to optimize the ordering based on cost and/or time. Thus, a decision was made to complement the knowledge base approach in DeMAID with a GA. This GA examines a large number of orderings of processes in each iterative subcycle and optimizes the ordering based on cost, time, and iteration requirements.

5. Genetic Algorithm

The use of GA's has been instrumental in achieving good solutions to discrete optimization problems that have not been satisfactorily addressed by other methods (Goldberg 1989). Because of the discrete nature of the sequencing problem, this solution technique has proved useful in solving this problem (Syswerda, 1990). A population of design points that are coded as finite-length, finite-alphabet strings is searched by the GA. Successive populations are produced primarily by the operations of selection, crossover, and mutation. The selection operator determines those members of the population that survive to participate in the production of members of the next population. Selection is based on the value of the fitness function, or the fitness of the individual members, such that members with greater fitness levels tend to survive. Crossover is the recombination of traits of the selected members, called the mating pool, in the hope of producing a child with better fitness levels than its parents. Crossover is accomplished by swapping parts of the string into which these design points have been coded. The final operation, mutation, prevents the search of the space from becoming too narrow. After the production of a child population, this operator randomizes small parts of the resulting strings, with a very low probability that any given string position will be affected.

Frequently, a binary coding is used with the GA; the values of the design variables are coded as binary numbers and then concatenated. While this approach works well with numerical problems, it is not efficient for the sequencing problem (Altus et al 1995; McCulley and Bloebaum 1994). The GA portion of DeMAID uses a direct representation of the order as a coding of an $\underline{n}$-process system, with each integer 1 through $\underline{n}$ used only once. For example, the string

$$[5\ 3\ 4\ 2\ 1]$$

represents the five-process DSM shown in Figure 3, in which the order from the top left corner of the DSM to the bottom right corner is 5, 3, 4, 2, and 1.

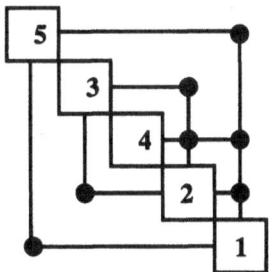

Figure 3. Five-process design structure matrix.

Selection, which only requires the use of the fitness function, is unaltered by this choice of coding. However, special operators for crossover and mutation must be used because these operators operate directly on the strings. The concern is that the result after a GA crossover or mutation operation must be a valid order (i.e. no repeated or missing processes). Valid orders cannot always be guaranteed with arbitrary switching of string information between or within strings.

Selection is accomplished by the *tournament selection* operator. To fill the mating pool, two strings are randomly selected from the parent pool and compared; the one with greater fitness is included in the mating pool. Crossover is accomplished by *position-based* (Syswerda, 1990) crossover as shown in Figure 4.

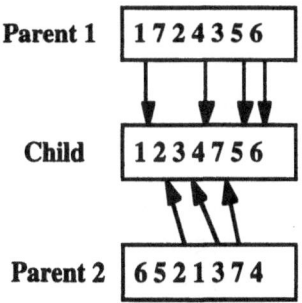

Figure 4. Position-based crossover.

Several processes (i.e. 1, 4, 5, and 6) are chosen from the first parent and placed in the same positions in the child string. Then, the processes (i.e. 2, 3, and 7) that were not taken from the first parent are taken from the second parent to fill the holes in the child string in the order in which they appear in the second parent. The result is a complete string with one and only one copy of each process number.

Mutation is accomplished through the *order-based* (Syswerda, 1990) mutation operator, as shown in Figure 5. Each string position is polled; if a given string position (i.e. position 2) is selected to undergo mutation, then its content is swapped with a randomly selected position (i.e. position 4) in the same string.

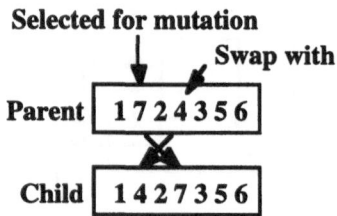

Figure 5. Order-based mutation.

In addition to minimizing feedbacks and crossovers, the fitness function for the GA in DeMAID can be used to determine the minimum cost and time required for convergence of each circuit. The GA sums the time and cost of each process contained in a feedback loop and multiplies those sums by the iteration factor for the feedback to obtain the total cost and time to converge a circuit. The user-definable weights determine the relative importance of each of the major components of the fitness function. The fitness function is:

$$\text{fitness} = 1.0/((\underline{wf}*\underline{f} + \underline{wc}*\underline{c} + \underline{wtime}*\underline{time} + \underline{wcost}*\underline{cost})**4)$$

where $\underline{f}$ is the number of feedbacks, $\underline{c}$ is the number of crossovers, $\underline{time}$ is the total time required to converge the circuit, $\underline{cost}$ is the total cost to converge the circuit; and $\underline{wf}$, $\underline{wc}$, $\underline{wtime}$, and $\underline{wcost}$ are user-definable weights. For the simple tournament selection, the relative scale of this fitness function is not important. Only the relation of the values (i.e. whether one fitness function is larger than the other) matters.

Each circuit is passed to the GA to optimize individually. A window (Figure 6) is displayed for each circuit. The window indicates the default values for the GA. The GA begins with a randomly generated initial population of a size determined by the user and proceeds from generation to generation by applying the three previously described operations.

The following parameters, shown in Figure 6, are available with their defaults in parentheses:

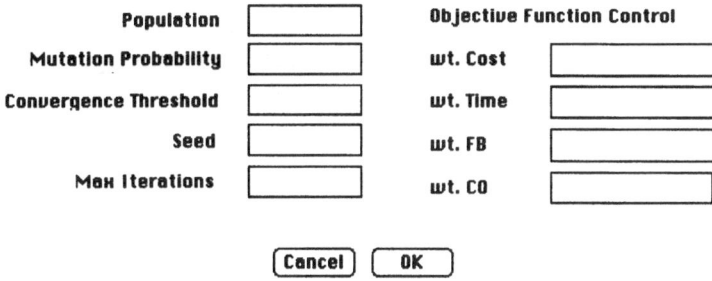

Figure 6. Window for setting genetic algorithm parameters.

- Population (100)—population size
- Mutation Probability (1.0)—mutation probability in percent, default is 1%
- Convergence Threshold (0.9)—a converged population is one for which the average fitness is at least convThresh of the best fitness, with the best fitness seen so far (default is 90%)
- Seed (3818969)—seed for random number generator
- Max Iterations (500)—maximum number of iterations to find the best sequence
- wt. Cost (1.0)—cost weight
- wt. Time (1.0)—time weight
- wt. FB (1.0)—feedback weight
- wt. CO (1.0)—crossover weight

Convergence is achieved when the average fitness of a population rises above some user-defined percentage (convergence threshold) of the best fitness for that population. At that point, the member of the population with the best fitness is chosen as the optimal. After the GA has completed ordering all the circuits, a new DSM can be displayed to demonstrate the changes.

6. Sample Cases

The two examples below indicate the savings that can be obtained by reordering the sequence of modules. In the figures, each process is assigned a cost and a time (units depend on the user). The numbers in the left-hand column correspond to the original process numbers assigned by the user. The sequence of processes has been reordered by DeMAID. Each table displays the modules coupled by feedbacks (iterative loops) for the corresponding DSM with the number of iterations for the feedback coupling along with the total time and cost to converge each iterative loop.

The DSM in Figure 7 is a circuit taken from a conceptual design project. This circuit contains 24 feedbacks and 16 crossovers. Coupling strengths are used to estimate the number of iterations required for convergence.

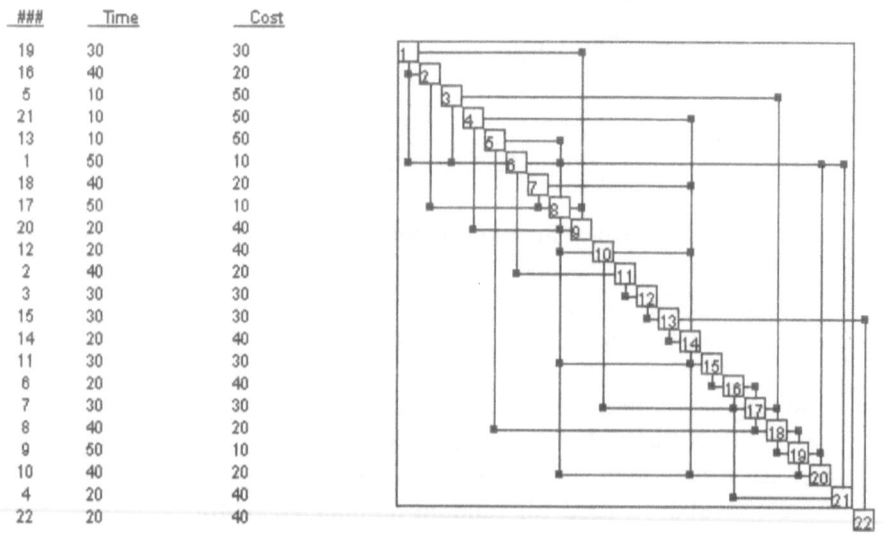

###	Time	Cost
19	30	30
16	40	20
5	10	50
21	10	50
13	10	50
1	50	10
18	40	20
17	50	10
20	20	40
12	20	40
2	40	20
3	30	30
15	30	30
14	20	40
11	30	30
6	20	40
7	30	30
8	40	20
9	50	10
10	40	20
4	20	40
22	20	40

Figure 7. A design structure matrix for example 1.

The DSM in Figure 8 contains the same set of processes with the same times, costs, and coupling strengths that are shown in Figure 7.

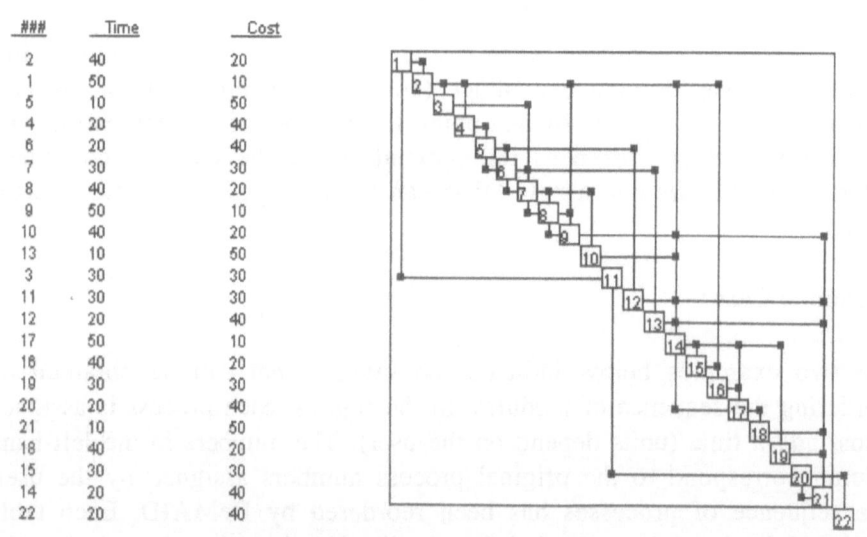

###	Time	Cost
2	40	20
1	50	10
5	10	50
4	20	40
6	20	40
7	30	30
8	40	20
9	50	10
10	40	20
13	10	50
3	30	30
11	30	30
12	20	40
17	50	10
16	40	20
19	30	30
20	20	40
21	10	50
18	40	20
15	30	30
14	20	40
22	20	40

Figure 8. Reordering of the design structure matrix for example 1

However, the sequence of processes has been reordered and optimized by the GA and are different from those in Figure 7 as shown by the numbers in the left-hand column. This DSM contains eight feedbacks and no crossovers.

Table 2 contains the data corresponding to Figure 7. The total design cycle for this DSM requires 21,340 time units and 19,640 cost units for completion.

TABLE 2. Time and cost for iterations in unordered design cycle for example 1.

To module	From module	Iterations	Time	Cost
1	2	8	560	400
1	6	4	600	840
2	8	8	1680	1680
3	6	2	160	320
4	9	7	1260	1260
5	18	6	2580	2460
6	11	8	1760	1120
7	8	6	540	180
8	9	2	140	100
8	10	8	720	720
8	15	4	960	960
8	20	7	2940	2520
10	17	8	1760	2080
11	12	5	350	250
12	13	3	180	180
13	14	6	300	420
14	15	8	400	560
14	20	4	920	760
15	16	6	300	420
16	17	7	350	490
16	21	8	1600	1280
17	18	8	560	400
18	19	6	540	180
19	20	2	180	60

Table 3 contains the data corresponding to Figure 8. The number of processes contained in the iterative loops has been reduced by reordering the sequence with the modified GA. With the same summing method described before, the total cost to complete the design cycle with this optimized ordering sequence is reduced from 19,640 to 3,950 units and the total time is reduced from 21,340 to 4,570 units.

TABLE 3. Time and cost for iteration in ordered design cycle for example 1.

To module	From module	Iterations	Time	Cost
1	11	5	1700	1600
5	6	7	350	490
6	7	8	560	400
7	8	6	540	180
8	9	2	180	400
11	21	3	960	1020
14	17	2	280	200

The DSM in Figure 2 is a circuit taken from another design project. The sequence of processes has been reordered by DeMAID. This circuit contains 8 feedbacks and no crossovers. Coupling strengths are not available therefore, the number of iterations required for convergence is set to 1.

The DSM in Figure 9 contains the same set of processes with the same times and costs, that are shown in Figure 2. However, the sequence of processes has been reordered and optimized by the GA and are different from those in Figure 2 as shown by the numbers in the left-hand column. This DSM also contains 8 feedbacks and no crossovers.

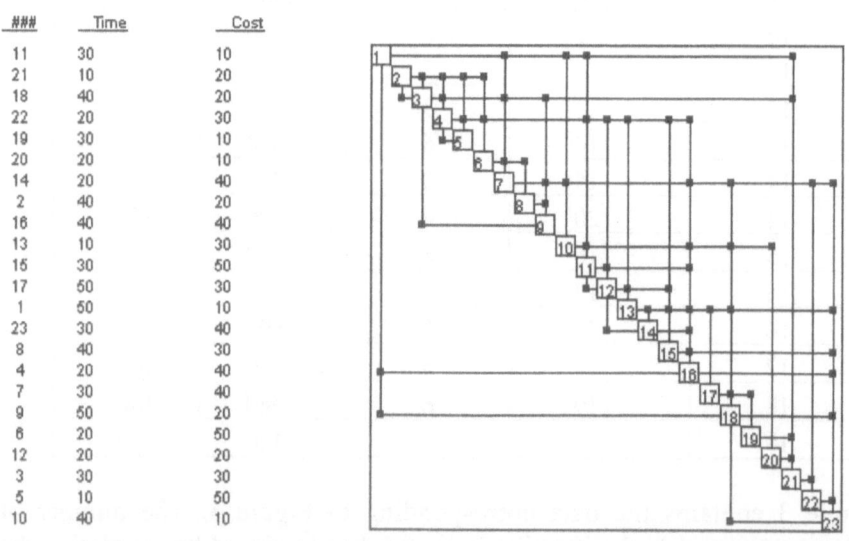

###	Time	Cost
11	30	10
21	10	20
18	40	20
22	20	30
19	30	10
20	20	10
14	20	40
2	40	20
16	40	40
13	10	30
15	30	50
17	50	30
1	50	10
23	30	40
8	40	30
4	20	40
7	30	40
9	50	20
6	20	50
12	20	20
3	30	30
5	10	50
10	40	10

Figure 9. Reordering of the design structure matrix for example 2.

Table 4 contains the data corresponding to Figure 2. The total design cycle for this DSM requires 2,430 time units and 2,330 cost units for completion.

TABLE 4. Time and cost for iteration in unordered design cycle in example 2.

To module	From module	Iterations	Time	Cost
1	21	1	590	620
1	23	1	680	650
2	3	1	50	40
2	19	1	530	520
4	6	1	70	50
7	9	1	130	80
9	18	1	290	340
22	23	1	90	30

Table 5 contains the data corresponding to Figure 9. The number of processes contained in the iterative loops has been reduced by reordering the sequence with the modified GA. The total cost to complete the design cycle with this optimized ordering sequence is reduced from 2,330 to 1,510 cost units and from 2,430 to 1,730 time units.

TABLE 5. Time and cost for iteration in ordered design cycle for example 2.

To module	From module	Iterations	Time	Cost
1	16	1	480	430
1	18	1	560	490
2	3	1	50	40
3	9	1	210	170
4	5	1	50	40
11	12	1	80	80
12	14	1	130	80
18	23	1	170	180

In the above examples, the number of processes contained in the iterative loops has been reduced by reordering the sequence with the modified GA. This reordering requires about 1 minute on a Macintosh Quadra 700. In each case, the total cost and time in the design cycle are substantially reduced by reordering the sequence of the design processes.

7. Concluding Remarks

The Design Manager's Aid for Intelligent Decomposition (DeMAID) is a knowledge-based software tool for ordering the sequence of complex design processes, grouping iterative subcycles, and identifying a possible multilevel structure for a design cycle. The DeMAID software displays the processes in a design structure matrix format in which an element on the diagonal is any process that requires input and generates output. Off-diagonal elements indicate a coupling between two processes. The knowledge base in DeMAID attempts to eliminate all feedbacks in the design cycle. If all feedbacks cannot be eliminated, iterative subcycles are identified. If sensitivity analysis results are available, the DeMAID software can be used to examine the ordering within a subcycle to determine the strengths of the couplings between any two processes. These coupling strengths, when input to the knowledge base, determine those processes and couplings might be removed or temporarily suspended without sacrificing system solution accuracy. In addition, a relation is formed between the coupling strengths and the number of iterations required to converge the iterative processes that are created by a feedback coupling.

In the original version of DeMAID, the optimal ordering of processes in an iterative subcycle was generated with a knowledge base, and only minimized the number of feedbacks. The primary drawback to the original method was that only a single ordering sequence could be examined at a time. Changes to the sequence were made interactively and then the costs and times were re-evaluated. This process was extremely slow with no guarantee that a reasonable optimum sequence would be found.

To remedy this problem, a genetic algorithm has been added to DeMAID to examine many possible orderings of the design processes in a design cycle. Each process can now have a time and/or cost associated with it. The GA in DeMAID examines the iterative subcycles to determine their time and cost. The GA fitness function is computed by summing the time and cost of each process contained in an iterative loop and multiplying the totals by the number of iterations required for convergence based on the coupling strength of the feedback coupling forming the loop. The GA determines the best ordering of each iterative subcycle by minimizing the total cost and time requirements, in addition to minimizing the number of feedbacks and crossovers for a particular ordering. This modification increases the likelihood that an optimal or near optimal sequence will be found.

Acknowledgments

The authors wish to acknowledge William J. LaMarsh II of the Computer Sciences Corporation for initially suggesting a genetic algorithm approach be incorporated into DeMAID. In addition, the second and third authors would like to acknowledge partial support of this work under NASA Grant NAG - 11599.

References

Altus, S. S.; Kroo, I. M.; and Gage, P. J.: 1995, A genetic algorithm for scheduling and decomposition of multidisciplinary design problems, *ASME Paper 95-141*.

Bloebaum, C. L.: 1992, An intelligent decomposition approach for coupled engineering systems, *Proceedings of the Fourth AIAA/AF/ NASA/OAI Symposium on Multidisciplinary Analysis and Optimization*, Cleveland, OH.

Giarranto, J. and Riley, G.: 1989, *Expert Systems Principles and Programming*, PWS–Kent Publishing Company, Boston.

Goldberg, D.: 1989, *Genetic Algorithms in Search, Optimization, and Machine Learning*, Addison-Wesley, New York.

McCulley, C. M.; and Bloebaum, C. L.: 1994, Optimal sequencing for complex engineering systems using genetic algorithms, *Fifth AIAA/USAF/NASA/OAI Symposium on Multidisciplinary Analysis and Optimization*, Panama City, FL.

Rogers, J. L.: 1989, A knowledge-based tool for multilevel decomposition of a complex design problem, *NASA TP-2903*.

Rogers, J. L. and Bloebaum, C. L.: 1994, Ordering design tasks based on coupling strengths, *AIAA Paper No. 94-4326*.

Steward, D. V.: 1981, *Systems Analysis and Management: Structure, Strategy and Design*, Petrocelli Books Inc.

Syswerda, G.: 1990, Schedule optimization using genetic algorithms, *Handbook of Genetic Algorithms*, Van Nostran Reinhold, New York.

J. S. Gero and F. Sudweeks (eds), Artificial Intelligence in Design '96, 135-150.
© 1996 *Kluwer Academic Publishers.*

AI IN CONTROL SYSTEM DESIGN USING A NEW PARADIGM FOR DESIGN REPRESENTATION

SOURAV KUNDU AND SEIICHI KAWATA
Control Engg. Laboratory, Department of Precision Engineering,
Tokyo Metropolitan University,1 - 1 Minami Ohsawa, Hachioji Shi,
Tokyo 192 - 03, Japan

Abstract. A new design representation paradigm different from traditional control system design is proposed. This representation of the control system design problem necessitates an Artificial Intelligence (AI) based search strategy to arrive at solutions. The search is performed by a multicriteria Genetic Algorithm (GA) to achieve Pareto optimal design solutions. The new design representation paradigm is used to implement both linear and non-linear state feedback. We also demonstrate with experimental results how non-linear state feedback expands the search space for the design. As an illustrative example an application of this new representation paradigm to control system design is presented.

1. Introduction

Design can be considered as a purposeful, constrained, decision making, exploration and learning activity (Gero *et al.*, 1993). Search is an important component of design, and is the process used in decision making and other design activities. Design, Artificial Intelligence (AI), Optimization and Decision Making (DM) are sometimes viewed as search problems. For a given design representation and state transition operators any search algorithm has to dynamically control the following (Kargupta and Goldberg, 1994):

- generating a feasible design solution within the desired portions of the search space; and
- assigning a goodness or usefulness evaluation to the generated design solution.

Thus for a given representation, the efficacy of any search algorithm in design eventually depends on how well it can control the two above aspects. An optimal search algorithm needs to sample from the universe of admissible design solutions and store the minimum required information about the sampling in its state and based on this information it should be able to decide correctly which direction to search next. The usefulness assigned to a generated design solution reflects the deterministic search efforts towards a certain search direction. Part of the AI in design research covers automated decision-making and search processes. These research efforts are aimed not at the replacement of the designer with machines but at the emulation of designer's cognitive activity to allow the transfer of routine and repetitive tasks to computers and to ensure more efficient decision-making. Human cognitive processes involved in design learning, are therefore studied in the subject of AI in design with a view to formalizing them and expressing them in an algorithmic form.

One formalized model of learning that has been extensively studied within the scientific community of AI and computer science is the Genetic Algorithms (GAs) (Holland, 1975; Goldberg, 1989). The genetic algorithms are distinguished by their parallel investigation of several areas of a search space simultaneously by manipulating a population, members of which are coded problem solutions. The task environment for these applications is modeled as an exclusive evaluation function, which in most cases is called a fitness function that maps an individual of the population to a real scalar. When a search is performed to locate the global extrema of the evaluation function over the search space with an extremely large number of solutions, the search is typically called optimization. The search algorithm merely sees the result of the fitness function as a feedback for the members of the population. Here each solution to the given problem is one member of the population. GAs have been used to design intelligent control systems as in Passino (1995), Porter (1995), Goldberg (1985), Huang and Fogarty (1992) and others, but usually the selection mechanism of the GA is dependent on a single valued scalar fitness function. For control system design the fitness function for the GA is usually taken to be the performance index function of the control system. In this paper a GA is used to design control systems using a new design representation technique, using a mathematical re-formulation. This representation is different to the traditional control system design representation as it uses a multiple-criteria (multiple-objective) formulation consequently with a multiple-criteria GA to perform the search for Pareto optimal design solutions. The proposed paradigm for the control system design representation calls for an improved search strategy which is implemented by the multiple-criteria GA. This improved search strategy produces better design solutions. We also show here that the new representation paradigm allows us to implement nonlinear state feedback for control which considerably expands the design space and a multicriteria GA is able to find better design solution in the expanded design space.

2. Control System Design

Control system design is a specific example of engineering design. The design goal here is three-fold. First is to obtain the configuration of the control system. Second is to achieve the specifications and the third is to identify and optimize the key parameters for the control task that meets an actual need. Control system designers first attempt to configure a system that will result in the desired control performance. The system configuration determines the topology of the sensors, process under control, actuators and controllers. Then the comes the second aspect of the design where the task is to specify proper candidates (mechanical, electrical, chemical or other hybrid components) which will serve as the sensor/s, actuator/s and controller/s for the given specific process under control. This is mainly routine design or a catalog lookup task. The third important aspect of control system design deals with the identification of key parameters and to optimize their values to achieve the desired result of the control action.

The key factor that affects the design of control systems is the `performance index' function that is selected to make the control system exhibit a desired type of performance. This performance index is generally denoted by $J(u(t))$ in control terminology. Some different possibilities of constructing the $J(u(t))$ index is discussed in the following. While formulating the generalized performance index for designing control systems, there are cases where we are concerned with both the expenditure of the control energy as well as the state variable performance. In these cases the performance index of the control system is the integral (over time) of a linear combination of the quadratic forms of the control input $\vec{u}$ ($\vec{u}^T R \vec{u}$) and the state variable description $\vec{x}$ ($\vec{x}^T Q \vec{x}$). The weighting matrices Q and R are chosen so that the relative importance of the state variable performance is contrasted with the importance of the expenditure of the system's energy resource. The matrices Q and R are called the `state' weighting matrix and the `control' weighting matrix respectively.

The issue of considerable concern to the control system designer is the selection of these weighting matrices Q and R. In practical cases only the minimization of the performance index is often not a true design objective. The problem, however, is that the true design objective can seldom be expressed in mathematical terms. In some instances when it can be described in mathematical terms it is usually impossible to solve for an optimal control input. Expression of the design objective in the form of a quadratic integrals is a prudent compromise between formulating the real problem that *cannot* be solved and formulating an artificial problem that *can* be solved. The quadratic form $\vec{x}^T Q \vec{x}$ in the performance index represents a penalty on the deviation of the state $\vec{x}$ from the origin and the term $\vec{u}^T R \vec{u}$ attempts to limit the magnitude of the control signal $\vec{u}$. It is intuitive to the understanding here that *if* it is possible to re-formulate the traditional performance index equation in a way so as to avoid the implicit necessity of the use of weighting matrices Q and R, that is, to consider the problem as one of multiple criteria, we can arrive at a control design technique which eliminates the difficulties in choice of the Q and R but still retains their versatility in providing for a suitable control design solution. This technique is outlined in sections 5 and 6.

The time integrals of the performance index expressions (taken from Rowland, 1986, p. 459) can be as follows: $J(u(t)) = \int_{t_0}^{t_f} e^2 dt$ (ISE) or $J(u(t)) = \int_{t_0}^{t_f} |e| dt$ (IAE) or $J(u(t)) = \int_{t_0}^{t_f} (t - t_0) e^2 dt$ (ITSE) or $J(u(t)) = \int_{t_0}^{t_f} (t - t_0) |e| dt$ (ITAE) or $J(u(t)) = \int_{t_0}^{t_f} u^2 dt$ (Minimum Energy) or $J(u(t)) = \int_{t_0}^{t_f} |u| dt$ (Minimum Fuel) or $J(u(t)) = \int_{t_0}^{t_f} dt$ (Minimum Time) or $J(u(t)) = \int_{t_0}^{t_f} (\vec{x}^T Q \vec{x} + \vec{u}^T R \vec{u}) dt$ (Quadratic in State and Control). Here $J(u(t))$ is the single criterion performance index for a given control input u which is a continuous or discrete function of the time t, t_0 and t_f being the starting and finishing times respectively, e is the system error or the error from the desired value of the system's state.

3. New Representation Paradigm

This paper principally deals with the third aspect of control system design outlined in the last section, utilizing an AI method (GA) to solve the design problem. Given a system configuration and the specification of the components used for the control, it is shown here how we can mathematically re-formulate the traditional optimization problem representation technique in optimal control design theory and how, with this new representation it is possible to consider a nonlinear feedback to control a nonlinear process. This is regarded as a new representation paradigm for control system design. The research presented here covers three main aspects of this new representation paradigm.

1. A new design representation via mathematical re-formulation of the traditional design formulation.
2. A new search algorithm to solve multiple criteria design problems using Pareto sets and a genetic algorithm.
3. Using (1) and (2) it is shown that a *nonlinear* state feedback can be considered which *expands* the search space for the design and the GA based search algorithm produces better design solutions, compared to the linear state feedback which constricts the search space.

In this paper the integral expressions shown in section 2 are re-formulated to avoid the use of weighting matrices. This allows us to deal with the design as a multicriteria problem and also enables us to use a concept of nonlinear state feedback instead of the traditional linear state feedback, when we use the GA based heuristic search method for the solution.

For analysis of the multiple-criteria problem we use the concept of Pareto sets. The Pareto optimal set of control design solutions is defined as :

"Let X be a set of feasible control design solutions. The Pareto optimum is defined as: A design solution (control input) $\vec{u}^*(t) \in U$, with N performance indices, is Pareto optimal if and only if there exists no $\vec{u}(t) \in U$ such that $J_i(\vec{u}(t)) \leq J_i(\vec{u}^*(t))$ for i = 1,2,...,N with $J_i(\vec{u}(t)) < J_i(\vec{u}^*(t))$ for at least one i. Here $U \in \Re^n$ is the universe of admissible control inputs. This definition is based upon the intuitive assumption that the control input $\vec{u}^*(t)$ is chosen as an optimal (or nondominated) one if none of performance indices $J_i(\vec{u}(t))$, can be improved without worsening at least one other performance index."

The Pareto set is plotted to find the set of non-dominated solutions that are generated considering all the N separate performance criteria simultaneously without placing any relative importance (weights) on any of them. With reference to this Pareto set, a fitness value is awarded to each control design solution. The technique proposed in this paper generates control inputs and compares the control output considering this Pareto optimality condition. At the end of the GA run, the control system designer chooses one of the solutions in the final Pareto optimal set, to suit his/her design goals and requirements.

4. Previous Work

4.1. PREVIOUS WORK ON LINEAR STATE FEEDBACK DESIGN

Considering a given plant and a desired reference input most of the design problems in control are to find a control input or an actuating signal so that the output of the plant will be as close possible to the reference input. If a control input depends on actual output of the system (plant), it is called a `feedback' or a closed-loop control. Since the state of a system contains all the essential information of the system, if the control input is designed to be a function of the state and the reference input, in most cases a reasonably good control can be achieved. The effects of introducing a linear state feedback of the form $\vec{u} = \vec{r} + k\vec{x}$ where $\vec{u}$ is a control input vector, $\vec{r}$ is the reference input vector, $k \in \Re^{r \times n}$ is the feedback gain matrix which is required to be a real and $\vec{x}$ is the state vector, have been extensively studied in control literature. Under the assumptions of controllability, linear state feedback can achieve stabilization in most of the cases. The design goal here is to find an optimal set of parameters or the optimal value of the matrix k in the linear state feedback equation above using a certain performance index or objective function of the control system, which is the basic function whose optimum (maximum or minimum) is sought subject to constraints on the values of some variables. A performance index to be useful, must be a number that is always a positive or zero. Then the best design of the control system is defined as the one which minimizes this index. Optimal control system design has been traditionally based on minimization of a quadratic performance index mentioned in section 2, which typically is a time integral over a function of $\vec{x}^T Q \vec{x}$ and $\vec{u}^T R \vec{u}$ (Lewis, 1986). Choice of the linear weighting factors Q and R are generally based on heuristics and some experiments are required to ascertain some satisfactory optimal value.

4.2. PREVIOUS WORK ON GA AND MULTICRITERIA OPTIMIZATION

Schaffer (1984, 1985) did some of the pioneering experiments in this area and proposed `Vector Evaluated Genetic Algorithm (VEGA)'. His method attempted to avoid the problems associated with scaling (or weighting) the different criteria of a multicriteria optimization problem. In the progressive generations VEGA found and maintained multiple solutions each one favoring one criterion. Thus it tried to optimize the multiple criteria by finding solutions in the neighborhood of the extreme points of the Pareto optimal frontier. This ultimately results in an equivalent weighting mechanism where one of the criterion weights is one and the rest are zero, while selection is being performed. The problem remained in finding solutions all along the Pareto optimal frontier. Husbands (1994) did some work on a distributed population co-evolving parallely to solve a scheduling problem.

In his book Goldberg (1989) suggests the use of a non-domination ranking scheme which moves the population towards the Pareto frontier in a multiple criteria problem. Horn et al. (1993) presents a new algorithm called the `Niched Pareto' genetic algorithm. This method uses a widely known selection scheme called the tournament selection (Goldberg, 1989) using Pareto domination within the tournament for selection. They also implement a technique called the fitness sharing which distributes the populations over a number of different peaks in the search space with each peak receiving a fraction of the population in proportion to the height of that peak. This in effect prevents the genetic drift in multimodal function optimization. In this sort of approach an important factor affecting the GA's success is controlling the selection pressure. This is determined by the size of the domination tournament which is mostly empirical now. Another parameter called the niche size or the σ_{share} exists, which has to be set beforehand. To do this setting again there is an implicit assumption that the solution set has a finite known number of peaks. Ritzel et al. (1994) uses a form of a Pareto optimality based rank evaluation scheme, consequently with a deduced form of VEGA and compares the results with the best trade-off surface found by a domain specific algorithm MICCP to judge the performance of the GA. This work does not report any experiments by using some form of Horn's `niching' mechanism or others.

Cieniawski *et al.* (1995) presents similar research and results.

Fonseca and Fleming (1993) and Belegundu *et al.* (1994) use a different form of Pareto optimality based ranking mechanism. The basic rank assignment in Fonseca and Fleming (1993) is simply a measure of the number of solutions that dominates the particular solution. Thus for the solution s_i, if it is dominated by d_i solutions in generation t, then its rank at t^{th} generation is $rank(s_i, t) = 1 + d_i$. Thus all the non-dominated individuals are assigned rank 1, and rest according to the given formula. They extend this rank assignment method by redefining the way in which comparisons are made between two solutions allowing one to prefer one solution to another even when they are both on Pareto optimal frontier. The algorithm then concentrates on specific region of the Pareto frontier and evolves towards better solutions. Louis and Rawlins (1993) uses a variant of binary tournament selection to incorporate Pareto optimality in the genetic algorithm. The algorithm selects two individuals at random from the current population, mates them, and produces the Pareto optimal set of parents and offsprings. Two random individuals from this Pareto optimal set form part of the next population. The procedure repeats until the new population fills up, thus becoming subsequent current population. Belegundu *et al.* (1994) uses just two ranks. All non-dominated points (Pareto frontier) are given rank 1 and all dominated points are given rank 2. Also any solution that violates a certain constraint is given rank 2. Then all rank 2 solutions are discarded from the population and only rank 1 members are bred. At every generation a lot of randomly generated solutions are added to the population to keep its size a constant. In this approach a lot of otherwise useful genetic material is discarded at every generation and this easily localizes the solutions to a specific region of the Pareto frontier. This method can be regarded equivalent to generating a huge number of random solutions and determining the Pareto frontier which is tantamount to an exhaustive search method.

4.3. PREVIOUS WORK ON GENETIC ALGORITHMS THEORY

Genetic Algorithms (GAs) (Holland, 1975) are commonly used for single criterion optimization and learning problems. The goodness or usefulness metric for a certain solution is the fitness function of the solution. By using a coded representation of the search space, every problem is presented to the GA in terms of the bit-string representation. For example, in a 5-bit problem any deterministic function-coding combination may be ultimately reduced to a list of fitness values associated with each $2^5 = 32$ strings. The fitness-lookup which is used to specify a problem in every GA application disguises the implied choice of `basis' functions. The basis being a linearly independent set of vector functions which span the underlying search space. As a result the GAs have the same fundamental sampling problem as any other search algorithm. In any traditional design problem the competing design solutions which are evaluated for a fitness function are quite explicitly represented. But as GAs view the search space via a coded representation of it which is often a bit-string, the search process with GAs is not an easy one. More often than not the existence of large hamming distances in the coded representation of the search space make the search with GAs very brittle due to the `hamming cliffs' (Kargupta and Goldberg, 1994). The motivational idea behind GAs is natural selection. Operators like selection, crossover and mutation are implemented to emulate the process of natural evolution. A population of `organisms'

Genetic Algorithm : (The basic algorithm in program form)
Procedure GA
begin

    $t = 0$
    initialize at random *P(t)*
    evaluate *P(t)*
    while termination is not valid;

        begin
            selection *P(t)*;
            crossover *P(t)*;
            mutation *P(t)*;
            evaluate *P(t)*;
            $t = t + 1$
        end
    endwhile
end

Figure 1. The simple genetic algorithm.

(usually represented as bit strings) is modified by the probabilistic application of the genetic operators from one generation to the next. The basic algorithm where $P(t)$ is the population of strings at generation t, is given in Figure 1. A more detailed explanation of the theory and working of the GA can be found in Goldberg (1989).

5. Proposed Mathematical Reformulation

5.1. TRADITIONAL DESIGN APPROACH

Let us consider a multiple input linear time-invariant system described by the state space equations :

$$\dot{\vec{x}}(t) = A\vec{x}(t) + B\vec{u}(t) \tag{1}$$

$$\vec{y}(t) = C\vec{x}(t) \tag{2}$$

with given initial conditions:

$$\vec{x}(t_0) = \vec{x}_0 \tag{3}$$

Where $\vec{x} \in \Re^n$, $\vec{u} \in \Re^r$, $A \in \Re^{m \times n}$, $B \in \Re^{n \times r}$ and $C \in \Re^{m \times n}$, and the pair (AB) is controllable given the initial and terminal state and the given performance index. $\vec{x}(t)$ is a n-dimensional state vector, $\vec{u}(t)$ is a r-dimensional control input and $\vec{x}_0$ is a constant n-dimensional vector. Associated with this system we have a performance index the minimization of which, is the goal of the control system design task. The conventional description of the performance index is :

$$J(u(t)) = \int_0^\infty [\vec{x}^T(t)Q\vec{x}(t) + \vec{u}^T(t)R\vec{u}(t)]dt \tag{4}$$

subject to equation (1) where Q and R are both positive definite and symmetric matrices. The control system design problem is to find a control input $u(t)$ such that the given performance index in equation (4) is minimized. Deducing Q and R depends on the experience of the control system designer and a number of trial and error strategies are generally required for satisfactory deduction. According to the linear control theory we can express the algebraic Riccati equation as:

$$A^T P + PA - PBR^{-1}B^T P + Q = 0 \tag{5}$$

The solution of equation (5) being the matrix P. By deducing P, optimal control input can be had as:

$$u(t) = -R^{-1}B^T P\vec{x} \tag{6}$$

and thereof we can have the minimal $J(u(t))$ as :

$$J(u(t))_{min} = \vec{x}_0^T P\vec{x}_0 \tag{7}$$

5.2. PROPOSED NEW DESIGN APPROACH

We resolve equation (4) for implementing a multicriteria design strategy. This new paradigm for the design representation fundamentally changes the search spaces for the design. We propose a new representation of the design spaces by a re-formulation of equation (4) with its decomposition, where $J_1(u(t))$ and $J_2(u(t))$ are defined as follows:

$$J_1(u(t)) = \int_0^\infty [\vec{x}^T(t)Q\vec{x}(t)]dt \tag{8}$$

$$J_2(u(t)) = \int_0^\infty [\vec{u}^T(t)R\vec{u}(t)]dt \tag{9}$$

Now we set $Q \triangleq I$ and $R \triangleq I$, (I being the identity matrix) and we treat the problem as one of multicriteria design by *avoiding* the use of the state and control weighting matrices Q and R as follows:

$$J_1(u(t)) = \int_0^\infty [\vec{x}^T(t)I\vec{x}(t)]dt \tag{10}$$

$$J_q(u(t)) = \int_0^\infty [\vec{u}^T(t)I\vec{u}(t)]dt \tag{11}$$

We thus redefine the goal of the control system design task as to find a control input $\vec{u}(t)$ such that the set of Pareto optimal solutions in the space of $J_1(u(t))$ and $J_2(u(t))$ minimize the multiple performance index functions mentioned in (10) and (11), simultaneously.

From the theory of dynamical control systems the control input $u(t)$ for the optimal control system design using *linear* state feedback is described as:

$$\vec{u}(t) = \vec{r} + k_1 x_1 + k_2 x_2 + ... + k_n x_n \tag{12}$$

where $k = k_i (i = 1, 2, ..., n) \in \Re^{r \times n}$, with $\vec{x}(t)$ being a n-dimensional state vector, $\vec{u}(t)$ being a r-dimensional control input and $\vec{r}$ being the reference input vector. By the use of the re-formulation described in equation (10) and (11) it is possible to construct a *non-linear* state feedback with combination of linear and non-linear terms in the feedback for the optimal control input $u(t)$ shown in (12). The introduction of non-linear term *expands* the design space. This expanded search space helps to find much better design solutions which are exemplified by the numerical example in section 7.

6. The Multicriteria GA Search Algorithm

The specifics of the GA search algorithm for multicriteria design which is presented in this section lie in the assignment of a fitness measure to a design solution based on reference to the dynamically updated Pareto optimal set generated during the progressing GA runs (Osyczka and Kundu, 1995; Kundu et al., 1996; Kundu, 1996). The GA uses this fitness measure to perform the selection operation. The fitness is the *proximity* value of a feasible design solution added to the niching value of the Pareto optimal solutions produced during the previous GA evaluation run. The fitness thus estimates how far the new design solution is, from most recent Pareto frontier. This Pareto frontier gets updated dynamically during the GA runs. An explanation of the notion of proximity in the solution space would help to understand this better. Every feasible design solution occupies a definite position in the n-dimensional universal space of feasible solutions. This position will have in some metrical proximity value (length) measured from each of Pareto solution (vectors) found in the previous run. We measure each of these proximity value. The value we return as the `fitness' of a design solution is *least* one of all these different proximity values. Analogous to an ordinary (scalar) GA where we measure the fitness value of a solution as its metrical distance (a straight line) from zero on a linear scale, our method is equivalent remembering that the the `shortest' distance between any two given points is a straight line.

6.1. PROXIMITY VALUES IN THE SOLUTION SPACE

Figure 2 shows the equivalent ways an ordinary GA and our multiple criteria GA assigns fitness to a new solution. In a single criterion GA the fitness d_f is the least distance (a straight line joining two given points) between the solution and some datum measure which is [0,0] in Figure 2. For a multiple criteria problem the fitness is the minimum of d_1, d_2, d_3, d_4 and d_5 in Figure 2. The minimum is d_3 in this case. The reference measure for d_1, d_2, d_3, d_4 and d_5 is the cumulative Pareto set of all the solutions found till the previous call to the GA evaluation routine. Thus the fitness is essentially equivalent to that of the working of a one-dimensional GA. A solution will always have a *proximity* value regardless of where is lies: whether in the negative or positive spaces of the given Pareto frontier. When it lies in the positive Pareto space (see figure 2) this proximity value should be greater than the case when it lies in the negative Pareto space, as the former is a better solution (Osyczka and Kundu, 1996). This is taken care by changing the signs in the proximity values. In the case when a solution lies in the negative Pareto space of the present Pareto frontier the method to assign the fitness is different to that when it lies in the positive Pareto space and this method is described in detail in section 6.2.

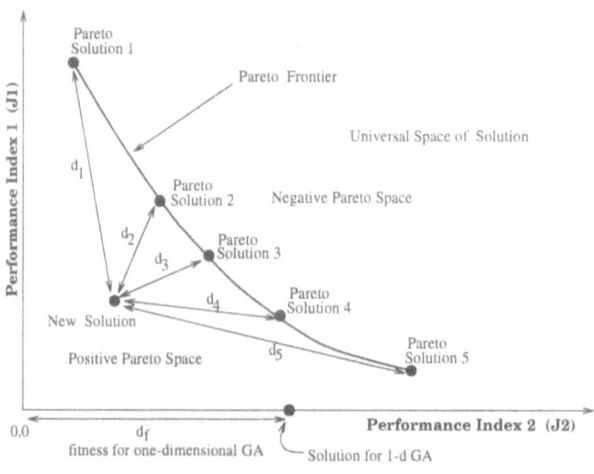

Figure 2. Fitness calculation method in GA based design of control system.

6.2. UPDATING OF THE PARETO SET

Throughout each generation we maintain a set of Pareto optimal solutions, all of which are assigned the same fitness value. This is equivalent to a assigning a `niche' as described in (Horn *et al.*, 1993) to the set of Pareto solutions. All members of a Pareto set have the same niching value. A new solution in a certain generation can fall in any of the three, and only the three of the following categories:

a. It is a new Pareto optimal solution and it *dominates* one, some (or all) of the Pareto optimal solutions found till the immediately preceding call to the GA evaluation routine.

b. Although it is a new Pareto optimal solution it *does not dominate* any of the Pareto optimal solutions found up till the immediately preceding call to the GA evaluation routine.

c. It is *not* a Pareto optimal solution.

For every new solution in a certain generation we first assign a proximity value to it. Then we deal with the three different categories mentioned above in three separate ways whereby a fitness value is returned for the GA roulette wheel selection mechanism to work. Note that for category [c.] solutions above we do not remove the solutions as in Belegundu *et al.* (1994) but keep them in the population with awarding them a lower fitness value. This is done as a measure to induce some form of atavism in the evolutionary process and to insure against the loss of any otherwise useful genetic material.

1. For category [a] solutions the fitness returned is the proximity value *added* to the niching value of the Pareto optimal solutions found till the immediately preceding call to the GA evaluation routine and the Pareto set is updated by **removing** those old solutions that this new Pareto solution dominates.

2. For category [b] solutions the fitness returned is the proximity value *added* to the niching value of the Pareto optimal solutions found till the immediately preceding call to the GA evaluation routine and the Pareto set is updated by **adding** this new Pareto solution to the old Pareto set.
3. For category [c] solutions the fitness returned is simply the proximity value *sub-tracted* from the niching value of the Pareto optimal solutions found till the immediately preceding GA evaluation run. There is no change in the Pareto set.

7. Application to a Design Example Using Proposed Method

This section presents the simulation results of a control problem taken from (Dorf and Bishop, 1995) [pp. 610] which has been coded and run in a Sun 4 machine using the proposed design representation paradigm and the multiple criteria search technique described in sections 5 and 6. Here the design goal here can be considered as a minimization task. The system considered can be represented by state space equation:

$$\vec{x}(t) = A\vec{x}(t) + B\vec{u}(t) \tag{13}$$

and a state feedback controller is selected so that:

- For a *linear* state feedback the $\vec{u}(t)$ is a linear function of the measured state variable $\vec{x}$ such that:

$$\vec{u}(t) = \vec{r} + k_1 x_1 + k_2 x_2 + \ldots + k_n x_n \tag{14}$$

- For a *non-linear* state feedback the $\vec{u}(t)$ is a combination of a linear and non-linear terms in the function of the measured state variable $\vec{x}$ such that:

$$\vec{u}(t) = \vec{r} + k_1 x_1 + k_2 x_2 + \ldots + k_n x_n + \begin{bmatrix} x^T K_1 x \\ x^T K_2 x \\ \vdots \\ x^T K_n x \end{bmatrix} \tag{15}$$

As an illustrative example a simple system with 2 state variables and 1 input variable is considered as follows:

$$\frac{d}{dt} \begin{bmatrix} x_1 \\ x_2 \end{bmatrix} = \begin{bmatrix} 0 & 1 \\ 0 & 0 \end{bmatrix} \begin{bmatrix} x_1 \\ x_2 \end{bmatrix} + \begin{bmatrix} 0 \\ 1 \end{bmatrix} u(t) \tag{16}$$

By setting $A = \begin{bmatrix} 0 & 1 \\ 0 & 0 \end{bmatrix}$, $B = \begin{bmatrix} 0 \\ 1 \end{bmatrix}$ and $\vec{x} = \begin{bmatrix} x_1 \\ x_2 \end{bmatrix}$ in equation (13). First a linear state feedback control system design is chosen such that:

$$u(t) = -k_1 x_1 - k_2 x_2 \tag{17}$$

Next a non-linear state feedback control system design is chosen such that:

$$u(t) = -k_1 x_1 - k_2 x_2 + \begin{bmatrix} x_1 & x_2 \end{bmatrix} \begin{bmatrix} k_3 & k_4 \\ k_5 & k_6 \end{bmatrix} \begin{bmatrix} x_1 \\ x_2 \end{bmatrix} \tag{18}$$

with the control $u(t)$ as a combination of the linear and non-linear terms of the two state variables. For nonlinear state feedback equation (16) produces:

$$\dot{x}_1 = x_2 \tag{19}$$

$$\dot{x}_2 = -k_1 x_1 - k_2 x_2 + \begin{bmatrix} x_1 & x_2 \end{bmatrix} \begin{bmatrix} k_3 & k_4 \\ k_5 & k_6 \end{bmatrix} \begin{bmatrix} x_1 \\ x_2 \end{bmatrix} \tag{20}$$

By application of the reformulation shown in equations (10) and (11) we can have two performance indices:

$$J_1(u(t)) = \int_0^{4\pi} (x_1^2 + x_2^2) \, dt \tag{21}$$

$$J_2(u(t)) = \int_0^{4\pi} (u^2) \, dt \tag{22}$$

both of which are to be minimized simultaneously.

For using linear state feedback the goal of the control system design task is to find the optimal value of the decision variables k_1, k_2 (see equation 17) such that when the system is simulated by using those optimal values equations (21) and (22) are minimized simultaneously. We consider the eigenvalues of the system as $\pm j$ and so we take the upper and lower limits of k_1, k_2 as ± 20.47. The GA genotype bit length is 24 which takes into account the values of k_1 (12 bits) and k_2 (12 bits). We note here that $(2^{11} - 1)/100 = 20.47$ and the 12th bit is for handling the negative values of k_1 and k_2.

For using nonlinear state feedback the goal of the control system design task is to find the optimal value of the decision variables k_1, k_2, k_3, k_4, k_5 and k_6, (see equation 18) such that when the system is simulated by using those optimal values equation (21) and (22) are minimized simultaneously. We consider the eigenvalues of the system as $\pm j$ and so we take the upper and lower limits of k_1, k_2, k_3, k_4, k_5 and k_6, as ± 20.47. The GA genotype bit length is 72 which takes into account the values of k_1 (12 bits) through k_6 (12 bits). We note here that $(2^{11} - 1)/100 = 20.47$ and the 12th bit is for handling the negative values of k_1, k_2, k_3, k_4, k_5 and k_6. Note here that the introduction of the extra variables k_3, k_4, k_5 and k_6 in the case of nonlinear state feedback expands the design search space.

7.1. DISCUSSION ON RESULTS

Figures 3 and 4 show the results from the computer simulation of the system described by equation (16) using linear state feedback. Figure 3 plots the error response of the two state variables of the system x_1 and x_2. Our design goal is to stabilize the system as soon as possible with minimum *overshoot*. The starting values of x_1 and x_2 are 1 and 0 respectively. By stabilization of the system we mean bringing down both the state variables x_1 and x_2 to zero (desired value), irrespective of their starting values. To bring x_1 to 0, x_2 has to increase from 0 towards the negative direction. At generation 1 of the GA we notice that there is a large amount of *overshoot* of both x_1 and x_2 variables before the damping off to 0 in about 6 seconds. Figure 4 plots the system's error response at the 50th generation of the GA when the final design solution was achieved (using linear state feedback). We readily notice the smoothness of the error

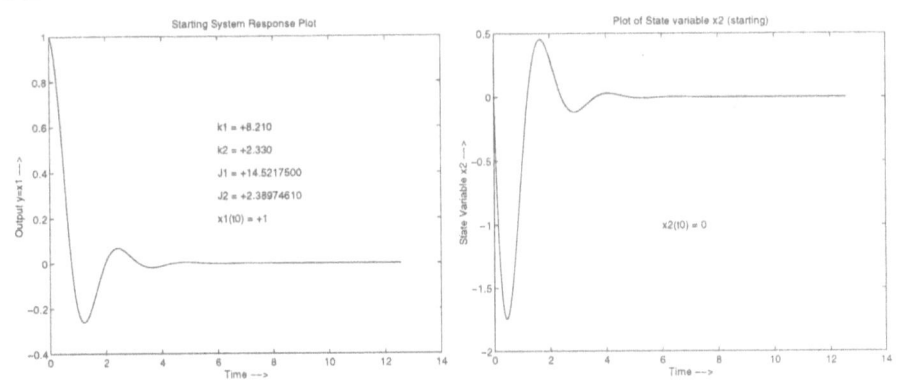

Figure 3. Starting solution for GA generation number 1 by *Linear State Feedback.*

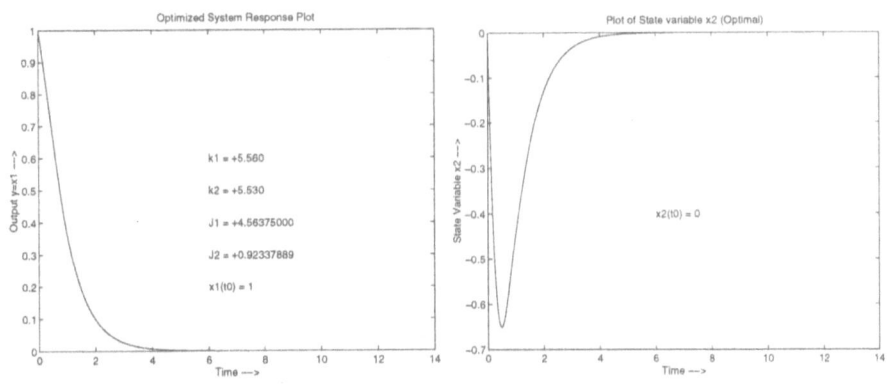

Figure 4. Finishing solution for GA generation number 50 by *Linear State Feedback.*

response with no overshoot. Both x_1 and x_2 state variables stabilize in a smooth way and in a shorter time which is about 4.6 seconds. We plot Figure 3 using one of the only 2 Pareto solutions found in generation 1 of the GA. Figure 4 was plotted using one of the 27 Pareto solutions found in the 50th generation of the GA.

Results of simulation of the system described by equation (16) using nonlinear state feedback are presented in figures (5) and (6). Comparing these results with plots in figures (3) and (4) shows what has been achieved by *expanding* the design search space via utilization of equation non-linear state feedback instead of linear state feedback. Figure 5 presents the error response curve of one of the only 2 Pareto solutions found in the GA generation number 1. We notice that there is some amount of overshoot present in both the response curves of the x_1 and x_2 state variables. But still the system stabilizes earlier (3.2 seconds) than the best solution found by linear state feedback. Figure 6 presents the system's error response of one of the 14 Pareto solutions found in the 50th generation of the GA. We readily notice that with absolutely no overshoot, both the x_1 and x_2 state variables of the system smoothly stabilizes in about 2.2 seconds.

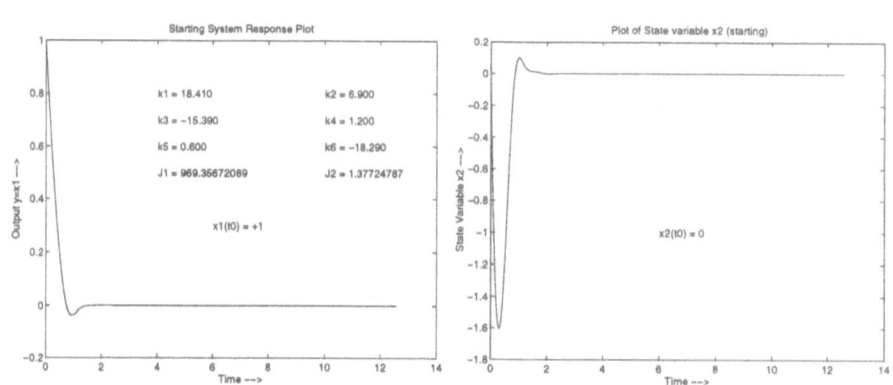

Figure 5. Starting solution for GA generation number 1 by *Non-Linear State Feedback*.

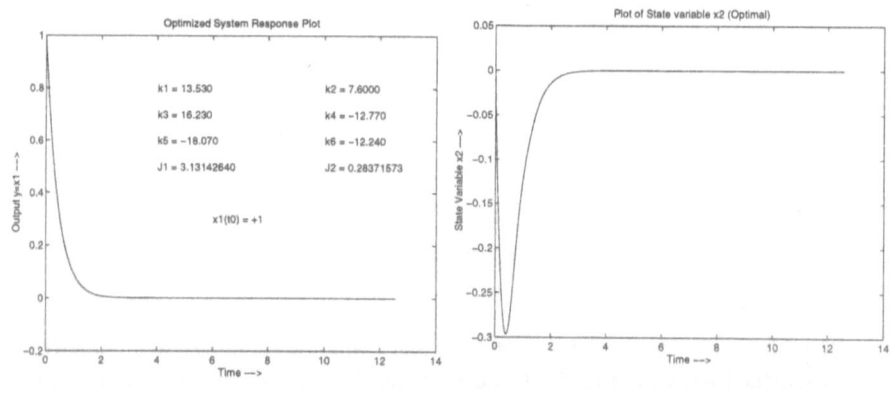

Figure 6. Finishing solution for GA generation number 50 by *Non-Linear State Feedback*.

Figure 7 shows the progress of the Pareto optimal frontier during the course of the 50 generations of the GA. In generation 1 we have only 2 solutions. In generation 50 we see that 14 Pareto solutions have been found out by the GA. The designer can now choose any of these 14 Pareto solutions according to his or her preference, the design considerations and requirements. One of these 14 solutions is used to plot figure (6). This solution is marked by "*" in figure (7) to show its position in the final Pareto frontier. For our simulations, the GA population size was 70. The GA crossover probability was 0.5. The GA mutation probability was 0.09. The number of generations required to produce all the results presented was 50.

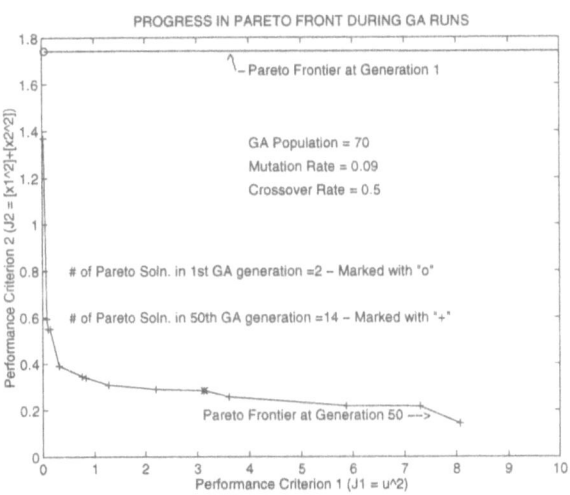

Figure 7. Pareto frontier progression as GA generations proceed.

8. Conclusions

Using any of the methods based on heuristic weighting of control performance criteria we obtain a single best solution which clearly reflects the choice of weights. In practical cases weights are not always easy to ascertain during the earlier stages of the control system design. Using our proposed method, usually we obtain a reasonably large set of Pareto optimal solutions which are well distributed all along the Pareto frontier, from which the control system designer can choose a design solution according to his/her preferences. Avoiding the existing mathematical methods to find a single optimal solution for the control system design problem we show that the Pareto approach combined with a GA gives a rich set of Pareto optimal solutions that could not have been otherwise produced by traditional methods from the dynamical systems control theory. It is apparent from the computer simulations that for the new design representation proposed, *nonlinear* state feedback produces better results by expanding the design search space. The multicriteria design formulation is essentially required to implement this nonlinear state feedback approach as there exists no other mathematical apparatus to solve it. The GA based search algorithm for multicriteria design proves very effective to generate and evaluate feasible design solutions and perform the search for Pareto optimal designs.

Acknowledgments

Sourav Kundu acknowledges the Ministry of Education, Japan for financial support on this project. He also wishes to acknowledge the research support of Prof. John S. Gero of the Key Center of Design Computing in University of Sydney Australia, Dr. Sushil J. Louis of the Department of Computer Science, University of Nevada, Reno and Prof. Andrzej Osyczka of the Department of Precision Engineering, Tokyo Metropolitan University, Japan.

References

Belegundu, A. D., Murthy, D. V., Salagame, R. R. and Constans, E. W.: 1994, Multi-objective optimization of laminated ceramic composites using genetic algorithms, *Proceedings of the 5th AIAA/NASA/USAF/ISSMO Symposium on Multidisciplinary Analysis and Optimization*, AIAA, Inc., pp. 1015–1022.

Cieniawski, S. E., Eheart, J. W. and Ranjithan, S.: 1995, Using genetic algorithms to solve a multiobjective groundwater monitoring problem, *Water Resource Research*, 31(2), 399–409.

Dorf, R. C. and Bishop, R. H.: 1989, *Modern Control Systems*, Addison-Wesley, Reading, Massachusetts.

Fonseca, C. M. and Fleming, P. J.: 1993, Genetic algorithms for multiobjective optimization: Formulation, discussion and generalization, *Proceedings of the Fifth International Conference on Genetic Algorithms*, Morgan-Kaufmann, pp. 416–423.

Gero, J. S., Louis, S. J. and Kundu, S.: 1993, Evolutionary learning of novel grammars for design improvement, *Artificial Intelligence in Engineering Design, Analysis and Manufacturing (AIEDAM)*, 8(2), 83–94.

Goldberg, D. E.: 1989, *Genetic Algorithms in Search, Optimization, and Machine Learning*, Addison-Wesley, Reading, Massachusetts.

Goldberg, D. E.: 1985, Genetic algorithms and rule-learning in dynamic system control, *in* J. J. Grefenstette (ed.), *Proceedings of the First International Conference on Genetic Algorithms and Their Applications*, Lawrence Erlbaum Associates, Hillsdale, New Jersey, pp. 8–15.

Holland, J. H.: 1975, *Adaptation in Natural and Artificial Systems*, University of Michigan Press, Ann Arbor, Michigan.

Horn, J., Nafpliotis, N. and Goldberg, D. E.: 1993, Multiobjective optimization using the niched pareto genetic algorithm, *IlliGAL Tech Report no 93005*, Department of General Engineering, University of Illinois at Urbana Champaign, Urbana, IL 61801-2996.

Huang, R. and Fogarty, T. C.: 1992, Learning prototype control rules for combustion control with genetic algorithm, *Journal of Modeling, Measurement and Control*, 38(4), 55–64.

Husbands, P.: 1994, Distributed coevolutionary genetic algorithms for multi-criteria and multi-constraint optimization, *in* T. C. Fogarty (ed.), *Evolutionary Computing* (selected papers from AISB Workshop Leeds, UK), Springer-Verlag, Berlin, Heidelberg, pp. 150–165.

Kargupta, H. and Goldberg, D. E.: 1994, Decision making in genetic algorithms: a signal-to-noise perspective, *IlliGAL Tech Report no 94004*, Department of General Engineering, University of Illinois at Urbana Champaign, Urbana, IL 61801-2996.

Kundu, S.: 1996, A multicriteria genetic algorithm to solve optimization problems in structural engineering design, *Proceedings of International Conference on Information Technology in Civil and Structural Engineering Design - Taking Stock and Future Directions*, 14th - 16th August 1996, Glasgow, Scotland (accepted for publication).

Kundu, S., Kawata, S. and Watanabe, A.: 1996, A multicriteria approach to control system design with genetic algorithm, *Proceedings of IFAC '96 - 13th World Congress, June 30th - July 5th, 1996*, International Federation of Automatic Control, Elsevier Science, Kidlington, Oxford, U.K. (in print).

Lewis, F. L.: 1986, *Optimal Control*, Wiley, New York.

Louis, S. J. and Rawlins, J. E.: 1993, Pareto optimality, GA-easiness and deception, *in* S. Forrest (ed.), *Proceedings of the Fifth International Conference on Genetic Algorithms*, Morgan Kaufmann Publishers, San Mateo, California, pp. 118–123.

Osyczka, A. and Kundu, S.: 1995, A new method to solve generalized multicriteria optimization problems using the simple genetic algorithm, *Structural Optimization*, 10(2), 94–99.

Osyczka, A. and Kundu, S.: 1996, A modified distance method for multicriteria optimization, using genetic algorithms, *Computers & Industrial Engineering Journal* - Special Issue on Genetic Algorithms, 30(2) (in print).

Passino, K. M.: 1995, Intelligent Control For Autonomous Systems, *IEEE Spectrum*, June, 55–62.

Porter, B.: 1995, Genetic design of control systems, *Transactions of the Society of Instrument and Control Engineers*, 34(5), 393–402.

Rowland, J. R.: 1986, *Linear Control Systems: Modeling, Analysis, and Design*, Wiley, New York.

Ritzel, B. J.; Eheart, W. and Ranjithan, S.: 1994, Using genetic algorithms to solve a multiple objective groundwater pollution containment problem, *Water Resources Research*, 30(5), 1589–1603.

Schaffer, J. D.: 1984, *Some Experiments in Machine Learning Using Vector Evaluated Genetic Algorithms*, Doctoral Dissertation, Vanderbilt University, Nashville, Tennessee.

Schaffer, J. D.: 1985, Multiple objective optimization with vector evaluated genetic algorithms, *in* J. Grefenstette (ed.), *Proceedings of an International Conference on Genetic Algorithms and their Applications*, pp. 93–100.

J. S. Gero and F. Sudweeks (eds), Artificial Intelligence in Design '96, 151-170
© 1996 Kluwer Academic Publishers.

AUTOMATED DESIGN OF BOTH THE TOPOLOGY AND SIZING OF ANALOG ELECTRICAL CIRCUITS USING GENETIC PROGRAMMING

JOHN R. KOZA, FORREST H BENNETT III, DAVID ANDRE
Department of Computer Science
Stanford University, Stanford, California

AND

MARTIN A. KEANE
Econometrics Inc., Chicago, Illinois USA

Abstract: This paper describes an automated process for designing analog electrical circuits based on the principles of natural selection, sexual recombination, and developmental biology. The design process starts with the random creation of a large population of program trees composed of circuit-constructing functions. Each program tree specifies the steps by which a fully developed circuit is to be progressively developed from a common embryonic circuit appropriate for the type of circuit that the user wishes to design. The fitness measure is a user-written computer program that may incorporate any calculable characteristic or combination of characteristics of the circuit. The population of program trees is genetically bred over a series of many generations using genetic programming. Genetic programming is driven by a fitness measure and employs genetic operations such as Darwinian reproduction, sexual recombination (crossover), and occasional mutation to create offspring. This automated evolutionary process produces both the topology of the circuit and the numerical values for each component. This paper describes how genetic programming can evolve the circuit for a difficult-to-design low-pass filter.

1. The Problem of Circuit Design

The design of an electrical circuit with specified operating characteristics is a complex task. Electrical circuits consist of a wide variety of different types of components, including wires, resistors, capacitors, inductors, diodes, transistors, transformers, and energy sources. The individual components are arranged in a particular *topology* to form a closed circuit. In addition, each component is further specified (*sized*) by a set of component values. Circuits typically receive input signals from one or more input sources and produce output signals at one or more output ports. A complete specification of an electrical circuit includes both its topology and the sizing of all of its components.

Considerable progress has been made in automating the design of certain categories of purely digital circuits; however, the design of analog circuits and mixed analog-digital circuits has not proved to be as amenable to automation (Rutenbar, 1993). In discussing "the analog dilemma," Aaserud and Nielsen (1995) observe,

> Analog designers are few and far between. In contrast to digital design, most of the analog circuits are still handcrafted by the experts or so-called 'zahs' of analog design. The design process is characterized by a combination of experience and intuition and requires a thorough knowledge of the process characteristics and the detailed specifications of the actual product.
>
> Analog circuit design is known to be a knowledge-intensive, multiphase, iterative task, which usually stretches over a significant period of time and is performed by designers with a large portfolio of skills. It is therefore considered by many to be a form of art rather than a science.

2. Previous Work

Numerous efforts have been made to automate the design process for analog and mixed analog-digital circuits. In an interactive design tool called IDAC for analog integrated circuits (Degrauwe ,1987), the user selects various possible topologies for the circuit; IDAC determines the values of the components in each circuit (in relation to the desired behavioral characteristics); and, the user chooses the best sized circuit.

In OASYS (Harjani, Rutenbar and Carley, 1989) and OPASYN (Koh, Sequin and Gray, 1990), a topology is chosen beforehand based on heuristic rules and the synthesis tool attempts to size the circuit. If the synthesis tool cannot size the chosen topology correctly, the tool creates a new topology using other heuristic rules and the process continues. The success of these systems depends on the effectiveness of the knowledge base of heuristic rules.

In SEAS (Ning, Kole, Mouthaan and Wallings,, 1992), evolution is used to modify the topology and simulated annealing is used to size the circuit. Maulik, Carley and Rutenbar (1992) attempt to handle topology selection and circuit sizing simultaneously using expert design knowledge. Higuchi et al. (1993) have employed genetic methods to the design of digital circuits using a hardware description language (HDL).

In DARWIN (Kruiskamp and Leenaerts, 1995), opamp circuits are designed using the genetic algorithm (Holland, 1975). In creating the initial population in DARWIN, the topology of each opamp in the population is picked randomly from a preestablished hand-designed set of 24 topologies in order to ensure that each circuit behaves as an opamp. In addition, a set of problem-specific constraints are solved to ensure that all transistors operate in their proper range and that all transistor sizes are between maximal and minimal values. The behavior of each opamp is evaluated using a small signal equivalent circuit and analytical calculations specialized to opamp circuits. The fitness of each opamp is computed using a combination of factors, including the deviation between the actual behavior of the circuit and the desired behavior and the power dissipation of the circuit. A crossover operation and mutation operation for the chromosome strings describing the opamps is used to create offspring chromosomes.

3. Background of Genetic Programming

John Holland's pioneering *Adaptation in Natural and Artificial Systems* (1975) described how an analog of the naturally-occurring evolutionary process can be applied to solving scientific and engineering problems using what is now called the *genetic algorithm*. The problem of automatic programming is one of the central questions in computer science. Paraphrasing Arthur Samuel (1959), the question is:

> How can computers learn to solve problems without being explicitly programmed? In other words, how can computers be made to do what needs to be done, without being told exactly how to do it?

Genetic Programming II: Automatic Discovery of Reusable Programs (Koza, 1994) demonstrates that genetic programming can evolve multi-part programs consisting of a main program and one or more reusable, parameterized, hierarchically-called subprograms (called *automatically defined functions* or *ADFs*).

4. Background on Cellular Encoding of Neural Networks

A feedforward neural network is a complex structure that can be represented by line-labeled, point-labeled, directed graph. The points of the graph are either neural processing units within the network, input points, or output points. The lines are labeled with weights to represent the weighted connections between two points. The neural processing units are labeled with numbers indicating both the threshold and the bias of the processing unit.

In his seminal *Cellular Encoding of Genetic Neural Networks*, Frederic Gruau (1992) described an innovative and clever technique, called *cellular encoding*, in which genetic programming is used to concurrently evolve the architecture of a neural network, along with all weights, thresholds, and biases. In cellular encoding, each individual program tree in the population is a specification for developing a complete neural network from a very simple embryonic neural network (consisting of a single neuron). Genetic programming is applied to populations of these network-constructing program trees in order to evolve a neural network capable of solving the problem at hand (see also Gruau, 1994).

Each program tree is a composition of network-constructing, neuron-creating, and neuron-adjusting functions and terminals. The program tree is the genotype and the neural network constructed in accordance with the tree's instructions is the phenotype. The fitness of an individual program tree in the population is measured by how well the neural network that is constructed in accordance with the instructions contained in the program tree performs the desired task. Genetic programming then breeds the population of program trees in the usual manner using Darwinian reproduction, crossover, and mutation.

5. Background on SPICE

SPICE (an acronym for Simulation Program with Integrated Circuit Emphasis) is a massive family of programs written over several decades at the University of California at Berkeley for the simulation of analog, digital, and mixed analog/digital electrical circuits (Quarles et al., 1994). The input to a SPICE simulation consists of a netlist describing the circuit to be analyzed and certain commands that instruct SPICE as to the type of analysis to be performed and the nature of the output to be produced.

6. The Mapping between Circuits and Program Trees

Genetic programming breeds a population of rooted, point-labeled trees (i.e., graphs without cycles) with ordered branches. There is a considerable difference between the kind of trees bred in the world of genetic programming and the special kind of labeled graphs employed in the world of circuits. Genetic programming can be applied to circuits if a mapping is established between the kind of point-labeled trees found in the world of genetic programming and the line-labeled (often doubly labeled) cyclic graphs employed in the world of circuits. In our case, developmental biology provides the motivation for this mapping. The growth process used herein begins with a very simple embryonic electrical circuit. The circuit is developed as the functions in the program tree are progressively executed. The result is both the topology of the circuit and the sizing of all of its components.

Each program tree contains (1) circuit-constructing functions and terminals that create the topology of circuit from the embryonic circuit, (2) component-setting functions that convert wires (and other components) within the circuit into specified components, and (3) arithmetic-performing functions and numerical terminals that together specify the numerical value (sizing) for each component of the circuit.

Program trees conform to a constrained syntactic structure. Component-setting functions have arithmetic-performing argument subtrees and construction-continuing argument subtrees, while the circuit-constructing functions that manipulate the topology of the circuit have one or more construction-continuing argument subtrees. The left argument subtree of each component-setting function consists of a composition of arithmetic functions and numerical constant terminals that together yield the numerical value for the component. The right argument subtree of each component-setting function specifies how the construction of the circuit is to be continued. Both the random program trees in the initial population (generation 0) and any random subtrees created by the mutation operation in later generations are created so as to conform to this constrained syntactic structure. This constrained syntactic structure is preserved by the crossover operation using structure-preserving crossover with point typing.

7. The Embryonic Circuit

An electrical circuit is created by executing the program tree. Each program tree in the population creates one electrical circuit from the common embryonic circuit. The embryonic circuit used on a particular problem depends on the number of input signals and the number of output signals (probe points). It may also contain certain fixed components that are required or desired for the circuit being designed. The embryonic circuit used herein contains one input signal, one probe point, two modifiable wires, a fixed source resistor, and a fixed load resistor. In the embryonic circuit, the two modifiable wires each initially possess a writing head (i.e., are highlighted with a circle). A circuit is developed by modifying the component to which a writing head is pointing in accordance with the circuit-constructing functions in the program tree. Each circuit-constructing function in the program tree changes its associated highlighted component in the developing circuit in a particular way and specifies the future disposition of successor writing head(s), if any.

The bottom three quarters of figure 1 shows the embryonic circuit used for a one-input, one-output circuit. The energy source is a 2 volt sinusoidal voltage source **VSOURCE** whose negative (−) end is connected to node 0 (ground) and whose positive (+) end is connected to node 1. There is a fixed 1000-Ohm source resistor **RSOURCE** between nodes 1 and 2. There is a modifiable wire (i.e., a wire with a writing head) **Z1** between nodes 2 and 3 and another modifiable wire **Z0** between nodes 3 and 4. There are circles around modifiable wires **Z0** and **Z1** to indicate that the two writing heads (thick lines) point to them. There is a fixed isolating wire **ZOUT** between nodes 3 and 5, a voltage probe labeled **VOUT** at node 5, and a fixed 1000-Ohm load resistor **RLOAD** between nodes 5 and 0 (ground). There is an isolating wire **ZGND** between nodes 4 and 0 (ground). All of the above elements of this embryonic circuit (except **Z 0** and **Z 1**) are fixed forever; they are not subject to modification during the process of developing the circuit. All subsequent development of the circuit originates from writing heads.

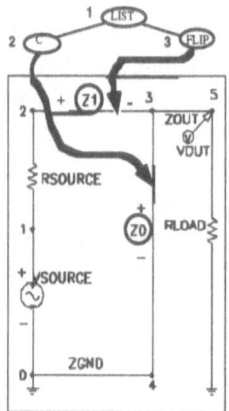

Figure 1 One-input, one-output embryonic electrical circuit.

A circuit is developed by modifying the component to which a writing head is pointing in accordance with the associated circuit-constructing function in the program tree. The figure shows L and FLIP functions just below the LIST and the two writing heads pointing to modifiable wires **Z 0** and **Z 1**. The L and FLIP functions will cause **Z 0** to be changed into a capacitor and the polarity of modifiable wire **Z 1** to be reversed.

The embryonic circuit is designed so that the number of lines impinging at any one node in the circuit is either two or three. This condition is maintained by all of the circuit-constructing functions. The isolating wire **ZOUT** protects the probe point **VOUT** from modification during the developmental process and the isolating wire **ZGND** protects the negative terminal of **VSOURCE**.

Note that little domain knowledge went into this embryonic circuit. Specifically, (1) the embryonic circuit is a circuit, (2) the embryonic circuit has one input and one output, and (3) there are modifable connections between the output and both source and ground. This embryonic circuit is applicable to any one-input, one-output circuit. It is the fitness measure that directs the evolutionary search process to the desired circuit.

8. Circuit-Constructing Functions

8.1. THE C AND L COMPONENT-SETTING FUNCTIONS

Each circuit-constructing function operates on a single component. Components are introduced into a circuit by the component-setting functions. The rightmost argument subtree of each component-setting function is a construction-continuing subtree that points to a successor function or terminal in the program tree. Upon completion, one writing head points to the new component. The left argument subtree of the component-setting functions is an arithmetic-performing subtree that contains a composition of arithmetic functions (addition and subtraction) and random constants (in the range −1.000 to +1.000). The arithmetic-performing subtree returns a floating-point value which is, in turn, interpreted as the value of the component using a logarithmic scale in the following way: If the return value is between −5.0 and +5.0, U is equated to the value returned by the argument subtree . If the return value is less than −100 or greater than +100, U is set to zero. If the return value is between −100.0 and −5.0, U is found from the straight line connecting the points (−100,0) and (−5, -5). If the return value is between +5.0 and +100, U is found from the straight line connecting (5,5) and (100, 0). The value of the component is 10^U in a unit that is appropriate for the type of component. This mapping gives the component a value within a range of 11 orders of magnitude centered on a certain value. This mapping gives the component a value within a range of 11 orders of magnitude that is centered on an appropriate value and that uses an appropriate unit of measurement that was settled upon after examining a large number of practical circuits in contemporary books.

If a component (e.g., a diode) has no numerical values, there is no left argument subtree. The two-argument C ("capacitor") function causes the highlighted component to be changed into a capacitor. The value of the capacitor is the antilogarithm (base 10) of the intermediate value U computed as above in nano-Farads (nF). This mapping gives the capacitor a value within a range of plus or minus 5 orders of magnitude centered on 1nF.

The two-argument L ("inductor") function causes the highlighted component to be changed into an inductor. The value of the inductor is the antilogarithm (base 10) of the intermediate value U in micro-Henrys (mH).

8.2. THE FLIP FUNCTION

All electrical components in SPICE have a designated positive (+) end and a designated negative (−) end. Polarity clearly matters for components such as diodes and transistors and it affects the course of the developmental process for all components. The one-argument FLIP function attaches the positive end of the highlighted component to the node to which its negative end is currently attached and vice versa. Upon completion, one writing head points to the now-flipped original component.

8.3. SERIES DIVISION

The three-argument SERIES ("series division") function operates on one highlighted component and creates a series composition consisting of the highlighted component, a copy of the highlighted component, one new modifable wire, and two new nodes. After execution of the SERIES function, there are three writing heads pointing to the original component, the new modifiable wire, and the copy of the original component.

Figure 2 shows a resistor **R1** connecting nodes 1 and 2 of a partial circuit containing various capacitors. **R1** is assumed to possess a writing head (i.e., is highlighted). Figure 3 illustrates the result of applying the SERIES division function to resistor **R1** from figure 2. First, the SERIES function creates two new nodes, 3 and 4. Second, SERIES relabels the positive (+) end of **R1** (currently labeled 2) as the first new node, 3. Third, SERIES creates a new wire **Z6** between the first new node, 3, and the second new node, 4. Fourth, SERIES inserts a duplicate (called **R7**) of the original component (including all its component values) between new node 4 and original node 2.

Note our convention of globally numbering components consecutively (rather than maintaining a different series of consecutive numbers for each type of component). Also, note that wires (such as **Z6**) are used only during the developmental process; all wires are edited out prior to the final creation of netlist for SPICE. Also, note that the SERIES function may be applied to a wire; in that event, the result is a series composition of three wires (each with its own writing head).

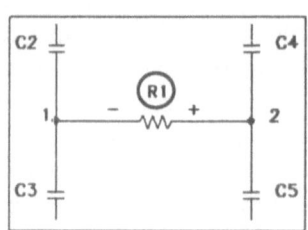

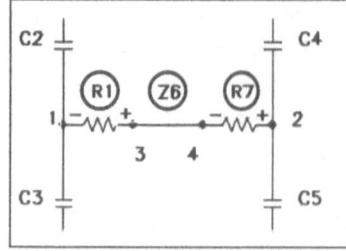

Figure 2. A circuit containing a resistor **R1**.   *Figure 3.* R esult after applying the series division function SERIES to resistor **R1**.

8.4. PARALLEL DIVISION FUNCTIONS

The two four-argument parallel division functions (PSS and PSL) each operate on one highlighted component to create a parallel composition consisting of the original highlighted component, a duplicate of the highlighted component, two new wires, and two new nodes. After execution of a parallel division, there are four writing heads. They point to the original component, the two new modifiable wires, and the copy of the original component. We describe (and use) only PSS herein.

First, the parallel division function PSS creates two new nodes, 3 and 4. Second, the parallel division function inserts a duplicate of the highlighted component (including all of its component values) between the new nodes 3 and 4 with the negative end of the duplicate connected to node 4 and the positive end of the duplicate connected to node 3. Third, the parallel division function creates a first new wire **Z6** between the positive (+) end of **R1** (which is at original node 2) and first new node, 3. Fourth, the parallel division function creates a second new wire **Z8** between the negative (–) end of **R1** (which is at original node 1) to second new node, 4.

The second character (i.e., the first S or L) of the name of the particular parallel division function indicates whether the positive end of the new component is connected to the smaller (S) or larger (L) numbered component of the two components that were originally connected to the positive end of the highlighted component. The third character (i.e., the second S or L) of the name of the particular parallel division function indicates whether the negative end of the new component is connected to the smaller (S) or larger (L) numbered component of the two components that were originally connected to the negative end of the highlighted component.

Figure 4 shows the results of applying the PSS function to resistor **R1** from figure 2. Since **C4** bears a smaller number than **C5**, new node 3 and new wire **Z6** are located between original node 2 and **C4**. Since **C2** bears a smaller number than **C3**, new node 4 and new wire **Z8** are located between original node 1 and **C2**.

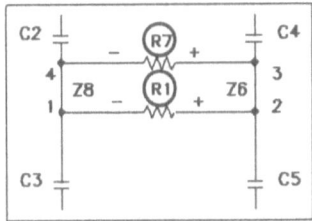

Figure 4. Result after applying PSS to resistor **R1**.

8.5. THE VIA AND GND FUNCTIONS

Eight two-argument functions (called VIA0, ..., VIA7) and the two-argument GND ("ground") function enable distant parts of a circuit to be connected together. The eight two-argument VIA0, ..., VIA7 functions create a series composition consisting of two wires that each possesses a successor writing head and a numbered port (called a *via*) that possesses no writing head. The port is connected to a designated one of eight imaginary layers (numbered from 0 to 7) in the wafer on which the circuit resides. If one or more other parts of the circuit connects to a particular layer, all such parts become electrically connected as if wires were running between them. If no other part of the circuit connects to a particular layer, then the one port connecting to the layer is useless (and this port is deleted when the netlist for the circuit is eventually created).

The two-argument GND ("ground") function is a special "via" function that connects directly to the electrical ground of the circuit. This direct connection to ground is made even if there is only one GND function calling for a connection to ground in the circuit. After execution of these functions, writing heads point to the two new wires.

8.6. THE NOP FUNCTION

The one-argument NOP function has no effect on the highlighted component; however, it delays activity on the developmental path on which it appears in relation to other developmental paths in the overall program tree – thereby (possibly) affecting the overall result produced by the construction process. After execution of NOP, one writing head points to the original highlighted component.

8.7. THE END FUNCTION

The zero-argument END function causes the highlighted component to lose its writing head – thereby ending that particular developmental path.

9. The Problem of Designing a Lowpass LC Filter

Consider a circuit design problem in which the goal is to design a filter using inductors and capacitors with an AC input signal with 2 volt amplitude. The filter is have a passband below 1,000 Hertz with voltage values between 970 millivolts and 1 volt and to have a stopband above 2,000 Hz with voltage values between 0 volts and 1 millivolts. This corresponds to a pass band ripple of at most 0.3 decibels and a stop band attenuation of at least 60 decibels. The circuit is to be driven from a source with an internal (source) resistance of 1,000 Ohms and terminated in a load of 1,000 Ohms.

A practising engineer would regard finding a circuit satisying the requirements as a non-trivial design problem. Using the terminology of Zverev (1967), these requirements can be satisfied by a Chebyshev-Cauer filter of complexity 5, with a relection coefficient of 20%, and modular angle of 30 degrees.

10. Preparatory Steps for Solving the Problem of Designing a Lowpass LC Filter

Before applying genetic programming to a circuit design problem, the user must perform seven major preparatory steps, namely (1) identifying the terminals of the to-be-evolved programs, (2) identifying the primitive functions contained in the to-be-evolved programs, (3) creating the fitness measure for evaluating how well a given program does at solving the problem at hand, (4) choosing certain control parameters (notably population size and the maximum number of generations to be run), (5) determining the termination criterion and method of result designation (typically the best-so-far individual from the populations produced during the run), (6) determining the architecture of the overall program, and (7) identifying the embryonic circuit that is suitable for the problem.

Since the problem of designing the lowpass LC filter calls for a one-input, one-output circuit with a source resistor and a load resistor, we use the embryonic circuit of figure 2 for this problem. Since the embryonic circuit starts with two writing heads, each program tree has two result-producing branches joined by a LIST function. There are no automatically defined functions. The terminal set and function set for both result-producing branches are the same. Each result-producing branch is created in accordance with the constrained syntactic structure that uses the leftmost (first) argument(s) of each component-creating function to specify the numerical value of the component. The numerical value is created by a composition of arithmetic functions and random constants in this arithmetic-performing subtree. Since the components involved in this problem (i.e., inductors and capacitors) each take exactly one component value, there is only one arithmetic-performing subtree. The rightmost (second) argument of each component-creating function is then used to continue the program tree.

In particular, the function set, $\mathcal{F}$aps for the arithmetic-performing subtree associated with each component-creating function contains the two-argument functions of addition and subtraction. That is,

$$\mathcal{F}\text{aps} = \{+, -\}.$$

The terminal set, $\mathcal{T}$aps, for the arithmetic-performing subtree consists of

$$\mathcal{T}\text{aps} = \{\mathfrak{R}\},$$

where $\mathfrak{R}$ represents floating-point random constants between −1.000 and +1.000.

The function set, $\mathcal{F}$ccs, for the construction-continuing subtree of each component-creating function is

$$\mathcal{F}\text{ccs} = \{C, L, SERIES, PSS, FLIP, NOP, GND, VIA0, VIA1, \\ VIA2, VIA3, VIA4, VIA5, VIA6, VIA7\},$$

taking 2, 2, 3, 4, 1, 1, 2, 2, 2, 2, 2, 2, 2, 2, and 2 arguments, respectively. The terminal set, $\mathcal{T}$ccs, for the construction-continuing subtree consists of

$$\mathcal{T}\text{ccs} = \{END\}.$$

The user provides a computer program to compute the fitness measure. The fitness measure drives the evolutionary process. For this problem, the voltage **VOUT** is probed at node 5 and the circuit is viewed in the frequency domain.

Note that the above is applicable to any one-input, one-output LC circuit. It is the fitness measure that directs the evolutionary process to the desired circuit.

Each circuit that is developed from the embryonic circuit is simulated using a modified version of the 217,000-line SPICE simulator that we modified to run as a submodule of our genetic programming system. The SPICE simulator is requested to perform an AC small signal analysis and to report the circuit's behavior for each of 101 frequency values chosen from the range between 101 frequency values chosen over five decades of frequency (from 1 Hz to 100,000 Hz). Each decade is divided into 20 parts (using a logarithmic scale).

Fitness is measured in terms of the sum, over these 101 fitness cases, of the absolute weighted deviation between the actual value of the voltage in the frequency domain) that is produced by the circuit at the probe point **UOUT** at node 5 and the target value for voltage. The smaller the value of fitness, the better. A fitness of zero is ideal. The fitness measure does not penalize ideal values; it slightly penalizes every acceptable deviation; and it heavily penalizes every unacceptable deviation.

The procedure for each of the 61 points in the 3-decade interval from 1 Hz to 1,000 Hz is as follows: If the voltage equals the ideal value of 1.0 volts in this interval, the deviation is 0.0. If the voltage is between 970 millivolts and 1,000 millivolts, the absolute value of the deviation from 1,000 millivolts is weighted by a factor of 1.0. If the voltage is less than 970 millivolts, the absolute value of the deviation from 1,000 millivolts is weighted by a factor of 10.0. This arrangement reflects the fact that the ideal voltage in the passband is 1.0 volt, the fact that a 30 millivolt shortfall is acceptable, and the fact that a voltage below 970 millivolts in the passband is not acceptable. It is not possible for the voltage to exceed 1.0 volts in an LC circuit of this kind, but if the voltage were to exceed the ideal, the deviation would be still be considered to be zero and there would still be no penalty for a filter design problem.

The procedure for each of the 35 points in the interval from 2,000 Hz to 100,000 Hz is as follows: If the voltage is between 0 millivolts and 1 millivolt, the absolute value of the deviation from 0 millivolts is weighted by a factor of 1.0. If if the voltage is more than 1 millvolt, the absolute value of the deviation from 0 millivolts is weighted by a factor of 10.0. This arrangement reflects the fact that the ideal voltage in the stopband is 0.0 volt, the fact that a 1 millivolt ripple above 0 millvolts is acceptable, and the fact that a voltage above 1 millivolt in the stopband is not acceptable.

We considered the number of fitness cases (61 and 35) in these two main bands to be sufficiently close that we did not attempt to equalize the weight given to the differing numbers of fitness cases in these two main bands. The deviation is considered to be zero for each of the 5 points in the interval above 1,000 Hz and below 2,000 Hz (i.e., the "don't care" band). Hits are defined as the number of fitness cases for which the voltage is acceptable or ideal or which lie in the "don't care" band. Thus, the number of hits ranges from a low of 5 to a high of 101 for this problem. Some of the bizarre circuits that are randomly created for the initial random population and that are created by the crossover operation and the mutation operation in later generations cannot be simulated by SPICE. Circuits that cannot be simulated by SPICE are assigned a high penalty value of fitness (10^8). These circuits become the worst-of-generation circuits for each generation. The practical effect of this high penalty value of fitness is that these individuals are rarely selected to participate in genetic operations and that they quickly disappear from the population.

The population size, M, is 320,000. Since this problem runs slowly, we set the maximum number of generations, G, to a large number and awaited developments. The percentage of genetic operations on each generation was 89% crossovers, 10% reproductions, and 1% mutations. A maximum size of 200 points was established for each of the two result-producing branches in each overall program. The other parameters for controlling the runs of genetic programming were the default values specified in Koza, 1994 (appendix D).

This problem was run on a medium-grained parallel Parystec computer system consisting of 64 Power PC 601 80 MHz processors arranged in a toroidal mesh with a host PC Pentium type computer. The so-called *distributed genetic algorithm* for parallelization was used with a population size of $Q = 10,000$ at each of the $D = 64$ demes. On each generation, four boatloads of emigrants, each consisting of $B = 2\%$ (the migration rate) of the node's subpopulation (selected on the basis of fitness) were dispatched to the four toroidally adjacent processing nodes. See Andre and Koza, 1996.

11. Results for the Problem of Designing a Lowpass LC Filter

We present the results of three different runs of genetic programming on the problem of designing the lowpass LC filter.

11.1. FIRST RUN

A run of genetic programming for this problem starts with the random creation of an initial population of 320,000 program trees (each consisting of two result-producing branches) composed of the functions and terminals identified above and in accordance with the syntactic constraints described above.

For each of the 320,000 program trees in the population, the sequence of circuit-constructing functions in the program tree is applied to the common embryonic circuit for this problem (figure 1) in order to create a circuit. The netlist for the resulting circuit is then determined. This netlist is wrapped inside an appropriate set of SPICE commands and the circuit is then simulated using our modified version of SPICE.

The initial random population of a run of genetic programming is a blind random search of the search space of the problem. As such, it provides a baseline for comparing the results of subsequent generations.

The best circuit of the 320,000 circuits from generation 0 had a fitness of 58.71 (on the scale of weighted volts described earlier) and scored 51 hits. The first result-producing branch of this program tree has 25 points (i.e., functions and terminals) and is shown below:

```
(C (- 0.963 (- (- -0.875 -0.113) 0.880)) (series (flip end) (series
(flip end) (L -0.277 end) end) (L (- -0.640 0.749) (L -0.123 end))))
```

The second result-producing branch has 5 points and is shown below:

```
(flip (nop (L -0.657 end))))
```

Figure 5 presents this best-of-generation program tree as a rooted, point-labeled tree with ordered branches. The first result-producing branch is rooted at the C function (labeled 2) and the second result-producing branch is rooted at the FLIP function (labeled 3).

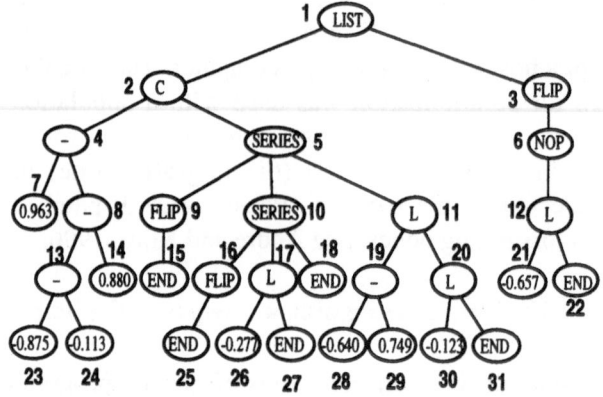

Figure 5. Program tree for best circuit of generation 0.

In executing the program tree, the connective LIST function (labeled 1) at the root of the tree is ignored. Most of the remainder of the tree is executed in a breadth-first order; however, arithmetic-performing subtrees (such as the 7-point subtree rooted at the point labeled 4) are executed in their entirety in a depth-first order immediately when its circuit-constructing function is first encountered. Thus, the C (capacitor) function (labeled 2) in figure 5 is executed first. Then, the 7-point arithmetic-performing subtree (labeled 4) is immediately executed in its entirety in a depth-first way so as to deliver the numerical component value needed by the capacitor function C. Then, the breadth-first order is resumed and the FLIP function (labeled 3) is executed.

Figure 6 shows the best circuit of generation 0 upon completion of the developmental process.

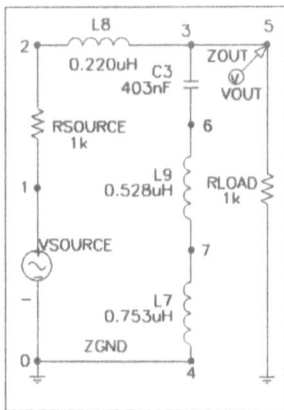

Figure 6. Best circuit of generation 0.

In the frequency domain, the voltages produced by this circuit in the interval between 1 Hz and 100 Hz are very close to the required 1 volt (accounting for most of the 51 hits scored by this individual). However, the voltages produced between 100 Hz and 1,000 Hz deviate considerably below the minimum of 970 millivolts required by the design specification (in fact, by hundreds of millivolts as one approaches 1,000 Hz). Moreover, the voltages produced above 2,000 Hz are, for the most part, considerably above the minimum of 1 millivolt required by the design specification (by hundreds of millivolts in most cases).

Generation 1 (and each subsequent generation of the run) is created from the population at the preceding generation by performing 142,400 crossover operations (producing 284,800 offspring or 89% of 320,000), 32,000 reproduction operations (10% of 320,000), and 3,200 mutation operations (1% of 320,000).

As the run proceeds from generation to generation, the fitness of the best-of-generation individual tends to improve. Figure 7 shows the standardized fitness and number of hits for the best-of-generation program of each generation of this run.

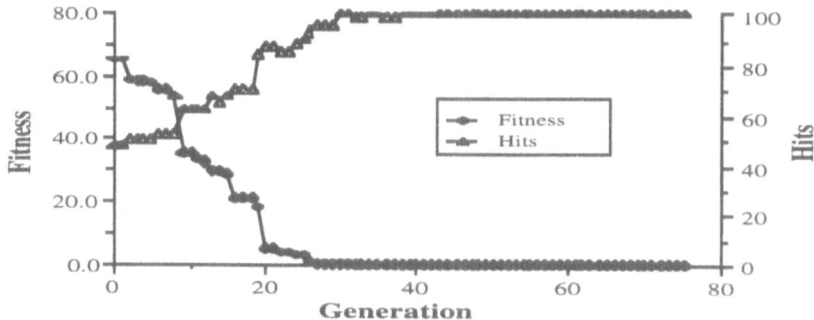

Figure 7. Fitness and hits for one run.

SPICE cannot simulate many of the bizarre circuits created by genetic programming. About two-thirds (65.3%) of the 320,000 programs of generation 0 for this problem produce circuits that cannot be simulated by SPICE. However, the percentage of unsimulatable circuits changes rapidly as new offspring are created by genetic programming using Darwinian selection, crossover, and mutation. The percentage of unsimulatable programs drops to 33% by generation 10, and 0.3% by generation 30. Figure 8 shows, by generation, the percentage of unsimulatable programs in this run.

In the genetic algorithm, the entire population generally improves from generation to generation. The hits histogram is a useful monitoring tool for visualizing the progressive learning of the population as a whole during a run. The horizontal axis of the hits histogram represents the number of hits (0 to 101 here) while the vertical axis represents the percentage of individuals in the population scoring that number of hits.

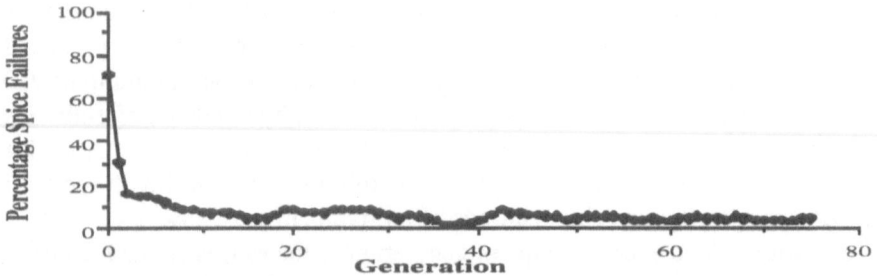

Figure 8. Percentage of unsimulatable programs

Figure 9 shows the hits histograms for generations 0, 20 and 40 of a typical run of this problem. The horizontal axis represents the number of hits (0 to 101 here) while the vertical axis represents the percentage of individuals in the population scoring that number of hits. Note the left-to-right undulating movement of both the high point and the center of mass of these histograms over the generations.

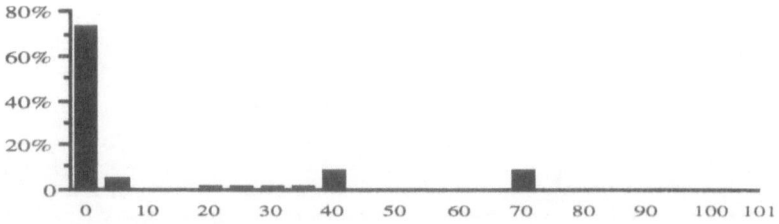

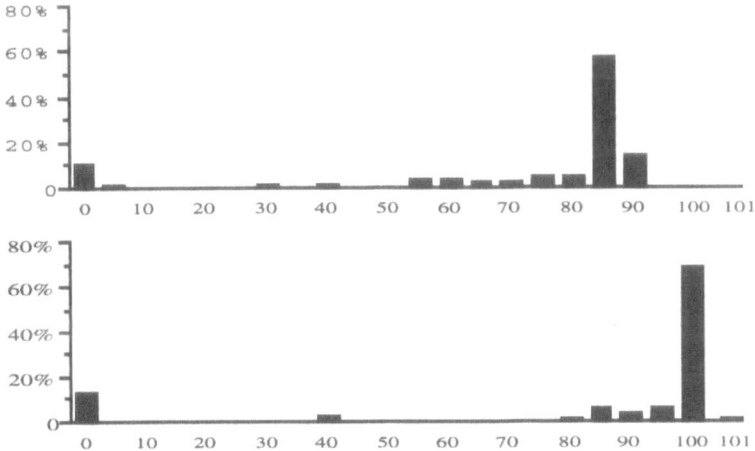

Figure 9. Hits histogram for generations 0, 20 and 40 of a run of this problem.

The improvement, from generation to generation, in the fitness of the population as a whole can also be seen by examining the average fitness of the population by generation. Figure 10 shows, by generation, the average fitness of the portion of the population that can be analyzed by SPICE (that is, after excluding individuals receiving the penalty value of fitness). As can be seen, the average fitness of the population as a whole is 1,054 for generation 0, 443 for generation 2, 213 for generation 5, 58.2 for generation 10, 38.0 for generation 20, and 16.5 by generation 30.

The best individual program tree of generation 32 has 306 points, has a fitness of 0.00781 and scores 101 hits. That is, by generation 32, all 101 sample points are in compliance with the design requirements for this problem.

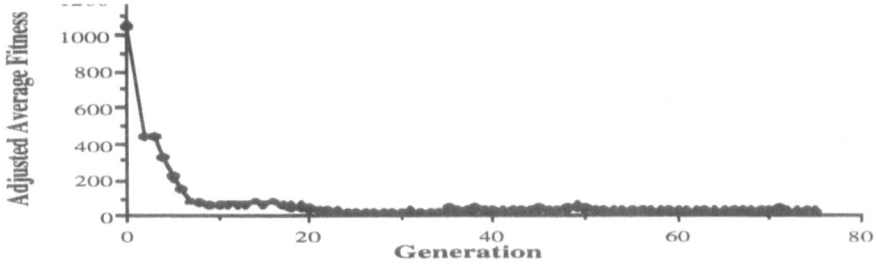

Figure 10. Average fitness of the simulatable circuits in the population.

Figure 11 shows the best-of-run circuit from generation 32. This circuit is a seven-rung ladder consisting of repeated values of various inductors and capacitors. Figure 12 shows the behavior in the frequency domain of the best-of-run circuit from generation 32.

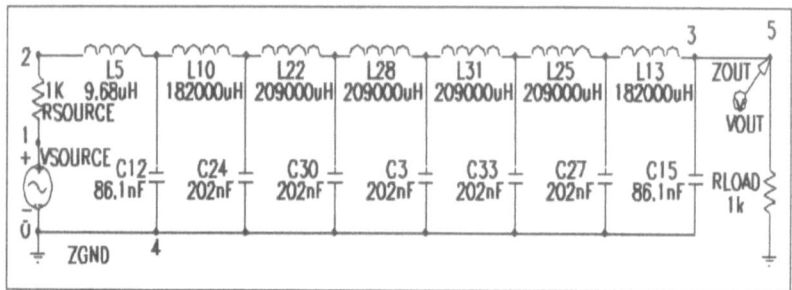

Figure 11. Best-of-run "seven-rung ladder" circuit from generation 32.

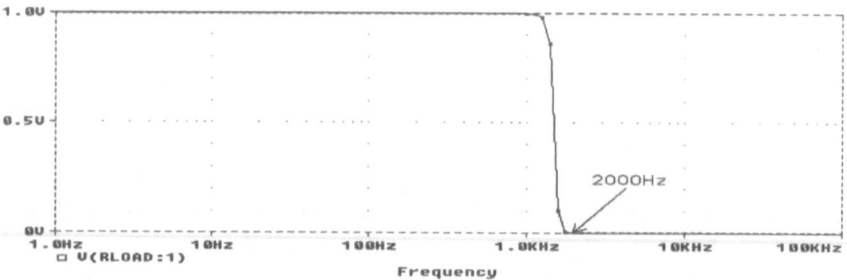

Figure 12. Frequency domain behavior of "seven-rung ladder" from generation 32.

As can be seen, the circuit delivers a voltage of virtually 1 volt in the entire passband from 1 Hz to 1,000 Hz and delivers a voltage of virtually 0 volts in the entire stopband starting at 2,000 Hz. The best individual from generation 76 has a fitness (0.000995) that is about an order of magnitude better than that of the fully compliant individual of generation 32.

11.2. A "BRIDGED T" CIRCUIT FROM ANOTHER RUN

Different runs of genetic programming produce different results. Moreover, when we continue the run of genetic programming after the emergence of the first 100%-compliant individual, additional 100%-compliant individuals often emerge. Figure 13 shows a fully compliant best-of-run circuit from generation 64 of another run. In this circuit (which has a fitness of 0.04224), inductor **L14** forms a "bridged T" subcircuit in conjunction with capacitors **C3** and **C15** and inductor **L11**. Of course, the parallel capacitors (the pair **C18** and **C33** as well as the triplet **C24**, **C21**, and **C12**) could be combined. This "bridged T" circuit is distinctly different in structure from the "ladder" circuit.

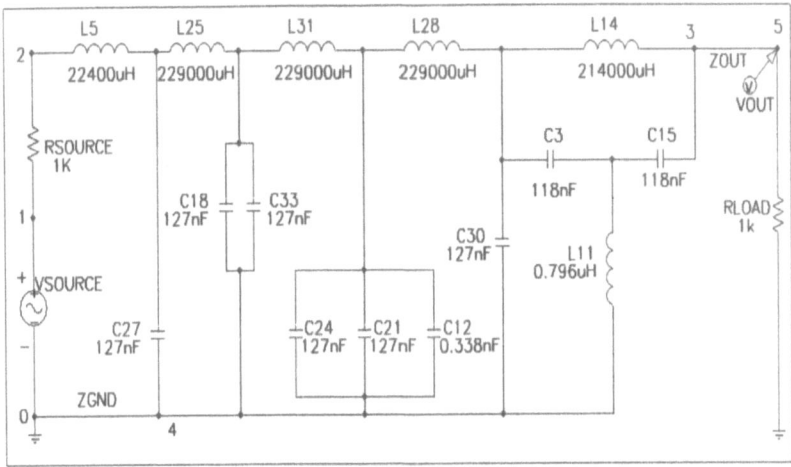

Figure 13. "Bridged T" circuit from generation 64.

12. Conclusions

We have also used this technique to design an asymmetric bandpass filter and a crossover (woofer and tweeter) filter. The latter requires a one-input, two-output embryonic circuit. We are currently working on circuits with active elements.

We have described an automated design process for designing analog electrical circuits based on the principles of natural selection, sexual recombination, and developmental biology. The design process starts with the random creation of a large population of program trees composed of circuit-constructing functions. Each program tree specifies the steps by which a fully developed circuit is to be progressively developed from a common embryonic circuit appropriate for the type of problem that the user wishes to solve. The population of program trees is genetically bred over a series of many generations using genetic programming that is driven by the fitness measure. Genetic programming employs genetic operations such as Darwinian reproduction, sexual recombination (crossover), and occasional mutation to create offspring. The paper described how genetic programming technique evolved the design of a low-pass filter.

Acknowledgements

Tom L. Quarles of Meta-Software of Campbell, California provided helpful advice concerning SPICE. Simon Handley made helpful comments on the above.

References

Aaserud, O. and Nielsen, I. R.: 1995. Trends in current analog design: A panel debate. *Analog Integrated Circuits and Signal Processing.* **7**(1) 5-9.

Andre, D. and Koza, J. R.: 1996, Parallel genetic programming: A scalable implementation using the transputer architecture, *in* P. J. Angeline and K. E. Kinnear Jr. (eds), *Advances in Genetic Programming 2*, MIT Press, Cambridge, MA.

Degrauwe, M.: 1987, IDAC: An interactive design tool for analog integrated circuits. *I1 Journal of Solid State Circuits*, **22**, 1106–1116.

Gruau, F.: 1992, Cellular Encoding of Genetic Neural Networks, *Technical report 92-21*, Laboratoire de l'Informatique du Parallélisme. Ecole Normale Supérieure de Lyon.

Gruau, F.: 1994, Genetic micro programming of neural networks, *in* K. E. Kinnear Jr. (ed.), *Advances in Genetic Programming.* MIT Press, Cambridge, MA, pp. 495-518.

Harjani, R., Rutenbar, R. A. and Carley, L. R.: 1989, OASYS: A framework for analog circuit synthesis. *I1 Transactions on Computer Aided Design*, **8**, 1247–1266.

Higuchi, T., Niwa, T., Tanaka, H., Iba, H., de Garis, H. and Furuya, T.: 1993, Evolvable hardware—Genetic-based generation of electric circuitry at gate and hardware description language (HDL) levels, *Electrotechnical Laboratory technical report 93-4*, Tsukuba, Ibaraki, Japan.

Holland, J. H.: 1975, *Adaptation in Natural and Artificial Systems: An Introductory Analysis with Applications to Biology, Control, and Artificial Intelligence*, University of Michigan Press, Ann Arbor, MI. Second edn MIT Press, Cambridge, MA, 1992.

Koh, H. Y., Sequin, C. H. and Gray, P. R.: 1990, OPASYN: A compiler for MOS operational amplifiers, *I1 Transactions on Computer Aided Design.* **9**, 113–125.

Koza, J. R.: 1992, *Genetic Programming: On the Programming of Computers by Means of Natural Selection,* MIT Press, Cambridge, MA.

Koza, J. R.: 1994, *Genetic Programming II: Automatic Discovery of Reusable Programs.* MIT Press, Cambridge, MA

Kruiskamp, W. and Leenaerts, D.: 1995, DARWIN: CMOS opamp synthesis by means of a genetic algorithm, *Proceedings of the 32nd Design Automation Conference*, Association for Computing Machinery, New York, NY, pp. 433–438.

Maulik, P. C. Carley, L. R., and Rutenbar, R. A.: 1992, A mixed-integer nonlinear programming approach to analog circuit synthesis, *Proceedings of the 29th Design Automation Conference*, I1 Press, Los Alamitos, CA, pp. 698–703.

Ning, Z., Kole, M., Mouthaan, T., and Wallings, H.: 1992, Analog circuit design automation for performance, *Proceedings of the 14th CICC*, I1 Press, New York, pp. 8.2.1–8.2.4.

Quarles, T., Newton, A. R., Pederson, D. O. and Sangiovanni-Vincentelli, A.: 1994, *SPICE 3 Version 3F5 User's Manual*, Department of Electrical Engineering and Computer Science, University of California, Berkeley, California.

Rutenbar, R. A:. 1993, Analog design automation: Where are we? Where are we going? *Proceedings of the l5th I1 CICC*, I1 Press, New York, pp. 13.1.1-13.1.8.

Samuel, A. L.: 1959, Some studies in machine learning using the game of checkers, *IBM Journal of Research and Development*, **3**(3), 210–229.

Zverev, A. I.: 1967, *Handbook of Filter Synthesis*, Wiley.

4

case-based design

J. S. Gero and F. Sudweeks (eds), Artificial Intelligence in Design '96, 173-189.
© 1996 *Kluwer Academic Publishers.*

A STUDY OF CASE ADAPTATION SYSTEMS

ANGI VOSS[1]
GMD, D-53754 Sankt Augustin, Germany
BRIGITTE BARTSCH-SPÖRL
BSR Consulting, Wirtstr. 38, D-81539 Munich, Germany

AND

RIVKA OXMAN
Faculty of Architecture and Town Planning,Technion,
Faculty of Civil Engineering, Technical University,
Delft, The Netherlands[2]

Abstract. This paper surveys ongoing research and implemented systems for a major step in the case-based reasoning (CBR) cycle which is the adaptation of similar cases to the current problem at hand. On the basis of the systems contained in the survey, we come up with several classifications by the type of task, by knowledge and methods, by characteristics of the solution space, and by the general strategy. The paper ends with the identification of important research issues for the future.

1. Introduction

Reasoning from past experience is widely recognized as a cognitive phenomenon in the creative process of design. In the course of exploring design solutions, designers appear to be able to browse freely among previous designs and adapt prior solutions. The computational paradigm known as Case-Based Reasoning (CBR) (Riesbeck and Schank, 1989; Kolodner, 1993) is relevant to design, since it supports reasoning from the specific knowledge which is associated with holistic designs. The CBR paradigm addresses issues of experience-based design where experience is strong, but the domain model is weak, or poorly formalized. It provides concepts and methods which reflect the cognitive way of designers.

[1]This research was supported by the German Ministry for Education and Research (BMBF) within the joint project FABEL under contract no. 01IW104. Project partners in FABEL are German National Research Center for Computer Science (GMD), Sankt Augustin, BSR Consulting GmbH, Muenchen, Technical University of Dresden, HTWK Leipzig, University of Freiburg, and University of Karlsruhe.
[2] This research was supported by the Informatics Group of Civil Engineering while Dr Oxman was a visiting professor, 1995.

Case-based reasoning has attracted many researchers in AI. CBR is a cognitively plausible theory in which previous cases are applied to current problems. It has been used in many domains such as classification, diagnosis, configuration, design and planning. CBR consists of the following main steps: problem specification, case retrieval, case adaptation and case storage. Stages of retrieval and storage are relatively understood and work in this field can be applied in many domains. In complex domains such as design, adaptation seems to play a major role, however, work on adaptation has not yet been developed both theoretically and computationally to provide useful methods across domains.

This paper describes our recent survey which was carried out as part of a workshop on adaptation in CBR at the First International CBR Conference (Veloso and Aamodt, 1995). Our goal was to make a survey of ongoing research in order to identify major issues, to classify current approaches, and to get a better understanding of adaptation approaches and their potential applicability to design.

To obtain an up-to-date survey, we circulated two questionnaires to active researchers in the field all over the world. Surprisingly, most answers concerned European systems, which sharply contrasts with a comparative but independent literature study, which mostly covers US American systems (Hanney et al., 1995). Does this indicate a geographic movement of active research on case adaptation? At the ICCBR conference we then had extended interviews with many system developers. As a result of this and by comparison with the study by Hanney et al. (1995), and with prior work by Kolodner (1993), this paper presents different classification schemes for case adaptation systems: by task, by methods and by knowledge, by the solution space, and by the global strategy. Each is devoted one of the following sections. We conlude with some open issues that need further elaboration.

Most systems in our study deal either directly with diverse design tasks, or generally with synthetic tasks. Therefore our findings are particularly relevant for CBR in design. In real-world design applications diverse kinds of tasks such as problem analysis and planning are involved and many sub-tasks include configurative problems. So according to our findings, it is possible to argue for a general theory of adaptation and to demonstrate its relevance to design tasks

2. Classification by Task

2.1 DEFINITION OF TASKS

In the expert systems community, there is a long tradition in dealing with problem classes and methods to solve them (Puppe, 1990). Case-based reasoning turned out not to fit very well into the existing classification scheme. In fact, CBR is a weak method that is suitable for a wide range of analytic and synthetic expert system tasks, with some exceptions like possibly simulation.

Analytic tasks typically analyse a situation and then try to map the description of the situation to something prefabricated like a class name, a diagnosis or a ready-made solution. Representative examples are object classification, medical diagnosis or help-desk support problems. Wess (1995) distinguishes the analytic CBR tasks classification, diagnosis and decision support. In *classification* problems all information necessary to do the job is known in advance and can be used when the problem solving process starts. In *diagnosis* problems the problem solving process typically starts with incomplete information and has to gather missing information on the fly. Both classification and diagnosis problems have a goal definition that is static which means clearly known in advance. For *decision support* problems, this last mentioned precondition is missing. Therefore they are explorative in nature and usually involve the user in making step-wise decisions about how to proceed and when to stop.

Synthetic tasks typically have to construct a solution from parts - mostly obeying to a set of domain-specific construction rules. Representative examples are route or transportation planning, configuration of technical equipment and design of complex artefacts ranging from production machines to buildings. Synthetic tasks can further be divided into planning, configuration and design. *Planning* problems deal with activities that have to reach a certain goal under given time constraints and where subsequent activities are based on prerequisites that former activities have generated. *Configuration* problems construct artefacts from a set of known parts and on the basis of reliable and complete knowledge about the compatibility constraints. In contrast, *design* tasks are typically less precisely defined and require loops and iteration steps because of incomplete knowledge about the interrelationships between functional, spatial, material and many other sorts of constraints (Bartsch-Spoerl, 1995).

Hanney et al. (1995), distinguish identification tasks, which can be combined with design or prediction. Their identification corresponds to our analysis, their design to our synthesis, while we have no representatives of their prediction tasks. That means, their distinction is essentially coarser than ours.

2.2 SELECTED SYSTEMS

Developers of case adaptation systems provided information about their current prototypes, altogether 22 systems. They cover a broad range of tasks, domains and approaches. In the rest of the paper we focus on selected representatives.

systems	analytic			synthetic		
	classification	diagnosis	decision support	planning	configuration	design
CaBaTa, INRECA, TUB-JANUS	X	X	X			
MoCAS, Torasso´s, AL		X				
CHESS	X	X	X			
DIAL, ROBBIE, PARIS, CAPlan/CbC, PRODIGY				X		
DÉJA VU				X	X	X
EADOCS						X
DOM,		X				X
AgentEX, AAAO, SYN, ToPo						X
IDIOM					X	X
Composer, Cunningham´s				X	X	

Figure 1. Our study comprises systems for many tasks. (The shaded area in the middle contains different tools from a single system, the FABEL system, FABEL-Report 35.)

CaBaTa

task: classification and diagnosis, domain: analytic

This is a CBR decision support system especially suitable for weak-theory domains (Lenz, 1994). It is comparable to INRECA (Bergmann et al., 1994), and to TUB-JANUS. The core inference engine relies on the CBR paradigm while additional knowledge can be added in model-based and rule-based form (the latter both for classification and adaptation rules). The main goal of CaBaTa was to use it as a test bench for various retrieval techniques. Adaptation itself occurs only as a "side effect". The Travel Agency Domain is the application cited in most papers: The task is to find an appropriate package holiday given some more or less vague specifications. However, CaBaTa has also been demonstrated for production planning support, financial decision aiding and is currently being applied to estate assessment.

MoCAS

task: diagnosis, domain: numerical control machine domain

This system deals with diagnosis applications of technical systems (Bergmann et al., 1994). Its main application is in the numerical control machine domain. It combines model-based and case-based approaches.

PARIS

task: planning, domain: manufacturing rotary-symmetric parts on a lathe

This system is used for plan abstraction and refinement in an integrated system (Bergmann and Wilke, 1995).

CAPlan/CbC

task: planning, domain: manufacturing mechanical workpieces

This is a case-based planning system supporting process planning for manufacturing mechanical workpieces (Munoz et al., 1995). The overall architecture is build on top of CAPlan (for: Computer Assisted Planning), a partial-order plan-space planner. To use episodic problem solving knowledge for both optimizing plan execution costs and minimizing search the case-based control component CAPlan/CbC has been realized in a way that allows incremental acquisition and reuse of strategical problem solving experience by storing solved problems as cases and reusing them in similar situations. For effective retrieval of cases CAPlan/CbC combines domain-independent and domain-specific retrieval mechanisms that are based on the domain model and problem representation.

PRODIGY

task: planning, domain: route planning

Since complete plans are too complex to be reusable, cases contain pieces (Haigh and Veloso, 1995). As a consequence, a problem must be covered by multiple cases. They are replayed, thereby eliminating inconsistencies and filling in gaps. Decomposition is implicitly achieved by the set of cases that cover the target.

DÉJÀVU

task: design, domain: plant control software design

This system employs a technique called hierarchical CBR which stores complex solutions as hierarchies of cases at varying levels of abstraction and which allows complex designs to be generated by decomposing them into simpler designs (using abstract cases) which can then be dealt with by using actual design cases (Smyth and Keane, 1995). The system also uses a novel retrieval technique called adaptation-guided retrieval which uses adaptation knowledge during retrieval to determine how easy it is to adapt a given case for a specific target problem, and hence ensures that the most "adaptable" case is retrieved.

EADOCS

task: design and optimisation, domain: aircraft design

This system will be used for conceptual subsonic civil aircraft design such as thin-walled fibre reinforced composite aircraft structures (more particular sandwich panels) (Netten et al., 1993). The system employs multiple techniques: qualitative constraint-based reasoning for evaluating feasible combinations of design components for prototype solutions, case-based reasoning to quantify prototype solutions and a rule-based heuristic modification and finally numerical optimisation to obtain a good conceptual design.

AAAO

task: design, domain: replacement of columns

This system uses knowledge which is formalized in constraints (Adami, 1995). These constraints depend on the statical requirements and architectural demands with respect to the layout of rooms and the kind of their use. This led to a model of active autonomous objects (columns being the objects) that behave according to simple heuristics trying to satisfy a set of applicable constraints in a concurrent way.

SYN

task: design, domain: layout of connections

SYN adds connections to a spatial layout that already contains the outlets (Boerner, 1995). It compares the layout with several prototypes, transfers the connections between matching outlets and generates them for the others by applying previous instantiations.

ToPo

task: design, domain: layout

This system compares the topological structure of a query and a case and matches identical substructures (Coulon, 1995). ToPo is able to transfer all structure of the case to the query, which is related to the matched substructure. The structure to be transferred may be selected by the user or any heuristic. The result is checked versus usual and unusual topologies occuring in the casebase. The matching is done by a modified graph-matching algorithm of Bron and Kerbosch.

DOM

task: design, domain: installation of pipes in buildings

This system establishes a connection between cases and generic domain knowledge (Bakhtari and Oertel, 1995). The generic domain knowledge can be used to evaluate, synthesize, and adapt parts of an actually handled case. Adaptation in this case means modifying a case in order to make it consistent with the underlying domain ontology.

Composer

task: planning, domain: mechanical design

This system formulates problems such as assembly sequence planning and configuration design as constraint satisfaction problems (Pu and Purvis,

1994). The advantage of this formulation is that it allows the use of a repair-based constraint resatisfaction algorithm to efficiently and systematically combine multiple cases and repair constraints in order to solve new problems. A very important result discovered in Composer is that using this method, one can assess the quality of adaptation before beginning the adaptation process.

IDIOM

task: design, domain: mechanical design

This is an interactive design system which employs intelligent objects and models, and is a result of a collaborative effort between architects and computer scientists (Smith et al., 1995). IDIOM uses parts of cases, domain models and user interaction to compose designs. Case adaptation is used to modify case parts according to the new design context. Spatial consistency is maintained through solving and propagation of constraints. Constraints are introduced through an analysis of topology, activation of domain models and through user interaction. Incremental dimensionality reduction simplifies constraint systems as case parts are added. Preference constraints are allowed to influence the design when they are not in conflict with fixed constraints.

3. Classification by Data Structures, Knowledge, and Methods

Given the domain and task of a system, the question is what approach to adaptation do they suggest? In the CBR community, adaptation is usually classified as being parametric, transformational or generative. Kolodner, (1993) introduces a finer distinction. Figure 2 compares the systems wrt. their task types, data structures for problems and solutions, knowledge, techniques, and the criteria from Kolodner (1993). (Please ignore for a moment the last column, it concerns the next section.)

We conclude that there is no one best way for representing data, adaptation knowledge and adaptation methods. The choice is essentially domain-dependent. Nevertheless, there are a few patterns that can be interpreted towards some tentative recommendations.

For analytic tasks, a good deal of the adaptation requirements are parametric and sometimes also transformational. Here, packages of rules can do the job - provided the necessary context dependencies can be expressed, e.g. in the form of preconditions. But keep in mind that the case has to be treated as a whole and known interrelationships e.g. between different slots must not be overlooked.

For synthetic tasks it is usually necessary to have at least a partial model of the underlying domain. They can range from complete models that are used by from scratch-problem solvers as in MOCAS, PARIS, CAPLan/CbC,

PRODIGY, to partial ones as the knowledge in DOM and EADOCS, or the constraints in AAAO, IDIOM, and Composer. SYN and ToPo form exceptions as they exclusively rely on structure matching operations.

Therefore, knowledge representation and methods have come to play only a minor role in our investigations. Adequacy for the problem and the application domain are much more important for this topic than any additional requirements that come from adaptation necessities.

systems	task type	data structures	knowledge	techniques	criteria from Kolodner	solution space level
CaBaTa	analytic	features	rules	rule interpretation	parameter adjustment	1
MoCAS	diagnosis	symptoms, diagnoses	behavior model	model-based diagnosis	replay	2
PARIS	planning	goals, plans	operators	hierarchical planner	replay	2
PRODIGY	planning	goals, plans	operators	planner	replay	3
CAPlan/CbC	planning	goals, plans	operators	planner	replay	2
DÉJÀ VU	synthesis	decomposition hierarchy	rules	rule interpretation	special adaptation	2
EADOCS	design	structure with qualitative parameters	heuristics + numerics	rule interpretation, calculation	parameter adjustment	2
AAAO	design	layout	constraints	constraint algorithms	model-guided repair	2
SYN	design	term	rules	algebraic	reinstantiation	2
ToPo	design	topology		graph algorithms	reinstantiation	3
DOM	design	layout	rules	rule interpretation	model-guided repair	3
Composer	configuration, planning	constraint network	constraints	constraint algorithms	model-guided repair	3
IDIOM	configuration, design	constraint network	constraints	constraint algorithms	model-guided repair	3

Figure 2. Adaptation systems, their data structures, knowledge, methods, classification according to Kolodner (1993) and by solution space characteristics.

4. Classification by Solution Space

As a more appropriate approach to classification, we propose to consider characteristics of the solution space. We distinguish five levels ranging from 'no adaptation at all' to 'no constructive system support for adaptation feasible'.

- *Level 0: No adaptation required:* This level includes all CBR systems solving pure classification problems, pure diagnosis problems without any therapy or repair, or pure artefact selection problems where e.g. a piece of technical equipment can be bought off the shelf without requiring any customer specific configuration.

- *Level 1: Only local and continuous adaptation required:* Here are all CBR problem solvers operating in a continuous solution space or having full knowledge about existing discontinuities. This is true of *most* analytic task systems in our survey, namely CaBaTa, and INRECA.

- *Level 2: Full adaptation in a closed world required and feasible:* The CBR problem solvers operate in well-structured solution spaces where reliable algorithms are known for de- and recomposing both the task and the solution. A typical prerequisite is a complete and consistent model of the application domain that allows to reason under the closed-world assumption. In our survey, this is clearly the case for *MoCAS*, PARIS, SYN and AAAO. They use generative problem solvers to transfer a whole case. DÉJÀVU, EADOCS, and CAPlan/CbC do a careful problem decomposition so that composition is guaranteed to work.

- *Level 3: Full adaptation in an open world required but only partially feasible:* This is the case for all synthetic task CBR systems operating in an open world domain on the basis of incomplete knowledge. This situation is typical for real world domains above a certain level of complexity. There are "holes" in the problem solving behaviour which means that there are (sub)problems the systems cannot solve fully automatically. Therefore most of them involve the user and call this interactive adaptation. In our survey, this is true of the FABEL modules DOM and ToPo, and of the design support systems IDIOM and Composer from Lausanne, and the planner PRODIGY.

- *Level 4: Full adaptation in an open world required but not feasible:* This level includes all synthetic task CBR systems that offer only browsing capabilities because there is not enough domain knowledge for meaningful adaptation. We did not get a questionnaire from such a system but we know that such systems have been built (Domeshek and Kolodner, 1993; Oxman, 1994).

We learned from this classification exercise that adaptation for analytic tasks usually is not a big problem - but for synthetic tasks this may change rather quickly, at least if the problems have to be decomposed and there are no secure de- and recomposition strategies available. As the last column in figure 2 shows, most synthetic systems from our study fall into levels 2 and 3. To obtain a finer-grained distinction, we have a closer look at the general strategy.

5. Classification by general strategy

In section 3 we concluded that the task and domain essentially influence the data structures, knowledge available for adaptation, and methods that can use this knowledge and that may be re-used for adaptation. What other decisions have to be taken for designing an adapting case-based reasoner?

5.1 GENERAL STRATEGIES

Crucial is the overall strategy. It has to address questions like the following ones:

With respect to the problem:
- Is the problem so simple that it can be solved in a single step?
- Or else, is a decomposition into subproblems available so that each subproblem can be input to retrieval?
- Or is the problem so intricate that it cannot be decomposed a priori?

With respect to the cases:
- Is it sufficient to retrieve a single case for each (sub)problem? - We do not consider alternative cases that are used for backtracking if the first one fails.
- Or are multiple cases required to cover the (sub)problem?

With respect to adaptation of individual cases:
- Can each case be adapted using heuristics?
- Or by using general structure transformations?
- Or by a from-scratch problem solver?

With respect to solution integration:
- Are there no partial solutions to be integrated?
- Or is there only one to be embedded into the global context of the problem?
- Or else, can the partial solutions be integrated incrementally according to a preceding decomposition?
- Or else, are there multiple cases that have to be adapted and integrated in a joint effort?

The answers to these questions are not independent. In our study, they lead to ten strategies, which are summarized in figure 3. They can be classified along each criterion: decomposition, cases, individual adaptation and integration. In the following, they are presented according to the numbering in the last column. It applies multiple criteria:

S1 is for level-1 systems, S1-3 need no decomposition, S1-4 deal with a single case, in S4 the case must be integrated into a context, S5-8 decompose the problem before retrieval, S5-7 compose the solutions incrementally, S9-10 have no a priori decomposition, they retrieve multiple cases for a single problem, S8-10 do integration in a joint effort.

systems	task type	problem decomposition	cases per (sub)problem	individual adaptation	integration	strategy	solution space level
CaBaTa	analytic	not required	1	heuristics	--	S1	1
MoCAS	diagnosis	not required	1	from-scratch problem solver	--	S2	2
PARIS	planning	not required	1	from-scratch problem solver	--	S2	2
CAPlan/ CbC	planning	specially designed	1	from-scratch problem solver	incremental	S6	2
PRODIGY	planning	not available	many		from-scratch problem solver	S9	3
DÉJÀ VU	synthesis	case-based	1	heuristics	incremental	S7	2
EADOCS	design	available	1	from-scratch problem solver	heuristics	S8	2
AAAO	design	not required	1	from-scratch problem solver	--	S2	2
SYN	design	not required	1	structural	--	S3	2
ToPo	design	not available	1	manual selection	structural	S4	3
DOM	design	available	1	from-scratch problem solver	incremental	S5	3
Composer	configuration planning	not available	many	--	from-scratch problem solver	S9	3
IDIOM	configuration design	manual	1 (manual)	from-scratch problem solver	from-scratch problem solver	S10	3

Figure 3. The systems demonstrate ten different strategies.

Strategy S1: The problem is simple, only one case needs to be retrieved and can be adapted adapted heuristically. S1 is used by the level-1 systems represented by CaBaTa.

Strategy S2: The problem is not simple. But the cases are as complex as the problems. Therefore a single case can be retrieved and adapted using a powerful from-scratch problem solver. S2 is applied by the level-2 systems MoCAS with a model-based diagnostic engine, by PARIS with a hierarchical planner, and by AAAO with a constraint-reasoner.

Strategy S3: The problem is simple, only one case needs to be retrieved and can be adapted by structure-matching and transfer. S3 is applied by the level-2 system SYN using term matching and transformation techniques.

Strategy S4: The problem is not simple and the cases may vary in complexity. A single case is retrieved and matched by structure. Parts for transfer are proposed and one is selected by the user. It is structurally integrated into the problem context. The level-3 system ToPo applies this strategy with graph matching and graph merging algorithms to layouts of buildings.

Strategy S5: The problem is a priori decomposable into simple subproblems. For each a single case is retrieved and adapted individually and incrementally. Level-3 system DOM applies S5 in an interactive mode to repair and extend pipe layouts in buildings.

Strategy S6: The problem is a priori decomposable into simple subproblems. For each a single case is retrieved and adapted individually and incrementally. S6 can be applied by the level-2 system CAPlan/CbC because its planner has a powerful dependency management. The decomposition strategy was specially designed.

Strategy S7: The problem is intricate, but by using cases with problem decompositions it can be decomposed in such a way that for each subproblem a single case can be retrieved and adapted incrementally. S7 is applied by level-2 system DÉJÀVU using heuristics for local adaptation and for incremental integration.

Strategy S8: The problem is decomposable and a single case is sufficient for each suproblem. But composition cannot be done incrementally. The cases are merged using heuristics and the solution is optimized using a from-scratch problem solver. S5 is applied by level-2 system EADOCS with numerical optimization algorithms.

Strategy S9: The problem is intricate, there is no apriori decomposition. Instead, several (complementary or overlapping) cases are retrieved and adapted jointly using from scratch problem solvers. S9 is applied by the level-3 systems Composer with a constraint problem solver and by PRODIGY with a planner. - Note that from the adaptation an a posteriori decomposition of the problem emerges: subproblems are those parts of the problem that were covered by a single case!

Strategy S10: The problem is intricate, there is no apriori decomposition. The user is asked to select a set of cases. They are integrated using a from-scratch problem-solver. S10 is applied by level-3 system IDIOM in connection with a constraint-problem solver.

In general, the more difficult the task and the higher the level of the solution space criterion, the more complex is the strategy. More specifically, the available methods (for decomposition, individual adaptation and integration) heavily dictate the choice of the strategy, and several tricks are employed to cover any discrepancies between available methods and the problems to be solved. Another factor not considered here is the complexity of cases. For instance, the systems applying S2 can only do so because they have complex cases and powerful reasoners. Otherwise, they would have to switch to more complex strategies. More detailed design guidelines are elaborated in Voss (1995).

5.2 RELATED WORK

As mentioned in the introduction, we got feedback mostly for European systems, though there are some from the USA and one from Israel. In contrast, the comparison by Hanney et al. (1995), was based on the literature and covered only systems from the USA, - apart from their own system, DÉJÀVU. It turned out that they came up similar criteria and consistent conclusions at a global level:
- Both consider the tasks of the systems, though using different task names and though our tasks are finer grained.
- Both are unsatisfied with the older classification as suggested (e.g. in Kolodner, 1993).
- Both consider complexity Hanney et al. focus on the solution and we consider both problem and solution.
- Both consider the number of cases.
- Both consider adaptation, though Hanney et al. only ask whether it is done or not, while we classify the methods employed.
- Both state a correlation between the difficulty of the task and the difficulty of the approach to adaptation. - In our paper it is refined to the correlation between tasks and level, and between level and strategy.

In general our paper applies finer distinctions and goes one step further: It identifies several strategies and relates them to the categorization by solution space and task.

6. Further issues

In this section we address several scientific and practical topics whose impact on case adaptation needs further attention.

Acquisition of adaptation knowledge
Adaptation is a knowledge-intensive task which relys on knowledge acquisition. As knowledge acquisition is always an effort-intensive task in itself, it is worthwhile to investigate the possibilities for enabling a system to learn and maintain its adaptation knowledge.

Acquisition of adaptation knowledge could be obtained manually with assistance of domain expert, manually without assistance of domain expert, or interactively supported by knowledge acquisition editors. This could serve as a starting point for machine learning support for both analytical and synthetic tasks.

For analytic adaptation problems the following steps are suggested:
Take a set of similar cases that are representative for a class of solutions and derive both, a prototypical case that can be used as a "solution pattern" for the whole class, and a package of adaptation rules that can be used in order to derive all members of the given set from the prototypical case.

For synthetic adaptation problems the following steps are suggested:
Take a set of cases that solve a similar problem situation. In contrast to the analytic domain where the learning task mostly deals with attributes and values, the learning task in synthetic domains is more likely to deal with pieces of plans or equipment and how they can be put together. From this set of cases try to abstract a common solution structure and packages of adaptation rules in order to instantiate the abstract cases by concrete ones which meet the functional requirements and fit together.

Structuring of cases
A central issue seems to be the structuring of cases. It is dependent on the domain. Concrete cases seems to be useful in analytical tasks. Here, most of the decomposition techniques are achieved by chunking the case into pieces. However, in complex tasks such as design the abstraction of cases plays a major role. In such cases, decomposition is achieved by different perspectives (Oxman, 1994). Another approach could be storing cases for refinement relations.

User interaction in the adapation process
Another major issue in the development of CBR systems is the role of the user. In some domains, the user has to make some difficult decisions and would like to interact with the process of adaptation. In other domains the

user might be restricted to selection and confirmation of each modification stage of a case.

Towards a methodology for adaptation
From a knowledge engineering point of view it would be desirable to have a procedure that helps first to diagnose the adaptation needs for a certain application and secondly to derive some recommendations for further steps to take like underlying knowledge needs, knowledge representation, and generic conceptual structures for carrying out the adaptation task itself.

Decreasing the risk and increasing the efficiency in the development of adaptation capabilities
The most urgent practical demand is how can we minimise the risk and maximise the efficiency for the delivery of adaptation capabilities for real world problems. As solving from scratch is usually the most risky and effort intensive alternative we propose instead to reuse the experience that is available.

7. Conclusions

After our look at very different approaches of doing adaptation in CBR - for a great variety of tasks, relying on very different prerequisites, using nearly all available knowledge representation mechanisms and many different problem solving methods up to involvement of the user, we have to come up with a common generalisation of what adaptation in CBR is like or what all systems investigated have in common.

A first, general conclusion would be: *Adaptation in CBR is knowledge-based problem solving - and not a well-defined subset thereof.* This means that essentially there is no limitation of the complexity in general, only for certain classes of problems with a set of common characteristics.

More precisely, the task and the domain will determine the most adequate data structures for problems and solutions, the knowledge to be used for adaptation, and what methods possibly can be re-used during adaptation. Besides, the general adaptation strategy is an important design decision. It concerns the decomposition of the problem, the number of cases, the complexity of cases, and the composition of a global solution from individual adaptations. The strategy is influenced by several factors. Here we proposed characteristics of the solution space. As another factor the discrepancy between the grainsize of problems and that of locally adaptable units is proposed in Voss (1995). Further influences may come from issues we did not consider in our study.

As a result of our research we now believe that we are beginning to understand and develop the theoretical foundations of adaptation and the relevance of these foundations to design tasks. We hope that Ralph Barletta's claim at the last European Conference on CBR: "adaptation should be avoided at all costs" is to be yet considered again.

Acknowledgements

Without the help of all the system developers who filled in our questionnaires this survey could not have been conducted:

CABATA: Mario Lenz, Berlin - INRECA: Ralph Bergmann, Wolfgang Wilke, Klaus-Dieter Althoff, Kaiserslautern - MoCAS: Ralph Bergmann, Gerd Pews, Kaiserslautern - PARIS: Ralph Bergmann and Wolfgang Wilke, Kaiserslautern - CAPLan/CbC: Hector Munoz-Avila, Kaiserslautern - PRODIGY: Mauela Veloso, Pittsburgh - CHESS: Yaakov Kerner - Composer: Pearl Pu, Lausanne - EADOCS: Bart Netten, Delft - DOM: Wolfgang Oertel, Dresden - AAAO: Parivash Adami/ Barbara Schmidt-Belz, Sankt Augustin - SYN: Katy Boerner, Freiburg, - ToPo: Carl Helmut Coulon, Sankt Augustin - IDIOM: Ian Smith, Lausanne - DÉJÀVU: Barry Smyth, Dublin.

Though their systems were not included in this paper, we thank Padraig Cunningham, Dublin - Raghu Bat, Karlsruhe - Ansgar Woltering, Berlin - David Leake, Indiana - Susan Fox, Indiana - Piero Torasso, Torino for answering our questionnaires. Boi Faltings made some critical comments which need further discussion.

References

Adami, P: 1995, Adaptation by active autonomous objects (AAAO), *in* K. Boerner (ed.), *Modules Supporting Design*, FABEL-REPORT 35, GMD, Sankt Augustin, June 1995.

Bakhtari, S. and Oertel, W.: 1995, DOM-ArC: An active decision support system for quality assessment of cases, *Veloso and Aamodt*.

Bartsch-Spoerl, B.: 1995, Towards the integration of case-based, schema-based and model-based reasoning for supporting complex design tasks, *Veloso and Aamodt*.

Bergmann, R., Wess, S., Traphoener, R. and Breen, S.: 1994, Using background knowledge in the integrated system: Specification and approach, *Deliverable D29*, Esprit-Project INRECA (P6322).

Bergmann, R., Pews, G. and Wilke, W.: 1994, Explanation-based similarity: A unifying approach for integrating domain knowledge into case-based reasoning for diagnosis and planning tasks, *Veloso and Aamodt*.

Bergmann, R. and Wilke, W.: 1995, Building and refining abstract planning cases by change of representation language, *Journal of AI Research*, 3, 53-118.

Boerner, K.: 1995, Analogical design layout, *in* K, Boerner (ed.), *Modules Supporting Design*, FABEL-REPORT 35, GMD, Sankt Augustin.

Coulon, C. H.: 1995, Automatic indexing, retrieval and reuse of topologies in complex designs, *Computing in Civil and Building Engineering*, Proceedings of the Sixth International Conference on Computing in Civil and Building Engineering, A. A. Balkema, Rotterdam, pp. 749-754.

Domeshek, E.and Kolodner, J.: 1993, Finding the points of large cases, *AIEDAM*, 7(2), 87-96.

FABEL-Report 35: 1995, Boerner, K. (ed.), *Modules Supporting Design*, GMD, Sankt Augustin.

Haigh, K. and Veloso, M.: 1995, Route planning by analogy, *Veloso and Aamodt*.

Hanney, K., Keane, M., Smyth, B. and Cunningham, P.: 1995, What kind of adaptation do CBR systems need? A review of current practice, *Adaptation of Knowledge for Reuse*, AAAI Fall Symposium, Working Notes, MIT, Cambridge, MA.

Kolodner, J.: 1993, *Case-Based Reasoning*, Morgan Kaufmann, San Mateo.

Lenz, M.: 1994, Case-based reasoning for holiday planning, *in* W/ Schertler, B. Schmid, A. M. Tjoa, H. Werthner (eds), *Information and Communications Technologies in Tourism*, Springer Verlag.

Munoz, H., and Huellen, J.: 1995, Retrieving cases in structured domains by using goal dependencies, *Veloso and Aamodt*.

Netten, B. D., Vingerhoeds, R. A., Koppelaar, H., Boullart, L.: 1993, Expert assisted discrete optimization of composite structures, *in* A. Verbraeck, E. J. H. Kerckhoffs (eds), *SCS European Simulation Symp. ESS'93*, Delft, pp. 143-148.

Pu, P. and Purvis, L.: 1994, Formalizing case adaptation in a case-based design system, *in* J. S. Gero and F. Sudweeks (eds), *Artificial Intelligence in Design '94*, Kluwer Academic Publishers, Dordrecht, pp. 77-91.

Oxman, R. E.: 1994, Precedents in design: A computational model for the organization of precedent knowledge, *Design Studies*, **15**(2), 141-157.

Puppe, F.: 1990, *Problemloesungsmethoden in Expertensystemen*, Springer-Verlag, Berlin.

Riesbeck, C. K. and Schank, R. C.: 1989, *Inside Case-based Reasoning*. Lawrence Erlbaum Associates, Hillsdale New Jersey.

Smith, I., Lottaz, C. and Faltings, B.: 1995, Spatial composition using cases: IDIOM, *Veloso and Aamodt*.

Smyth, B. and Keane, M.: 1995, Experiments on adaptation-guided retrieval in case-based design, *Veloso and Aamodt*.

Veloso, M. and Aamodt, A. (eds): 1995, *Topics in Case-Based Reasoning Proceedings of the International Conference on Case-Based Reasoning*, LNAI series, Springer-Verlag.

Voss, A.: 1995, Exploiting previous cases - made easy, ftp://ftp.gmd.de//GMD/ai-research/Publications/Fabel/Prev-sol-voss.ps.gz.

Wess, S.: 1995, *Fallbasiertes Schliessen in wissensbasierten Systemen zur Entscheidungsunterstuertzung und Diagnose*, Doctoral Dissertation, University of Kaiserslautern.

J. S. Gero and F. Sudweeks (eds), Artificial Intelligence in Design '96, 191-210.
© 1996 Kluwer Academic Publishers.

APPLYING FORMAL METHODS TO CASE BASED DESIGN AIDS

MARIO DE GRASSI, ALBERTO GIRETTI
Laboratorio di Progettazione Assistita e Intelligenza Artificiale
IDAU - Università degli Studi di Ancona
via Brecce Bianche 60131 Ancona Italy

AND

LUCA SPALAZZI
Istituto di Informatica, Università degli Studi di Ancona
via Brecce Bianche 60131 Ancona, Italy

Abstract. Case based reasoners have been applied to design support in order to overcome the lack of complete domain theories and the complexity of the design space. Most of the case based design aiding systems make use of large unstructured chunks of knowledge. The use of unstructured knowledge has been proved quite effective in the early phases of design. Moreover a complete design support needs the integration of autonomous problem solvers (e.g. for case adaptation and design evaluation) which, in turn, require a structured knowledge representation. In this paper we propose some noteworthy aspects of the application of a formally defined knowledge representation schema to the development of a Case Based Design Aiding system. For this purpose we propose a computational framework for case based design support, that uses a terminological language as the basic knowledge representation tool. The terminological language is used to represent domain objects, relations and design intentions. On the basis of the formal features of the knowledge representation schema we define a set of domain independent procedures for searching, retrieving and adapting cases.

1. Introduction

The operational representation of knowledge is a main issue in Case Based Reasoning (CBR) (Kolodner, 1993). In CBR the operational requirement means that knowledge should be used by every inference process without any additional cost due to knowledge reformulation. Research on CBR conducted so far has interpreted the operational requirement mostly as the issue of explicitly representing all the information involved into the problem solving tasks within the case (i.e. what particular strategies to accomplish a goal were used, what pieces of knowledge were processed, etc.). For this purpose a number of domain analysis have been carried out in order to define what are the relevant domain information and how they are related (e.g. Universal Index Frame (Schank, 1990), Structure Behaviour Function Models (Goel, 1991)). We believe that a complementary investigation about knowledge representation at machine level needs more attention. The main issue is the definition of the data structures that are suitable for every CBR inferential process without any structural translation. The development of a Case Based Design Aid (CBDA) inherits the problem of the definition of suitable knowledge structures, because in a CBDA the insights and techniques developed in the Case Based Reasoner paradigm are used to have a real effect on the quality of design. In order to overcome both the incompleteness of domain models and the design space complexity, works on CBDA have used unstructured chunks of knowledge carrying relevant design information organised by means of an index frame (Domeshek, 1994). Even if this approach have been proved quite effective during the early phases of design conception, it can not support other CBR inferences like case adaptation and design evaluation. In fact the development of systems with limited but autonomous problem solving abilities necessary requires a structured knowledge representation (e.g. Battha, 1994).

In this paper we examine some noteworthy aspects of the application of a formally defined knowledge representation schema in the development of a Case Based Design Aiding system. For this purpose we propose a computational framework for case based design aids. Section 2 introduces the formal background of the paper. The main topic of that section is the ASA Concept Language (ASA-CL) that has been designed to support the CBDA framework. Section 3 introduces a sample knowledge base. Section 4 introduces the computational framework for CBDA. Section 4.1 describes the memory organisation, sections 4.2, 4.3 and 4.4 contain respectively the algorithms for searching, retrieving and adapting cases.

2. Terminological Languages

The first contribution to the terminological representation paradigm has been given by Brachman with his PhD thesis in 1977. Previous works on semantic networks and frames proposed representation schemata with significant semantic ambiguities (Woods, 1975). Brachman (1979) introduces a noteworthy rationalisation of structured knowledge representation. He identifies five distinct representation levels and assigns to each level well defined properties:

Implementational   *Atoms, pointers*
Logical   *Propositions, predicates, logical operators*

Epistemological	*Concept types, conceptual subpieces, inheritance and structuring relations*
Conceptual	*Semantic or conceptual relations, primitive objects and actions*
Linguistic	*Arbitrary concepts, words, expression*

Every level is based on the structure and on the procedures of the previous one. The definition of a knowledge representation schema limited to the epistemological level produces a knowledge representation tool that is not committed to any conceptual schema. In other words the expressive and computational properties of the schema are totally general and applicable to a variety of domains[1]. The first and most representative system that uses an epistemologically defined knowledge representation schema is KL-ONE (Brachman, 1985). Even if KL-ONE represents a significant step forward to avoid the ambiguity of the structured knowledge representation schemata, it still lacks a formal semantics. *Concept based KR systems* (or *terminological systems*, or *description logics*) have been recently proposed as KL-ONE successors. Terminological systems are usually defined as a subset of the KL-ONE framework plus the Tarskian semantics of First Order Logic (Rick, 1991). Terminological languages are based on a representation primitive called *structured conceptual object* or *concept*. A concept description asserts necessary and sufficient conditions for the inclusion of individuals in the class denoted by the concept. The primary structural component of a concept is the *role*. A role represents a binary relationship between one individual of the type denoted by the concept and individuals of other types. In a concept definition schema, role fillers can be constrained by means of structural descriptions. Role Value Map is the most common structural description between a two roles, it constraints the role fillers to be the same set of individuals. When the number of possible fillers of a role is limited to one, the role is called *Attribute*. Concepts are organised into a generalisation-specialisation hierarchy defined by means of the *subsumption* relationship. A concept C_1 is said to *subsume* a concept C_2 if every individual that belongs to the class denoted by C_2 belongs also to the class denoted by C_1.

2.1. KNOWLEDGE REPRESENTATION IN TERMINOLOGICAL SYSTEMS

A terminological knowledge base is made up of two components:
- a general schema describing classes of individuals (i.e. concepts), their properties and mutual relationships (T-Box)
- a set of individuals defined as concept instances by means of assertions which relate individuals to classes and individuals to each other (A-Box)

---

[1] As far as we know the current research on CBR has been conducted at conceptual level and extensive investigation at epistemological level are still lacking.

Concepts are defined through the system concept language. For example the formula *(And (All X NUMBER) (All Y NUMBER) (All Z NUMBER))* denotes the set of points in 3D space. The formula is built by means of the *And* constructor that conjoins three sub-concept expressions. Every sub-expression has the form *(All <Attribute-id> NUMBER)* and denotes the set of individuals in which every *<Attribute-id>* belongs to the class denoted by the *NUMBER* concept. The above concept can be inserted into the T-Box hierarchy using a concept definition construct as follows: *3D-POINT* := *(And (All X NUMBER) (All Y NUMBER) (All Z NUMBER))*. The A-Box appears as a set of predicates over a set of individuals. In this case a possible A-Box is the following *{X(a,1),   Y(a,2),   Z(a,0),   3D-POINT(a)}*. The semantics of terminological systems adopts the open world assumption: when an information is missing no restriction is imposed on the possible interpretation of the knowledge base.

2.2. ASA CONCEPT LANGUAGE

ASA Concept Language (ASA-CL) is a terminological language designed to fulfil the representational and computational requirements of the CBDA paradigm. The language syntax and semantics are outlined below.

Syntax

Given three distinct sets of symbols: derivate concept names, host concept names and attribute names, the sets of concept and attribute terms are inductively defined as follow:

<u>Base</u>:

      Every derivate concept name is a concept term
      Every host concept name is a concept term
      Every attribute name is a attribute term

<u>Step</u>:

Let C be a concept term. Let $D, D_1, ..., D_k$ be derivate concept terms. Let $A, A_1, ..., A_k$ be attribute terms. Let $h_1, ... , h_n$ be host domain individuals. Let k_1, k_2 be host numbers. Let n be a non negative integer and let λ be a function name. Then:

HOST-THING
NUMBER
SYMBOL,
STRING
(ONE-OF $h_1, ... , h_n$)
(RANGE $k_1 k_2$)
are host concept terms
ASA-THING
(And $D_1 D_2 ... D_k$)
(Or $D_1 D_2 ... D_k$)
(All A C)
(Rule R D $(A_{11} A_{21})(A_{12} A_{22})...(A_{1n} A_{2n})$)
(Test R λ $(A_{11} A_{12} ... A_{1n})$)

THING
BOTTOM *are concept terms*
Every host concept term is a concept term.
Every derivate concept term is a concept term.
$(A_1,...,A_n)$
is an attribute term

Semantics

The ASA-CL semantics is defined relative to an interpretation I. An interpretation I consists of a domain, D, and of an interpretation function $(\cdot)^I$. D is divided into two sub-domains: D_A related to ASA-CL individuals and D_H related to host individuals. The interpretation function is recursively defined as follow:

Base

Let E be either an host concept name or a derivate concept name, let A be an attribute name, let f be a total function over $D_{A'}$. Then:

1. $\text{THING}^I := \Delta$

2. $\text{ASA} - \text{THING}^I := \Delta_A$

3. $\text{HOST} - \text{THING}^I := \Delta_H$

4. $\text{BOTTOM}^I := \emptyset$

5. $E^I \subseteq \Delta$

6. $A^I = f : \Delta_A \rightarrow \Delta$

7. $(\text{ONE-OF } h_1,...,h_n) := \{ h_1,...,h_n \}$

8. $(\text{RANGE } k_1\ k_2) := \{x \mid x \in \text{NUMBER} , x > k_1 , x < k_2\}$

9. $\lambda^I : \Delta \rightarrow \{True, False\}$

Step

Let C, D be concept terms, let $A_1,...,A_n$ $B_1,...,B_m$ be attribute terms, let n be a positive integer. Then:

1. $(\text{And C D})^I := C^I \cap D^I$

2. $(\text{Or C D})^I := C^I \cup D^I$

3. $\left(\text{All A C}\right)^I := \left\{a \in \Delta_A{}^I \mid A^I(a) \in C^I\right\}$

4. $\left(\text{All } (A_1...A_n)\ C\right)^I := \left\{a \in \Delta_A{}^I \mid A_n{}^I(... A_1{}^I(a)) \in C^I\right\}$

5. $(\text{Rule A C } ((A_{11}...A_{n1})(B_{11}...B_{m1}))...((A_{1k}...A_{nk})(B_{1k}...B_{mk})))^I :=$
$\left\{a \in \Delta_A \mid \exists b \in C^I : A_{n1}^I(... A_{11}^I(a)) = B_{m1}^I(... B_{11}^I(a)) \wedge ... \wedge A_{nk}^I(... A_{1k}^I(a)) = B_{mk}^I(... B_{1k}^I(a))\right\}$

6. $(\text{Test R } \lambda\ (A_{11}...A_{1k})...(A_{n1}...A_{nk}))^I :=$
$\left\{a \in \Delta_A \mid \lambda^I((A'_{1k}(... A'_{11}(a)))...(A'_{nk}(... A'_{n1}(a)))) = True\right\}$

2.3. PROCESSING CONCEPT DESCRIPTIONS

In this section we propose the subsumption and closure algorithms for ASA-CL. They are based on the ASA-CL formal features, thus they are domain independent. The algorithms operate on a normalised internal concept description structure which has the form:

$$(\text{Or } D_1 \ D_2 \ ... \ D_n)$$

where every D_j has the form:

$$(\text{And } (\text{All } R_1 \ C_1) \ ... \ (\text{All } R_n \ C_n) \ (\text{Rule } C'_1 \ (R_i \ R_k)...) \ ... \ (\text{Rule } C'_n \ (R_j \ R_l) \ ...))$$

and where $C_1,...,C_n$, $C'_1,...,C'_n$ are concept names and $R_1,..., R_n, R_i, R_k, R_j, R_l$ are attribute names[2].

2.3.1. Subsumption algorithm for ASA-CL

Given the ASA-CL feature language, it is possible to define a procedure that verifies if the subsumption relation holds between two concept descriptions. Given two concept expressions D and G, the relation *subsumes(D,G)* holds if and only if one of the following condition is verified:

1. G is BOTTOM
2. D is THING
3. D is ASA-THING and G is a derivate concept term
4. D is HOST-THING and G is a host concept term
5. D and G are both atomic concept names and D is equal to G
6. D and G are both NUMBER or STRING or SYMBOL
7. D is (RANGE k1 k2) and G is (RANGE k3 k4) and k1£k3 and k2≥k4
8. D is (ONE-OF h1 h2 h3 ... hn) and G is (ONE-OF k1 k2 k3 .... km) and the set {k1 k2 k3 .... km} is a subset of {h1 h2 h3 ... hn}
9. D is (All A C) and G is (All A E) and *subsumes(C,E)* is true.
10. D is (Rule R1 C1 Map1) and G is (Rule R2 C2 Map2) and *subsumes (C1,C2)* is true and Map1 is equal to Map2.
11. D is (Not-Rule R1 C1 Map1) and G is (Not-Rule R2 C2 Map2) and *subsumes (C2,C1)* is true and Map1 is equal to Map2.
12. D is (Test R1 E1 Arg1) and G is (Test R2 E2 Map2) and *E1* is equal to *E2* and Arg1 is equal to Arg2.
13. D is (And D1 D2 ... Dn) and G is (And G1 G2 ... Gm) and *subsumes(D1,Gi)*, *subsumes(D2,Gi)*, ..., *subsumes (Dn,Gi)* are true for at least one Gi in {G1, G2, ..., Gm}
14. D is (And D1 D2 ... Dn) and *subsumes(D1,G)*, *subsumes(D2,G)*, ..., *subsumes (Dn,G)* are true
15. G is (And G1 G2 ... Gm) and at least one of *subsumes(D,Gi)*, *subsumes(D,Gi)*, ..., *subsumes (D,Gi)* is true for Gi in {G1, G2, ..., Gm}
16. D is (Or D1 D2 ... Dn) and G is (Or G1 G2 ... Gm) and *subsumes(Di,G1)*, *subsumes(Di,G2)*, ..., *subsumes (Di,Gm)* are true for at least one Di in {D1, D2, ..., Dn}

---

[2] The normalised form is obtained by means of a translation of the linear form into an internal representation structure called descritpion graph whose discussion is out of the scope of the article. The normalised form is logically equivalent to the original one. The reader can find further information in Borgida (1994).

16. D is (Or D1 D2 ... Dn) and G is (Or G1 G2 ... Gm) and *subsumes(Di,G1)* , *subsumes(Di,G2)*, ..., *subsumes (Di,Gm)* are true for at least one Di in {D1, D2, ..., Dn}

17. G is (And G1 G2 ... Gm) and *subsumes(D,G1)* , *subsumes(D,G2)*, ..., *subsumes (D,Gm)* are true

18. D is (Or D1 D2 ... Dn) and at least one of *subsumes(D1,G)*, *subsumes(D2,G)*, ..., *subsumes(Dn,G)* is true.

The algorithm outlined above is correct but incomplete. Incompleteness derives from points 12, 16 and 18[3].

2.4.2. *Closure algorithm for ASA-CL*

Given a knowledge base S = <T-BOX, A-BOX>, it is possible to define an algorithm that produces the instance set of a concept. The iteration of this procedure over the entire T-BOX produces a deductive closure of the knowledge base S. Before introducing the algorithm it is necessary to build a minimum of glossary. We say that a concept description is *structurally complete* if every attribute takes part at least in one rule, that is, if it appears at least in a *rule-map* expression. We call *rule-maps* the final part of the ASA-CL rule expression, where pair of attributes are enclosed in brackets. Every pair of attributes means that attribute fillers must be equal. When an attribute does not takes part in any rule, it is in principle impossible to deduce its filler value from the concept structural description. Thus the algorithm returns an instance set only for descriptions that are structurally complete and that, recursively, have attribute fillers and rule fillers that are structurally complete.

```
Procedure closure(Σ)
  Input a knowledge base Σ =<T-BOX,A-BOX>
  Output a knowledge base Σ'
  Begin
    A-BOX'={}
    For every concept C in T-BOX
      A-BOX'=A-BOX' ∪ concept_intepretation(C);
      Return Σ'=<A-BOX ∪ A-BOX',T-BOX >
  End;

Procedure concept_interpretation(C)
  Input a concept C
  Output an instance set I
  Begin
    If the concept has already been interpreted
      Then return the concept instance set
      Else If the description of C is not structurally complete
        Then Return Ø
    Else
      Begin
        For every attribute Ao
```

---

[3] Even if the algorithm is incomplete, in the applications of the formalism to CBDA experienced so far, incompleteness has not caused great problems.

> let C_{Ao} be the attribute filler of Ao, call *concept_intepretation*(C_{Ao});
> **For** every rule Ru, let C_{Ru} be the rule filler of Ru, call *concept_intepretation*(C_{Ru});
> Solve the Constraint Satisfaction Problem where the set of attribute names
> and rule names of C are the variables. Variable domains are the filler instance sets
> computed in the previous steps. Problem constraints are both the *mappings* that
> links rule attributes to the concept attributes and the *tests* of the description.
> **Return** the computed instance set.
> **End**
>
> **End**

The formal features of the representation schema offer further possibilities to define noteworthy algorithms. It is possible to define a procedure that computes the differences between two given descriptions, producing a new concept built by attributes and rules that first description has not in common with the second one. The computed description can be used to explain the transition between two concepts, and, more generally, to produce explanations of complex T-BOX navigation. The development of the algorithm is a work in progress.

3. Design Representation

In this section we analyse the application of ASA-CL to the representation of design. For that purpose it is sufficient to assume a simplified model of design. Thus design representation consists of the representation of domain object structure and the representation of design intentions[4].

3.1. THE REPRESENTATION OF DOMAIN OBJECTS

ASA-CL contains a lot of constructs that can be fruitfully used in the description of the structural features of domain elements. Figure 1 shows the graphical representation of a sample knowledge base whose formal description is given in Table 1. A domain object description (e.g. window) is obtained by means of a T-BOX concept. Its expression contains the attributes and the compositional rules that individuals belonging to the object class must satisfy. Concept attributes are used to represent both components (e.g. sash, ledge of window, etc.) and features (e.g. colour, length, width, etc.). For example, the *WINDOW* concept in Table 1 is described by means of three attributes: the *upper-sash* and the *lower-sash* which must belong to the *SASH* concept, and a *ledge* which must belong to the *WINDOW-LEDGE* concept. A *WINDOW* is thus represented by three components. Two of them are in class *SASH* and are respectively the *upper-sash* and the *lower-sash* of the window. The third component is in class

---

[4]The extension to a more complete model, embracing SBF models, design critics, design failures etc., can be obtained at the cost of an expansion of the proposed knowledge base without any noteworthy extension of the expressive power of the language.

WINDOW-LEDGE and is the *ledge* of the window. Language rules are used to express the compositional constraints of the *WINDOW* class of object. Rule *R1* asserts that the *upper-sash* must be *UPPER-PLACED* (i.e. placed to upper side and adjacent) to the *lower-sash*. Rule *R2* asserts that the *lower-sash* must be *UPPER-PLACED* to the *ledge*. Test R3 verifies if the width of the lower window sash is equal to the width of the window ledge. Test R4 verifies if the width of the lower window sash is equal to the width of the upper window sash.

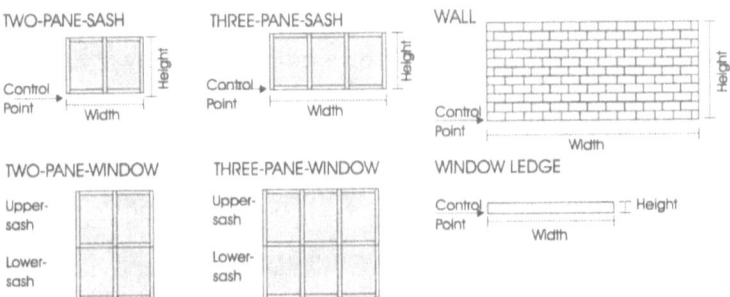

Figure 1. Graphical representation of a sample knowledge base.

POINT := (**And** (**All** X NUMBER) (**All** Y NUMBER) (**All** Z NUMBER))

GEOMETRIC-ELEMENT := (**And** (**All** Control-Point POINT) (**All** Height NUMBER) (**All** Width NUMBER))

TWO-PANE-SASH := (**And** GEOMETRIC-ELEMENT (**All** Glass-Colour (**ONE-OF** Dark Light) (**Test** = Width 90))

THREE-PANE-SASH := (**And** GEOMETRIC-ELEMENT (**All** Glass-Colour (**ONE-OF** Dark Light))(**Test** = Width 120)))

SASH := (**Or** TWO-PANE-SASH THREE-PANE-SASH)

WINDOW-LEDGE := (**And** GEOMETRIC-ELEMENT(**All** Material (**ONE-OF** Marble Stone Wood))

WALL := (**And** GEOMETRIC-ELEMENT (**All** Material (**ONE-OF** Bricks Concrete Wood))

UPPER-PLACED := (**And** (**All** Element-1 GEOMETRIC-ELEMENT) (**All** Element-2 GEOMETRIC-ELEMENT)
          (**Test** R1 at-y-distance (Element-1 control-point X) (Element-2 control-point X) (Element-2 Height))
          (**Test** R2 = (Element-1 control-point X) (Element-2 control-point X)))

WINDOW := (**And** (**All** Upper-Sash SASH) (**All** Lower-Sash SASH)   (**All** Ledge WINDOW-LEDGE)
          (**Rule** R1 UPPER-PLACED (Element-1 Upper-Sash)(Element-2 Lower-Sash))
          (**Rule** R2 UPPER-PLACED (Element-1 Lower-Sash)(Element-2 Ledge))
          (**Test** R3 = (Lower-Sash Width) (Ledge Width))
          (**Test** R4 = (Lower-Sash Width) (Upper-Sash Width)))

TWO-PANE-WINDOW := (**And** WINDOW (**All** Upper-Sash TWO-PANE-SASH) (**All** Lower-Sash TWO-PANE-SASH))

THREE-PANE-WINDOW := (**And** WINDOW (**All** Upper-Sash THREE-PANE-SASH)
          (**All** Lower-Sash THREE-PANE-SASH))

WALL-WITH-OPENING := (**And** (**All** window WINDOW) (**All** wall WALL)
          (**Test** T1 at-y-distance (window ledge control-point) (wall control-point) 1,20)
          (**Test** T2 > (window ledge control-point X) (wall control-point X)))

TABLE 1. Formal representation of the sample knowledge base.

The reader should notice that even the *UPPER-PLACED* relation, used to build the *WINDOW* concept description, is expressed as a concept. Therefore a domain relation is represented with the same expressive power as a domain object. In a domain relation the attributes of the language are used to represent the arguments of the relation and the rules are used to represent the conditions that should be satisfied in order to verify the relation. The language also allows to define relations in an implicit form, by means of the *Test* construct.

In fact the *Test* construct verifies if an implicitly defined relation (i.e. some function defined in the host language) with the attributes of the compound description as arguments holds. An example of the application of the *test* construct is given in the description of the *RIGHT-PLACED* concept. Here the *Test* constructs has attribute chains as its arguments. An attribute chain allows us address a component of a component down to the desired depth.

In the *TWO-PANE-WINDOW* concept definition the *And* construct conjoins the concept name *WINDOW* with other constructs. When some language expressions are conjoined by the *And* construct, the resulting concept inherits all the expression attributes and rules. For example, the resulting description of the *TWO-PANE-WINDOW* concept has the *Upper-Sash, Lower-Sash, Ledge* attributes and the rules R1, R2, R3 and R4 inherited by the *WINDOW* description. If an inherited attribute has the same name of an attribute in the current expression, then the two attributes are merged (i.e. it is generated an attribute with a filler built of the conjunction of the fillers). Finally the *SASH* class of objects is defined as a collection (i.e. a disjunction) of *TWO-PANE-SASH* and *THREE-PANE-SASH*[5].

The language expressiveness outlined above can be reached by means of a lot of different structured knowledge representation schemata. Moreover the use of a formally defined language offers a set of unique computational properties that depend exclusively on the formal features of the knowledge representation schema. Therefore they are domain independent. In section 4 we will show the application of those properties in a CBDA system.

3.2. THE REPRESENTATION OF DESIGN INTENTIONS AND STRATEGIES

Design is basically a purposeful activity in which a lot of strategies are applied in order to accomplish a design goal. The application of a strategy in a design domain produces a transformation of the current domain state into a new state. In a terminological representation schema a design domain state can be represented by a set of concept instances in the A-BOX. A state transition consists of a variation of the A-BOX content. An A-BOX content

---

[5] Our experience in the definition of knowledge bases for architectural design support has identified a set of standard primitive components. In our example the *SASH*, the *WINDOW-LEDGE* and the *WALL* concepts represent primitive graphical components. The shape of every primitive component is implicitly defined by a lot of parameters and control points as we usually find in variational CAD systems.

variation can be obtained by the addition, the deletion or the substitution of an instance. Instance addition and deletion are basic A-BOX operation and they are straightforward, except for the problem of the knowledge base consistency maintenance. In a terminological schema the substitution of an instance a with an instance β of the A-BOX can be structurally explained as the transition between the concept Cα, whose α is an instance of, and the concept Cβ, whose β is an instance of. In fact, the transition between two concepts outlines what changes and what remains when an instance of the concept Cα is substituted with an instance of the concept Cβ. In the proposed design model a transition between two concepts is called *action*. In terminological knowledge base, actions are represented as a concept sub-hierarchy which starts from the following basic description:

ACTION := (**And** (**All** from ASA-THING) (**All** to ASA-THING))

in which the *from* attribute represents the concept from which the transition starts and the *to* attribute the target concept.

Even if actions are meaningful tools to represent domain state changes, the representation of complex design strategies needs more expressive tools to combine actions into sequences and to link them to design goals. For this purpose we have developed the ASA Planning Language (ASA-PL) that can be used to define complex design strategies. ASA-PL is based on the *operator* structure which links a planning strategy to a design goal. The operator is represented as a concept with the following structure:

OPERATOR := (**And** (**All** goal STRING)(**All** strategy STRING))

A strategy can be defined according to the following syntax and semantics of the ASA-PL language.

Syntax
Given the set of *concept terms* and two disjoint alphabets of symbols: *primitive actions* and *goals*. The set of *planning expressions* is inductively defined as follows:
Base:
  Every primitive action is a planning expression.
  Every goal is a planning expression.
Step:
  Let $P,P_1,...,P_k$ be planning expressions. Let C be a concept term. Then:
  NOP
  (And P$_1$ P$_2$ ... P$_k$)
  (Or P$_1$ P$_2$ ... P$_k$)
  (Not P)
  (Inclass P C)
  (Seq P$_1$ P$_2$ ... P$_k$)

(Iffail P_1 P_2 P_3)
are planning expressions.
And, Or, Not and **Inclass** are called *algebraic operators*, **Seq** and **Iffail** are called *control structures* (Giunchiglia, 1994).

Semantics

The semantics of ASA-PL is defined relative to an interpretation $J = <X^J, \cdot^J>$ where the set $X^J \subseteq \Delta^I$ is the *domain* of J (X^J is the set of individuals such that they are linked to elements of the data base), and the function $\cdot^J$ is the *interpretation function* of J.

Base
Let A be an action. Let G be a goal. Then
$$A \subseteq X^J \times X^J$$
$$G \subseteq X^J \times X^J$$

Step
Let $S_0 \subseteq X^J$ be a starting situation. Let $P, P_1, ..., P_k$ be planning expressions. Let C be a concept term. Then

$$NOP^J := \left\{ (a,a) \big| a \in S_0 \right\}$$

$$(\text{And } P_1 P_2)^J := \left\{ (a,b) \big| a \in S_0 \wedge (a,b) \in P_1^J \cup P_2^J \wedge b \in im(P_1^J) \cap im(P_2^J) \right\}$$

$$(\text{Or } P_1 P_2)^J := \left\{ (a,b) \big| a \in S_0 \wedge (a,b) \in P_1^J \cup P_2^J \wedge b \in im(P_1^J) \cup im(P_2^J) \right\}$$

$$(\text{Not } P)^J := \left\{ (a,b) \big| a \in S_0 \wedge (a,b) \notin P^J \right\}$$

$$(\text{Inclass } P\ C)^J := \left\{ (a,b) \big| a \in S_0 \wedge (a,b) \in P^J \wedge b \in C^J \right\}$$

$$(\text{Seq } P_1 P_2 .... P_n)^J := \left\{ (c_0,c_n) \big| c_0 \in S_0 \wedge \exists c_1,...,c_{n-1}.(c_0,c_1) \in P_1^J \wedge ... \wedge (c_{n-1},c_n) \in P_n^J \right\}$$

$$(\text{Iffail } P_1 P_2 P_3)^J := \left\{ (a,b) \big| a \in S_0 \wedge (a,b) \notin P_1^J \wedge (a,b) \in P_2^J \right\} \cup$$
$$\left\{ (a,b) \big| a \in S_0 \wedge \exists c.\left[(a,c) \in P_1^J \wedge (c,b) \in P_3^J \right] \right\}$$

The most simple strategy that can be expressed by means of ASA-PL is the primitive action. Given an initial instance set (i.e. subset of the A-BOX), the execution of an action over the initial set is essentially a searching operation that retrieves the instance set of the target concept (i.e. the one that appears in the *to* attribute of the action). The retrieved instance set of an action execution may be either the target concept instance set of the current A-BOX or, much more interesting for CBDA, an instance set of a long term memory related to the target concept. For example the target concept instance set of a particular case in a case memory. The execution of an action may fail. A failure consists of the retrieval of an empty instance set.

More complex searching strategies can be obtained by using the ASA-PL constructs. Algebraic operators combine the results (i.e. the final instance sets) of the executions of their arguments expressions. The **(Inclass P C)** expression restricts the result of the execution of the planning expression P to the instances of concept C. The *Seq* control structure chains the execution of its argument planning expressions. When two expressions are chained, the resulting instance set of the first one becomes the initial set of the second. The **(Iffail P_1 P_2 P_3)** expression has the following meaning: if the execution of P_1 results in an empty set then execute P_2, execute P_3 starting form the result of P_1 otherwise. The planning algorithm is essentially an interpreter of the ASA-PL language. It builds a plan node for every planning expression it encounters. A plan node contains the initial and the final instance set of the planning expression execution. When a *goal* expression is encountered the interpreter searches in the knowledge base for operators with the *goal* attribute filled with the same expression. The set of planning expressions appearing in the *plot* attribute is retrieved and the first of them is executed. If the execution fails then the second one is executed until either the execution of one of the expressions succeeds or there are no more expressions, in which case the goal is not satisfied. Finally the NOP expression is the action that does nothing but succeeds.

4. A Computational Framework for Case Based Design Aids

In this section we propose a framework in which the representational and computational features of ASA-CL are combined in order to obtain the functionality of a case based design aiding system. The framework is defined at conceptual level, using the terminological schema as the underlying epistemological layer (see section 2). The framework has a long term conceptual memory, a short term working memory and a long term episodic memory. The conceptual memory stores partial domain models. The working memory contains the current design status represented as a set of conceptual model instances. The episodic memory contains past design. The adoption of a terminological language as the knowledge representation schema specialises the above framework as follows. The long term conceptual memory corresponds to the terminological T-BOX and the short term working memory to the A-BOX. The episodic memory corresponds to a set of tables, one for each concept of the T-BOX. For example Figure 2 shows a snapshot of the knowledge base, when the designer has just drawn with a CAD interface a set of domain objects corresponding to a *TWO-PANE-WINDOW* and to a *WALL*. The A-BOX contains the instances corresponding to the various window components and to the relationship that regulates the *TWO-PANE-WINDOW* concept (i.e. *UPPER-PLACED*) plus an instance for the *WALL* and the *WALL-WITH-OPENING* concepts.

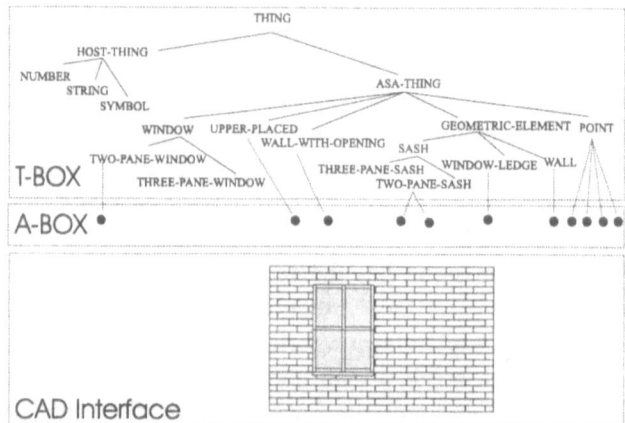

Figure 2. A snapshot of the terminological knowledge base during the design process.

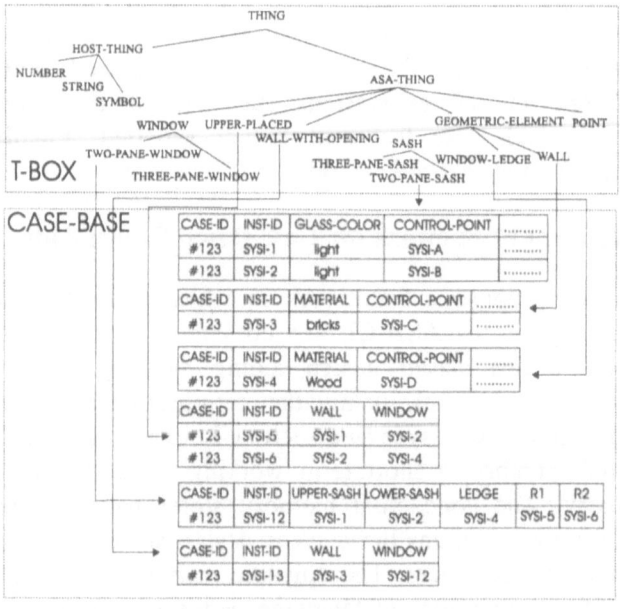

Figure 3. A snapshot of a case memory.

The instances of the *UPPER-PLACED, WALL-WITH-OPENING* concepts have been automatically asserted using the *closure* algorithm. Figure 3 shows the instances of the *TWO-PANE-WINDOW* of Figure 2 stored as a case into the case base memory tables. The case memory contains a set of tables, one for each concept of the T-BOX. Table columns correspond to the concept attributes and rules, plus one column for the instance identifier and one for the case identifier. Instance attributes and rules are filled with concept instance identifiers[6] that logically link the compound object with its components.This framework adopts a distributed case representation. Notice that cases are stored as a collection of descriptions that can be independently recalled and processed.

This feature partially avoids the *grain size* problem often encountered in the design of case representations in traditional case based systems (Kolodner, 1993). Thus, in the following, we will not make any distinction between a case or one of its components.

4.1. SEARCHING AND RETRIEVING CASES

A basic operation of a case based design aiding system is the retrieval of a helpful piece of situated knowledge. In the proposed CBDA framework, we can access cases by addressing the concept table they are inserted in. In fact, every table is directly linked to a concept and inherits the structure of the concept description. Thus case retrieval can be guided both by its contents, as in usual data-base queries, and by the structure of the knowledge description schema. In the following we investigate the possibilities offered by the ASA-CL language to access a case on the basis of the description structure. ASA-CL offers two computational tools that can be used for searching cases. The first one is the subsumption algorithm that can be used to find out structural similarities between a concept description and a searching key. The second tool is the planning sub-system that can be used to traverse the T-BOX and to find out a concept on the basis of its structural differences form the current situation. We call the first strategy *searching for similarities* and the second *searching for differences*.

Once the desired case has been identified, the case is retrieved from the case memory. In the previous section we pointed out that the proposed CBDA model uses a distribute case representation. Case retrieval is thus a matter of traversing the parthonomic hierarchy. Starting from the current concept instance, the case retrieval algorithm loads recursively the instance components and rules until it finds host concept values. The retrieval algorithm has an elementary structure, thus we do not discuss it further.

4.1.1. Searching for similarities
The schema of the searching for similarities procedure is defined as follows:
1. Define a searching key as a concept description.
2. Associate the key description to a default concept name.
3. Place the key concept into the T-BOX hierarchy using the subsumption algorithm.
4. If there are subsumed concept, then retrieve the cases contained into the tables pointed by the subsumed concepts
5. Delete the key concept from the T-BOX.

---

[6]Host concept instance places are filled with values.

The specification of a searching key is accomplished with the same technique used to describe concepts. In fact a searching key is basically a partial description of the desired case, so it can be represented as a concept and defined by means of the ASA-CL language. Once the key concept is placed into the T-BOX hierarchy, it is possible to access the subsumed concept. That is to access concepts that contain the description of the key concept. The procedure usually retrieves a multiplicity of cases. Retrieved cases can be subsequently ranked[7] on the basis of their information content.

For example the expression:

(**And** (**All** Upper-Sash SASH)(**All** Lower-Sash SASH))

used as a searching key produces a key concept that subsumes the *WINDOW* concept. Thus the set of windows stored into the case base can be retrieved.

4.1.2. Searching for differences

Designers sometime express retrieval requirements as a variation of a given situation. In section 3.4 we have introduced *actions* and *strategies* as representations of this kind of design intentions. Strategies are represented by means of transitions between concepts. The final concept of a transition path represents the structure that satisfies the goal in the current situation. The set of instances contained into the case-base table linked to the concept is the result of the searching procedure.

For example consider the knowledge base of Table 1 augmented with the following definitions:

```
OP1 := (And OPERATOR
              (All goal (ONE-OF "increase-lighting"))
              (All plot (Iffail TWO-THREE TWO-FOUR NOP)))
TWO-THREE := (And ACTION
                        (All from TWO-PANE-WINDOW)
                        (All to THREE-PANE-WINDOW))
THREE-FOUR := (And ACTION
                        (All from THREE-PANE-WINDOW)
                        (All to FOUR-PANE-WINDOW))
TWO-FOUR := (And ACTION
                        (All from TWO-PANE-WINDOW)
                        (All to FOUR-PANE-WINDOW))
```

Suppose that the designers activate the *"increase-lighting"* goal. Planning for the *"increase-lighting"* goal produces the planning tree of Figure 4. The plan execution results in following the path from the instance of the *TWO-PANE-WINDOW* concept, contained into the A-BOX, to the set of instances of the *THREE-PANE-WINDOW* concept, contained into the concept table of the case memory. Figure 5 depicts the situation.

---

[7] Ranking cases usually needs domain dependent heuristics. The discussion of this issue is out of the scope of this article.

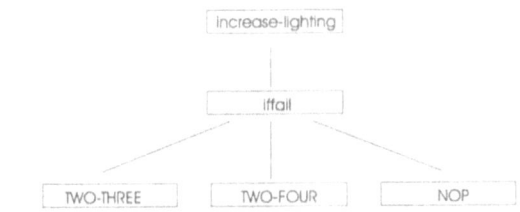

Figure 4. Planning tree for the "increase-lighting" goal.

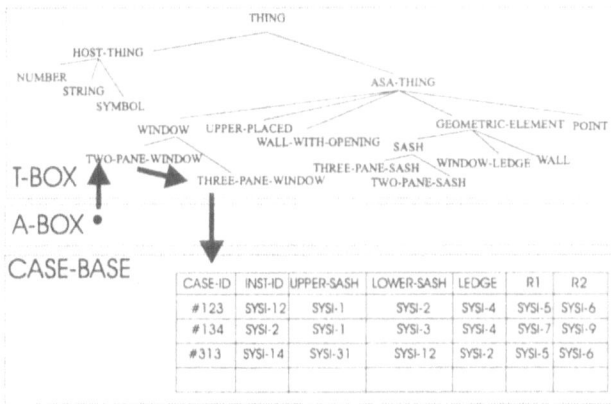

Figure 5. An example of case base navigation related to the "increase-lighting" goal.

4.2. MODEL GUIDED CONTEXT INSERTION

Reusing past experience is strongly connected to the ability of relating that experience to the situation one is working on. In case based reasoning, this process is usually called *adaptation* (e.g. Hua et al., 1992). Thus adaptation is the process that, given a previously retrieved case, tries to insert it in the current context in order to obtain a coherent situation. In the proposed paradigm the insertion basically consists of a juxtaposition of two sets of concept instances, the first one related to the current context and the second one to the retrieved case. Cases are juxtaposed by finding out the appropriate compositional rules.

The context insertion algorithm has two phases: *context definition* and *case insertion*. Context definition identifies the set of instances of the current situation that are directly related to the case to insert. Case insertion builds the set of rules that regulates the interaction between the retrieved case and the given context. The context definition process strongly depends on the type of search. If a case is the result of a search for similarities, it is hard to automatically recognise the context. On the contrary, if a case is the result of a search for differences, the system has enough information to go back to the initial instance set of the transition path. The initial instance set contains the instances that should be substituted for the retrieved set. For every instance i in the initial instance set, the context definition procedure determines the set $\mathfrak{S}$ of instances in the A-BOX such that any instance contains i as one of its components. The set $\mathfrak{S}$ is called the *context* of i. Once the context of an instance has been calculated, the case insertion algorithm replaces every reference to i with a reference to one of the instances resulting from the searching phase. In the current release, the choice is arbitrary. Nevertheless the system performs a consistency checking. It verifies that the instance to replace belongs to a superclass of the chosen instance. Then the procedure updates every instance attribute in the A-BOX which is filled with host values. For this purpose, case insertion builds a Constraint Satisfaction Problem (CSP) in which the *Test* constructs are used as constraints and the host value filled attributes are the variables.

The CSP problem may fail. In this case a new instance is selected until either the insertion succeeds or there are no more instances. In the latter case the whole context insertion fails. Suppose, for example, to be in the design status depicted in Figure 2 and to activate the *"increase-lighting"* goal (see Figure 4). The searching for differences procedure retrieves an instance of a *THREE-PANE-WINDOW* concept. The procedure determines the initial instance set composed by the *TWO-PANE-WINDOW* instance. The context definition process determines the context of the *TWO-PANE-WINDOW* instance and finds the instance of the *WALL-WITH-OPENING* concept. The case insertion procedure first updates every logical reference to the *TWO-PANE-WINDOW* instance with the identifier of *THREE-PANE-WINDOW* instance, then uses the tests T1 and T2 of *WALL-WITH-OPENING* concept as constraints of a Constraint Satisfaction Problem. The CSP variables are the arguments of the tests T1 and T2.

5. Related Work

The ASA concept language has been designed following the principles outlined in the design of the CLASSIC concept language (Brachman, 1991) with some restrictions (i.e. the limitation to attributive relationships) and some minor extensions (i.e. the RANGE and the RULE constructs). In fact the RULE construct is a syntactic facility that can be reduced to a composition of attributes and role value maps[8], and the RANGE construct can be viewed semantically as the ONE-OF construct. The relationships between CLASSIC like languages and other terminological languages are discussed in Rich (1991). The representation of domain n-ary relationships by means of concepts is a well established technique for terminological

---

[8] The expression (Rule R C ((A11 A21)...(A1n A2n)) is equivalent to the expression (AND (All A21 THING) .... (All A2n THING) (All R C) (RVM (R A11) A21) .... (RVM (R A1n) A2n))

languages (e.g. Bergamaschi, 1992). Moreover the use of n-ary relationships for knowledge base closure is, as far as we know, original. The episodic memory of the proposed framework has been built by coupling the conceptual schema with a set of tables that reside on persistent storage media. Coupling concept languages to data-base systems is an active research area in the Description Logic community (e.g. Borgida, 1994; 1993). In the proposed framework the data base schema strictly follows the concept structure defined in the knowledge base, allowing the inference procedures to efficiently access case data. That requirement is substantially different from standard data-base specifications in which storage space optimisation requires only a partial storage of the information related to views. An attempt to formalise Case Based Reasoning can be found in Koton (1989), where the discrimination net presented in Kolodner (1983) has been reformulated in terms of default logic. The work proposes a well founded knowledge representation schema but it does not provide any attempt to support Case Based inferences like adaptation. More recently a work on the application of description logics to the modelling of Case Based Reasoning has been proposed in Yen (1994). In that work the LOOM concept language has been used to support the classification and storage of cases and their semantic based analysis. The work focuses on the retrieval problem and proposes a measure of similarity based on the semantic properties of the LOOM terminological language. The analysis introduced in Yen (1994) can be ported to the ASA-CL environment with some restrictions, allowing the definition of a more refined case retrieval procedure. Moreover the work is limited to similarity measures and thus it does not face the adaptation problem.

Acknowledgements This work has been conducted as part of ASA, a project under development at the University of Ancona and has been done with the support of the Italian National Research Council (CNR), Special Project on Building (Progetto Finalizzato Edilizia). Mauro Di Manzo have supervised the project and provided many useful intuitions. Andrea Fornarelli, Massimo Lemma, Berardo Naticchia have participated to the development of ASA.

References

Bergamaschi, S., Lodi, S. and Sartori, C.: 1992, The ES Knowledge Representation System in AI*IA Notizie Anno V, **2**, June, 31-40.
Bhatta, S., Goel, A. and Prabhakar, S.: 1994, Innovation in analogical design: A model-based approach, *in* J. S. Gero and F. Sudweeks (eds), *Artificial Intelligence in Design '94*, Kluwer, Dordrecht, pp. 57-73.
Patel-Schneider, P. F.: 1994, A semantics and complete algorithm for subsumption in the CLASSIC description logic, *Journal of Artificial Intelligence Research*, **1**, 277,308.

Borgida, A: 1994, Description logics for quering databases, *in* F. Bader, M. Lenzerini, W. Nutt, P. F. Patel-Schneider (eds), *Prooceedings of International Workshop on Description Logics,* DFKI Dokument D-10, 95-96.

Borgida, A. and Brachman, R.J.: 1993, Loading data into description reasoners, *ACM SIGMOD Conference on Data Management,* Washington, DC, pp. 217-226

Brachman, R. J., Schmolze, J. G.: 1985, An overview of the KL-ONE knowledge representation system, *Cognitive Science* 9, 217-260.

Brachman, R. J.: 1979, On the epistemological status of semantic networks, *in* N. V. Findler (eds), *Associative Networks: Representation and Use of Knowledge by Computers,* Academic Press, New York, pp. 3-50.

Brachman, R. J., McGuinness, D., Patel-Schneider, P. F., Resnik, L. A., Borgida, A.: 1991, Living with classic: When and how to use A KL-ONE like language, *in* J. F. Sowa (ed.), *Principles of Semantic Networks,* Morgan Kaufman, San Matteo, pp. 157-190

Domeshek, E. A., Kolodner, J. L. and Zimring, C. M.: 1994, The design of a tool kit for case-based design aid, *in* J. S. Gero and F. Sudweeks (eds), *Artificial Intelligence in Design '94,* Kluwer, Dordrecht, pp. 109-126

Giunchiglia, F., Spalazzi, L. and Traverso, P.: 1994, Planning with failure, *Proceedings of AIPS 94,* Chicago.

Goel, A.: 1991, A model-based approach to case adaptation, *Proceedings of the Thirteenth Annual Conference of the Congitive Science Society,* Chicago, IL, pp. 143-148

Hua, K., Smith, I., Faltings, B., Shih, S. and Schmitt, G.: 1992, Adaptation of spatial design cases, *in* J. S. Gero (ed.), *Artificial Intelligence in Design '92,* Kluwer, Dordrecht, pp. 559-575.

Kolodner, J. L.: 1983, Maintaining organization in a dynamic long-term memory, *Cognitive Science,* 7, 243-280.

Kolodner, J. L.: 1993, *Case Based Reasoning,* Morgan Kaufmann.

Koton, P. and Chase, M. P.: 1989, Knowledge representation in a case-based reasoning system: Default and exceptions, *in* R. J. Brachman, H. J. Levesque and R. Reiter (eds), *Proceedings of KR '89,* Morgan Kaufman, pp. 203-211.

Rich, C. (ed.): 1991, Special issue on implemiented knowledge representation and reasoning systems, *SIGART Bulletin,* 2(3).

Schank, R. and Osgood, R.: 1990, A content theory of memory indexing, *Technical Report no.2,* Northwestern University, Institute for the Learning Sciences.

Woods, W. A.: 1975, What's in a link: Foundations for semantic networks, *in* D. G. Bobrow and A. M. Collins (eds), *Representation and Understanding: Studies in Cognitive Science,* Academic Press, pp. 35-82.

Yen, J., Hor Teh, S., Liu, X.: 1994, Using description logics for software reuse and case-based reasoning, *in* F. Bader, M. Lenzerini, W. Nutt and P. F. Patel-Schneider (eds), *Proceedings International Workshop on Description Logics,* DFKI Dokument D-10, pp. 51-54.

J. S. Gero and F. Sudweeks (eds), Artificial Intelligence in Design '96, 211-227.
© 1996 *Kluwer Academic Publishers.*

DESIGNING NUTRITIONAL MENUS USING CASE-BASED AND RULE-BASED REASONING

CYNTHIA R. MARLING AND LEON S. STERLING
Department of Computer Engineering and Science
Case Western Reserve University
Cleveland, Ohio 44106, USA[†]

Abstract. Case-based reasoning (CBR) and rule-based reasoning (RBR) are two paradigms for building knowledge-based systems. They represent both distinct approaches to knowledge-based systems development and distinct cognitive models of human problem solving. They are usually viewed as competing, rather than complementary, paradigms. However, our investigation shows that in combination, they can provide both a stronger approach to knowledge-based systems development and a broader cognitive model. The domain of our investigation is the design of nutritious, yet appetizing, menus. Both logic and experience play roles in this domain. Our approach is to construct two expert systems, one case-based and one rule-based, to perform the same task. We compare and contrast our two systems, to identify the strengths and weaknesses of each.

1. Introduction

For those who accept the definition,

"Artificial intelligence (AI) is the study of how to make computers do things which, at the moment, people do better," (Rich and Knight, 1991)

menu planning[1] is a rich domain. Ordinary people plan menus for themselves, their families and friends, as a matter of course. Food service professionals plan specialized menus for use in restaurants, catering, school cafeterias, hospitals, military bases, prisons, and other institutions. The food service industry makes extensive use of computer applications, for functions ranging from payroll to controlling the amount of Coca Cola released from beverage dispensers. However,

---

[†]Leon Sterling's new address is: Department of Computer Science, University of Melbourne, Parkville, Victoria, Australia.

[1]Menu *planning* is the unfortunate vernacular term for a task which is quintessentially one of *design*.

"Computer-assisted menu planning is not widely used today because of the difficulty in quantifying the many variables involved in menu planning, such as flavor, color, and texture." (Spears, 1995)

Both case-based reasoning (CBR) and rule-based reasoning (RBR) systems have been built in this domain (Galotra *et. al.*, 1991; Hinrichs, 1992; Kovacic *et. al.*, 1992; Yang, 1989). Yet, human capabilities still outstrip the computer. Unsuccessful attempts to build computer-assisted menu planners date from the 1960's (Balintfy, 1964; Eckstein, 1967). The high levels of both interest and difficulty in this domain may stem from the observation that

"Menu planning is both an art and a science." (Eckstein, 1978)

We have chosen this domain to investigate opportunities for CBR/RBR hybridization. While some believe,

"It is probably an axiom of artificial intelligence that intelligent behaviour is rule-governed." (Jackson, 1990)

and others counter,

"Real thinking has nothing to do with logic at all. Real thinking means retrieval of the right information at the right time." (Riesbeck and Schank, 1989)

we suspect the truth lies somewhere in between. To test our supposition, we have built two expert systems, one rule-based and one case-based. Both systems have identical problem statements and the same domain experts. However, the implementations are independent, to minimize any effect of learning in one mode on performance in the other. We compare and contrast our systems, to identify the strengths and weaknesses of each.

2. CBR, RBR and Hybridization

A CBR system solves new problems by finding, adapting, and reusing the solutions to similar problems encountered in the past. An RBR system solves new problems by drawing inferences from rules which embody problem-solving knowledge. The following definitions and examples are offered to clarify the distinction.

A *case* is a knowledge representation comprising a past problem and its solution. It contains an approximate solution that can be found and modified to obtain a good solution to a new problem.

A *rule* is a knowledge representation expressing a relationship between objects. It contains a piece of knowledge that can be combined, or chained together, with other pieces of knowledge to build a good solution to a problem.

Figures 1 and 2 show examples of cases and rules in the menu planning domain. Simple examples are provided for pedagogic purposes; details follow in Section 3. Given the cases shown in Figure 1, we just need methods for finding and adapting them. For example, if we wanted an easy to prepare, brown bag[2] lunch meal for a vegetarian, we might find Case 2 and adapt it by substituting a cheese sandwich for the turkey sandwich. Given the rules shown in Figure 2, we just need some facts like:

```
protein(steak).
beverage(apple_juice).
starch(potato_chips).
```

plus definitions for includes_meat and calorie_count, and an inference engine, like Prolog, to deduce a solution.

A hybrid system has both CBR and RBR components or modules. Note that this is not the same as having both cases and rules. CBR systems have long used rules in a supportive role, for indexing, matching and adaptation. The few hybrid systems which have been built to date have taken two approaches to hybridization. One approach is to have independent CBR and RBR modules, each of which can solve the problem independently of the other. The other approach is to take an essentially RBR system, and add a CBR module to provide some portion of the system's overall functionality.

Rissland and Skalak's CABARET is a system with independent CBR and RBR problem solvers (Rissland and Skalak, 1991). CABARET grew out of an effort to extend the HYPO system (Ashley and Rissland, 1988), which operates in the trade secrets law domain, to another legal domain, that of tax law. While HYPO's domain was primarily case-based, tax law has statutes, or rules. These rules are vague, because they include words that are open to interpretation. The current interpretation is determined by past interpretations made in courts of law for similar cases. In effect, cases determine whether or not rules apply. CABARET interleaves CBR and RBR, using heuristics to post CBR and RBR tasks to an agenda.

Golding's ANAPRON is a system which uses a CBR module to provide one portion of an essentially RBR system's functionality (Golding, 1991). ANAPRON is a speech synthesizer which pronounces surnames aloud. ANAPRON uses rules to generate a probable pronunciation, and then uses cases to handle exceptions to the rules. Golding found that rules were good for capturing large trends,

---

[2]This Americanism stems from the brown paper bags in which lunches were commonly carried to work or school. Americans still "brown bag it," even when carrying reusable lunch containers today.

```
Case 1:  Lunch Meal                    ; Features of the problem
         Fast Food                     ; useful for later retrieval
         Fun for Kids
         High Calorie

         We had six kids over for      ; Old problem statement,
         lunch, and we needed a        ; useful for explanations
         meal that was quick, fun
         and filling.

         Pizza Hut Pizza               ; Old solution:
         Coke                          ; Meal actually served

Case 2:  Lunch Meal
         Brown Bag
         Easy to Prepare
         Cheap

         I packed my own lunch to take to school. I'm on a
         student budget, both in terms of time and money.

         Turkey Sandwich
         Carrot Sticks
         Potato Chips
         Apple Juice

Case 3:  Lunch Meal
         Sit Down
         Suitable for Executives

         We hosted some visiting managers from GM at work.
         We needed to cater a suitable lunch in the company
         dining room.

         Steak
         Baked Potato with Sour Cream
         Green Beans Almondine
         Strawberry Cheese Cake
         Coffee
```

Figure 1. Example cases.

and cases were good for filling in small pockets where there were exceptions to the rules.

A third approach looks promising for designing menus. This is to take an essentially CBR system and to enhance it with an RBR module. Here, CBR provides a menu to meet a user's nutritional and personal preference requirements. RBR allows the user to customize the menu in creative ways, adding flair, while tracking

```
/*      lunch(Meal,Constraints) :-
        lunch plans a meal given a list of constraints.
        A typical use would be:
        lunch(X,[vegetarian,low_cal,easy_to_prepare]). */

lunch(meal(A,B,C,D),Constraints) :-
        protein(A), starch(B), fruit_or_veg(C), beverage(D),
        satisfies_constraints(meal(A,B,C,D),Constraints).

satisfies_constraints(Meal,[C|Constraints]) :-
        meets_constraint(Meal,C),
        satisfies_constraints(Meal,Constraints).

satisfies_constraints(Meal,[]).

meets_constraint(meal(A,B,C,D),vegetarian) :-
        \+(includes_meat(A)), \+(includes_meat(B)),
        \+(includes_meat(C)), \+(includes_meat(D)).

meets_constraint(meal(A,B,C,D),low_cal) :-
        calorie_count(meal(A,B,C,D),Calories),
        Calories < 700.
```

Figure 2. Example rules.

the nutritional effects. The customized cases can then be stored in the case base to strengthen the CBR component of the system.

3. System Comparison and Contrast

We have built two menu planners, the CAse-Based Menu Planner (CAMP) and the rule-based Pattern Regulator for the Intelligent Selection of Menus (PRISM). While both menu planners share the same problem statement and the same domain experts, the implementations are independent. This allows us to objectively compare the effort involved in constructing the systems, as well as the quality of system output.

There are several related, but distinct, forms of menu planning. A caterer, like JULIA, plans a single meal for the enjoyment of many eaters (Hinrichs, 1992). A restaurant owner plans a multi-choice menu, allowing each customer to choose his own favorites, while ensuring that kitchen capacity, supplies and personnel are adequate to implement the plan. A dietitian, in a hospital or community outreach program, designs a daily menu for a single individual, taking their dietary requirements and personal preferences into account. Our menu planners are of the last type. Our experts are nutrition professors in Case Western Reserve University's School of Medicine. We aim to provide practical assistance to those who, for medical reasons, must adjust their daily diets.

3.1. CAMP

CAMP's approach to the design process is that of case-based design. In case-based design, cases serve as examples showing how multiple constraints have been successfully met in the past. A case may suggest a design or a design frame-work (Kolodner, 1993). A system exemplifying case-based design in another domain is CADET, which designs small mechanical devices (Sycara *et. al*, 1992).

CAMP is a "pure" case-based reasoner. As much knowledge as possible was kept directly in cases, so as not to overlook opportunities or gloss over shortcomings of CBR in our domain. Like any canonical CBR system, CAMP operates by storing, retrieving and adapting cases.

A representative case in CAMP is shown in Figure 3. A solution in CAMP is a single day's menu. The features that indicate the usefulness of a case are:

- the nutrient vector for the menu, including: calories, protein, fat, carbohydrate, alcohol, fiber, cholesterol, Vitamin C, thiamin, niacin, riboflavin, Vitamin B6, Vitamin B12, folic acid, Vitamin A, Vitamin E, iron, calcium, phosphorus, sodium, potassium, magnesium, copper and zinc
- the types of meals and number of snacks included
- foods on the menu

The menu shown in Figure 3 was obtained from the U.S. Department of Agriculture (USDA, 1982). Each of the 81 menus used in CAMP was obtained from a recognized nutritional source and reviewed by our experts. Our experts modified menus as needed to ensure that each one conforms to the Recommended Dietary Allowances (RDA's) (Food and Nutrition Board, 1989) and the Dietary Guidelines (USDA, 1990), while meeting aesthetic standards for color, texture, temperature, taste and variety. Each case contains a "good" menu, at least for some individuals. Because individuals vary in their tastes and nutritional needs, not every menu is good for every individual. The retrieval process selects the menu that best suits a given individual's requirements. Then, CAMP adapts that menu to meet any unmet constraints. It uses "snippets," or parts, of other "almost-right" menus to aid in adaptation.

CAMP's cases are stored in a flat memory structure, for both methodological and domain specific reasons. Historically, CBR systems used hierarchically indexed case libraries to facilitate efficient case retrieval (Kolodner, 1993). Current research indicates that flat memories offer greater flexibility, and are amenable to parallel implementation should efficiency become a problem (Kettler *et. al.*, 1994). In the menu planning domain, cases are not naturally ordered into meaningful categories. The same menu may be suitable, with adaptation, for many different eaters. This is illustrated by the family dinner, where all family members

```
Breakfast:
   3/4 cup orange juice
   1 poached egg
   2 medium bran muffins with
   2 tsp margarine
   1 cup skim milk

Lunch:
   Sandwich
      2 slices rye bread
      1/2 cup chicken salad
   1 cup split pea soup
   2 halves pears, canned in light syrup
   1 cup water, tea, or coffee

Dinner:
   3 oz. pork chop
   1/2 cup cooked broccoli
   1 medium baked sweet potato
   1 medium whole wheat roll
   2/3 cup canned fruit salad
   1 cup water, tea, or coffee

Snack:
   4 Triscuit whole wheat crackers
   1 cup skim milk

Breakfast Type: Egg Breakfast
Lunch Type: Soup and Sandwich Meal
Dinner Type: Meat and Vegetable Meal
Snack 1 Type: Salty Snack
Snack 2 Type: None
Snack 3 Type: None

Calories:       1754.07 kc, Protein:        95.21 gm, Fat:             53.76 gm
Carbohydrate:    234.38 gm, Alcohol:         0.00 gm, Fiber:           21.40 gm
Cholesterol:     425.25 mg, Vitamin C:     193.05 mg, Thiamin:          2.13 mg
Niacin:           17.83 mg, Riboflavin:      2.04 mg, Vitamin B6:    1423.58 ug
Vitamin B12:       3.54 ug, Folic Acid:      0.21 mg, Vitamin A:    29463.56 IU
Vitamin E:         4.40 mg, Iron:           12.87 mg, Calcium:        975.60 mg
Phosphorus:     1749.27 mg, Sodium:       2928.06 mg, Potassium:     3670.08 mg
Magnesium:       276.73 mg, Copper:          1.26 mg, Zinc:             8.15 mg

Source of Menu: USDA
```

Figure 3. A representative case in CAMP.

are served essentially the same menu. Adaptation may be made so that children receive smaller portion sizes and/or milk instead of beer. In contrast, the CBR recipe generator CHEF could make effective use of hierarchical indexing in its domain (Hammond, 1989). A stir-fry recipe is of no use at all in generating a recipe for a desert souffle, so recipe type effectively partitioned CHEF's case base.

CAMP uses an adaptation-oriented retrieval technique, which chooses a case based on the ease of adapting it to meet current goals. In CAMP, a case must be adapted until it meets all user-specified constraints, plus additional constraints imposed as minimum RDA's. To find the best case, CAMP checks each case against all constraints. Any case meeting all constraints constitutes an exact match and is retrieved. When a case does not comply with a constraint, a penalty score is assigned based on how difficult it would be to bring the case into compliance. CAMP finds the case that is easiest to adapt, striking a balance between the number and severity of constraint violations. It uses this case as a starting point, and uses the next best cases to aid in adaptation.

The adaptation framework, based on our expert's approach to adapting menus, is:

1. Check the number of snacks. Adjust, if necessary.
2. Check meal types. Swap meals to accommodate preferences, if necessary.
3. Eliminate any forbidden food items.
4. Check calorie level. Adjust serving sizes, if necessary.
5. Fix any nutrient specific deficiencies.

Many adaptations use snippets from other cases. Snippets may be whole meals, or parts of meals, such as main dishes or side dishes. Changes are made at the largest granularity possible, to maintain aesthetic qualities such as color combinations, textures, temperatures, shapes and compatible flavors.

A menu designed by CAMP is shown in Figure 4. The menu shown was constrained by the user to include one snack, between 1800 and 2200 calories, at least 800 mg of calcium, and no more than 30% of calories from fat.

3.2. PRISM

PRISM's design process is one of hierarchical refinement. In this more traditional design approach, skeletal designs, or design patterns, are instantiated and refined. PRISM is based on an earlier RBR system, the Expert System on Menu Planning (ESOMP) (Yang, 1989). ESOMP planned menus for patients on a severely restricted low-protein diet. PRISM expands on ESOMP by planning menus for a wide range of dietary requirements. To do this, PRISM relies on menu and meal patterns. A daily menu pattern takes the form:

breakfast optional-snack lunch optional-snack dinner optional-snack

Each meal within the menu pattern may fit one of several patterns. PRISM's algorithm for planning a meal, given a pattern, is shown in Figure 5. A *go-with food*, as used in Figure 5, is something a person normally expects to eat with an-

```
Breakfast:
    1/2 cup orange juice
    1/2 cup bran flakes
    1/2 cup skim milk
    1/4 cup omelette, made from egg substitute
    1 English muffin with
    1 Tbsp. cream cheese
    1 cup coffee

Lunch:
    1 cup tuna-noodle casserole
    1/2 cup spinach
    1/2 cup steamed squash
    1 medium whole wheat roll with
    2 tsp. reduced-calorie margarine
    2 pear halves, canned in light syrup
    1 cup iced tea

Dinner:
    3 oz. roast beef
    1/2 cup cooked broccoli
    1/2 cup mashed potatoes
    1/2 cup glazed carrots
    1 medium whole wheat roll with
    2 tsp. reduced-calorie margarine
    1 baked apple
    1 cup skim milk

Snack:
    4 graham crackers
    1 oz. low-fat American cheese
    1 cup skim milk
```

Figure 4. Menu designed by CAMP.

other food. For example, butter and jelly are go-with foods for bread, at least in the American heartland.

PRISM's approach to menu creation is one of generate, test, and repair. A daily menu is initially generated by successively refining patterns for meals, dishes, and foods, filling general pattern slots, such as *breakfast bread dish* with specific foods, such as *1 slice of cinnamon raisin toast with 1 teaspoon of margarine*. A multi-layered hierarchical structure, relating meal parts to each other, was implemented to ensure that each meal conforms to common sense expectations for the form of a Western meal. At the implementation level, this structure consists of four databases, containing meal types, dish types, food types and foods. At the conceptual level, the structure can be viewed as a four-layered network of nodes connected by arcs defining relationships. Arcs are unidirectional, and may con-

```
While there are dish slots in the meal pattern to fill
    Select a dish type from the meal pattern
    Select a dish of that type as follows:
        If there is a client-preferred food of that type
            Select the client-preferred food
        Else
            Randomly select a food type from the possibility list
            Repeat
                Choose a food of that food type
                Check that food against constraints
            Until a food satisfies constraints or no foods are left
            If no foods are left
                Randomly choose any food of that type
            End if
        End if
        Calculate serving amount for the selected food
    Choose go-with foods for the selected food
End while
```

Figure 5. PRISM's meal planning algorithm.

nect two nodes within a layer, or a node in one layer to a node in an adjacent, more specific, level. An example relationship within a layer is: a *continental breakfast* is one type of *light breakfast*. An example interlayer relationship is: a *continental breakfast* includes a *breakfast bread dish*.

Another view of PRISM's initial menu generation process is that the multilayered network implements a context free grammar for the production of well-formed menus. Example production rules of this grammar are:

```
<breakfast> -> <light_breakfast> |
               <hearty_breakfast>
<light_breakfast> -> <continental_breakfast> |
                     <cereal_breakfast>
<continental_breakfast> ->
  <breakfast_bread_dish> <breakfast_beverage_dish> |
  <juice_dish> <breakfast_bread_dish> <breakfast_beverage_dish>
<breakfast_bread_dish> -> <muffin_dish> |
                          <quick_bread_dish> |
                          <toast_dish>
<muffin_dish> -> <muffin_food> |
                 <muffin_food> <muffin_spread>
<muffin_food> -> corn_muffin |
                 bran_muffin |
                 blueberry_muffin
```

After a menu is generated in compliance with both user specifications and common sense expectation as to form, it is tested to see if it meets nutritional constraints. Because many of these constraints can not be built into the menu up front, repair is usually necessary. Repair, in PRISM, is a backtracking process, in which new foods, dishes or meals are substituted for those found to be nutritionally lacking. The PRISM implementor has noted that repair is most likely to be successful when the original menu comes close to meeting constraints. When the original menu did not meet constraints, early PRISM could churn nonproductively, correcting one nutritional deficiency only to create another. PRISM now always produces a menu within reasonable time limits, but not always one which meets all constraints.

After PRISM generates a menu, it displays it, and then allows the user to perform "what if" analysis. The user can choose to delete foods from and/or add foods to the menu. PRISM keeps a running total of the effects on the nutritional value of the menu. This allows a user to evaluate tradeoffs: if he wants a chocolate milkshake, then he can learn what else needs to change in his daily menu to accommodate it. In practice, a nutritionist can use this analysis for educational purposes and/or to better satisfy individual preferences.

A menu designed by PRISM, to meet the same constraints as the menu shown in Figure 4, is shown in Figure 6.

3.3. STRENGTHS AND WEAKNESSES

Both systems successfully generate useful menus, as judged by expert nutritionists. However, they have different strengths and weaknesses.

3.3.1. *Meeting Nutritional and Preference Constraints*

While PRISM handles aesthetic and preference constraints well, it is limited in its ability to handle nutritional constraints. Typically, it can satisfy only three or four nutritional constraints at a time. One difficulty is that whereas we can use context-free rules to form aesthetically pleasing menus, we can not determine the nutritional validity of a menu before it is fully designed. A nutritionist does not think in terms of "bad" foods or meals, only in terms of bad menus. Eggs are good for breakfast, when the remaining meals are low in cholesterol. Given a large steak dinner, eggs are a poor breakfast choice for anyone trying to limit cholesterol. Much effort was expended on backtracking strategies, with limited success.

CAMP meets all constraints that PRISM meets and more. In addition to user-specified constraints, CAMP also constrains all menus to meet the RDA's. Unspecified nutrient levels are treated as don't care's in PRISM. Stricter standards were imposed on CAMP when it became apparent they could be met without much additional effort. It is an easier task to find a menu that nearly meets all constraints and to modify it than to create such a menu from scratch.

3.3.2. *Creative Design*

Menu planning is a domain in which creativity is valued. Today's ideal menu is monotonous tomorrow. New foods, served in new ways and combinations, provide variety and appeal. PRISM has over 1200 different foods in its database, and it

```
Breakfast:
    1 cup "Coffee brewed"
    1 cup "Malt-o-meal flavored ckd w/salt"
        2 tablespoons "Cream light coffee or table 19%"
        1 teaspoon "Sugar brown"

Lunch:
    1 cup "Tomato soup canned mw/ milk"
    1 cup "Grapefruit juice unsw froz w/ 3 pts water"
    1 piece "White cake w/ unckd white icing"

Dinner:
    0.5 medium "Lettuce romaine/cos raw leaf"
        3 number "Tomato raw cherry"
        1 tablespoon "Bean sprouts/alfalfa raw"
        1 tablespoon "Pepper red sweet raw chopped"
        0.125 cup "Ham cured L rst chopped or diced"
        0.125 cup "Cheddar cheese shredded"
        0.5 number "Egg whole hard cooked"
        1 tablespoon "Croutons Croutettes Kellogg's"
        0.125 cup "Thousand Island dressing 8 kcal/tsp"
    0.5 average "Baking powder biscuit"
    0.5 teaspoon "Honey"
    0.5 cup "Hot chocolate mix powder w/ sugar"
    0.25 cup "Chocolate ice cream Baskin-Robbins"

Snack:
    1 weight-ounce "Muenster cheese"
        1 slice "Low protein bread Sherwin"
```

Figure 6. Menu designed by PRISM.

can combine them in a wide variety of ways. PRISM also allows users to propose and evaluate their own creative food combinations, using "what if" analysis. "What if" analysis is a useful thinking process, not easily supported by CBR, with its alternate emphasis on "what did."

A case-based reasoner's strength lies in remembering old solutions which can be reused, not in considering new possibilities. CAMP's innovation is limited by the possibilities stored in its case base. CAMP generates new menus by combining parts of old menus and/or by making minor changes to them.

The number of different menus CAMP can output is significantly less than the number PRISM can produce. It may be noted that human innovation in menu planning also has limitations. Experience shapes and limits the range of menus any individual can realistically plan.

3.3.3. *Knowledge Engineering*

While it is often claimed that CBR eliminates the knowledge acquisition bottleneck, this was not our experience. We found both CBR and RBR knowledge engineering to be difficult, albeit interesting, despite the efforts of exceptionally articulate and cooperative experts.

The major challenge for CAMP was to find cases. Locating cases was difficult because:

— Few publications contain daily menus.
— Publications which do contain sample menus do not ordinarily give quantities for menu items, and these are needed for nutritional analysis.
— Nutritional knowledge is continually evolving, making older references out-of-date.
— There are many considerations to juggle in planning a menu, including aesthetic considerations. No nutritional benefit is derived unless a person *eats* what's on the menu.
— Human experts find it difficult to plan menus that meet all desired criteria. A recent study found that only 11% of menus prepared by qualified nutritionists met both the RDAs and the Dietary Guidelines (Dollahite, 1995).
— Experts do not always agree on what is a good menu. Personal preference is involved.

It took over three months of full-time effort to acquire the first forty cases for CAMP. An early idea that weeklong menus might make good cases could not even be explored.

The challenge for PRISM was of a different nature. At the beginning, the knowledge engineer knew little about nutrition and the nutritionists knew little about AI. There was a feeling that "they weren't speaking the same language." Much mutual education ensued, and sessions became more productive. Knowledge acquired early on had to be revised in light of new understanding, and PRISM's initial design had to change. Ultimately, rules approved by both the experts and the knowledge engineer were acquired and incorporated into PRISM.

The processes involved in implementing the two systems were by and large *different*, rather than better or worse. However, some functionalities required less engineering to provide with CBR than RBR. PRISM devoted many rules to serving go-with foods together. For example, rules were needed to serve butter with bread, catsup with French fries, and cranberry sauce with turkey. In contrast, these relationships are implicit in CAMP, as go-with foods are already together in cases. Another task requiring a complex rule set in PRISM is calculating serving sizes. In CAMP, serving sizes are also stored in cases. They may be adapted to accommodate individual eaters. Implementing the adaptation strategy required less time than developing the rule set.

3.3.4. *Task Complexity*

Designing a menu from scratch proved to be a more complex task than retrieving and adapting one. This is primarily because of the amount of common sense knowledge involved in menu planning (Kovacic, 1995). The expert provided a rule, for example, that breakfast could contain two fruit exchanges, and expected PRISM to *know* that these should not be half a cup of orange juice and half a cup of apple juice. There's a *sense* that some meals appeal and others do not. PRISM tackled this problem in the tradition of CYC (Lenat and Guha, 1990), which still represents a grand challenge for AI.

On the other hand, the common sense of a human nutritionist is already embedded in each of CAMP's menus. CAMP can never retrieve an implausible menu; bad combinations can only be introduced through adaptation. Guarding against this possibility was a trivial task, in comparison.

Menu repair is an aspect of menu design which is more complex in PRISM than in CAMP. PRISM creates menus via a generate-test-repair process. Repair rules are designed to bring any initially generated menu into compliance with nutritional constraints. Repair works well in PRISM for menus which come close to satisfying their constraints. However, not all menus do come close, and not all are effectively repaired. In contrast, CAMP's process is one of retrieve-adapt, which can also be viewed as retrieve-test-repair. Because the retrieval mechanism selects a menu which needs as little adaptation as possible, CAMP's repair process is simpler and more effective than PRISM's.

3.3.5. *Cognitive Aspects*

Our experts tell us they use both case-based and rule-based reasoning as they design menus. Because CAMP and PRISM use single-reasoning approaches, neither fully captures the cognitive processes of our human experts. In studying how human experts plan menus, we have found evidence of both CBR and RBR at work. The best example of CBR is provided by a system for planning healthful, well-balanced lunches for school children. This is a manual CBR system, developed by the American Heart Association, to reduce the amount of fat and salt children eat, to prevent future heart disease (American Heart Association, 1992). Over 12,000 kits have been distributed for use in schools. The kit consists of two thick binders, containing:

- over 100 complete school lunch menus (*cases*)
- extensive lists of foods that can be substituted in the menus for local customization (*adaptation rules*)
- criteria for determining that a resultant menu meets the standard for healthful meals (*evaluation criteria*)

Examples of RBR are provided by texts used in college courses to train future menu planning professionals (Eckstein, 1978; Shugart and Molt, 1989; Spears, 1995). Rules given in (Spears, 1995) are:

1. Plan the dinner entree first
2. Plan the lunch entree or main dish, avoiding that served for dinner
3. Select starch dishes appropriate to serve with the entrees
4. Select salads, accompaniments and appetizers next
5. Plan deserts for both lunch and dinner
6. After dinner and lunch are planned, plan breakfast and snacks
7. Evaluate the entire daily menu as a unit

We also found evidence of hybridization. One rule given in (Shugart and Molt, 1989) is that you should have previous menus handy while planning new ones. This evidence confirms what our experts have told us, that both CBR and RBR play essential roles in menu design.

3.4. SCOPE FOR HYBRIDIZATION

Our goal is to capitalize on the strengths and to mitigate the weaknesses of CBR and RBR through hybridization. Our system comparison elucidates how we can accomplish this goal in the menu planning domain. A hybrid system should combine CAMP's ability to satisfy contraints with PRISM's flair for creative design.

A CBR module to store, retrieve and adapt potential menus contributes toward the design of menus which meet multiple nutritional and personal preference constraints. It reduces system complexity by embedding common sense knowledge in cases, rather than representing it explicitly. It simplifies the menu repair process by retrieving menus which already meet constraints as closely as possible.

An RBR module to perform "what if" analysis and to introduce new foods into menus contributes creativity in design. It facilitates keeping the system up-to-date as new foods become popular. It allows the user to interact with the system, evaluating trade-offs and personalizing menus. These personalized menus can become new cases, which are otherwise difficult to acquire. The two modules then function symbiotically, to design better menus in concert than either CAMP or PRISM can design alone.

4. Related Work in Menu Planning

Computer-assisted menu planning systems have been built since the 1960's. Using linear programming techniques to build the first of these, Balintfy optimized a menu for nutritional adequacy, cost, and palatability (Balintfy, 1964). Shortly thereafter, Eckstein adopted a "random" approach to satisfice, rather than optimize, menus (Eckstein, 1967). Using a simple meal pattern, she composed each menu of a meat, starchy food, vegetable, salad, desert, bread and beverage. Within each category, a food item was selected randomly and evaluated with respect to constraints. The program would iterate until satisfactory items were found.

Two decades later, AI approaches to menu-planning were first tried. Yang built ESOMP to plan nutritionally sound menus for patients on a severely restricted low-protein diet (Yang, 1989). Galotra *et. al.* developed a Prolog expert system to plan therapeutic menus for patients in India (Galotra *et. al.*, 1991). They used Operations Research methods to match nutritional requirements to specific food items and heuristic rules and reasoning to convert the food items into complete menus. Hinrichs combined CBR with constraint propagation techniques to build JULIA, an interactive menu planner (Hinrichs, 1992). JULIA plans meals for dinner parties, functioning in the role of caterer. It plans a meal to satisfy a group of guests, despite conflicting food preferences and evolving constraints. Ganeshan and Farmer have implemented an RBR catering system for a large Australian catering corporation (Ganeshan and Farmer, 1995).

5. Summary and Conclusions

Two expert systems, one case-based and one rule-based, were built in the domain of nutritional menu planning. The systems shared the same problem statement and the same domain experts, but were implemented independently by different knowledge engineers. The systems were compared and contrasted to identify their strengths and weaknesses.

In our experience, the CBR system was better at constraint handling, and the RBR system was better at creative design. The task of designing a menu from scratch proved to be more complex than the task of retrieving and adapting one. The added complexity in the RBR system stemmed from the need to explicitly represent common sense knowledge. Neither CBR nor RBR provided an edge in knowledge acquisition. Both CBR and RBR were necessary in order to model the cognitive processes our human experts use in designing nutritional menus.

Menu planning has proven to be a fertile domain for exploring issues of AI in design. In future work, we will incorporate our findings in a hybrid CBR/RBR system. The hybrid menu planner will combine CAMP's ability to meet nutritional and personal preference contraints with PRISM's creative flair. This will move us one step closer toward automating the design of nutritionally balanced, yet appetizing, menus.

6. Acknowledgements

This research was partially supported by the National Science Foundation under NSF Grant CCR-9303484. The authors would like to thank Grace Petot and Karen Fiedler, whose expertise makes this work possible. Special thanks go to Kathy Kovacic, implementor of PRISM.

References

American Dietetic Association and the U.S. Department of Agriculture: 1982, *FOOD 2*, Chicago, IL.

American Heart Association: 1992, *The Hearty School Lunch Menus*, Dallas, TX.

Ashley, K. D. and Rissland, E. L. : 1988, A case-based approach to modeling legal expertise, *IEEE Expert*, 3(3), 70–77.

Balintfy, J. L.: 1964, Menu planning by computer, *Communications of the ACM*, 7(4), 255–259.

Dollahite, J., Franklin, D. and McNew, R.: 1995, Problems encountered in meeting the recommended dietary allowances for menus designed according to the dietary guidelines for Americans, *Journal of the American Dietetic Association*, **95**(3), 341–347.

Eckstein, E. F.: 1967, Menu planning by computer: The random approach, *Journal of the American Dietetic Association*, **51**, 529–533.

Eckstein, E. F.: 1978, *Menu Planning*, 2nd edn, AVI Pub., Westport, CT.

Food and Nutrition Board: 1989, *Recommended Dietary Allowances*, 10th edn, National Academy Press, Washington, DC.

Galotra, V., Ramachandran, S., Singh, H. and Bajaj, K. K.: 1991, Nutrition diet programme - An expert system, Unpublished Report, Artificial Intelligence Division, National Informatics Centre, New Delhi, India.

Ganeshan, K. and Farmer, J.: 1995, Menu planning system for a large catering corporation, *Proceedings of the Third International Conference on the Practical Application of Prolog*, Paris, France, pp. 262–265.

Golding, A. R.: 1991, *Pronouncing Names by a Combination of Case-Based and Rule-Based Reasoning*, PhD Dissertation, Stanford University, CA.

Hammond, K. J.: 1989, *Case-Based Planning: Viewing Planning as a Memory Task*, Academic Press, San Diego, CA.

Hinrichs, T. R.: 1992, *Problem Solving in Open Worlds: A Case Study in Design*, Lawrence Erlbaum, Northvale, NJ.

Jackson, P.: 1990, *Introduction to Expert Systems*, 2nd edn, Addison-Wesley, Reading, MA.

Kettler, B. P., Hendler, J. A., Andersen, W. A. and Evett, M. P.: 1994, Massively parallel support for case-based planning, *IEEE Expert*, 9(1), 8–14.

Kolodner, J.: 1993, *Case-Based Reasoning*, Morgan Kaufmann, San Mateo, CA.

Kovacic, K. J.: 1995, *Using Common Sense Knowledge for Computer Menu Planning*, PhD Dissertation, Department of Computer Engineering and Science, Case Western Reserve University, Cleveland, OH.

Kovacic, K., Sterling, L., Petot, G., Ernst, G. and Yang, N.: 1992, Towards an intelligent nutrition manager, *Proceedings of the ACM/SIGAPP Symposium on Computer Applications*, ACM Press, New York, NY, pp. 1293–1296.

Lenat, D. and Guha, R. V.: 1990, *Building Large Knowledge-Base Systems: Representation and Inference in the Cyc Project*, Addison-Wesley, Reading, MA.

Rich, E. and Knight, K.: 1991, *Artificial Intelligence*, 2nd edn, McGraw-Hill, New York, NY.

Rissland, E. L. and Skalak, D. B.: 1991, CABARET: Rule interpretation in a hybrid architecture, *International Journal of Man-Machine Studies*, **34**, 839–887.

Riesbeck, C. K. and Schank, R. C.: 1989, *Inside Case-Based Reasoning*, Lawrence Erlbaum, Hillsdale, NJ.

Shugart, G. and Molt, M.: 1989, *Food for Fifty*, 8th edn, Macmillan, New York, NY.

Spears, M. C.: 1995, *Foodservice Organizations: A Managerial and Systems Approach*, 3rd edn, Macmillan, New York, NY.

Sycara, K., Guttal, R., Konig, J., Navasimhan, S. and Navinchandra, D.: 1992, CADET: A case-based synthesis tool for engineering design, *International Journal of Expert Systems*, 4(2).

U.S. Department of Agriculture: 1990, *Nutrition and Your Health: Dietary Guidelines for Americans*, 3rd. edn, U.S. Government Printing Office, Washington, DC.

Yang, N.: 1989, *An Expert System on Menu Planning*, M.S. Thesis, Department of Computer Engineering and Science, Case Western Reserve University, Cleveland, OH.

5

reuse of designs

On design formalization and retrieval of reuse candidates
Joachim Altmeyer, Bernd Schürmann
Design rationale and design patterns in reusable software design
Feniosky Peña-Mora, Sanjeev Vadhavkar
Constraint-based retrieval of engineering design cases
Taner Bilgic, Mark Fox

J. S. Gero and F. Sudweeks (eds), Artificial Intelligence in Design '96, 231-250
© 1996 *Kluwer Academic Publishers*.

ON DESIGN FORMALIZATION AND RETRIEVAL OF REUSE CANDIDATES

JOACHIM ALTMEYER AND BERND SCHÜRMANN
University of Kaiserslautern, Germany

Abstract. To enable reuse in design, a formal model of design artifacts and design processes is necessary. In this contribution, we present a feature-based formalization of design and the exploitation of this formalism for retrieving suitable reuse candidates. Using a given requirement specification, we describe designs as goal-oriented refinement processes, and we show how we use this formalism within a case-based and rule-based retrieval approach. The main goal of this paper is to formalize generic structures of the design space and to show how these structures can be used to improve the retrieval.

1. Introduction

Reuse has one of the largest saving potential in design. It can be supported in different ways: for instance, building libraries, creating parameterized descriptions or templates, or providing reuse tools which base on case-based reasoning techniques (Kolodner, 1993). An important question which should be discussed is: "Are there generic reuse techniques in design possible?" or, with other words, "Can we identify mechanisms which are typical for design but independent of a specific design system, and how can we make use of these mechanisms during the reuse process?"

During the recent years, many design theories have been developed to describe and classify designs (Yoshikawa, 1981; Gero, 1990; Takeda, Veerkamp and Tomiyama, 1990). One criteria to classify designs is the number of abstraction levels within the design process. A design can be realized in one or many refinement steps. For example, in *computer-aided architectural design* (CAAD), a building can be drawn from scratch in one step, or the design process can be divided in different phases: formulating constraints, creating topological descriptions, building floorplans, and constructing the final detail drawings. Another criteria to classify designs is the structure of the design artifacts which are mostly complex objects. For instance, a design artifact can be an aggregate composed of a set of subobjects (e.g. Siepmann and Zimmermann, 1989, or Katz, 1990). All in all, we see that the *design space*, i.e. the space where the design takes place, is structured.

This paper gives a characterization of designs based on a refinement process, and it works out how such characterized designs can be supported by a framework retrieving suitable reuse candidates. Using the domain of *electronic computer-aided design* (ECAD), we summarize and formalize properties of the majority of design systems. Considering these properties and their formalization, we discuss two retrieval techniques. The first is more in the sense of indexing in case-based reasoning, whereas the second is rule-based.

This paper is organized as follows: In section 2, properties of designs are described and different reuse strategies are addressed. Section 3 shows a formal design model based on a refinement process. Using this formalism two different strategies to find useful reuse candidates are proposed in section 4. In section 5, the application and integration of these two strategies is shown by presenting a prototype implementation of a retrieval system in ECAD. At the end, in section 6, we summarize interesting future works.

2. Design Properties and Reuse in Design

PROPERTIES OF DESIGNS

In the following, we give an overview of design properties as they are valid for many ECAD processes. Therefore, this property list does not claim to characterize all possible design situations but it is valid for a large and interesting set of design processes, for example, the software engineering processes.

1.   One difference between design and general computer-aided tasks is the presence of a *requirement specification* in form of a list of constraints which has to be satisfied by the final result. Considering this requirement specification, two problems typically occur during the design. First, the list of requirements does not determine the final design artifact in its entirety. Therefore, there may exist alternative *realizations* which all fulfill the given requirements (*ambiguousness of the specification*). And second, the design often cannot fulfill all of the given requirements (*contradictoriness of the specification*). Descriptions of the concepts of *requirement specification* and *realization* can be found in Brazier et. al. (1994) or Schürmann, Altmeyer, and Schütze (1994).

2.   Requirement specifications can be represented by a list of *properties* (*features*) which should be fulfilled by the final design object. Concerning a category of features (e.g. the category Function), we distinguish different types of features, for instance, single features (e.g. 'Function is alarm'), compound features (e.g. 'Function is alarm and

report'), and complex features (e.g. the function is described by a state diagram).

3. It is obvious that it often does not cause the same costs to adapt a property A of an artifact to a property B as vice versa. For example, if we look for the topology of a design artifact which can be abstracted by a graph, it is generally easier to transform a planar graph to a non-planar graph (nothing to do) than to transform a non-planar graph to a planar one.

4. The amount of data in design is immense. To handle this amount, adequate data clusters are defined as complex objects which divide the data into disjunct sets. For example, within ECAD systems, a whole netlist or a complex layout are handled as complex objects. Although these data are necessary for the design tools, they are unwieldy to handle. Therefore, abstract concepts are introduced to simplify the treatment of these complex objects. For example, an ECAD management system may handle an adder without knowing the exact boolean equation. Instead, the management uses the abstraction that the adder fulfills the requirements of the 'IEEE floating-point standard'.

5. Defining data clusters is not sufficient to handle the large amount of data. In addition, the creation of types which can be instantiated within designs is necessary to avoid redundant descriptions of design artifacts. The recursive application of this type concept results in the concept of *configuration hierarchy* (Siepmann and Zimmermann, 1989).

6. One important characterization of design systems is the number of abstraction levels during the design. For example, in software engineering, one could start with creating a (mostly informal) problem description, and during the phases analysis, design, and implementation this description is refined on different abstraction levels to the final programming code. Or within the ECAD domain, the design space can be divided into three refinement domains: *behavioral domain*, *structural domain*, and *physical domain* (Gajski, 1988). Within the behavioral domain, an integrated circuit may be represented by an abstract algorithm or by a boolean equation. Within the structural domain, the design artifacts may be represented by modules and nets connecting these modules. And within the physical domain, the circuit may be represented by a layout description language. These levels defines a hierarchy which we call *design hierarchy*, and we call the different abstraction levels *refinement levels*.

7. The most important observation within a design system with more than one refinement level is that a more precise artifact should fit to its more abstract representations. For example, in software engineering,

the final application program has to fulfill the given problem description, or, within ECAD, the final mask layout has to fit to its representation in form of a boolean equation. Many design management systems do not support this requirement.

8. If the more precise description does not fulfill the requirement specification, and the decision which cause the problems was made in a more abstract domain, it should not be allowed to revise the precise descriptions without correcting the abstract one's accordingly. For example, Figure 1 shows the design hierarchy and the handling of such a *revision step*. The design space spans four refinement levels. The design process starts with the design steps φ_0, φ_1, and φ_2. Then, two alternative design steps φ_3 and φ_4 try to fulfill the requirement specification but both fail. However, the designer detects that the decision which causes the problem was made in step φ_1 and, therefore, within φ_5, the decision is revised to get the final result d_7 which fulfills the given requirements.

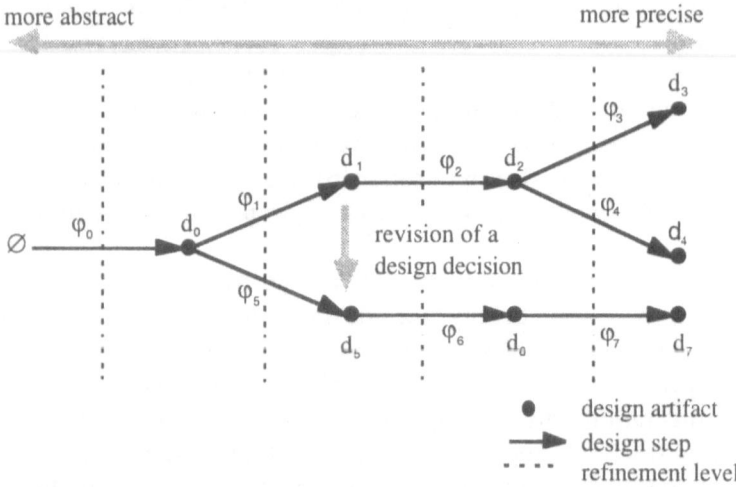

Figure 1. Example of a design hierarchy.

All in all, we see that the database of the design artifacts has a complex structure. The design artifacts build a network of different semantic relationships. A general, domain-independent reuse framework has to exploit these relations during the selection of suitable reuse candidates.

REUSE TYPES

Before we formalize the design, we present different reuse types which should be supported by a reuse framework. Following Riesbeck and Schank

(1989), we discern three modes of reuse in ECAD (Altmeyer, Ohnsorge, and Schürmann, 1994):

❑ *Reuse by instantiation*

The main idea is to reuse often used components, e.g. standard cells, instead of designing them always from scratch. Here, candidates for the reuse are well-tested and/or frequently-used components or components which implement standards. Often, these modules are arranged in libraries and serve as types which can be instantiated within new designs.

❑ *Reuse by parameterization*

A parameterized object is instantiated with fixed values. For example, an existing n-bit adder can be instantiated with a given fixed bit-width. Parameterized objects build equivalence classes over a set of concrete objects. Before realizing the final layout, a generator or a compiler has to concretize the parameterized objects. Examples of possible parameter classes in ECAD are the bit-width of circuits, the circuit function, the module shape, and the microprogram memory.

❑ *Reuse by adaptation*

Previously designed objects are adapted with regard to a given requirement specification. For example, it is much easier to write a high level language program of a 32-bit adder by using the code of a 16-bit adder than designing the adder from scratch.

So far, we informally presented a set of design properties, and we addressed the different reuse modes which should be supported by a reuse framework. In the following sections, we present a formal model of design, and we demonstrate how we exploit this model to retrieve suitable reuse candidates.

3. A formal design model

REQUIREMENTS AND REALIZATIONS

We first have to introduce a set of definitions which are used to characterize requirement specifications and their realizations objects.

Definition 1 (feature): An object o can be described by its *features* (properties). A feature p is a one-place predicate with p (o) is true.

To support a facet classification, we build sets of features concerning the same field. We call these sets *feature sets* (*property sets*) (Onosato and Yoshikawa, 1987). Examples of feature sets are Function, Size, and Technology.

Definition 2 (specialization-generalization relation; Onosato and Yoshikawa, 1987): A feature p of a feature set $\mathcal{F}$ is more specific than a feature q of $\mathcal{F}$, written as q > p, iff q holds always if p holds, i.e. p $\rightarrow$ q. For each feature set a partial order is defined by the *specialization-generalization relation* >.

Definition 3 (empty feature): We define the *empty feature* $\perp_{\mathcal{F}}$ of a feature set $\mathcal{F}$ as p > $\perp_{\mathcal{F}}$ for all features p of $\mathcal{F}$.

Definition 4 (universal feature): We define the *universal feature* $\top_{\mathcal{F}}$ of a feature set $\mathcal{F}$ as $\top_{\mathcal{F}}$ > p for all features p of $\mathcal{F}$.

Definition 5 (single feature): A feature α is a *single feature* of the feature set $\mathcal{F}$ iff $\alpha \neq \perp_{\mathcal{F}}$ and there is no p in $\mathcal{F}$ with α > p.

Definition 6 (generic feature): A feature γ is a *generic feature* of the feature set $\mathcal{F}$ iff there are two features p and q of $\mathcal{F}$ with γ > p and γ > q.

For each feature set the specialization-generalization relation builds a mathematical lattice structure (Onosato and Yoshikawa, 1987). Figure 2 depicts this structure.

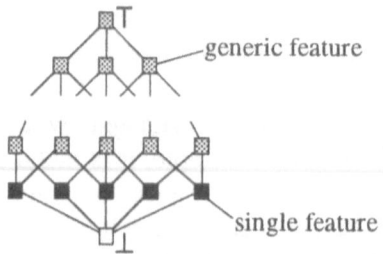

Figure 2. Lattice structure of a feature set

Definition 7 (generic feature set): A *generic feature* γ of the feature set $\mathcal{F}$ defines the generic feature set G_γ with $G_\gamma ::= \{p \mid \gamma > p \wedge p$ is a single feature of the feature set $\mathcal{F}\}$.

Definition 8 (feature set of an object): The *feature set* $\mathcal{P}_o$ *of an object* o is defined as $\mathcal{P}_o ::= \{p \mid p\,(o)\}$.

Definition 9 (design object): A *design object* o is described by its feature set $\mathcal{P}_o$.

Every feature p of a design object o is also an instance of a feature set $\mathcal{F}$.

Definition 10 (set of possible designs): We denote *the set of all possible design objects* as $\mathcal{T}$.

The definitions above allows us to characterize requirement specifications (see also section 2):

Definition 11 (specification): A *(requirement) specification* s is described by the set of its features $\mathcal{P}_s$. A requirement specification s is an *ambiguous specification* iff there is a set of possible design objects $O \subseteq \mathcal{T}$ with $\mathcal{P}_s \subseteq \mathcal{P}_o$ for all $o \in O$ and $|O| > 1$. Iff $|O| = 1$, we call this requirement specification *ideal specification*, and iff $O = \varnothing$ *contradictory specification*.

Definition 12 (successful design): Let s be a requirement specification. A design is *successful* iff the design results in a design object x with $\mathcal{P}_s \subseteq \mathcal{P}_x$.

So far, a design is described as one monolithic process with a specification as input and a design object as result. In the following subsection, we divide this process in several design steps.

SYSTEM SPECIFICATION VERSUS TOOL SPECIFICATION

The input of a design process is a *system specification*. The features of this specification span different refinement levels in the design space. For example, in the ECAD specification {'Function is Multiplier', 'Bit-Width is 16-Bit', 'Technology is CMOS', 'Size is smaller than 0.15 mm^2', 'Aspect Ratio is 1'}, the function belongs to the behavioral domain, whereas the aspect ratio belongs to the physical domain. Only some of the features of the system specification represent the input specification of the first design step in the behavioral domain. Together with other features of the system specification, the result of the first design step is the specification of the second design step. The specification of a single design step is called a *tool specification*. Note that this concept is recursive: The system specification is a tool specification at a meta level.

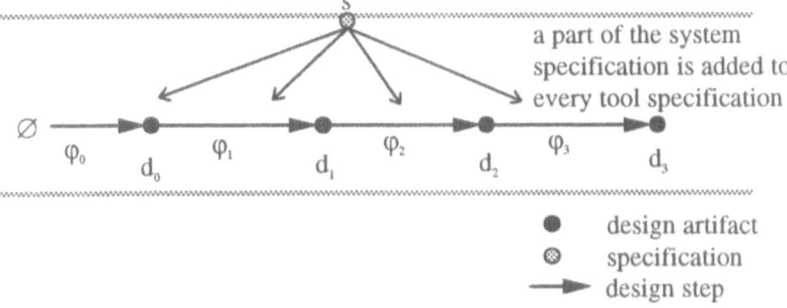

Figure 3. Data flow at one level of the configuration hierarchy

In the example of figure 3, the problem description of a design step φ_i is composed by the design object d_{i-1} (for the design step φ_0, this input is empty) together with the corresponding part of the requirement specification s (see also the definition of the *design problem description* in Brazier et. al. (1994)). Of course, during the design, other design objects, for example design objects as submodules, can be included into the input data of a design step.

FORMALIZATION OF THE REFINEMENT PROCESS

In this subsection, we formalize the design process by a *refinement model*. Definitions of a refinement design process can be found in Brazier et. al. (1994)

or Schürmann, Altmeyer, and Schütze (1994). In Brazier et. al. (1994), a refinement relation is introduced which orders the design objects based on truth values (*true*, *false*, and *undefined*) of *their ground atoms* (ground atoms can be constants, instances of object types, functions on objects, or relations between objects). In Schürmann, Altmeyer, and Schütze (1994), the iterative refinement process is defined by using attribute states *unknown*, *default*, *predicted*, *preliminary*, and *final* with the partial order {unknown, default} ➞ predicted ➞ preliminary ➞ final. Here 's1 ➞ s2' means that s2 is a refinement of s1, i.e. s2 is more precise than s1. For simplification, this paper is restricted to a set oriented definition without additional attribute states.

Design as described by the properties 6 and 7 in section 2 is a goal-oriented *convergence process* (Yoshikawa, 1981). With every design step in a convergence, we try to get closer to fulfill the given requirement specification s, i.e. every design step φ that is applied on a design object x resulting in a new design object y has to hold $(\mathcal{P}_s \cap \mathcal{P}_x) \subset (\mathcal{P}_s \cap \mathcal{P}_y)$.

Definition 13 (refinement): Let x and y be design objects. y is a *refinement* of x iff $\mathcal{P}_x \subset \mathcal{P}_y$.

Definition 14 (refinement design step): Let x and y be design objects and s be a design specification. φ is a *refinement design step* iff φ applied on x results in y, y is a refinement of x, and $(\mathcal{P}_s \cap \mathcal{P}_x) \subset (\mathcal{P}_s \cap \mathcal{P}_y)$.[1]

Figure 4 shows the refinement process of a successful design using an ambiguous requirement specification s. $\mathcal{P}_s$ is the set of all features of s, φ_i are refinement steps, and $\mathcal{P}_i$ is the set of all features of the design object d_i.

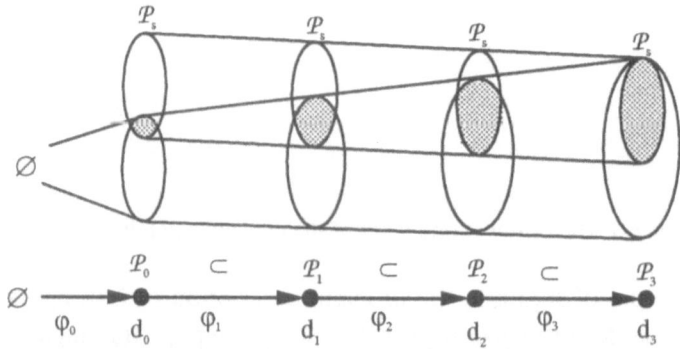

Figure 4. Venn diagrams of design refinement steps

If we change parameters of a design step or the user makes different design decisions within a design step, we get alternative results using the same input

---

[1]   In terms of the General Design Theory (Yoshikawa, 1981), the refinement model is an alternative convergence model to the *paradigm model* in which $\mathcal{P}_x \subset \mathcal{P}_y$ need not hold.

specification. Design objects x and y are alternatives with respect to a specification s iff $\mathcal{P}_s \subseteq \mathcal{P}_x$ and $\mathcal{P}_s \subseteq \mathcal{P}_y$.

If we forbid two or more design objects in the problem description of a design step at the same level of the configuration hierarchy (merge operation), the design process can be described by a tree (see for example figure 1). We call this tree *refinement tree*.

In real design processes, the given requirement specification cannot always be fulfilled, and the requirement specification has to be revised. Therefore, the development of an adequate requirement specification describes a process similar to the design process itself, and this process can also be described by refinement design steps and revision steps (Brazier et. al., 1994; Schürmann, Altmeyer, and Schütze, 1994).

4. Retrieval of design knowledge

So far, we described design objects and requirement specifications, and we characterized designs as goal-oriented refinement processes. This characterization of designs is useful to understand design processes and helps to classify designs. Beside this, a formalization of design can serve as the basis of a design data model (see for example the data model based on a refinement model in Schürmann, Altmeyer, and Schütze, (1994)). In this contribution, we show how this formalization could be a basis for the development of retrieval techniques which help the designer to find useful reuse candidates. To do this, we present two retrieval techniques, a case-based approach and a rule-based approach, which use this design knowledge when retrieving suitable reuse candidates from a design database.

On the one hand, case-based approaches use former situations to solve the current problem (Kolodner, 1993). These (selected) situations are stored as concrete information in the working memory, and they are selected by a partial matching. Within case-based reasoning, the problem of recalling suitable former cases is called *indexing problem*. On the other hand, rules are based on generalized design knowledge. Only if the precondition matches exactly, a rule is activated. The link between these two approaches lies in the fact that rules can be built by expert knowledge or by generalization of former cases. For a discussion of these techniques and their relations between each other, we refer to Kolodner (1993).

Our case-based approach does not focus on efficient indexing techniques. Our main focus is the exploitation of design properties and the combination of the case-based and the rule-based approach. With our retrieval approach, we only focus on the first step of case retrieval: finding a set of good´ cases as a starting point for case selection (Kolodner, 1993). In section 5, we will show

how we combine our two approaches in a way that the case-based approach is complemented by the rule-based one.

4.1 CASE-BASED APPROACH

A priori determination of adaptation costs
Within the case-based approach for design, the goal is to find design objects and to reuse these objects for solving the current design problem described by a requirement specification. At reuse by instantiation or parameterization, an exact matching between the specification and the reuse object is necessary. But if we allow an adaptation of the old situation considering the given requirement specification, we must search for a measure which expresses the costs for this adaptation a priori.

The relations between the adaptation costs and an a priori determination of these costs are motivated by figure 5. The costs for adapting a retrieved design object o to a design object o* which fulfills the specification s are not known in advance. Therefore, it is necessary to choose a *reusability value* which corresponds to the adaptation efforts as well as possible. In this subsection, we provide a model to determine this reusability value by a function REUSE which estimates the reusability comparing the design object o with the current requirement specification s. This function reduces the reusability to the fitness for each feature set separately by a function FIT which uses the similarity of single features of the feature set expressed by a function SIM.

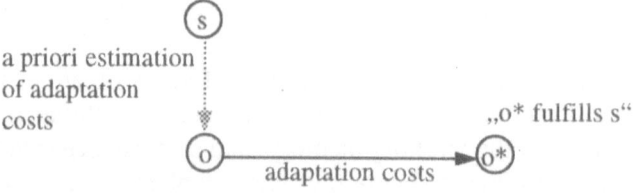

Figure 5. A priori estimation of adaptation costs

Definition 15 (similarity function of a feature set): Let $\mathcal{F}$ be a feature set. The *similarity function of a feature set* $\mathcal{F}$ SIM$_{\mathcal{F}}$ is defined as SIM$_{\mathcal{F}}$: $\mathcal{F} \times \mathcal{F} \rightarrow [0, 1]$. If SIM$_{\mathcal{F}}$ (p, q) = 1, then the feature p is equivalent to the feature q. In the case of SIM$_{\mathcal{F}}$ (p, q) = 0, p is totally different to q.

The feature set similarity function SIM$_{\mathcal{F}}$ (p, q) roughly expresses the expense of converting the feature q to the feature p, and this function has to be individually computed or defined for each pair of features of the feature set $\mathcal{F}$.
A function FIT$_{\mathcal{F}}$ enables us to handle generic features by substituting these by their generic feature sets (see definition 7).

Definition 16 (fitness function of a feature set): Let s be a specification and x be a design object. If the feature sets $\mathcal{P}_s^{\mathcal{F}} ::= \{p \mid p \in \mathcal{P}_s \wedge p \in \mathcal{F}\}$ and $\mathcal{P}_x^{\mathcal{F}} ::= \{p \mid p \in \mathcal{P}_x \wedge p \in \mathcal{F}\}$, with $\mathcal{F}$ is feature set, contain only one element each, the fitness function of the feature set $\mathcal{F}$ $\text{FIT}_{\mathcal{F}}(s, x)$ is $\text{SIM}_{\mathcal{F}}$ with the instances of $\mathcal{P}_s^{\mathcal{F}}$ and $\mathcal{P}_x^{\mathcal{F}}$, respectively. $[0, 1]$ is the range of $\text{FIT}_{\mathcal{F}}$. If it is allowed that the feature sets $\mathcal{P}_s^{\mathcal{F}}$ and $\mathcal{P}_x^{\mathcal{F}}$ contain more than one element, this function has to be defined individually for the feature set $\mathcal{F}$.

In the next example, we propose a fitness function for the feature set Function. This fitness function allows us to handle *multi-functional design objects*:

Example 1 (fitness function of a feature set Function): If the feature sets $\mathcal{P}_s^{\mathcal{F}} ::= \{p \mid p \in \mathcal{P}_s \wedge p \in \mathcal{F}\}$ and $\mathcal{P}_x^{\mathcal{F}} ::= \{p \mid p \in \mathcal{P}_x \wedge p \in \mathcal{F}\}$ with $\mathcal{F}$ is the feature set Function, the fitness function of the feature set $\mathcal{F}$ $\text{FIT}_{\mathcal{F}}$ is defined as:

$$\text{FIT}_{\mathcal{F}}(s,x) = \frac{\sum_{p_s \in \mathcal{P}_s^{\mathcal{F}}} \max_{p_x \in \mathcal{P}_x^{\mathcal{F}}} \left(\bigcup \{ \text{SIM}_{\mathcal{F}}(p_s, p_x) \} \right)}{\left| \mathcal{P}_s^{\mathcal{F}} \right|}$$

with $\text{SIM}_{\mathcal{F}}$ is the feature set similarity function of the feature set Function.

Definition 17 (reusability function): Let s be a design specification, x be a design object, and Γ be the set of all feature sets of which the corresponding feature set has instances in $\mathcal{P}_s$. The reusability function REUSE (s, x) is defined as:

$$\text{REUSE}(s,x) = \sum_{\mathcal{F} \in \Gamma} \gamma_{\mathcal{F}} \cdot \text{FIT}_{\mathcal{F}}(s,x)$$

$[0, 1]$ is the range of REUSE. $\gamma_{\mathcal{F}} \in [0, 1]$ is a *relevance factor* (weight factor) expressing the importance of the corresponding feature set for the comparison (the sum of all relevance factors has to be 1).

Using this function, we are able to determine the fitness of a design object compared to the requirement specification. But two cases have to be considered (see also the *contrast model* of Tversky (1977)). First, if the specification contains an additional feature set $\mathcal{F}$ which is not available for the current object x. In this case, we do not know the influence (positive or negative) on the fitness, and we therefore use the value 1/2 for the function $\text{FIT}_{\mathcal{F}}$. Second, if the design object x contains additional features or feature sets which are not available at the specification. In this case, theses features have no effect on the reusability, and they are ignored. We use this view because we explicitly model costs for these additional features which are evaluated by the reusability function. For example, if a design object realizes additional functions which are not included in the requirement specification, the additional area consumption for these needless functions is considered by the feature set Size.

Other approaches to express similarity can be found in Tversky (1977) or Richter (1992). But we choose this formalization because our reusability function considers generic features (design property 2 of section 2), and it is asymmetric (design property 3 of section 2).

Consideration of the design hierarchy

Regarding the design process, we make the following observation: the abstraction of the design decreases during the design process, i.e. the design artifacts become more and more precise. The degree of abstraction expressed by the refinement level can be regarded as a rough measure for the design progress. Therefore, the expense of changing early decisions increases with each design step. In the example of figure 6, it is less complex to change an n-bit adder to an m-bit adder in the behavioral domain (adaptation step ψ_1) and then to synthesize (semi-)automatically the mask layout of the m-bit adder (refinement step φ_2) than to change the adder in the physical domain (adaptation step ψ_2).

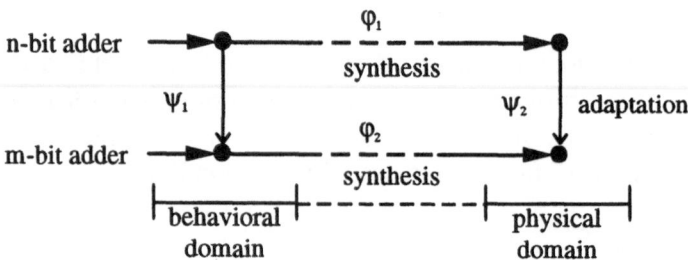

Figure 6: Example of changing an n-bit adder to an m-bit adder

Due to this observation, we define the similarity function of the ordinal feature set RefinementLevel as follows: Assume a refinement tree with the refinement levels 0 (root) to n (leaves). The feature p_k represents the refinement level k. The similarity function $SIM_{RefinementLevel}$ (or short: SIM_r) has to meet three conditions:

(1) $SIM_r(p_i, p_k) < SIM_r(p_i, p_j)$, $0 \leq i \leq j < k \leq n$

(2) $SIM_r(p_k, p_i) < SIM_r(p_j, p_i)$, $0 \leq i \leq j < k \leq n$

(3) $SIM_r(p_j, p_k) < SIM_r(p_j, p_i)$, $0 \leq i \leq j < k \leq n$.

Conditions (1) and (2) state that the similarity of two design objects decreases with increasing distance of the refinement levels. For instance, design object d_2 in figure 2 is more similar to d_1 than to d_0. We need two conditions to describe this behavior because the similarity function of a feature set is not symmetrical. Condition (3) reflects the observation described above. For a given design object at the refinement level i, all objects at a smaller level j are more similar than the objects at a larger refinement level k.

A typical shape of the similarity function $SIM_{RefinementLevel}$ is shown in figure 7. The feature p_j is more similar to p_i in the case that the refinement level j is smaller than i for the case that j is larger than i. The similarity decreases with increasing distance of the refinement levels. The shape of the two partial functions may be any strong monotone curve.

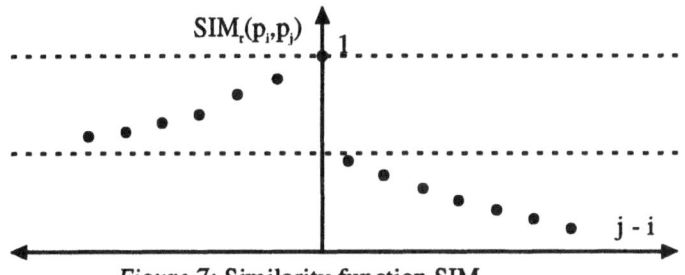

Figure 7: Similarity function $SIM_{RefinementLevel}$

4.2 RULE-BASED APPROACH

One of the basic problems of the formalism described so far is that features of each feature set $\mathcal{F}$ defined as $\mathcal{P}^{\mathcal{F}} ::= \{p \mid p \in \mathcal{P} \wedge p \in \mathcal{F}\}$ are considered within the reusability function independent of their context, defined as $\overline{\mathcal{P}}^{\mathcal{F}} ::= \mathcal{P} - \mathcal{P}^{\mathcal{F}}$. In the following subsection, we introduce an approach which helps us to overcome this problem.

Design goal substitution
We begin this subsection introducing some definitions which are the basis for a rule-based reasoning using backward-chaining similar to PROLOG.

Definition 18 (fact): A *fact* f is described as 'f: $\mathcal{P}$.' with $\mathcal{P}$ is a set of features.

Definition 19 (goal): A *goal* g is described as 'g: ?-$\mathcal{P}$.' with $\mathcal{P}$ is a set of features.

Definition 20 (rule): A *rule* r is described as 'r: $\mathcal{P}_0$:- $\mathcal{P}_1$, $\mathcal{P}_2$, ..., $\mathcal{P}_n$' with $\mathcal{P}_0$, $\mathcal{P}_1$, $\mathcal{P}_2$, ..., $\mathcal{P}_n$ are sets of features. We call '$\mathcal{P}_0$' the *rule header* and '$\mathcal{P}_1$, $\mathcal{P}_2$, ..., $\mathcal{P}_n$' the *rule body*.

Definition 21 (goal substitution by rules): Let 'g: ?- $\mathcal{P}$.' be a goal and 'r: $\mathcal{P}_0$:- $\mathcal{P}_1$, $\mathcal{P}_2$, ..., $\mathcal{P}_n$.' be a rule. Then, r can be applied to g iff $\mathcal{P}_0 \subseteq \mathcal{P}$. If r is applied to g new subgoals g_1 to g_n are created as 'g_i: ?- $\mathcal{P}_i^*$' with $\mathcal{P}_i^* ::= (\mathcal{P} - \mathcal{P}_0) \cup \mathcal{P}_i$ for all i = 1 to n. The goal g is substituted by the subgoals g_0 to g_n in which the features $\mathcal{P}_0$ of $\mathcal{P}$ are *substituted* by features $\mathcal{P}_1$, $\mathcal{P}_2$, ..., $\mathcal{P}_n$.[2]

---

[2]   The semantics of this goal substitution base on the semantics of tasks in the planning system STRIPS (Lifschitz, 1986).

To handle these rules we use backward chaining and unification. A fact 'f: P_0.' represents a design object with the feature set P_0. The requirement specification s is the initial goal 'g: ?- P_s.'. If the *conflict set* (the set of goal matching rules) contains more than one element, different alternative rules or facts are suitable. As we will see in section 5, the preference of rules from the conflict set is handled by using additional design-specific information.

During the inference process goals are substituted by subgoals of matching rules. Similar to parallel PROLOG, we identify two types of parallelism: OR parallelism for alternative rules or facts and AND parallelism for the subgoals of a rule. In design, OR indicates design alternatives whereas AND shows that a design object can be built by a set of subobjects.

Now, we are able to describe design objects and specifications by facts and goals, and we described an inference mechanism. In the following subsection, we discuss the application of rules within our retrieval approach.

Rule-based consideration of the design and the configuration hierarchy
The design hierarchy can also be exploited by the rule-based retrieval. For example, we can represent a design tool t by a rule 'r_t: P_{out} :- P_{in}.' in the following way: P_{in} contains the precondition (necessary input of t) and P_{out} the expected output of the tool t. Of course, we do not know the exact output but we know which feature sets are influenced by a tool. Using the trace of the inference process, an explanation facility can create a plan which describes the tools which have to run to get the desired data.

As well as the design hierarchy, the configuration hierarchy can be considered by the rule-based approach. For example, it is possible that the configuration of a design object is represented by an abstract rule in which the rule header represents the aggregate and the subgoals of the rule body the parts. As described by definition 21, the rule 'r: P_0 :- P_1, P_2, ..., P_n.' substitutes the goal 'g: ?- P.' by new subgoals 'g_i: ?- P_i^*.' with P_i^* ::= $(P - P_0) \cup P_i$, i = 1 to n. This definition does not only allow the substitution of features. It allows addition, deletion, and forwarding of features and feature sets, respectively. If we consider the subgoal g_i the feature set P_i is added, the feature set P_0 deleted, and the feature set P_i' ::= $P - P_0$ is forwarded to the feature set of the subgoal g_i. An example of a feature set which can be forwarded is the feature set Technology: if we search for the parts of an aggregate, we are interested in parts using the same technology as the aggregate. But if we look for a feature set Size which expresses the size of the artifact, we see that features of this feature set cannot forwarded directly. Therefore, the header of the rule has to be expanded by a construct which prevent forwarding, or the size of the parts of the aggregate has to be considered explicitly. In example 6 of the following section, we will exemplify these relations by using a rule from our prototype implementation.

5. Application in ECAD

So far, we described a formal design model and an indexing model which supports the retrieval of reuse candidates using two different reasoning techniques. Within this section, we demonstrate the application of these models within an existing ECAD system, and we show how the rule-based and the case-based approach can be combined within a retrieval system. To validate our approach, we implemented the prototype design retrieval system **RODEO** (**RODEO** is an acronym for reuse of design objects) which is integrated in the ECAD system PLAYOUT (Zimmermann, 1989). PLAYOUT allows the computer-aided design of very large scale integrated circuits (VLSI circuits). The most important difference of PLAYOUT to other VLSI design systems is the well-developed top-down floorplanning approach. **RODEO** is implemented in C++ and uses the data of the PLAYOUT design database (Siepmann and Zimmermann, 1989). Currently, there are about 5.000 complex design objects within the PLAYOUT database.

FEATURE SETS IN **RODEO**

As described in section 2, only a subset of all available data of a complex design object is considered as feature sets during the retrieval process. Within **RODEO**, we abstract feature sets by classes, called *feature classes*. These data types encapsulate variables, for example the features and their states, and functions, for example the similarity and the fitness function of the corresponding feature set. Examples of **RODEO** feature classes are Function, Technology, Bit-Width, Arity, Size, Aspect Ratio, and RefinementLevel. Using the fitness function of the feature set Function of definition 1, we are able to handle multi-functional integrated circuits:

Example 2 (multi-functional design object): If we search for a multi-functional design object with specification s with $\mathcal{P}_s$ = { 'Function is multiply', 'Function is divide', 'Bit-Width is 32', 'Arity is 2'}, and we find a design object x with the feature set { 'Function is multiply', 'Bit-Width is 32', 'Arity is 2'} then $\text{FIT}_{\text{Function}}$ (s, x) = 1/2 if $\text{SIM}_{\text{Function}}$ ('Function is multiply', 'Function is divide') = 0.

DESIGN HIERARCHY IN **RODEO**

The design hierarchy of the ECAD system PLAYOUT is divided into four different domains: *behavioral domain*, *structural domain*, *floorplan domain*, and *physical domain*. In addition to most other ECAD systems, PLAYOUT performs a floorplanning of the VLSI chips based on estimated shape information.

Example 3 (modeling tools by rules): A tool t which assembles the mask layout (Glasmacher and Zimmermann, 1992) needs a floorplan as input, and it synthesizes mask layout data, i.e. data of the physical domain. Therefore, such a tool can be described by a rule: r_t: {'Domain is physical', 'Size is smaller than x'}:- {'Domain is floorplan', 'Size is smaller than x - 10% tolerance'}. If the current specification is s ::= {'Function is multiplex', 'Domain is physical', 'Bit is 4', 'Arity is 4', 'Size is smaller than 0.3 mm^2'} the new specification s* built by goal substitution using the rule r_t is s* = {'Function is multiplex', 'Domain is floorplan', 'Bit is 4', 'Arity is 4', 'Size is smaller than 0.27 mm^2'}. A tolerance of 10% is introduced because the floorplan based on an area estimation process with a precision of ± 10%.

CONFIGURATION HIERARCHY IN RODEO

The number of hierarchy levels within the configuration hierarchy of PLAYOUT is not fixed. The design tools can be recursively applied to each hierarchy level. The following three examples demonstrate how the configuration hierarchy can be modeled by rules. In example 4, a rule codes the part-of relationship. In example 5, we see how a feature of the requirement specification are passed along the configuration hierarchy. Example 6 shows the relations between a rule header and a rule body considering the feature class Size.

Example 4 (modeling the configuration hierarchy by rules): If we search for a 2^n-bit multiplier, and we know that it can be constructed by a 2^n-bit adder, 2^n-bit register, and a 2^{n+1}-bit shifter, we use the following rule during the retrieval: 'r_1: {'Function is multiply', 'Bit-Width is 2^n', 'Arity is 2'} :- {'Function is add', 'Bit-Width is 2^n','Arity is 2'}, {'Function is store', 'Bit-Width is 2^n', 'Arity is 1'}, {'Function is store and shift-right', 'Bit-Width is 2^{n+1}','Arity is 1'}.'. Explanations and hints how the assembly of the parts should annotate the rule.

Only abstracted informations are coded by the rules. If the complete structure of the aggregate is known, the information of the configuration hierarchy is explicitly modeled by a design object instead of a rule.

Example 5 (forwarding of features): Regarding definition 21, we see that features which are not in $\mathcal{P}_0$ are forwarded to the subgoals. For example, if we use rule r_1 of example 4 and the goal 'g: ?- {'Function is multiply', 'Bit-Width is 16', 'Arity is 2', 'Technology is CMOS'}.', we get 'g_1: ?- {'Function is add', 'Bit-Width is 16', 'Arity is 2', 'Technology is CMOS'}.' as the first subgoal. In this example, the feature 'Technology is CMOS' is forwarded because the parts of the module should have the same technology as the module itself.

Example 6 (assembling of features): Considering the feature set Size, we know that the sum of all areas of the subcircuits should not exceed a given total area. Using this information, the rule r_1 of example 4 is expanded to 'r_1*:

{ 'Function is multiply', 'Bit-Width is 2^n', 'Arity is 2', 'Size is smaller than m' }
:- { 'Function is add', 'Bit-Width is 2^n', 'Arity is 2', 'Size is smaller than m_1' },
{ 'Function is store', 'Bit-Width is 2^n', 'Arity is 1', 'Size is smaller than m_2' },
{ 'Function is store and shift-right', 'Bit-Width is 2^{n+1}', 'Arity is 1', 'Size is smaller than m_3' }, $m \geq m_1 + m_2 + m_3$.'.

CASE BASE OF **RODEO**

Besides design objects, the working memory of the retrieval algorithm contains requirement specifications and rule headers as described above. The requirement specifications are integrated to exploit the knowledge whether a design with a similar specification to the current one has already been tried before. The advantage of the integration of rules is obvious: in addition of finding similar design objects, we allow to decompose the current design problem or to go to a more abstract refinement level within the design hierarchy. Therefore, the domain of x within the reusability function REUSE (s, x) is extended by specifications and rule headers. The representation of design data - among other things design objects, specifications, and rule headers - is frame-based.

RETRIEVAL STRATEGY OF **RODEO**

RODEO is a hybrid retrieval system using the case-based as well as the rule-based approach described in this contribution. Each goal is realized by a lightweight process which looks for design objects, specifications, and rule headers based on the reusability function REUSE defined in definition 17. Currently, this search is realized by a best first strategy but other retrieval strategies are investigated. The user has the possibility to define a threshold value which defines a maximum value for the best first search.

If a lightweight process finds a rule-header, new processes with new subgoals are created as described in definition 21. But the old search process remains active because it searches for alternative objects. As described before, rules consider the design and the configuration hierarchy, respectively. For the design hierarchy, the rules correspond to tool executions, and for the configuration hierarchy rules correspond to assembly steps. Both activities cause costs which can be expressed by a roughly estimated cost factor assigned to each rule. For instance, this cost factor can base on the experiences of previous designs. These factors are used to schedule the lightweight processes: if the estimated costs are low, the new lightweight processes should be prioritized compared to processes which correspond to rules with higher costs. In **RODEO**, these cost values are currently hand-coded but complex formulas which estimate synthesis or assembly costs based on the involved data can be used as well.

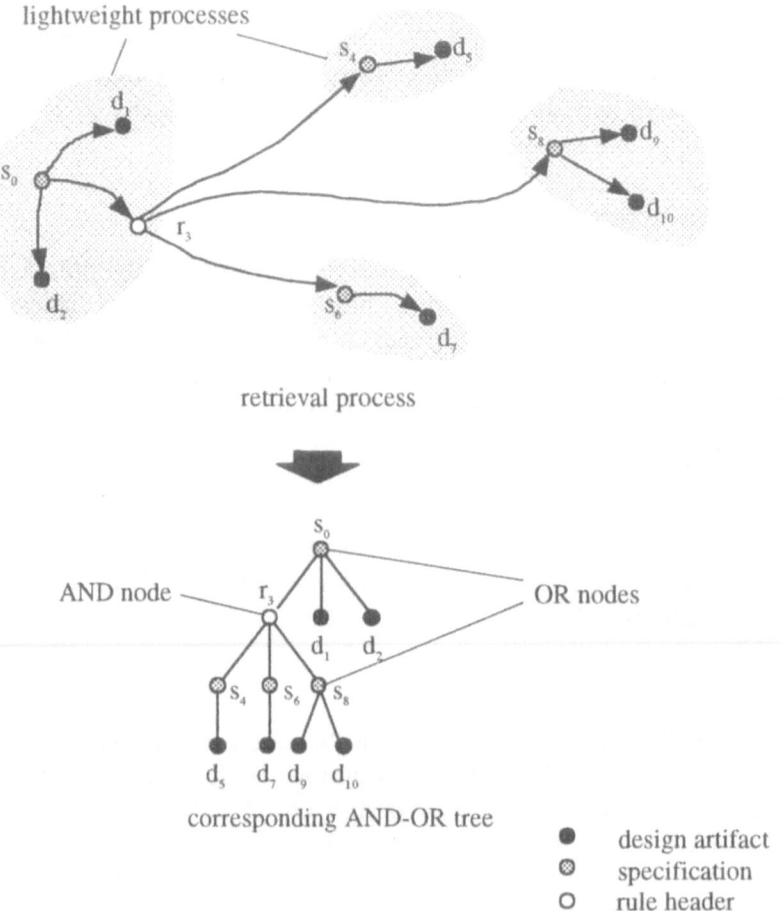

Figure 8: Example of a **RODEO** retrieval process

Figure 8   shows an example of a retrieval process based on the following rule base:

f_1: $\mathcal{P}_1$. f_9: $\mathcal{P}_9$.

f_2: $\mathcal{P}_2$. f_{10}: $\mathcal{P}_{10}$.

f_5: $\mathcal{P}_5$. r_3: $\mathcal{P}_3$:- $\mathcal{P}_4$, $\mathcal{P}_6$, $\mathcal{P}_8$.

f_7: $\mathcal{P}_7$. g_0:   ?- $\mathcal{P}_0$.

The retrieval is shown at an arbitrary state (e.g. when stopped by a threshold value). The design objects d_i are represented by the facts f_i with the feature sets $\mathcal{P}_i$. The goal g_0 represents the initial specification s_0 with the feature set $\mathcal{P}_0$. The retrieval has found two reuse candidates d_1 and d_2 and the header of rule r_3. Therefore, rule r_3 fired. The body of this rule consists of three parts which results in three new lightweight processes with the specifications s_4, s_6, and s_8 using the feature sets $\mathcal{P}_4$, $\mathcal{P}_6$, and $\mathcal{P}_8$. The user now has the following reuse possibilities: trying to adapt d_1 or d_2 or to assemble the sub-objects d_5, d_7, (d_9 or d_{10}) to fulfill the given requirement specification s_0. If he chooses the assembling, he has the possibilities to use d_9 or d_{10} together with d_5 and d_7 to fulfill the requirements. Finally, figure 8 illustrates the inference process, i.e. the transformation of goals into several subgoals, by an equivalent AND-OR tree (Nilsson, 1971; Barr and Feigenbaum, 1981).

As described in section 2, we distinguish three reuse types: reuse by instantiation, reuse by generation, and reuse by adaptation. Since our approach focuses on the third reuse type, the first two are supported automatically. There, we only allow a total matching of the design object with the given specification, i.e. we choose a threshold value of 1 for the reusability function.

The retrieval algorithm of **RODEO** is examined by many experiments. The retrieval times of **RODEO** to find first reuse candidates are lower than one second (CPU time on a HP730 workstation).

6. Conclusions and Future Works

In this contribution, we introduced a formalism describing design processes based on a refinement model. Based on this convergence model, we presented a hybrid, i.e. a case-based and rule-based, approach which supports the retrieval of adequate reuse candidates. Our approach considers most of the design properties listed in section 2. We exemplified and validated our approach by the prototype retrieval system **RODEO** which is integrated into the ECAD system PLAYOUT.

Future works are the integration of complex requirement specifications and design data which currently cannot be compared fast enough by using the feature set similarity function. Another problem is the determination of suitable feature set similarity functions. For example, in ECAD, the similarity function of the feature set Function can be defined by using logical or structural representations (for example truth tables and netlists). However, the change of only one net within a netlist may result in a behavior which bases on a totally different truth table. Finding adequate and domain-dependent similarity functions that overcomes these problems remains an interesting problem of case-based reasoning approaches in different application domains. Another open problem is the exploitation of the feature context. Although the rule-based approach considers this context, our approach cannot cover all dependencies.

References

Allen, J.: 1990, Performance-directed synthesis of VLSI systems, *Proceedings of the IEEE*, **78**(2).

Altmeyer, J., Ohnsorge, S. and Schürmann, B.: 1994, Reuse of design objects in CAD frameworks, *Proceedings of the IEEE/ACM International Conference of Computer Aided Design*, San Jose, California

Barr, A. and Feigenbaum, E. A.: 1981, *The Handbook of Artificial Intelligence*, Volume I, William Kaufmann, Los Altos, California.

Brazier, F. M. T., van Langen P. H. G., Ruttkay, Z. and Treur, J.: 1994, On formal specification of design tasks, *in* J. S. Gero, and F. Sudweeks (eds), *Artificial Intelligence in Design '94*, Kluwer , Dordrecht.

Gajski, D. D. (ed.): 1988, *Silicon Compilation*, Addison-Wesley.

Gero, J. S.: 1990, Design prototypes: a knowledge representation schema for design, *AI Magazine*, **11**(4).

Glasmacher, K. and Zimmermann, G.: 1992, Chip assembly in the PLAYOUT VLSI design system, *Proceedings of the European Design Automation Conference*, Hamburg, Germany.

Katz, R. H.: 1990, Towards a unified framework for version modeling in engineering databases, *ACM Computing Surveys*, **22**(4).

Kolodner, J.: 1993, *Case-Based Reasoning*, Morgan Kaufmann Publishers.

Lifschitz, V.: 1986, On the semantics of STRIPS, *Proceedings of the Workshop on Reasoning about Actions and Plans*, Los Altos.

Nilsson, N.: 1971, *Problem-Solving Methods in Artificial Intelligence*, McGraw-Hill, New York.

Onosato M. and Yoshikawa H.: 1987, A framework on formalization of design objects for inelligent CAD, *Proceedings of IFIP TC 5/WG 5.2 Workshop on Intelligent CAD*, Boston, MA.

Richter, M. M.: 1992, Classification and learning of similarity measures, *Studies in Classification, Data Analysis and Knowledge Organization*, Springer-Verlag, Berlin.

Riesbeck, C. K. and Schank, C. E.: 1989, *Inside Case-Based Reasoning*, Lawrence Erlbaum Associates, Hillsdale, New Jersey.

Schürmann, B., Altmeyer, J. and Schütze, M.: 1994, On modeling top-down VLSI design, *Proceedings of the IEEE/ACM International Conference of Computer Aided Design*, San Jose, California.

Siepmann, E. and Zimmermann, G.: 1989, An object-oriented datamodel for the VLSI design system PLAYOUT, *Proceedings of the 26th Design Automation Conference'89*.

Tversky, A.: 1977, Features of similarity, *Psychological Review*, **84**.

Takeda, H., Veerkamp, P., Tomiyama, T. and Yoshikawa, H.: 1990, Modeling design processes, *AI Magazine*, **11**(4).

Yoshikawa, H.: 1981, General design theory and a CAD system, *in* T. Sata, and E. Warman (eds), *Man-Machine Communication in CAD/CAM*, Proceedings of IFIP WG 5.2/5.3 Working Conference (Tokyo), North-Holland, Amsterdam.

Zimmermann, G.: 1989: PLAYOUT - a hierarchical design system, *in* G. X. Ritter (ed.), *Information Processing 89*, Elsevier Science Publishers B.V.

J. S. Gero and F. Sudweeks (eds), Artificial Intelligence in Design '96, 251-268.

DESIGN RATIONALE AND DESIGN PATTERNS IN REUSABLE SOFTWARE DESIGN

FENIOSKY PEÑA-MORA AND SANJEEV VADHAVKAR
Intelligent Engineering Systems Laboratory
Department of Civil and Environmental Engineering
Massachusetts Institute of Technology, Room 1-253
Cambridge, MA 02139, USA

Abstract. This paper presents an in-progress development of a framework for using design rationale and design patterns for developing reusable software systems. The proposed framework will be used as an integrated design environment for reusable software design, to support collaborative development of software applications by a group of software specialists from a library of building block cases. These goals translate into the effort of exploring the use of Artificial Intelligence in better management of software development and maintenance process by providing faster, less costly, smarter and on-time decisions. The paper details the use of an explicit software development process to capture and disseminate specialized knowledge that augments the description of the cases in a library during the development of software applications by heterogeneous groups. This specialized knowledge constitutes an important part of a software organization's memory, that is, the sharing of information and it's common interpretations as a result of conceiving and implementing the combination of cases from a library when making software design decisions. The importance of preserving and using this specialized knowledge has become apparent with the recent trend of combining both the software development process and product. It has become essential to capture the design rationale to develop and design software systems efficiently and reliably.

1. Introduction

Design of software reuse involves the application of a variety of kinds of knowledge about one software system to another software system in order to reduce both time and cost to develop, run and maintain that software system. The reused knowledge includes concepts such as domain/context knowledge, development experience, design decisions, design history, code and documentation. Until recently most research in providing computer support for software design has focused on issues concerned with the synthesis and development of reusable software components (Smith, 1990). It is now being realized that, effective software reuse requires more than building an easy to browse, well cataloged, convenient software components (Shaw, 1990; Lubars, 1991). Methodologies combining catalogs of standardized software components and corresponding retrieval tools with models that capture and retrieve relevant design rationale need to be formalised. The goal of reuse research should be to establish a software engineering discipline based on such methodologies. Reusable software can best be accomplished with knowledge not only of what the software system is and what it does, but also why it was put together that way and why other approaches were discarded (Kim and Lochovsky, 1989).

The proposed framework will allow the conception, development and testing of a new methodology that adhere to a software development process, allow the recording and easy retrieval of valuable design rationale information and are able to record and present the knowledge gained during the collaboration. To address the issue of retrieval of project information, the framework will use Truth Maintenance Systems (Doyle, 1979), Case-Based Reasoning (Kolodner, 1993) and C4.5 (Quinlan, 1993). To test the framework, software developed from MIT Intelligent Engineering Systems Laboratory as well as Air Traffic Control Software, Satellite Design Software and Hostile Missile Counter Attack Software developed at Charles Stark Draper Laboratory will be used.

The following sections cover in details the ideas put forward to address these challenges. Section 2 provides a survey and comparison of research efforts on the capture of design rationale in various domains. The Design Recommendation and Intent Model (DRIM) is presented in Section 3. In Section 4, design patterns are described in details. The use of DRIM and design patterns for software reusability is explored in Section 5. Finally, a brief presentation on reasoning mechanisms, i.e., Truth Maintenance Systems, Case-Based Reasoning and C4.5, is given in Section 6.

2. Survey of Current Models or Systems that Capture Design Rationale

Any large scale software engineering system involves the expertise and knowledge of numerous software developers, engineers and programmers. A large scale involvement of such a nature results in an interaction of different ideas and views, which invariably leads to a conflict. The conflict arises from one group's lack of information about the previous group's thinking behind accepting or rejecting a particular proposal, i.e. the design rationale is not carried forward as the design process goes on.

Capturing rationale has been a research topic for several decades (Pena-Mora et al., 1995). There have been a number of models and systems developed by researchers in different application areas ranging from discourse (Toumlin, 1958), to engineering design (Garcia and Howard, 1992). Figure 1 shows a classification of these research efforts.

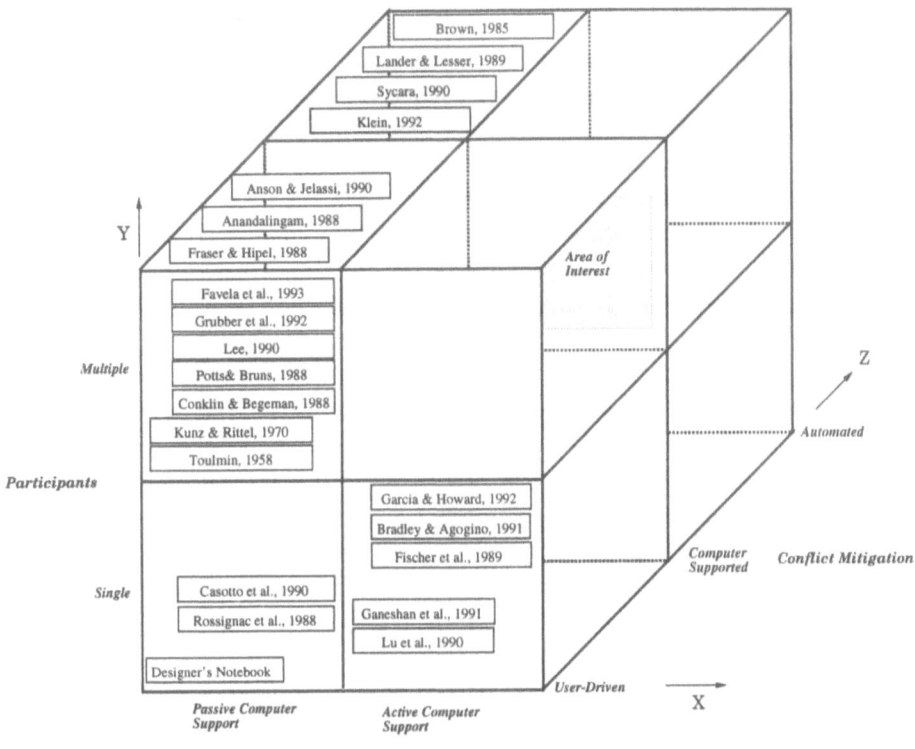

Figure 1. Comparison of design rationale research efforts.

In Figure 1, the **Y** coordinate represents the number of designers who are able to record their interacting rationale and are able to participate in the mitigation of the conflicts. The scale is divided into *single* and *multiple participants*. In other words, this parameter represents how the different models or systems handle different designers close relationship on generating a product. The **X** coordinate represents the computer support for recording and using the rationale for conflict mitigation. The scale is divided into *passive* and *active computer support*. *Passive computer support* indicates that the computer helps the designer to store the rationale. The designer inputs the rationale in the computer and the system creates some links among the different components of the rationale. *Active computer support* indicates that the computer helps in recording the rationale by providing part of it. The **Z** coordinate represents the support provided by the computer during conflict mitigation. The scale is divided into *user-driven, computer supported*, and *automated*. *User-driven* indicates that the user inputs most of the intents (preferences) and recommendations (options) into the system and the computer uses some general strategy like game and bargaining theories to evaluate recommendations with respect to the intents. *Computer supported* indicates that the computer provides some of the intents and recommendations to be analyzed, and it provides some domain dependent knowledge (i.e., heuristics, cases, first principles, etc.) for mitigating the conflicts. Of course, this does not preclude user interaction and application of general strategies, as available in *user-driven* systems. *Automated* indicates that the computer provides solutions to the conflict with very little interaction with the user, where intents and recommendations are implicit in the conflicts and the solutions presented.

The **X** scale in Figure 1 is a continuous measurement with more computer support as the boxes get farther away from the origin. The **Y** scale is discrete and there is no relation among the distances of the boxes to the origin. The **Z** scale is a continuous measurement ranging from mostly user-driven mitigation to mostly computer automated mitigation with a middle balance where an interactive user-computer mitigation is achieved.

Most of the research in design rationale has focused on capturing design rationale without concern for its later use. The use has been limited to maintaining the design history. In that case, the design rationale models or systems fall into the plane *participants-design rationale* without going into the *conflict mitigation* direction.

In Figure 1, the *single participant-passive computer support* quadrant has the designer's notebook which represents the notes taken by the designer during the design process. This document is usually private and manually developed. It also has Rossignac *et al.*'s (1988) MAMOUR and Cassotto *et al.*'s (1990) VOV which keep a trace of the design as it evolves, but leave the design intent implicit in the trace. The idea behind these systems is that a sequence of transformations represents the design and captures some of the designer's intent. Here, the transformations are operations performed on a model, and the sequence of these operations give the final product. Thus, it is believed that by recording that sequence, the product could be reproduced, if needed. One important point is that design rationale is defined as the operations that can re-create the product while intent is believed to be the operations performed. Intents are more than operations. They also refer to objectives to be achieved which are not related to a specific task but to the comparison among design alternatives.

The *multiple participants-passive computer support* quadrant has a series of research efforts from academia and industry: Toulmin's (1958) Model; Kunz and Rittel's (1970) Issue Based Information System (IBIS); Potts and Bruns' (1988) Model; Conklin and Begeman's (1988) Graphical Issue Based Information System (gIBIS); Lee's (1990) Design Representation Language (DRL); Gruber *et al.*'s (1992) SHADE; and Favela *et al.*'s (1993) CADS. It is important to note in this quadrant the ontology used by these systems. Their ontology lacks a representation and a structure for the process and the product as they evolve. Missing is the notion of artifact evolution. Most of them concentrate on the decisions made but without any underlying model of the artifact. The artifact model is important because that is the product developed which connects the design to the physical entity. This in turn guides all the subsequent design decisions. Also missing is the notion of classification of the intents (i.e., objectives, constraints, function, and goals), as well as the classification of the justifications for a proposal (i.e., rules, catalog entry, first principles, etc) since they have different characteristics and are used different by the designers. Section 2 explains in more detail these classifications. In addition, these systems do not really attempt to perform any conflict mitigation. This is due to the lack of structure of the models. It is difficult to assert that an intent can only be satisfied after comparison among different alternatives when there is no control mechanism to enforce that.

Models or systems in the area of conflict mitigation have focused primarily on the resolution of conflicts. To that end, they have provided support in terms of evaluating participants' options (*user-driven* systems) or in terms of providing solution to the conflict based on some domain-dependent knowledge (*Automated* systems). However, they have lacked support in the area of rationale capture, conflict causes, and conflict prevention. In addition, a balance is needed in terms of *user-driven* and *automated* support. Some solutions will be available on domain-dependent knowledge (i.e., heuristics, rules, first principles, etc.). However, some novel solutions will come from the users/designers experience in dealing with similar problems. Thus, support needs to be provided such that both user resolution and computer solution can co-exist.

The *multiple participant-user driven- passive computer support* quadrant has a series of research efforts: Fraser and Hipel's (1988) Conflict Analysis; Anandalingan and Apprey's (1988) use of bi-level linear programming; and Anson and Jelassi's (1990) use of integrative bargaining. These systems take designers' options, evaluate them, and help the designers select the best option. However, the computer does not provide any support in generating some of these options and their accompanying preferences.

There is little or no documentation of research in the *multiple participants-computer supported conflict mitigation-active computer support design rationale* quadrant. However, this quadrant is the one need in order to satisfy the conflict mitigation requirements. Thus, a model and system for capturing the rationale of negotiating participants in which the computer provides support for providing rationale and mitigating the conflicts is necessary.

3. An Overview of DRIM

To capture the design experience in a form that others can use effectively at a later stage and to use the concept of design rationale in a collaborative environment, the Design Recommendation and Intent Model is suggested. The Design Recommendation and Intent Management System provides a method by which

design rationale information from multiple participants can be partially generated, stored and later retrieved by a computer system (Pena-Mora, 1994). It uses
domain knowledge, design experiences from past occasions and interaction with
designers to capture design rationale.

In Figure 2, the Design Recommendation and Intent Model (DRIM) is presented. The DRIM uses the Object-Oriented Modeling Technique (OMT) described
in Rumbaugh and Blaha (1991). The DRIM represents a software designer who
can be either a human expert or a specific computer program. The software designer after negotiating and collaborating with other designers, presents project
proposals based on a design intent. The design intent refers to the objective of
the software project, the constraints involved, the function considered or the goal
of the project. The software designer can present a number of different proposals satisfying a common design intent. The proposals presented can be either a
version of a proposal or completely alternative proposals. A given proposal may
consist of sub-proposals. A proposal may react to an existing proposal by either
supporting, contradicting or changing the ideas put in the existing proposal. A
project proposal includes the designer's recommendation and the justification of
why that particular proposal is recommended. The design recommendation can
either introduce or modify a design intent, a plan or an artifact. When a design
intent is recommended, it refers to more entities that need to be satisfied in order to achieve the design intent. With a plan, more goals are brought into picture.
The artifact denotes the product in a design process. An artifact has both behavioral and structual properties. The artifact comprises the system as well as the
components in the system. Justification explains why the recommendation satisfies the proposed design intent. A justification can be either a rule, e.g. a suggestion from a past experience, or a case, e.g. pertaining to similar performance in an
existing software system, or a catalog, e.g. from a standard library of classes, or
a principle, e.g a set of equations, or a trade-off, i.e the best design considering
trade-off between two constraints, or a constraint-network, e.g. satisfying all the
systems constraint considered in proposing the design intent, or a pareto optimal
surface, e.g the design falls on the surface of best possible design after considering many factors. A justification reacts to other justifications by either suppoort-
ing or contradicting their claims. A context is the data generated during the entire
design process and consists of evidence and assumptions.

4. Design Patterns

Design Patterns can be considered as descriptions of communicating objects and
classes that are customized to solve a general design problem in a particular context. A design pattern names, abstracts and identifies the key aspects of a common design structure that make it useful for creating a reusable object-oriented
design. The term *design patterns* has been established in the lexicon of software

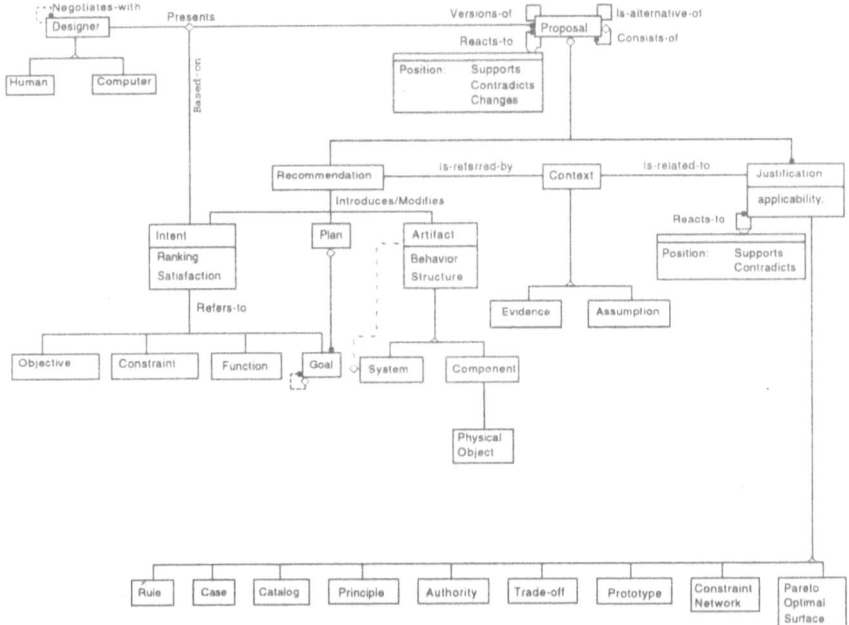

Figure 2. Design recommendation and intent model.

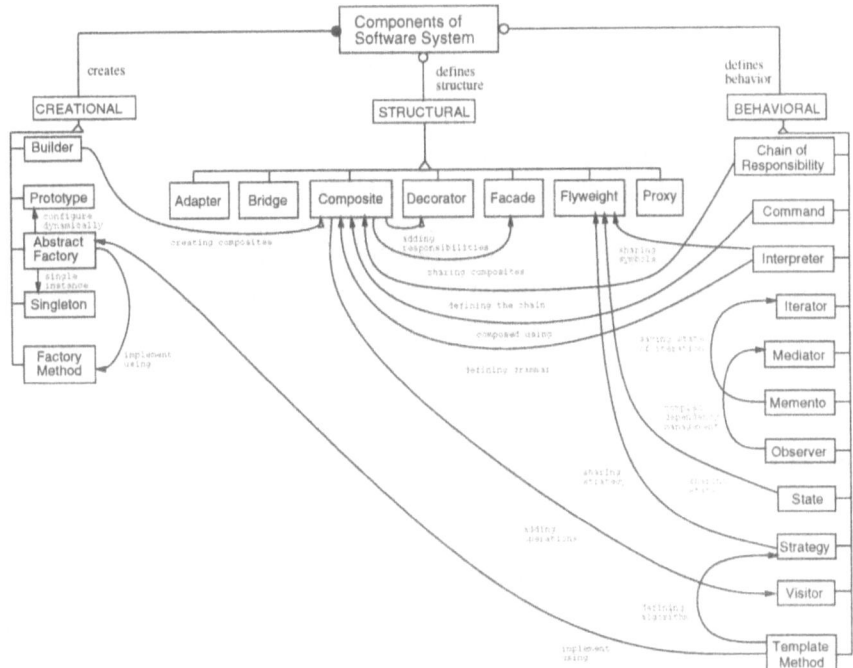

Figure 3. Design patterns relationships.

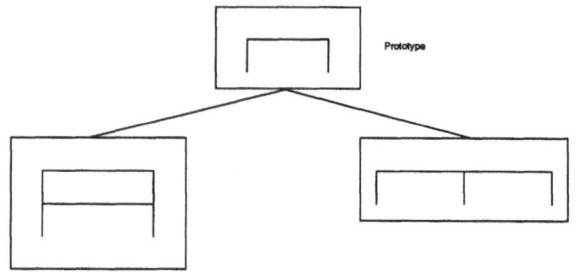

Figure 4. Prototype design pattern for drawing a structural frame.

design by Gamma, Helm, Johnson and Vlissides' (1994) pioneering book Design Patterns: Elements of Reusable Object- Oriented Software. Design pattern can be used as mechanisms for matching information with knowledge from previously developed projects. Design patterns in software can be considered similar to the architectural patterns that exist in building complexes and communities.

Design pattern relationships are shown in Figure 3. The catalog of design patterns contains 23 design patterns. Two criteria have been considered for classifying the design patterns. Design patterns have been classified into three types, based on their purpose or the function they play in object-oriented design. Creational patterns concern the process of object creation. Structural patterns deal with the composition of classes or objects. Behavioral patterns characterize the ways in which classes or objects interact and distribute responsibility. Design patterns can also be classified according to their scope, specifying whether the pattern applies primarily to classes or objects. Class patterns deal with relationships between classes and their subclasses. These relationships are established through inheritance, so they are static. Object patterns deal with object relationships which are dynamic as they can be changed at run-time. Creational design patterns are described in Section 4.1. In Section 4.2, structural design patterns are reviewed. Behavioral design patterns are discussed in Section 4.3.

4.1. CREATIONAL DESIGN PATTERN

The creational design patterns help to make a system independent of how its objects are created, composed and represented. They give a flexibility in what gets created and when it is created.

Five types of creational design patterns have been defined: Prototype, Builder, Abstract Factory, Factory Method, Singleton. *Prototype* specifies the kinds of objects to create using a prototypical instance and creating new objects by copying this prototype. Consider a graphic tool that draws frame structures. As shown in Figure 4, the basic structure of the frame gets repeated successively in some form or the other. The idea is to create a new graphic, by cloning an instance of

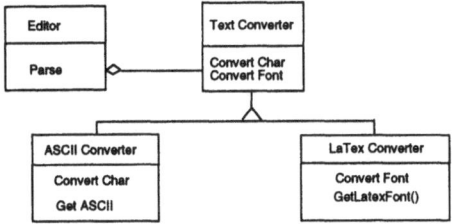

Figure 5. Builder design pattern for a text converter.

a graphic prototype. *Builder* separates the construction of a complex object from it's representation so that the same construction process creates different representations. As shown in Figure 5, the *builder* pattern can be used to convert a text format from a typical document editor, to other formats. The *builder* pattern separates the algorithm for interpreting a textual format from how a converted format gets created and represented. *Abstract Factory* provides an interface to create families of related or dependent objects without specifying their concrete classes. *Abstract factory* pattern can be used in generating windows under different types of operating systems like XWindows and SunView. An abstract base class can be created that defines the interface for creating objects that represent the various parameters such as size, location, font and color of the window. Concrete subclasses implement the interfaces for a specific system. *Singleton* ensures a class has only a single instance and provides a global point of access to it. For example, there may be many windows but only one window manager. The *singleton* pattern ensures that no other instance of the window manager is created and it also provides a way to access the window created by the window manager.

4.2. STRUCTURAL DESIGN PATTERNS

Structural patterns are concerned with how classes and objects are composed to form larger structures. There are two types of structural patterns depending on whether they are applied to objects or classes. Structural class patterns use inheritance to compose interfaces or implementations. Structural object patterns describe ways to compose objects to realize new functionality. The design pattern catalog contains 7 structural patterns. Some of the structural design patterns are described below.

Adapter converts the interface of a class into another interface thereby allowing classes with incompatible interfaces to work together. Adapter design pattern can be used effectively in a drawing application to combine interface elements such as menus, scroll bars with graphic objects such as line, circle, polygon. *Bridge* allows abstraction and implementation to vary individually. Consider the implementation of a window tool kit in different operating systems. An abstract class window can be defined with subclasses for different operating systems. For every kind of window, different subclasses will have to be generated to account for the operating systems. Using the *bridge* pattern, the window abstraction and it's implementation are placed in separate class hierarchies. There is a class hierarchy for window interfaces and a separate hierarchy for operating system specific implementation. *Decorator* attaches an additional responsibility to an object dynamically. For example, a document editor should add properties like borders and scrolling facilities to the user's document. Instead of inheriting these properties, the *decorator* design pattern encloses it as an object and for·vards requests to the object. *Facade* provides an unified interface to a set of interfaces in a subsystem. Clients communicate with a subsystem by sending requests to the facade design pattern, which in turn forwards the requests to the appropriate subsystem objects.

4.3. BEHAVIORAL DESIGN PATTERNS

Behavioral design patterns are concerned with the algorithms and the assignment of responsibilities between objects. Behavioral patterns describe patterns of objects and also patterns of communication between them. Behavioral class patterns use inheritance to distribute behavior between classes. Behavioral object patterns use object composition. The design pattern catalog contains 11 behavioral patterns. Some of the behavioral patterns are described below.

Chain of responsibility decouples the sending and receiving objects by allowing multiple objects a chance to handle the request. The request gets passed along a chain until one of the objects handles it. For example, in a menu driven help system, the user can obtain help on a specific topic. If the information is not available on that topic then the request is passed on to higher levels, until a more general help is available. *Iterator* provides a way to access the elements of an aggregate object sequentially without exposing its underlying representation. The *iterator* pattern provides the means to access a list of employees without going through the internal structure of the list. *Memento* captures and externalizes an object's internal state so that the object can be restored to it's state later. *Memento* design pattern proves useful in implementing undo mechanisms. In such cases, the object can be restored to it's earlier state. *Observer* defines a one-to-many dependency between objects so that when one object changes state, all it's dependents are notified and updated automatically. For example, in a spreadsheet program, when the data is changed, the bar chart and pie chart diagrams change automatically.

5. Combination of Design Patterns and Design Recommendation Intent Model for Software Reusability

Figure 6 shows the entire software development scheme. The *problem domain* includes the requirements of the software system. The *design process* is concerned

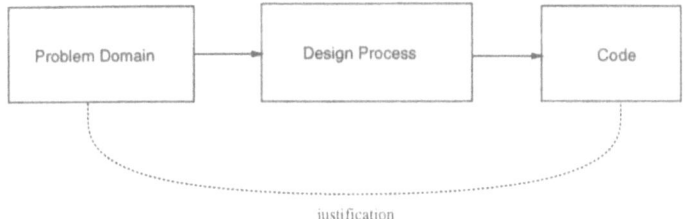

Figure 6. Software development scheme.

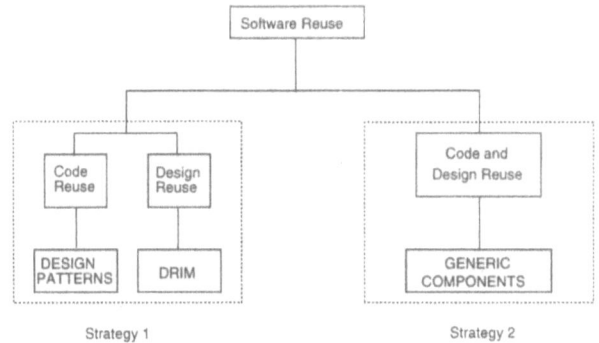

Figure 7. Strategies involved in software reusability.

with the design of the software system from the understanding of the problem to the code generation. The *code* is the final end-product of the *design process*.

As shown in Figure 7, there are two strategies to achieve software reusability. In strategy 1, the code and design reuse components are treated separately. In this strategy, the key issues in the role of software development as described in Coplien (1994) are:

1. Whether design patterns define the components of the *design process*.
2. Whether design patterns define the components in a software development scheme.
3. Whether the design patterns justify that the *code* generated satisfies the requirements of the *problem domain*.

The position of this research is that, the design patterns handle the code reuse, while DRIM captures the design rationale necessary for design reuse. In strategy 2, the code and design reuse components are not separated. Components are defined where all the features necessary for their effective reuse are stored within the component. Features to be included are design rationale, formal specifications and code history.

The following sections cover in details the possibility of using design patterns and DRIM for software reusability (strategy 1). First, an overview of how design

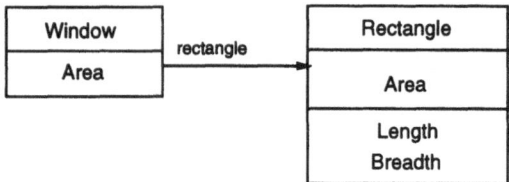

Figure 8. Delegation.

patterns help in code reuse is presented in Section 5.1. In Section 5.2, the use of DRIM in design reuse is documented. The combined design patterns and DRIM model is outlined as a means for achieving software reusability in Section 5.3.

5.1. HOW DESIGN PATTERNS HELP IN PUTTING REUSE MECHANISMS TO WORK

There are basically three types of reuse. Reuse by subclassing is often referred to as **white-box reuse**. The white-box reuse technique uses class inheritance for reusing functionality in object-oriented systems. As an alternative, new functionality can be achieved by assembling or composing objects, i.e. by object composition. This style is known as **black-box reuse**. Most design patterns use **delegation** as a means of achieving reuse. In delegation, two objects are involved in handling a request, a receiving object delegating operations to it's delegates. For example, in Figure 8, instead of making *Window* a class of *Rectangle*, the Window class might reuse the behavior of *Rectangle* by keeping a *Rectangle* instance variable and delegating *Rectangle* specific behavior to the *Window*. That is the area of the *Window* can be obtained by delegating it's area operation to a *Rectangle* instance. The main advantage of delegation is that it is easy to compose behavior at run-time. *Windows* can be made circular, by replacing the *Rectangle* instance with *Circle* instance.

Maximizing reuse lies in anticipating new requirements and changes to an existing model. Design Patterns let some aspects of a model vary independently of other objects. In this way the model is made more robust to a change and redesigning is rendered unnecessary. Some cause of redesigning along with the design patterns that address them are described below.

1. Creating an object by specifying the class explicitly: This commits the designer to a particular implementation, creational design patterns like Abstract Factory, Factory Method and Prototype create object indirectly.
2. Dependence on specific operations: Instead of specifying a particular operation, behavioral design patterns such as Chain of Responsibility and Command, make it more easier the way a request is satisfied. These design patterns either chain the receiving objects or encapsulate a request as an object.

3. Algorithmic dependencies: Algorithms are often extended or replaced during reuse. Objects depending on an algorithm need to be changed when that happens. Creational design patterns such as Builder, avoid that by separating the construction of an object from it's representation. Behavioral patterns such as Strategy, define and encapsulate a family of algorithms. This makes the algorithms vary independently of the clients that use it.

4. Tight coupling: Tight coupling between classes leads to a systems which are hard to reuse in isolation. Design patterns such as Abstract Factory, Facade, Mediator use techniques like abstract coupling and layering to promote loosely coupled systems.

5. Extending functionality: Object composition and delegation offer flexible alternatives to inheritance for combining behavior. Design patterns such as Bridge, Chain of Responsibility and Composite introduce functionality by defining a subclass and composing it's instances with existing ones. Design pattern Observer, extends functionality by defining a one-to-many dependency between objects.

6. Inability to change classes conveniently: In some cases a class cannot be modified because the code is not available or may involve altering many subclasses. Design pattern Adapter, converts the interface of class into the interface that clients want.

While design patterns are useful in utilizing code reuse, they have some limitations for achieving reusability. For instance, design patterns fail to keep a track of what objects have been created and the reason why the objects had to be created, Analysis patterns are not accounted for in the catalog. Reusable software will require user-directed viewing of formal and informal information of the software. However the catalog does not include design patterns for user interface design. Design patterns without explicit documentation fail to provide the software designer with clear requirements and design alternatives that can help in solving the problem and make the reuse effort worthwhile. There is a growing consensus, that simply providing a library of reusable software artifacts is insufficient for supporting software reuse. To make reuse worthwhile, the library of components should be used within well-defined and well-understood domains.

5.2. USING DRIM FOR SOFTWARE REUSABILITY

To capture the design experience in a form that others can use effectively at a later stage and to use the concept of design rationale in a collaborative environment, the Design Recommendation and Intent Model is suggested. The Design Recommendation and Intent Management System provides a method by which design rationale information from multiple participants can be partially generated, stored and later retrieved by a computer system. It uses domain knowledge, design experiences from past occasions and interaction with designers to capture design rationale.

Design Recommendation and Intent Model can be considered as a framework of complimentary classes that make up a reusable design for a specific class of reusable software. Specific objectives in design reuse as described in Bhansali and Nii (1992) are:

1. Identifies good design that maps the solution to the implementation.
2. Explicitly specify how reusable software modules relate to the design.
3. Defines the context in which software components or systems are valid.
4. Explicitly specifies key issues, assumptions, constraints and dependencies in prior designs.

Design Recommendation and Intent Model can be used for supporting the capture, modular structuring and effective access of design rationale information needed for building reusable software.

Design patterns need validation by experience rather than by conventional testing. Design patterns are usually validated by periodic patterns reviews (Schmidt, 1995). DRIM can provide better documentation of such reviews by capturing the strength and weakness of each pattern from past experience. Thus DRIM provides a way to solve the shortcoming of design patterns in capturing the rationale of choosing the objects.

5.3. COMBINED DRIM-DESIGN PATTERNS MODEL FOR SOFTWARE REUSABILITY

Figure 9 represents the combined DRIM-Design Pattern Model. In DRIM, the *artifact* component in a software design context represents the components in a software system. A *software designer* presents a *proposal* which includes a *recommendation* and a *justification*. The *recommendation* introduces or modifies the components in a software system. The design patterns either create these components (Creational Patterns) or define their structure (Structural Patterns) or define their behavior (Behavioral Patterns).

DRIM allows for the explicit capture of design rationale during a software development process. If this design rationale is not captured explicitly, it may be lost over time. This loss deprives the maintenance teams of critical design information, and makes it difficult to motivate strategic design choices to other groups within a project or organization (Schmidt, 1995). Identifying, documenting and reusing useful design patterns requires concrete experience in domain. By capturing the past experience, the combined DRIM-Design Patterns model offers essentially a mechanism to leverage patterns effectively. The combined DRIM-Design Patterns model integrates the concepts of design and code reuse. The combined approach leads to the "patterns-by-intent" approach. The power of design patterns or any library aprroach derives from the reuse of components. The combined

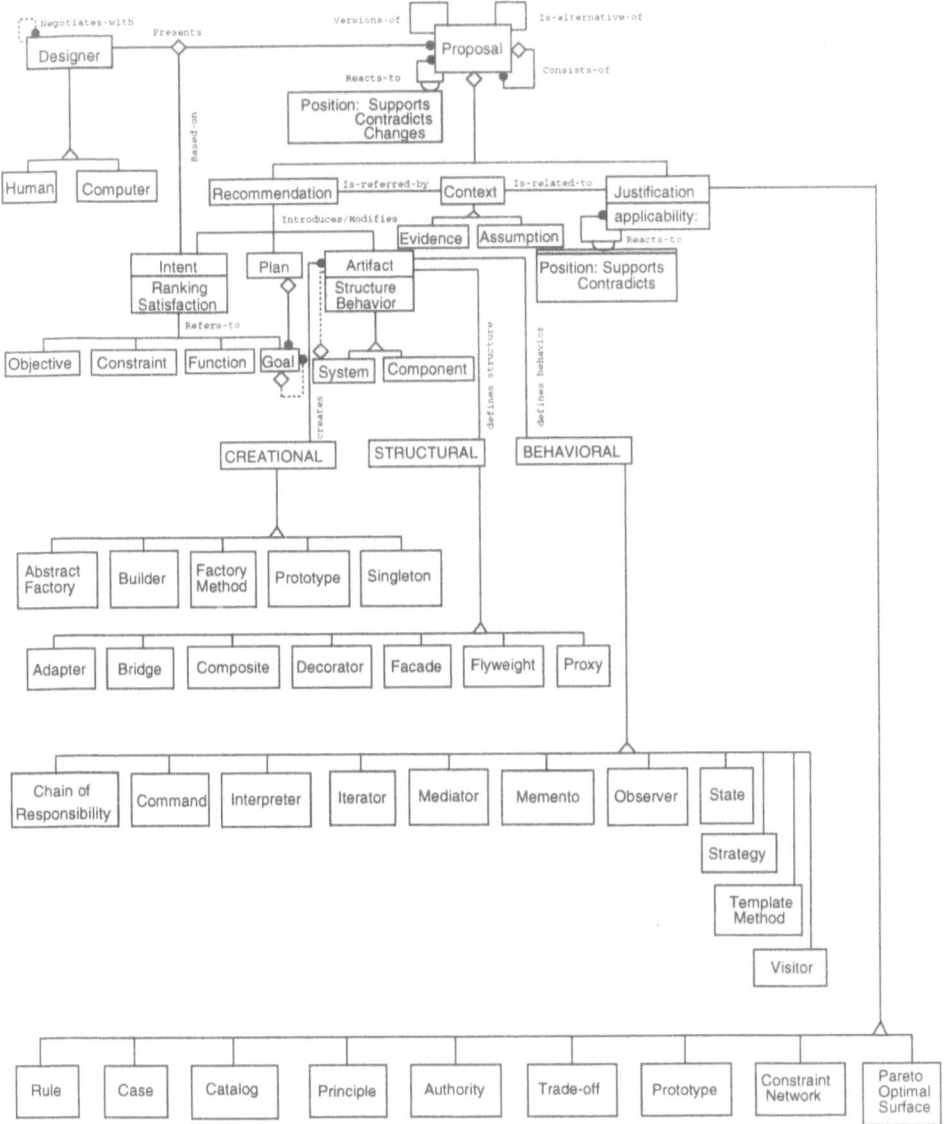

Figure 9. Combined DRIM-Design Patterns Model.

DRIM-Design Patterns model will achieve the same advantages of knowledge reuse and automation, but for a more general class of domains and for multiple modeling purposes.

6. Reasoning Mechanisms

To choose their actions, reasoning programs must be able to make assumptions and subsequently revise their beliefs when discoveries contradict their assumptions. The Truth Maintenance System (TMS) is a problem solver sub-system for performing these functions by recording and maintaining the reasons for program beliefs. Such recorded reasons are useful in constructing explanations of program actions and in guiding the course of actions of a problem solver. The TMS serves as a powerful tool for automated problem solvers and has been used to support several model-based reasoning tasks such as prediction and diagnosis. It provides an efficient mechanism for maintaining consistent set of beliefs and recording the assumptions underlying them. This enables the problem solver to switch rapidly between contexts and compare them. The application of this capability of TMS for designing and planning, needs to be reviewed.

Case-based reasoning is one of the fastest growing areas in the field of knowledge based systems. Case-based reasoning systems are systems that store information about situations in their memory. As new problems arise, similar situations are searched out to help solve these problems. Problems are understood and inferences are made by finding the closest cases in memory, comparing and contrasting the problem with those cases, making inferences based on those comparisons and asking questions when those inferences can't be made. Learning occurs a natural consequence of reasoning where novel procedures applied to problems are indexed in memory.

C4.5 is a machine-learning algorithm that consists of sets of computer programs that examine numerous recorded classifications and construct classification models by discovering and analyzing patterns in these records. An initial decision tree is generated from a set of training cases. After the tree definition, C4.5 creates the production rules and prunes the original decision tree by removing its parts that do not contribute to classification accuracy on unseen cases.

7. Conclusions

The proposed framework combining Design Recommendation and Intent Model with design patterns, offers active assistance to software designers in designing reusable software systems. Although the framework emphasises the importance of documenting the software process, instead of laying extra burden on the code writers, it assists the code writers by providing active computer assistance in recording the key design decisions. The framework acts a software design tool that facilitates software reuse by: 1] Using an excellent library of tested software components. 2] Recording and allowing easy retrieval of decisions made during the software design process. 3] Providing economic methods for systems by providing a context for design modifications when the requirements change over their life time.

The research described herein has significant technological and economical benefits to the modern software design process. The project envisions a paradigm shift from the specify-build-then-maintain life cycle assumed in the past to one of reusable software. Reusable software offers an economic relief to the change activity associated with modern software development wherein costs are incurred disproportionate to the size of the change. By supporting the capture as well as effective access of design rationale information, a strong information base for software understanding can be provided.

Acknowledgments

The authors would like to acknowledge the support received from the Charles Stark Draper Laboratory (CSDL). Funding for this project comes from CSDL award, No. DL-H-4847757.

References

Anandalingam, G. and Apprey, V.: 1988, *Multi-Level Programming and Conflict Resolution in International River Management.*
Anson, R. and Jelassi, M.: 1990, A development framework for computer-supported conflict resolution, *European Journal of Operational Research*, **46**, 181–189.
Bhansali, S. and Nii, H.: 1992, Software design by reusing architectures, *The 2nd International Conference on Artificial Intelligence in Design.*
Coplien, J.: 1994, Progress on patterns: Highlights of PLoP/94, *Proceedings of the Object Expo Europe.*
Casotto, A., Newton, A. and Sangiovanni-Vincentelli, A.: 1990, Design management based on design traces, *27th ACM/IEEE Design Automation Conference*, IEEE, pp. 136–141.
Conklin, J. and Begeman, M.: 1988, gIBIS: A hypertext tool for exploratory policy discussion, *ACM Transactions on Office Information System*, **6**(4), pp. 303–331.
Doyle, J.: 1979, A Truth maintenance system, *Artificial Intelligence*, **12**, 231–272.
Favela, J., Wong, A. and Chakravarthy, A. S.: 1993, Supporting collaborative engineering design, *Engineering with Computers*, **9**(4), 125–132.
Fraser, S.: 1991, Reuse by design—A team approach, *Proceedings of the WISR - 5.*
Fraser, N. and Hipel, K.: 1988, Using the decision maker computer program for analyzing environmental conflicts, *Journal of Environmental Management*, **7**, 213–228.
Gamma, E., Helm, R., Johnson, R. and Vlissides, J.: 1994, *Design Patterns*. Addison Wesley.
Garcia, A. and Howard, H.: 1992, Acquiring design knowledge through design decision justification, *AI EDAM*, 59–71.
Graves, H.: 1991, Lockheed environment for automatic programming, *Proceedings of the 6th Annual Knowledge-Based Software Engineering Conference*, pp. 78–91.
Gruber, T., Tenenbaum, J. and Webber, J.: 1992, Toward a knowledge medium for collaborative product development, *in* J. S. Gero (ed.), *Artificial Intelligence in Design '92*, Kluwer, Dordrecht, pp. 413–432.
Harris, K.: 1991, Increasing reusability through architectural design, *Proceedings of the WISR - 4.*
Hislop, G.: 1994, Evaluating a software reuse tool, *Proc. Third Symposium on Assessment of Quality Software Development Tools*, IEEE, pp. 184–190.
Kim, W. and Lochovsky, F.: 1989, *Object-Oriented Concepts, Databases and Applications*, ACM Press.
Kolodner, J.: 1993, *Case-Based Reasoning*, Morgan-Kaufmann.
Kunz, W. and Rittel, H.: 1970, *Issues as Elements of Information System*

Lee, J.: 1990, SIBYL: A qualitative decision management system, *in* P. Winston and S. Shellard (eds), *Artificial Intelligence at MIT: Expanding Frontiers*, MIT Press, pp. 104–133.

Lubars, M.: 1991, The ROSE-2 strategies for supporting high-level software design reuse, *Automating Software Design*. AAAI/MIT Press.

Pena-Mora, F.: 1994. *Design Rationale for Computer Supported Conflict Mitigation during the Design-Construction Process of Large-Scale Civil Engineering Systems*. ScD Thesis, Massachusetts Institute of Technology.

Pena-Mora, F. and Sriram, R. and Logcher, R.: 1995, Conflict mitigation system for collaborative engineering, *AI EDAM*.

Potts, C. and Bruns, G.: 1988, Recording the reasons for design decisions, *Proceedings of the 10th International Conference on Software Engineering*, IEEE, pp. 418–427.

Pree, W.: 1994, *Design Patterns for Object-Oriented Software Development*, Addison-Wesley.

Quinlan, J.: 1993, *C4.5 Programs for Machine Learning*, Morgan-Kaufmann.

Rossignac, J., Borrel, P. and Nackman, L.: 1988, Interactive design with sequences of parameterized transformations, *Intelligent CAD Systems 2: Implementational Issues*, Springer-Verlag.

Rumbaugh, J. and Blaha, M.: 1991, *Object-Oriented Modeling and Design*, Prentice Hall.

Schmidt, D.: 1995, Experience using design patterns to develop reusable object-oriented communication software, *Communications of the ACM* 38(10), 65–74.

Shaw, M.: 1990, Towards higher-level abstractions for software systems, *Data and Knowledge Engineering*, 5, 119–128.

Smith, D.: 1990, KIDS: A semi-automatic program development system, *IEEE Transactions on Software Engineering*, 16, 1024–1043.

Toumlin, S.: 1958, *The Uses of Arguement*. Cambridge University Press.

J. S. Gero and F. Sudweeks (eds), Artificial Intelligence in Design '96, 269-288.
© 1996 *Kluwer Academic Publishers.*

CONSTRAINT-BASED RETRIEVAL OF ENGINEERING DESIGN CASES

Context as constraints

TANER BILGIC AND MARK S. FOX
Enterprise Integration Laboratory
University of Toronto
Toronto, Ontario M5S 1A4, Canada

Abstract. The case-based retrieval is frequently reported as a valuable tool for engineering design. We discuss similarity based retrieval in the engineering design domain when the context is given as a set of constraints. This approach comprises the lowest level with which we support case-based retrieval from our Integrated Knowledge-Base. The characterization of the retrieval process yields a robust compliance measure and a similarity measure for the cases in a given context. The problematic concept of context is taken up front by making it an explicit part of the query.

1. Introduction

Engineering design involves usage of domain specific technical knowledge together with creative problem solving skills to come up with a properly functioning artifact that complies with a set of requirements — performance goals, physical constraints, etc. It is a creative process that relies heavily upon associations to past experiences and similar designs (Goel, 1994). Consequently case-based reasoning (CBR) (Kolodner, 1993) has been the focus of the design research community (Maher *et al.*, 1995).

Our interest in case-based design arises from the design of complex artifacts for the aerospace industry where design is requirements driven. It is initiated with a "high level" set of requirements, i.e., goals, functional requirements and constraints, that trigger the retrieval of one or more "high level" design cases. These cases are used by engineers to guide their construction of an abstract design that in turn provides a set of requirements for the next design level. Design is therefore a process of successive refinement, when each level iterates between requirements specification/analysis, design case retrieval and design decision-making.

Indexing, case retrieval and case modification are key issues in case-based design. In many case-based reasoning systems, case retrieval is performed based on the *similarity* between the new problem *context* and cases represented in the case memory. An *indexing* scheme which defines the situations under which the new context is similar to the ones in the case-base drives the retrieval process. A case-based retrieval system is effective to the extent its indexing scheme covers *all possible contexts* since similarity is known to change from context to context.

Our work focuses on the indexing/case retrieval problem and leaves the problem of adaptation to the engineer. In particular, given that design is requirements driven, we are interested in how requirements, i.e., goals and constraints, can be used to dynamically retrieve relevant cases from a case library, and how cases in the library should be represented to support this style of dynamic indexing.

The rest of the paper is organized as follows: we review the relevant literature in Section 2 and then propose a view of the design process that is consistent with Fox and Salustri (1994) for one-off, high-tech artifacts. We then discuss what needs to be represented to support this particular view of the design for the *whole design life cycle* in Section 3. Particularly, we mention the framework we are working in, which is a broad scope project to support the concurrent and collaborative engineering design projects using knowledge-based technologies. In Section 4 we elaborate on the requirement-driven retrieval. Our retrieval strategy employs a dynamic indexing mechanism that is based on a compliance measure and resolves the problem of context dependency by providing an explicit representation for the context as a set of constraints. We also define a similarity measure between cases and briefly discuss the properties of the measures defined relevant to case-based retrieval. In Section 5, we give the implementation details and a small example to illustrate the system. We conclude the paper with a summary and further research directions.

2. Previous Related Research

Serrano and Gossard (1988) discuss a constraint-based approach to conceptual design. They build on Serrano's earlier work on constraint management in the context of computational design. Their constraint representation is parameters on nodes and constraints on the arcs of a graph. They discuss graph theoretic methods to handle constraints efficiently.

Sycara and Navinchandra (1992) consider retrieval strategies in a case-based design system. Their representation not only includes physical attributes but *function* and *behaviour* as well. They represent the behaviour as an influence graph where the nodes correspond to parts and the arcs to causal relations between them. The input to their case-based retrieval system is a similar graph which depicts the new design situation. The input is matched to other graphs or parts of graphs stored in the case-base.

Nakatani *et al.* (1992) describe a case-based engineering design support system called SUPPORT which is an interactive system for supporting various phases of engineering design. Their case-based retrieval module uses a three-level representation based on features, functions, and parts hierarchies. They use a constraint-based search to select parts from line-ups.

Maher and Zhang (1993) represent design cases using two indexes: design problem specifications and design solutions. They construct hierarchies of design cases in this manner. Retrieval is done by finding the closest match for given specifications. If a match does not occur at one level in the hierarchy the process is repeated for each of the lower levels.

Wood and Agogino (1993) discuss an architecture to support case-based conceptual design. Their architecture relies heavily on the emerging Internet protocols like WWW and WAIS. They propose to store the design cases in several different multimedia formats (hypertext, CAD drawings, audio, video etc.) and then to search the case-base as guided by the design engineer.

Domeshek *et al.* (1994) discuss MIDAS (a Memory for Initial Design of Aircraft Subsystems). The authors use a repository of *design stories* and discuss ways of creating and indexing those stories. They state that the most developed part of CBR technology is the retrieval. However, the major challenge in the retrieval is building a comprehensive *indexing vocabulary*. They suggest that creating a design story requires two types of information: *presentation* and *connections*.

Maher and Balachandran (1994) explicitly mention the *iterative* nature of the case-based retrieval in the engineering design process. They model the retrieval process as *exploration* rather than a one-shot search. They represent only function, behaviour, and structure for case-based retrieval and propose two index elaboration methods for iterative retrieval.

Kumar and Krishnamoorty (1995) argue that the indexing process is highly *context dependent* and must be carried out for each domain separately.

In the case-based design systems mentioned above the retrieval is implemented as a memory search task and the system's ability is directly proportional to the "richness" of the indexing scheme. One has to foresee and provide indices for most of the query contexts that may arise (i.e. one should be able to abstract apples and oranges as similar in the context of edible items but be able to differentiate between them in the context of fruits). However, we observed that in the process of case-based retrieval:

- a person begins an information interaction with only a vague understanding of the design problem,
- her knowledge, constraints, and goals change over time.

This tends to suggest that, *(i)* case-base retrieval should be *iterative*, and *(ii)* instead of trying to foresee the context in which the retrieval is to be performed, the indexing mechanism has to be *dynamic* and similarity of one case to the context has to be computed on-the-fly.

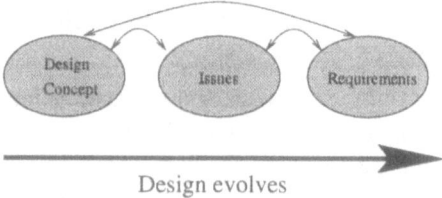

Design evolves

Figure 1. A particular view of design that emphasizes the iterative nature of the process.

3. Case Representation

At the heart of the case-based reasoning paradigm lies case *representation, indexing* and *matching*.

For engineering design purposes, comprehensive case libraries can be built in-house or distributed libraries available on the Internet can be used (e.g. PARTNET (http://part.net/), INDUSTRY NET (http://www.industry.net/)).

We have found the following three concepts crucial in the process of designing one-off, high-tech artifacts (cf. Figure 1):

- *Concepts*: The first thing that the design engineers come up with are concepts which provide a solution to the (design) problem at hand. These can either be competing or complementary design alternatives (e.g. Let's build a remote manipulator arm with six joints to solve the problem).
- *Issues*: Then the design team raises issues and deals with them until a compromise closure is attained. The issues can be from anywhere within the life cycle of the design. There can be issues of risk, cost, schedule, control, stability, manufacturability, quality, fit, form, function etc. (e.g. How are we going to stabilize the system? What is the power consumption of a particular joint? How did we handle the manufacturability issues for past designs? etc.)
- *Requirements*: The design objectives together with design concepts and issues yield functional, structural and performance requirements. Design is satisfaction of these requirements. (e.g. The remote manipulator should have six degrees of freedom. The remote manipulator should be able to handle payloads upto 1000 kg. The shoulder joint should provide inclination and travel to the arm, while elbow and wrist joints should provide travel to the end effector etc.). The requirements are iteratively decomposed and elaborated on (e.g. Providing travel decomposes to provide rotation for all joints.)

Although there seems to be a natural hierarchy between concepts, issues, and requirements, one should bear in mind that the concurrent engineering practices allow for a concept's requirements to be refined, while issues arising from another concept are investigated at the same time.

The knowledge to support the *full life cycle* of this particular view of the design is captured in the TOronto Virtual Enterprise (TOVE) (Fox, 1992; Fox *et al.*, 1993; Fox and Gruninger, 1994). Particularly TOVE *(i)* provides a shared terminology for the enterprise that each agent can jointly understand and use, *(ii)* defines the meaning of each term in a precise and as unambiguous manner as possible, *(iii)* implements the semantics in a set of axioms that will enable TOVE to perform *deductive* query processing to answer "common sense" questions about the enterprise. TOVE represents both generic concepts (time, causality, activity, and constraints) as well as enterprise specific entities (products, requirements, activities, organisation, cost, and quality).

The framework we operate in is a complex engineering design project which requires the services of many engineers and their efficient collaboration. Reusing existing designs, which we address in this paper, is *one of* the objectives of the system we are developing.

The product, parameter, requirement, constraint and function representations in TOVE are closely related to the case-based retrieval of engineering design cases (cf. Figure 2) since TOVE provides a sophisticated representation of the design. In this paper, we do not utilize TOVE's activity, organisation, and cost ontologies.

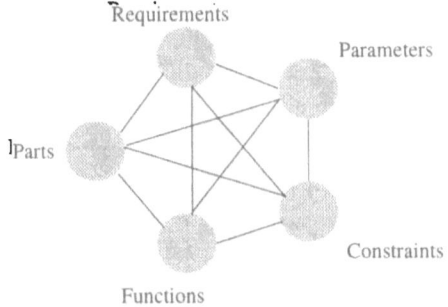

Figure 2. Representations in TOVE that are used to represent engineering design cases.

We adopt the *repository* view of TOVE and consider it as a repository of knowledge relevant for engineering design. We use the term "Integrated Knowledge-Base" to refer to the repository. This view induces a "monolithic" representation of design cases (as contrasted with the "snippet" representation) (Kolodner, 1993, Section 5.4.1) from which sub-cases need to be extracted. We make that extraction via functions. The functional representation is the higher level indexing mechanism of design cases and it complements the dynamic indexing based on requirements that will be discussed in detail in Section 4.

Furthermore, we assume that the case-based retrieval process is *iterative*, a view shared by others (Domeshek *et al.*, 1994; Maher and Balachandran, 1994),

which is consistent with the V-model of systems engineering view (Fox and Salustri, 1994). The user is not confined to any level of representation at any time
and can ask a question of arbitrary generality at will.

We briefly identify what needs to be represented to support case-based retrieval for the particular view of design put forth in this section and then discuss
the details of requirement-driven retrieval in Section 4.

3.1. DESIGN CONCEPTS

In order to reason about *design concepts* one has to consider notions of:

- *fit*: how do the parts of the design fit together?
- *form*: the structure of the design as captured by the parts hierarchy.
- *function*: the intended behaviour of the artifact that is designed as might be
  found in a functional classification of parts.
- *behaviour*: the causal relationships between different parts of the artifact.
- *working principle* hydraulic, electro-mechanic etc.

Therefore any case-based design tool should be able to represent and *reason
on* the above items. TOVE provides explicit representations for fit, form, and function. Behaviour and working principle is simply implemented as a classification
of design parameters. A simplified schematic representation of an engineering
design case is shown in Figure 3.

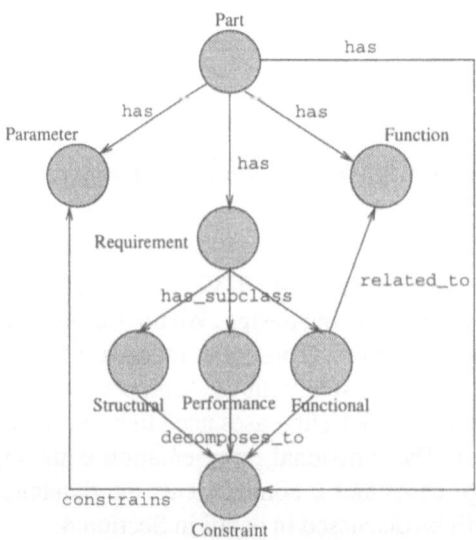

Figure 3. Representation of engineering design cases using TOVE.

Furthermore, each category has its internal classification (e.g. allocated parameters, estimated parameters, actual parameters, basic functions, non-basic functions, unary functions, binary functions etc.).

We briefly mention the representation of functions since that is usually the starting point of case-based retrieval in engineering design domains (Goel and Chandraskaran, 1989; Sycara and Navinchandra, 1989). The functional representation is as described in (Pahl and Beitz, 1988) who identifies five *generally valid* functions (change, vary, connect, channel, and store). These functions take energy, materials, and signals as their arguments. We distinguish user-defined non-basic functions from the generally valid basic functions (e.g. the non-basic function "provide rotation" is related to joints of the manipulator arm as well as to the basic function "connect(energy,matter)"). Generally valid functions are useful when the system cannot retrieve any prototype for a given non-basic function. In that case, if there is another level, the retrieval algorithm moves one level up in the non-basic function hierarchy. If there is still not one item retrieved, the algorithm moves to the related generally valid basic function and retrieves prototypes related to that function with the hope of retrieving something relevant. The retrieved cases are pruned by the designer with respect to their relevance.

3.2. ISSUES

In a sense, issues define the solution context for which more detailed questions can be asked. What issues have been dealt with in the previous cases and how they were dealt with is an important piece of knowledge. Eventually one can discover recurring issues in "similar" situations (e.g. how risk was reduced in a previous project when stabilization issue was raised can be a valuable piece of information in the current context.)

Hence a case-based design tool should be able to represent and reason on issues. Issue-based retrieval is used to retrieve those cases in which the same issues have been dealt with. In the current prototype, issues are simply indexed by their names and no further abstraction is available. The extension of issue-management and categorizing issues lie in out further research agenda.

3.3. REQUIREMENTS

Design is highly requirement-driven in the engineering design domain we are concentrating on. Usually the customer comes in with a set of higher level requirements which get decomposed and elaborated on during the design process and find their way to every detail of the design, usually in the form of a constraint on design parameters.

TOVE supports the requirement management process in various ways. The requirements are elaborated on and decomposed into sub-requirements until they are represented (internally) as *constraints* in the knowledge network (cf. Figure 3).

We do not impose a way of managing requirements but provide a rich representation which can support many requirement management schemes.

We differentiate between functional, structural, and performance requirements. Functional requirements dictate "how" to achieve a desired behaviour whereas performance requirements dictate "how well" a behaviour must be achieved. Structural requirements are usually physical laws that are required for the design to achieve its goals.

Requirements are the basic means to describe the design to our system. The higher level requirements (which are usually functional) retrieve candidate design prototypes which come with their own requirements and constraints. The designer modifies and prioritize the new requirements and continue retrieving in an iterative manner.

The requirement-driven retrieval is elaborated fully in Section 4. An example is given is Section 5.

4. A Characterization of the Requirement Driven Retrieval

In this section we outline the retrieval mechanism in the presence of constraints. The formal treatment in this section should not turn the reader off. What we are saying is really simple: when context is given as a set of constraints, individual cases comply with the context to the degree they satisfy the constraints. The case that satisfies most of the constraints (or more than a predetermined number of constraints) is retrieved. To yield more flexibility, the constraints can be weighed as to their importance, in which case our compliance measure is the weighted average of the number of constraints satisfied. Furthermore, when a case does not contain some attribute mentioned in the constraints, we propose to solve for it with the purpose of aiding the designer in *selecting* an appropriate design case. We do that by allowing to solve for multiple, weighted objectives.

The formalism will allow us to discuss the properties of the compliance measure and the similarity of cases which, we believe, are too important to be overlooked.

Basically we define individual *cases* with finite number of attribute-value pairs and a *retrieval context* with finite number of constraints on the attributes. Then a case satisfies the retrieval context to the degree it satisfies the constraints.

Formally, the situation is as follows: an individual case, S, is assumed to be comprised of a finite list of attribute-value pairs:

$$S = \{\langle a_1, v_1 \rangle, \langle a_2, v_2 \rangle, \cdots, \langle a_k, v_k \rangle\}.$$

Then the *case-base*, CB, is a finite collection of individual cases:

$$CB = \{S_1, S_2, \ldots, S_\ell\}.$$

The context of retrieval is explicitly defined by a set of constraints [1] :

$$X = \{\langle a_1, C_1 \rangle, \langle a_2, C_2 \rangle, \ldots, \langle a_{m'}, C_{m'} \rangle\}.$$

These can either be explicit constraints on the particular design or constraints related to functional or structural requirements as well as constraints from anywhere in the life cycle of the design.

We use the characteristic function, χ, to denote satisfaction of a constraint by a case: for any case S_j and constraint C_i:

$$\chi_i(S_j) \begin{cases} 1 & \text{if } S_j \text{ satisfies } C_i \\ 0 & \text{otherwise} \end{cases}$$

Then we define another relation for a case which satisfies a given *context*:

$$sat(S, X) \text{ iff } \forall C_i \in X, \exists S \in CB, \chi_i(S) = 1.$$

However, this is not flexible enough for retrieval purposes: a case either satisfies a context or not! We are interested in cases which *almost* satisfy the context as well. To achieve this flexibility we can define a measure which shows *how much* a case satisfies a given context:

$$\mu_X(S) = \frac{\sum_{i=1}^{m} \chi_i(S)}{m}.$$

Clearly, $\mu_X(S) \in [0, 1]$ with $\mu_X(S) = 0$ showing no compliance with the given context, $\mu_X(S) = 1$ denoting full compliance and $\mu_X(S) \in (0, 1)$ denoting partial compliance.

Consider the situation where one objective dominates all the others in the sense that if it is not fulfilled then the satisfaction of the rest is not that important (when an envelope objective is not satisfied it really may not matter whether a weight objective is fulfilled or not). This situation is typical in engineering design.

To be able to provide more flexibility to the user in terms of retrieval the constraints can be *weighed* by the user as to their importance.

Therefore the context, X, is now given as:

$$X = \{\langle a_1, w_1 C_1 \rangle, \cdots, \langle a_{m'}, w_{m'} C_{m'} \rangle\}$$

---

[1] Since a single constraint can apply to several attributes the list contains more tuples than the number of constraints. We assume that there are m constraints but the context is given by m' attribute-constraint pairs where $m' \geq m$.

where w_i are the weights which denote importance of each constraint[2]. We assume that weights are positive real numbers. The situation is as depicted in Figure 4.

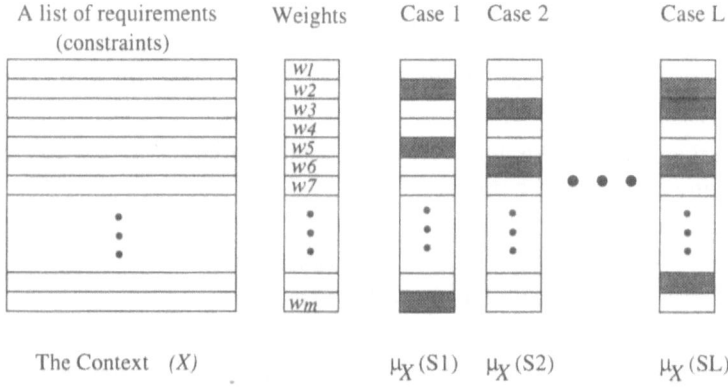

Figure 4. Constraint-based retrieval: case of weighted compliance measure. The shaded areas of the cases represent unconforming parameters.

The retrieval with weighted constraints is based on the extended compliance measure, $\mu_X(S)$, which is given as:

$$\mu_X(S) = \frac{\sum_{i=1}^m w_i \chi_i(S)}{\sum_{i=1}^m w_i}.$$

Some other properties of μ_X are as follows:

- Two $\mu_X(\cdot)$ values are *commensurate* as long as they denote the same context X. On the other hand, μ_X and μ_Y are incommensurate if the relation of X to Y is not known.
- $\mu_X(S)$ is a *summary* measure in the sense that it gives an average compliance measure. It does not tell anything about the *similarity* of one case to another (i.e., two cases that have the same compliance measure can be totally dissimilar simply because they satisfy different constraints but end up satisfying the same number of constraints!).

As far as the retrieval is concerned for a given context X, one can choose the case(s) with:

$$\max_i \{\mu_X(S_i)\}$$

---

[2]There is a major assumption here about the weights from a measurement-theoretic point of view. It has to be the case that the weights attributed by the user must be on a ratio scale (Krantz *et al.*, 1971) (i.e., if the user is assigning 10 and 20 to two different constraints she means not only the latter is a more important constraint but it is *twice* as important.). This cognitive task is usually fulfilled when aided with proper visual tools like sliding scales. This issue is should not be overlooked. If the designer is not able to fulfill the cognitive requirement that the weights are on a ratio scale, the averaging operation (and the retrieval based on it) is simply meaningless!

or cases with compliance measure greater than a user defined threshold:

$$\mu_X(S_i) \geq \tau.$$

However, μ_X is not monotonic as are most of the retrieval measures on which retrieval is based (i.e., if one adds a constraint to context X to construct context Y the relation between μ_X and μ_Y is undetermined).

4.1. RETRIEVAL BY SOLVING CONSTRAINTS

Although the representation of the context by weighted constraints and using the weighted compliance measure for retrieval of cases is a flexible way of retrieving design cases, it is not sufficient to retrieve a case which does not have a particular attribute that the constraint requires. In such a case, the unknown parameter required by the constraint(s) must be *solved for*.

Furthermore, the engineering design usually has *performance requirements* set as goals to achieve.

To account for the two concepts above we extend the definition of a context to include *objectives* to be optimized as well as constraints. Hence, the context, X, is now given as:

$$X = \{\langle a_1, w_1^o O_1 \rangle, \cdots, \langle a_{n'}, w_{n'}^o O_{n'} \rangle, \langle a_1, w_1^c C_1 \rangle, \cdots, \langle a_{m'}, w_{m'}^c C_{m'} \rangle\}$$

where the tuple $\langle a_i, w_i^o O_i \rangle$ (footnote 1 applies here as well) denotes a weighted objective on the attribute (e.g. maximize torque, minimize risk, maximize power output etc.) and w^o and w^c denote weights on objectives and constraints, respectively. Note that we allow for *multiple* objective functions and both the objectives and constraints can be assigned weights by the user.

During the interaction of the designer with the system we are assuming that she defines the context using equations to be optimized with respect to constraints to be satisfied. This is not an unreasonable assumption, since the design proceeds by posting requirements (of which performance requirements are the goals, and structural and functional requirements are constraints) and trying to fulfill them.

This complicates the problem particularly when the constraints are not linear. From an implementation point of view such a system of constraints can be handled either by constraint logic programming techniques (e.g. CLP($\mathcal{R}$) (Jaffar *et al.*, 1992; Holzbaur, 1995)) or mathematical programming techniques (e.g. many implementations of the Simplex algorithm or interior point methods for linear constraints or non-linear search algorithms). In this paper we tackle the case where the objective functions and constraints are linear.

When the the objective function and the constraints are linear the problem can be solved by using numerous Multiple Objective Linear Programming (MOLP) techniques. In fact, multiple criteria optimization techniques have been employed in *detailed* engineering design (Statnikov and Matusov, 1995). We are employing similar techniques for the full life cycle of the design process. The situation with unknowns in constraints and objectives to be optimized is depicted in Figure 5.

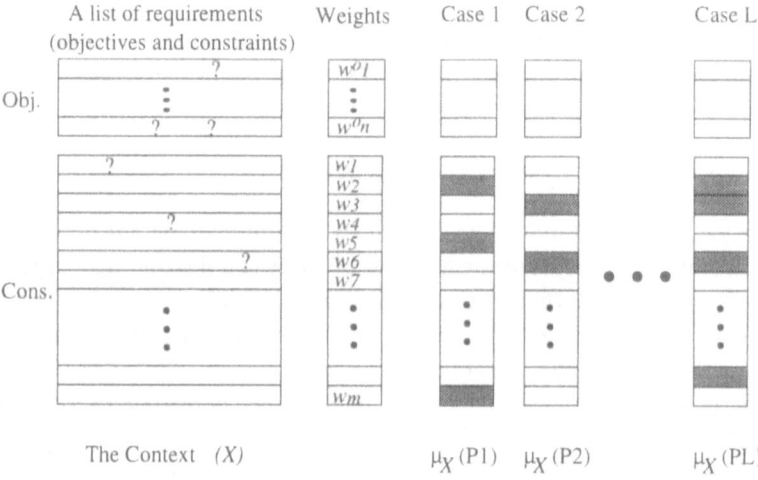

Figure 5. Constraint-based retrieval: the general case. There are multiple objectives to be optimized with respect to the given constraints.

The measure of compliance can be extended to the multiple objective case in a natural manner retaining all the desired properties of the measure for retrieval process:

$$\mu_X(S) = \frac{\sum_{i=1}^{m} w_i^c \chi_i(S) + \sum_{i=1}^{n} w_i^o \chi_i(S)}{\sum_{i=1}^{m} w_i^c + \sum_{i=1}^{n} w_i^o}.$$

The characteristic function for the objectives is evaluated by the designer and only then the compliance measure can be calculated (cf. Section 5).

The retrieval is again based on $\mu_X(S)$ but the process of retrieval requires *solving* a MOLP.

In fact the cognitive task of coming up with weights for objectives and constraints can be quite a hard task in large domains. In order to eliminate the need for an *intrusive* acquisition of weights from the user we assume further generality in the sense that the weights w_i are given as *intervals* rather than a single number: $w_i \in [\ell_i, u_i]$.

When this is the case the problem is still tractable as long as the objectives and the constraints are linear. This formulation gives rise to the family of weighted sum problems (Steuer, 1986).

4.2. SIMILARITY OF CASES

If one is interested in the *similarity* of one case to the other the compliance measure μ_X is useless. For similarity one can define another measure, s_X, which measures the similarity of two cases for a given context X as (for the most general case with weights):

$$s_X(S_1, S_2) = \frac{\sum_{i=1}^{m}[w_i^c\chi_i(S_1)\chi_i(S_2)] + \sum_{i=1}^{n}[w_i^o\chi_i(O_1)\chi_i(O_2)]}{\sum_{i=1}^{m}w_i^c + \sum_{i=1}^{n}w_i^o}.$$

$s_X(S_1, S_2)$ measures the similarity of case S_1 to case S_2 in context X (It simply counts the occurrences where the two cases satisfy the same constraints and normalizes it using the weights). This is a particularly novel definition of similarity since context is explicitly taken care of. Cases that are similar in one context may be totally dissimilar in others. This effect of context on the similarity measure is a well known problem in the research and practice of similarity measures (Tversky, 1977).

As defined here s_X is a *valued relation*[3] (Ovchinnikov, 1991; Bilgiç, 1995). It has the following desired properties:

- $s_X \in [0, 1]$,
- s_X is reflexive (i.e., $\forall S, s_X(S, S) = 1$), a design case is totally similar to itself.
- s_X is symmetric (i.e., $\forall S_i, S_j, s_X(S_i, S_j) = s_X(S_j, S_i)$), if a design case, S_i, is similar to another, S_j, to some extent then S_j must also be similar to S_i to the *same* extent.
- a more robust definition of transitivity holds:

$$\forall S_i, S_j, S_k, s_X(S_i, S_k) \geq \min\{s_X(S_i, S_j), s_X(S_j, S_k)\}.$$

If a design case S_i is similar to S_j to an extent and S_j is similar to S_k then S_i must also be similar to S_k to *some* extent. Note that this reduces to the usual definition of transitivity when $s_X \in \{0, 1\}$.

Therefore s_X is a *bona fide* valued similarity relation. It has a robust transitivity condition which avoids heap paradoxes[4] The transitivity condition of the similarity measure we define will make the retrieval algorithm stop at a point when the similarities of the two items diminish to zero (or when it is under a predefined threshold).

Similarity of cases is important if one needs to index cases according to their similarity and *store it that way*. However this can result in inadvertent effects since similarity is dependent on the context. Therefore computing similarity on-the-fly for a particular context at the time of query seems to be the superior alternative. However note that for retrieval purposes the compliance measure μ_X is sufficient.

---

[3] A valued relation, R, is an extension of the concept of classical relation which takes on either 0 or 1 as its values. Valued relations take on values in the unit interval $[0, 1]$.

[4] A cup of coffee without any sugar and another one with just one grain sugar added are similar in terms of sweetness. The transitivity condition entails that the first cup and a cup with thousands of grains of sugar added are still similar in terms of sweetness.

The valued similarity relation can be a basis for *soft classification* of cases in which every case belongs to a cluster to a degree (for a given context). Such a framework provides flexible retrieval strategies and it reduces to crisp clustering methods similarities are an all-or-nothing matter.

5. Implementation and an Example

In this section, we briefly discuss the implementation of the techniques we described and give a small example.

The implementation of the design (and design case) representation has been carried out in Prolog in an object-oriented manner. An object-oriented layer built on top of Prolog (together with inheritance mechanisms) contains the design representations.

The case retrieval module is also implemented in Prolog. We use ECRC's constraint logic programming system ECLiPSe (Wallace and Veron, 1993) and the CLP(Q,R) constraint solver (Holzbaur, 1995) that comes with it on a Sun SPARC workstation running SunOS version 4.1. The user interface to the system is via World Wide Web (WWW) and it requires a web browser.

We have the representations of several manipulator arms in the system complete with their part, parameter, requirement, constraint, and function hierarchies. We assume that we are faced with the situation of designing another manipulator arm for a totally different task. The aim is to be able to reuse some of the past designs to reduce development costs.

For purposes of illustration we assume that the design has come to a stage where it is decided that another arm with three joints is going to be designed. The designer is faced with the problem of selecting brakes for each joint. Each joint has its own requirements from the brakes to be used (i.e, each joint is a different context for brake selection) and there are several types of brakes used in the past designs.

The functional requirement for a particular joint (context) is represented in the system as shown in Figure 6.

This particular functional requirement (REQ202) is related to two functions that are already defined in the system (FUN30 and FUN31). The representation of FUN30 is shown in Figure 7.

Since the functional requirement is related to the function "stop-motion", a first retrieval on the basis of this requirement retrieves eight parts (design prototypes) that are functionally related to that requirement (cf. Figure 7). In this example, all the prototypes that are retrieved are brakes. It could have been the case that there were other prototypes (e.g. motors with inverse drive capabilities) retrieved for the same function. Pruning of design prototypes can be done by the designer at this moment or the designer may choose to continue, with everything retrieved so far, to requirement-based retrieval where prototypes are evaluated for

```
Class Frame    REQ202

Subclass of:   REQ190
Instances:
Relations:     related-to-function [FUN30,FUN31]

Attributes:    token [slow-stop]
               name [rms.slow-stop.req]
               req-type [functional-req]
               req-title [Functional requirement for slowing and stopping]
               req-id [3.2.7.11b]
               req-documentation [RMS-SG-1944A]
               req-short-description [The artifact should slow down and
               stop the motion produced by the joint motors]
Messages:
```

Figure 6. Representation of the functional requirement in the system.

```
Class Frame    FUN30

Subclass of:
Instance of:   non-basic-function
Relations:     related-to-part [PRT138,PRT139,PRT140,PRT141
               PRT142,PRT143,PRT144,PRT145]
               generalizes-to [FUN3]

Attributes:    name [stop-motion]
Messages:
```

Figure 7. Representation of a function in the system.

their compliance (e.g. motors would have been eliminated from further consideration because they would violate weight and envelope constraints of the context).

In our example, the designer continues with the eight brakes retrieved and
evaluates their compliance for the given context. Since the design solution seems
to be the concept of a brake, the detailed requirements for the brakes can either
be entered explicitly at this point or the requirement tree of one of the retrieved
brakes can be adopted and modified.

There are fourteen requirements for the brake in the current context and the
representation of one of the requirements (REQ142) is shown in Figure 8 as an
example.

The designer weighs each of the fourteen requirements as to their import-

Class Frame	REQ142
Subclass of:	REQ90
Instances:	
Relations:	requirement-of [PRT133,PRT134,PRT135]
	has-expression [CON53]
	has-document ['slip-torque.req.html']
Attributes:	name [brake.slip-torque.req]
	req-derivation-type [derived-req]
	req-title [Breakaway torque of the brake]
	req-id [3.2.1.2.3]
	req-documentation [RMS-SG-1954A]
	req-short-description [The peak torque level required
	to induce brake slip shall not exceed 12- oz-inches]
Messages:	

Figure 8.   Representation of a derived requirement in the system.

ance and starts the requirement-based retrieval. The results of such a transaction is shown in Figure 9.

The results indicate that PRT138 has the maximum compliance for this particular context. However, the designer selects the top four of the retrieved prototypes (PRT138, PRT139, PRT144, PRT143) and adds the following goals (performance requirements) to the system:

− maximize brake actuation life (10), and
− minimize cost (7).

The numbers by the objectives denote their relative importance. The designer submits this new query to the system which solves for actuation life and cost parameters of three parts (PRT144 did not yield a solution to the required parameters due to insufficient information). Then, the designer evaluates the objectives for each part as to their acceptance and the system returns the new compliance measures: (PRT138:0.88, PRT139:0.89, PRT143:0.92).

The designer selects PRT143 as the new brake of the joint with the knowledge that it has to be modified to meet requirements REQ125, REQ128, and REQ142.

The designer also has the option of classifying retrieved items on the basis of their similarity. To illustrate that, assume that the designer wanted to classify the eight brakes retrieved before the introduction of the objectives. Figure 10 shows the clustering of the eight brakes at two levels of similarity, 0.55 and 0.70.

Note that this clustering is only valid for the particular context (joint).

PRT138 (0.971429)	PRT139 (0.914286)
satisfies 13 out of 14 requirements.	satisfies 11 out of 14 requirements.
Requirements that are NOT satisfied:	Requirements that are NOT satisfied:
REQ128	REQ125
	REQ128
	REQ130
PRT140 (0.6)	PRT141 (0.619048)
satisfies 8 out of 14 requirements.	satisfies 8 out of 14 requirements.
Requirements that are NOT satisfied:	Requirements that are NOT satisfied:
REQ123	REQ125
REQ125	REQ128
REQ130	REQ130
REQ137	REQ136
REQ140	REQ137
REQ142	REQ142
PRT142 (0.752381)	PRT143 (0.819048)
satisfies 9 out of 14 requirements.	satisfies 11 out of 14 requirements.
Requirements that are NOT satisfied:	Requirements that are NOT satisfied:
REQ123	REQ125
REQ125	REQ128
REQ128	REQ142
REQ136	
REQ140	
PRT144 (0.914286)	PRT145 (0.628571)
satisfies 11 out of 14 requirements.	satisfies 9 out of 14 requirements.
Requirements that are NOT satisfied:	Requirements that are NOT satisfied:
REQ125	REQ125
REQ128	REQ128
REQ130	REQ138
	REQ139
	REQ141

Figure 9.   Results of requirement-based retrieval.

6. Summary

In this paper we describe a certain view of design for one-off, high-tech artifacts and outline how that design process can be supported in concurrent engineering environments. Our aim is to be able to support all phases of the design life cycle. We briefly mention the type of information we have in TOVE to represent knowledge necessary for the design task. We suggest that fit, form, function, behaviour, working principle, issues and requirements need to be explicitly rep-

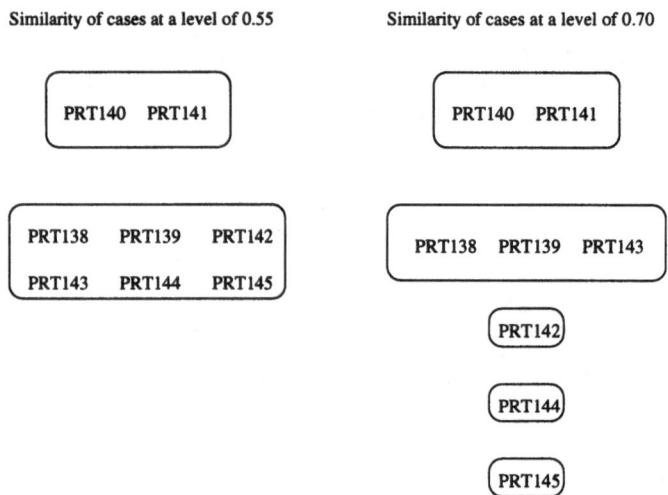

Figure 10. Clustering of the retrieved cases for two similarity levels.

resented. We formalize the retrieval process on the basis of constraints in which we make the constraints, goals, and their appropriate weights an explicit part of the query rather than part of the knowledge-base. This results in a flexible way of retrieving and selecting design cases. This formalization leads to:

— A definition of *context* in terms of constraints on the design. Therefore context becomes an explicit part of the case-based query itself.
— A concept of *compliance* measure, $\mu_X(S)$ for the given context. Each case, S, can be evaluated on the basis of this measure as to its compliance with the context, X.
— An extension to the context such that it not only contains constraints but goals (objectives) to be satisfied as well. This approach is more realistic in the engineering design domain where constraints are decompositions of structural and functional requirements and objectives stem from performance requirements.
— A further extension to the context in the sense that not all objectives and constraints are weighed equal. The designer has the flexibility to choose weights for each objective which denote the importance of that particular objective. The weights can be given as *intervals* if there is any doubt about their validity.
— A definition of similarity of two cases, $s_X(S_i, S_j)$. This measure is based on the context X and can change from one context to the next. (e.g. apples and oranges are not similar in the context of fruits that contain starch but they are similar in the context of edible things).

We are planning on extending the approach provided here in several respects. One immediate concern is the measurement units used in parameters and constraints. A retrieval mechanism should be able to distinguish between different units that are used and should be able to convert from one unit system to another for correct retrieval.

Solving for non-linear objectives and constraints are computationally expensive. Case-based design systems must have well defined protocols to communicate with commercially available symbolic mathematics and detailed engineering design software to care for non-linear objectives and constraints.

Acknowledgments

We would like to thank JinXin Lin for fruitful discussions on an earlier version of this paper. Anonymous referee reports contributed to the clarity of the presentation.

References

Bilgiç, T.: 1995, *Measurement-Theoretic Frameworks for Fuzzy Set Theory with Applications to Preference Modelling*. PhD Thesis, University of Toronto, Department of Industrial Engineering Toronto Ontario M5S 1A4 Canada.

Domeshek, E. A., Herndon, M. F., Bennett, A. W. and Kolodner, J. L.: 1994, Case-based design aid for conceptual design of aircraft subsystems, *Proceedings of the 10th Conference on Artificial Intelligence for Applications*, IEEE, Piscataway, NJ, pp. 63–69.

Fox, M. S. and Gruninger, M.: 1994, Ontologies for enterprise integration, *Proceedings of the 2nd Conference on Cooperative Information Systems*, Toronto, Ontario.

Fox, M. S. and Salustri, F.: 1994, A model for one-off systems engineering, *Proceedings of the AI and Systems Engineering Workshop, AAAI '94*, Seattle, Washington.

Fox, M. S., Chionglo, J. F. and Fadel, F. G.: 1993, A common sense model of the enterprise, *Proceedings of the 2nd Industrial Engineering Research Conference*, Institute for Industrial Engineers, Norcsross, GA, pp. 425–429.

Fox, M. S.: 1992, The TOVE project: Towards a common sense model of the enterprise, *in* C. Petrie (ed.), *Enterprise Integration*. MIT Press, Cambridge, MA.

Goel, A. and Chandraskaran, B.: 1989, Use of device models in adaptation of design cases, In *Proceedingsof the DARPA Workshops on Case-Based Reasoning*, Morgan Kaufmann, New York, pp. 109-120.

Goel, V.: 1994, A comparison of design and nondesign problem spaces, *Artificial Intelligence in Engineering*, 9(1), 53–72.

Holzbaur, C.: 1995, OFAI clp(q,r), manual, edn 1.3.3, *Technical Report TR-95-09*, Austrian Research Institute for Artificial Intelligence, Vienna.

Jaffar, J., Michayov, S., Stuckey, P. and Yap, R.: 1992, The CLP($\mathcal{R}$) language and system, *ACM Transactions on Programming Languages and Systems*, 14(3), 339–395.

Kolodner, J.: 1993, *Case-Based Reasoning*. Morgan Kaufmann, New York.

Krantz, D. H., Luce, R. D., Suppes, P. and Tversky, A.: 1971, *Foundations of Measurement*, Vol. 1, Academic Press, San Diego.

Kumar, H. S. and Krishnamoorty, C. S.: 1995, A framework for case-based reasoning in engineering design. *Artificial Intelligence for Engineering Design, Analysis and Manufacturing (AI EDAM)*, 9, 161–182.

Maher, M. L. and Balachandran, M. B.: 1994, Flexible retrieval strategies for case-based design, *in* J. S. Gero and F. Sudweeks (eds), *Artificial Intelligence in Design '94*, Kluwer, Dordrecht, pp. 163–180.

Maher, M. L. and Zhang, D. M.: 1993, CADSYN: A case-based design process model, *Artificial Intelligence for Engineering Design, Analysis and Manufacturing (AI EDAM)*, 7(2), 97–110.

Maher, M. L., Balachandran, M. B. and Zhang, D. M.: 1995, *Case-Based Reasoning in Design*, Lawrence Erlbaum, Hillsdale, New Jersey, USA.

Nakatani, Y., Tsukiyama, M. and Fukuda, T.: 1992, Engineering design support framework by case-based reasoning, *ISA Transactions*, 31(2), 165–180.

Ovchinnikov, S.: 1991, Similarity relations, fuzzy partitions, and fuzzy orderings, *Fuzzy Sets and Systems*, 40, 107–126

Pahl, G. and Beitz, W.: 1988, *Engineering Design: A systematic approach*, Springer-Verlag, Berlin, trans. A. Pomerans and K. Wallace.

Serrano, D. and Gossard, D.: 1988, Constraint management in MCAE, *in* J. S. Gero (ed.), *Artificial Intelligence in Engineering: Design*, Elsevier/CMP, Boston/Southampton, pp. 217–240.

Statnikov, R. B. and Matusov, J. B.: 1995, *Multicriteria Optimization and Engineering*, Chapman and Hall, New York.

Steuer, R. E.: 1986, *Multiple Criteria Optimization: Theory, Computation, and Application*, John Wiley, New York.

Sycara, K. P. and Navinchandra, D.: 1992, Retrieval strategies in a case-based design system, *in* C. Tong and D. Sriram (eds), *Artificial Intelligence in Engineering Design. Vol. II*, Academic Press, New York, NY, pp. 145–164.

Sycara, K. P. and Navinchandra, D.: 1989, Integrated case-based reasoning and qualitative reasoning in engineering design, *in* J. S. Gero (ed.), *Artificial Intelligence in Design*, CMP/Springer-Verlag, Berlin, pp. 231–250.

Tversky, A.: 1977, Features of similarity, *Psychological Review*, 84(4), 327–352.

Wallace, M. and Veron, A.: 1993, Two problems – two solutions: One system – ECLiPSe. *Proceedings IEE Colloquium on Advanced Software Technologies for Scheduling*, London.

Wood III, W. H. and Agogino, A. M. 1993, A case-based conceptual design information server for concurrent engineering, *Technical Report 93-1104-1*, University of California, Berkeley, Department of Mechanical Engineering, BEST Laboratory, Berkeley , CA 94706, USA.

6

grammars in design

A networks approach for representation and evolution of shape
grammars
Sourav Kundu, Michael Hellgardt
Variable-complexity evolution of shape grammars for
engineering design
Peter J. Gage
Grammars for machine design
Linda C. Schmidt, Jonathan Cagan

J. S. Gero and F. Sudweeks (eds), Artificial Intelligence in Design '96, 291-310.
© 1996 *Kluwer Academic Publishers.*

A NETWORKS APPROACH FOR REPRESENTATION AND EVOLUTION OF SHAPE GRAMMARS

Networks and Shape Grammars

SOURAV KUNDU
Control Engg. Laboratory, Department of Precision Engineering
Tokyo Metropolitan University,1 - 1 Minami Ohsawa, Hachioji Shi
Tokyo 192–03, Japan

AND

MICHAEL HELLGARDT
Architect, Prinsengracht 151
1015 DR Amsterdam, The Netherlands

Abstract. The way in which meaningful shapes are put together to form meaningful designs is the subject of *shape-syntax*, which forms the basis of shape-grammars. In this paper we describe some experiments with a network model to represent and evolve shape grammars. A theory of space configuration in built environment - the *Space-Between* theory is first presented along with aspects of the theory of shape grammars. Experiments, that have demonstrated that the Augmented Transition Network (ATN)-frame is a reliable tool to simulate the idea of space-between and generate real instances of non-bisymmetric Palladio villas via a shape grammar, in the background of cultural expressions, are described. It is also shown how a Genetic Algorithm (GA) model can be used for implementing a directed search through the network, avoiding the combinatorial explosion. The concept of "evolving" shape-grammars is also presented, as opposed to a fixed grammar.

1. Introduction

The common structure of all design machines should provide the basis for a future science of design (Stiny, 1980b) pp. 461.

A series of pioneer-experiments from the late 1970s to early 1980s (for example on Palladio, Frank Lloyd Wright, Terragni, American building-types and others) were based on a *Shape Grammar* formalism developed and defined by Stiny (1980a). This formalism soon became a generally accepted standard. This was a successful, promising and a challenging start but today a certain stagnation can and has been observed, as evident from the comments by many authors among which (Coyne and Snodgrass, 1993) is noteworthy - "AI in design does not appear to be producing useful results. Claims to its success are usually couched in terms of future promise". This applies certainly to shape grammars too, particularly in architecture.

The shape grammar formalism, in some way seems to be too narrow or insufficient to serve as a *common structure* in terms of Stiny's self declared intentions (Stiny, 1980a). This conclusion is precisely the point from where our paper commences. We therefore proceed to construct a formal method to represent shape grammars in a more universal and technical framework, which is the *Networks*. We present the formal mathematical notations of a network and show how this can be used to represent shape grammars and how the *Genetic Algorithm*, which is a restricted but computable model of natural evolution, can be used to evolve these networks and consequently the shape grammars. Issues relating to network based representation technique and the genetic algorithm based search and evolution technique, in the context of cultural expressions, will be discussed in this paper.

But nevertheless we do not claim to provide an exhaustive and consistent idea of a "common structure of all design machines". Instead we will present a proposal for an essentially experimental approach highlighting two main aspects, which seem to be less explored with respect to design mechanisms (design-machines) producing shape configurations:

- Rrepresentation of the design knowledge. Even if we can claim that we understand, to some extent, what knowledge is, or what its model should contain, we still do not have any universal mechanism or concrete framework to represent it in computable terms. Networks seem to be a plausible approach as they can be easily used for distributed storage and handling of data. We represent the rules of a shape grammar using a class of networks. We then take the "Pittsburgh Approach", where knowledge is said to be an implicit quality embedded in the rules, and therefore we attempt to evolve the rules as a result of the interaction of the system with the task environment, to gain the domain specific knowledge. We show how a genetic algorithm model can possibly be used to evolve rules of a shape grammar.
- Design or shape generating formalisms (shape grammars) in the field of architecture. Some experimentation with the translation of a theory of space configuration in built environment - the *Space-Between Theory* - into a formal mechanism, will be presented. This will be used as a reference to discuss aspects of the representation of design knowledge in the light of an application. The target is to give an example of how the development of abstract and applied formal mechanisms can proceed mutually.

Cagan and Mitchell (1993) were the first to combine shape grammars with optimizing search and they showed how shape grammars can be used to model a class of engineering problems to network flow problems in (Cagan and Mitchell, 1994). This paper is aimed not at engineering problems but at questions of how to simulate cultural expressions. We start with a short explanation of that.

2. Shape Grammars

A *Grammar* is a tool to describe and understand language, the faculty of exchanging thought by speaking (or writing) and hearing (or reading). Technical expressions are related to objectifiable goals. They are not necessarily the same as language-expressions in their functioning as carriers of thought. The application of grammar in non-language and above all in technical fields require some clarification of this difference. *Shape Grammar* - as coined predominantly by Stiny (1980a) - is such a field where the borderline between technical and non-technical expressions are insufficiently elaborated, if not ignored. But some minimum basis of theoretical assumptions must be clarified here.

2.1. THEORETICAL ASPECTS

According to Stiny (1980a) a *Shape Grammar* has four components: (**1**) S is a finite set of shapes; (**2**) L is a finite set of symbols; (**3**) R is a finite set of shape rules; and (**4**) I is a labeled shape called the initial shape. In Chomsky (1957), we read:

> *The grammar of L (a language) will be thus a device that generates all of the grammatical sentences of L...* p.13.

Various alternative grammars can apply within a same language and refer to the same vocabulary. For this reason it is more appropriate not to subsume language (a given set of all possible expressions or shapes) under grammar but to associate sets of possible grammars and vocabularies with languages. It was a discovery by Humboldt (1795/96, 1973) that certain grammatical principles must be of universal validity underlying all languages. This, as Chomsky (1965) emphasizes, belongs to the fundamentals of a generative grammar. Furthermore it belongs to the fundamentals of semiology, a science which claims that certain *linguistic universals* apply not only in all languages, but also in all fields of cultural, or "discursive" (not to confuse with technical) expression, such as music, painting, architecture etc. On the other hand universality with respect to all languages and even all fields of expression, is a necessary but not sufficient condition of a grammar. A grammar cannot be defined without reference to a particular language like English or German etc. Consequently the idea of some kind of a universal shape grammar at least implicitly contained in Stiny's shape grammar formalism (Stiny, 1980a) is not tenable. Shape grammars can be approached only on the basis of some theory of shape configuration containing:

- A particular component dealing with a particular field of shape evolution, an architectural period for instance; and
- An universal component dealing with features and mechanisms of universal validity, as for instance space configuration in the built environment.

2.2. READING AND SPEAKING

Grammars in modern linguistics are primarily designed to *read*, or parse sentences and expressions, rather than to *speak*. But Shape Grammars described in the literature are primarily aimed at production systems or "speaking" systems. Yet speaking is no one-way traffic. Fleisher (1992), in his critique of shape grammars seen in the light of linguistics expands on that. He argues that shape grammars are "traveling at random" as long as mutual relations between parsing and generating are ignored. Two questions with respect to the reading aspect and one conclusion with respect to the speaking aspect, which are of basic importance for the approach to work presented here, arise with that:

1. Is it possible to establish as a hypothesis a theory of shape evolution, with which given sets of historical shape-evolution (an architectural or a vernacular building style) can be *parsed* with the aid of possible architectural grammars with an assumption of conventions of configurative mechanisms?
2. Given a positive answer to 1), is it possible to simulate knowledge acquisition, comparable with mechanisms of language learning discussed by Chomsky (1965), as a process resulting in the evolution of new rules and rule-sets in an environment of already given rule-sets and procedures of evaluation. Gero (Gero, 1992; Gero *et al.*, 1994) presents some work in this direction and shows that this is a possibility, though the shape grammar used by them is very rudimentary.
3. (with respect to speaking) The child learns a language through practice, speaking included. We assume that artificial knowledge acquisition—certainly not in the technical fields but cultural, discursive or language expressions—can proceed similarly only when it is related to both parsing and generating expressions.

A reading act results in some meaning. A speaking-, or *speech-act* (Searle, 1969) starts with one. But meanings are not communicated without some knowledge or some reflection or intention about how this meaning may be read and which formal grammatical device may be the most appropriate tool to achieve certain goals of communication. Some conclusions about the opening, the *initial state* of a system producing expressions and its relation to language and grammar result from that. We can think of an initial state of an expression or shape configuration as a speech-act to be initialized within a given language or shape-language, using some kind of grammar or shape-grammar. But we can not think of an initial state of a grammar or shape grammar. An initial state of an expression (a proposition) reflects goals and intentions of an imaginary speaker or designer. It sets a generative grammar in motion. This means that an initial state must at least satisfy three requirements:

1. It contains certain goal parameters, which can be zero ;

2. In a spatial or built environment it must contain certain environmental data which can be zero as well (a tabula rasa); and
3. It is related to a cultural context defined by a variety of alternative ways of how to express meaning and intention, a variety of tools of selectable grammars.

2.3. THE QUESTION OF REPRESENTATION

Graphic representation
We do not know how a shape is represented in the brain, neither do we know how is it linked to perception and reasoning. Stiny (1980b) defines a shape in terms of lines. Some architects certainly do not only design with lines, but also by marking a kind of *hatched patches* representing space partitions emerging with the design process. Probably the mental elaboration of space-units is nearer to a hatched-patches-technique than to line drawing.

Knowledge Representation
Some experimentation with one particular method of a *Network* as a means of representation (to be discussed in sub-section 4.2) has demonstrated that networks are an excellent, if not indispensable, means to represent shape grammars. Partially this was inspired by an example showing how networks are used in a field obviously fundamental for shape grammars - parsing natural language expressions.

2.4. GENERATIVITY AND CREATIVITY

Stiny claimed that the set of expressions (shapes) defining a grammar is finite (see sub-section 2.1). This does not apply in discursive expressions. Language *makes infinite use of finite means*, as in Humboldt (1795/96, 1973). A grammar must describe the processes that make this possible. At least to some extent its rules are *1)* context-sensitive and *2)* applied recursively, possibly creating embedded sub-expressions. Such a grammar is generative, which provokes a remark on creativity: in discursive expressions there is no fundamental difference between generativity and creativity. Any discursive expression, seen in its generative context, is creative even if it is not a new one. Uniqueness of technical inventions is not necessarily the same.

3. Computational Shape Grammars - Palladio Villas

As said in the introduction we do not claim to present an elaborated theory but rather an essentially inductive approach towards one. In this section we present results based on some preliminary experimentation with computational shape grammars, mainly with *Palladio Villas* (Palladio, 1570/1983), using Allegro-Lisp and

CLOS, which will be discussed in more detail in sub-section 4.2. We begin with an enumeration of main components to be subsumed under shape language.

Shape Language:

1. an infinite set of designs (sentences);
2. a vocabulary of room-types (category symbols);
3. a shape grammar (a generative device):

    3.1 a set of *space-between* generators;

    3.2 a set of *extentio* generators;

    3.3 a set of *spatium* generators;

4. Two continuation-functions:

    4.1 controlling the *extensio* generator;

    4.2 controlling the *spatium* generator.

We proceed with some explanation of these components.

3.1. THE *SPACE-BETWEEN* MECHANISM

It is shown that the *space-between* mechanism is at least of relative universality in the built environment (Hellgardt, 1993, 1994). In various texts by Heidegger (1954, 1985) the etymological roots of "Raum" (space) are discussed. In *Building Dwelling Thinking*, (Heidegger, 1954, 1985) this is confronted with *extensio*. An extensio "fills" a quantified section of a Cartesian grid, and it is a solid. (Lovejoy, 1936). *Extensio* seems to be an artificial term (we hardly encounter it in colloquial language) in contradiction to space, the root of which is *Spatium*.

The emphasis of many connotations of the Latin "spatium", lies on movement and duration. In terms of geometric descriptions this can be translated as follows - "space in the function of *spatium* is an empty and to some extent amorphous region, partially defined by surrounding *extensio's*". Translating this in terms of the direction of time of the proceeding of a design, we get the following fragment of a serial computer algorithm:

```
WHILE extensio
   spatium
```

We start with an extensio(n) and we continue in the direction of time with another extensio creating a space-between these two extensio's, to be evaluated as a spatium. This is continued ad infinitum, or until: (1) All possible and feasible evaluable variants of the extensio and spatium functions are evaluated; and (2) A desired configuration or a set of configurations is achieved and that, no more continuation conditions for either extensio's or spatium's are given.

Consequently we have to add continuation functions for extensio and spatium. Start and termination of a project is controlled by continuation-spatium on the basis of a given design-context and design-brief (requirements). The resulting schematic summary of an algorithm (where the | prefix marks arguments) is:

```
define-procedure  space-between-generator  |context  |brief
  WHILE  continuation-spatium  |brief
   WHILE  continuation-extensio  |brief-section
    WHILE  extensio  |brief-section
     WHILE  continuation-spatium  |brief-section
      spatium  |brief-section
```

The space-between mechanism was discovered by one of the authors in his experience in the office of Scharoun, an architect whose work is obviously dominated by the space-between mechanism, documented in texts by Scharoun himself (refer Figure 1).

The space-between mechanism can be observed in anonymous building (medieval town development, Arab building cultures etc.) as well as in professional architecture. The central parts of Palladio villas for instance, can be interpreted as a space, mainly a hall and a loggia, between confining mirrored wings (refer Figure 2). This was a professional assimilation of the wing-centralspace-wing type, dominating Venetian popular building culture (Ackerman, 1977).

Finally, the space-between mechanism appears also, though rather as an undesired by-product, in the literature, specially addressed as: *non-trivial holes* in the *rectangular-dissections* (or LOOS) approach by Flemming *et al.* (1992). These are resulting "space-betweens" actually.

3.2. THE VOCABULARY

The space vocabulary *villa* is largely described in (Palladio, 1570/1983): essentially *chambers*, *halls* and *loggias*. Less mentioned are auxiliary rooms as staircases and a species of by-rooms obviously resulting from floor-plan articulation and serving as niches or fill-blocks (see Figure 2, Pisani). Important in the terms of typology (sub-section 4.3.2) is the degree of universality of such collections of room-types which we call *categories* referring to Chomsky's *category symbols*.

3.3. THE *EXTENSIO* GENERATOR

The wing of Palladio villas can be interpreted as linear arrays of rooms. They can also be read as grid-configurations. This underlies the *Palladian Grammar* by Stiny and Mitchell (1978) which however ignores the space-between structure. This array of rooms is a method of *extensio* which is simple but widely found in building history. The wing-configuration of Palladio villas, however, is not just a simple array but a special kind of an array, geared by Palladio's system of harmonic proportions. This can be simulated fairly easily by a filtered Cartesian product over sets of pairs of $n \times Vicentian$-feet, representing harmonic shaped rooms. This rather peculiar method can be presumably marked as a particular shape grammar mechanism. Various kinds of *extensio*-generators can

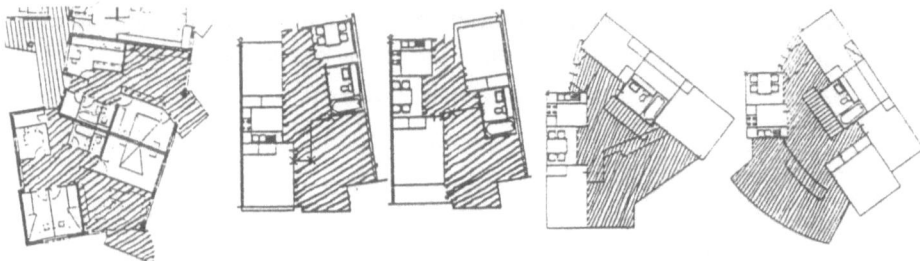

Figure 1. Hans Scharoun, SALUTE estate (Stuttgart), detail and computer generated paraphrases (Hellgardt, 1993).

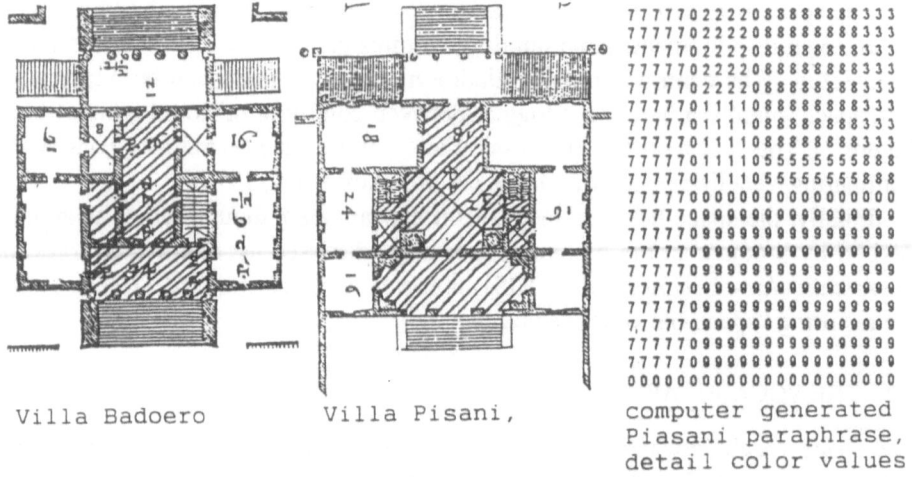

Villa Badoero   Villa Pisani,   computer generated
                                              Piasani paraphrase,
                                              detail color values

Figure 2. Palladio, Space-Between interpretation.

be imagined, and all of the published floor-plan generators can be seen in this light.

Finally, the *space-between* generator itself can serve as an *extensio*-generator. We then get an embedded structure, as mentioned in *2)* of sub-section 2.4 (and also in sub-section 4.3.3). The *spatium*-generator will be discussed in some detail in this paper.

3.4. CONTINUATION FUNCTIONS

Evaluation of available input data: Continuation-extensio controls the elaboration of all brief-sections. If "Continuation-spatium" detects no evaluable space-between, the next extensio brief-section is called.

A final test: A final test may be added to the last evaluated spatium-rule.

Allocation: A temporal coordinate system allocates any additional extensio in an either initial or emerging context.

Selection of possible grammars: Influenced by Heidegger, Scharoun called this *Gestaltanweisung* (shape/gestalt-indication): the assignment of a particular device of shape configuration (a grammar), on the basis of an evaluation of a given context, a kind of reading (mentioned in the opening of section 2), or recognition performance resulting in a strategy of how to proceed in a design process.

3.5. ENCODING RULES OF THE *SPATIUM*-GENERATOR

3.5.1. *Delimitation rules*

Delimitation rules draw a borderline to complete a space between extensios.

Productive substitution rules
A study of all available Palladio Villas confirms sub-section 2.3. Staircases, loggias, etc. can be interpreted as *hatched patches* gradually filling the central space between wings. The corresponding representation and implementation technique obviously is pixel-maps. The space-between generator starts with an *empty* pixel-map. Certain patches of the color-value representing emptiness are substituted then. Translated into a prescription this means that zones or zone-regions have to be specified to be filled entirely or, provided $xy-$ values are given, partially. If no $x-$ or $y-$value is given, the entire corresponding zone or zone-region-section is filled. This very simple and fundamental by definition context-sensitive mechanism imitates the presumably partially subconscious exploration and reading of thoroughly ambiguous contexts in practice; a kind of active space perception. It can be expressed by one single anonymous substitution-rule to be called "on behalf of" various room-categories. If lists of possible parameter-values are provided all rules associated with room-types can be encoded as strings of numbers defining a position in the corresponding list. This possibility intuitively suggests experiments with crossing-over of rules, which in effect produces *new* rules that did not exist before, using a Genetic Algorithm (see section 5).

Evaluative substitution rules
These rules quantify features as topology, surface or shape of an expression (room) created. They are context-free or sensitive. In terms of typology they do not necessarily coincide with productive rules. All rooms must be evaluated, but not necessarily produced, because they can simply result as remaining patches, as in our Palladio simulation the hall (see Figure 4). The selection and concatenation of rules is organized by means of a network, which is probably an indispensable tool which will be discussed in the following section.

4. Networks

Networks are powerful problem representation models. In network search, analysis and optimization, the goal is to search for a path through the network that optimizes certain given objective functions, satisfying the given constraints. Thus each path traversed through a network should have an unique utility measure which will allow us to award a certain fitness measure to that path, to compare it with other (grammatically) feasible paths. Searching for an optimal path through the network can be *depth-first* search or *breadth-first* search, both of which involve blind groping. Another blind way to search for a path through the network is *exhaustive-search*. This method often results in a combinatorial explosion. In this section we describe a formal network notation and then transcribe an Augmented Transition Network (ATN-frame) with this formal notation which is used to simulate all real instances of non-bisymmetric Palladio villas, via a shape grammar. The solution of this network-representation involves search for a feasible path through the network that results in a grammatically correct Palladio Villa. This searching problem can be performed by heuristic search algorithm, which is a Genetic Algorithm (GA), the encoding for which is described in section 5.

4.1. NETWORKS - FORMAL NOTATION

Here we briefly outline a formal mathematical framework adopted from Osyczka (1980) and Osyczka (1984), to encode a network model which can be manipulated using a computer.

Let $S = <X, R, \vec{f}>$ denote a network whose graph $G = <X, R>$ belongs to the class of directed and acyclic graphs, where X is a set of nodes, R is a two argument relation defined on the set X, $\vec{f}$ a vector function defined on the arcs of the graph G. The two argument relation R defines which nodes are connected by arcs and in which direction. For example $x_n R x_m$ denotes that node x_n is connected with node x_m and the direction is from the node x_n to the node x_m.

We assume that the set of all the nodes X can be divided into subsets $X_1, X_2, ..., X_n$ and that the nodes can be connected only between neighboring subsets X_1 with X_2, X_2 with X_3, X_3 with X_4 and so on. Each subset X_n contains no empty set of nodes $X_n = x_n^1, x_n^2, ... x_n^{M_n}$. The two argument relation R can be defined only for neighboring subsets i.e. $x_n^{m_n} R x_{n+1}^{m_{n+1}}$ for $n = 1, 2, ..., N - 1, m_n = 1, 2, ..., M_n, m_{n+1} = 1, 2, ..., M_{n+1}$. The vector function $\vec{f}$ is defined as follows. A vector:

$$\vec{f}\left(x_n^{m_n}, x_{n+1}^{m_{n+1}}\right) = \left[f_1\left(x_n^{m_n}, x_{n+1}^{m_{n+1}}\right), ..., f_i\left(x_n^{m_n}, x_{n+1}^{m_{n+1}}\right), ..., f_I\left(x_n^{m_n}, x_{n+1}^{m_{n+1}}\right)\right]^T$$

$$(1)$$

is associated with any pair of nodes $x_n^{m_n}, x_{n+1}^{m_{n+1}} \in X$ for which $x_n^{m_n} R x_{n+1}^{m_{n+1}}$. The I components of the vector $\vec{f}\left(x_n^{m_n}, x_{n+1}^{m_{n+1}}\right)$ which are given as the *weights* assigned to the arcs connecting the pair of nodes $x_n^{m_n}, x_{n+1}^{m_{n+1}}$ represent the objective functions. To comprehend the matter presented in this paper the following set of nodes would be relevant:

• : Γ_{x_n} - a set of the nodes $x_n \in X$ for which $x_n R x_{n+1}$.

In other words the set Γ_{x_n} contains those nodes for which the arcs lead out of the node x_n. We denote: *1)* $p_j = \{x_1, ..., x_n, ..., x_N\}$ the $j - th$ path in the network joining the node from the set X_1 to the node from set X_N and *2)* $P = \{p_j\}$ a set of all the p_j paths in the networks where $j = 1, 2, ..., J$. The network analysis and search problem of the network $S = < X, R, \vec{f} >$ can now be formulated as: Find the path $p^* = \{x_1^*, ..., x_n^*, ..., x_N^*\}$ in the network S which optimize a vector function : $\vec{f}(p_j) = [f_1(p_j), ..., f_i(p_j), ..., f_I(p_j)]^T$ for a general multiple criteria case. The i-th component of the vector $\vec{f}(p_j)$ is evaluated as follows:

$$f_i(p_j) = \sum_{n=1}^{N-1} f_i(x_n, x_{n+1}) \qquad (2)$$

In other words the search problem is to find a path $p^* = \{x_1^*, ..., x_n^*, ..., x_N^*\}$ for which

$$f_i(p^*) = \min_{p_j \in P} \sum_{n=1}^{N-1} f_i(x_n, x_{n+1}) \ for \ i = 1, 2, ..., I \qquad (3)$$

We assume that all the functions are to be minimized. For a multiple criteria model there is no unique solution which satisfies (3). Thus Pareto optimality concept which is of particular interest in the context of this paper, is introduced. The path $p^* = \{x_1^*, ..., x_n^*, ..., x_N^*\}$ is Pareto optimal if and only if there exists no path $p_j \in P$ such that $f_i(p_j) \le f_i(p^*)$ for $i = 1, 2, ..., I$ with $f_i(p_j) < f_i(p^*)$ for at least one i. This definition is intuitively based upon the fact that the path p^* is chosen as the optimal if no criterion can be improved without worsening at least one other criterion. For most of the network search models, there exists a set of Pareto optimal paths and the problem is to find this set.

4.2. DESCRIPTION OF AN ATN FRAME

When we adopt the space-between-generator described in sub-section 3.1 to the main mechanisms underlying Palladio Villas described in section 3, we get the following reduced version of the algorithm (again the | prefix marks arguments):

```
defin-procedure  palladio-villa  |scale
    FORALL  wing-instances  |scale
      WHILE  continue-spatium-mirror
        spatium
```

scale is a list of $n \times Vicentinian$-feet, and is the input of the wing-configuration function (wing-instances, refer sub-section 3.3). In a realistic simulation at least some hundred add wing-instances are resulting from this function. In the real cases we can observe various underlying scales, which is not important here. The continue-spatium control-function is reduced to the mirroring of the wings in various widths of the space-between. Symmetry is a main feature of all Palladio villas. The bi-symmetric cases have not yet been dealt with.

In the formal notation presented in the sub-section 4.1, we described that the set of all the nodes X can be divided into subsets $X_1, X_2, ..., X_n$ and that the nodes can be connected only between neighboring subsets X_1 with X_2, X_2 with X_3, X_3 with X_4 and so on. Thus $X_1, X_2, ..., X_n$ are the *stages* (or levels) of the network and n is the index of the stage of the network. The notation $x_n^{m_n} R x_{n+1}^{m_{n+1}}$ denotes that node $x_n^{m_n}$ is connected with node $x_{n+1}^{m_{n+1}}$ where n denotes the stage of the network and m_n denotes the index of nodes at stage n of the network. In fact m_n is a pair of indices as described below. There are a total of N stages in the network and the total number of nodes in each stage is M. The function $f_i \left(x_n^{m_n}, x_{n+1}^{m_{n+1}} \right)$ in equation (1) will then refer to $rule_i$ (a rule with index i) that can be applied at node $x_n^{m_n}$, to traverse to node $x_{n+1}^{m_{n+1}}$. The vector function

$$\vec{f} \left(x_n^{m_n}, x_{n+1}^{m_{n+1}} \right)$$

will then be a *rule vector*.

On this background we transcribe the algorithm above into the form of a network $S =< X, R, \vec{f} >$ (refer sub-section 4.1), represented by an array of stages of nodes (set X):

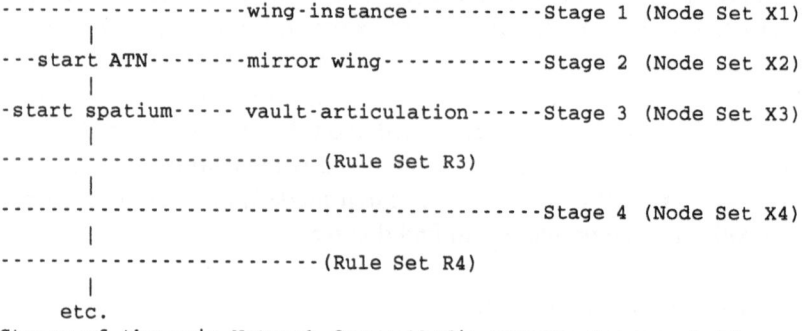

```
-------------------wing-instance-----------Stage 1 (Node Set X1)
        |
---start ATN--------mirror wing------------Stage 2 (Node Set X2)
        |
-start spatium----- vault-articulation------Stage 3 (Node Set X3)
        |
-------------------------------- (Rule Set R3)
        |
-------------------------------------------Stage 4 (Node Set X4)
        |
------------------------- (Rule Set R4)
        |
     etc.
Stages of the main Network for Palladio Grammar Representation
```

At Stage 2 starts an ATN (Augmented Transition Network) with the mirroring of the current wing-instance, thus creating a *space-between*. At Stage 3 starts the spatium-network with a subdivision of the space-between in order to prepare for the articulation of vaults in them: tunnel or cross-vaults, possibly constructed in various ways (defined by *vault rules*). The spatium-network, in our Palladio-interpretation consists of 5 category-levels, corresponding to a Palladio-villa-space-between-vocabulary consisting of 5 categories (halls, loggias, staircases, niches and fill-blocks) with which 5 substitution-rule-sets are associated (substitution-rules are explained in sub-section 3.5). At any node there are two options:

1. either proceed to the next stage in case the previous function is satisfied, or if it isn't;
2. go back to the previous level in order to choose another arc.

Thus the arcs in our network are bi-directional. The nodes of this network are encoded and written in the following form:

```
define-node mirror-wing
       start vault-articulation mirror(1) wing
       ....
       start vault-articulation mirror(n) wing

define-node vault-articulation
       continue room-category(c) <vault-articulations>
       ....
       continue room-category(c) <vault-articulations>

define-node room-category(c)
       continue room-category(c) <substitution-rule-set(r)>
       ....
       continue room-category(c) <substitution-rule-set(r)>
       resume <substitution-rule-set(1)>
       ....
       resume <substitution-rule-set(n)>
```

Here (c) marks the c-th category-node, and (r) the r-th substitution- rule-set. The method underlying this notation and its implementation in Lisp is described by Graham (1994) with the demonstration of the functioning of an ATN (a grammar), parsing simple natural language expressions.

Define-node defines a macro with a name of a node (x) and a body of code of outgoing arcs ($x_n^{mn} R x_{n+1}^{mn+1}$). When a node is defined these arcs are macro-expanded, that is, translated into the form defined by the (macro) definitions of their first expressions which are functions f (refer sub-section 4.1), here: the start, continue and resume-functions. The arcs evaluate a register of information stored away as closures with the proceeding of the network. The ATN starts with a start function with the argument of the first register, here: the result of the mirroring of the current wing input. Continue-arcs are of the form: $f(g(x))$, where x is the current register, g the current function to be evaluated (a substitution-rule from the given rule-set) and f the next function to be evaluated, the next node. The terminal arcs are the resume-arcs, evaluating all substitution-rules as possible final operations. For the ease of clarity the resume-arcs can be subsumed under one 'coda'-node. This is shown in Figure 3.

The ATN-arc notation above shows that within the spatium part of the network the vector function $\vec{f}$ described above assigns: (1) the next node to address and (2) a rule to be chosen from the rule set in the arc of the current node. The general form of this is:

$$f(n, node - index, rule - index)$$

with n as the stage index and thus x_n^{mn} corresponding to $x_n^{node-index,rule-index}$.

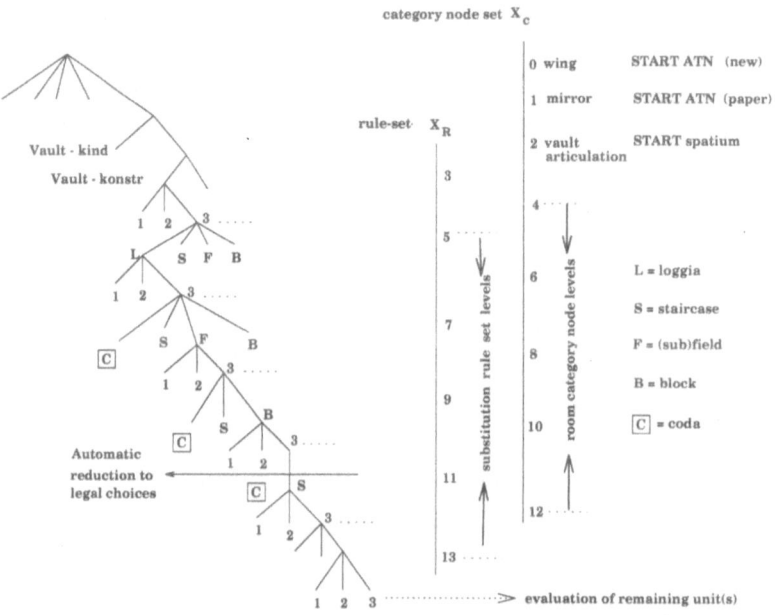

Figure 3. The evaluation tree.

Figure 3 shows how one fictitious path (L-3, F-3, B-3, S-3) traverses through the network. This figure unfolds visually, the interplay between these node and rule-indices, which correspond to the interplay between the selection of sequences of (room) categories and of the rules. Figure 4 shows the visual surface (not as pixel but as line representation, for reasons of screen printing) of three examples of real *successful* paths. The cases 3) and 4) demonstrate that the inversion of sequences, here of loggia and staircase, can result in different appearances. Such expressions can be compared with structural homonymities discussed in language theory (Chomsky, 1957) pp. 28. Case *1* shows the working on an empty rule, the one of L2 (*loggia*$_2$): under particular contextual constraints, no loggia is required within the the central villa-corpus, it might be added outside in some way.

ATNs can be run for undirected exhaustive or various ways of directed search. One way of directed search is that when a node is called, its first arc is chosen to be evaluated and the remaining (macro-expanded) arcs are stored as closures in a paths-list. These can be addressed to be evaluated by a fail-operator which can be called as long as no feasible path, or no path satisfying some objective function is found. In case of exhaustive search (refer also sub-section 4.3.4) the fill-operator subsequently calls all available arcs. Another way of directed search is that a choose-operator, which is addressed when a node is called, assigns an arc and a rule within it to be chosen, using node and rule-indices as described above. These pairs of indices are to be extracted from the result of the heuristically direc-

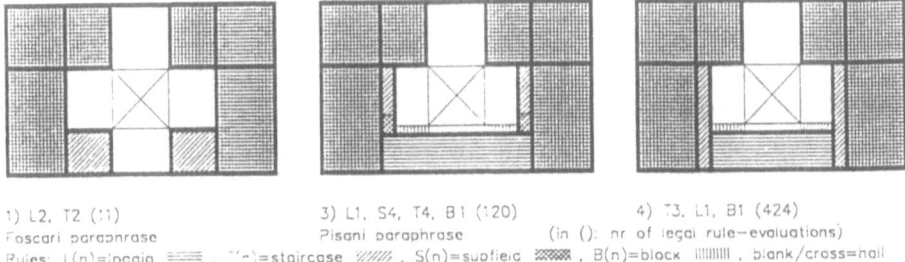

1) L2, T2 (11)
Foscari paraphrase

3) L1, S4, T4, B1 (120)
Pisani paraphrase

4) T3, L1, B1 (424)
(in (): nr of legal rule—evaluations)

Rules: L(n)=loggia ▦ , (n)=staircase ▨ , S(n)=subfield ▦ , B(n)=block ▥ , blank/cross=nall

Figure 4. Simulating Palladio villas - 3 of 5 syntactically correct, but not equally ranking instances, resulting from exhaustive search for 1 out of 208 wing-instance in 1 mirror-width and vault-articulation

ted search mechanism, for instance the Genetic Algorithm (GA) to be described in section 5.

4.3. DISCUSSION PART 1: ASPECTS OF KNOWLEDGE REPRESENTATION

4.3.1. *ATNs as frames of knowledge*

Neither shape grammars nor networks are just tools to model technical problems (refer section 1). ATNs, which are independent from particular grammars, but designed to represent grammatical principles, can synthesize chunks of knowledge for special scientific or practical aims. In the notation above they are transparent enough to serve as a flexible tool for the definition of grammatical structures as for their experimental adjustment.

4.3.2. *Typification*

No genesis of design can be explained beyond typification. The *hatched-patches* mechanism described in subsection 3.5.2.1 is obviously a universal, partially subconsciously working one, so is the space-between mechanism and the type underlying Palladio villas (see subsection 3.1). Both the Palladio as the Scharoun paraphrases (Figures 1-3) can be produced with these tools, the same generator and the same rule-mechanism, using the same ATN- frame. Palladio-specific was his consciously applied method of "spatial chords" (see sub-section 3.3). Scharoun-specific, also provably consciously, was his way to generate space-betweens in inclined and "deconstructed" environments. Culturally determined design obviously combines subconsciously and intentionally working performances. It has a corresponding profile in terms of typification. Using object systems, ATNs allow extensive recourse to typologies. Particular sequences and types of operations and material to be explored can be programmed connectedly. All rule-evaluations within arcs create instances of category (super)classes and are linked to them as a kind of intelligent objects by the virtue of knowledge stored in them. Connectedly all rule operations are related by means of method-inheritance to features of the objects they are transforming (in our case for example, special substitution rules are provided for corresponding *wing-types* - Columns, L or Grid fragmented-shaped). The link to typification is ensured by inbuilt prescriptions and procedures of the ATN described.

4.3.3. *Recursion*

ATNs can address sub-networks, or shape-configurations within shape-configurations. Until now we have experimented with embedded structures only on a primitive level: a staircase can be inserted in a subfield, possibly leaving an unused niche (case 3, Figure 4). For more complicated structures it is presumably necessary to define special continue-subfield functions identifying subfields in a given space-between configuration. The way how sub-networks are addressed in language (ATN) parsers will be a major reference for that.

4.3.4. *Exhaustive search*

The system has been applied on the level of exhaustive search to simulate Scharoun (Figure 1), to elaborate on which is impossible here, and Palladio (Figure 4). Random search has been applied too and it arbitrarily comes up with the same result, though very inefficiently. The price of exhaustive search is a combinatorial explosion. But enabling visual control is - particularly when related to strategies of typology - a useful, if not indispensable tool in the field of cultural expression where we cannot trust our abstract assumptions, and where quantifiable standards of evaluation do not simply correspond to technical *problem-solving*.

5. Genetic Search and Evolution

While using the shape grammar paradigm within a shape language for generating architectural designs (speaking aspect) or for parsing architectural expressions (reading aspect), we are restricted by the choice of a *fixed* grammar. Any design activity, has two principal components that the designer consciously or subconsciously performs. These are *learning* and *reasoning*. Reasoning is done via previously learned design knowledge and the inherent belief system in the designer, and learning is done via a feedback from the design output. If the design is what we might call a *routine* or technically oriented design task, learning will involve finding a feasible, perhaps optimal, sequence of shape-rule applications resulting in (shape) syntactically correct structures. This is equivalent to searching for paths satisfying Pareto Optimality condition outlined in sub-section 4.1. In technically oriented design we might call this as an equivalent to search for an optimal path through the network. This will eventually result in a desired shape configuration (final design). The problem here is one of search. A Genetic Algorithm (GA) can be easily used to perform a directed search and avoid the combinatorial explosion.

In the cases of *non-routine* or culturally oriented design, a fixed grammar often limits expressions or ability of the designer. In that case learning will involve "restructuring" of the design knowledge embedded in the shape grammar itself, or in other words, evolution of the shape grammar (Gero *et al.*, 1994). We thus come to the aspect of - evolving a shape grammars - represented by means of a network. An approach - called the Pittsburgh approach (De Jong, 1987) - can be used to evolve the shape rules. In this case, the individual subjected to evolution (a shape grammar) encodes a whole tentative solution to the learning problem, such as a whole set of shape rules of one particular shape grammar, or a set of paths through the network. The feedback to be used for learning is then some evaluation of the performance of the intermediate solution produced by the shape rules, with respect to the goal of the learning task. The outcome of such evolution task would be new shape grammar rules, or new paths (set of arcs) in the network which never existed before. Thus we discover new shape grammar which helps us to either generate or parse new architectural designs. We explain the computational evolution model (the GA) in the following.

5.1. GENETIC ALGORITHMS

Genetic Algorithms (GAs) (Holland, 1975) are distinguished by their parallel investigation of several areas of a search space simultaneously by manipulating a *population*, members of which are coded problem solutions. The task environment for these applications, is modeled as an exclusive evaluation function which, in most cases is called a *fitness function* that maps an individual of the population into a real scalar. The motivational idea behind GAs is natural selection. Randomized genetic operators like *selection, crossover* and *mutation* are implemented to emulate the process of natural evolution. A population of "organisms" (usually represented as bit strings) is modified by the probabilistic application of the genetic operators from one generation to the next. A more detailed explanation of the theory and working of the GA can be found in Goldberg (1989).

5.2. ENCODING NETWORKS FOR GENETIC ALGORITHM APPLICATION

In this sub-section we describe a technique using which, the mathematical network formulation described in sections 4.1 and 4.2 can be easily transformed into a coding, that is compatible with the genetic algorithm evolution model. We can also have the matrix notation of the network $S = < X, R, \vec{f} >$. For the i-th objective value (rule evaluation value) and for the arcs which connect the nodes from the set $X_n = \{x_n^1, x_n^2, ..., x_{n+1}^{M_n}\}$ to the set $X_{n+1} = \{x_{n+1}^1, x_{n+1}^2, ..., x_{n+1}^{M_{n+1}}\}$ we have:

$$\vec{F}_{in} = \begin{bmatrix} f_i(x_n^1, x_{n+1}^1) & f_i(x_n^1, x_{n+1}^2) & \cdots & f_i(x_n^1, x_{n+1}^{M_{n+1}}) \\ f_i(x_n^2, x_{n+1}^1) & f_i(x_n^2, x_{n+1}^2) & \cdots & f_i(x_n^2, x_{n+1}^{M_{n+1}}) \\ \cdots & & & \\ f_i(x_n^{M_n}, x_{n+1}^1) & f_i(x_n^{M_n}, x_{n+1}^2) & \cdots & f_i(x_n^{M_n}, x_{n+1}^{M_{n+1}}) \end{bmatrix} \quad (4)$$

The full description of the network is obtained when we write the matrices $\vec{F}_{in}$ for $i = 1, 2, ..., I$ and $n = 1, 2, ..., N - 1$. In the matrix notation if the node x_n^{mn}

is not connected with the node $x_{n+1}^{m_{n+1}}$, then the value of $f_i(x_n^{m_n}, x_{n+1}^{m_{n+1}}) = A$, for $n = 1, 2, ..., N - 1$, where A is an arbitrarily chosen great number so the path going through these nodes cannot be taken as the feasible one in the shape grammar search and evolution task. In the process of evolution some new arcs are generated while some exiting arcs perish (get the value A). Thus giving rise to new shape grammar rules.

Consider the network presented in Figure 5, which we may consider as a part of a reduced version of the ATN, described in sub-section 4.2, connecting only rule-sets. The two numbers designated to each arc of the network are the example fictitious values resulting from a rule evaluation. (a bi-criterion function $\vec{f}(x_n^{m_n}, x_{n+1}^{m_{n+1}})$ as described in sub-section 4.1). This means they are the evaluation values contributing to the fitness that is awarded to a shape-grammar if that particular arc is traversed. The matrix notation of this network is:

$$\vec{F}_{11} = \begin{bmatrix} 20 & 18 & 21 & 1000 \\ 28 & 23 & 1000 & 20 \\ 29 & 27 & 19 & 22 \end{bmatrix} \quad \vec{F}_{21} = \begin{bmatrix} 5 & 7 & 6 & 1000 \\ 3 & 8 & 1000 & 6 \\ 10 & 9 & 6 & 5 \end{bmatrix}$$

$$\vec{F}_{12} = \begin{bmatrix} 41 & 45 & 39 \\ 1000 & 42 & 44 \\ 46 & 38 & 40 \\ 1000 & 42 & 47 \end{bmatrix} \quad \vec{F}_{22} = \begin{bmatrix} 12 & 11 & 14 \\ 1000 & 13 & 10 \\ 9 & 15 & 10 \\ 1000 & 16 & 8 \end{bmatrix}$$

Here we have assumed that $A = 1000$. We use the matrices shown above to encode the whole network as the genetic algorithm *genotype* string. Our genotype string is a concatenation of each rows of the matrices above. Here we make a very brief note on the encoding method of the Network for use by a GA, as this is out of scope of this paper. We serially concatenate the binary coding of each of these matrix entries and make up the GA usable genotype strings. This genotype string is a member of the GA population, which undergoes the evolution process. These strings are subjected to the genetic operators that produce *new* arcs of the network, thus giving rise to new rules.

5.3. DISCUSSION PART 2: ASPECTS OF DIRECTED SEARCH

We have shown how the GA described is linked to the ATN-frame. How this formal template can be applied to a real field, *Palladio Villas* or some other, have not been shown yet. As said (in subsection 4.3.4), the difficulty lies in quantifiable selectional rules. Thus, it is not plausible that we could simply adopt some technical "problem-solving" and allegedly apply harmless quantitative standards of use-value in terms of surface, topology (accessibility) and similar aspects, to a phenomenon such as Palladio. The result will not make much sense in terms of architectural and environmental development. We can never grasp some quantifiable aesthetic standards in this way. How such standards and objective, or tech-

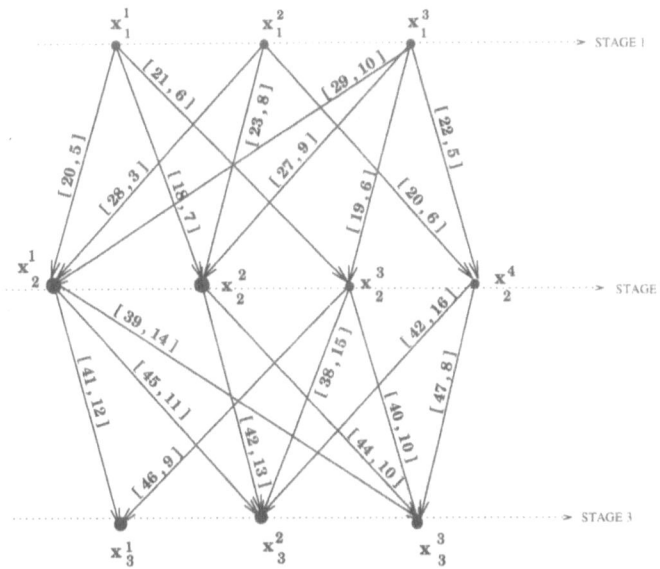

Figure 5. Example of a multicriteria network having fictitious rule evaluation (bi-criterion) values assigned to it.

nical properties are interlinked, might be connected to laws of grammaticalness in related fields of configurative expression, such as architecture. We have presented the Network and GA models to compute shape grammars, not as technical panacea but as a tool to pursue such questions.

Acknowledgments

One of the authors wishes of acknowledge the research support of Prof. John S. Gero, of the Design Computing Laboratory, University of Sydney.

References

Ackerman, J. S.: 1977, *Palladio*, Penguin Books, Harmandsworth, Great Britain.
Cagan, J. and Mitchell, W. J.: 1993, Optimally directed shape generation by shape annealing, *Environment and Planning B*, **20**, 5–12.
Cagan, J. and Mitchell, W. J.: 1994. A grammatical approach to network flow synthesis, *in* J. S. Gero and E. Tyugu (eds), *Formal Design Methods for CAD*. Elsevier Science, Amsterdam, pp. 173–189.
Chomsky, N.: 1957, *Syntactic Structures*, Mouton, The Hague/Paris.
Chomsky, N.: 1965, *Aspects of the Theory of Syntax*, The MIT Press.
Chomsky, N.: 1994, *Language and Thought*, 3rd Anshen Transdisciplinary lecture in Art, Science and the Philosophy of Culture, Emeryville CA.
Coyne, R. and Snodgrass, A.: 1993, Rescuing CAD from rationalism, *Design Studies* **14**.
De Jong, K. A.: 1987, Learning with genetic algorithms: An overview, *Machine Learning*. **3**, 121–138.

Fleisher, A.: 1992, Grammatical architecture, *Environment and Planning B.*

Flemming, U.; Baycan, C. A.; Coyne, R. F. and Fox, M. S.: 1992, Hierarcical generate-and-test vs. Constraint-directed search, *in* J. S. Gero (ed.) *Artificial Intelligence '92*, Kluwer, Dordrecht.

Gero, J. S.: 1992, Creativity, emergence and evolution in design, *in* J. S. Gero, and F. Sudweeks (eds), *Preprints 2nd International Round Table Conference on Computational Models of Creative Design*, Department of Architectrual and design Science, University of Sydney. pp. 1–28.

Gero, J. S.; Louis, S. J. and Kundu, S.: 1994, Evolutionary learning of novel grammars for design improvement. *Artificial Intelligence in Engineering Design, Analysis and Manufacturing (AIEDAM).* **8**(2), 83–94.

Goldberg, D. E.: 1989, *Genetic Algorithms in Search, Optimization, and Machine Learning.* Addison-Wesley, Reading, Massachusetts.

Graham, P.: 1994, *On Lisp, Advanced Techniques for Common Lisp,*, Prentice Hall, Englewood Cliffs, New Jersey.

Heidegger, M.: 1954 and 1985, Bauen Wohnen Denken (Bulding Dwelling Thinking), *Vorträge und Aufsätze*, Verlag Günther Neske.

Hellgardt, M.: 1993, Syntaktische Aspekte der Arbeit von Hans Scharoun, Begründungen und Erläuterungen zu den im Begleitprogramm der Ausstellung "Werkschau Hans Scharoun" vorgeführten Fragmenten einer Scharoun-Syntax, Akademie der Künste, Abteilung Baukunst, Berlin.

Hellgardt, M.: 1994, Dentro l' architettura di Scharoun, *Housing 6*, Etaslibri, Milano.

Holland, J. H.: 1975, *Adaptation in Natural and Artificial Systems.* University of Michigan Press, Ann Arbor, Michigan.

Humboldt, W. von: 1795/96 and 1973, Über Denken und Sprechen (On Thinking and Speaking), *Schriften zur Sprache*, Reclam, Stuttgart.

Lovejoy, A. O.: 1936, *The Great Chain of Being*, Harvard University Press.

Osyczka, A.: 1984, *Multicriteria Optimization in Engineering with FORTRAN programs*, Ellis Horwood Limited, Halsted Press: a division of John Wiley & Sons, Chichester, England.

Osyczka, A.: 1980, Multicriteria network optimization, *Computing*, **25**, 363–368.

Palladio, A.: 1570 and 1983, Die vier Bücher zur Baukunst, Artemis Verlag, Zürich/München, (I Quatro Libri Dell' Architettura, Venice 1570)

Searle, J. R.: 1969, *Speech Acts*, Cambridge University Press.

Stiny, G.: 1980a, Introduction to shape and shape grammars, Environment and Planning B, **7**, 343–351

Stiny, G.: 1980b, Kindergarten grammars: Designing with Froebel's buildings gifts. *Environment and Planning B*, **7**, 409–462.

Stiny, G. and Mitchell, W. J.: 1978, The Palladian Grammar, *Environment and Planning B.*

J. S. Gero and F. Sudweeks (eds), Artificial Intelligence in Design '96, 311-324.
© 1996 *Kluwer Academic Publishers.*

VARIABLE-COMPLEXITY EVOLUTION OF SHAPE GRAMMARS FOR ENGINEERING DESIGN

PETER J. GAGE
School of Aerospace and Mechanical Engineering
Australian Defence Force Academy
Canberra ACT 2600 Australia

Abstract. Shape grammars provide a formal method for efficient description of engineering designs, by listing a sequence of modifications to a baseline design. A genetic algorithm can be used to identify helpful elements of the grammar and to search for optimal combinations of grammatical elements. A variable complexity genetic algorithm, which permits modification sequences of varying length, can identify useful elements in short sequences and subsequently exploit them in longer sequences. Application to multicriteria optimization of a beam section, to maximize stiffness-to-weight ratio and minimize perimeter, demonstrates the benefits of the variable-complexity algorithm.

1. Introduction

Genetic algorithms are designed to mimic evolutionary selection (Goldberg, 1989). They provide a robust search method which permits the use of discrete-valued variables, and is effective in multi-modal and non-smooth domains. Initially, a population of candidate designs is distributed throughout the global design space. New populations are produced by recombination of the descriptions of existing designs, and the algorithm learns to concentrate the search in promising subspaces. A variable-complexity genetic algorithm can operate on alternative designs described by different numbers of variables (Gage, 1994). This is particularly appropriate for design studies, where it is common to start with a simple representation and progress to more detailed descriptions which use more variables.

Each individual in a population of the genetic algorithm is represented by a `genetic' string, which is a coded listing of the values of the design variables. The entire string is analogous to a chromosome, with genes for the different features (or variables). The high fitness individuals do not actually survive across generations, but the description of their features is propagated. Genetic algorithm performance is strongly influenced by the encoding scheme, which defines the extent of the search space, and influences the identification of promising subspaces (Liepins and Vose, 1991; Davidor, 1991).

Shape grammars efficiently describe complex shapes as an assembly of simple components. Originally developed by Stiny to formally prescribe the elements of particular architectural styles (Stiny, 1980), they have been used by several researchers to describe the search space in optimization tasks (Cagan, 1993; Gage et al., 1994; Gero et al., 1994). Gero et al. (Gero et al., 1994) showed that a genetic algorithm could be used to evolve a shape grammar, while simultaneously evolving sequences of shape transitions which use the rules of the grammar being evolved. In this paper, that work is extended by using a variable-complexity genetic algorithm for the evolution. An application to the multicriteria optimization of beam cross sections demonstrates that the standard and variable-complexity algorithms are effective for beam shapes of moderate complexity, but the greater flexibility of the new method produces superior designs for very simple and very complex beams.

2. Genetic Algorithms for Evolutionary Search

Genetic algorithms are global search methods which use operators modelled on biological reproductive mechanisms observed in the natural world. A population of candidate designs is randomly generated, to provide a statistically meaningful sample of the global search space. Descriptions of these designs are encoded in `genetic' strings, and strings of relatively fit designs are selected to contribute to the production of new strings describing new designs.

The effectiveness of genetic algorithms depends on useful correlations between parts of the genetic string (genotype) and the performance of the individual it represents (phenotype). Substrings, or building blocks, which appear in the description of above-average phenotypes are likely to survive into the next generation, even if the genotype is broken up by the action of crossover and mutation. Short, low-order building blocks are retained and combined to form higher-order building blocks, with the process repeating over many generations until the best design is found. Promising features of different candidates can be recombined to produce improvements in complete designs.

A wide variety of exotic genetic operators have been devised, but a few fundamental features are common to most genetic algorithms: encoding, selection, crossover and mutation. Each of these characteristics is described in the following paragraphs. Details of the particular implementations used in the current study are also discussed.

2.1. ENCODING

Genetic algorithms do not operate on design variables directly, but manipulate a genetic string, which encodes the variables. Algorithm efficiency depends directly on the recombination of low-order building blocks into higher-order assemblies, so encodings which promote the recognition of promising building blocks should be carefully chosen for each application (Liepins and Vose, 1991). Although Holland's schema theory (Holland, 1975) suggests that binary representations are to be favored because they maximize the number of potential building blocks in the string, they are not always appropriate. This is particularly true when a binary encoding produces a large proportion of infeasible candidates. In a sequencing task, for example, a binary encoding of a precedence matrix allows representation of orderings that are logically inconsistent (e.g. A precedes B, B precedes C, C precedes A) (McCulley and Bloebaum, 1994). With only 8 items to be placed in sequence, 99.98% of all possible strings describe impossible orderings, and a randomly-generated initial population is unlikely to contain any feasible candidates. Permutation encodings used in conjunction with re-ordering operators are much more successful in problems of this type (Kroo et al., 1994).

In some situations, identification of promising building blocks is simplified if the encoding is not limited to a fixed length. Messy genetic algorithms (Goldberg et al., 1989; Goldberg et al., 1990) use variable-length strings to avoid deception that can arise when promising building blocks have a large defining length in a chosen (fixed) encoding. These encodings refer to a fixed set of parameters, but may include several references to each parameter. It is also possible to use a variable-length encoding to refer to a varying number of parameters, as in genetic programming (Koza, 1992) and in the variable-complexity genetic algorithm used in this study. The schema theorem was extended by Smith (1980), to show that promising building blocks are appropriately retained and recombined when string length is not constant. Koza cites the empirical evidence of successful applications of genetic programming in a variety of fields as proof that adaptation of variable-length strings is a valid search mechanism.

A shape grammar encoding of a complex shape is typically efficient, because it need refer only to component shapes that appear in the design. This approach compares favorably with the exhaustive representation schemes commonly employed in genetic encodings, which refer to the existence or absence of all possible components (Sakamoto and Oda, 1993; Grierson and Pak, 1993; Grierson and Pak, 1993b; Yang, 1993; Hajela et al., 1993). More detailed discussion of the advantages of shape grammar encodings is provided in the next section of this paper.

2.2. SELECTION

Selection is the operation which rewards high-fitness designs, because there is a relatively high probability that they will be chosen for reproduction. The selection scheme should be chosen to balance the competing desires to exploit promising features contained in the existing population and to explore the design space for

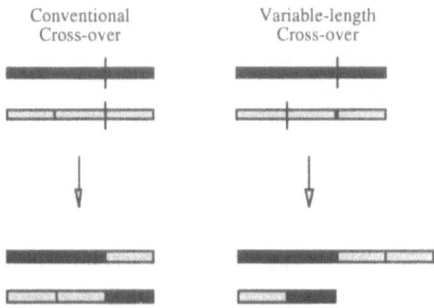

Figure 1. Standard and variable-complexity crossover operations.

new possibilities. If selection pressure is too great, diversity can be quickly lost, and the population will converge to a sub-optimal design. If selection pressure is too weak, the algorithm is reduced to random search.

Roulette-wheel selection is commonly used in simple genetic algorithms (Goldberg, 1989), but it is susceptible to premature convergence when a poorly scaled fitness domain allows one individual to dominate reproduction, by occupying most of the space on the wheel. Ranking schemes prevent such dominance, because they are not affected by the margin of superiority of higher-fitness individuals. Tournament methods perform a local ranking at each selection operation, without ever requiring the entire population to be sorted. Each time a parent is needed, k members of the current population are selected at random, where k is the tournament size. Their fitness is compared, and the highest fitness individual becomes the parent. With this scheme, it is expected that the best individual will be a parent k times per generation (it will participate in k tournaments and win them all), with linear decline in expectation of reproduction to the worst individual, which cannot win a tournament.

When a genetic algorithm is used for multicriteria optimization, the selection method must produce a scalar fitness function from the objective function vector (Fonseca and Fleming, 1995). In this paper, the notion of dominance (one individual dominates another if it has a superior value for each component of the objective vector) is used to select the winner of a tournament of size two ($k = 2$). If neither candidate is dominant, selection is random. This weak selection scheme preserves variety in the population, which is useful when searching for a Pareto optimal set of alternative designs.

2.3. CROSSOVER

Crossover operators generate new designs composed from elements of two earlier designs, thus exploiting features already present in the population. Figure 1 shows how parts of two parent strings are recombined, both for standard and variable-complexity operators. The position of the crossover point along the string is chosen at random. If variable-length encodings are permitted, the crossover point may be different in each parent, so offspring of different length are produced.

2.4. MUTATION

Standard genetic algorithms generally include a pointwise mutation operation, which modifies individual bits of the string at random, and produces corresponding changes to a design variable. This operation can introduce features not present in either parent, so it helps to maintain diversity in the population.

Variable-complexity encodings permit the introduction of deletion and insertion mutations (Fig. 2). In these operations, a segment of the existing string is selected at random, and either removed (deletion) or duplicated (insertion).

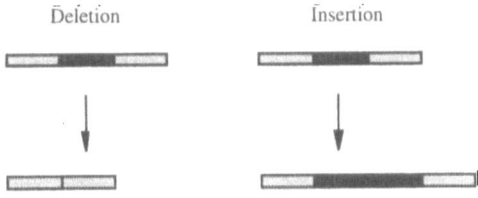

Figure 2.   Deletion and insertion mutations.

3. Shape Grammars in Engineering Design

Shape grammars describe complex topologies as assemblies of simple components. They include four basic features: an initial shape, a set (or alphabet) of component shapes, a set of labels, and a set of shape transition rules. The shape transition rules describe all legal modifications to an existing shape, and have the form: $X \rightarrow Y$, where X and Y are instantiations of component shapes (or assemblies of component shapes). The initial shape is transformed by application of a shape transition rule (so at least one rule must have the initial shape as its left hand side). The resulting shape can be further modified by subsequent application of more transition rules. Labels are used to instantiate special features of each shape, such as size, material density or sites where additional shapes may be added.

By starting with a basic shape, and adding complexity only as necessary, shape grammars bias the search in favor of simplicity, which is beneficial for learning (Mitchell, 1990). The rules of a shape grammar can also incorporate factual knowledge of the domain. For example, when components of a complex design must be connected, the rules might only permit transitions which add components adjacent to existing components. This further simplifies learning about superior designs, by biasing search in favor of feasible candidates.

Reddy and Cagan employed a shape grammar to represent topologies of structural trusses (Cagan, 1993). They sought the optimal topology by starting with the simplest shape that would support the applied loads, and then using shape transition rules to produce more complex topologies. Search was performed by a simulated annealing algorithm, and at each iteration of the algorithm, a single shape transition rule (randomly selected from all rules in the shape grammar) was appended to the sequence describing the current design. The length of the rule sequence is variable in this application, but the set of rules is not refined.

This `shape annealing' is an evolutionary process, but the only operator is random mutation. A genetic method, which also includes the powerful crossover operator, can provide the capability for more efficient search, but a standard algorithm restricts the rule sequence to be of fixed length. However, a variable-complexity algorithm can be used to combine the power of crossover with the flexibility of variable-length encoding. Complex truss geometries have been successfully developed using this approach (Gage et al., 1994).

Gero et al. (1994) have used a shape grammar in conjunction with a standard genetic algorithm to design efficient structural beam sections. Their encoding implicitly includes the set of shape transition rules in the genetic string, which permits evolution of the grammar itself. Beam section shapes are evolved simultaneously, because the genetic operators also affect the sequence of shape transition rules encoded in the string.

A variable-complexity genetic algorithm can also be used to refine the shape grammar for a given design domain. It should be more efficient than a standard genetic algorithm, because promising rules can be identified in short sequences which describe relatively simple compound shapes, and these sequences can subsequently be recombined in extended sequences that produce more complex shapes. These conjectures are examined in the next section of this paper, which describes the application of a variable-complexity genetic algorithm to the design of beam sections, and compares its performance with a standard algorithm.

4. Multicriteria Design of Beam Sections

A standard genetic algorithm with shape grammar encoding was used by Gero et al. (1994) for beam section design, with the objective of maximizing moment of inertia while minimizing perimeter. In this paper, The goal of maximizing moment of inertia is replaced with the requirement to maximize stiffness to weight ratio. This modified objective requires a more complex grammar to represent the family of optimal designs (because various densities are required), so the task is more difficult than that solved in the earlier work. The perimeter is still minimized (to minimize surface area of the beam), and the multicriteria objective is handled by Pareto optimization, which produces a family of non-dominated solutions (i.e. the set of solutions which are better than all other solutions in at least one component of the objective function).

Gero et al. used an example problem where sequences of 9 transition rules were used to describe candidate shapes. When the task involves simultaneous optimization of the representation language and the optimal shapes described by that grammar, the space of candidate designs increases rapidly. Consequently, transition sequences of up to 25 rules are considered here, to demonstrate that the genetic method performs well in larger search spaces. Each of the computer runs used to generate the results presented here took less than 5 minutes on a Sun SparcStation, so combinatorial explosion is not a serious concern for problemms of this type.

The shape grammar encoding used in this example is based on the one developed by Gero et al (1994). Beam sections are represented by a number of square elements, each of which may have one of four density values. Candidate designs are described using a shape grammar with the following features:

Initial shape Each candidate design starts with a single element, placed at the centerline of the beam.

Component shapes All component shapes are square elements of unit side length.

Labels Each component square is labelled with a density (there are four alternative values: A, B, C, or D). The most recently introduced element is designated as the site for new elements to be added.

Transition rules Transition rules have 3 parts:

1. Modify density of element labelled as the building site

2. Define a direction to move (up, down, left or right)

3. Add a new element of specified density adjacent to the building site (on the side specified by the move direction)

There are sixty-four possible rules, because there are four alternative values for each of the three parts of the rule. The restriction to add elements only adjacent to existing elements ensures that candidate cross-sections are simply connected, and hence are able to support bending loads.

The decoding of a short sequence of genetic string is shown in Figure 3. Three transition rules are produced. The cross-section resulting from the sequential application of these rules is also shown.

Longer rule sequences typically produce beam sections with more elements, but there is not a direct correspondence between genotype and phenotype complexity. Figure 4 shows that long genetic strings can produce beams with few elements. (The solid line marks the maximum number of elements in the phenotype for a given length of genotype. If each rule produced a new element in the phenotype, all population members would lie on this line. The squares represent individuals in an example population of candidate designs. Most designs have lower complexity than their genotype length might produce.) Low complexity

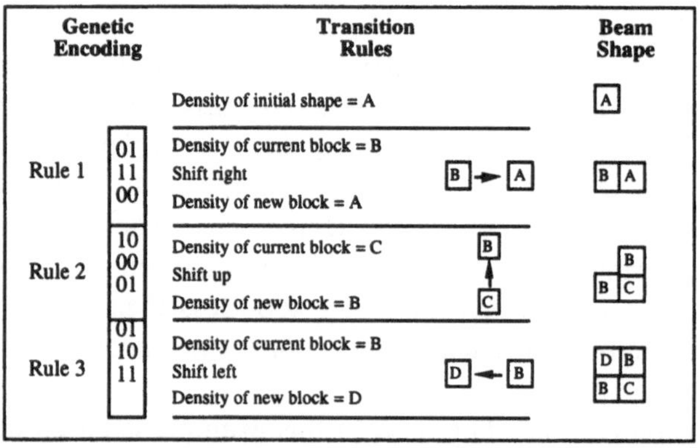

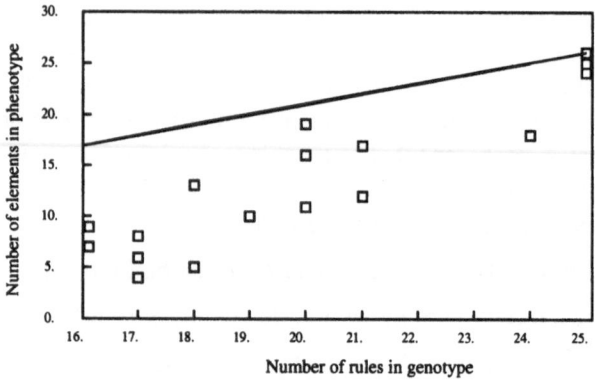

Figure 3. Construction of beam section by decoding and application of shape transition rule sequence.

Figure 4. Long genotypes can produce phenotypes with few elements.

phenotypes are produced when later rules in the string refer to elements defined by earlier rules. In these cases, the density of the element might be changed, but the number of elements is not increased. A fixed length genetic encoding, corresponding to a fixed number of rule applications in the shape transition sequence, is consequently able to describe cross-sectional shapes of varying complexity, by referring to some elements several times. A variable-length encoding can also refer to each element several times, but it also has the freedom to describe low-complexity phenotypes more simply and naturally, by using shorter rule sequences.

The operation of the rules in this grammar may cause changes to an existing element in the phenotype, or may cause the introduction of a new element. The decoding of each rule does not depend on its absolute position in the genetic string, but the relative position within the sequence of rules is important. The ef-

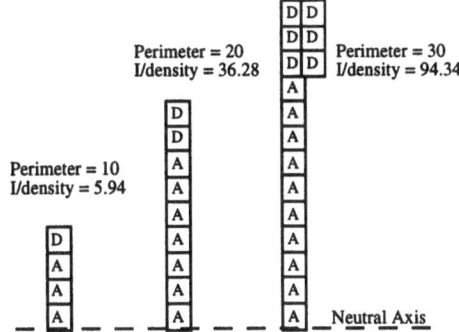

Figure 5. Three optimal cross-sections from the Pareto set.

fect of each rule depends on the phenotype constructed by the action of all prior rules in the string. This limited dependence on position increases the flexibility of the genetic algorithm, and permits the evolution of complexity by recombination of short rule sequences.

Three of the optimal beam sections in the Pareto set are shown in Figure 5. In this design task, it is advantageous to have elements of low density close to the neutral axis (where they make a small contribution to moment of inertia), with high density elements at the outer edge of the beam.

For simple beams, a single column of elements is appropriate, because it is better to add a single element at the end of the beam than to add an element to the side. Only rules which prescribe addition of an element above the current building site are needed in the transition sequence. As the height of the beam increases, however, it becomes attractive to add several dense elements beside existing elements. This changes the nature of the best sequence of transition rules, because rules prescribing a horizontal shift must be included.

Figure 6 indicates that the standard algorithm and the variable- complexity algorithm each fail to locate one of the optimal cross-section types. The standard algorithm, which must use twenty-five rules to describe even the simplest shapes, fails to produce a sequence that excludes any horizontal shifts. On the other hand, the variable-complexity algorithm generates a population that is quickly dominated by short rule sequences which efficiently describe optimal simple shapes. When these are recombined to form complex shapes, they only produce a longer single column of elements.

The difference in results produced by the two genetic algorithms indicates that their search procedure is quite different. The two aspects of the search are evolution of the shape grammar, by modification of the set of available transition rules, and evolution of transition sequences composed of the rules included in the new grammar. Both search processes contribute to the development of optimal cross-sections.

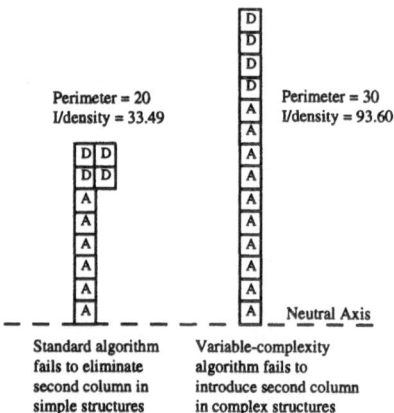

Figure 6. The standard and variable-complexity algorithms fail to locate different types of optimal cross-section.

4.1. EVOLUTION OF THE SHAPE GRAMMAR

The initial shape grammar includes all 64 transition rules that can describe modification of the density of the building site, and addition of a new element adjacent to that site. They are almost certainly all represented in the initial population which has 200 members, each encoding a sequence of up to 25 rules (exactly 25 rules for the standard, fixed-length, encoding). With 2500 to 5000 randomly generated rules, each member of the grammar is expected to be represented 40 to 80 times.

The shape grammar evolves by modifying the variety of rules used to describe candidate sections. Gero et al described this as a substitutive learning process (the learned grammar searches a space partially disjoint from the original space) when they developed beam sections to maximize moment of inertia and minimize perimeter (and found that rules referring to high-density elements quickly dominated the population). Although these rules replace others in the population, new rules (outside the original 64) are not produced. The learning process might be described as subtractive, because a subset of the original set of transition rules is retained, and the learned grammar searches a subspace of the original space. The genetic algorithm restructures the original knowledge to localize its search in promising regions, thereby increasing the likelihood of identifying the best designs.

In the current application, rules must be retained for both low density and high density elements, but reference to intermediate densities is unnecessary. Both genetic algorithms achieve this refinement effectively. The evolved grammar should also include rules that add elements in each of the four possible directions, and it is in this regard that the variable- complexity algorithm is unsatisfactory. It is too aggressive in the refinement of the rule set, because short rule sequences describing optimal simple shapes quickly dominate the population, and rules producing horizontal addition of material are lost from the grammar.

This dominance of short rule sequences which add material only in the vertical direction can be remedied in two ways: short sequences which include horizontal addition of material can be favoured, or short sequences can be avoided entirely. In the first case, the objective function must be modified to extend the Pareto set. (If the second component of the objective, minimization of perimeter, is replaced by minimization of perimeter + height, the variable-complexity algorithm successfully retains the necessary rules, and more complex optima are subsequently found.) Such adjustments are difficult to make *a priori*, and the second remedy is therefore preferred. It simply involves the imposition of an explicit lower limit on the length of rule sequences developed by the variable-complexity algorithm, instead of implicitly using a lower limit of one rule in the sequence. This increases the total number of rules in the population, and increases the proportion of complex shapes being described.

4.2. EVOLUTION OF THE RULE SEQUENCES

For the standard genetic algorithm, the evolution of rule sequences is inextricably linked to evolution of rules in the grammar. In fact, the standard algorithm essentially evolves a grammar for each position in the sequence, independently (and in parallel). There is no indication that a good rule for position 4, say, would also be useful at position 5. The set of transition rules for the entire sequence is the union of sets for each position. There is no opportunity to generalize by creating new rule sequences using the global shape grammar, because the grammar is not defined until the set of transition rules for each point in the sequence has evolved.

In contrast, the variable-complexity algorithm does develop a global set of shape transition rules, each of which can be applied at any point in the sequence of rules. It readily describes cross-sections composed of radically different numbers of elements, and broad generalizations are possible. The full range of shapes describable by the rules in the evolved shape grammar is efficiently explored by this method.

4.3. RESULTS WITH LOWER BOUND ON LENGTH OF RULE SEQUENCE

The lower limit on the length of transition sequences is now set at 10 rules, instead of implicitly permitting sequences as short as one rule. The variable-complexity implementation no longer eliminates horizontal transitions from the rule set. Consequently, it can identify optimal shapes of both types, and is thus effective for a wide range of perimeters. Performance comparisons for the standard and variable-complexity encodings are presented in Figures 7, 8 and 9. Data for three separate runs of each algorithm are included, to indicate that the observed trends are repeatable.

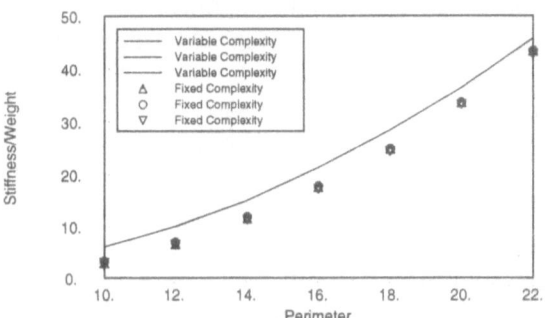

Figure 7. Pareto curve for designs of low complexity.

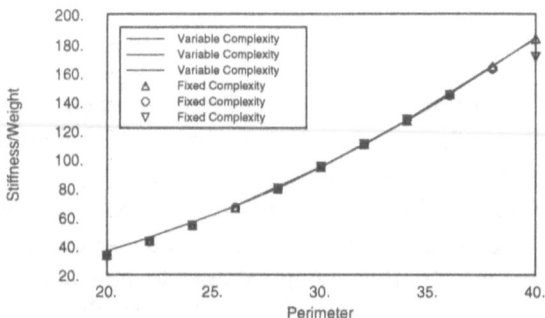

Figure 8. Pareto curve for designs of moderate complexity.

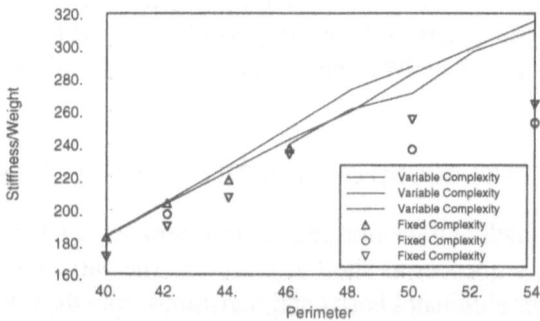

Figure 9. Pareto curve for designs of high complexity.

These results clearly demonstrate that the variable-complexity encoding generalizes more widely than the standard algorithm. The standard algorithm effectively locates shapes of moderate complexity. Sequences of 25 rules do well in finding shapes composed of 10 to 20 elements, which means that some elements are referenced more than once. The variable complexity algorithm uses shorter sequences to locate optimal simple shapes (Figure 7). It recombines short sequences to identify highly complex designs (Figure 9).

5. Conclusions

Genetic algorithms are able to identify optimal beam cross-sections, with the twin objectives of maximizing stiffness-to-weight ratio, and minimizing perimeter. A shape grammar encoding is efficient for this application, because it refers only to elements which actually appear in the design. It also limits search to feasible candidates, by restricting the rule set to produce only simply connected shapes.

During optimization, the knowledge contained in the shape grammar is restructured. The set of shape transition rules is refined, to include only those rules consistently associated with high performance cross sections. The learned grammar concentrates search in promising subspaces of the original design space, thereby increasing the likelihood of identifying optimal rule sequences. The variable-complexity algorithm can reduce the rule set too greatly by excessively exploiting short rule sequences that describe optimal simple shapes. If the nature of optimal solutions changes as complexity increases (as it does by switching from a single column of elements to two columns), the second type of solution can be missed. This defect is remedied by imposing a lower limit on the length of transition sequences, so that more complex shapes also influence the development of the refined rule set.

Rule sequences, corresponding to particular beam cross-sections, are also evolved during optimization. For the standard algorithm, this occurs simultaneously with grammar refinement, and there is no opportunity to generalize using the evolved grammar. The variable-complexity algorithm can use rules at any location in the transition sequence, so that genuinely new sequences are produced by recombination of general rules. This capacity for generalization permits the identification of optimal shapes for a wide range of perimeters. Thus, the variable-complexity algorithm is able to produce better combinations of shape grammar and transition sequence than the standard genetic algorithm.

References

Davidor, Y.: 1991, Epistasis variance: A viewpoint on GA-hardness, *in* G. Rawlins (ed.), *Foundations of Genetic Algorithms*, Morgan Kaufmann.
Fonseca, C. and Fleming, P.: 1995, An overview of evolutionary algorithms in multiobjective optimization, *Evolutionary Computation*, 3(1).
Gage, P.: 1994, *New Approaches to Optimization in Aerospace Conceptual Design*, PhD Thesis, Stanford University.
Gage, P., Kroo, I. and Sobieski, I.: 1994, A variable-complexity genetic algorithm for topological

design, *AIAA 94-4413 AIAA/NASA/USAF/ISSMO Symposium on Multidisciplinary Analysis and Optimization*, Panama City, FL.

Gero, J. S., Louis, S. J. and Kundu, S.: 1994, Evolutionary learning of novel grammars for design improvement, *Artificial Intelligence for Engineering Design, Analysis and Manufacturing (AIEDAM)*, **8**, 83–94.

Goldberg, D.: 1989, *Genetic Algorithms in Search, Optimization, and Machine Learning*, Addison Wesley.

Goldberg, D. E., Deb, K. and Korb, B.: 1990, An investigation of messy genetic algorithms", *TCGA Report 90005*, May.

Goldberg, D. E., Korb, B. and Deb, K.: 1989, Messy genetic algorithms: Motivation, analysis and first results, *TCGA Report 89003*, May.

Grierson, D. and Pak, W.: 1993a, Optimal sizing, geometrical and topological design using a genetic algorithm, *Structural Optimization 6*, 151–159.

Grierson, D. and Pak, W.: 1993b, Discrete optimal design using a genetic algorithm, *in* M. P. Bendsoe *Topology Design of Structures*, Kluwer, Mota Soares, C.A.

Hajela, P., Lee, E. and Lin, C.-Y.: 1993, Genetic algorithms in structural topology optimization, *in* M. P. Bendsoe, *Topology Design of Structures*, Kluwer, Mota Soares, C.A.

Holland, J.: 1975, *Adaptation in Natural and Artificial Systems* University of Michigan Press.

Koumousis, V.: 1993, Layout and sizing design of civil engineering structures in accordance with the eurocodes, *in* M. P. Bendsoe, *Topology Design of Structures*, Kluwer, Mota Soares, C.A.

Koza, J. R.: 1992, *Genetic Programming: On the Programming of Computers by Means of Natural Selection*, MIT Press, Cambridge, MA.

Kroo, I., Altus, S., Braun, R., Gage, P. and Sobieski, I.: 1994, Multidisciplinary optimization methods for aircraft preliminary design, *AIAA 94-4325 AIAA/NASA/USAF/ISSMO Symposium on Multidisciplinary Analysis and Optimization*, Panama City, FL.

Liepins, G. and Vose, M.: 1991, Deceptiveness and genetic algorithm dynamics, *in* G. Rawlins (ed.), *Foundations of Genetic Algorithms*, Morgan Kaufmann.

McCulley, C. and Bloebaum, C. L.: 1994, Optimal sequencing for complex engineering systems using genetic algorithms, *AIAA 94-4325 AIAA/NASA/USAF/ISSMO Symposium on Multidisciplinary Analysis and Optimization*, Panama City, FL.

Mitchell, T. M.: 1990, The need for biases in learning generalizations, *in* J. W. Shavlik and T. G. Dietterich (eds), *Readings in Machine Learning*, Morgan Kaufmann.

Reddy, G. M. and Cagan, J.: 1993, Optimally directed truss topology generation using shape annealing, *DE-Vol. 65-1, Advances in Design Automation - Volume 1*, ASME, pp. 749–759.

Sakamoto, J. and Oda, J.: 1993, A technique of optimal layout design for truss structures using genetic algorithm, *AIAA 93-1582, SDM 93*, La Jolla.

Smith, S. F.: 1980, *A Learning System Based On Genetic Adaptive Algorithms*, PhD Thesis, University of Pittsburgh.

Stiny, G.: 1980, Introduction to shape and shape grammars, *Environment and Planning B*, **7**, 343–351.

J. S. Gero and F. Sudweeks (eds), Artificial Intelligence in Design '96, 325–344.
© 1996 Kluwer Academic Publishers.

GRAMMARS FOR MACHINE DESIGN

LINDA C. SCHMDT
University of Maryland, College Park, Maryland, USA

AND

JONATHAN CAGAN
Carnegie Mellon University, Pittsburgh, Pennsylvania, USA

Abstract. The use of grammars in mechanical design research is growing in popularity, largely due to the ability of a grammar to concisely express a language of designs. It is natural to attempt to achieve the level of success at writing descriptive languages for mechanical devices as is seen with spatial grammars in describing architectural styles. However, the differences in representing form and function between the fields and the mechanical designer's focus on function must impact the type of grammars that can be used in describing mechanical designs. Two grammars for mechanical configuration design are briefly described: a string grammar for the design of cordless power drills and a graph grammar for the design of rolling carts. The ability of the grammars to generate a space of machine designs is discussed. How a mechanical design grammar can provide a platform for a designer assistance tool and the strengths and weaknesses of such a tool are presented.

1. Introduction to Grammar-Based Design

At the heart of the mechanical design problem is a desire to find a best design amidst a host of designs. A number of optimization techniques exist to search through a space of designs for the best one. We are concerned with the challenge of finding a means to generate the space of designs for the search. Gips and Stiny (1980) list four methods to specify a space of designs:
1. compiling a catalog of all members of the space;
2. describing one element of the space and the transformations to create all other elements from it;
3. providing a computer program that generates all designs in the space; and
4. writing a grammar.

A grammar is also the means later suggested by Stiny and March (1981) to define a language of designs in the creation of a machine for design. An exploration of the potential of grammars to express a language of mechanical designs and, in this role, become an integral part of a mechanical design tool is undertaken here.

Grammars are enjoying growing popularity in mechanical design research and their success in describing languages of design in architecture tempts the mechanical design researcher. Stiny (1988) proposed that a common concern with the design of artifacts described by function and form allows the same formal devices to be used in mechanical design that are used in architectural design. Mitchell's (1991) work on functional grammars for architecture provides a glimpse of how grammars can be used to design forms from mechanical functional intent. His functional grammar introduces the idea of the function template that is instantiated by one or more forms that can fulfill the structural function. Mitchell foreshadowed the spirit of this work when he ended his paper with the following quote, "It seems reasonable to suggest (though this has not yet been tested) that large, dynamically-extensible function grammars, allied with fast search engines, would be capable of producing unexpected, and perhaps, valuably innovative, design solutions." This paper presents two proof-of-concept examples of grammar-based, mechanical design algorithms and considers their potential usefulness as designer assistance tools.

2. Background on Grammars in Design

The term *grammar* was coined by Chomsky (1957) who applied it in analyzing natural language. Chomsky observed that a set of grammatical transformations used to produce meaningful sentences had formal properties and the study of these transformations would lead to insight in how language is used and understood. A grammar is a formal device consisting of a set of productions or rules, a set of symbols, and an initial symbol or symbol set. The grammar rules manipulate an initial symbol into a set of symbols which together create a meaningful expression. Grammars exist in many forms and are classified according to their productions (e.g., context-free or context-sensitive) and according to the symbols they manipulate (e.g., linguistic, symbolic, and spatial). A particular grammar defines a language of meaningful expressions such as an English-language sentence or a floor plan of a bungalow.

2.1. SHAPE GRAMMARS IN ARCHITECTURE

Shape grammars (Gips and Stiny, 1972; Stiny, 1980a, 19980b) have been used to describe a complete language of spatial designs. Shape grammars capture a style of design and can be used to create both existing examples and new instances of that style. Sample languages include: Palladian-style villas (Stiny and Mitchell, 1978); Hepplewhite-style chair backs (Knight, 1980); bungalows of Buffalo, New York (Downing and Flemming, 1981); Frank Lloyd Wright prairie houses (Koning and Eizenberg, 1981); and Queen Anne houses in Shadyside, Pittsburgh (Flemming, 1987). A few authors have explored computational issues of implementing shape grammars, building momentum toward, perhaps, a new CAD modeling paradigm (Krishnamurti, 1980; 1981, 1992a, 1992b; Chase, 1989; Krishnamurti and Stouffs, 1993.)

2.2. GRAMMARS IN MECHANICAL DESIGN

The exploration of formal grammars for engineering design purposes was officially acknowledged in a 1991 issue of *Research in Engineering Design* devoted to papers on the subject. Early applications of grammars in engineering appeared in solid model representations (Fitzhorn, 1986; Pinilla, et al., 1989; Longenecker and Fitzhorn, 1991; Fu et al., 1993). This grammar work on solids and features may have sparked some of the current interest in expanding grammar applications to increasingly higher levels of description -- the physical component (Sthanusubramonian et al., 1992), the mechanical device (Hoover and Rinderle, 1989; Mullins and Rinderle, 1991; Rinderle, 1991; Deng, 1994), the structure (Reddy and Cagan, 1995; Shea et al., 1995), and the manufacturing process plan (Brown et al., 1994).

Implementations of grammars coupled with an optimizing technique to control and guide design generation are the newest additions to the field of grammar research in mechanical design. This type of inquiry begins with Cagan and Mitchell (1993) who combine a shape grammar with simulated annealing to create shape annealing, a controlled method of generating an optimal shape. Reddy and Cagan (1995) and Shea et al., (1995) developed structural design applications of shape annealing in the generation of optimal trusses.

Schmidt and Cagan (1992 and 1995) apply the shape annealing idea in a new direction, proposing a recursive grammar approach to highly idealized machine design. Schmidt (1995) proposes string and graph, grammar-based design algorithms that generate designs from a library of components using a grammar. FFREADA (Function-to-Form REcursive Annealing Design Algorithm) uses a string grammar to design hand-held, cordless power drills. GGREADA (Graph Grammar Recursive Annealing Design Algorithm) uses a rudimentary graph-grammar to design carts from a limited subset of Meccano Set components. Both algorithms were able to demonstrate success in using grammars to generate design solutions state spaces and in exploring the generated spaces. The success of these applications is owed to the generative power of their grammar-based design mechanisms.

3. Machine Design by Grammars

A machine design process transforms initial specifications of machine function into an arrangement of machine components. Transformational models make up one popular class of design models for designer assistance tools. Transformational models lend themselves well to automation because they are descriptive in nature, providing structured methods for mapping function to form, often simplifying the problem by decomposing it into a hierarchical one. A transformational model of design lends itself to implementation by a grammar, which is itself transformational.

3.1. MODEL FOR MACHINE DESIGN BY GRAMMARS

A machine design model created for implementation with a grammar-based design mechanism is the recursive design model (Figure 1) proposed by Schmidt and Cagan (1995). In this model, machine design is accomplished by converting specifications into increasingly detailed descriptions of machine functioning (called function structures by Pahl and Beitz (1988)) until the function descriptions can be replaced by single components. At each level of detail on which the designer considers the machine (i.e., at each level of abstraction used in the design process) many alternative designs exist and must be developed in great physical detail in order to predict their suitability.

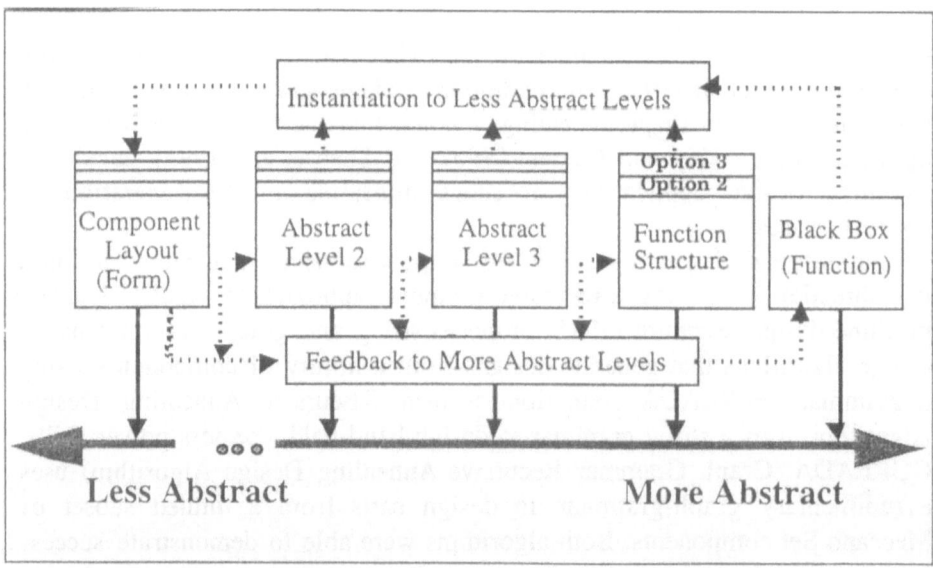

Figure 1. Conceptual design model on abstraction continuum.

The recursive design model assumes that a meaningful design process can proceed hierarchically along a sequence of levels of abstraction chosen to be relevant for a particular design problem. Each level of abstraction has associated with it a set of parameters that describe some aspect of machine functioning or machine form (i.e., the machine components it contains). Lower level designs are instantiations of higher level designs. Higher level designs are generalizations of lower level designs and can serve as patterns for designs on lower levels of abstraction. To implement this recursive design model, a mechanism for generating designs in a hierarchical fashion is necessary. Then, since many designs can be generated from an abstract design, the means of assuring that all possible designs are generated and that good designs can be recognized are needed to build a complete design generation algorithm.

To implement the recursive design model with grammars, hierarchical procedures must be built into the grammar's rule system that direct the design process to proceed sequentially through the levels of abstraction. "Abstraction grammar" (Schmidt and Cagan, 1995) is the term we apply to grammars written for this type of recursive design, although the abstraction grammar concept is more general and is not limited to recursive design. We add to the abstraction grammar a stochastic mechanism that randomizes the selection of machine components for rule applications. What results is a generative machine design tool that can, in theory, design all machines possible from the given set of machine component symbols. The grammar with its symbols and rules has, in fact, defined a space of design solutions to a given machine design problem. Each state in the space represents a different design.

3.2. MACHINE DESIGN ABSTRACTION GRAMMAR

In an abstraction grammar each level of abstraction fits within an ordered hierarchy describing a machine in function and form terms. The knowledge and detail expressed by a design accumulates from levels of less to more abstraction (left to right) along the abstraction continuum. Each level of abstraction has symbols that can be manipulated into a machine design as it appears on that abstraction level from a pattern design created at the next higher level of abstraction. The design process on any level is guided both by design specifications for that level of abstraction and the need to satisfy the design pattern provided by the design from the next higher level of abstraction.

Figure 2 holds a description of a generic abstraction grammar's design process. The process begins with a set of machine specifications. Inputs passed to Level J of design includes the grammar's initial symbol signaling the start of the design process, the symbol indicating design will occur on

Level J of abstraction, and machine specifications for designs on all levels of abstraction. After design on Level J is complete, inputs to of design Level J-1 include the design created on Level J, the symbol indicating design will now occur on Level J-1 of abstraction, and machine specifications for designs on levels of abstraction J-1 to 0. Completed designs, design start symbols, and appropriate machine specifications are passed to each successive level of abstraction as inputs to the design process on that level of abstraction.

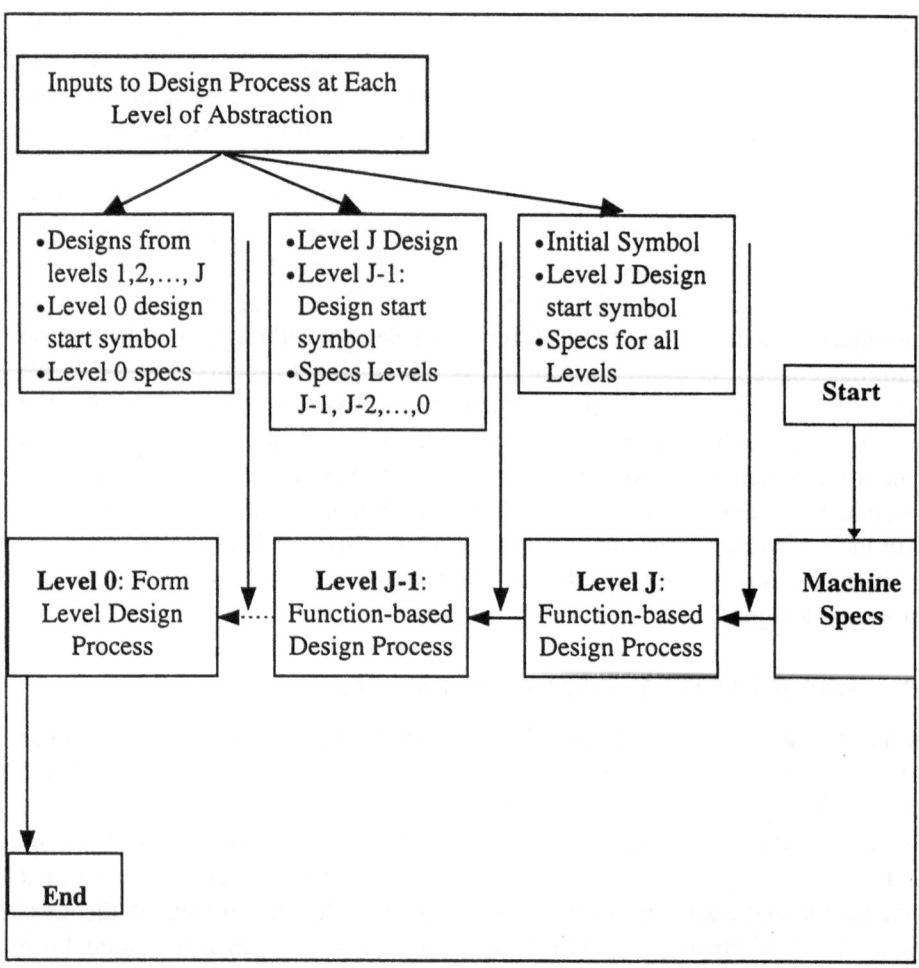

Figure 2. Generic abstraction grammar design process.

A complete design produced by an abstraction grammar is an arrangement of the symbols of the grammar that represent machine functions and machine components. A complete design is labeled "D" and is comprised of the designs created on each level of abstraction as follows:

$$D = \mathcal{D}_0 \# \mathcal{D}_1 \# \mathcal{D}_2 \# \ldots \# \mathcal{D}_J.$$

In this notation, there are J+1 levels of abstraction in the grammar. The symbol "#" indicates an arrangement depicting hierarchical functional dependence. A partial design, $\mathcal{D}_j$, is the machine design created on level j of abstraction, represented as an arrangement of symbols defined to have meaning on level j. This design, $\mathcal{D}_j$, is an abstract description of the machine in the function and form characteristics defined for level j.

The string and graph abstraction grammars used for machine design here are both spatial grammars and can be described by a four-tuple G = (N, T, R, I) (Krishnamurti and Stouffs, 1993):

N = $\{\mathcal{N}_0 \cup \mathcal{N}_1 \cup \ldots \cup \mathcal{N}_j\}$, a collection of sets of non-terminal symbols, those which will not appear in a final design (e.g., design start symbols on each level of abstraction). Each set $\mathcal{N}_j$ contains the non-terminal symbols, design procedure symbols, and machine specifications for the jth level of abstraction.

T = $\{\mathcal{T}_0 \cup \mathcal{T}_1 \cup \ldots \cup \mathcal{T}_j\}$, a collection of sets of terminal symbols, those of which a final design is comprised (e.g., on a form level of abstraction, terminal symbols represent machine components or sub-assemblies). Each set $\mathcal{T}_j$ contains the terminal symbols for the jth level of abstraction. Note that sets N and T are disjoint (i.e., N∩T = ∅).

R = $\{R_m = (a, b)\}$, m = 0, 1, 2, ..., M. M is the number of rules in the grammar. Each rule transforms a design, a, into a new design, b. For a rule to apply to design a, it must contain at least one non-terminal symbol.

I = an initial symbol.

Designs are created from the set of vocabulary elements, V, where V = N ∪ T and T ⊂ U. U is the power set of designs that can be created from members of T, and U includes ε, the empty design. The language of designs generated by the grammar is L(G) = $\{D \mid D \in T^*\}$. When the abstraction grammar is a string grammar, T^* is the least set of terminal vocabulary elements closed under string concatenation and the rules of the grammar. Typical transformations on strings, including addition, subtraction, and the substring relation, are outlined by Krishnamurti and Stouffs (1993). Analogous relationships exist for graph grammars.

3.3. MACHINE COMPONENT REPRESENTATION

An abstraction grammar for machine design requires a set of symbols that represent machine components (e.g., gears, motors, linkages and belts) and machine functions (e.g., convert electrical energy to rotational energy). Alone or in combination, symbols must be able express machine functions to allow reasoning about a machine at a high level of abstraction. Symbols must also express machine component forms to provide representation of component layout. An initial symbol or set of symbols that encode specifications into a functional description of the machine to be designed is also necessary.

There are numerous function and form representation systems that exist. Many different approaches can be successful with grammar-based design. The FFREADA drill design example uses the Pahl and Beitz (1988) concept of energy, material, and signal flows through machine components as a basis for representing functions and forms. The governing principle for combining function and form symbols is that the flows match.

The GGREADA cart design example uses an even simpler representation system. Functions, sub-functions, Meccano set components, and component sub-assemblies are declared to be capable of satisfying specific higher-level functions. The Meccano set components and component sub-assemblies are defined to allow joining to others if physically feasible and if the necessary number of joining cites (joints) are open.

4. String Grammar Design Application: Cordless Drills

Cordless, hand-held power drills are a common machine used to create holes for driving wood or metal screws. The drills of interest in this design problem are those that a homeowner might choose to purchase for occasional use. The drill is activated by pulling a trigger, sending a signal to the power supply. The output of the drill is a constant, high speed rotation of the bit placed in the drill chuck. In the string grammar, a completed string represents a machine design. A drill design can be described by a string of symbols, each symbol representing either a function necessary to the operation of the drill or a component of the drill. One such drill design is as follows:

$$S_{1,0}S_{3,0}S_{52,1}S_{7,0}S_{20,0}S_{35,0}S_{4,0}S_{53,1}S_{61,2.}$$

This drill representation is described in more detail in this section.

4.1. STRING ABSTRACTION GRAMMAR

FFREADA uses a string abstraction grammar and three levels of abstraction ($J = 2$) to design hand-held power drills. The grammar acts on levels 2 and 1 to combine vocabulary symbols representing simple functions into more complex expressions of machine behavior called function structures (Pahl and Beitz, 1988). For example, symbol $S_{61,2}$ on Level 2 represents the production of rotational energy from a non-energy input like a signal, and symbol $S_{53,1}$ on Level 1 represents the conversion of continuous electrical energy into continuous rotational energy. Level 0 of this abstraction grammar transforms a string of Level 2 and Level 1 vocabulary symbols and machine specifications into a string of machine component symbols arranged to mimic placement in a drill. The set of terminal symbols on Level 0 of this grammar (7_0) represents triggers (2), power supplies (1), chuck attachments (1), motors (11), shafts (9), and gears (25), for a total of 49 represented components.

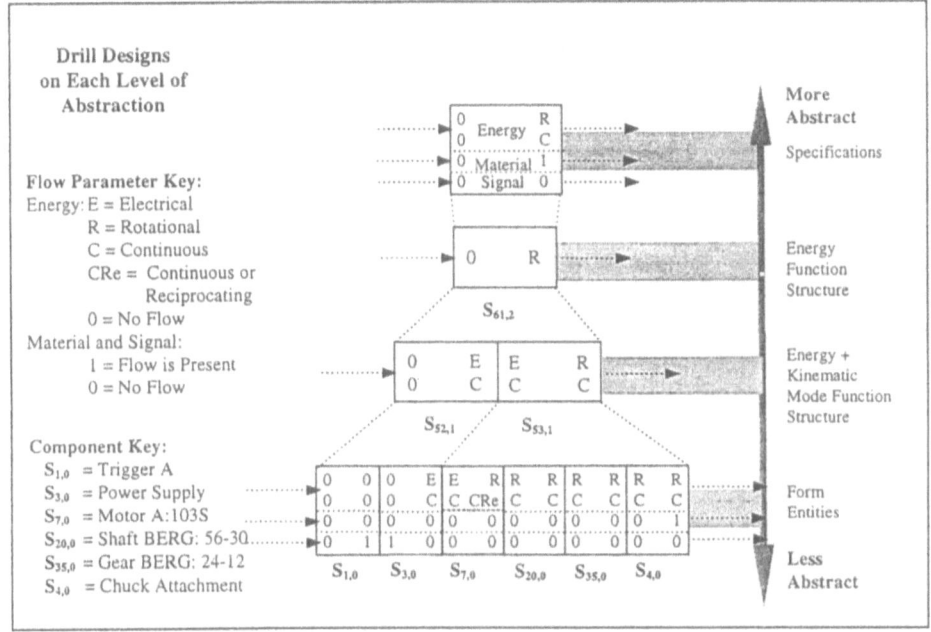

Figure 3. A FFREADA drill design.

The drill design introducing this section can also be described by a tree diagram (Figure 3). The string is a complete design (D) made up of three different designs of the same drill, each valid for one level of abstraction. The design on Level 1 is as follows:

$$D_1 = S_{52,1} S_{53,1}.$$

The Level 1 design, a function structure, depicts a machine that receives no energy input but produces continuous electrical energy that is then converted into continuous rotational energy. This Level 1 design satisfies the pattern created by the Level 2 design (a single symbol, $S_{61,2}$) and satisfies the specifications on energy input. The abstraction grammar uses the Level 1 design and specifications as a pattern for creating the Level 0 design of symbols representing form components as seen at the bottom of the figure.

The ability to generate a single drill design is not sufficient to declare the creation of a designer assistance tool. The goal is to generate all possible designs and, by combining some form of search mechanism with the grammar, to generate only a fraction of those possible designs in order to efficiently present the designer with a selection of very good designs. FFREADA integrates a recursive simulated annealing optimization process with design on each level of abstraction (Figure 4). When FFREADA is run in optimization mode, the algorithm converges to near-optimal designs.

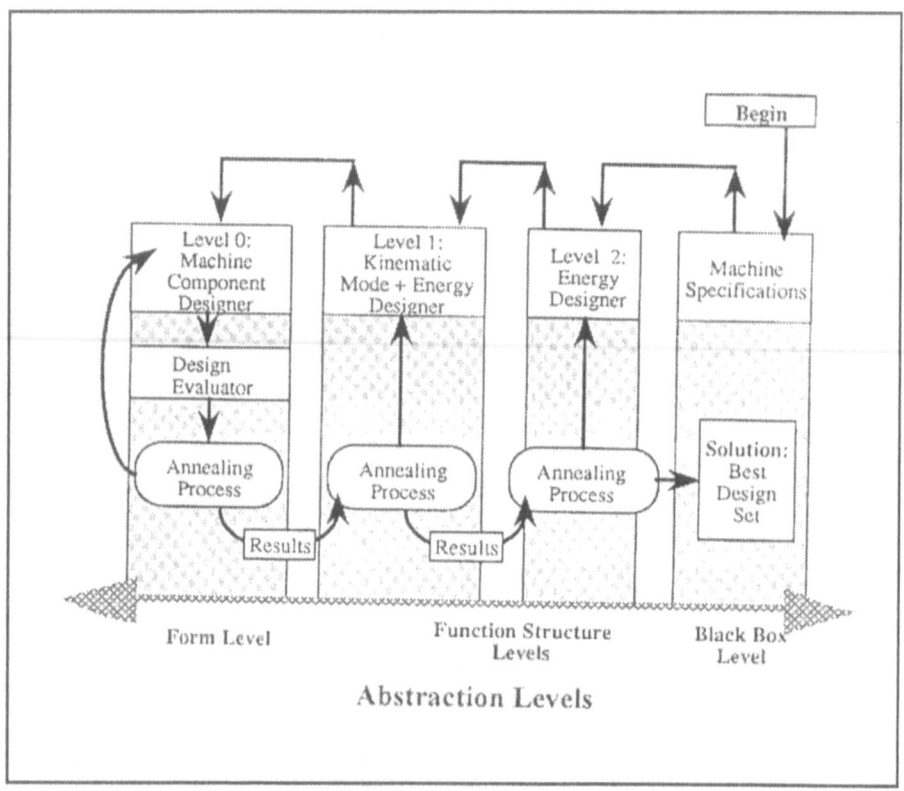

Figure 4. The FFREADA design algorithm.

4.2. ABSTRACTION GRAMMAR DESIGN RESULTS

The design with the global minimum cost for a drill generating at least 1 ft-lb of torque is easily found during a set of 10 optimization runs by FFREADA. This design was confirmed to be the optimal solution for this objective function by three, independent, random generation runs of 2,000,000 designs each. The design, represented as a string of entities, is as follows:

$$S_{1,0}S_{3,0}S_{10,0}S_{16,0}S_{45,0}S_{47,0}S_{4,0}S_{50,1}S_{61,2}.$$

The design can also be described in words as follows (refer to the flow parameter key in Figure 3):

Level 2:	0->R ($S_{61,2}$)	Fulfills:	Specifications
Level 1:	0->R 0->C ($S_{50,1}$)	Fulfills:	0->R
Level 0:			
	trigger B ($S_{1,0}$)	Fulfills:	0->R 0->C
	power supply ($S_{3,0}$)	Fulfills:	0->R 0->C
	motor HL:2100 ($S_{10,0}$)	Fulfills:	0->R 0->C
	shaft BERG: 54-25 ($S_{16,0}$)	Fulfills:	0->R 0->C
	gear BERG: 64-12 ($S_{45,0}$)	Fulfills:	0->R 0->C
	gear BERG: 64-30 ($S_{47,0}$)	Fulfills:	0->R 0->C
	chuck attachment ($S_{4,0}$)	Fulfills:	0->R 0->C

Design torque:	1.02 ft-lbs.
Number of form level designs generated in run:	271,611
Run time:	3.25 Minutes

In the 10-run set, FFREADA converges to designs with costs within 1.010% of the minimum generated in the run and within 1.014% of the global minimum. The average number of form level designs generated per run is 301,391. This represents less than 0.15% of the design state space when a 10-component limit is imposed on the drill designs. Limiting FFREADA to designing drills of 10 machine components or less results in a design space of roughly 200 million design states. However, a 250-component limit was used in the annealing runs. Assume only a 10-fold increase in designs solutions with each additional component and the resulting design state space would exceed $(200,000,000)10^{240}$ in size. The percentage of visited states prior to FFREADA's convergence is impressively small. The algorithms of this paper are run on a DEC 3000 workstation.

4.3. STRING ABSTRACTION GRAMMAR LIMITATIONS

The string abstraction grammar is successful in the FFREADA algorithm for drill designs because drill components are arranged in series. A series arrangement is naturally expressed as a string. Many machines can be designed as an arrangement of components in series, especially those that have a process orientation. Overall, however, the serial design limitation is restrictive.

In addition to being unable to express non-serial arrangements of machine components, a string grammar is unable to express function sharing, an essential characteristic of many good designs. Function sharing occurs when a component fulfills multiple functions in a device. Reviewing Figure 3 again, it is clear that there is no mechanism in the string abstraction grammar for expressing shared functionality. Additional design steps applied after the design grammar is finished or other types of grammars must be considered for this purpose. In the next section, a graph grammar is described that is well-suited for implementing function sharing during the design process.

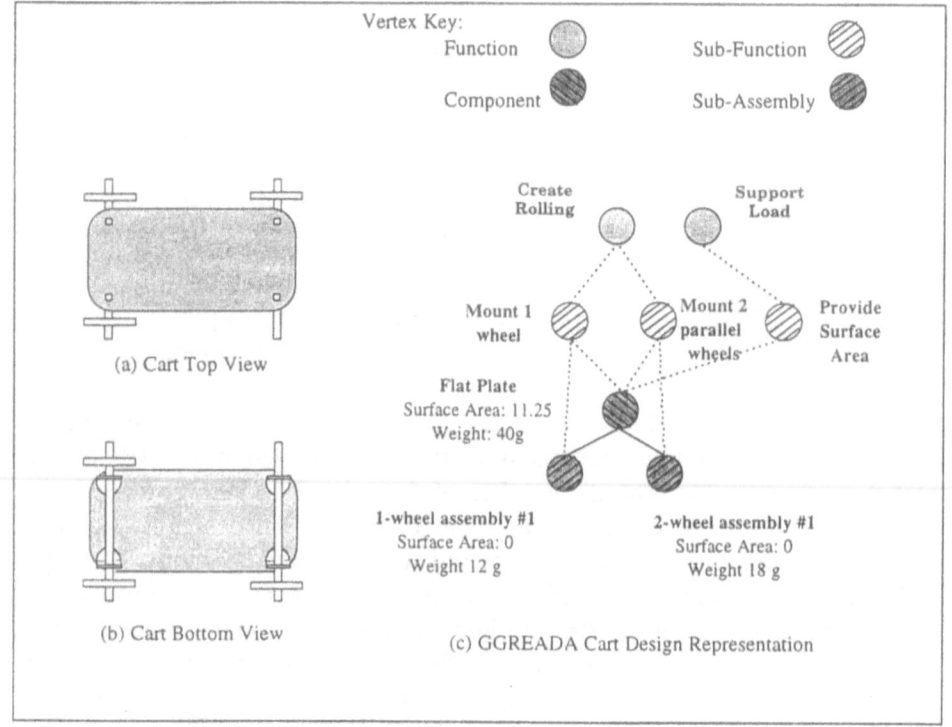

Figure 5. A typical GGREADA Meccano Set cart design.

5. Graph Grammar Design Application: Meccano Set Carts

A cart is a machine that moves by rolling and provides space to carry cargo. An automobile would satisfy the functional definition of a cart, as would a horse-drawn or child-propelled wagon. The GGREADA algorithm generates rolling carts meeting user specifications from scale model Meccano Set mechanism components (Figure 5). GGREADA (Graph Grammar REcursive Annealing Design Algorithm), like FFREADA, uses the design process described by the recursive design model (Figure 1) where design occurs on hierarchically ordered levels of abstraction. GGREADA designs on two levels of abstraction below the black box machine specifications level. During the search for an optimal design, simulated annealing is applied in a recursive fashion to control the design generation process. Like FFREADA, GGREADA's abstraction grammar generates designs on each level of abstraction from a pattern design generated on the next higher level of abstraction. In the cart of Figure 5, the abstract design passed to the component level of abstraction is "mount 1 wheel, mount 2 parallel wheels, and provide surface area." GGREADA's grammar is a graph grammar, allowing GGREADA's designs to have complex and non-serial component arrangements.

5.1. GRAPH ABSTRACTION GRAMMAR

A graph abstraction grammar is used to create cart designs from a vocabulary of vertices. Each terminal vertex in the grammar's vocabulary represents a function, sub-function, Meccano Set component, or component sub-assembly. These are denoted by $t_{i,j}$, where the first subscript is an identification number and the second indicates the level of abstraction to which the vertex is assigned. Non-terminal vertices are labeled $n_{\varnothing, j}$. Figure 6 holds the Mecanno Set pieces in the grammar's component vocabulary and two of the sub-component assemblies relevant to the examples shown in this paper.

The edges of GGREADA's graphs represent either a physical relationship between the vertices or a functional relationship. A physical edge between two components indicates that they are connected. Functional relationships exist between vertices on different levels of abstraction. As with FFREADA, a vertex is selected to participate in a design only if it satisfies, in full or in part, functionality required by a vertex in the design pattern from the next higher level of abstraction. For example, in the cart of Figure 5, two sub-function vertices, "mount 1 wheel" and "mount 2 parallel wheels", are used to instantiate the function "create rolling."

GGREADA's graph-based abstraction grammar represents function sharing explicitly. The Meccano component named "flat plate" is satisfying the functionality required by all three of the sub-function components. The "flat plate" component acts as a mounting piece for the wheel sub-assemblies and it provides surface area for the cart.

GGREADA's graph grammar uses rules to transform a graph g_a, which includes at least one non-terminal vertex, into a graph g_b. GGREADA's rules allow addition and subtraction of symbols, in this case vertices representing vocabulary symbols. The non-terminal vertices mark the spot where the design will be changing. Figure 7 shows a GGREADA cart design at four different points during the design process. For example, Figure 7 (a) is the graph after the 5-sided plate, represented as terminal vertex $t_{3,1}$, is chosen as a base piece. Figure 6 holds the key for component and sub-assemblies used in Figure 7.

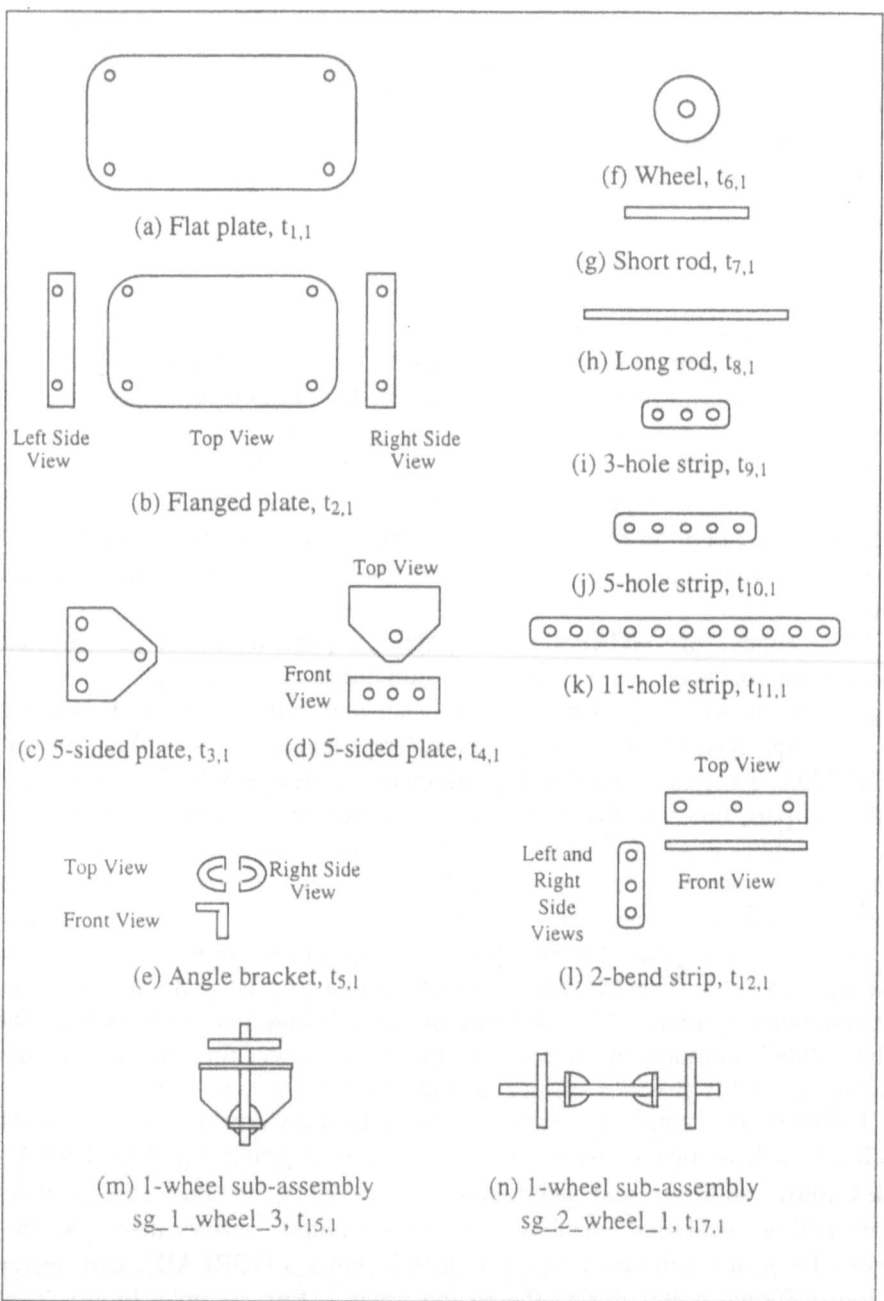

(a) Flat plate, $t_{1,1}$

(b) Flanged plate, $t_{2,1}$

(c) 5-sided plate, $t_{3,1}$

(d) 5-sided plate, $t_{4,1}$

(e) Angle bracket, $t_{5,1}$

(f) Wheel, $t_{6,1}$

(g) Short rod, $t_{7,1}$

(h) Long rod, $t_{8,1}$

(i) 3-hole strip, $t_{9,1}$

(j) 5-hole strip, $t_{10,1}$

(k) 11-hole strip, $t_{11,1}$

(l) 2-bend strip, $t_{12,1}$

(m) 1-wheel sub-assembly
sg_1_wheel_3, $t_{15,1}$

(n) 1-wheel sub-assembly
sg_2_wheel_1, $t_{17,1}$

Figure 6. Meccano Mechanism Set pieces used for cart design.

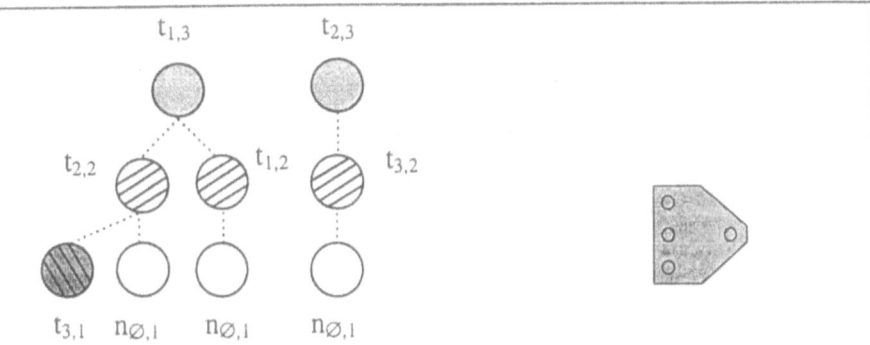

(a) Graph and component version of cart design with base component.

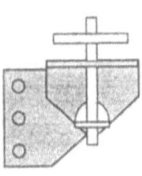

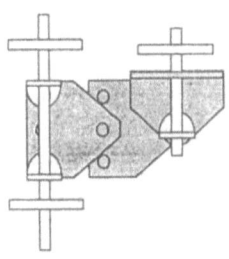

(b) Cart design after $t_{15,1}$ wheel sub-assembly is selected to satisfy $t_{2,2}$.

(c) Cart design after $t_{17,1}$ and an additional flanged plate are selected to satisfy $t_{1,2}$.

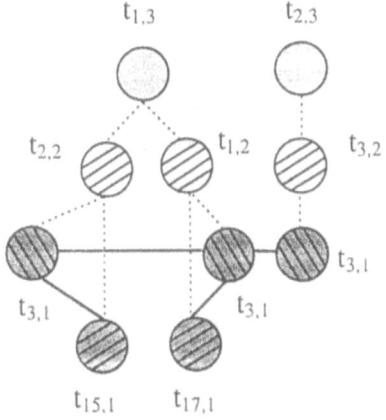

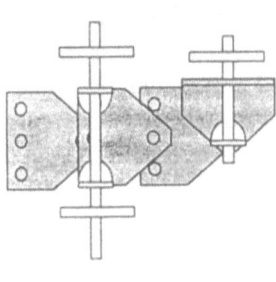

(d) Graph and component versions of a complete cart design.

Figure 7. Interim GGREADA cart designs.

5.2. ABSTRACTION GRAMMAR DESIGN RESULTS

GGREADA's Meccano Set cart design problem has a large space of solutions relative to the number of unique component pieces (12) of which the solutions are comprised. Multiple occurrences of the same component in a single design are possible. To derive the size of the design state space, consider first the number of options for mounting wheels to satisfy the "create rolling" function. This function requires the mounting of 3 or 4 wheels. There are 32 unique ways to mount either 1 or 2 wheels (8 base component pieces with 2 mounting joints and 4 sub-assemblies for each mounting scenario). There are 4 different ways in which the "mounting" components can be combined to produce the mounting of 3 or 4 wheels. These options produce 67,584 different cart designs before a base piece is selected for the mounting and before the load carrying requirement is addressed.

If there is no minimum surface area requirement for the cart design, there are 2(67,584) designs available from the library, using one Meccano base piece (with 4 or more joints). The designs begin their combinatorial explosion from this point as new pieces are selected for addition to the designs to provide surface area. For example, to provide 8 in^2 of surface area, there are over 600,000 designs with 3 base pieces and over 3.6 million designs with 4 base pieces. The design space grows in size as designs using more base pieces are included.

GGREADA is able to generate optimal cart designs using the amount of surface provided as a constraint and minimizing the weight of the cart as the objective function. The optimal design from GGREADA's vocabulary of Mecanno Set components for a cart that provides 4 in^2 of surface area is the cart depicted in Figure 7. This design weighs 68 grams and provides 4.125 in^2 of surface area. A near-optimal solution is the cart depicted in Figure 5. It weighs 70 grams and provides 8 in^2 of surface area. GGREADA was able to converge to one or the other of these solutions in each of a set of 5 annealing runs, averaging 11,400 design generations and about 45 seconds per run.

5.3. GRAPH ABSTRACTION GRAMMAR LIMITATIONS

GGREADA shows that it is possible to write an abstraction graph-grammar for the design of carts. GGREADA's limitations in this example stem from the representation of the Meccano Set component pieces. The amount of geometric knowledge available to GGREADA via the representation is minimal. The rules assure that the connection between the Meccano pieces can be made but they do not specify at exactly which joints of the pieces they are being made. Full connectivity information is determined by

GGREADA, yet some interpretation of results by the designer is required to assemble the final designs. To increase the power of the algorithm, more geometric knowledge about each component is needed as is a means to reason about the knowledge during the rule applications.

6. Strengths and Weaknesses of Grammar-Based Machine Design

Designing abstraction grammar-based design algorithms involves two categories of issues, those pertaining to the grammar and those pertaining to the search mechanism used to select good designs. This discussion focuses exclusively on the grammar issues.

The strength of the abstraction grammar-based algorithms presented here is that an entire space of designs can be generated automatically. This provides a designer with access to considerable knowledge about the design problem. The weakness of this approach is the amount of work it takes to write a grammar, the representational complexities involved, and the limitations placed on designs that can be generated from a well-defined system like a grammar.

6.1. STRENGTHS

A design state space generated by a grammar-based algorithm can be explored in several ways with several goals. One form of exploration already demonstrated is to search for an optimal solution using one or more measures of goodness expressed in an evaluation function. Exploration of a space of designs can have other goals such as:

1. To search for designs using a particular machine component, or component assembly.
2. To catalog all designs of a given nature.
3. To assess the impact of altered design rules on the language.
4. To explore changes in the design state space as the result of adding entities to the library of design building blocks or changing performance parameters of existing entities.
5. To compare different design models and optimization strategies in the search for best designs for a class of problems.
6. To observe the effects of different mappings of the design state space by different evaluation functions.
7. To determine the response of design generation to imposition of design constraints.

6.2. WEAKNESSES

Practical problems in implementing this type of algorithm center on establishing a grammar for the design of realistic machines and devising efficient optimization procedures to search the resultant space. Developing representation systems sensitive enough to express the behavior of single machine components in order that they can be used to their full potential is difficult. Yet more challenging is the task of developing a means to represent component interactions so that "the whole is greater than the sum of the parts", as is true in examples of good designs. The two sample design grammars presented here indicate that the concept of a grammar-based designer assistance tool is possible. Successful implementation of more complex grammars with richer function and form representation is necessary to prove that this type of algorithm is practical.

Of theoretical concern is the question, "What possibility does this type of algorithm provide for generating novel, innovative, or creative designs?" It can be argued that once the design problem is so well-defined that it is possible to implement a grammar-based, generative design tool the design problem has long been solved. On the other hand, the task of establishing a grammar vocabulary so sufficient for designing a particular type of device that a complete design space is defined is, in a sense, impossible. A new functional idea or machine component can always be identified that will expand the space of design solutions. However, it is highly probable that, in the combinatoric design spaces under consideration, machine components will be combined in novel ways by a generative, grammar-based algorithm, and some of these combinations are bound to be innovative. The question of the creativity of the paradigm depends entirely on the definition of creativity and remains an open question.

7. Conclusion

Grammars are a useful formalism for mechanical design because they provide the means of representing an entire space of design solutions to a given problem. An automated grammar-based design generation system coupled with a stochastic vocabulary selection processes in the grammar rule applications can generate a wide variety of designs in the solution space. This generative capacity will allow the designer to survey that solution space in general and glean information about the design problem and the solutions. Also, a search engine can be integrated with the design generation process, guiding it to converge on good solutions for the designer's review. The major challenge to creating designer assistance tools using grammars is developing representations for the grammar that are adequate to express machine component functionality on any number of levels of abstraction meaningful to a given design problem.

Acknowledgements

Linda Schmidt gratefully acknowledges the support of the National Defense Science and Engineering Graduate (NDSEG) Fellowship Program. The authors acknowledge the National Science Foundation for providing support for this research under grants DDM-9258090 and DDM-9301096. Both authors also thank Dr. Robert Sturges for helpful discussions on the cart design problem.

References

Brown, K. N., McMahon, C. A. and Sims Williams, J. H.: 1994, A formal language for the design of manufacturable objects, *in* J. S. Gero and E. Tyugu (eds), *Formal Design Methods for CAD*, North Holland, Amsterdam, pp. 135-155.

Cagan, J. and Mitchell, W. J. (1993), Optimally directed shape generation by shape annealing, *Environment and Planning B*, **20**, 5-12.

Chase, S. C.: 1989, Shapes and shape grammars: from mathematical model to computer implementation, *Environment and Planning B*, **16**, 215-242.

Chomsky, N.: 1957, *Syntactic Structures*, The Hague: Mouton.

Deng, Y-S.: 1994, *Feature Based Design: Synthesizing Structure from Behavior*, PhD Thesis, Department of Industrial Engineering, University of Pittsburgh.

Downing F. and Flemming, U.: 1981, The bungalows of Buffalo, *Environment and Planning B*, **8**, 269-293.

Fitzhorn, P.: 1986, A linguistic formalism for engineering solid modeling, *in* H. Ehrig, M. Nagl and A. Rosenfeld (eds), *Graph-Grammars and Their Application to Computer Science*, Springer-Verlag, Berlin, pp. 202-215.

Flemming, U.: 1987, More that the sum of the parts: the grammar of Queen Anne houses, *Environment and Planning B*, **14**, 323-350.

Fu, Z., De Pennington, A. and Saia, A.: 1993, A graph grammar approach to feature representation and transformation, *International Journal of Computer Integrated Manufacturing*, **6**(102), 137-151.

Gips, J. and Stiny, G.: 1972, Shape grammars and the generative specification of painting and sculpture, *in* C. V. Freiman (ed.), *Information Processing 71*, North-Holland, Amsterdam, pp. 1460-1465.

Gips, J. and Stiny, G.: 1980, Production systems and grammars: a uniform characterization, *Environment and Planning B*, **7**, 399-408.

Hoover, S. P. and Rinderle, J. R.: 1989, A synthesis strategy for mechanical devices, *Research in Engineering Design*, **1**, 87-103.

Knight, T. W.: 1980, The generation of Hepplewhite-style chair back designs, *Environment and Planning B*, **7**, 227-238.

Koning, H. and Eizenberg, J.: 1981, The language of the prairie: Frank Lloyd Wright's prairie houses, *Environment and Planning B*, **8**, 295-323.

Krishnamurti, R.: 1980, The arithmetic of shapes, *Environment and Planning B*, **7**, 463-484.

Krishnamurti, R.: 1981, The construction of shapes, *Environment and Planning B*, **8**, 5-40.

Krishnamurti, R.: 1992a, The maximal representation of a shape, *Environment and Planning B: Planning and Design*, **19**, 267-288.

Krishnamurti, R.: 1992b, The arithmetic of maximal planes, *Environment and Planning B: Planning and Design*, **19**, 431-464.

Krishnamurti, R. and Stouffs, R.: 1993, Spatial grammars: motivation, comparison, and new results, *in* U. Flemming and S. Van Wyk (eds), *CAAD Futures '93*, Elsevier Science, pp. 57-74.

Longenecker, S. N. and Fitzhorn, P. A.: 1991, A shape grammar for non-manifold modeling, *Research in Engineering Design*, **2**, 159-170.

Mitchell, W.: 1991, Functional grammars: an introduction, *in* Goldman and Zdepski (eds), *Proceedings of Association for Computer Aided Design in Architecture '91*, Reality and Virtual Reality, Los Angeles, California.

Mullins, S. and Rinderle, J. R.: 1991, Grammatical approaches to engineering design, Part I: an introduction and commentary, *Research in Engineering Design*, **2**, 121-135.

Pahl, G. and Beitz, W.: 1988, *Engineering Design—A Systematic Approach*, Springer-Verlag, New York.

Pinilla, J. M., Finger, S. and Prinz, F. B.: 1989, Shape feature description using an augmented topology graph grammar, *Preprints: NSF Engineering Design Research Conference*, Amherst, MA, pp. 285-300.

Reddy, G., and Cagan, J.: 1995, An improved shape annealing algorithm for truss topology generation, *ASME Journal of Mechanical Design*, **117**(2), 315-321.

Rinderle, J.: 1991, Grammatical approaches to engineering design, Part II: Melding configuration and parametric design using attribute grammars, *Research in Engineering Design*, **2**, 137-146.

Schmidt, L. and Cagan, J.: 1992, A recursive shape annealing approach to machine design, *in* J. S. Gero and F. Sudweeks (eds), *Preprints of the Second International Round-Table Conference on Computational Models of Creative Design*, Key Centre of Design Computing, University of Sydney, pp. 145-171.

Schmidt, L. C. and Cagan, J.: 1995, Recursive annealing: A computational model for machine design, *Research in Engineering Design*, **7**, 102-125.

Schmidt, L.: 1995, *An Implementation using Grammars of an Abstraction-Based Model of Mechanical Design for Design Optimization and Design Space Characterization*, PhD Thesis, Carnegie Mellon University, Pittsburgh, PA.

Shea, K., Cagan, J. and Fenves, S.: 1995, A shape annealing approach to optimal truss design with dynamic grouping of members, *ASME 21st Design Automation Conference*, Boston, MA, DE-Vol. 82, pp. 377-384.

Sthanusubramonian, T., Finger, S and Rinderle, J. R.: 1992, A transformational approach to configuration design, *Proceedings of the 1992 NSF Design and Manufacturing Systems Conference*, Atlanta, Georgia, pp. 419-424.

Stiny, G.: 1980a, Introduction to shape and shape grammars, *Environment and Planning B*, **7**, 343-351.

Stiny, G.: 1980b, Kindergarten grammars: designing with Froebel's building gifts, *Environment and Planning B*, **7**, 409-462.

Stiny, G.: 1988, Formal devices for design, *Design Theory 88*, Springer-Verlag, New York.

Stiny, G. and March, L.: 1981, Design machines, *Environment and Planning B*, **8**, 245-255.

Stiny, G. and Mitchell, W. J.: 1978, The Palladian grammar, *Environment and Planning B*, **5**, 5-18.

7

design spaces

J. S. Gero and F. Sudweeks (eds), Artificial Intelligence in Design '96, 347-366.

DESIGN SHEET: A SYSTEM FOR EXPLORING DESIGN SPACE

Application to automotive drive train life analysis

SUDHAKAR Y. REDDY AND KENNETH W. FERTIG
Rockwell Science Center, Palo Alto Laboratory
444 High Street, Suite 400, Palo Alto, CA 94301, USA

Abstract. This paper describes Design Sheet, an advanced software system which facilitates the conceptual design of complex engineering systems. Design Sheet enables the designer to quickly explore large areas of design space and study how the different performance and cost criteria tradeoff with respect to one another. It provides an interactive interface for building analysis models in terms of systems of nonlinear algebraic equations, and automatically writes computational procedures for solving these equations based on user-specified tradeoff criteria. The paper briefly describes the principles and the methodology behind the software, and showcases some of its capabilities. To demonstrate its practical applicability, a system-level fatigue life analysis model of an automotive drive train has been developed. The paper discusses the model and how it is used in performing design tradeoff studies.

1. Introduction

The design of complex engineering systems is a hierarchical process. In current industry practice, computer-aided design tools are widely used during the later stages of this process for determining detailed design specifications. Though design decisions made during the early stages have a far greater impact on the final design quality and cost, tools for supporting conceptual design are limited in availability as well as scope. In contrast to detailed design, where the goal is to modify nominal design specifications such that an optimal design is obtained, the critical need during conceptual design is to quickly search the entire design space in order to provide good specifications for the detailed design phase. Therefore, the approaches used for detailed design analysis and optimization are not suitable for use during the early conceptual stages.

During conceptual system-level design, a candidate design needs to be evaluated with respect to multiple performance criteria. This will ensure that the designs generated in the later stages are optimal with respect to the overall system criteria, rather than being locally optimal solutions from individual perspectives. Additionally, as cost is often a crucial criterion in system design, multiple attribute tradeoffs with combined performance and cost models are needed for exploring the large design space. Which tradeoffs are important at any stage during design, however, are not known a priori and depend on the results of other tradeoff analyses. Conceptual design tools, therefore, require an ability to use integrated performance and cost models and a capability to flexibly define tradeoff studies.

The research community has recognized the importance of integrated models in design; several efforts are aimed at developing collaborative design environments that integrate multiple models and tools used by engineers (e.g., Herman and Lu, 1992; Cutkosky, 1993). However, the type of model integration facilitated by these efforts is not quite appropriate for conceptual design, where models based on multiple, incomplete and quickly evolving system descriptions need to be used together for tradeoff analyses. Conceptual design research, on the other hand, has mainly focused on the role of optimization and robust design (e.g., Dixon, et al., 1993), without any emphasis on how to use integrated models. Further, most design tools either do not provide integrated support for tradeoff studies or hard-code a limited set of them.

The difficulties discussed above can be overcome if the representation of knowledge is in a declarative form, and separated from the mechanisms that control the use of the knowledge. This has been the thrust of Artificial Intelligence, and the application of appropriate techniques from this realm is bound to make a significant impact on the development of tools for system design. In many system design problems, the natural representation of knowledge is in the form of algebraic equations. Using the declarative knowledge for tradeoff analysis requires the solution of systems of nonlinear equations. Constraint propagation techniques provide a powerful mechanism for accomplishing this task; they represent the equations as constraints between variables, and propagate the changes in variable values across the constraint network.

Constraint propagation approaches have been used by several researchers in creating conceptual design systems. The main advantage of these approaches is in decomposing large systems of equations into subsets of more manageable size, which are solved individually before being combined to obtain the overall solution. Bouchard et al. (1988) use directed constraints between design variables and numerical solution approaches to allow rapid production of tradeoff studies. This approach, however, forces the designer to decide a priori which variables are input and which are output. Serrano (1987) has developed a constraint management approach based on bipartite matching for efficiently decomposing large systems of algebraic equations, and strong component identification for determining subsets of equations that need to be solved simultaneously. Fromont and Sriram (1992) use planning techniques to add flexibility to such systems, such as allowing constraints to be added incrementally. In these approaches, the simultaneous subsets are solved either symbolically or numerically, without further decomposition. In practical applications, where such subsets can include several tens of simultaneous nonlinear equations, neither symbolic nor numerical solution techniques are feasible. Krishnan, et al. (1990) discuss issues in user-directed constraints as well as suggestions for further decomposing strong components representing the equations that need to be solved simultaneously. However, their treatment is not comprehensive in either aspect. Ward (1989) has extended constraint management for propagating interval values. However, this is only practical for linear systems. Ramaswamy and Ulrich (1993) have extended Serrano's work by developing a constraint system with a spread-sheet interface.

In summary, current constraint-based design systems suffer from one or more of the following shortcomings: do not solve large nonlinear systems of simultaneous equations that are often present in practical system analysis models; fail to scale up to real-life analysis problems, which require from a few hundreds to a few thousands of equations to be solved; and do not provide flexible interfaces needed for performing what-if analysis and defining tradeoff studies, which is essential during conceptual design. We have developed a constraint-based system, called Design Sheet, for solving large systems of simultaneous nonlinear equations. It uses graph-theoretic algorithms for breaking down large systems of equations into sub-systems of more manageable size, before solving them. It further integrates this with a graphical user-interface, which allows the system designer to quickly and easily define and perform new tradeoff analyses.

The next section (Section 2) briefly describes the principles and method-ology underlying Design Sheet, and discusses its unique features and capabilities. The rest of the paper demonstrates the application of Design Sheet to the fatigue life analysis of an automotive drive train. Section 3 describes the problem, derives the governing equations, and discusses the important issues in developing the Design Sheet model for drive train life prediction. Section 4 presents sample analyses and interesting tradeoff plots obtained using the Design Sheet model. Finally, Section 5 concludes with a discussion of the current status and the future direction of this research.

2. Design Sheet: A Tool For System Design

Design Sheet is an engineering design and analysis tool for conceptual design. Design Sheet allows the user to input design equations in their natural mathematical representation and solves these equations without requiring the user to provide computational procedures. It represents algebraic equations as constraints between variables, and uses a constraint propagation approach for determining which variables are dependent on which others. It also finds a computational sequence for evaluating the values of dependent variables, given the values of independent variables. In this process, it solves the set of nonlinear equations which make up the constraint network using symbolic methods when possible and resorting to numerical methods otherwise.

Design Sheet uses graph-theoretic algorithms to decompose the constraint set into tractable subsets, solves them individually and combines them to get the overall solution. The constraint network is represented as a directed bipartite graph; a bipartite graph has two types of nodes with arcs only between nodes of different types. In this case, the variables and the relations are the two types of nodes and the arcs between them signify that a variable is in a particular relation. A variable can be either independent, meaning that its value can be set and freely varied by the user; dependent, meaning that its value is determined by a relation, whose other variables are all either dependent or independent; or undetermined, meaning that its value cannot be determined by the current state of the constraint network.

Figure 1 shows a simple set of equations and the corresponding bipartite graph. The arc directions in the graph signify what variables are inputs to, and outputs of, a particular relation. In this example, the variables m1 and f are independent; therefore, the arcs from these variables are directed outward. When a relation is used to compute a variable, the arc between them is directed towards the variable; the variable is then determined and the rest of the arcs from it are directed outward. When all but one of the arcs on a relation node are directed inward, the relation can be used to calculate the remaining variable.

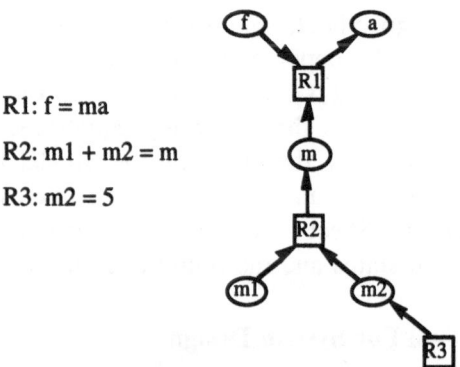

R1: f = ma

R2: m1 + m2 = m

R3: m2 = 5

Figure 1. Bipartite graph of a simple set of equations.

As equations are added or variables made independent, Design Sheet assigns appropriate arc directions and propagates them to determine all the dependent variables. This process essentially decomposes the constraint network into a sequence of subsystems which are solved individually, thus improving the robustness of the overall solution procedure. In the example shown in Figure 1, m1 and f are independent, and the three equations need to be solved for the remaining variables, m2, m and a. Instead of solving the system as a 3x3 system, Design Sheet uses equation R3 to solve for m2, equation R2 to solve for m and equation R1 to solve for a.

Unlike in the above example, it is not always possible to decompose the solution of a constraint network into steps which use a single equation to calculate the value of a single variable. When a set of equations need to be solved simultaneously, they form a directed cycle in the graph, called a strongly connected component (SCC). The graph-theoretic approach is essential for decomposing a non-trivial constraint network. The decomposition process involves several steps—directing the graph, variable determination, plan construction and component decomposition. Directing is accomplished using a variant of the Ford-Fulkerson algorithm for finding maximal matchings on bipartite graphs (Cormen et al., 1991). SCCs are identified using a standard backward search and marking algorithm. Once Design Sheet figures out which variables are determined in the graph, it can construct a plan for computing the value of any or all of the determined variables. This is accomplished by topologically sorting the appropriate portion of the constraint graph. In the resulting plan, a single evaluation step corresponds to each SCC. A detailed description of the decomposition and the plan construction mechanisms in Design Sheet is described in Buckley, Fertig and Smith (1992).

A SCC can be solved symbolically when feasible, or numerically by guessing at initial values for all the variables in the SCC. However, the robustness of such procedures deteriorates rapidly as the number of equations and variables increases. A unique feature of Design Sheet is in further decomposing a SCC by judiciously choosing a set of iteration variables, whose values when known unravel the cycle, so that the system can be solved sequentially. Once an appropriate set of iteration variables is determined, Design Sheet uses constraint propagation to derive an error relation corresponding to each iteration variable. The system of simultaneous equations is then solved by iterating over values of these variables so as to make the residual errors in the error relations negligible. All possible combinations of variables in the component are possible candidates for selection as iteration variables. Design Sheet uses a heuristically guided branch and bound search together with smart pruning of the set of candidate iteration variables, to efficiently find the minimal set from among all these possibilities.

With nonlinear systems, it is sometimes easier to use an equation in one direction than another, and in other cases, an equation cannot be solved in a particular direction. For example, the equation, $Sin(x) + y = z * Log(z)$, is difficult to solve for z and the equation, $z = If (x>y, y, y^2)$, cannot be solved for x. Design Sheet uses direction penalties to account for the former and forced directions to account for the latter.

2.1. CAPABILITIES OF DESIGN SHEET

Conceptual design requires capabilities for what-if analysis and tradeoff studies. Design Sheet provides a spread-sheet like interface for carrying out what-if analysis, and a menu-driven graphical interface for performing design tradeoff studies.

2.1.1. What-if analysis

Design Sheet allows the designer to type in new values for independent variables and see their effect on dependent variables. It allows the user to decide what the input and output parameters are at run time. It automatically restricts such choices to those that are mathematically feasible. When the user changes the choice of inputs, Design Sheet automatically produces the computational procedures needed to invert the model. Design Sheet can also use Lisp or FORTRAN programs to define constraints among design variables, and can reverse flow through such programs by iterating on input parameters.

2.1.2. Tradeoff studies

The main purpose of Design Sheet is to perform tradeoff studies, which involve determining the effect of different values of independent variables on the values of dependent variables. The user defines new tradeoff studies simply by specifying the independent and dependent parameters of interest and the ranges in which to vary the independent variables. Design Sheet displays the results of tradeoff studies using trade tables and plots. These facilities have proved very useful for studying performance-cost and other multiple-objective tradeoffs, and for quickly exploring large areas of design space.

2.1.3. Error propagation

During conceptual design, estimated values are often used for some of the independent parameters, but the designer would still like to know how errors in these estimates affect the various tradeoffs. Design Sheet allows standard deviations to be specified for the independent variables. It uses the same constraint network that defines the variables, for propagating the effects of errors in the independent variables. It assumes that the errors in the independent parameters are not correlated and performs a first order error analysis.

2.1.4. Constrained optimization

Design Sheet has a limited capability for constrained optimization. It allows any dependent variable to be minimized, subject to user-defined inequality constraints. The basic algorithm iteratively determines if a given constraint is active (or inactive), and if so adds (or removes) it as an equality constraint, before solving the minimization problem using a modified gradient-search algorithm.

3. Developing a Model for Drive Train Life Prediction

System-level drive train life prediction is an important analysis activity at Rockwell Automotive's engineering department. Currently, the major tool for drive train life prediction is DTL, a large FORTRAN program (Shih and Keeney, 1992). Though it provides extensive predictive capability for wide range of drive train configurations, it is difficult to provide true flexibility with this tool as new tradeoff studies require extensive re-programming. Design Sheet, on the other hand, is ideal for performing new tradeoff studies rapidly. In order to demonstrate the applicability of Design Sheet to analysis problems as complex as drive train life prediction, we have developed a Design Sheet model for the critical portions of the DTL program.

The model reported here is for a single axle automotive drive train. The goal is to predict the life of the drive train subject to eight critical component failures, which include the bending failure of the ring and the pinion gears, the gear surface contact failure, and the fatigue failure of five bearings. The life of each of these components depends on the stresses at the components, and can be determined from the stresses by using the corresponding S-N fatigue curves. The stresses at the components depend on the effective torque at these components, which are determined by the output or pinion torque and the corresponding gear ratios. The output torque is obtained by solving the governing equations for drive train operation.

3.1. GOVERNING EQUATIONS FOR DRIVE TRAIN OPERATION

A bond graph[1] representation has been used to derive the governing equations for use in Design Sheet. Figure 2 shows the bond graph for the drive train.

The engine can provide a maximum torque of τ_{ea} at full throttle, for a given angular velocity, ω_e. The actual torque output, τ_e, depends on the engine throttle conditions. The effect of a partial throttle is modeled using a variable resistor, which produces a torque loss, $\Delta\tau$, from the available full-throttle engine torque of τ_{ea}. There is an inertial element, I, which models the storage of rotational inertial energy. The angular velocity and torque are transferred through a clutch, a transmission, and a differential to the wheel. The wheel converts angular velocity to linear velocity. The labels "0" and

---

[1] See Rosenberg and Karnopp (1983) for an introduction to bond graphs.

"1" refer to the two types of junctions in the bond graph. The angular (or linear) velocities are summed across the "0" junctions, whereas the torques (or forces) are transmitted unchanged. On the other hand, the torques (or forces) are summed across the "1" junctions, whereas the angular (or linear) velocities are transmitted unchanged.

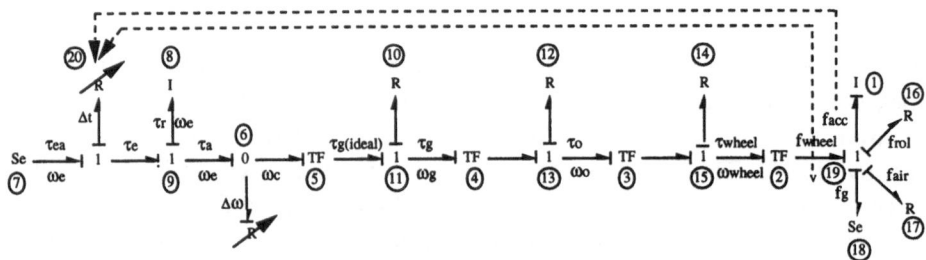

Figure 2. The bond graph of the drive train.

The clutch is modeled with a variable resistor. Effectively, an incremental angular velocity ($\Delta\omega$) is created which assures that ω_e is at least a specified idling value. The bulk of the drive train is modeled with transformer elements, with the addition of resistive elements to account for non-ideal effects which reduce the available torque. Finally, there is a transformer element modeling the wheel, which converts rotational velocity to linear velocity. The forces due to wind resistance (f_{air}), rolling friction (f_{rol}), gravity (f_g) and acceleration (f_{acc}) are considered. The "1" junction at the end balances all the relevant forces: $f_{wheel} = f_{acc}+f_{rol}+f_{air}+f_g$.

The set of all equations for all the junctions and elements of the bond graph constitute the system of governing equations for the drive train. Table 1 shows the equations derived from the bond graph. The equation labels in Table 1 correspond to the numbers beside the junctions and elements in Figure 1. Labels with a "b" suffix refer to equations on the torque side of the "TF" junctions, whereas the non-suffixed labels refer to equations on the speed side of the same junctions. The first six speed equations are differentiated to determine the derivatives needed for calculating the rotational inertia in the engine. The last differential equation (label 19) in Table 1 is merely an equivalence relationship. Maximum limits on velocity and acceleration require the partial-load operation of the engine. The dotted lines in Figure 1 represent this as a variable torque loss, which depends non-linearly on the velocity and the acceleration, and is modeled using equations 20 and 21 in Table 1.

3.2. DESIGN SHEET MODEL FOR DRIVE TRAIN LIFE PREDICTION

The equations for calculating the output torques, derived from the bond graph, are input to Design Sheet along with additional relationships for deriving the gear and bearing lives from the torques. The overall Design Sheet model for drive train life is made up of several modules, namely *params, grade, acceleration, grade life, acceleration life, bearing* and *damage* modules, which are described below.

TABLE 1. The governing equations for the drive train.

Label	Equation	Differential equation
1	$v = p/m$	$dv/dt = (dp/dt)/m$
2	$\omega_{wheel} = v/(2*\pi*trad)$	$d(\omega_{wheel})/dt = dv/dt/(2*p*trad)$
2b	$f_{wheel} = \tau_{wheel}*(2*\pi*trad)$	
3	$\omega_o = \omega_{wheel}*gra$	$d(\omega_O)/dt = d(\omega_{wheel})/dt*gra$
3b	$\tau_{wheel}(ideal) = \tau_o*gpin$	
4	$\omega_g = \omega_o*srg$	$d(\omega_g)/dt = d(\omega_o)/dt*srg$
4b	$\tau_o(ideal) = \tau_g/trg$	
5	$\omega_c = \omega_g*ig$	$d(\omega_c)/dt = d(\omega_{trans})/dt*ig$
5b	$\tau_g(ideal) = \tau_a/ig$	
6	$\omega_e = \Delta\omega + \omega_c$	$d(\omega_e)/dt = 0, \Delta\omega>0; d(\omega_e)/dt = d(\omega_c)/dt, \Delta\omega=0$
7	$\tau_{ea} = EngineCurve(\omega_e)$	
8	$\tau_r = I_{rot}*d(\omega_e)/dt$	
9	$\tau_a = \tau_e - \tau_r$	
10	$\Delta\tau_g = \omega_g*geff$	
11	$\tau_g = \tau_g (ideal) - \Delta\tau_g$	
12	$\Delta\tau_o = \omega_o*deff$	
13	$\tau_o = \tau_o(ideal) - \Delta\tau_o$	
14	$\Delta\tau_{wheel} = \omega_{wheel}*weff$	
15	$\tau_{wheel} = \tau_{wheel}(ideal) - \Delta\tau_{wheel}$	
16	$f_{rol} = wtn*res$	
17	$f_{air} = car*v^2$	
18	$f_g = sin(a)*m*g$	
19	$f_{wheel} = f_{acc} + f_{rol} + f_{air} + f_g$	$dp/dt = f_{acc}$
20	$\tau_{ea} = \tau_e + \Delta\tau(v, f_{acc})$	
21	$\Delta\tau = Function(v, f_{acc})$	

3.2.1. The params module

The *params* module is the repository for data tables, relations and variables, which are relevant across the different modules. Data tables are used for modeling the engine torque-speed curve, gear ratios for different gears, and grade distribution over the route. Table 2 shows the table used to represent the engine torque-speed curve; Design Sheet uses linear interpolation to obtain the values of torque for intermediate values of engine speed.

The primary independent variables, which are varied during tradeoff studies, are also stored in this module. Table 3 shows a partial list of these variables, along with their default values, units of measurement and brief

TABLE 2. Engine torque-speed curve data.

speed, rpm	torque, lb-ft
0.0	0.0
1200.0	711.9907
1285.714	732.681
1371.429	731.1246
1457.143	719.921
1542.857	707.7778
1628.571	694.8433
1714.286	681.2361
1800.0	667.0525
3000.0	667.0525

TABLE 3. The primary independent variables.

Name	Value	Units	Description
ar	94.0	ft^2	Vehicle frontal projected area
car	0.188	-	Aerodynamic resistance parameter
dvdtmax	5.0	ft/s^2	Maximum vehicle acceleration
ediff	0.96	-	Differential efficiency
effg	0.99	-	Transmission gear efficiency
ird	1	-	Parameter specifying road type
irotat	0.2	lb-ft^2	Moment of inertia
nax	1	-	Number of axles
rag	3.5833	-	Hypoid gear ratio
rah	1	-	Helical gear ratio
rap	1	-	Gear reduction ratio at wheel end
res	0.012	-	Base value of rolling resistance
rlrs	0.0103	-	Coefficient of rolling resistance
rrm	0.2	-	Rolling resistance parameter for a tire type
srg	1	-	Differential speed ratio
trdd	20.0	in	Tire radius
trg	1	-	Differential torque ratio
vmaxmph	55	mph	Vehicle maximum speed
whi	1800	rpm	Maximum engine rpm controlled by the governor
widle	950	rpm	Idling speed/rpm of engine
wlo	1400	rpm	Minimum engine rpm when shifted into new gear
wt	40000	lb	Gross combined vehicle weight

descriptions. These variables represent the various design parameters, and include the vehicle parameters such as the gross weight and the frontal area, the engine parameters such as the idling speed and the maximum limit on the acceleration, the transmission and differential gear ratios, the tire radius, the coefficients of rolling and air resistance, the different efficiencies, and the operational parameter limits such as the maximum velocity.

3.2.2. The grade and acceleration modules

The *grade* and the *acceleration* modules model the two operating modes, namely constant velocity travel on a grade and acceleration from a stop, respectively. The relations in these modules are essentially those in Table 1, and can be used to calculate the output torque for a given grade or acceleration profile. The relations are represented independently in both the modules, because the drive train life depends on a given acceleration profile together with a given grade profile. Though both the modules have similar equations for the most part, the variables participating in the relations are different between the modules. The velocity in the grade module is the constant velocity at which the vehicle goes up the grade, whereas in the acceleration module, it is the velocity at a given instant during acceleration.

Vector variables and equations are used in both the modules. In the *grade* module, elements of the vectors correspond to different grades in the route, whereas in the *acceleration* module, they correspond to discrete time instances during acceleration. In addition to the equations in Table 1, these modules have relations for enforcing maximum limits on acceleration, velocity and engine speed, and for modeling the gear shift policy. These additional relations are shown in Table 4, for the grade module.

TABLE 4. Additional grade module relations.

Name	Form	Description
GGR01	{(wlo*60)/(trpm*gra*fgr(PG_igrlo[i])) = Max(vlo,PG_v[i])*mph l i<=PG_NMAX}	Use highest possible gear
GGR02	{PG_ig[i] = Max(1, Int(PG_igrlo[i])) l i<=PG_NMAX}	Lowest gear
GLM00	{PG_gr[i] = fgr(PG_ig[i])/fgr(PG_ig[i]+1) l i<=PG_NMAX}	
GLM01	PG_teloss1 = 0.0	Zero acceleration
GLM02	{PG_teloss2[i] = If(PG_v[i] < vmax, 0.0, (PG_teavail[i] * 100*(PG_v[i] - vmax))/vmax) l i<=PG_NMAX}	Maximum limit on velocity, vmax
GLM03	{PG_teloss3[i] = If(PG_we[i] < whi, 0.0, ((PG_gr[i] - 1) * PG_te[i] - (telo - tehi))*(1 - ((PG_gr[i]*wlo - PG_we[i]) / (PG_gr[i]*wlo - whi))^(2*exp))) l i<=PG_NMAX}	Maximum limit on engine rpm, we
GLM04	{PG_teloss[i] = Min(telossmax*PG_teavail[i], PG_teloss1 + PG_teloss2[i] + PG_teloss3[i]) l i<=PG_NMAX}	
GLM05	PG_te = PG_teavail - PG_teloss	

The maximum limits on acceleration, velocity and engine speed are enforced by a reduction in the available engine torque, when such limits are exceeded. The torque reduction (or loss) accounts for the partial load operation of the engine. In practice, this is done by reducing the throttle, but in Design Sheet this is modeled by smooth nonlinear functions which sharply increase the torque loss when the limits are exceeded (see Table 4).

The gear selection policy is modeled by the first two equations in Table 4 for the *grade* module, and is illustrated in Figure 3 for both the modules. The lower curve in Figure 3 corresponds to the constant velocity mode, where the highest gear at which the engine can operate above a minimum engine rpm (e.g., 1400) is selected. The upper curve in Figure 3 corresponds to the acceleration mode, where the gear is shifted up when the engine speed reaches a given value (e.g., 1800 rpm). According to this policy, from Figure 3, at velocity v1, gear 3 is selected in the constant velocity mode, whereas gear 2 is selected in the acceleration mode; at velocity v2, however, gear 4 is selected in both modes.

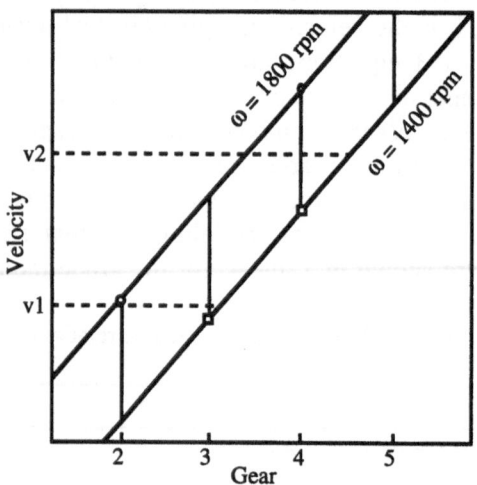

Figure 3. Gear shifting in different modes of operation.

The main difference between the *grade* and *acceleration* modules is the direction in which Design Sheet automatically chooses to use the different relations for calculating the output torque. To illustrate this point, the simplified directed graphs, which represent the systems of equations that need to be solved simultaneously are shown in Figure 4 and Figure 5, for the constant velocity mode and the acceleration mode, respectively. In the figures, the actual systems of simultaneous equations and the variables involved are connected by thick arcs, whereas the variables that are upstream to this system are connected by thin arcs.

As is clear from Figures 4 and 5, the same sets of variables and relations are involved in both the modules. The main difference between them is whether the acceleration force or the velocity is the independent variable. The *grade* module (Figure 4) sets f_{acc} to be equal to zero, and solves for the terminal velocity as well as the output torque for a particular grade. The

variable, and calculates the resulting f_{acc}, the force due to acceleration, and eventually the output torque by solving the force balance equations.

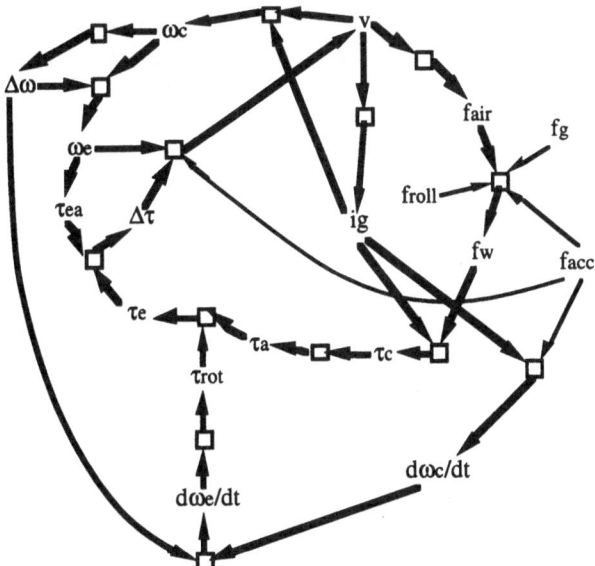

Figure 4. Simplified directed graph of the component in the constant velocity mode.

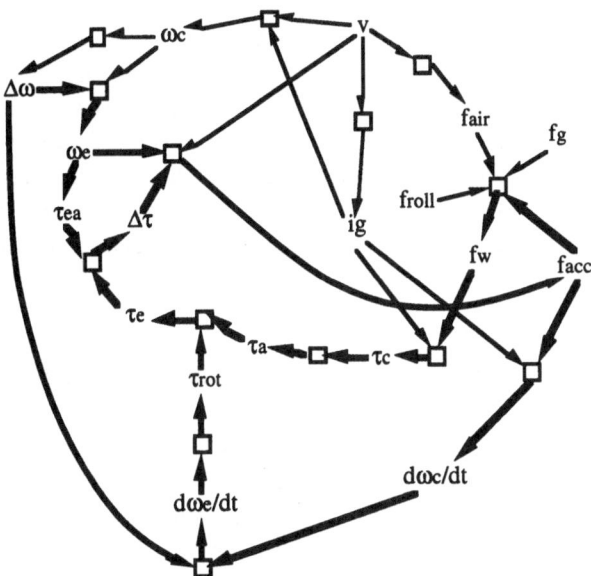

Figure 5. Simplified directed graph of the component in the acceleration mode.

3.2.3. The grade life and acceleration life modules

Corresponding to the two modes of operation, the two *life* modules have equations relating the output (or pinion) torque to the expected life in number of miles for the different components at that torque. Table 5 shows the relations in the *grade life* module; similar relations are defined in the *acceleration life* module.

TABLE 5. Grade life module relations.

Name	Form	Description
GLF11	{PG_spsi[1,i] = (PG_to[i]/gpbr)*cdam | i<=PG_NMAX}	Gear 1 bending stress
GLF12	{PG_spsi[2,i] = (PG_to[i]/ggbr)*cdam | i<=PG_NMAX}	Gear 2 bending stress
GLF13	{PG_spsi[3,i] = Sqrt(PG_to[i])*kpgr | i<=PG_NMAX}	Gear surface stress
GLF14	{PG_ncyc[j,i] = If(PG_to[i] = 0, 1.0e+12, (cgb/PG_spsi[j,i])^egb) | j={1,2}, i<=PG_NMAX}	Gear bending S-N curve
GLF15	{PG_ncyc[3,i] = If(PG_to[i] = 0, 1.0e+12, (cgs/PG_spsi[3,i])^(2*egs)) | i<=PG_NMAX}	Gear surface S-N curve
GLF16	{PG_ncyc[j,i] = If(PG_to[i] = 0, 1.0e+12, If(bd[j-3] = 0, 1.0e+8, cbr*(bd[j-3]/PG_to[i])^ebr)) | j = {4,5,6,7,8}, i<=PG_NMAX}	Bearing lives in terms of bearing factors and pinion torques
GLF17	{PG_nmile[j,i] = PG_ncyc[j,i]/trpm/rag | j={1,3,4,5,6}, i<=PG_NMAX}	Cycles to miles conversion
GLF18	{PG_nmile[j,i] = PG_ncyc[j,i]/trpm | j={2,7,8}, i<=PG_NMAX}	Cycles to miles conversion

The first three relations in Table 5 calculate the stresses in the gears from the pinion torque. The next two relations model the S-N curves for calculating the number of cycles to fatigue failure of the gears; the S-N curves are represented by empirical (exponential) functions. The next equation is used to calculate the bearing fatigue lives from the pinion torque and the bearing factors. Finally, the last two relations are used for calculating the number of miles to failure from the number of cycles to failure, the tire revolutions per mile and the appropriate gear ratios.

3.2.4. The bearing module

The *bearing* module calculates the bearing factors, which are needed to calculate the bearing lives for specific pinion torques. Essentially, the bearing factors are the stresses on the different bearings per unit pinion torque. The *bearing* module calculates these unit stresses by balancing the moments on the different bearings due to the pinion torque. In these equations, load factors are used to account for bearing locations and bearing ratings are used to account for the bearing type and material. These are independent variables that can be varied depending on the selected bearing locations and bearing types.

3.2.5. The damage module

The *damage* module has relations for calculating the total damage on the different components as a result of travel over a specified route and start-stop requirements. Table 6 shows the relations from this module. The first four relations calculate the distance traveled in different modes and the total distance traveled. The next three relations calculate the total damage by summing damage due to different grades as well as accelerations in a route. The last equation calculates life in miles, which is the reciprocal of damage.

TABLE 6. Damage module relations.

Name	Form	Description
DG01	GR_distsum = Sum(fdist(-0.09 + k * 0.01), k<=GR_NMAX)	Distance traveled in grade mode.
DG02	AC_distsum = AC_dist[AC_NMAX]/5280	Distance per acceleration.
DG03	numstarts = Int(GR_distsum/milesperstop)	Number of starts.
DG04	totaldist = (AC_distsum + DC_distsum) * numstarts + GR_distsum	Total distance traveled.
DG05	{AC_damage[j] = (numstarts/totaldist/5280) * Sum((AC_v[i]/AC_dvdt[i]/AC_nmile[j,i] + AC_v[i-1]/AC_dvdt[i-1]/AC_nmile[j,i-1])*0.5* (AC_v[i] - AC_v[i-1]), 2<=i, i<=AC_NMAX) | j<=8}	Damage due to acceleration
DG06	{PG_damage[j] = Sum(PG_dist[i]/PG_nmile[j,i], i<=PG_NMAX) / totaldist) | j<=8}	Damage due to constant velocity travel on a grade
DG07	{Damage[j] = AC_damage[j] + PG_damage[j] | j<=8}	Overall damage.
DG08	{Life[j] = 1/Damage[j] | j<=8}	Reciprocal of damage.

Damage in a component is expressed as the fraction of the distance traveled at a specific torque level with respect to the expected life of the component in number of miles at that torque level. Using Miner's rule, the damage due to constant velocity travel on a particular grade is summed for the different grades to obtain the total damage due to constant velocity travel on a positive grade. Similarly, the damage due to acceleration per each start is obtained by integrating the damage due to instantaneous torques during acceleration, and is computed by:

$$\int_0^x \frac{dx}{N_m(x)} = \int_0^v \frac{\left(v/a\right)dv}{N_m(v)} ,$$

where v is the velocity, x is the distance traveled and N_m is the life in miles at any given instance during acceleration. A numerical approximation to the right hand side integral is used in this module. The total damage in the acceleration mode is calculated by multiplying this damage due to an individual profile by the number of starts over the complete route.

4. Using Design Sheet Model For Tradeoff Studies

The drive train life model described in the previous section can be used for calculating the life of the components for a particular design specification and route description (in terms of grade distribution and the number of accelerations from stop). The interactive spread-sheet like interface of Design Sheet, shown in Figure 6, can be used for what-if analysis to study the effect of specific design and route changes.

Figure 6. Design Sheet user interface.

More importantly, the tradeoff study capability provided by Design Sheet can be used to explore a large region of the design space by studying the effect of various attributes on the drive train life. Figure 7 shows a trade table of the effect of gross vehicle weight (in lb.) on the eight component fatigue lives (in miles), Life[1] through Life[8]. As can be seen from the trade table, Life[5] and Life[6], corresponding to bearings 2 and 3, respectively, are critical.

wt	Life[1]	Life[2]	Life[3]	Life[4]	Life[5]	Life[6]	Life[7]	Life[8]
30000	3829781.	4084075.	2313997.	1397291.	79725.29	133584.4	2.809861e+7	1.802486e+8
40000	1092939.	1165509.	1331842.	544210.3	78286.73	52027.84	1.094371e+7	7.020238e+7
50000	462147.3	492833.5	891354.2	282537.5	76549.93	27011.28	5681645.	3.644695e+7
60000	222600.2	237380.7	642579.8	163654.9	74321.87	15645.81	3290992.	2.111125e+7

Figure 7. Trade table showing effect of gross vehicle weight on component lives.

The plotting interface to the tradeoff facility can be used to get a clearer view of the effect of gross vehicle weight on the lives of the two critical bearings. Obtaining such a tradeoff plot, shown in Figure 8, is trivial in Design Sheet and only takes a couple of minutes. As can be seen from the figure, change in weight has only a small effect on the life of bearing 2 relative to its effect on the life of bearing 3. Further, bearing 2 is critical below a gross vehicle weight of 36000 lb., whereas bearing 3 is critical for larger values of vehicle weight.

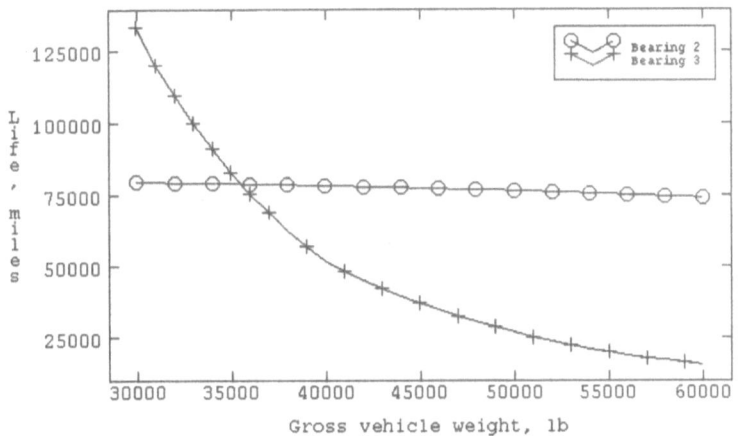

Figure 8. Effect of gross vehicle weight on bearing life.

Figure 9 shows a similar tradeoff for the bending and surface contact failure of the pinion gear. It shows the contribution of acceleration mode travel to the total damage (reciprocal of life). This tradeoff reveals that the majority of the damage is a result of acceleration. It further illustrates that increasing the vehicle weight has a far greater impact on the bending damage. Also evident is the fact that surface damage is more critical than, though comparable to, bending damage below a weight of about 36,000 lb., whereas bending damage becomes critical above that weight.

Though it is enticing to draw conclusions about the transition of criticality between components at specific weight values, one has to be cautious while interpreting the results. For example, the tradeoffs in Figure 8 and Figure 9 are obtained for a constant vehicle frontal area, which may not be valid considering the wide range of gross weights used. Accurate conclusions can only be drawn when the relationship between the gross weight and frontal area is established. The power of Design Sheet lies in here, in that such relations can be added incrementally at run time, without requiring time-consuming modifications to the software.

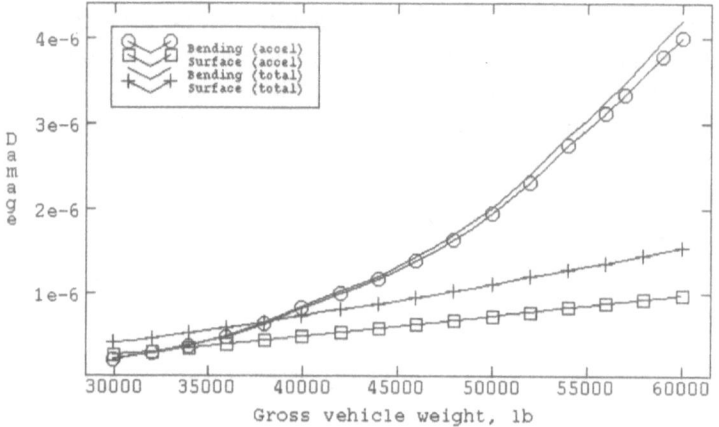

Figure 9. Effect of gross vehicle weight on pinion gear damage.

Let us now consider the tradeoff between the two bearing lives as the vehicle weight and frontal area are varied independently. Figure 10 shows this as a cross plot produced by Design Sheet. The tradeoff from Figure 10 clearly shows that there are regions in design space where the lives of both bearings 2 and 3 are high, whereas there are other regions where the life of bearing 3 falls dramatically. This information is similar to that obtained in Figure 8, but provides a more appropriate view of the design space to the engineer, by demonstrating the effect of both the vehicle gross weight and frontal area on a single plot. The tradeoff shows that gross weight has a much greater effect on bearing lives than the frontal area of the vehicle.

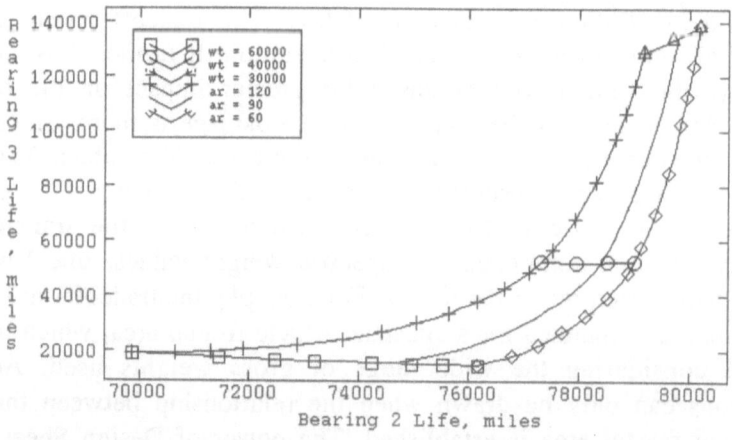

Figure 10. Effect of gross weight and frontal area on bearing life.

As a final example of Design Sheet's flexibility, Figure 11 shows the effect of maximum velocity (in mph) on the vehicle weight and fatigue life tradeoff for bearing 2, which shows that reducing the maximum velocity has a much larger impact on bearing life at higher values of gross weight.

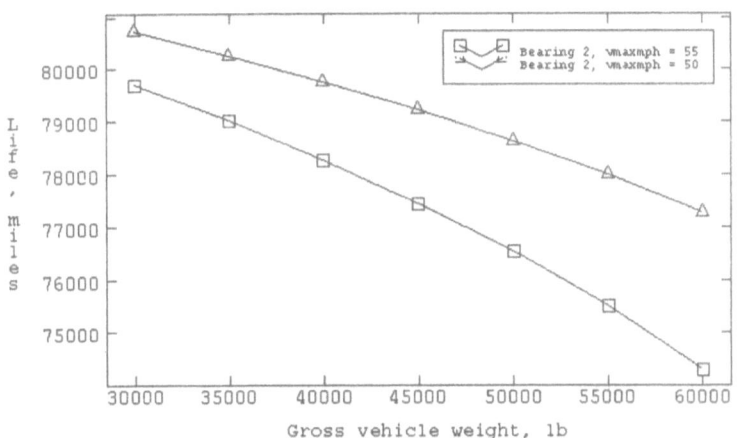

Figure 11. Effect of maximum velocity on the life of bearing 2.

5. Conclusions

Design Sheet has proved to be a highly flexible environment for building analysis models and using them for tradeoff studies during conceptual design. It derives its strength by integrating constraint management methods with symbolic mathematics, robust equation solving, and a specially designed software environment for supporting tradeoff studies. Like other constraint-based systems, Design Sheet decomposes the design constraints into manageable subsets, before solving them. However, in contrast to other systems, Design Sheet further decomposes simultaneous subsets of equations before applying numerical solution techniques, thus improving the robustness of the overall solution process. This is also the main factor behind the success of Design Sheet on applications with as many as five thousand constraints.

The drive train life application clearly demonstrates that Design Sheet can be used to quickly develop complex models and flexibly obtain interesting tradeoffs. The tradeoff plots illustrated in this paper were obtained in a matter of minutes, but more importantly, each successive tradeoff was decided upon only after reviewing the results of the previous tradeoff plots. It is this flexibility that sets Design Sheet apart from other design tools; generating similar tradeoff plots without recourse to Design Sheet would have taken hours, if not days. This application also illustrates how Design Sheet automatically uses the same equations in different directions, in the two operating modes, to calculate velocity and acceleration force. In certain situations, such as when one tries to specify a value for the life of a component and calculate the value of an upstream design parameter, the computational complexity of determining a cut set for large components could prohibit Design Sheet from propagating values backwards through the constraint network. In future, we plan to overcome such difficulties by adding to Design Sheet the capability to reason with function-valued attributes.

Acknowledgments. We wish to thank Dr. Shan Shih for help with the modeling of the automotive drive train life, and the current and previous members of the laboratory who have contributed to the Design Sheet effort.

References

Bouchard, E. E., Kidwall, G. H. and Rogan, J. E.: 1988, The application of artificial intelligence technology to aeronautical system design, *AIAA 88-4426*, AIAA Aircraft Design Systems and Operations Meeting, Atlanta, Georgia.

Buckley, M. J., Fertig, K. W. and Smith, D. E.: 1992, Design sheet: an environment for facilitating flexible trade studies during conceptual design, *AIAA 92-1191*, Aerospace Design Conference, Irvine, California.

Cormen, T., Leiserson, C. and Rivest, R.: 1991, *Introduction to Algorithms.*, McGraw-Hill, New York.

Cutkosky, M., R.: 1993, PACT: An experiment in integrating concurrent engineering systems, *IEEE Computer*, **26**(1), 28–37.

Dixon, J. R., Orelup, M. F. and Welch, R. V.: 1993, A research progress report: robust parametric designs and conceptual design models, *Proceedings of the NSF Design and Manufacturing Systems Conference*, SME, Dearborn, Michigan, pp. 499-506.

Fromont, B. and Sriram, D.: 1992, Constraint satisfaction as planning process, *in* J. S. Gero (ed.), *Artificial Intelligence in Design '92*, Kluwer, Dordrecht, pp. 97-117.

Herman, A. E. and Lu, S. C-Y.: 1992, Computer methods for distributed reasoning to support concurrent engineering, *Proceedings Prolamat '92 Conference*, Tokyo, Japan, pp. 1–19.

Krishnan, V., Navinchandra, D., Rane, P. and Rinderle, J. R.: 1990, Constraint reasoning and planning in concurrent design, *Technical Report CMU-RI-TR-90-03*, The Robotics Institute, Carnegie Mellon University, Pittsburgh, Pennsylvania.

Ramaswamy, R. and Ulrich, K.: 1993, A designer's spreadsheet, *in* T. K. Hight and L. A. Stauffer (eds), *Design Theory and Methodology*, **53**, pp. 105-113.

Rosenberg, R. C. and Karnopp, D. C.: 1983, *Introduction to Physical System Dynamics*, McGraw-Hill, New York, New York.

Serrano, D.: 1987, *Constraint management in conceptual design*, PhD dissertation, MIT, Department of Mechanical Engineering, Cambridge, Massachusetts.

Shih, S. and Keeney, C. S.: 1992, *DTL2 User's Manual*, Rockwell International Automotive Operations, Troy, Michigan.

Ward, A. C.: 1989, *A Theory of Quantitative Inference for Artificial Sets Applied to a Mechanical Design Compiler*, PhD Dissertation, MIT, Department of Mechanical Engineering, Cambridge, Massachusetts.

J. S. Gero and F. Sudweeks (eds), Artificial Intelligence in Design '96, 367-385.
© 1996 *Kluwer Academic Publishers.*

USING MODELING KNOWLEDGE TO GUIDE DESIGN SPACE SEARCH

ANDREW GELSEY, MARK SCHWABACHER AND DON SMITH
Computer Science Department
Rutgers University
New Brunswick, NJ 08903, USA

Abstract. Automated search of a space of candidate designs seems an attractive way to improve the traditional engineering design process. To make this approach work, however, the automated design system must include both knowledge of the modeling limitations of the method used to evaluate candidate designs and also an effective way to use this knowledge to influence the search process. We suggest that a productive approach is to include this knowledge by implementing a set of *model constraint* functions which measure how much each modeling assumptions is violated, and to influence the search by using the values of these model constraint functions as constraint inputs to a standard constrained nonlinear optimization numerical method. We test this idea in the domain of conceptual design of supersonic transport aircraft, and our experiments indicate that our model constraint communication strategy can decrease the cost of design space search by **one or more orders of magnitude.**

1. Introduction

Automated search of a space of candidate designs seems an attractive way to improve the traditional engineering design process. Each step of such automated search requires evaluating the quality of candidate designs, and for complex artifacts (e.g., aircraft, our main example), this evaluation must be done by computational simulation. However, computational simulation is based on a model of the physics of the artifact, and this model will generally make simplifying assumptions in order to be computationally tractable. Most existing computational simulators are intended to be used by human experts, and thus they typically include no explicit representation of their modeling assumptions. Instead, it is assumed that the experts know enough to stay away from portions of the design space that will violate the simulator's assumptions.

For example, a typical assumption for an aircraft simulator might be that the wings won't stall. Stall is a physical phenomenon that occurs when a wing is operated at too high an angle of attack and therefore ceases to generate lift. The physics of stall is understood, and there is in principle no reason not to model it in a simulator. However, a human expert aircraft designer doesn't want to design a plane that stalls during normal operation, so he doesn't need a detailed prediction of stall behavior. The designer is satisfied with an incomplete model as long as he can recognize "impossibly high" lift coefficients and realize that the design he is considering would actually stall and thus should be discarded.

However, if the simulator is invoked by another program such as an automated search procedure rather than by a human expert, it is quite likely that in exploring the design space, the automated search procedure will examine designs which violate the simulator's assumptions, and for those candidate designs, the evaluation of the design quality computed by the simulator may be meaningless. Furthermore, this meaningless value may appear better than the value for any physically realizable design, thus leading the search procedure to a worthless but apparently very good design.

In our earlier work (Gelsey, 1995b), we have investigated the types of modeling knowledge that are needed so that a simulator can be reliably invoked by another program, and we have described algorithms for detecting assumption violations and other problems that might lead to low-quality or unreliable simulation results. In the present paper, we address the question of how information about model assumption violations can be effectively communicated to an automated search procedure so that the search procedure can find candidate designs that don't violate model assumptions.

2. Communication Strategies

Strategies for communicating information about model violations to the search procedure include:

The Null Strategy: ignore the model violation — the search procedure uses whatever value happens to be computed by the inapplicable model for the quality of the candidate design.

The Boolean Strategy: when any model violation occurs, always give the search procedure a standard "very bad value" as the quality of the candidate design.

Model Constraints: when a candidate design is evaluated, give the search procedure not only a value for the quality of the candidate design, but also values for a set of "model constraint" functions which measure how much the various modeling assumptions are satisfied or violated.

Model Penalties: same as the model constraints strategy, except that only the value for the quality of the candidate design is returned to the search procedure, and that value is penalized in proportion to the amount by which the various modeling assumptions are violated.

In this paper we will focus primarily on the boolean strategy and model constraints. The null strategy is unlikely to be useful unless it coincidentally happens to be the same as either the boolean strategy or the model penalties strategy. The boolean strategy can be useful — its advantages include:

- easy to implement: as soon as a violation is detected, just return immediately with a standard "very bad" value for the objective function
- it can be used even with unconstrained search methods

The model constraints strategy is more complicated to implement than the boolean strategy, but our experimental results later in this paper show that when used with a search method that allows constraints, the performance of the model constraints strategy is considerably better than that of the boolean strategy. We don't investigate the model penalties strategy in this paper, but discuss possible uses for it in our Future Work section.

3. Aircraft Design

We have pursued our investigation in the domain of conceptual design of supersonic transport aircraft. Figure 1 shows a diagram of a typical airplane automatically designed by our software system to fly the mission shown in Figure 2, and Figure 3 shows a block diagram of the system's software architecture. The search controller attempts to find a good aircraft conceptual design for a particular mission by varying major aircraft parameters such as wing area, aspect ratio, engine size, etc. using a numerical optimization algorithm. The search controller evaluates candidate designs using a multidisciplinary simulator with which it communicates via the Model/Simulation Associate (MSA), which implements the various communication strategies described in the previous section. In our current implementation, the search controller's goal is to minimize the takeoff mass of the aircraft, a measure of merit commonly used in the aircraft industry at the conceptual design stage. Takeoff mass is the sum of fuel mass, which provides a rough approximation of the operating cost of the aircraft, and "dry" mass, which provides a rough approximation of the cost of building the aircraft. The simulator computes the takeoff mass of a particular aircraft design for a particular mission as follows:

1. Compute "dry" mass using historical data to estimate the weight of the aircraft as a function of the design parameters and passenger capacity required for the mission.
2. Compute the landing mass $m(t_{\text{final}})$ which is the sum of the fuel reserve plus the "dry" mass.
3. Compute the takeoff mass by numerically solving the ordinary differential equation

$$\frac{dm}{dt} = f(m, t)$$

which indicates that the rate at which the mass of the aircraft changes is equal to the rate of fuel consumption, which in turn is a function of the cur-

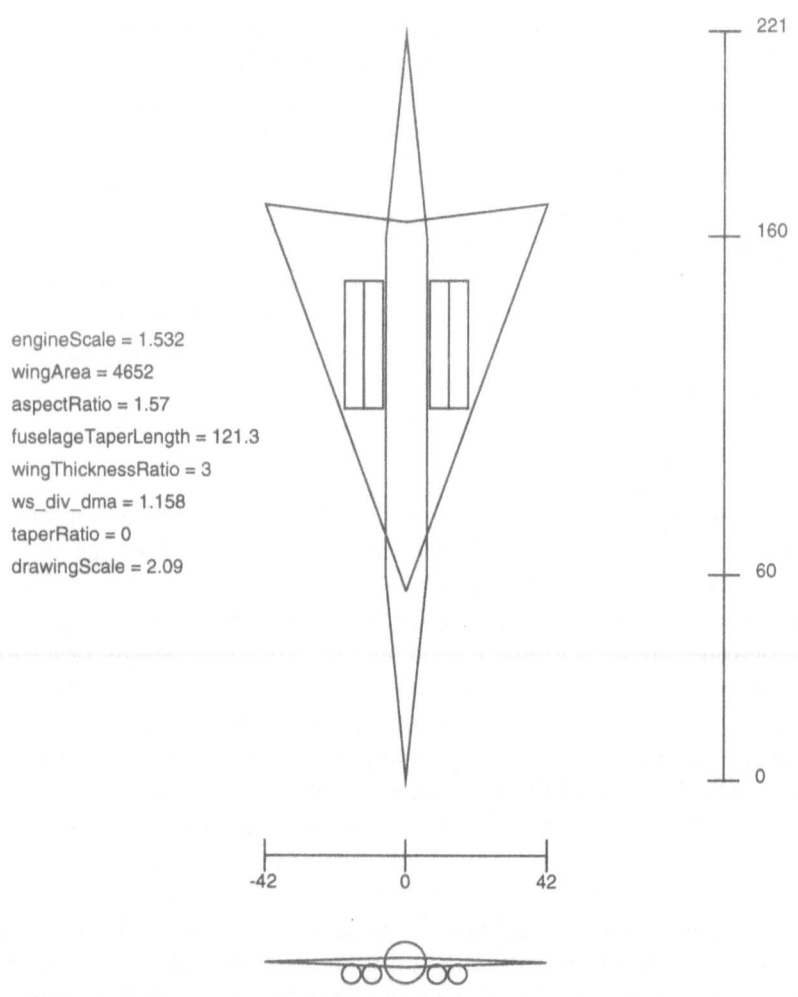

Figure 1. Supersonic transport aircraft designed by our system (dimensions in feet).

Phase	Mach	Altitude (ft.)	Duration (min.s)	comment
1	0.227	0	5	"takeoff"
2	0.85	40,000	85	subsonic cruise (over land)
3	2.0	60,000	180	supersonic cruise (over ocean)

capacity: 70 passengers.

Figure 2. Mission specification for aircraft in Figure 1.

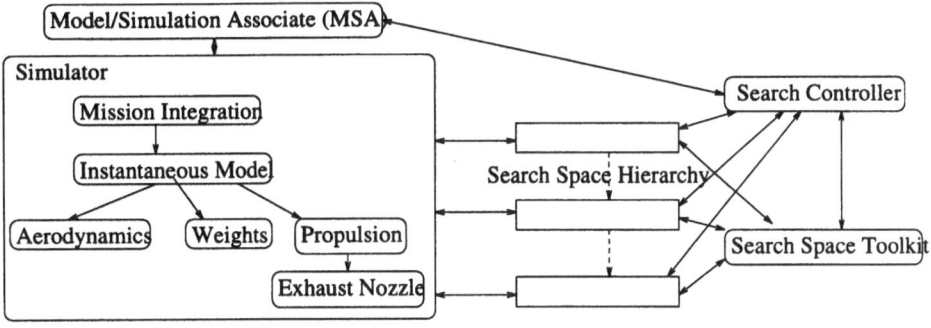

Figure 3. Software architecture block diagram.

rent mass of the aircraft and the current time in the mission. At each time step, the simulator's aerodynamic model is used to compute the current drag, and the simulator's propulsion model is used to compute the fuel consumption required to generate the thrust which will compensate for the current drag.

The software architecture in Figure 3 also includes a "search space toolkit" for determining the design space structure, which is described in Gelsey and Smith (1995) and Gelsey *et al.* (1996) and therefore will not be discussed further in this paper.

A complete mission simulation requires about 1/4 second of CPU time on a DEC Alpha 250 4/266 desktop workstation.

4. Search Procedure

In this paper we will focus on search of a space of candidate designs using numerical optimization methods which vary a set of continuous parameters to minimize[1] a nonlinear objective function subject to a set of nonlinear equality and inequality constraints. The numerical optimizer used in this paper is CFSQP (Lawrence *et al.*, 1995), a state-of-the-art implementation of the Sequential Quadratic Programming method. Sequential Quadratic Programming is a quasi-Newton method that solves a nonlinear constrained optimization problem by fitting a sequence of quadratic programs[2] to it, and then solving each of these problems using a quadratic programming method.

In order to handle unevaluable points (i.e., points whose objective function was assigned the boolean strategy standard "very bad" value), we have supplemented CFSQP with *knowledge-based gradients*. Knowledge-based gradients are

---

[1] To instead *maximize* the objective function, just multiply it by −1 and minimize.

[2] A quadratic program consists of a quadratic objective function to be optimized, and a set of linear constraints.

computed by using a set of rules that specify how to compute gradients with reasonable accuracy in the presence of unevaluable points. In addition, we have arranged for the line searches in CFSQP to terminate when they encounter unevaluable points. These enhancements to CFSQP are crucial when using the boolean communication strategy, which results in numerous unevaluable points. They can also be helpful when using the model constraints communication strategy, since some limitations of the simulator are not modeled in the model constraints, so some unevaluable points exist even when using model constraints. In the experiments reported in this paper, with the boolean strategy, 76% of the points encountered were unevaluable, and with the model constraints strategy, 4% of the points encountered were unevaluable. (Note: the optimizer tends to avoid unevaluable points, so these percentages are considerably lower than the average density of unevaluable points in the search spaces, as indicated by the data presented later in this paper.) Knowledge-based gradients are further described in Schwabacher and Gelsey (1996).

5. Model Constraints

For the experiments in this paper, the MSA module in Figure 3 computes the following model constraint functions, which are ≤ 0 if a constraint is satisfied and positive otherwise:

ETUB = <maximum throttle required during mission simulation> − <maximum throttle setting allowed for engine>. If an impossibly high throttle is required to fly the mission, the simulation will continue using extrapolation, but the value of ETUB will indicate the extent to which the engine model assumptions are violated.

ETLB = <minimum throttle setting allowed for engine> − <minimum throttle required during mission simulation>.

A1LB, A1UB, A2LB, A2UB: Similar to above — violation of bounds for a two-dimensional table of experimental data on supersonic drag.

WLUB = <maximum wing loading during mission simulation> − <maximum wing loading simulator can validly model>.

FM = <fuel mass that current candidate design requires to complete mission> − <fuel mass that can be stored in available volume for current candidate design>.

STALL = <maximum lift coefficient during mission simulation> − <maximum lift coefficient simulator can validly model>. The simulator assumes wings won't stall, and this constraint function computes how well that assumption is satisfied.

These model constraint functions are continuous and usually smooth with respect to the design parameters as their values change sign, which is very important so that when MSA is using the model constraint communication strategy, CF-

SQP (the numerical optimizer) can follow constraint boundaries if necessary as it searches for an aircraft design which can fly the given mission with minimal takeoff mass. If MSA is following the "boolean" communication strategy, it does not give the values of the model constraint functions to CFSQP: instead, any candidate design for which some model constraint function is positive will be evaluated to have a standard "very large" takeoff mass.

In addition to these model constraints, MSA computes the following design constraint:

PASS = <passenger capacity required for the mission> − <passenger capacity available with current design parameters>.

Note: differences between model constraints and design constraints include:

— Design constraints can be extracted directly from design goals, while formulating model constraints requires carefully examining the underlying assumptions of the model which the simulator is based on.

— Design constraints can be violated without reducing the quality (i.e., correctness) of the objective function computed by the simulator, but when a model constraint is violated, the value of the objective function computed by the simulator cannot be trusted. For example, even if the PASS constraint is violated, the simulator can still correctly compute the takeoff mass needed to fly though the mission carrying whatever number of passengers the aircraft is actually able to hold. However, if a model constraint is violated, then the takeoff mass computed by the simulator may be wildly wrong. For example, if the simulator is allowed to violate the STALL constraint, the optimizer may design an aircraft with very small wings operated at a very high angle of attack which may appear to be a very efficient aircraft, much better than the best physically plausible design, but which in fact is not capable of flying at all.

— If a design constraint happens to be inactive at the optimal design (i.e., the constraint is satisfied for all designs near the optimal design, so the optimum does not lie on a constraint boundary), then the "null" communication strategy will be effective when applied to this constraint — i.e., the constraint may safely be ignored without a detrimental effect on the optimization. However, the null communication strategy will not in general be effective when applied to model constraints, even if they are inactive at the optimal design. In the region where a model constraint is violated, the value of the objective function computed by the simulator may include random meaningless values, so therefore if the model constraint violations are ignored by the null strategy, the region where the model constraint is violated may include local optima of the objective function or, even worse, points having (spurious) values of the objective function better than the best value for any design satisfying all the model assumptions. Either of these conditions can "trap" the op-

| | Small Box | | Big Box | |
Design Parameter	low	high	low	high
engine size	0.5	3	0.1	5
wing area (sq. ft.)	1500	13500	500	20000
wing aspect ratio	1	2	0.5	3
fuselage taper length (ft.)	100	200	50	300
effective structural thickness over chord	1	5	0.5	10
wing sweep over design mach angle	1	1.45	0	1.45
wing taper ratio	0	0.1	0	0.1
fuel annulus width (ft.)	0	4	0	8

Figure 4. Subsets of design space explored.

timizer and keep if from getting to the true optimum, even though the model constraint in question is inactive at the true optimum.

6. Experimental Results

To experimentally test MSA communication strategies, we used a seven-dimensional design space in which the optimizer varied the following aircraft conceptual design parameters over a continuous range of values:

1. engine size
2. wing area
3. wing aspect ratio
4. fuselage taper length (how "pointed" the fuselage is)
5. effective structural thickness over chord (a nondimensionalized measure of wing thickness)
6. wing sweep over design mach angle (a nondimensionalized measure of wing sweep)
7. wing taper ratio (wing tip chord divided by wing root chord)
8. fuel annulus width (space available in fuselage for fuel storage)

Figure 4 shows the two subsets we explored in the design space defined by these seven design parameters.

To test the effect of the MSA communication strategy on the design process, we considered the following strategy combinations:

1. Return values of all model constraint functions to the optimizer as nonlinear inequality constraints.
2. Return values of all model constraint functions except ETLB and ETUB to the optimizer, but for candidate designs where the engine table constraints were violated (ETLB or ETUB positive), use the "boolean" strategy and return a standard "very large" value for takeoff mass.

Design Parameters:

engine size	1.532
wing area	4652 sq. ft.
wing aspect ratio	1.570
fuselage taper length	121.3 ft.
effective structural thickness over chord	3.002
wing sweep over design mach angle	1.158
wing taper ratio	0
fuel annulus width	0

Objective Function:

Takeoff Mass	167.4 tonnes

Model Constraints:

ETUB	−41.57
ETLB	−0.76
A1LB	−2.2
A1UB	−1.8
A2LB	−1.5
A2UB	−8.5
WLUB	−149.8
FM	−0.0011 tonnes
STALL	0

Design Constraint:

PASS	−2

Figure 5. Best design found for mission of Figure 2.

3. Return values of all model constraint functions except A1LB, A1UB, A2LB, and A2UB to the optimizer; use "boolean" strategy for points which required extrapolation outside the aerodynamics table bounds.
4. Return values of all model constraint functions except FM, which is "boolean".
5. Return values of all model constraint functions except STALL, which is "boolean".
6. Return values of all model constraint functions except WLUB, which is "boolean".
7. Use the "boolean" communication strategy for all model constraint functions.
8. A two-level approach in which the "boolean" communication strategy is used to find a feasible point, and then all model constraints are used to find an optimum.

Strategy Combination	Success	Start Cost	Opt. Cost	Est. 99% Cost
All model constraints returned	65/74	16	42375	1252
ETLB and ETUB "boolean"	52/74	3203	67158	3609
FM "boolean"	0/74	603	99215	≫ 456565
STALL "boolean"	18/74	5441	81566	19427
A1LB/A1UB/A2LB/A2UB "boolean"	67/74	57	47042	1242
WLUB "boolean"	62/74	721	39404	1372
All model constraints "boolean"	0/74	21946	75804	≫ 447106
Two level	72/74	18098	39389	990

Figure 6. Performance of the various strategy combinations.

For each strategy combination, our system randomly choose points in the "small box" until it found 74 "evaluable" points (i.e., points whose objective function was not assigned the boolean strategy standard "very bad" value).[3] Each of these 74 points was then used as a starting point for a design optimization using CF-SQP to try to find an optimal aircraft design for the mission shown in Figure 2. (We required the starting points to be evaluable because if CFSQP happened to be started in an unevaluable region, then all components of the gradient would be zero and the optimization would terminate immediately.) The best design found for this mission in all the experiments is shown in Figure 5, and a diagram of this aircraft appears in Figure 1.

The performance of the strategy combinations is shown in a table in Figure 6. The "Success" column for each strategy combination shows what fraction of the 74 optimizations found aircraft designs having takeoff masses within 1% of the takeoff mass of the apparent "global optimum" — the best design we found for this mission (Figure 5). The "Start Cost" column shows how many simulations had to be run on unevaluable points while finding the 74 optimization starting points, and "Opt. Cost" shows the total number of simulations that were run during each set of 74 optimizations.[4] The "Est. 99% Cost" column in Figure 6 gives the estimated cost with each strategy combination to have a 99% chance of finding the global optimum, which is computed by multiplying the average cost per optimization times $\log(1 - P_{\text{desired}})/\log(1 - P_{\text{success}})$, where P_{desired} is the desired probability of finding the global optimum (99% in this case) and P_{success} is the probability of any single optimization finding the global optimum (which we estimate with the value in the "Success" column). Figure 7 shows graphically the "Est. 99% Cost" to achieve a range of different design qualities. (Note that the curves for some of the strategies are so bad that they are above the largest vertical

---

[3] 74 is not a "magic" number; it was just a convenient choice given available disk space.

[4] As mentioned earlier, a complete mission simulation requires about 1/4 second of CPU time on a DEC Alpha 250 4/266 workstation.

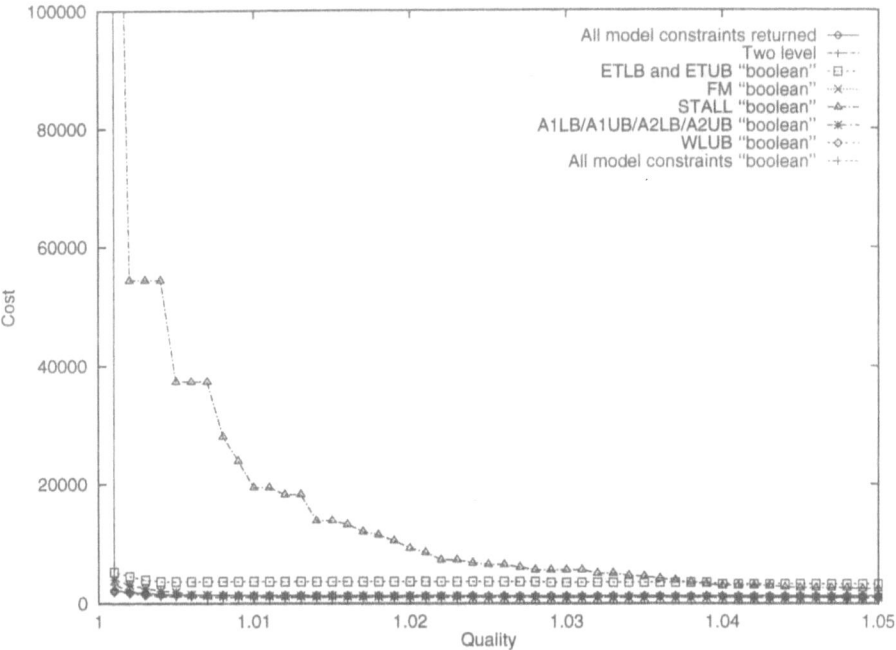

Figure 7. Cost to achieve a range of design qualities with 99% confidence. Quality is takeoff mass, normalized by the best takeoff mass found (Figure 5), so quality = 1.01 corresponds to Figure 6.

axis value shown and therefore they do not appear in the plot. See the "Est. 99% Cost" column in Figure 6 for their values at Quality = 1.01.)

The data in Figure 6 indicates that the model constraints communication strategy can find the global optimum with a 99% confidence at a cost which is **one or more orders of magnitude smaller** than the cost to achieve comparable results with the boolean communication strategy. Examination of the different strategy combinations indicates that the model constraints which contribute most to this performance difference are the constraints active at the global optimum (constraint values ≈ 0 in Figure 5), but that even the constraints which are inactive at the global optimum may give a factor of two to three speedup when handled using model constraints rather than the boolean strategy. If a constraint is active at the global optimum, then CFSQP must "navigate" along the constraint boundary when searching for the optimum. This navigation is easy when the boundary is defined by a smooth model constraint, but much more difficult when the boundary is marked only by a sudden jump in the objective function from a reasonable value to the boolean "very bad" value. Model constraints which are inactive at the optimum may still be active during some parts of the search and thus can help guide the search and prevent the optimizer from getting stuck.

Strategy Combination	Success	Start Cost	Opt. Cost	Est. 99% Cost
All model constraints returned	55/74	48	58376	2674
ETLB and ETUB "boolean"	14/74	6979	124796	39102
FM "boolean"	0/74	879	128618	≫ 592317
STALL "boolean"	15/74	21669	78508	27520
A1LB/A1UB/A2LB/A2UB "boolean"	40/74	280	176113	14114
WLUB "boolean"	60/74	2181	58976	2285
All model constraints "boolean"	0/74	354761	80835	≫ 1992408
Two level	54/74	301917	56198	17034

Figure 8. Performance of the various strategy combinations in a bigger box.

Model constraints which are active at the global optimum are more critical, but it is important to note that there will typically be no reliable *a priori* way to determine which model constraints will be active at the global optimum. This fact suggests that the model constraints communication strategy should be used to handle all model assumptions, even though implementing smooth model constraint functions may require more work than implementing the simpler boolean communication strategy.

An issue that should be considered is the question of why any model constraints are active for the globally optimal design. Does this situation indicate that there are actually better designs on the other side of the constraint boundary which the optimizer would be able to find if only we had a more sophisticated model that didn't need as many constraints? Not necessarily. For example, lift initially rises as a function of angle of attack and later begins falling rapidly as stall occurs for higher angles of attack. The STALL constraint, which is active at our global optimum (see Figure 5) cuts off this function at its peak so that the lift function is monotonic where the constraint is satisfied. A more sophisticated simulator which modeled stall would not find better designs on the other side of the STALL constraint boundary — it would just find that the lift function ceased to be monotonic when the boundary was crossed. The ETLB constraint is also active at our global optimum (see Figure 5). In this case, the engine stops running when the throttle is too low. Modifying the engine model to correctly predict the sudden low temperatures and pressure produced by the engine when it stops running would not uncover better designs.

To test the effect of box size on our conclusions, we repeated our experiments in a larger box. Figure 4 shows the two "boxes" in the design space used in our experiments. The bigger box contains the smaller box, and the volume of the larger box is about 300 times greater than the volume of the smaller box. Figure 8 shows the performance of the various communication strategies in the larger box, and Figure 9 shows graphically the "Est. 99% Cost" to achieve a range of dif-

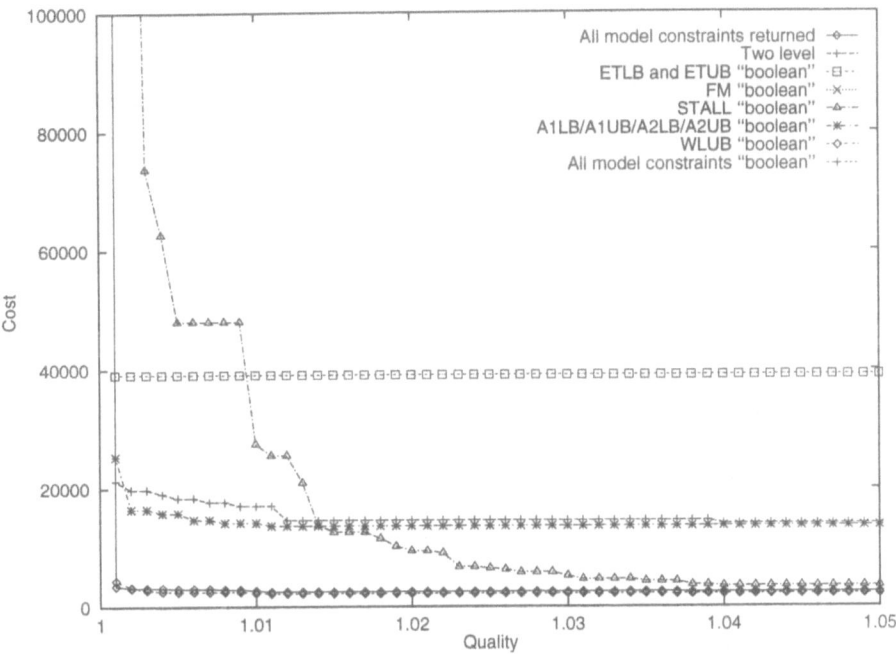

Figure 9. Cost to achieve a range of design qualities with 99% confidence. Quality is takeoff mass, normalized by the best takeoff mass found (Figure 5), so quality = 1.01 corresponds to Figure 8.

ferent design qualities. Search cost increases in the larger box, as expected, but model constraints still cost orders of magnitude less than the boolean strategy.

It is important to compare the performance of the "two-level" strategy combination for the two boxes. In the "small" box, the two-level approach was actually superior to the pure model constraints approach: it was slightly better to use a boolean strategy to find a feasible point before starting to use model constraints to find the optimum. The reverse was true in the big box, however: the pure model constraints approach was a factor of six less expensive then the two-level approach. These results are quite plausible, because the "start cost" data for "all boolean" combination indicates that the density of feasible points in the small box is about 1/300 while in the big box it is only 1/4800. The big box has such a small feasible region that the benefit of using model constraints to search for the feasible region outweighs the model constraints overhead, while in the smaller box random probes can find the feasible region cheaply enough that the overhead of using model constraint to find the feasible region is not justified. However, even in the small box model constraints are still extremely useful for searching within the feasible region in order to find an optimum.

To test the effect of the design goal on our conclusions, we repeated our experiments with a different goal. We used the same boxes as for the previous exper-

Phase	Mach	Altitude (ft.)	Duration (min.s)	comment
1	0.227	0	5	"takeoff"
2	0.85	40,000	50	subsonic cruise (over land)
3	2.0	60,000	225	supersonic cruise (over ocean)

capacity: 70 passengers.

Figure 10. Another mission specification.

Design Parameters:

engine size	1.146
wing area	3690 sq. ft.
wing aspect ratio	1.089
fuselage taper length	130.1 ft.
effective structural thickness over chord	2.728
wing sweep over design mach angle	1.235
wing taper ratio	0
fuel annulus width	0

Objective Function:

Takeoff Mass	134.8 tonnes

Model Constraints:

ETUB	−2.89
ETLB	−18.19
A1LB	−1.83
A1UB	−2.17
A2LB	−2.03
A2UB	−7.97
WLUB	−143.8
FM	−0.00038 tonnes
STALL	0

Design Constraint:

PASS	−2

Figure 11. Best design found for the 2nd mission (Figure 10).

Strategy Combination	Success	Start Cost	Opt. Cost	Est. 99% Cost
All model constraints returned	62/74	13	36204	1238
ETLB and ETUB "boolean"	57/74	1275	65556	2827
FM "boolean"	1/74	31	65105	297930
STALL "boolean"	36/74	1681	50847	4904
A1LB/A1UB/A2LB/A2UB "boolean"	65/74	55	40377	1194
WLUB "boolean"	64/74	227	34899	1092
All model constraints "boolean"	0/74	6576	67046	≫ 336745
Two level	64/74	4307	34477	1205

Figure 12. Performance of the various strategy combinations for the 2nd mission.

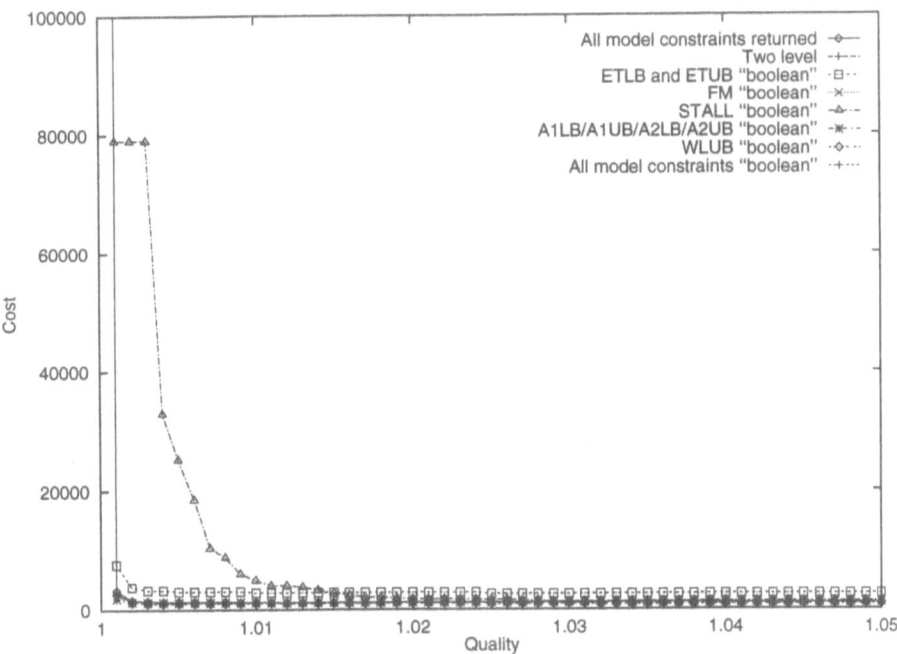

Figure 13. Cost to achieve a range of design qualities with 99% confidence. Quality is takeoff mass, normalized by the best takeoff mass found (Figure 11), so quality = 1.01 corresponds to Figure 12.

iments, but instead the goal was to design the best aircraft for the mission shown in Figure 10. Figure 11 shows the best design found, which differs considerably from the optimal design for the previous mission. We performed the same set of experiments for this case, and the experimental data which appears in Figures 12, 13, 14, and 15 supports our previous conclusion that the model constraint communication strategy can cut search cost by an order of magnitude or more.

Strategy Combination	Success	Start Cost	Opt. Cost	Est. 99% Cost
All model constraints returned	48/74	41	57607	3429
ETLB and ETUB "boolean"	13/74	5616	145770	48765
FM "boolean"	0/74	162	93146	≫ 426789
STALL "boolean"	39/74	7602	64000	5951
A1LB/A1UB/A2LB/A2UB "boolean"	34/74	342	131652	13352
WLUB "boolean"	51/74	1420	56171	3066
All model constraints "boolean"	0/74	133265	117224	≫ 1145732
Two level	49/74	231396	48049	16025

Figure 14. Performance of the various strategy combinations for the 2nd mission in the bigger box.

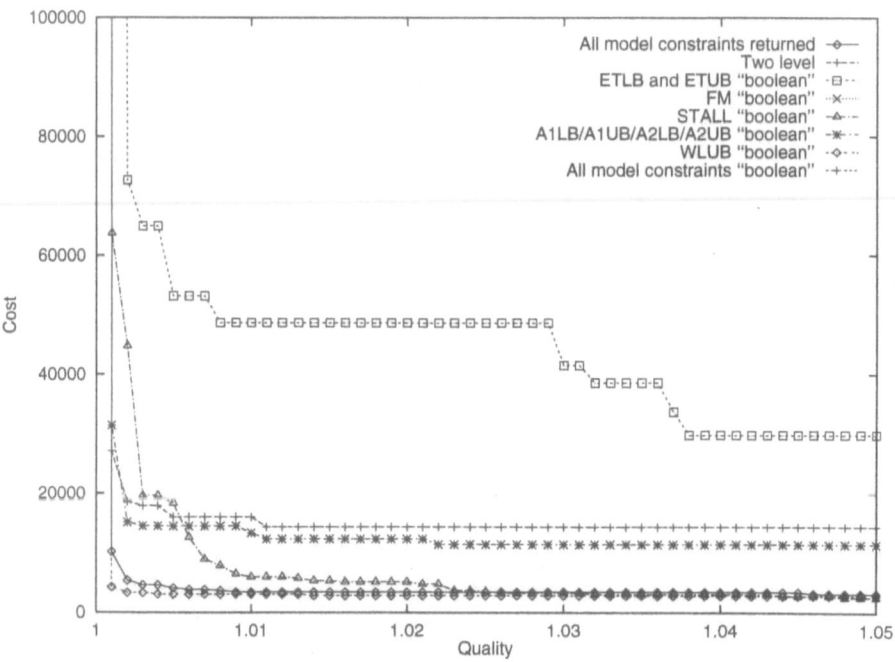

Figure 15. Cost to achieve a range of design qualities with 99% confidence. Quality is takeoff mass, normalized by the best takeoff mass found (Figure 11), so quality = 1.01 corresponds to Figure 14.

7. Related Work

Gelsey (1995b) examines the types of modeling knowledge that are needed so that a simulator can be reliably invoked by another program and describes algorithms for detecting assumption violations and other problems that might lead to low-quality or unreliable simulation results, but strategies for communicating information about modeling failures to an automated design systems are not discussed. Forbus and Falkenhainer (1990; 1992; 1995) discuss the use of qualitative simulation to check the quality of numerical simulation results, but here strategies for communicating information about modeling failures to an automated design systems are also not discussed.

Other automated intelligent controllers for numerical simulators are described in Gelsey (1991; 1995a), Sacks (1991), Yip (1991) and Zhao (1994), but these do not address the issue of model and simulation quality assurance.

Intelligent monitoring for complex systems has received considerable attention (e.g., Dvorak and Kuipers (1991)), but this work has focused on diagnosis of problems in dynamically changing physical systems as opposed to problems in the execution of computational algorithms which are attempting to simulate the behavior of physical systems.

A great deal of work has been done in the area of numerical optimization algorithms (Gill *et al.*, 1981; Vanderplaats, 1984; Peressini *et al.*, 1988; Moré and Wright, 1993; Papalambros and Wilde, 1988), though not much has been published about the particular difficulties of attempting to optimize functions defined by large "real-world" numerical simulators. A number of researchers have combined AI techniques with numerical optimization (Ellman *et al.*, 1993; Schwabacher *et al.*, 1994; Schwabacher *et al.*, 1996; Tong *et al.*, 1992; Powell, 1990; Bouchard *et al.*, 1988; Bouchard, 1992; Sobieszczanski-Sobieski *et al.*, 1985; Agogino and Almgren, 1987; Williams and Cagan, 1994; Hoeltzel and Chieng, 1987; Cerbone, 1992), but have not addressed the issue of model and simulation quality assurance.

8. Limitations and Future Work

A limitation of the model constraints communication strategy is the need to implement fairly well-behaved model constraint functions for all model assumptions. Implementing the model constraint functions was not too difficult for our conceptual design of aircraft domain, but investigating the difficulty of implementing model constraints in other domains is an important area for future work.

Our experiments have been performed in a domain in which the global optimum has a fairly large "basin of attraction", so that a local optimization method like Sequential Quadratic Programming will give a high confidence of finding the global optimum if started from a small number of random starting points. For domains in which this property fails to hold, global optimization methods such as Simulated Annealing will often be preferable. Such methods would not typically be able to make direct use of model constraint functions, so for such a domain investigating the "model penalties" communication strategy described in Section 2 might be worthwhile area for future work.

9. Conclusion

Automated search of a space of candidate designs seems an attractive way to improve the traditional engineering design process. To make this approach work, however, the automated design system must include both knowledge of the modeling limitations of the method used to evaluate candidate designs and also an effective way to use this knowledge to influence the search process. We suggest that a productive approach is to include this knowledge by implementing a set of *model constraint* functions which measure how much each modeling assumptions is violated, and to influence the search by using the values of these model constraint functions as constraint inputs to a standard constrained nonlinear optimization numerical method. Our experiments indicate that our model constraint communication strategy can decrease the cost of design space search by one or more orders of magnitude.

10. Acknowledgments

We thank our aircraft design expert, Gene Bouchard of Lockheed, for his invaluable assistance in this research. We thank our programmer, Keith Miyake, for his effort in implementing large parts of the software described in this paper. This research was partially supported by NASA under grant NAG2-817 and is also part of the Rutgers-based HPCD (Hypercomputing and Design) project supported by the Advanced Research Projects Agency of the Department of Defense through contract ARPA-DABT 63-93-C-0064.

References

Agogino, A. M. and Almgren, A. S.: 1987, Techniques for integrating qualitative reasoning and symbolic computing, *Engineering Optimization*, **12**, 117–135.

Bouchard, E. E., Kidwell, G. H. and Rogan, J. E.: 1988, The application of artificial intelligence technology to aeronautical system design, *AIAA/AHS/ASEE Aircraft Design Systems and Operations Meeting*, Atlanta, Georgia, AIAA-88-4426.

Bouchard, E. E.: 1992, Concepts for a future aircraft design environment, *Proceedings, 1992 Aerospace Design Conference*, Irvine, CA, AIAA-92-1188.

Cerbone, G.: 1992, Machine learning in engineering: Techniques to speed up numerical optimization, PhD Thesis, *Technical Report 92-30-09*, Oregon State University Department of Computer Science.

Dvorak, D. and Kuipers, B.: 1991, Process monitoring and diagnosis, *IEEE Expert*, **6**(3), 67–74.

Ellman, T., Keane, J. and Schwabacher, M.: 1993, Intelligent model selection for hillclimbing search in computer-aided design. *Proceedings of the Eleventh National Conference on Artificial Intelligence*, Washington, DC.

Forbus, K. D. and Falkenhainer, B.: 1990, Self-explanatory simulations: An integration of qualitative and quantitative knowledge, *Proceedings, Eighth National Conference on Artificial Intelligence*, Boston, MA, pp. 380–387.

Forbus, K. D. and Falkenhainer, B.: 1992, Self-explanatory simulations: Scaling up to large models, *Proceedings, Tenth National Conference on Artificial Intelligence*, San Jose, CA.

Forbus, K. D. and Falkenhainer, B.: 1995, Scaling up self-explanatory simulations: Polynomial-time compilation, *Proceedings, Fourteenth Internationnal Joint Conference on Artificial Intel-

ligence, Montreal, Quebec, Canada.

Gelsey, A. and Smith, D.: 1995, A search space toolkit, *Proceedings, 11th IEEE Conference on Artificial Intelligence Applications*, Los Angeles, CA, pp. 117–123.

Gelsey, A., Smith, D., Schwabacher, M., Rasheed, K. and Miyake, K.: 1996, A search space toolkit, *Decision Support Systems, special issue on Unification of Artificial Intelligence with Optimization* (to appear).

Gelsey, A.: 1991, Using intelligently controlled simulation to predict a machine's long-term behavior, *Proceedings, Ninth National Conference on Artificial Intelligence*, Cambridge, MA, pp. 880–887.

Gelsey, A.: 1995, Automated reasoning about machines, *Artificial Intelligence*, **74**(1), 1–53.

Gelsey, A.: 1995, Intelligent automated quality control for computational simulation, *Artificial Intelligence for Engineering Design, Analysis and Manufacturing (AI EDAM)*, **9**(5), 387–400.

Gill, P. E., Murray, W. and Wright, M. H.: 1981, *Practical Optimization*, Academic Press, London/New York.

Hoeltzel, D. and Chieng, W.: 1987, Statistical machine learning for the cognitive selection of nonlinear programming algorithms in engineering design optimization, *Advances in Design Automation*, Boston, MA.

Lawrence, C., Zhou, J. and Tits, A: 1995, User's guide for CFSQP version 2.3: A C code for solving (large scale) constrained nonlinear (minimax) optimization problems, generating iterates satisfying all inequality constraints, *Technical Report TR-94-16r1*, Institute for Systems Research, University of Maryland, August.

Moré, J. J. and Wright, S. J.: 1993, *Optimization Software Guide*, SIAM, Philadelphia.

Papalambros, P. and Wilde, J.: 1988, *Principles of Optimal Design*, Cambridge University Press, New York, NY.

Peressini, A. L., Sullivan, F. E. and Uhl, Jr, J. J.: 1988, *The Mathematics of Nonlinear Programming*, Springer-Verlag, New York.

Powell, D.: 1990, Inter-GEN: A hybrid approach to engineering design optimization, *Technical Report*, PhD Thesis, Rensselaer Polytechnic Institute Department of Computer Science.

Sacks, E. P.: 1991, Automatic analysis of one-parameter ordinary differential equations by intelligent numeric simulation, *Artificial Intelligence*, **48**(1), 27–56.

Schwabacher, M. and Gelsey, A.: 1996, Intelligent gradient-based search of incompletely defined design spaces, *Technical Report HPCD-TR-38*, Department of Computer Science, Rutgers University, New Brunswick, NJ. ftp://ftp.cs.rutgers.edu/pub/technical-reports/hpcd-tr-38.ps.Z.

Schwabacher, M., Hirsh, H. and Ellman, T.: 1994, Learning prototype-selection rules for case-based iterative design, *Proceedings of the Tenth IEEE Conference on Artificial Intelligence for Applications*, San Antonio, Texas.

Schwabacher, M., Ellman, T., Hirsh, H. and Richter, G.: 1996, Learning to choose a reformulation for numerical optimization of engineering designs, *in* J. S. Gero and F. Sudweeks (eds), *Artificial Intelligence in Design '96*, Kluwer, Dordrecht (this volume).

Sobieszczanski-Sobieski, J., James, B. B. and Dovi, A. R.: 1985, Structural optimization by multilevel decomposition, *AIAA Journal*, **23**(11), 1775–1782.

Tong, S. S., Powell, D. and Goel, S.: 1992, Integration of artificial intelligence and numerical optimization techniques for the design of complex aerospace systems, *Proceedings, 1992 Aerospace Design Conference*, Irvine, CA.

Vanderplaats, G. N.: 1984, *Numerical Optimization Techniques for Engineering Design: With Applications*, McGraw-Hill, New York.

Williams, B. C. and Cagan, J.: 1994, Activity analysis: The qualitative analysis of stationary points for optimal reasoning, *Proceedings, 12th National Conference on Artificial Intelligence*, Seattle, Washington, pp. 1224–1230.

Yip, K.: 1991, Understanding complex dynamics by visual and symbolic reasoning, *Artificial Intelligence*, **51**(1–3), 179–221.

Zhao, F.: 1994, Extracting and representing qualitative behaviors of complex systems in phase space, *Artificial Intelligence*, **69**(1-2), 51–92.

J. S. Gero and F. Sudweeks (eds), Artificial Intelligence in Design '96, 387-405.

EXPLANATORY INTERFACE IN INTERACTIVE DESIGN ENVIRONMENTS

ASHOK GOEL, ANDRÉS GÓMEZ DE SILVER GARZA, NATHALIE
GRUÉ, J. WILLIAM MURDOCK AND MARGARET RECKER
College of Computing
Georgia Institute of Technology
Atlanta, Georgia 30332, USA

AND

T. GOVINDARAJ
School of Industrial and Systems Engineering
Georgia Institute of Technology
Atlanta, Georgia 30332, USA

Abstract. Explanation is an important issue in building computer-based interactive design environments in which a human designer and a knowledge system may cooperatively solve a design problem. We consider the two related problems of explaining the system's reasoning and the design generated by the system. In particular, we analyze the content of explanations of design reasoning and design solutions in the domain of physical devices. We describe two complementary languages: task-method-knowledge models for explaining design reasoning, and structure-behavior-function models for explaining device designs. INTERACTIVE KRITIK is a computer program that uses these representations to visually illustrate the system's reasoning and the result of a design episode. The explanation of design reasoning in INTERACTIVE KRITIK is in the context of the evolving design solution, and, similarly, the explanation of the design solution is in the context of the design reasoning.

1. Background, Motivations and Goals

Effective communication of both the design process and the design product is critical in collaborative design. Communicating the process of design and the evidence that the product satisfies its requirements can help build confidence in the design. When members of a design team work on different parts of a design problem, this kind of communication about one part of the problem can help in constraining other parts of the problem. In addition, explanation of design reasoning and its result can enable reuse of parts of the reasoning/result in subsequent design projects. Within the course of a design project, the explanation can enable reflection, support the detection of flaws, and suggest remedies for fixing them.

This is no less true of collaboration between a human designer and a knowledge system in the context of computer-based interactive design environments. When a human designer and a knowledge system are cooperatively addressing a design problem, the system must be able to explain to the designer precisely what it is doing, how and why. In addition, the system must be able to justify why the design solution it has proposed is acceptable for the given problem. Without this the user will have little confidence in the design and may be unable to detect potential flaws in it. Building usable interactive design environments thus requires both a theory of design explanations and the creation of explanatory interfaces.

The issue then becomes how may a knowledge system explain both its reasoning and the design solutions it proposes. This issue has several related but distinct facets pertaining to the content, generation, and presentation of explanations. To illustrate, let us consider the problem of explaining the design of a gyroscope. The explanation may specify how the design works, how its structure delivers its functions, how its design satisfies its requirements. Within a knowledge system, knowledge of the gyroscope's behaviors may be represented as a causal network, or generated at run-time from a representation of its design structure. To the user, the system may present the explanation in text form, or as graphics, or in some other modality such as animation. Our research on design explanations centers on the content of explanations presented to the user, and the content and representation of design knowledge and reasoning needed for generating the explanations.

The content of explanation and justification of design solutions, such as that of a gyroscope, depends both on the design phase and the design domain. For example, the explanation of the result of preliminary design is different from that of the result of configuration design: the former pertains to the function and structure of the design while the latter refers to its geometry. Similarly, the content of a justification for the design of gyroscope is different from that of an office building or a software interface. This is because the relationships between the function and the structure of the gyroscope design are fundamentally different from the function-structure relations in the design of an office building or a software interface. Our work focuses on the preliminary (conceptual, qualitative) design of physical devices such as electrical circuits, heat exchangers, and angular momentum controllers. The input to this task is a specification of the desired functions, and the output is a specification of a structure that can deliver the desired functions.

We are developing an interactive design and learning environment called INTERACTIVE KRITIK. When complete, INTERACTIVE KRITIK is intended to serve as an interactive constructive design environment. At present, when asked by a human user INTERACTIVE KRITIK can invoke a knowledge-based design system called KRITIK3 to address specific kinds of design problems. KRITIK3 evolves from KRITIK, which has been extensively described elsewhere (e.g., Goel (1991, 1992); Goel and Chandrasekaran (1989, 1992)). INTERACTIVE KRITIK provides an explanatory interface to KRITIK3. In particular, it provides visual explanations and justifications of both KRITIK3's reasoning and the solutions it proposes. In addition, it enables the user to explore the system's design knowledge and also the design of the device generated by the system. A key feature of INTERACTIVE KRITIK is that explanation of the design reasoning is presented in the context of the evolving design solution, and, similarly, explanation of the design solution is presented in the context of the reasoning that led to it.

2. INTERACTIVE KRITIK

INTERACTIVE KRITIK's architecture consists of two agents: a design agent in the form of KRITIK3[1] and an interface agent[2]. Figure 1 illustrates INTERACT-IVE KRITIK's architecture. The solid lines in the figure represent data flow while dotted lines represent control flow.

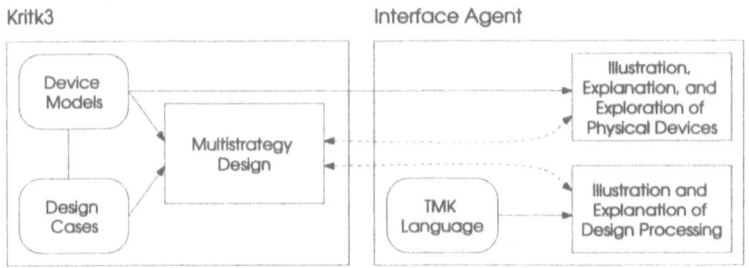

Figure 1. INTERACTIVE KRITIK's architecture.

2.1. STRUCTURE-BEHAVIOR-FUNCTION MODELS IN INTERACTIVE KRITIK

We use structure-behavior-function models (SBF models) (Chandrasekaran et al., 1993; Goel, 1991, 1992) for explaining and justifying designs of physical devices. The SBF model of a device provides a functional and causal explanation of how the device works, how its structure delivers its functions. This explanation makes explicit the functional and causal roles played by each structural element in the device design. Since KRITIK3 addresses the function-to-structure design task, and because the SBF model of a design created by the system explains how the pro-posed structure delivers the desired functions, the SBF model provides a justific-ation for the design.

---

[1] KRITIK3 runs under Common Lisp using CLOS.
[2] The interface is built using the Garnet tool (Myers and Zanden, 1992).

The SBF model of a device explicitly represents (i) the function(s) of the device, (ii) the structure of the device, and (iii) the internal causal behaviors of the device. The internal causal behaviors specify how the functions of the structural elements of the device are composed into the device functions. As a simple (almost trivial) example, let us consider the SBF model of an electrical circuit that produces light of intensity 9 lumens.

Structure: The structure of a device in the SBF language is expressed in terms of its constituent components and substances and the interactions between them. Components and substances can interact both *structurally* and *behaviorally*. For example, electricity can flow from battery to bulb only if they are structurally connected, and only if supported by the function allow electricity of switch that connects the battery and the bulb.

Function: The function of a device in the SBF language is represented as a schema that specifies the input behavioral state of the device, the behavioral state it produces as output, and a pointer to the internal causal behavior of the design that achieves this transformation. Both the input state and the output state are represented as *substance schemas*. The input state specifies that the substance electricity has the property *voltage* and the corresponding parameter, *10 volts*. The output state specifies the property *intensity* and the corresponding parameter, *9 lumens*, of a different substance, light. Finally, the slot by-behavior points to the causal behavior that achieves the function of producing light. Figure 2 illustrates INTER-ACTIVE KRITIK's visual representation of the function of the electrical circuit.

Behavior: The SBF model of a device also specifies the internal causal behaviors that compose the functions of device substructures into the functions of the device as a whole. In the SBF language, the internal causal behaviors of a device are represented as sequences of *transitions* between *behavioral states*. The annotations on the state transitions express the *causal, structural,* and *functional contexts* in which the state transitions occur and the state variables get transformed. The causal context specifies *causal relations* between the variables in preceding and succeeding states. The structural context specifies different *structural relations* among the components, the substances, and the different spatial locations of the device. The functional context indicates which functions of which components in the device are responsible for the transition. The behaviors are organized along the flow of specific substances through the device.

Figure 3 illustrates INTERACTIVE KRITIK's visual representation of *Light-Behavior*, the causal behavior that explains how light is generated. The state transition in this behavior has three annotations Using Function, Under Condition Transition, and Parametric Equation as indicated in the side bar on the top right of the figure. In the screen shot depicted, the description of one of these annotations, Using Function, is displayed in the pop-up dialog box in the right center of the figure. This description explains that the transition occurs due to the function *Bulb-Function-Light* component *Bulb*. Although not shown in Figure 3, the

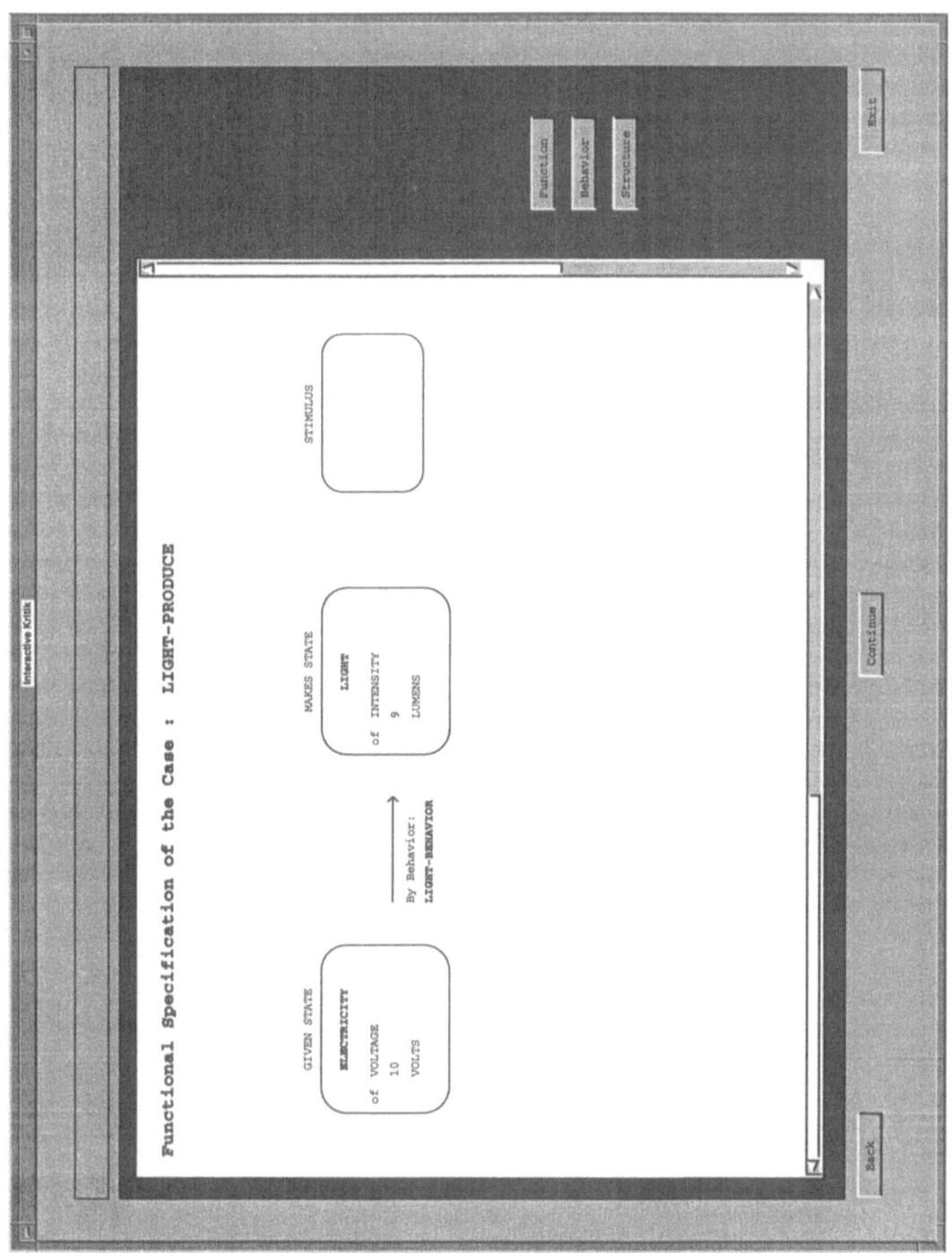

Figure 2. The function of an electrical circuit.

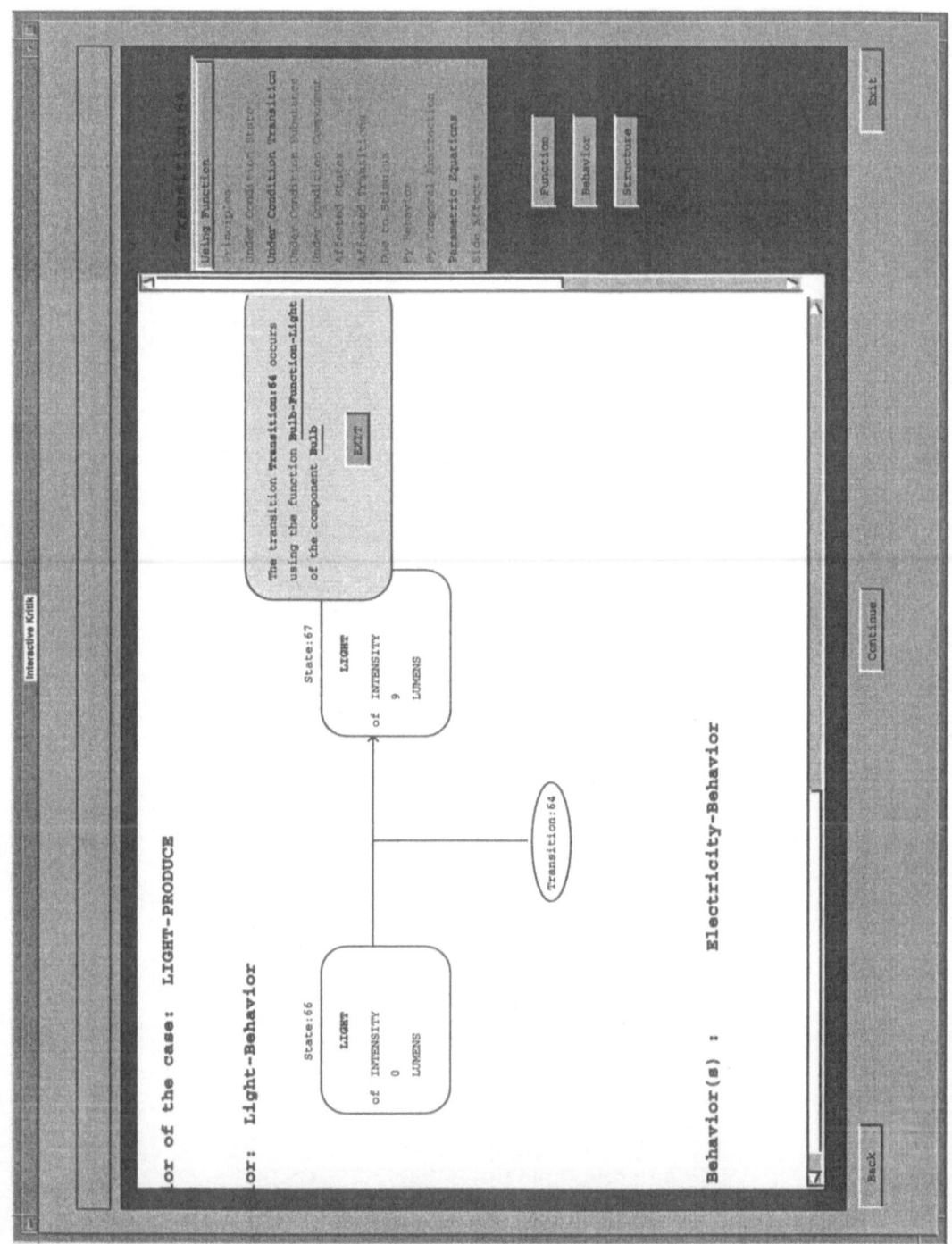

Figure 3. A behavioral transition within an electrical circuit.

description for *Under Condition Transition* specifies that the transition is contingent on the flow of electricity through the bulb as detailed in a separate behavior labeled *Electricity-Behavior*. Similarly, the description for **Parametric Equation** specifies the specific equation relating the state variables.

The use of SBF models for explanation of designs is consistent with Simon's (1981) notion of functional explanations of artifacts. He has argued that explanations of artifacts pertain to, and are referenced by, the purpose of the artifact. This leads us to hypothesize that *SBF models capture the content of explanation of a device design at the "right" level of abstraction for comprehension by human designers.*

2.2. TASK-METHOD-KNOWLEDGE MODELS IN INTERACTIVE KRITIK

We use task-method-knowledge models (TMK models) (Chandrasekaran, 1989, 1990; Goel and Chandrasekaran, 1992) for explaining and justifying reasoning about a design problem. The TMK model provides a functional and strategic explanation of design reasoning in terms of the task, the methods used to accomplish the task, the subtasks spawned by the methods, and the knowledge used by the methods. Since subtasks are spawned by the methods available to the reasoner, the TMK model also provides a justification of specific tasks addressed by the reasoner in terms of the methods that spawn the tasks. Similarly, since methods serve tasks and are afforded by the available knowledge, the TMK model provides a justification of the use of specific methods by the reasoner in terms of the tasks being addressed and the knowledge that affords the methods.

The TMK model of design reasoning has three main elements. The first element, the task, is characterized by the types of information it takes as input and gives as output. KRITIK3 addresses the functions-to-structure design task in the domain of physical devices. This task takes as input a specification of the functions of the desired design. It has the goal of giving as output the specification of a structure that delivers the desired functions. The second element in the TMK model is the method. A method is characterized by (i) the type of knowledge it uses, (ii) the subtasks (if any) it sets up, and (iii) the control it exercises over the processing of subtasks. KRITIK3 uses the method of case-based reasoning for addressing the function-to-structure design task. Figure 4 illustrates INTERACTIVE KRITIK's visual representation of this method. The figure shows that the method sets up the subtasks of problem elaboration, case retrieval, design adaptation, and case storage. It also shows the order in which these tasks are executed. In addition, it shows the input-output specification of these tasks. For example, the task of design adaptation takes as input the specification of the desired functions and the best matching case retrieved from the case memory. It gives as output an SBF model for a candidate design as indicated in Figure 4.

The third element in the TMK model is knowledge. A specific type of domain

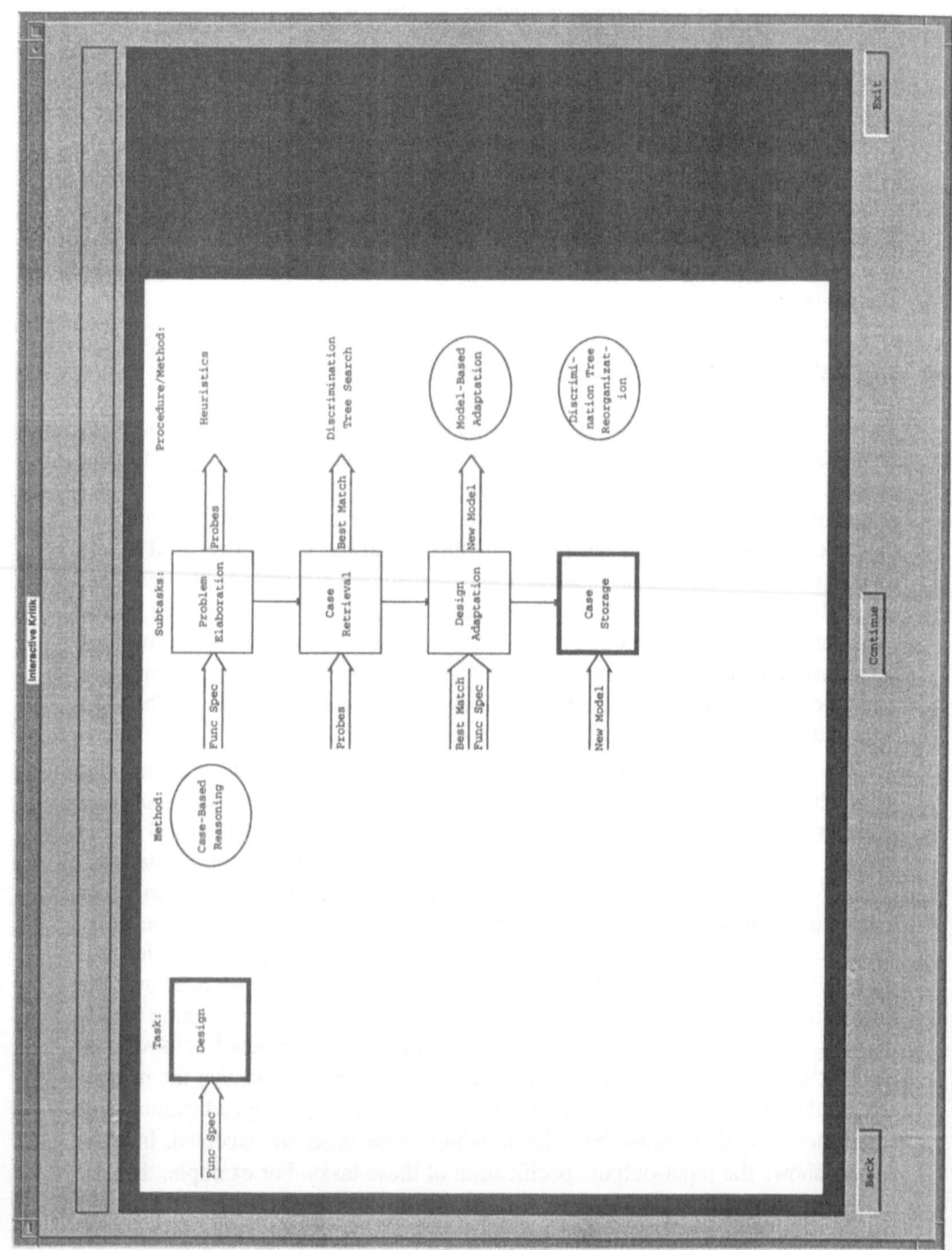

Figure 4. The overall design task.

knowledge is characterized by its content, by its form of representation, and by its organization. Consider the example of diagnostic knowledge. In some domains, heuristic associations that directly map signs and symptoms into fault categories may be available. In a knowledge system, this associative knowledge might be represented in the form of production rules and organized as an unordered list. KRITIK3 contains two kinds of domain knowledge: past design cases and case-specific SBF models. We already have briefly described the representation and organization of the SBF models. Design cases are indexed by the functions delivered by the stored designs, and organized as leaf nodes of a discrimination tree.

The design of the KRITIK family of systems embodies a TMK model of of function-to-structure design of common physical devices (Goel and Chandrasekaran, 1992). We derived this model by analysis of the above task domain using the following methodology (Chandrasekaran, 1989, 1990):

Task Identification: First, the task is specified in terms of the generic types of information it takes as input and the generic types of information desired as its output.

Knowledge Identification: Next, the domain is analyzed in terms of the kinds of knowledge available in it.

Method Identification: Then, the different methods afforded by the different kinds of available knowledge are identified. This step also involves the identification of the subtasks that each method may set up.

Method Selection: Next, since more than one method may be feasible, the criteria for selecting a specific method is specified. These criteria may include factors such as properties of the desired solution and computational properties of the methods.

Recursive Task-Domain Analysis: Finally, the above steps are repeated for each of the subtasks that the selected method sets up.

This recursive decomposition of the given task continues up to an "elementary" level at which the domain affords knowledge that can "directly" map the input to the (sub)task into its desired output. At this level, no method is needed; instead, a procedure directly applies the relevant knowledge to solve the task. The recursive task decomposition results in a task-method-subtask tree. For example, design adaptation is a subtask of the design task set up by the case-based method as illustrated in Figure 4. KRITIK3 uses a model-based method for addressing the task of adaptation as Figure 5 illustrates. The model-based method sets up its own subtasks of the design adaptation task. The first of these subtasks is the computation of differences between the desired function and the function delivered by the design retrieved from the case memory. KRITIK3 uses a simple pattern matching procedure for this task.

The TMK language for describing a knowledge system's reasoning is consistent with Marr's (1977) task-level and Newell's (1982) knowledge-level analyses of intelligent agents. Marr proposed that the reasoning of an intelligent agent can

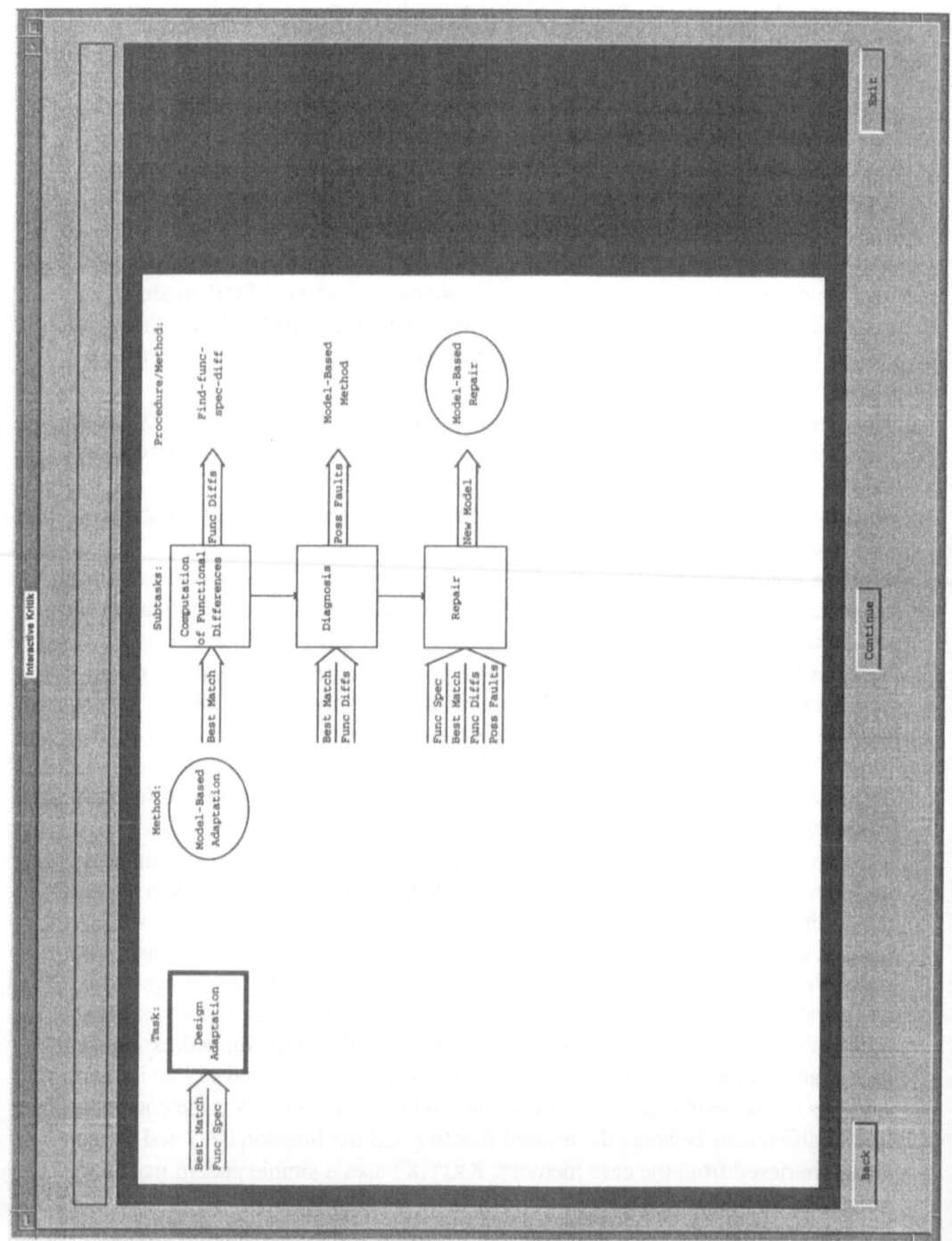

Figure 5. The design adaptation task.

be analyzed at three levels. At the highest level is a specification of the tasks addressed and the mechanisms used by the agent. At the next level are the specific algorithms and data structures that the mechanism uses. At the lowest level is the architecture (or language) of implementation. Similarly, Newell proposed several levels of analysis of intelligent agents. The highest level in his scheme pertains to the agent's goals and the knowledge that enables the accomplishment of the goals. The next level concerns the symbolic structures that implement the mechanisms of the higher level. The next lower level specifies the physical devices that implement the symbolic structures, and so on. Marr suggested that the highest level in his scheme, the task-level, constituted the computational theory of the agent. Similarly, Newell suggested that the highest level in his scheme, the knowledge level, constituted the computational theory of the agent. This leads us to the hypothesis that *TMK models capture the content of explanation of design reasoning at the "right" level of abstraction for communication with human designers.*

3. The Explanatory Interface in INTERACTIVE KRITIK

The explanatory interface in INTERACTIVE KRITIK not only explains and justifies design reasoning and device designs, but also enables the user to explore the device designs and to reflect on the reasoning.

3.1. DESIGN EXPLANATION IN INTERACTIVE KRITIK

The interface agent in INTERACTIVE KRITIK has access to all the knowledge of KRITIK3 including its design cases and device models. It uses KRITIK3's SBF models of physical devices to graphically illustrate and explain the functioning of the devices to the users. It also graphically illustrates and explains the reasoning of the system in generating a new design. Within the context of a design episode, INTERACTIVE KRITIK provides graphical representations of both the designs retrieved from the case memory and the new designs created. Thus it provides representations of intermediate designs in addition to the final designs. The different design versions are presented as the design reasoning unfolds, i.e., in the context of the design subtask at hand.

The working of a device is illustrated to the user on several interrelated screens. One screen represents the device function; Figure 2 is an example of INTERACTIVE KRITIK's screen illustrating the function of an electrical circuit. The means by which the function of a device is achieved is explained by the internal causal behaviors in the SBF device model. Figure 3 shows an illustration by INTERACTIVE KRITIK of the main behavior, *Light-Behavior*, of the electrical circuit that produces light. A different screen shows the secondary behavior, *Electricity-Behavior*, of this device: the behavior of the electricity in this circuit.

KRITIK3's reasoning is illustrated on multiple screens identifying the tasks that the system performs while solving a problem and the methods it uses, as in-

dicated in Figures 4 and 5. For each (sub)task, INTERACTIVE KRITIK illustrates
the *reasoning state* both before and after the accomplishment of the (sub)task.
The reasoning state specifies the task context and the method context. In addition,
when appropriate, INTERACTIVE KRITIK illustrates the design knowledge available to KRITIK3. For example, in explaining the task of case retrieval, it graphically illustrates the case memory.

3.2. DESIGN EXPLORATION IN INTERACTIVE KRITIK

INTERACTIVE KRITIK enables the user to browse through different facets of a
device design. This includes not only the final design proposed by KRITIK3 but
also the intermediate designs it may have generated, for example, the design retrieved from the case memory. Exploration of a given design through browsing is
enabled by the SBF model for the design.

As we explained in Section 2.1, different parts of an SBF model are closely
interrelated. For example, the specification of a function in the SBF model acts as
an index to the causal behaviors that accomplish the function. Also, the specifications of the state transitions in a causal behavior act as indices into the functional
specifications of the structural components of the device. In addition, the description of a device component contains a specification of its functions, and points to
the causal behaviors of the device in which the component plays a functional role.
This indexing scheme enables the user to browse through the SBF model of the
design.

The initial view of an SBF model is a representation of the device's functional specification, as in Figure 2. From here the user can push interface but-
tons to move among the functional, behavioral, and structural representations of
the device. Additionally, the user can click on the name of the behavior by which
the function is achieved (e.g., *Light-Behavior* in Figure 2) and "jump" directly to
that behavior. Figure 3 illustrates the *Light-Behavior* screen. This screen presents
Light-Behavior and labels all other behaviors (in this case, just the *Electricity-
Behavior*) which the user can select to jump to a different behavior. When a user
clicks on a particular transition a menu pops up allowing the user access to a variety of options relating to that transition, as indicated in Figure 3. This allows direct access to structural and behavioral information relating to that transition. For
example, if the transition selected is dependent on another behavior, the user can
jump directly to that behavior. The structure screen provides similar capabilities
for looking at the components of a device and the connections between them.

3.3. DESIGN REFLECTION IN INTERACTIVE KRITIK

The explicit SBF representation of a design enables the user to inspect each element and aspect of the device design. Similarly, the explicit TMK representation
of the trace of design reasoning enables the user to inspect each task, method,

knowledge source, and reasoning state. This enables the user to reflect on the design reasoning. For example, the user can examine the TMK reasoning trace and detect potential flaws in it.

As we mentioned in Section 2.1, the SBF model of a device design not only explains how the device works but also justifies the design by showing how its structure delivers the desired functions. And as we mentioned in Section 2.2., the TMK model not only explains the reasoning of KRITIK3 but also justifies the tasks it sets up and the methods it uses. In addition, the user can also ask IN-TERACTIVE KRITIK for a justification for specific reasoning choices. As an example, consider the situation in which KRITIK3 retrieves a design case from its case memory. The TMK trace shows the user the probe KRITIK3 had prepared to retrieve a case and the case the system actually retrieved from its case memory. The user can now ask why did KRITIK3 retrieve this particular design case. Since the reasoning trace explicitly specifies the probe prepared by KRITIK3, and how the system's retrieval method probed the case memory - the branches it followed, the matches it made, and their results - the trace provides a justification for why the particular case best matches the given problem.

3.4. CRITIQUE

There is still a great deal of work to be done on INTERACTIVE KRITIK's user interface. Some issues which would need to be addressed before the system could be used as a practical tool include the improved display of the structure of a device, the building of better graphical representations, and provision of additional interaction capabilities. More importantly, INTERACTIVE KRITIK needs to be formally evaluated in a real world setting. But this kind of evaluation also requires additional work on the user interface.

4. Discussion

This research builds on earlier work on three topics at the intersection of AI and Design: design methods and process models, design knowledge and device models, and interactive design environments.

Design Methods and Process Models: A major goal of AI research on design has been to develop computational methods and process models for design. This has led to the development of several computational methods for design; examples include heuristic search (Stallman and Sussman, 1977), heuristic association (Mc-Dermott, 1982), and plan instantiation and expansion (Brown and Chandrasekaran, 1989; Mittal, Dym and Morjaria, 1986). Recent research on case-based design (e.g., Goel and Chandrasekaran, 1992; Maher, Balachandran and Zhang, 1995; Navinchandra, 1991) has led to the development of multi-strategy process models for design. KRITIK3 is a multi-strategy process model of design in two senses. First, while the high-level design process in KRITIK3 is case-based, the reasoning about individual subtasks in the case-based process is model-based. For example, KRITIK3 uses SBF device models for adapting a past design and for evaluating a candidate design. Second, design adaptation in KRITIK3 involves multiple modification methods. While all modification methods make use of SBF device models, different methods are applicable to different kinds of adaptation tasks.

A closely related research direction concerns the language for specifying the computational methods and process models for design. McDermott (1982) describes R1's method for configuration design in the language of constraints of a design problem, components available in the design domain, heuristic associations pertaining to the constraints and the components, and selection and activation of the associations. But this language is much too specific to R1's method. This method-specificness of the language becomes a major problem for describing and explaining multi-strategy process models such as KRITIK3.

Task-level (Marr, 1977) (or, equivalently, knowledge-level (Newell, 1982)) accounts make a clearer separation between knowledge-based reasoning and its implementation in a knowledge system. In the mid-eighties, Chandrasekaran (1988) proposed the language of Generic Tasks for analyzing and modeling knowledge-based problem solving, and showed that this language enables more perspicuous explanations (Chandrasekaran, Tanner, and Josephson, 1989). In the late eighties, Chandrasekaran (1990) related Generic Tasks with task structures: Chandrasekaran (1989) describes a high-level task structure for design; Goel and Chandrasekaran (1992) describe a fine-grained task structure for case-based design. In their work on the elevator design project called VT, McDermott and his colleagues (McDermott, 1988; Marcus et al., 1988) described a similar task-oriented language for analyzing knowledge-based design.

Our TMK models represent a generalization of task structures based on Generic Tasks. Also, our hypothesis that TMK models provide the "right" level of abstraction for explaining knowledge-based reasoning is based in part on earlier work on explanation in the Generic Task framework. But TMK models make the specific role played by a particular type of knowledge more explicit than earlier models. Consider, for example, the functional role of an SBF model of a past design in KRITIK3. Since the SBF model is associated with the past case, it affords a method for adapting the past design. The TMK model makes this affordance explicit. Thus, while task structures are useful for explaining the control of reasoning in terms of task-method interactions, TMK models are also useful for explaining knowledge-method interactions. In particular, they enable the explanation of the organization and indexing of different kinds of knowledge, the kinds of knowledge available for addressing a task, and the methods that become feasible because of the available knowledge.

Design Knowledge and Device Models: Explanation of physical devices has been a major topic of research not only in AI and in Design but also in Cognitive Engineering. AI research on device modeling and explanation can be traced as far back as Hayes (1979) work on "naive physics" in which he described a component-substance ontology. At about the same time, de Kleer developed the method of qualitative simulation for diagnosing electrical circuits (de Kleer, 1984). This work led to the no-function-in-structure principle (de Kleer and Brown, 1984) which states that the behaviors of each structural component must be represented in a manner independent of their functional contexts.

In contrast, in the early eighties, Chandrasekaran and his colleagues developed the Functional Representation (FR) scheme (Sembugamorthy and Chandrasekaran, 1986; Chandrasekaran et al., 1993) in which the functions are not only represented explicitly, but also used to reference the causal behaviors responsible for their accomplishment. The causal behaviors in turn reference the functions of the device substructures. Since the function of a substructure refers to the causal behaviors that result in it, this gives rise to a hierarchical organization of the device model. Also in the mid-eighties, Bylander proposed a taxonomy of primitive behaviors (Bylander, 1991) based in part on Hayes' component-substance ontology. He also described a method of composing the primitive behaviors into more complex behaviors. Our SBF models evolve from Chandrasekaran's Functional Representation scheme and Bylander's ontology of behaviors. In particular, they use FR's organizational scheme in which the device functions act as indices to the causal behaviors and the causal behaviors index the functions of device substructures. The specification of the functions, behaviors and structure in SBF models, however, is based on Bylander's well-defined behavioral ontology.

In Cognitive Engineering, Rasmussen (1985) proposed a hierarchical organization for presenting device knowledge to human users. His device models also specify the structure, the behaviors, and the functions at each level in the hierarchy. Our hypothesis that SBF models provide the "right" level of abstraction for explaining the working of a device to a human user is supported by Rasmussen's empirical work. Govindaraj (1987) has used similar hierarchical organization schemes for enabling engineering students to explore the design of complex devices containing hundreds of components. Following his device models, the causal behaviors in our SBF models too are organized along the flow of specific substances in the device.

In Design research, Gero et al. (1991) and Umeda et al. (1990) have also described FBS models (for function-behavior-structure). While the details of the representation schemes differ, in both their FBS models and in our SBF models, behavior mediates between function and structure. Indeed, a major theme of our work on the KRITIK family of systems has been that while the design task takes a functional specification as input and gives a structural specification as output, much of the design reasoning is at the intermediate behavioral level.

Interactive Design Environments: A core issue in interactive design environments is how human designers and knowledge systems may share design responsibilities. AI research on interactive design environments covers a broad range of human/system responsibility sharing. At one extreme, the system acts as a knowledge source but leaves almost all reasoning to the human designer. Traditionally, knowledge bases for design have contained knowledge of design components and materials. But recent work on design knowledge bases has focused on providing human designers with access to libraries of design cases; examples include CADET (Sycara et al., 1991), CADRE (Hua and Faltings, 1992), CASECAD (Maher, Balachandran and Zhang, 1995), FABEL (Voss et al., 1994), Archie (Pearce et al., 1992), AskJef (Barber et al., 1992), and ArchieTutor (Goel et al., 1993). At the other extreme of this spectrum are autonomous knowledge systems that perform almost all design reasoning by themselves. Human interaction with these systems is limited to formulating design problems, supplying the problems to the system, and receiving the solutions generated by the system; examples include R1, AIR-CYL, and the original KRITIK system.

In between these two extremes lies a large range of potential sharing of responsibility between the system and the user. An important goal of design environments in the middle of this spectrum is to enable humans to construct new designs. Fischer et al.'s (1992) JANUS and Steinberg (1987) VEXED are two examples of constructive design environments. The goal of the INTERACTIVE KRITIK project is also the building of a constructive design environment. We will not describe here how, when completed, INTERACTIVE KRITIK may enable a human to construct new designs (but see Grué, 1994). Instead, we focus the rest of this discussion on the issue of explanatory interface in the current version of INTERACTIVE KRITIK since this is already operational.

Mostow (1989) has argued that when a knowledge system in an interactive design environment proposes a solution to a design problem, then the system should also provide the human designer with an explanation of the reasoning that led to the solution. His BOGART system uses derivational analogy (Carbonell et al, 1989) for generating solutions to design problems. Following the theory of derivational analogy, BOGART provides the human designer with an explanation of its reasoning in the form of a derivational record. The derivational record contains a trace of the system's reasoning in the language of design goals, operators, and heuristics for goal decomposition and operator selection.

We share the premise that in any interactive design environment, the knowledge system must be able to explain its reasoning. However, we believe that the language of goals, operators and heuristics is too low level to be accessible and comprehensible to human designers, especially novice designers. Instead, we hypothesize that the TMK language is at the "right" level of abstraction. More importantly, we believe that in addition to explaining its reasoning, the knowledge system must also be able to justify the design solution it proposes. INTERACTIVE KRITIK uses SBF models for justifying its design solutions.

Further, we believe that it is critical that the explanation of design reasoning should be grounded in the context of the evolving design solution, and, similarly, the explanation of the evolving design should be grounded in the context of the design reasoning that led to it. The advantages of situating design explanations in this way are two fold. First, situating the explanation of design reasoning in the context of the evolving design solution makes the explanation more meaningful. This is because the explanatory terms can now get their meaning from the specific parts of the design to which they refer. Second, situating the explanation of the design solution in the context of the design reasoning makes the explanation more complete because of the availability of previous versions in the evolution of the design solution.

5. Conclusions

Interactive design environments typically contain knowledge systems as major components. A human designer may use the interactive environment for design construction and experimentation. The knowledge systems may help automate specific and selected portions of this process, leading to human-system cooperative design. This raises the issues of usability and learnability of the knowledge systems. Human designers are unlikely to work with these systems if they cannot easily use them and also easily learn how to use them. Designers are more likely to use these systems if they can form a mental model of how the system works, how it reasons about problems, and if they can develop some confidence in the solutions generated by the system.

So the issue becomes how might a knowledge system enable the user to form a mental model of its reasoning, how might it explain its reasoning and justify its answers. Our work on INTERACTIVE KRITIK is based on three related ideas:

1. Explanations of a knowledge system need to capture the functional and strategic content of reasoning in addition to its knowledge content. Task-method-knowledge models enable this kind of task-level and knowledge-level explanation, which facilitates effective communication between the system and the user.

2. Explanations of physical systems need to capture the functionality and causality of the systems in addition to their structure. Structure-behavior-function models enable this kind of explanation at a level of abstraction that facilitates effective communication between the system and the user.

3. Explanation of design reasoning needs to be grounded in the context of the evolving design solution, and, similarly, the explanation of the evolving design needs to be grounded in the context of the design reasoning that led to it.

INTERACTIVE KRITIK demonstrates the computational feasibility of these ideas.

Acknowledgments

Much of this research was done during 1993-94 when all the authors were with Georgia Institute of Technology in Atlanta, Georgia, USA. Andrés Gómez is now with the Key

Centre of Design Computing, University of Sydney, Sydney, Australia; Nathalie Grué is now with the Institute for Learning Sciences, Northwestern University, Evanston, Illinois, USA; and Margaret Recker is now with Victoria University, Wellington, New Zealand. This work has benefited from contributions by Sambasiva Bhatta, Michael Donahoo, Vinay Pandey, and Eleni Stroulia. It has been funded in part by a grant from the Advanced Research Projects Agency and partly by internal seed grants from Georgia Tech's Educational Technology Institute, College of Computing, Cognitive Science Program, and Graphics, Visualization and Usability Center.

References

Barber, J., Jacobson, M., Penberthy, L., Simpson, R., Bhatta, S., Goel, A., Pearce, M., Shankar, M. and Stroulia, E.: 1992, Integrating artificial intelligence and multimedia technologies for interface design advising, *NCR Journal of Research and Development*, **6**(1), 75–85.

Brown, D. and Chandrasekaran, B.: 1989, *Design Problem Solving: Knowledge Structures and Control Strategies*, Pitman, London, UK.

Bylander, T.: 1991, A Theory of consolidation for reasoning about devices, *Man-Machine Studies*, **35**, 467–489.

Carbonell, J., Knoblock, C. and Minton, S.: 1989, PRODIGY: An integrated architecture for planning and learning, *in* Van Lehn (ed.), *Architectures for Intelligence*, Lawrence Erlbaum.

Chandrasekaran, B.: 1988, Generic tasks as building blocks for knowledge-based systems: The diagnosis and routine design examples, *Knowledge Engineering Review*, **3**(3), 183–219.

Chandrasekaran, B.: 1989, Task structures, knowledge acquisition and machine learning, *Machine Learning*, **4**, 341–347.

Chandrasekaran, B.: 1990, Design problem solving: A task analysis, *AI Magazine*, **Winter**, 59–71.

B. Chandrasekaran, M. Tanner, and J. Josephson. Explaining control strategies in problem solving. *IEEE Expert*. 4(1):9-24, 1989.

B. Chandrasekaran, A. Goel, and Y. Iwasaki. Functional Representation as Design Rationale. *IEEE Computer*, 48-56, January 1993.

de Kleer, J.: 1984, How circuits work, *Artificial Intelligence*, **24**, 205–280.

de Kleer, J. and Brown, J.: 1984, A qualitative physics based on confluences, *Artificial Intelligence*, **24**, 7–83.

Fischer, G., Grudin, J., Lemke, A., McCall, R., Ostwald, J., Reeves B. and Shipman, F.: 1992, Supporting indirect collaborative design with integrated knowledge-based design environment, *Human-Computer Interactions*, **7**(3), 281–314.

Gero, J. S., Lee H. and Tham, K.: 1991, Behavior: A link between function and structure in design, *Proceedings IFIP WG 5.2 Working Conference on Intelligent CAD*, Columbus, Ohio, pp. 201–230.

Goel, A.: 1991, A model-based approach to case adaptation, *Proceedings Thirteenth Annual Conference of the Cognitive Science Society*, Lawrence Erlbaum, pp. 143–148.

Goel, A.: 1992, Representation of design functions in experience-based design, *in* D. Brown, M. Waldron and H. Yoshikawa (eds), *Intelligent Computer Aided Design*, North-Holland, pp. 283–308.

Goel, A. and Chandrasekaran, B.: 1989, Functional representation of designs and redesign problem solving, *Proceedings Eleventh International Joint Conference on Artificial Intelligence*, Morgan Kaufmann, pp. 1388–1394.

Goel, A. and Chandrasekaran, B.: 1992, Case-based design: A task analysis, *in* C. Tong and D. Sriram (eds), *Artificial Intelligence Approaches to Engineering Design, Volume II: Innovative Design*, Academic Press, pp. 165–184.

Goel, A., Pearce, M., Malkawi, A. and Liu, K.: 1993, A cross-domain experiment in case-based design support: ARCHIETUTOR, *Proceedings AAAI Workshop on Case-Based Reasoning*, pp. 111–117.

Govindaraj, T.: 1987, Qualitative approximation methodology for modeling and simulation of large dynamic systems: Applications to a marine power plant, *IEEE Transactions on Systems, Man and Cybernetics*, **SMC-17**(6), 937–955.

Grué, N.: 1994, *Illustration, Explanation and Navigation of Physical Devices and Design Processes*, MS Thesis, College of Computing, Georgia Institute of Technology.

Hayes, P.: 1979, Naive physics manifesto, *Expert Systems in the Microelectronics Age*, Edinburgh University Press, Edinbugh, UK, pp. 242–270.

Hua, K. and Faltings, B.: 1993, Exploring case-based building design - CADRE. *AI(EDAM)*, **7**(2), 135–143.

McDermott, J.: 1982, R1: A rule-based configurer of computer systems, *Artificial Intelligence*, **19**, 39–88.

McDermott, J.: 1988, Preliminary steps towards a taxonomy of problem solving methods, *in* S. Marcus (ed.), *Automating Knowledge Acquisition for Expert Systems*, Kluwer, Boston, MA.

Maher, M. L., Balachandran, M. B. and Zhang, D.: 1995, *Case-Based Reasoning in Design*, Erlbaum, Hillsdale, NJ.

Marcus, S. Stout, J. and McDermott, J.: 1988, VT: An expert elevator designer that uses knowledge-based backtracking, *AI Magazine*, **9**(1), 95–112.

Marr, D.: 1977, Artificial intelligence—A personal view, *Artificial Intelligence*, **9**(1).

Mittal, S., Dym, C. and Morjaria, M.: 1986, PRIDE: An expert system for the design of paper handling systems, *Computer*, **19**(7), 102–114.

Mostow, J.: 1989, Design by derivational analogy: Issues in the automated replay of design plans, *Artificial Intelligence*.

Myers, B. and Zanden, B.: 1992, Environment for rapidly creating interactive design tools, *Visual Computer*, **8**, 94–116.

Navinchandra, D.: 1991, *Exploration and Innovation in Design: Towards a Computational Model*, Springer-Verlag, New York.

Newell, A.: 1982, The knowledge level, *Artificial Intelligence*, **18**(1), 87–127.

Pearce, M., Goel, A., Kolodner, J., Zimring, C., Sentosa, L. and Billington, R.: 1992, Case-based design support: A case study in architectural design, *IEEE Expert*, **7**(5), 14–20.

Rasmussen, J.: 1985, The role of hierarchical knowledge representation in decision making and system management, *IEEE Trans. Systems, Man and Cybernetics*, **15**, 234–243.

Sembugamoorthy, V. and Chandrasekaran, B.: 1986, Functional representation of devices and compilation of diagnostic problem solving systems, *in* J. Kolodner and C. Riesbeck (eds), *Experience, Memory and Reasoning*, Lawrence Erlbaum, Hillsdale, NJ, pp. 47–73.

Simon, H.: 1981, *The Sciences of the Artificial*, 2nd edn, MIT Press.

Stallman, R. and Sussman, G.: 1977, Forward reasoning and dependency-directed backtracking in a system for computer-aided circuit analysis, *Artificial Intelligence*, **9**, 135–196.

Steinberg, L.: 1987, Design as refinement plus constraint propagation: The VEXED experience, *Proceedings Sixth National Conference on Artificial Intelligence*, pp. 830–835.

Sycara, K., Navinchandra, D., Guttal, R., Koning, J. and Narsimhan, S.: 1991, CADET: A case-based synthesis tool for engineering design, *Expert Systems*, **4**(2), 157–188.

Umeda, Y., Takeda, H., Tomiyama, T. and Yoshikawa, H.: 1990, Function, behavior and structure, *Proceedings Fifth International Conference on Applications of AI in Engineering*, Vol. 1, pp. 177–193.

Voss, A. Coulon, C.-H., Grather, W. Linowski, B., Schaaf, J., Barstsch-Spörl, B., Borner, K., Tammer, E. Durscke, H. and Knauff, M.: 1994, Retrieval of similar layouts - About a very hybrid approach in FABEL, *in* J. S. Gero (ed.), *Artificial Intelligence in Design '94*, Kluwer, Dordrecht, pp. 625–640.

8

learning in design

J. S. Gero and F. Sudweeks (eds), Artificial Intelligence in Design '96, 409-428.
© 1996 *Kluwer Academic Publishers.*

LEARNING BY SINGLE FUNCTION AGENTS DURING SPRING DESIGN

DAN L. GRECU AND DAVID C. BROWN
AI in Design Group, Computer Science Department
Worcester Polytechnic Institute, Worcester, MA 01609, USA

1. Introduction

This paper reports on some initial experiments on learning in multi-agent design systems. These experiments have several goals. The first is to study the ease with which simple learning techniques fit into the multi-agent paradigm we are using. The second is to determine the performance of these techniques. The third is to study the application of the multi-agent paradigm we use to "real" problems, as its development has mostly been concerned with a more theoretical view.

The design system we use is built from small knowledge-based (expert) systems that we call Single Function Agents (SiFAs) (Victor and Brown, 1995; Dunskus et al., 1995; Berker and Brown, 1995). There are a small set of types of agents, each of which has restricted capabilities. SiFAs will be explained in more detail below.

1.1 THE TYPE OF DESIGN PROBLEM

The type of design problem addressed is Parametric design. This occurs when the topology of the artifact being designed is already decided, and the design is to be completed by determining values for a set of parameters that specify the remaining details, such as color, surface finish, or geometry. This is a knowledge-based process, where the amount of available theoretical and experiential knowledge (e.g., heuristics or relevant design history) can vary from problem to problem. We are concerned with non-routine or near-routine situations that have many constraints present, and perhaps tangled dependencies between parameters. Parametric design is not necessarily routine or simple.

Deciding a parameter's value can require various types of reasoning using different kinds of knowledge. At least three factors make the decision difficult, even though the range of choices may not be that large. These factors are the different design requirements, a multitude of attributes that characterize the parameter, and the dependencies between parameters.

The lack of a fixed, known order of parameters for which to decide values makes the search for a good set of values even more complex. Under these circumstances it is virtually impossible to avoid failures due to constraint violations. This leads to conflicts between the agent that decided a value and the constraint that rejected it.

To reflect the complexity involved in parameter decisions, we allow multiple design agents to have the same "job" (i.e., to provide a value for parameter X), but to have different points of view (e.g., cost, strength). The agents will often produce different values for X, leading to a conflict. To resolve conflicts, trade-offs have to be made via negotiations, with an exchange of information between the agents involved.

1.2 WHY USE LEARNING?

A multi-agent design system includes many agents with lots of potential interactions between them. Conflicts considerably increase the number of interactions needed during design. The more serious the conflict (i.e., the more difficult it is to resolve), the more messages are exchanged to find an acceptable solution. As the ratio of messages per design decision increases, the efficiency of the system decreases. High conflict situations, such as the ones we are addressing, provide a strong motivation for any technique that leads to reduced overhead. Learning can reduce this kind of overhead, by reducing the number of conflicts and/or by reducing the number of messages during interactions.

In this paper, after briefly describing other related research we will present Single Function Agents. The domain of material selection in Spring design is described next, along with its mapping to SiFAs. A discussion of conflicts in material selection motivates the description of the method of learning and its implementation. The paper concludes after a presentation of the results of our experiments.

2. Related Work

Agents are becoming increasingly important in Artificial Intelligence and Computer Science (Wooldridge and Jennings, 1994). Design, as well as other complex tasks, have parallel decompositions that remove problems resulting from forced pre-execution serialization. Parallel decompositions allow opportunistic collaboration. These map well to agents. This has led to a slow but steady growth in the amount of research into multi-agent design systems. Examples include Klein (1991), Kuokka et al. (1993), Lander and Lesser (1991), Sycara (1990), Taleb-Bendiab and Oh, (1993), Victor et al. (1993), Werkman and Barone (1991).

The study of Conflict Resolution and Negotiation is also growing. Klein's (1991) model of conflict resolution uses a hierarchy of conflicts with the most abstract conflicts at the top and most concrete conflicts at the leaves. A corresponding hierarchy of resolution strategies allows conflicts to be mapped to resolution strategies. Klein's agents have both design and conflict resolution knowledge.

Negotiation is a common approach to conflict resolution in the design domain, and is the process by which resolution of inconsistencies is achieved in order to arrive at a coherent set of design decisions (Sycara, 1990). Negotiation proceeds with generation of a proposal, then a counter proposal based on feedback from dissenting agents, and communication of justifications and supporting evidence.

As the Single Function Agent approach is relatively new only three systems have been developed so far. I3D is a system that integrates part design and manufacturing plan production for Powder Processing applications (Victor et al., 1993). In I3D the agents were not allowed to conflict, but were in I3D+ (Victor and Brown, 1994). The SNEAKERS system was built to train users in Concurrent Engineering. The user interacted with agents that had different functions and points of view (Douglas et al., 1993).

Recently, researchers have started to work on the application of Machine Learning algorithms to multi-agent systems (Sen, 1995), and design researchers have investigated learning in design systems (Maher, Brown and Duffy, 1994). However, there is still relatively little work on learning in multi-agent design systems (NagendraPrasad, 1995).

3. Single Function Agents

A SiFA is a small knowledge-based, expert agent. It performs a single Function, on a single Target, from a single Point of view. For example, the selection of a material from the point of view of reliability.

A SiFA's type is determined by its *function*. The function describes what kind of information it processes and what kind of result it produces. There are a limited number of types currently allowed. The types are intended to be combinable to have the problem-solving power to do (at least) Parametric design.

The SiFA types used in previous work are:

1. *Selector:* selects a value for a parameter, by picking a value from prestored or calculated possible values, according to some preferences.
2. *Estimator:* produces an estimate of the value of a parameter. Unlike selectors, estimators can work quickly and with insufficient information, so that the values they produce are just estimates of what the final value should be. They may also be imprecise, producing a range of possible values.

3.   *Evaluator:* gives a quality rating for the value of a parameter, producing a measure of goodness for that value, usually represented as a percentage or as a symbol (e.g., "good").
4.   *Critic:* criticizes the value of a parameter by pointing out constraints or quality requirements that are not met by the current value.
5.   *Praiser:* praises values of parameters by pointing out why the value is desirable.

Each SiFA has a single *target*. The main type of target of a SiFA is a value of a single parameter of the design. But as the critiques, praises, estimations and evaluations are all treated as "first class objects", any of these entities can be the target of an agent (Berker and Brown, 1995).

The *point of view* of an agent is some direction or aspect of the design that the agent considers while doing its work – often a goal that the agent is trying to achieve. Examples of points of view are cost, strength, style, weight, reliability and availability. The points of view partition the knowledge about the target.

SiFAs communicate by sending each other messages. Each agent can communicate directly with any other agent. The communication language used is based on KQML (Knowledge Query and Manipulation Language) (Finin et al., 1993). The current state of the design is accessible to all agents. Although a scheduling mechanism is required for any implementation of SiFA-based systems, no particular method is assumed by the SiFA model. SiFAs are assumed to be able to be "triggered" by satisfied preconditions. It is likely that Selectors should be given priority, and that conflict detection and resolution is more important than carrying on with the design.

3.1 WHY STUDY SIFAS?

SiFAs were originally conceived from experiences with building expert systems for Concurrent Engineering support systems (for example, see Douglas et al., 1993; Victor et al., 1993). Since that time they have gradually been refined. They are not principally seen as a new way of building design systems. Rather, they are considered to be, and have been used as, a tool for investigating knowledge, problem-solving and conflict resolution. They provide a context for precisely conceptualizing, generating, clarifying and categorizing:
• types of knowledge e.g., conflict identification (Berker and Brown, 1995);
• types of reasoning, e.g., conflict detection (Berker and Brown, 1995);
• types of conflicts, e.g., Estimator-Critic (Dunskus et al., 1995);
• trade-offs in the design process, e.g., material processing vs. cost);
• what might be learned in multi-agent design systems, e.g., selection preferences (Grecu and Brown, 1995);
• knowledge for knowledge acquisition, e.g., selection list (Currier, 1995).

3.2 WHAT SIFAS AREN'T

SiFAs are not intended to be a realistic simulation of a team of designers (such as in Concurrent Engineering). They are too fine-grained for that. At present, SiFA research has little to offer the study of multi-agent systems built from legacy code (such as Kuokka et al., 1993), or other large-grained design systems (such as Lander and Lesser, 1991; Werkman and Barone, 1991). However, the detailed study of conflict management that SiFAs facilitate should yield general results (Dunskus et al., 1995).

SiFAs are also not "overhead free". For the same overall functionality, having many small agents, rather than fewer large agents, leads to increased overhead from inter-agent communication. Currently, the research benefits of SiFAs outweigh this disadvantage. Learning, however, can help reduce this overhead. The small grain size increases the number of potential locations where learning can occur. However, it remains to be seen if SiFAs will ever become a fully viable design system building tool, in addition to being a research tool.

4. Material Selection in Spring Design

A variety of types of springs are used in mechanical engineering design. Each type corresponds to a basic spring configuration. After a specific configuration is chosen, the designer has to reason about the values of the design parameters which describe the configuration. The parameter set is not necessarily the same for all the spring configurations.

Our experiments were limited to helical compression springs. The most important parameters which define these springs are the spring material, the wire diameter, the mean coil diameter and the number of coils. These are primary design parameters, meaning that they are not computed based on other spring parameters. Note that the primary design parameters are not totally independent, as they are related through design constraints. The non-primary spring parameters, such as the spring index (the ratio of the mean coil diameter of a spring to the wire diameter), are derived from the primary design parameters.

All the single function agents we used to test agent interactions targeted one single design parameter: the spring material. Material selection requires the designer to take into account a wide diversity of attributes, such as stress, electrical conductivity and cost, and to decide which of the possible materials satisfies the design requirements. The choice influences the decisions on the other spring parameters. For example, the deflection of a spring coil depends on material elasticity and will be used in deciding the number of coils. This is an example of parameter dependency.

There are about 30 materials most commonly used in spring design (Machinery, 1982). Based on their composition they are grouped in 7 categories with related physical properties. The selection of a material is determined by these factors: *temperature range,* where the material has its normal physical properties; *tensile strength,* dependent on the manufacturing process; *resistance under various shock loading conditions*; *resistance under various impact loading conditions*; *allowable working stress* given the intended service of the spring; *modulus of elasticity; fatigue life* – the time after which the spring fails, far below its normal elastic limit, due to continuous deflection; *endurance limit*—the highest stress, or range of stress, that can be repeated indefinitely without causing spring failure; *hardness* value that can be achieved through treatments; *electrical conductivity; magnetic properties; corrosion resistance; shape* and *diameter* of the wire section, from the manufacturer; and *cost* of manufacturing the material.

The selection of the spring material is usually the first step in parametric spring design. As such, it influences many of the subsequent design decisions. Poor choices can lead to the need to reconsider the entire design process. Therefore, it is desirable to consider as many of the design requirements as possible when deciding the spring material.

A spring design problem does not necessarily have requirements given for all of the previously enumerated factors. Requirements impose thresholds for the admissible values of these factors. Ideally one would like to achieve an optimal value for each factor. However, this is rarely the case. The goal of the design will be to find materials which satisfy all the requirements. Optimality is a different issue as it is relative to various criteria. A global measure of optimality is not available. If the user needs higher standards of quality for some of the material properties, (s)he adds the corresponding constraints to the problem specification.

The material properties are not independent. For example, additional hardness can be obtained by special treatments, but these procedures raise the cost of the spring material. In order to find satisfactory solutions, designers use only some of the properties as selection guides. The rest of the material properties are used to verify that design requirements are not violated. The knowledge about these properties is neither complete, nor uniform across the entire range of materials. Some materials simply do not exhibit one property or another (e.g., electrical conductivity, or magnetic properties). Sometimes, even when the property is known to exist, the knowledge describing it can be available in different amounts and under various representations (e.g., physical laws, graphs, or experimental data stored in tables). Therefore, given varying degrees of completeness and uniformity, analyzing a material's suitability for a design is a matter of expertise.

5. Material Selection with SiFAs

We approached the problem of material selection for spring design by defining a set of selectors, critics, and praisers. All of them have the same target – a material value. The points of view were chosen from the criteria enumerated in the previous section. Not all of them were included in these experiments. But for every point of view we decided to use, there was at least one agent defined.

The important questions we had to address at this point were:
- What type of agents should be defined for each point of view?
- How do we partition the available knowledge among the agents?
- Do we need to use every type of SiFA?

Since we intend to choose a material value we need to define at least one selector. Selectors are the only agents allowed to propose parameter values. Selectors are defined only for those points of view for which there exists knowledge about acceptability, as well as preferences for all parameter values. In other words, a selector should be able to distinguish, from its point of view, between *any* two materials in the range of the application.

For example, every material has a cost. Therefore, we can order all the material values from the point of view of cost, and select the most preferable value. But not all materials have magnetic properties. It isn't possible to prefer one of those materials over another from that point of view, as they don't have any.

Our experiments used two selectors:
1. The first selector proposes materials from the point of view of the working temperature range. Materials are ordered by how well they cover the temperature interval in which the spring is required to work.
2. The second selector chooses materials from the point of view of cost.

The critics are used to express objections to the choices made by the selectors. A critic is not supposed to have knowledge about all the possible choices for the parameter value. It is only supposed to point out those material values that are not acceptable given its point of view. The acceptability of parameter values may depend not only on the design knowledge of the critic, but also on the design requirements. In fact, from its point of view, a critic is capable of detecting the entire set of unacceptable values in the current design context.

We are currently using critics from the following points of view: tensile strength, resistance to shock loading, resistance to impact loading, fatigue life, and electrical conductivity.

These critics use points of view independent of the other spring parameters. As the design experiments will be extended to include all the spring parameters, critics will also be responsible for the dependencies between parameters. For example, an availability critic for wire diameters will restrict the materials to those for which the desired wire diameters are provided.

Praisers are the complementary agents to critics. Their role is to highlight every selection which has a very good rating from their point of view. They are not required to know the entire set of high rating selections in the current design context – as opposed to critics, which do have to know all the 'bad' choices. A praiser's duty is just to mention whether the current proposal for a parameter value surpasses its internal standard. The information provided by the praiser is not critical. That means that neglecting a praiser's observation will not lead to constraint violations and failures. Praisers are useful in negotiations. Whether a proposal is praised or not can influence which of the agents in conflict should reconsider (e.g., relax) its proposal. The current version of the system uses material praisers from the following points of view: stress, endurance, corrosion resistance, and hardness.

Advisers and estimators have also been implemented. Advisers help selectors decide between values which have very similar support. The adviser that assists will have the same point of view as the selector. Since they are not critical for the illustration of the following experiments, advisers are not further mentioned.

Estimators are used in the selection process for parameter values that are connected by constraints with other parameters. Assuming that the other parameter values in the constraint have not been decided yet, they can be estimated in order to make a more informed selection for the current design parameter. As this paper focuses on the decisions associated with one single parameter, estimators are not used either.

Agents from the same functional class (e.g., selectors, critics, etc.) have heterogeneous domain knowledge representation and reasoning. The tensile strength critic uses functions extracted from plotted graphs on which to base its decisions, while the electrical conductivity critic uses data available in tables. The temperature selector uses qualitative interval matching as one of its techniques, while the cost selector simply orders the current costs.

As the target and point of view of each agent are domain dependent, they provide the interface of the agents with the design domain. The functionality of the agents (e.g., selection, critique, praising, etc.) creates a general framework of interaction among the agents independent of the underlying domain.

Finally, another important distinction which differentiates agents with the same functionality is how context sensitive they are. A material selector from the point of view of cost will use a list of preferences which is independent of the current design problem, since the costs of the materials are given and can be viewed as independent of the design specifications. In contrast, the material selector from the point of view of the temperature range is context sensitive. Its preferences vary depending on the working temperature range which needs to be covered. The working temperature range of each available material will overlap the required range differently, thus producing a different preference rating for each material. The issue of context sensitivity will prove to be important when we discuss what agents can learn about each other.

6. System Design and Implementation

The spring design system is composed of several modules: the single function agents and the design board (Figure 1). The design board is a module which is visible to all the agents participating in design. It is subdivided in three parts:

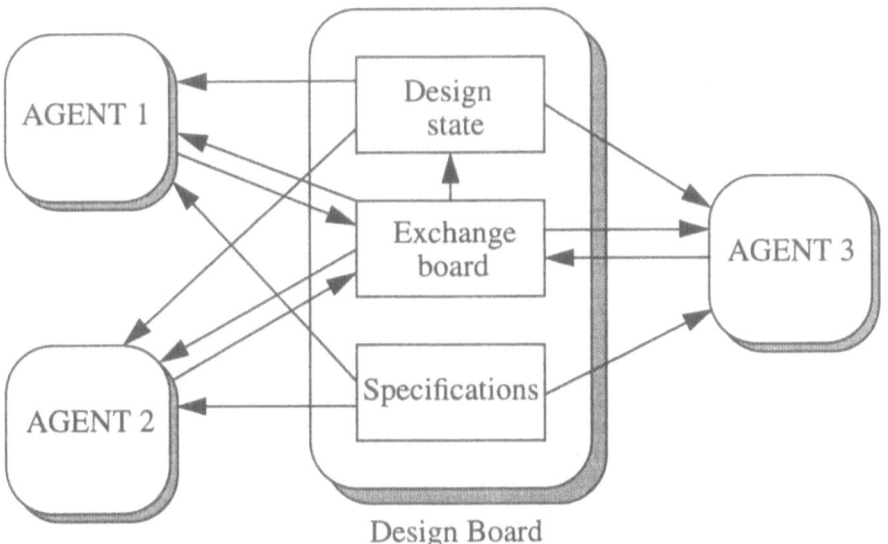

Figure 1. Architecture of the spring design system.

- *design specifications*, including all the requirements provided by the user for the current design;
- *design state*, which describes the design and records the parameter values decided so far;
- *exchange board*, where agents make their proposals, engage in negotiations and reach agreements.

The agents encapsulate the domain knowledge about the corresponding target and point of view. Most of it is represented as rules and facts. Additional specialized routines carry out numerical computations associated with reasoning based on equations and physical laws. Selectors use such routines to establish their lists of preferences, and critics use them to compute the ranges of admissibility.

Every agent can see the information describing the design state and the specifications. The exchange board is used by selectors to make proposals, by critics to post objections, and by praisers to announce the superiority of a proposal, as seen from their point of view. In addition, agents can communicate directly.

An agent is activated by a set of preconditions, confirming that the information for carrying out its task is available and that the result of the task is needed. The control of its activity belongs to each agent and is not imposed from the outside. The way in which design agents work together can be easily altered through these preconditions.

The system is implemented in CLIPS (Giarratano and Riley, 1994). Each agent is a rule-based system, with the addition of specialized C functions. The design state is described in an object oriented manner, allowing uniform parameter handling.

7. Agent Conflicts in Material Selection

Choosing an acceptable parameter value involves the following stages:

1. *The selectors negotiate a common acceptable value for the design parameter.* Each selector computes a range of possible values and an order of preferences for these values. Selectors start negotiation with the most preferred value from their point of view.

    One of the selectors, let's call it A, makes a first proposal for the parameter value. The other selector (B) accepts the value only if it is currently the highest ranking value in its ordering of values. Otherwise, a conflict is detected and a negotiation session starts. B will make a counterproposal. A will proceed in the same manner as B did. The process continues until one agent makes a proposal that is also the best one for the other agent at this stage. Negotiation, also stops if the counterproposal to be sent by one selector has been previously posted by the other agent. Given the following material preferences of the two selectors (temperature and cost):

SELECTOR	PREFERENCES
Temperature	J I D E F B
Cost	B C A F E D G I

the following negotiation would take place:

SELECTOR	PROPOSALS				
Temperature	J	I	D	E	F *agreement on F*
Cost		B	C	A	F

The negotiation sequence can be altered by a praiser. Assume a praiser P praises a proposal v made by an agent (say A), and B does not consider the value as being the best one. Considering that v is claimed as a benefit from more than one point of view (A and P), B will match v against *several* preferences before making a counterproposal. Currently an agent is required to match a proposal against its 3 next best values, if the proposal is reinforced by a praiser.

For example, consider the same preferences of the same two agents. Assume that the praiser from the point of view of hardness praises material I, while the praiser from the point of view of stress praises material E. Praised values are marked in bold face. The additional values considered after a praised proposal are included in parentheses.

SELECTOR	PROPOSALS			
Temperature	J I		D	E
Cost	B	C (A, F)	A	F (E, D) *agreement on E*

The outcome is different than in the first case. Proposal I is not accepted, since the selector from the point of view of cost cannot find it among its top 3 preferences at that moment. However, proposal E is accepted, as it is found in the required range. In this situation, one selector relaxes its preferences in the light of evidence from other agents.

2. *The critics post their objections to the agreement reached by the selectors.* A critic which rejects a parameter value, will post the entire set of values that are not acceptable under the current conditions. This operation is computationally expensive and is carried out only by those critics which are not satisfied with the current decision. If at least one objection arises, a critic-selector conflict is signalled and a third stage becomes necessary.

3. *The selectors start a new negotiation round, during which they avoid use of the values considered not to be acceptable by the critics.*

If, for example, the design problem requires high resistance to shock loading, the corresponding critic would object to the material value E, and point out that the values A, B and C are also unacceptable. The new preference table for the selectors would be:

SELECTOR	PREFERENCES
Temperature	J I D F
Cost	F D G I

Knowing, that I is a praised value, the new negotiation round would run as follows:

SELECTOR	PROPOSALS	
Temperature	J I	
Cost	F	D (G, I) agreement

Steps 2 and 3 are repeated until the system reaches a solution agreed upon by all the agents. One of the advantages of the method proposed is that it allows multi-partite negotiation. Considering the number of agents, any method approaching conflicts in a pairwise manner would have to cope with a serious overhead and with convergence problems.

In designing our experiments we have considered two types of negotiation, as possible versions of the previous strategy:

I) *Point-to-point negotiations:* This technique is the method described above. Each agent proposes a single value at a given moment. When analyzing a proposal made by another agent, the current agent matches it against a single value – the most preferred value from the local point of view. Proposals are made and analyzed one value (point) at a time.

II) *Range-to-range negotiations:* This strategy generalizes point-to-point negotiations in two directions. An agent (A) proposes its preferred value at that moment and also posts a set of alternative options. The other agent (B) matches the proposal against its best value. In case of mismatch, B compares the alternatives with its own next best set (range) of values. If the two sets intersect, a value is chosen as an agreement. Otherwise, B will prepare its own counterproposal as a new preferred value followed by a set of alternatives. Considering the initial preferences of the two selectors:

SELECTOR	PREFERENCES
Temperature	J I D E F B
Cost	B C A F E D G I

a negotiation session (without praisers) in which an agent posts two alternatives to its best proposal and compares an incoming proposal with its first three best values, would run as follows:

SELECTOR	NEGOTIATION
Temperature	prop: J (I, D)
Cost	comp to: B (C, A) prop: F (E, D) *agreement on D*

While this strategy is computationally more costly for both agents, it requires fewer interactions and fewer proposals by each side.

8. Learning in SiFAs

SiFAs generate a significant number of interactions while deciding the value of a parameter. This overhead arises because each single function agent is responsible for only for a portion of a decision, and the final decision has to gain the approval from all points of view. The number of single function agents involved in a parameter decision is not trivial. We have used 11 agents just for the material selection. This number can easily be increased if we take additional points of view into account.

The thorough exploration of the choices for the design parameter's value justifies this overhead. Designers usually prune the large number of comparisons necessary for an informed choice, by using subranges for their evaluations. Due to their specialization, SiFAs are much more powerful, provided they quickly learn to cope with situations which occur frequently.

As interactions consume a large portion of the computational effort expended on value selection, the primary goal of learning is to improve the knowledge which the agents have about each other. The learning experiments investigate:

- how difficult it is to learn about the other agents;
- how good the prediction of the behavior of the other agents will be;
- how much learning contributes to reducing the interaction overhead.

The learning results from agents interacting, and is aimed at predicting the future behavior of the interaction partners. Since agents act based on their functionality, learning can take advantage of this knowledge. Therefore, in our experiments we have made the learning strategies dependent on the type of the agent whose behavior is to be predicted.

The agents that learn are the selectors. They search in the space of values to find an acceptable solution. The other agents will 'encourage' or 'discourage' their search. It is to a selector's advantage if it can predict the feedback, and already take it into account when it evaluates alternatives, without exploring unproductive search paths. Critics' and praisers' design opinions will not be influenced by the actions of the other agents. The generic strategy used by an agent A to learn about another agent B is:

1. *Create a case from the interaction with agent B.* Cases are indexed by the design requirements relevant for the point of view of the agent one learns about. A case records the sequence of responses of the other agent. For example, assume the learning refers to the selector from the point of view of temperature. The case will be indexed by the working temperature range required for the spring and will reflect a decreasing preference sequence of material values of the selector from the point of temperature. The sequence is recorded only up to the point where an agreement is reached in that negotiation session.

2. Integrate the case in the knowledge already available about agent B. The goal is to create a mapping of the options and/or preferences of agent B under as many conditions as possible.

The learning strategy attempts to create a model of another agent's behavior that is closely tied to the specific design conditions rather than to the details of the other agent's domain reasoning. Two arguments favor this approach:

- Single function agents are very specialized. It is virtually impossible to assume that an agent can understand another agent's domain reasoning.
- An agent is much better off quickly learning small things about the other agents. There are many agents to learn about, therefore it makes sense to

look for knowledge that can be applied easy and which reduces interactions.

It is important to mention that agents only discover the other agents through the "signed" proposals, criticisms and praises posted on the design board.

8.1 WHAT CAN BE LEARNED ABOUT EACH AGENT TYPE?

The learning approach used is concept formation (Gennari et al., 1990; Michalski, 1983; Mitchell, 1982). Concepts reflect the connection between design conditions and the decisions based on those conditions. Whenever an agent posts a design proposal, criticism or praise, it also posts the elements from the design context on which the decision was based. If no conditions are posted, it is assumed that the decision is valid independent of any particular circumstances.

Each case recorded about an agent A's behavior represents a training instance in developing a conceptual description of A. The concept features are the design conditions which led to A's proposal, expressed as allowable design ranges and thresholds. The contents of the proposal determines the class partitioning of the concept instances.

As the heterogeneity of the domains which have to be covered by the inductive learning is extremely diverse, no assumption can be made about the continuity of the concept representations in the feature domain. For generality we have used disjunctive induction methods (Michalski et al., 1986). The dependency of the learned data on the learning of another agent raises additional issues which are as important as the accuracy of the classification. When learning about selectors the learning agent has to learn an evolving description of the other agent's behavior, as the training instances get refined in time. Even though the internal preferences of an agent repeat themselves under identical conditions, the way they are perceived from the outside varies depending on the other agents. The external perception is used to construct the model of the agent.

The classification features are the same for all the agent types, since they originate in the design. However, what is learned about an agent depends on its type:

1. Learning about praisers. Praisers point out parameter values which are particularly suitable from their point of view. Whenever a praiser praises a parameter value, the information is recorded by selectors. In the current experiments, only context-insensitive praisers were used. For example, a material is considered excellent from the point of view of hardness regardless of any other conditions or design values. Thus learning is merely storing praised values associated with praisers. If context-sensitive praisers are used the inductive technique used for critics applies to praisers too.

Praisers effectively provide only one learnable item at a time, and can only act when a selector proposes a value. This can occur during a negotiation. Consequently, knowledge about the praisers is acquired only to the extent to which proposals are praised during negotiations.

2. Learning about critics. Critics can be context-insensitive, or context-sensitive. The values they do not accept can depend on the design requirements; e.g., the resistance of a spring material to impact loading can be good or bad, depending on the conditions under which the spring is required to work. The information recorded about a critic is the set of values it considered unacceptable and the design conditions that the critic used to make those decisions. The classes correspond, in this case, to the individual materials a critic objects to. For any particular requirement(s), the information learned about the critic is complete (i.e. the critic makes all its objections known).

3. Learning about selectors. Selector behavior is the most difficult to predict, since, for a given design context/state, a selector can use different methods or sets of prestored values. Even if the set of values remains constant, the preferences among those values can be variable. The classes that partition a selector's behavior are defined by sets of parameter value preferences.

Assume that selector A encounters selector B several times, with the same design requirements for B each time. A still has reason to record B's responses every time, as they might offer sparse sequences of B's preferences. The sparsity can be due to B's knowledge that some values will be unacceptable to critics and are therefore left out of its proposals. However, the critics may have different requirements each time and thus the 'holes' in B's sequence of proposals will be different. In addition, the recording is incomplete because it stops when an agreement is reached. A will integrate the current response sequence of B with the sequence compiled from the previous encounters under the same conditions.

The main issue is that, while being refined, concepts can 'migrate'. Different sequences of responses can become identical after several interaction sessions, merging into a single concept. Alternatively, concepts can be split if under some particular subconditions the refined responses become different.

8.2 HOW DO AGENTS USE THE LEARNED KNOWLEDGE?

The knowledge learned by an agent about the other agents is used differently, depending on the type of the agent to which it refers:

Knowledge about selectors. During a negotiation a selector will decide on its next proposal. Before posting its proposal the selector will see whether it has knowledge about the behavior of the other selectors under these design conditions. If so, the selector will anticipate their proposals and prepare a new proposal. A new evaluation of the next-best proposals of the other selectors will be used to determine their responses. The predictive reasoning continues until either one of the following conditions is fulfilled:

- a value is found which is likely to be agreed on by all the selectors;
- for one of the other selectors no further preferences are known.

The selector will do predictive reasoning only if it knows how *all* the other selectors will respond. This condition is not imposed on other agent types.

If the proposal a selector submits after predictive reasoning is not known to be an agreement, it will be accompanied by the last value taken into account for each of the other selectors. This will allow the other selectors to know where to continue with their proposal evaluation.

The incompleteness of the knowledge about a selector can cause an undesired phenomenon: A will assume that the responses learned from B are consecutive preferences. This may cause possible solutions to be overlooked, as A will unknowingly skip them in its prediction of B's behavior. We have assumed that selectors can always reach an agreement, unless either A or B have insufficient knowledge about each other. Lack of agreement initiates a new negotiation in which they will not use the knowledge they have about each other. The new sequence of responses is used by each agent to correct its knowledge about the other selector.

Knowledge about critics. Knowledge about the critics is used to eliminate proposals known to be unacceptable. This holds true for the proposals which the current selector intends to submit, as well as for the proposals it anticipates from the other agents. Even if there is no knowledge about how a critic will respond, the selector continues its anticipatory reasoning. The critic will eventually 'protest' about an unacceptable agreement of the selectors.

Knowledge about praisers. The knowledge about praisers is used in a very similar manner to that of the critics. During the anticipation of the other selectors' proposals, praiser opinions will be taken into account to find out which agent has to revise its counterproposal, assuming that no agreement is anticipated.

These methods attempt to ensure that what has been learned is used effectively, and that information is exchanged in a cooperative fashion.

9. Experimental Results

Our experiments used eleven SiFAs (see Section 5). The two selectors, five critics and four praisers encoded knowledge about the 20 materials that are considered representative for helical compression springs. This makes the material choice problem non-trivial, due to the many comparisons from various points of view.

The design problems used included design constraints for each selector and critic. This made their proposals and responses context-sensitive. In these experiments the measure used for evaluation was the number of interactions needed to generate a material value accepted by the selectors as well as the critics. By an "interaction" we mean a proposal posted by a selector, an objection posted by a critic, or a praise posted by a praiser.

The first type of analysis required running the system through a sequence of 22 generated design problems. One of the problems was considered a reference problem, while each of the other 21 problems introduced a change in the requirements affecting the reasoning of one selector or one critic, relative to the preceding problem. A run through all the 22 problems is called an "experiment". Several experiments were carried out, with the same changes in the requirements ordered differently. During each experiment an agent encountered three changes in the design conditions affecting its proposals. At first, the changes in the design problems in an experiment were made in random order. There were 22 experiments in all. Each experiment was run with and without learning capabilities. Table 1 summarizes the average results for the set of experiments.

TABLE 1. Decrease of interactions in point-to-point negotiations (random problem ordering)

Type of analysis	Average number of interactions (rounded to closest integer) after			
	6 expts.	12 expts.	17 expts	22 expts
without learning	34	36	33	34
with learning	31	29	23	19

The slow initial decrease in the number of interactions is explained by the fact that the selectors do not use predictions if they have no information about the behavior of the other selector in the new situations. The initial decrease in the number of interactions is due mainly to the information learned about the critics. Another type of analysis involved scheduling the changes affecting selectors first, and then making the other changes in random order. The results of this set of runs are summarized in Table 2. The numbers in parentheses represent the interactions due to selectors.

The major improvement can be seen towards the end of this run. The learning of selectors about selectors happened mostly during the first six experiments and without any changes in the behavior of critics and praisers, as their requirements did not change during that time. As a consequence, interactions during the next five experiments were reduced mostly due to shorter negotiations between selectors. Interactions due to praisers are reduced only to the extent that the negotiations between selectors become shorter and the praised material values occur less often on the average.

TABLE 2. Decrease of interactions in point-to-point negotiations (design problems were ordered such that requirements for selectors were changed first)

Type of experiment	Average number (rounded to closest integer) of interactions after			
	6 expts	12 expts	17 expts	22 expts
without learning	34	35	33	34
with learning	31 (25)	22 (18)	19 (16)	15 (13)

For all of the previous experiments, the changes in requirements were close to the ones in the reference design problem. The primary goal was to test the decrease in the number of interactions under the assumption that the variations can be captured relatively fast. Another type of analysis assumed that the changes in the design requirements affected only one agent, but on a large scale. The goal was to investigate how a selector would build a reliable model of the corresponding agent for a large number of situations. Even though the reduction of the amount of interaction was smaller than in the case where learning extended to all the agents, accuracy of prediction proved to be better.

Having a selector learn without the critics and then running the system on the same (or similar) cases, but with the critics, generated the most accurate learning. Accuracy was measured in terms of distance between the concepts developed and the actual preferences of the targeted agent. The explanation relies on the fact that learning without interference of critics generates an accurate partitioning of the design requirements into concepts. Adding the critics does not change the conceptual partitioning, but causes the concepts to be refined.

The experiments described so far used point-to-point negotiations. The first type of experiment was also done using range-to-range negotiations. An agent considered two additional options besides its most preferred one. Table 3 reflects the learning results for the range-to-range interactions.

TABLE 3. Decrease of interactions in range-to-range negotiations (random problem ordering)

Type of analysis	Average number of interactions (rounded to closest integer) after			
	6 expts	12 expts	17 expts	22 expts
without learning	21	22	21	21
with learning	19	17	16	15

The design problems were randomly ordered. The learning rate was slower than in the point-to-point negotiations, as a much larger amount of negotiation was needed to capture correct information. This is mostly due to the 'hidden' preferences of a selector (due to the look-ahead), which were not seen by the other selector. The number of failures (no agreement reached and renegotiation without using the learned knowledge) was higher

(5 cases vs. 2 cases in point-to-point negotiations). However, rerunning the system over a total of 50 experiments, with the same changes in the requirements affecting the agents, but combined in different ways, led to interaction reductions that compared well with the point-to-point case.

10. Conclusions

We want a system to learn to reduce the design overhead in the application range needed by a specific user. Designers will most often use the system only in a particular region of the design space. Pre-training the system is not really possible, as it is hard to foresee the class of problems explored by a particular user. Therefore it is important to have the system learn from applications. Although the order of presentation of design problems can make the learning faster, note that similar final results were obtained by a random ordering of the design problems seen by the system. This is important, and a positive result. A factor that makes experimentation hard, and clear conclusions difficult, is that the learning processes are not independent. One agent learn something, can delay the occurrence of situations leading to learning by another agent. The range of phenomena of this type has yet to be sufficiently explored.

We have dealt with interactions between agents having the same target. There will be many additional aspects to be considered for the entire parameter set for spring design. For example, the knowledge representation will have to be extended to handle domain types specific to various parameters. Advisors and estimators will also require the model to be enhanced. Advisors raise the interesting issue that their behavior is modelled in conjunction with the behavior of the agent they advise. Predicting estimator behavior will be more complex as it is an example of a situation where an agent attached to one parameter is needed in decisions related to another design parameter.

The general conclusion of this work is that SiFAs, coupled with relatively simple negotiation schemes, provide a basis for interesting experiments in learning in multi-agent design systems, but that more remains to be done.

References

Berker, I. and Brown, D.C.: 1996, Conflicts and negotiations in single function agent based design systems, *Concurrent Engineering: Research and Applications*, special issue on Multi-Agent Systems in Concurrent Engineering, D. C. Brown, S. Lander and C. Petrie (eds) (submitted).

Currier, P. M.: 1995, *SiFAKA: Knowledge Acquisition for Single Function Agents*, Major Qualifying Project, Computer Science Department, WPI, Worcester, MA.

Douglas, R. E., Brown, D. C. and Zenger, D. C.: 1993, A concurrent engineering demonstration and training system for engineers and managers, *International Journal of*

CADCAM and Computer Graphics, **8**(3), special issue on AI and Computer Graphics, I. Costea (ed.), pp. 263-301.

Dunskus, B. V., Grecu, D. L., Brown, D. C. and Berker, I.: 1995, Using single function agents to investigate conflicts, *AI EDAM,* **9**(4), special issue on Conflict Management in Design, I. Smith (ed.), pp. 299-312.

Finin, T., Weber, J., Wiederhold, G., Genesereth, M., Fritzon, R., McKay, D., McGuire, J., Pelavin, R., Shapiro, S. and Beck, C.: 1993, *DRAFT Specification of the KQML Agent-Communication Language,* DARPA Knowledge Sharing Initiative External Interfaces Working Group.

Gennari, J.C., Langley, P. and Fisher, D.: 1990, Models of Incremental Concept Formation, *in* J. Carbonell (ed.), Machine Learning – Paradigms and Methods, MIT Press, pp.11-61

Giarratano, J. C. and Riley, G.: 1994, *Expert Systems: Principles and Programming,* 2nd edn, PWSx Publishing Co., Boston, MA.

Grecu, D. L. and Brown, D. C.: 1995, Design agents that learn, *AI EDAM,* special issue on Machine Learning in Design, A. Duffy, M. L. Maher and D. C. Brown (eds) (to appear).

Klein, M.: 1991: Supporting conflict resolution in cooperative design systems, *IEEE Transactions on Systems, Man, and Cybernetics,* **21**(6), 1379-1390.

Kuokka, D. R., McGuire, J. G., Pelavin, R. N., Weber, H. C., Tenenbaum, H. M., Gruber, T. and Olsen, G.: 1993, SHADE: Technology for knowledge-based collaborative engineering, *Concurrent Engineering: Research and Applications,* **1**, 137-146.

Lander, S. E. and Lesser, V. R.: 1991, Customizing distributed search among agents with heterogeneous knowledge, *Proceedings 5th International Symposium on AI Applications in Manufacturing and Robotics,* Cancun, Mexico.

Machinery's Handbook, 1982, revised 21st ed., Industrial Press Inc., New York, NY.

Maher, M. L., Brown, D. C. and Duffy, A.H.B. (eds): 1994, *AI EDAM,* **8**(2), special issue on Machine Learning in Design.

Michalski, R. S.: 1983, A theory and methodology of inductive learning, *Artificial Intelligence,* **20**, 110-156.

Michalski, R. S., Mozetic, I., Hong, J. and Lavrac, N.: 1986, The multi-purpose incremental learning system AQ15 and its testing application to three medical domains, *Proceedings of AAAI-86,* Morgan Kaufmann, Los Altos, CA, pp. 1041-1045.

Mitchell, T. M.: 1982, Generalization as search, *Artificial Intelligence,* **18**, 203-266.

NagendraPrasad, M. V., Lesser, V. and Lander, S. E.: 1995, Learning experiments in a heterogeneous multi-agent system, *in* S. Sen (ed.), *Working Notes of the IJCAI-95 Workshop on Adaptation and Learning in Multiagent Systems,* Montréal, pp.59-64,

Sen, S. (ed.): 1995, *Working Notes of the IJCAI-95 Workshop on Adaptation and Learning in Multiagent Systems,* Montréal, Canada.

Sycara, K. P.: 1990, Cooperative negotiation in concurrent engineering design, *Cooperative Engineering Design,* Springer-Verlag, pp. 269-297.

Taleb-Bendiab, A. and Oh, V.: 1993, Speech-act based communication protocol to support multi-agent cooperative design systems, *Proceedings 1993 AI in Engineering Conference,* CMI Publ.

Victor, S. K., Brown, D. C., Bausch, J. J., Zenger, D. C., Ludwig, R. and Sisson, R. D.: 1993, Using multiple expert systems with distinct roles in a concurrent engineering system for powder ceramic components, *in* G. Rzevski, J. Pastor and R. A. Adey (eds), *Applications of AI in Engineering VIII,* Vol. 1, *Design, Methods and Techniques,* Elsevier Press, pp. 83-96.

Victor, S. K. and Brown, D. C.: 1994, Designing with negotiation using single function agents, *in* G. Rzevski, R. A. Adey and D. W. Russell (eds), *Applications of Artificial Intelligence in Engineering IX,* CMI Publ., pp. 173-179.

Werkman, K. J. and Barone, M.: 1991, Evaluating alternative connection designs through multiagent negotiation, *in* D. Sriram, R. Logcher and S. Fukuda (eds), *Computer Aided Cooperative Product Development,* Springer Verlag, pp. 298-333.

Wooldridge, M. and Jennings, N. R.: 1995, Intelligent agents: Theory and practice, *Knowledge Engineering Review,* **10**(2), 115-152.

J. S. Gero and F. Sudweeks (eds), Artificial Intelligence in Design '96, 429-445.
© 1996 *Kluwer Academic Publishers.*

A MACHINE LEARNING APPROACH TO AUTOMATED DESIGN CLASSIFICATION, ASSOCIATION AND RETRIEVAL

ANIL VARMA, WILLIAM H. WOOD III AND ALICE AGOGINO
University of California at Berkeley
Department of Mechanical Engineering
5136 Etcheverry Hall
Berkeley, CA 94720-1740

Abstract. Acquisition and recall of associations between problem descriptions and solutions is a critical task of case based design systems. The organization of design knowledge impacts the quality of inference and support a designer may derive from a case based system. Machine learning over case data may be used to create an intelligent interface between designer requirements and available design knowledge. Such an interface assists the designer in navigating the case base for effective case based retrieval. This paper explores two neural architectures based upon the Adaptive Resonance Theory for automated generation of design representations useful during the preliminary stages of case based retrieval. A standard bridge design case base is used to demonstrate the approach.

1. Introduction

Remembering past experiences and abstracting broad relationships from complex data helps a designer conserve precious cognitive and computational resources during decision making. Computer support for the direct application of experience, cast in an artificial intelligence framework, is called case-based reasoning (CBR).

A fundamental problem in CBR is identifying the stored case data that is appropriate for use in a new situation; the mappings between a new problem and stored case descriptions are usually complex and context dependent. As the case base grows in size, it is desirable to invest some effort in clarifying and matching the information requirements of the designer to the information available in the case base. Machine learning techniques can aid in this process by providing "compiled" representations of the design knowledge contained in the case base to the designer. Such summarized knowledge can help the designer clarify, expand or otherwise revise a problem description to place it into an appropriate context so that case based reasoning can be applied. In this paper, we describe the application of neural architectures based upon the Adaptive Resonance Theory (ART) paradigm (Grossberg, 1976a, 1976b) to the problem of flexibly organizing design data for effective case based retrieval. The paper is structured as follows. In section 2 and 3 we present out motivation and discuss background research. Section 4 and 5 introduces the ART1 and ARTMAP architectures respectively. Section 6 contains some results from performing ART1 based classification over a bridge design case database. Section 7 illustrates the performance of the ARTMAP network over our test case database. Finally, Section 8 contains our discussions and conclusions.

2. Motivation

The organization of design knowledge within a system has a great impact upon the quality of inference and support a designer may derive from it. The thrust towards adding more evaluative information to design descriptions means that both individual case size as well as overall case base size will continue to expand. This suggests that the following two issues will be important in the design of effective case based systems:
1. Case retrieval with complex stored design records and tentative new design problem specifications is better modeled as exploration and assessment of context leading to search rather that a single step retrieval (Maher and Balachandran, 1994).
2. Not all aspects of a case are necessarily interesting or useful in a new design situation. Segmenting cases into chunks of appropriate size and content is desirable from the standpoint of storage efficiency as well context appropriate retrieval (Domeshek and Kolodner, 1993). The segment of the case that encodes context need not necessarily contain the solution details that are the final objective of the designer. The mappings between different case segments also need to be stored for navigation from one segment of the case to another. This can be an essential requirement in case of a heterogenous case base requiring diverse representations for different segments.

Supporting exploration implies that the system must incorporate flexible retrieval strategies that do not assume that the mapping between a new problem situation and stored case information is necessarily straightforward or predefined. Our motivation in this paper is to look at how a machine learning approach can be used to aid flexible case retrieval strategies by creating compressed, approximate characterizations of underlying case data with which the designer can interact during the case retrieval process. Specifically we examine the performance of Adaptive Resonance Theory

(ART) networks at this task. ART networks are chosen because they possess the following desirable characteristics:

1. The networks create compressed representations of design data composed of attribute-value pairs in a simple, rapid fashion. The mode of learning is incremental so each case need be encountered only once. Clusters of design cases are created based on a simple criterion that looks at the number of shared features between a cluster prototype and incoming case. Both these aspects allow the algorithm to perform its tasks in a rapid, real time fashion. As a result, the algorithm can be embedded as part of the case exploration procedure that allows the designer to survey different characterizations of case data. Tedious learning times associated with backpropagation type networks or more sophisticated learning schemes are avoided.

2. ART networks incorporate architectures that can learn in both a supervised and unsupervised fashion. Supervised learning in the context of case based design means that the network is exposed to predefined mappings between case segments for learning the associations among them. One example would be to learn to associate a functional description of requirements to structural and behavioral characteristics of the completed design. Supervised learning supports flexibility in navigation from one segment of case data to another. Unsupervised learning in the context of case based design means that the system autonomously creates categories of "similar" cases and provides summarized representations of these categories for examination by the designer. Such an approach permits the designer to interact with a manageable view of the case base and understand the broad trends present in case data. This better understanding of case data provides an opportunity for more efficient and effective case retrieval.

3. Background

A fundamental mapping in engineering design is that from problem specifications to a realizable artifact description. A number of researchers have developed strategies for achieving this mapping through a two-step process. The first step is to identify frequently occurring or "typical" design experiences and associate them with standardized solution procedures or inference mechanisms. Next, a mechanism to map new problem specifications to these standard solution paths is implemented. Such approaches include concepts like Design prototypes (Gero, 1990; Tham et al., 1992), Computational prototypes (Donaldson and Maccallum, 1994) and Generic components (Alberts et al., 1992). In contrast to CBR, the structure of prototypical design experience in such approaches is predefined. CBR is

interesting as a paradigm for knowledge support since it requires little by
way of background knowledge and allows integration of computerized
experience into the design process in a manner that has resisted traditional
model based approaches. However, in absence of hand-crafted
representation schemes, determining relevance from stored case data has
become an increasingly critical task for large case bases.

Machine learning techniques for compiling knowledge in the
engineering design domain have been adopted by a variety of researchers.
One approach has been to allow the design description to form categories in
an unsupervised fashion and use this structure for prediction and
classification. This approach is illustrated by the inductive clustering
methods developed by the machine learning community (Fisher, 1987;
Quinlan, 1986) and applied to engineering design (Reich, 1991;Reich and
Fenves, 1992). In CADSYN (Maher and Zhang,, 1993) a pure case based
resoning approach is complemented with a generalized decomposition
approach to provide a hybrid strategy for case based design. Maher and Li
(1994) extend this approach toward creating useful partitions of the design
space called design concepts. ART networks have been implemented as
underlying mechanisms for engineering design retrieval (Caudell et al.,
1994) and evaluated for the task of forming an association model between
design problem and solution information (Kamarthi and Kumara, 1993).
The need for flexible characterization of mappings in case based retrieval is
highlighted in Maher and Balachandran (1994). This paper proposes that
certain classification and association functions of ART networks can be
effectively utilized to allow iterative, exploratory and flexible retrieval
strategies needed in a case based design system.

4. The ART Paradigm

Adaptive resonance networks are a class of neural networks introduced by
Grossberg (1976a, 1976b). ART networks function as classifiers that display
properties of self organization and self stabilization. Self organization refers
to the ability of the network to carry out it's learning autonomously in
absence of an explicit teacher. Self stabilization is the property of a network
that keeps it's learned memory from degrading due to irrelevant inputs from
the environment.

ART1 is a member of the ART family designed for classification of
binary valued input patterns (Carpenter and Grossberg, 1991). Once
classification is completed, the network can be queried as a neural database
by presenting an input pattern. The network then returns the prototypical
representation of the pattern family to which the input belongs.

4.1. THE ART1 NETWORK

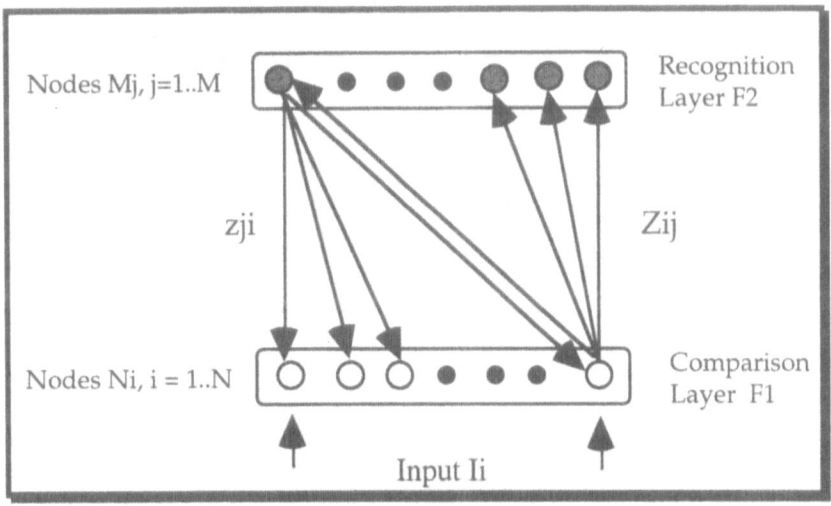

Figure 1. ART1 Architecture.

The ART I architecture consists of two layers of nodes F1 and F2 which are fully connected in both directions. Weights Z_{ij} represent the connection weights from node N_i to node M_j and take continuous values. The corresponding connection weights from node M_j to node N_i are represented as z_{ji}. These weights take binary values. F2 nodes store the different categories learned by the network. Each node M_J is associated with a weight vector z_{Ji}. This vector stores the prototypical features of the classification family represented by the F2 node J. The ART1 algorithm proceeds as follows:

1. On the first presentation of input **I**, node M_1 is selected as the first category node and input **I** is copied into the feature vector z_{1i}. The number of categories is initialized to 1.
2. On subsequent presentations of **I**, the system first hypothesizes a category that may be most suitable for classifying the current input . This is achieved by evaluating the bottom up input to each node M_j as

$$\sum_{i=1}^{N} I_i Z_{iJ} .$$

(1)

The node with the maximum input is the winner. A user specified parameter called vigilance ρ calibrates how close an input vector must be to a family prototype vector to be classified as part of that family. This condition may be expressed as

$$\frac{\displaystyle\sum_{i=1}^{N} z_{Ji}\, I_i}{\displaystyle\sum_{i=1}^{N} I_i} \geq \rho \ , \ \rho \in (0,1) \tag{2}$$

A value of 1 for the vigilance parameter indicates extreme selectivity of classification where the input vector I must be a subset of the category prototype vector zji to be classified together. This condition may be expressed as $z_{Ji} \cap I_i = I_i$. A network with a high vigilance value typically results in many clusters with feature rich family prototype vectors. On the other hand, a baseline $\rho = 0$ vigilance condition indicates that a match on *any* feature vector value between the input and the family prototype vector is a sufficient condition for classification of the input as a part of that family. A low vigilance value results in few categories with sparse category prototype vectors.

If (2) is satisfied, the input vector is classified as a member of the category represented by F2 node J. The feature vector associated with node M_j is updated as

$$z_{Ji\,New} \rightarrow z_{Ji} \cap I_i \tag{3}$$

where $\cap$ represents the logical AND operation.

The F1 -> F2 weights Z_{iJ} are updated to represent a normalized version of feature vector J as

$$Z_{iJ} = \frac{z_{Ji} \cap I_i}{\beta + \displaystyle\sum_{i=1}^{N} (z_{Ji} \cap I_i)} \tag{4}$$

Small values of β bias the system towards searching the prototypes in it's memory in preference to creating a new prototype while seeking to classify a new input. If (2) is not satisfied, the node with the next largest bottom up input (1) is selected and the procedure is repeated. If no previously learned feature prototype vector is found to satisfy (2), a new category is created with input I_i as its category prototype vector. The system is then ready for it's next input.

The user specified parameters are the vigilance threshold ρ and M, the maximum number of nodes in layer F2. This is a measure of the maximum number of clusters that may constitute the memory set of a particular ART network. Depending upon the ρ parameter, however, the actual number of categories may be less than or equal to M.

5. The ARTMAP architecture

The ARTMAP architecture extends the functionality of ART1 and incorporates supervised learning (Carpenter, Grossberg et al., 1991). It consists of two ART1 modules connected by a "map field" that associates the families created by the ART1 modules with each other. The operation of the ARTMAP architecture is illustrated in Figure 2.

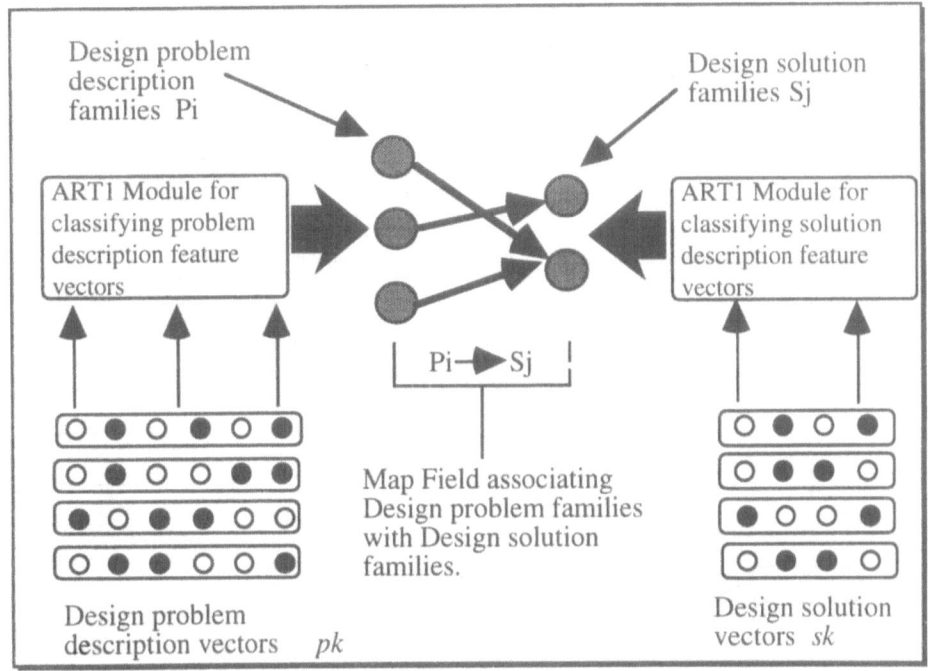

Figure 2 . The ARTMAP architecture.

The ARTMAP algorithm assumes that a corpus of design problem descriptions and their corresponding solutions are available for training in a supervised fashion. ARTMAP adaptively creates families from both problem and solution descriptions at user specified levels of the vigilance parameter and learns the associations between them. During operation, the ARTMAP algorithm classifies a new problem vector into a design problem family. It then identifies the associated design solution family and returns the feature vector representing it. The algorithm is shown in Figure 3.

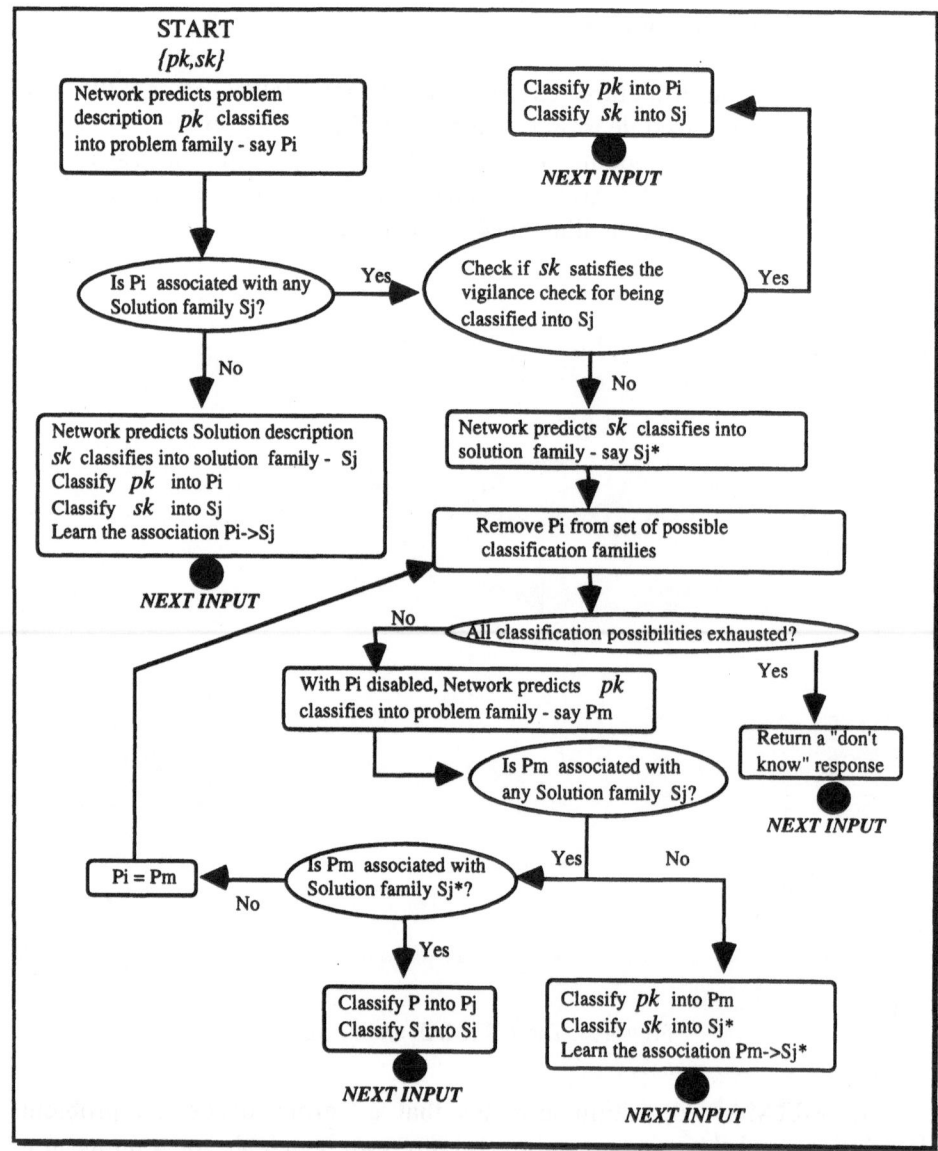

Figure 3. The ARTMAP algorithm.

6. Organizing Case Data using ART 1 Networks

Organizing case data with ART1 networks has utility when using hand-crafted coding and indexing schemes using features is cumbersome to create, maintain and use (Caudell et al., 1994). In such a case, the ART1 network is first used to cluster designs into families. Subsequently, the trained network can be used to retrieve a small set of similar families in response to new design case. The level of discrimination used by the network while clustering families is controlled by the vigilance parameter and is specified at training time. Network weights can be stored for several vigilance levels to provide multiple levels of abstraction in case recall. The next section provides an example of this type of implementation using some standard case data.

6.1 DESIGN REPRESENTATION

The set of design cases for the simulations in this paper was obtained from a machine learning database containing descriptions for Pittsburgh area bridges built since 1818 (Murphy and Aha, 1995). This database contains 108 examples of bridges, each described by 12 features as shown in Table 1.

TABLE 1. Values of 12 features constituting the bridge design description

Number	Feature	Possible Values
1	River Name	A,M,O
2	River Location	1..52
3	When Erected	Crafts, Emerging, Mature, Modern
4	Purpose	Walk, Aquaduct, RR, Highway
5	Length of Crossing	Short, Medium, Long
6	No. of Lanes	1, 2, 4, 6
7	Clear-G	N,G
8	Through or Deck	T,D
9	Material	Wood, Iron, Steel
10	Span	Short, Medium, Long
11	Rel-L: Relative length of main span to total crossing length	S, S-F,F
12	Type of bridge	Wood, Suspen, Simple-T, Arch, Cantilev, Cont-T

Each of these features is described by a set of discrete values. This allows each case to be represented as a binary vector suitable for processing by the ART1 network. The first 7 properties are listed as design specifications and the remaining 5 properties are design descriptions of bridges that resulted from those design specifications. The relatively small number of samples in the database (108) is typical of real life engineering design situations where only sparse and limited data may be available for characterizing a complex domain. This database has been previously used for learning by inductive clustering (Reich and Fenves, 1992).

There are 89 possible feature values for the 12 descriptive features of the bridge database. Each of the 108 examples available was represented by a 89 element binary feature vector. Each vector would ideally have 12 entries of "1" to denote the features associated with that particular example. Cases with missing or unknown feature values had fewer "1" entries in the feature vector and this was checked for by the program to prevent misinterpretation by the learning algorithm.

6.2 ORGANIZING CASE DATA WITH ART1 NETWORKS

Five simulations were carried out with increasing values of the vigilance parameter. The vigilance parameter is a measure of the similarity that needs to exist between a input vector and a cluster prototype vector to be classified together. The results are summarized in Table 2.

TABLE 2. Sample Simulation Results.

Simulation No.	Vigilance Factor	No. of Clusters	Average No. of Prototype Features representing each family
1	0.0	3	1.0
2	0.3	12	3.8
3	0.5	26	6.0
4	0.7	47	8.7
5	0.8	73	10.0

Table 2 illustrates how the network offers a variety of representational levels at which the user may interact with it. Low vigilance values provide data compression i.e. fewer clusters but the prototype of each cluster is relatively sparse. Trial 2 provides a reasonable mix of prototype attribute information and number of cases per design cluster and may be useful in the early stage of the case retrieval process. If the new design problem specifications are fairly well developed, a high vigilance value like trial 4 will map these specifications to a cluster of 2-3 cases. The specification revision process is illustrated in Figure 4.

After interacting with the trained ART module, a revised set of specifications may be submitted to the case database for retrieval of the actual cases meeting the similarity criteria. Two situations may arise. If the number of cases retrieved is sufficiently small that the user may gainfully examine each for guidance towards a solution, then the actual cases are delivered to the user. It may also happen that a number of cases, possibly each of considerable size may be retrieved, making individual examination impossible. In such a case, an untrained ART1 module can be used to autonomously and rapidly create structured categories from the retrieved case data. If the large number of cases returned is due to many similar cases in the database meeting the input specifications, the compressed ART1 representation will highlight the salient features of the cases retrieved. Alternately, the input specifications may be too abstract to retrieve a sufficiently small set of cases. This would indicate a need for a revision of specifications, possible by choosing a higher vigilance value for the trained ART1 module to retrieve case prototypes with a higher number of features.

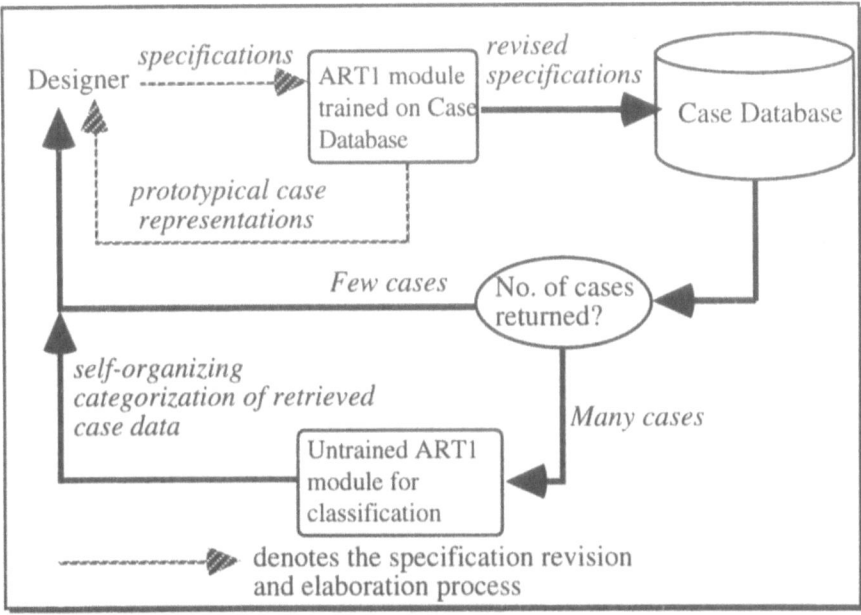

Figure 4. ART1 modules for elaborating specifications and organizing retrieved case data.

Since ART1 operates with the entire case data, it treats all case features alike. In actual design situations, complete cases may be distinguished by features relating to design requirements, design solutions etc. An efficient utilization of case memory is to partition the case base into segments and store the relationships between these segments. A case memory structure organized around mappings within case data is implemented by the ARTMAP network.

7. Organizing Case Data Using ARTMAP Networks

A natural way to structure case data is to exploit some basic categories that exist in the case structure itself. Once basic division is between the problem description Pi and the solution description Sj . Another may be between the functional requirements and their mappings to structures and behaviors. It is likely that these mappings are not unique. A more representative structure, then, is to view a case Ck as

$$Ck = \{Pi, Sj\} , Pi \in \mathbf{P} \; \& \; Sj \in \mathbf{S}$$

where **P** and **S** are sets of problem specification and solution descriptions respectively. Such a partition can be viewed as a first step towards creating flexible mappings between relevant categories of case memories (Maher and Balachandran, 1994).

A mapping between case fragments as described above may be learned by the ARTMAP architecture. The revised case retrieval procedure using the ARTMAP network is depicted in Figure 5.

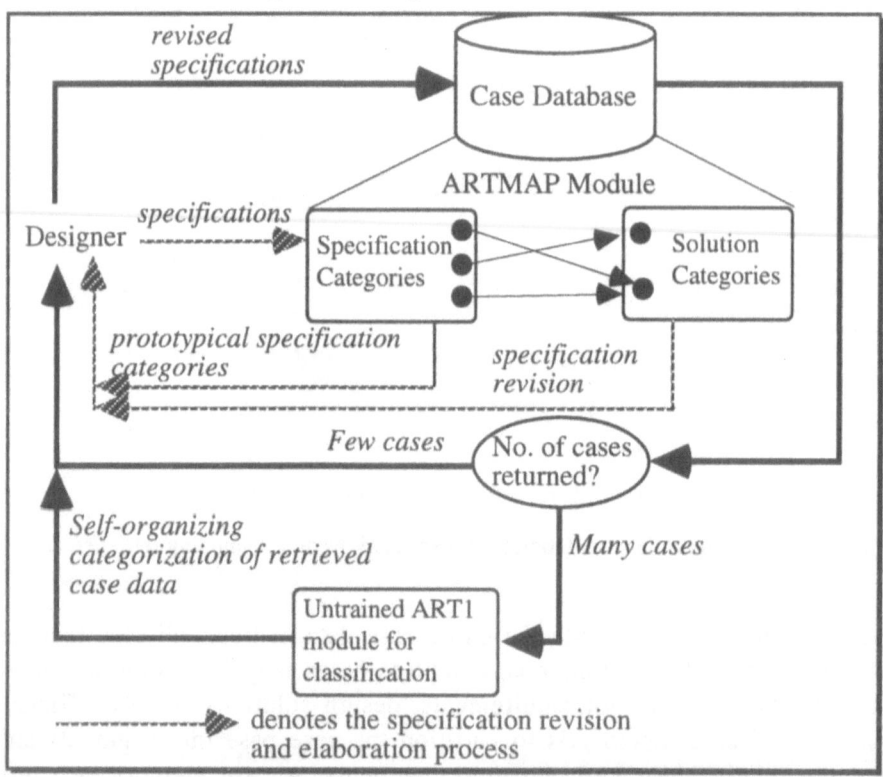

Figure 5. Mappings between categories within the case structure

Instead of an immediate retrieval of cases, the designer begins by classifying his/her initial set of specifications into a specification family. The characteristics of the retrieved family may lead to some revision of the initial specification set or the designer may choose to retrieve the solution family associated with the selected specification set. An examination of the features of the prototype solution set may again lead to specification revision resulting a solution set with desirable prototype features. This refined specification set is then submitted to the case database for retrieval of cases. The following section provides examples of using ARTMAP for learning mappings between case features using the bridge database.

7.1 LEARNING ASSOCIATIONS USING ARTMAP

The network was given the task of separately clustering the design specification attributes and design description attributes and learning the associations between them. The bridge design data identifies the first 7 attributes as design specification attributes. These are {*River Name, River Location, When Erected, Purpose of Bridge, Length of Bridge, Number of Lanes, Clear-G*}. The remaining five attributes describe the resulting designs. The design description attributes are {*Through or Deck, Material, Span, Rel-L, Type of Bridge*}. During learning, both the design specification attribute vector and the design description attribute vectors form families. There is a many to one mapping as specification attribute vectors are mapped to specification families and specification families are mapped onto solution families. Several design specification families may be associated with a particular design solution family.

7.2 DESIGN REPRESENTATION AND TESTING

For learning the associations between design specification and solution attribute vectors, the input to the design problem specification ART1 module was a feature vector of length 83. Each vector consisted of 7 ones to specify the feature values. The input to the design solution ART1 module was a feature vector of length 6, with exactly one value equal to 1 and the rest zero. Several features had missing values leading to vectors of less than standard length. Missing feature values were checked for in such vectors to ensure that the network prototypes were not erroneously degraded. The performance of ART1 networks is extremely sensitive to exposure of network prototypes to missing values in the incoming feature vector. This shortcoming is overcome to some extent in ART2 networks.

The network was allowed to train during exactly one pass on the entire data set. Each design instance was encountered only once by the network and classified before the next input.

For testing, the network was presented with a the design specification vector pi and associated design solution vector si from the training data . This practice differs from traditional machine learning approaches where the system's learning ability is tested using unseen data in a predictive mode. The purpose of the network here is to provide a faithful characterization of the learnt case data so that the designer can interact with the compressed network representation while retaining much of the accuracy available if he/she were interacting with the full case base. The success rate reported in the following trials thus should be interpreted as the "apparent success rate" since the training and test data are the same.

The network calculated the design specification family P* the test input belonged to and the solution family $S_{predicted}$ that was associated with it. $S_{predicted}$ was recorded. The network then calculated the actual design solution family S* it would classify *si* into. If $S_{predicted}$ = S* then the prediction was confirmed and the instance was recorded as a success. If $S_{predicted} \neq S*$, the instance was recorded as an error. Occasionally, the network was unable to classify *pi* or *si* into any corresponding families at the prescribed vigilance levels. In that case, the network returned a "don't know" response. Due to the real-time nature of ART learning, all simulations were completed in under a minute .

7.3 SIMULATION RESULTS FOR ARTMAP

Let **P** and **S** denote the ART1 modules used for classifying problem descriptions and solution descriptions respectively. For each trial the following information was recorded:

1. **P** vigilance and **S** vigilance: These refer to the vigilance levels at which the problem and solution classification ART1 modules operated.
2. Results: The number of correct, incorrect and don't know responses by the system during testing.
3. Number of **P** and **S** families: The number of clusters or families formed by the problem and solution classification ART1 modules was recorded.

This simulation attempted to map to a set of 5 properties, namely {*Through or Deck, Material, Span, Rel-L, Type*}. In such a case, the vigilance level at which the Solution ART1 module S operates becomes important.

TABLE 4. Results for ARTMAP simulation predicting bridge design description

Trial No.	P Vigilance S Vigilance	Correct Predictions	Incorrect Predictions	Don't know responses	Apparent Success Rate (%)	P families, S families	Average P features, S features
1	0.0,0.0	46	60	0	43.4	10 , 3	2.8 , 1
2	0.5,0.0	69	37	0	65.0	32 , 4	4.91 , 1
3	0.7,0.0	89	17	0	83.9	48 , 4	5.85 , 1
4	0.0,0.90	96	10	0	90.5	89 , 33	6.74,4.97
5	0.5,0.5	95	11	0	89.6	55 , 7	6.0 ,3.0

If retrieval of a set of related cases is important, relatively low vigilance values can provide mappings from specification families to solution families. Trial 5 illustrates how a good compromise between generalization and accuracy can be achieved and the system operates in a case retrieval mode where a solution prototype represents a set of cases.

Alternately, the system can be forced to provide predictions for all five design solution properties in trial 4 by raising the S vigilance value to 0.9. This virtually ensures that each solution family contains a unique solution vector with no generalization. Notice that even though P vigilance was set at the baseline value of zero, the extreme selectivity of the solution classification procedure caused a relatively large number of problem description families to be generated so that accuracy of prediction could be maintained. In this situation, the system recalls all five design solution description values 90% of the time. Varying the vigilance parameter provides a continuum of behavior by the ARTMAP system between learning by generalization at one end and simply storing each input for future reference at the other.

The ARTMAP network tries to balance both predictive capability and generalization. If significant similarities exist in the incoming data, the network tries to generalize them into specification and solution families. If a new specification being classified into the closest matching specification family leads to an incorrect prediction, then a classification into the next closest family is attempted where a correct association might be learned. Novel instances are stored by memorization by the network and families of instances are represented by a compressed prototype vector.

8. Discussion and Conclusions

In this paper we have examined how ART networks can form a mediating information support mechanism between a designer and a case base. Such a strategy promotes exploration. For example, a case may be retrieved based on a match on functional specifications. The behavioral data of that case can be used as the starting point for exploring similar behaviors without requiring a match on the other specifications. The flexibility such a scheme affords a designer in exploring case memory can be very important. ART networks allow for unsupervised formation of categories from case data. This can be useful in exploring a new domain. Alternately, if specific categories needed from data fragments are known beforehand, they can be assigned hand-crafted representations so that ART networks create the appropriate families. We plan to exploit both of these features in applying ART to the organization and interpretation of queries over a heterogeneous database of mechatronic components - the Concept Database. Here, components are represented by the values of pertinent performance parameters; indexing in the Concept Database also includes hierarchical classifications of component's type or family. For conceptual design, where preserving the ambiguity of component type can help toward making better design decision, this latter index can be ignored. By applying structure externally, the ART methodology laid out here can dynamically structure components and their performance specifications into sets that can help the designer navigate the design space toward arriving at a final design concept.

While the ART networks discussed in this paper are do not allow continuous valued attributes, this is not seen as a major obstacle toward more general use. It is relatively straightforward to discretize continuous design domains for application of the algorithms discussed here. ART2 (Carpenter and Grossberg, 1987) extends ART networks for processing both continuous and discrete inputs. A hierarchy of mapping layers is provided by ART3 (Carpenter and Grossberg, 1993).

In conclusion, this paper has examined how ART networks can be utilized to provide some of the functionality required to support the preliminary, iterative, phase of case based retrieval. ART's representation of cases as mappings between fragments rather than monolithic entities can aid design exploration, helping to set defaults by identifying similar design specification families and lending focus by illuminating relevant design solution families. In the initial phase of case retrieval, when the designer decides on what specifications to search the case base, ART networks can aid in the process of adapting input specifications for increased likelihood of retrieval of relevant cases. In the case retrieval phase, ART networks can autonomously structure incoming case data in real time so the designer may focus on broad similarities and variations in the retrieved case data. Where natural categories exist in case data, separate mappings may be learnt between each category to allow the designer to focus on the category of interest and what differentiates it from other design categories. Such mappings maybe learnt using the ARTMAP network that associates case fragments based upon maximizing predictive accuracy and generalization.

Acknowledgments. This work has been supported in part by NSF Grant #DDM-9300025, The Conceptual Design Database. In addition we would like to thank our industrial partners Rockwell International and Autodesk Inc.

References

Alberts, L. K., Wognum, P. M. and Mars, N. J. I.: 1992, Structuring design knowledge on the basis of generic components, *in* J. S. Gero (ed.), *Artificial Intelligence in Design '92*, Kluwer, Dordrecht, pp. 639-656.
Carpenter, G. A. and Grossberg, S.: 1991, A massively parallel architecture for a self-organizing neural pattern recognition machine, *in* G. Carpenter and S. Grossberg (eds),

Pattern Recognition by Self Organizing Neural Networks, MIT Press, London, pp. 316-382.

Carpenter, G. A., Grossberg, S. and Reynolds J. H.: 1991, ARTMAP: supervised real-time learning and classification of nonstationary data by a self-organizing neural network, *in* G. Carpenter and S. Grossberg (eds), *Pattern Recognition by Self Organizing Neural Networks*. MIT Press, London, pp. 503-544.

Carpenter, G. A. and Grossberg, S.: 1987, ART2: self-organization of stable category recognition codes for analog input patterns, *Applied Optics*, **26**, 4919-4930.

Carpenter, G. A. and Grossberg, S.: 1990, ART3: hierarchical search using chemical transmitters in self-organizing pattern recognition architectures, *Neural Networks*, **3**, 129-152.

Caudell, T. P., Smith, S. D. G., Escobedo, R. and Anderson, M.: 1994, NIRS: Large scale ART-1 neural architectures for engineering design retrieval, *Neural Networks*, **7**(9), 1339-1350.

Domeshek, E. and Kolodner, J.: 1993, Using the points of large cases, *Artificial Intelligence in Engineering Design, Analysis and Manufacturing*, **7**, 87-96.

Donaldson, I. and Maccallum, K.: 1994, The role of computational prototypes in conceptual models for engineering design, *in* J. S. Gero and F. Sudweeks (ed.), *Artificial Intelligence in Design '94*, Kluwer, Dordrecht, pp. 3-20.

Fisher, D. H.: 1987, Knowledge acquisition via incremental conceptual clustering, *Machine Learning*, **2**(2), 139-172.

Gero, J. S.: 1990, Design prototypes: a knowledge representation schema for design, *AI Magazine*, **11**(4), 6-36.

Grossberg, S.: 1976a, Adaptive pattern classification and universal recoding, I: Parallel development and coding of neural feature detectors, *Biological Cybernetics*, **23**, 121-134.

Grossberg, S.: 1976b, Adaptive pattern classification and universal recoding, II: Feedback, expectation, olfaction, and illusions, *Biological Cybernetics*, **23**, 187-202.

Kamarthi, S. V. and Kumara, S. R. T.: 1993, Neural networks in conceptual design, *in* J. Wang and Y. Takefuji (eds), *Neural Networks in Design and Manufacturing*, Singapore, World Scientific Publishing, pp. 99-120.

Maher, M. L. and Balachandran, B.: 1994, Flexible retrieval strategies for case based design, *in* J. S. Gero and F. Sudweeks (eds), *Artificial Intelligence in Design '94*, Kluwer, Dordrecht, pp. 163-180.

Maher, M. L. and Li, H.: 1994, Learning design concepts using machine learning techniques, *Artificial Intelligence in Engineering Design, Analysis and Manufacturing*, **8**, 95-111.

Maher, M. L. and Zhang, D. M.: 1993, CADSYN: a case-based design process model, *Artificial Intelligence in Engineering Design, Analysis and Manufacturing*, **7**, 97-110.

Murphy, P. M. and Aha, D. W.: 1995, *UCI Repository of machine learning databases* [http://www.ics.uci.edu/~mlearn/MLRepository.html]. Irvine, CA: University of California, Department of Information and Computer Science.

Quinlan, J. R.: 1986, Induction of decision trees, *Machine learning*, **1**(1), 81-106.

Reich, Y.: 1991, Constructive induction by incremental concept formation, *in* Y. A. Feldman and A. Bruckstein (eds), *Artificial Intelligence and Computer Vision*, Elsevier Science, Amsterdam, pp. 191-204.

Reich, Y. and Fenves, S. J.: 1992, Inductive learning of synthesis knowledge, *International Journal of Expert Systems*, **5**(4), 275-297.

Tham, K. W. and Gero, J. S.: 1992, PROBER - a design system based on design prototypes, *in* Gero, J. S. (ed), *Artificial Intelligence in Design '92*, Kluwer, Dordrecht, pp. 657-675.

J. S. Gero and F. Sudweeks (eds), Artificial Intelligence in Design '96, 447-462.

LEARNING TO CHOOSE A REFORMULATION
FOR NUMERICAL OPTIMIZATION OF ENGINEERING DESIGNS

MARK SCHWABACHER, THOMAS ELLMAN AND HAYM HIRSH
AND GERARD RICHTER
Computer Science Department
Rutgers University, New Brunswick, NJ 08903 USA

Abstract. It is well known that search-space reformulation can improve the speed and reliability of numerical optimization in engineering design. We argue that the best choice
of reformulation depends on the design goal, and present a technique for automatically
constructing rules that map the design goal into a reformulation chosen from a space
of possible reformulations. We tested our technique in the domain of racing-yacht-hull
design, where each reformulation corresponds to incorporating constraints into the search
space. We used a standard inductive-learning algorithm, C4.5, to learn rules from a set of
training data describing which constraints are active in the optimal design for each goal
encountered in a previous design session. We then used these rules to choose an appropriate reformulation for each of a set of test cases. Our experimental results show that using
these reformulations improves both the speed and the reliability of design optimization,
outperforming competing methods and approaching the best performance possible.

1. Introduction

In a simulation-based automated engineering design system that uses numerical
optimization, the decision on how to formulate the search space can dramatically affect the performance of the optimizer in two ways. First, using a lower-
dimensional formulation of the search space makes optimization faster, since each
gradient computation requires fewer runs of the simulator, and the distance in
design space from the starting point to the optimum is smaller. In design problems
where evaluating even just a single design can take tremendous amounts of time,
selecting an appropriate formulation can be the determining factor in the success
or failure of the design process. Second, different formulations of the search space
can result in different degrees of "smoothness" of the search space, which can impact not only the speed of the optimizer, but also the ability of the optimizer to
get to the optimum, and therefore the quality of the resulting designs. We present

a method of reformulation called "constraint incorporation," which reduces the dimensionality of the search space and increases its smoothness by incorporating constraints into the search space.

Traditionally, numerical optimization has dealt with explicit, "hard" constraints. The optimizer assumes that these constraints can never be violated. A hard constraint can be expressed as

$$f(x_1, x_2, \ldots, x_n) \leq k$$

(Here $x_1, x_2, \ldots, x_n$ are the *design parameters* that represent the design.) The constraint is said to be *inactive* if $f(x_1, x_2, \ldots, x_n) < k$, *active* if $f(x_1, x_2, \ldots, x_n) = k$, and *violated* if $f(x_1, x_2, \ldots, x_n) > k$. Hard constraints can result from the laws of physics, for example.

Another type of constraint is the "soft" constraint, for which there is some sort of known penalty for violating the constraint. A soft constraint can be expressed as

if $f(x_1, x_2, \ldots, x_n) > k$ then apply penalty $P(x_1, x_2, \ldots, x_n)$

These usually arise from human-written laws, such as regulations specifying a monetary penalty for exceeding a certain noise level. In either case, if it is known that the constraint will be active at the optimal design point, and the constraint function f is invertible, then the constraint can be incorporated into the search space by using the inverse of f to eliminate one of the design parameters. Papalambros and Wilde (1988) describe how monotonicity knowledge can be used to determine that certain constraints will be active at the optimum. Incorporating these constraints produces a new search space with lower dimensionality, since the incorporation eliminates a design parameter, and greater smoothness, since the incorporation eliminates the "ridge" (or nonsmoothness) in the search space caused by the "if" statement in the constraint. If there are n constraints that can be incorporated in this way, then there are 2^n possible reformulations that can be produced by incorporating different subsets of constraints.

Optimization can be done for a variety of *design goals*. A design goal consists of *environment parameters*, which are inputs to the simulator other than the design parameters (and which typically describe the environment in which the designed artifact will operate), and the thresholds on the various constraints.

Constraint activity depends on the goal (some constraints are active at the optimum for only some design goals), for two reasons: First, the constraint thresholds are part of the design goal. Second, different design goals will result in different optimal values of the design parameters on which the constraint functions depend.

Because constraint activity depends on the goal, different reformulated search spaces are appropriate for different design goals. We describe a way in which inductive learning can be used to map the design goal into the appropriate reformulation.

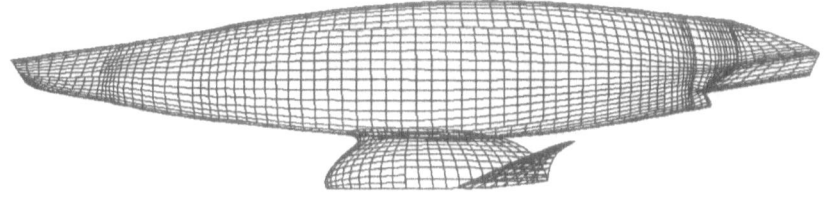

Figure 1. The Stars and Stripes '87.

2. Learning Reformulation Rules

The problem addressed by an inductive-learning system is to take a collection of labeled "training" data and produce rules that make accurate predictions on future data. To use inductive learning to form reformulation-selection rules, we take as training data a collection of design goals, each labeled with the set of constraints that are active at the optimal design point. We run the inductive learner once for each constraint, producing for each constraint a set of rules that can be used to predict whether the constraint will be active for new design goals.

Inductive learning is particularly suitable in the context of an automated design system because training data can be generated in an automated fashion. For example, one can choose a set of training goals and perform an optimization for each goal. One can then evaluate each constraint function for each optimal design, and then construct a table that records which constraints were active (within a threshold) for each training goal. This table can be used by the inductive-learning algorithm to generate a set of rules for each constraint, mapping the space of all possible goals into a prediction of whether or not that constraint will be active at the optimal design point for that goal. If learning is successful, these mappings extrapolate from the training data and can be used successfully in future design sessions to map a new goal into an appropriate reformulation.

The specific inductive-learning system used in this work is C4.5 (Quinlan, 1993) (release 6.0). The approach taken by C4.5 is to find a small decision tree that correctly classifies the training data, and to then remove lower portions of the tree that appear to fit noise in the data. The resulting tree is then used to assign labels to future, unlabeled data.

3. Yacht Design

Our reformulation-selection techniques have been developed as part of the "Design Associate," a system for assisting human experts in the design of complex physical engineering structures (Ellman *et al.*, 1992). One of the domains in which The Design Associate is currently being tested is the domain of 12-meter racing yachts, which until recently was the class of sailboats raced in America's Cup competitions. An example of a 12-meter yacht is the Stars and Stripes '87; its hull is shown in Figure 1. [1]

Racing yachts can be designed to meet a variety of objectives, such as course time or cost. In our work we have chosen to focus on a course-time goal, namely minimizing the time it takes for a yacht to traverse a given race course under given expected wind conditions. A particular course-time goal thus requires the specification of two environment parameters: (1) the race course, represented as a set of (*distance, heading*) pairs; and (2) the wind speed, represented as a scalar number, in knots. Our design system represents a yacht geometry by a set of design parameters, and evaluates course time using a "Velocity-Prediction Program" called "RUVPP,"(Schwabacher *et al.*, 1994) a somewhat simplified version of "AHVPP" from AeroHydro, Inc., which is a marketed product used in yacht design (Letcher, 1991).

Yacht designs are modified by operators that manipulate design parameters. A search space is thus specified by providing the parameters that define an initial prototype, and a set of operators for modifying that prototype. In the experiments described in this paper, the following design parameters were varied:

1. **Length.** The length of the yacht, as measured along the water line.
2. **Beam.** The maximum width of the yacht at the water line.
3. **Hull Depth.** The maximum vertical distance from the water line to the bottom of the "canoe body" of the hull.
4. **Keel Height.** The height of the keel.
5. **Keel Taper Ratio.** The tip chord of the keel divided by the root chord of the keel.
6. **Winglet Span.** The width of the winglets that are attached to the keel.

To find a yacht for a given design goal our system uses CFSQP, a state-of-the-art implementation of the Sequential Quadratic Programming method (Lawrence *et al.*, 1995). [2] Sequential Quadratic Programming is a quasi-Newton method that solves a nonlinear constrained optimization problem by fitting a sequence of quadratic programs[3] to it, and then solving each of these problems using a quadratic programming method. We have supplemented CFSQP with *knowledge-based gradients* (Schwabacher and Gelsey, 1996) to handle designs that cannot be evaluated due to limitations of the simulator.

In the experiments described in this paper, we ran CFSQP with *course-time* as the objective function, and with one explicit, nonlinear, "hard" constraint. This

---

[1] This is the boat that won the 1987 America's Cup competition, returning the trophy to the United States after an Australian win in 1983 (Letcher *et al.*, 1987).

[2] CFSQP stands for "C code for Feasible Sequential Quadratic Programming."

[3] A quadratic program consists of a quadratic objective function to be optimized, and a set of linear constraints.

constraint specifies that the mass of the yacht, before adding any ballast, must be less than or equal to the mass of the water that it displaces. (In other words, the boat must not sink.)

Although the program we use to compute course time (RUVPP) is a state-of-the-art simulator, it nevertheless suffers from a number of deficiencies that make optimization difficult. For example, it will sometimes return a spurious root of the balance-of-force equations that it solves. It may also exhibit discontinuities, due to numerical round-off error, or due to truncation error in the numerical solver used to solve the balance-of-force equations. These deficiencies can produce "noise" in the evaluation function surface over which the optimization algorithm is moving. The algorithm can therefore easily get stuck at a point that appears to be a local optimum, but is nevertheless not locally optimal in terms of the true physics of the yacht design space. There is also noise in the search space caused by the constraints of the 12-Meter Rule, which is discussed further in the next section.

4. The Reformulations

Yachts entered in the 1987 America's Cup race had to satisfy what is known as the 12-Meter Rule (IYRU, 1985). The basic formula in the rule is:

$$\frac{length - freeboard + \sqrt{sailarea}}{2.37} \leq 12m$$

In addition to the basic formula, the rule contains several other constraints, along with associated penalties for violating these constraints. These constraints are:

— draft constraint
— beam constraint
— displacement constraint
— winglet span constraint

For example, the *beam constraint* states

if *beam* < 3.6m, then add four times the difference to *length*

While constructing the simulator, we used a reasoning process similar to that described in Papalambros and Wilde (1988) to determine that the constraint described by the basic formula of the 12-Meter Rule, above, will always be active, since the objective function being minimized, *course-time*, is monotonically decreasing in *sail-area*, and the left-hand-side of the constraint is monotonically increasing in *sail-area*. We therefore *incorporated* this constraint into the simulator by solving for *sail-area* in terms of the other design parameters. So, for example, when the optimizer makes *length* bigger, *sail-area* is automatically made smaller. In addition, because we also incorporated the other constraints into the sim-

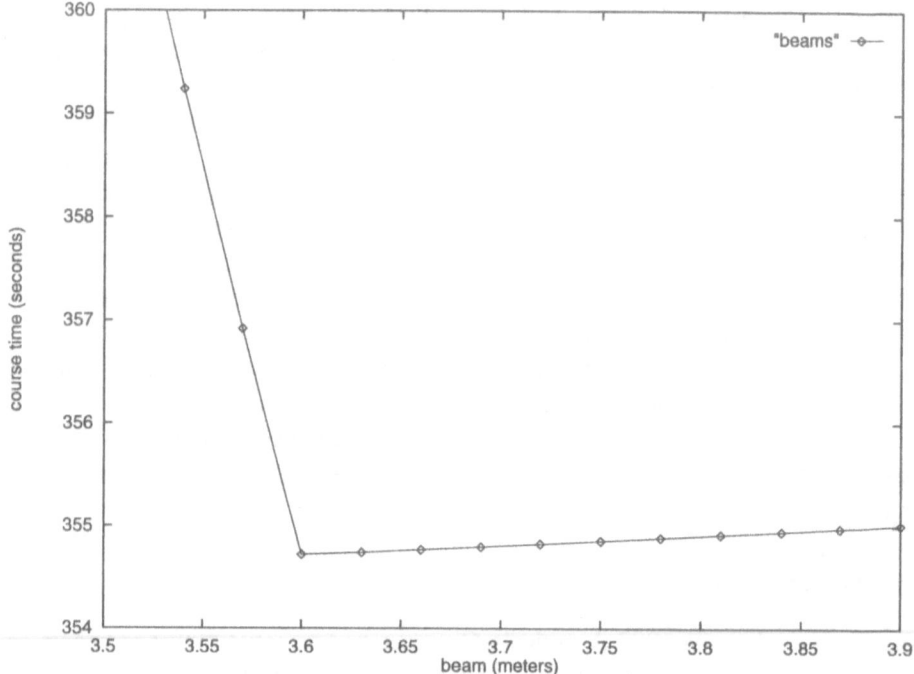

Figure 2. The nonsmoothness in the search space caused by the beam constraint.

ulator, reducing *beam* beyond 3.6*m* causes the quantity *length* in the formula to increase, which causes *sail-area* to decrease.[4]

Because the beam constraint contains an *if* statement, this incorporation causes a nonsmoothness in *course-time* as a function of *beam*. That is, there is a discontinuity in the first derivative of *course-time* with respect to *beam*. Figure 2 illustrates this nonsmoothness by showing the cross-section of the search space corresponding to the *beam* design parameter.[5] This nonsmoothness can cause a gradient-based optimizer such as CFSQP to get stuck, and to fail to get the optimum.

For many design goals, the optimal design is right on the constraint boundary. The optimal beam is often 3.6*m*. If we expect the optimal beam to be 3.6*m*, then we can incorporate the beam constraint into the operators. In the case of the beam constraint, this incorporation is trivial — we simply set *beam* to 3.6*m* and leave it there. For other constraints, the incorporation is more complicated. For example, there is a constraint that specifies a penalty if *displacement* does not vary with

---

[4] Because we incorporated each part of the 12-Meter rule into the simulator, we did not need to use it as an explicit constraint.

[5] Although this figure shows only a "snapshot" of the search space for specific values of the other design parameters, we believe that the trend shown in the figure is generally applicable.

a certain cubic polynomial in *length*. *Displacement* is not a design parameter; rather, it is a quantity computed from all of the design parameters. In order to incorporate the displacement constraint, we used Maple (Char *et al.*, 1992), a symbolic algebra package, to invert the displacement formula, and created a new set of operators that vary certain parameters while maintaining *displacement* at the minimum displacement allowed by the constraint. For still-more-complicated constraints, it might not be possible to invert the constraint function using Maple; it might therefore be necessary for the operators to contain numerical solvers that find the right values of the incorporated design parameters so as to put the design on the constraint boundary.[6]

We created operators to incorporate all four of the above-listed 12-Meter Rule constraints: the draft constraint, the beam constraint, the displacement constraint, and the winglet constraint. Using these operators, we are able to either incorporate or not incorporate each of these four constraints independently. We thus defined a set of sixteen (2^4) possible reformulations of the search space. From our initial experiments with these operators, we determined empirically that incorporating the draft constraint substantially improved the reliability and speed of optimization for any design goal. We therefore decided to always incorporate the draft constraint, leaving us with a space of eight possible reformulations that we used in the experiments described below.

5. Learning to Choose a Reformulation

Having defined eight reformulations of the search space, we used inductive learning to decide, based on the design goal, which reformulation to use. As training data, we used 100 previous optimizations.[7] For each previous optimization, we evaluated each 12-Meter Rule constraint function at the optimum, and determined if the constraint was active (within a tolerance). Each of these previous optimizations had as its design goal minimizing course time for a single-leg racecourse, which can be represented using two numbers: the wind speed, and the heading (the angle between the yacht's direction and the wind direction). The design goal can therefore be represented using these two numbers. We ran the inductive learner once for each of the three constraints. Each time, the inductive learner was provided with a set of triples: the wind speed, the heading, and a ternary value indicating whether the constraint was inactive, active, or violated. One of the constraints was violated at the optimum in 10 of these optimizations. Figure 3 gives an example of a decision tree output by C4.5. This decision tree predicts whether the displacement constraint will be active at the optimum, based

---

[6]Operators containing numerical solvers would probably be more computationally expensive than operators containing the algebraic solutions of the constraint functions, so the CPU time savings from reformulation would probably be smaller.

[7]The optimizer failed for one of these goals, so we used the remaining 99 goals as training data in the results that follow.

```
heading <= 109 :
|   windspeed <= 6.3 : active
|   windspeed > 6.3 :
|   |   windspeed > 8.2 : violated
|   |   windspeed <= 8.2 :
|   |   |   heading <= 65 : violated
|   |   |   heading > 65 : active
heading > 109 :
|   windspeed > 11.5 : active
|   windspeed <= 11.5 :
|   |   heading <= 135 : active
|   |   heading > 135 : inactive
```

Figure 3. Learned decision tree for the displacement constraint.

TABLE 1. Cross-validated error rates for selecting whether to incorporate each constraint.

method	Beam	Displacement	Winglet
C4.5 w/ pruning	11.1%	15.1%	7.0%
C4.5 w/o pruning	11.1%	15.1%	10.0%
C4.5rules	11.1%	15.1%	10.0%
MFC	33.3%	53.5%	13.1%
Random	66.7%	66.7%	66.7%

on the design goal. By running a new design goal down three decision trees, one for each of the three constraints that can be incorporated, the system can make predictions of whether each constraint will be active at the optimum. These three yes/no predictions directly map into one of the eight (2^3) reformulations of the search space.

We used C4.5 to perform tenfold cross-validation (Weiss and Kulikowski, 1991), and obtained the error rates shown in Table 1. Here we compare the error rates of C4.5 with and without pruning, and of C4.5rules, a variant of C4.5 that extracts rules from the trees, with the expected error rate of random guessing (which is two-thirds since there are three classes from which to guess), and the error rate of the Most Frequent Class (MFC) learning method. MFC always chooses the class that occurs most frequently in the training data. In this case, that means that it always chooses the same reformulation, namely the one that is most often the best reformulation in the training data.

As Table 1 shows, C4.5 with pruning performed slightly better than C4.5

without pruning or C4.5rules (and so in our further experiments reported below we use only C4.5 with pruning), and all three significantly outperformed MFC, which in turn significantly outperformed random guessing.

However, these results are for error rates, the proportion of cases where learning makes an incorrect guess, and more important in this domain is how learning affects the overall problem-solving task, namely how it improves the speed and reliability of the design optimization process. Does learning make the design process faster or slower? Are the resulting designs better or worse? To measure these effects, we performed optimizations for 25 new randomly generated goals using the reformulations suggested by each learning method. Table 2 shows the effect that C4.5 (with pruning) and MFC had on the average course time (the quality of the design), and average number of evaluations (the speed of the optimization), as compared with the "old way" of doing optimization without incorporating any of the three constraints into the operators. The first column in the table shows the percentage difference between the optimized course-time produced without reformulation, and the optimized course time produced with the specified reformulation. The second column shows the percentage difference between the cost of performing the optimization without reformulation, and the cost of performing it with the specified reformulation.

We also include in this table the performance of several other methods. A hypothetical "omniscient" problem solver always magically guesses the best possible choice. [8] No learning method will enable results superior to this. The "exhaustive" optimization method performs eight optimizations for each goal, using all eight possible reformulations, and then chooses the best resulting design. Incorporating "all" constraints all the time results in the fastest possible optimization within this set of reformulations (at the cost of quality loss).

C4.5 produced a significant speedup in optimization, with no quality loss. In fact, it produced a small quality increase. (This quality increase suggests that without any reformulation, the optimizer gets "stuck" on the "ridges" that the constraints cause the search space to have, and therefore sometimes fails to get the optimum.) MFC produced a slightly smaller speedup and a slightly smaller quality improvement. The difference between C4.5 and MFC was, however, statistically significant at the 99% confidence level, according to the paired t-test. Both learning methods performed substantially better than random guessing. C4.5 performed almost as well as the hypothetical omniscient learner, which means it performed almost as well as any learner could possibly do. [9]

---

[8] We simulated the omniscient learner by performing optimizations using all eight reformulations for each goal (as in the "exhaustive" method), and then ignoring the cost of the seven optimizations that turned out not to be best.

[9] Interestingly, according to the t-test, the difference between C4.5 and the omniscient method was not statistically significant, but this just illustrates a limitation of the t-test, since we know that the omniscient method really is better, on average, than C4.5.

TABLE 2. Effect of using reformulations chosen by learner on optimization performance.

method	quality change	time change
omniscient	+0.085%	-36%
exhaustive	+0.085%	+384%
C4.5	+0.080%	-35%
MFC	+0.029%	-32%
none	0	0
random	-0.276%	-40%
all	-0.599%	-74%

Incorporating all of the constraints all of the time resulted in a very large speedup, with a modest quality loss. This method may be appropriate if one wants a quick and approximate optimization. It might, for example, be used in the early stages of design when the engineer wants to get a feel for the search space by asking "what-if" questions.

One question that these results raise is how training-data quantity affects performance. If one does not have results from a large number of previous optimizations available, then one can either run some extra optimizations to generate training data (which is expensive), or do the learning with less training data (which is likely to produce higher error rates and lower optimization performance). To explore this issue we applied our learning approach to datasets of varying sizes, with the error rates shown in Figure 4. For each training-set size in the figure, we randomly chose 10 different subsets of our training data of that size, and performed 10-fold cross-validation on each subset. The figure shows the averages. The three symbols at the right side of the figure show MFC's performance on the full training set. C4.5 outperformed MFC for every training-set size, but C4.5's error rate on smaller training sets was significantly larger than C4.5's error rate for larger training sets (with performance reaching an asymptote for training sets of about 60 cases or more).

6. Related Work

Another way of selecting a search space is by selecting a starting point which, when combined with a set of operators, defines a search space. Previous results (Schwabacher et al., 1994) showed that machine learning can improve optimization performance by learning how to select an initial prototype from which to start the optimizer. Cerbone (1992) has reported work which applied machine-

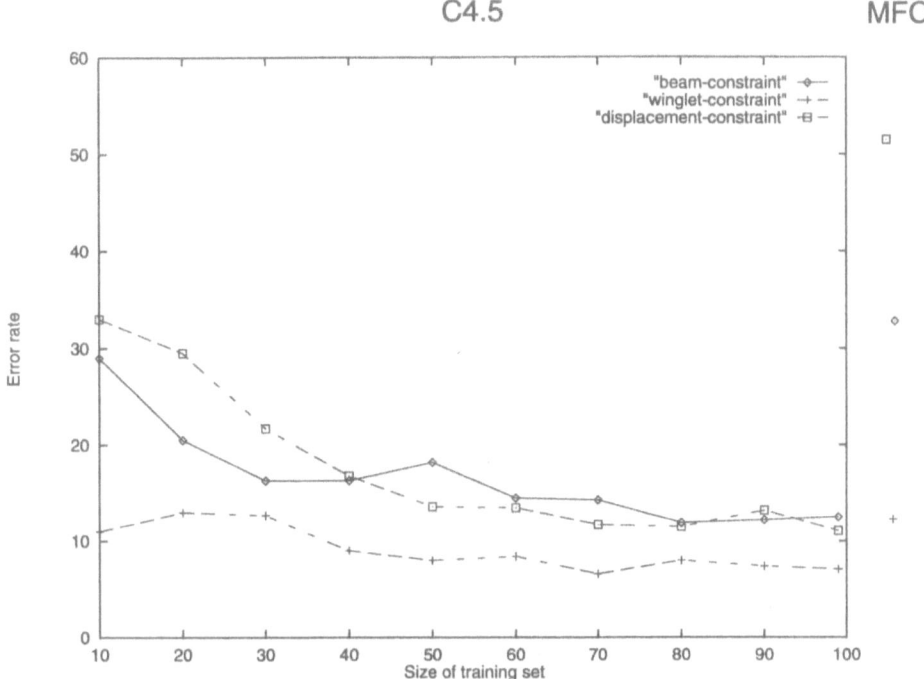

Figure 4. Effect of training set size on learner performance.

learning techniques to a problem similar to our prototype-selection problem. Cerbone's design space, in the domain of truss design, has an exponential number of disconnected search spaces. He uses inductive learning techniques to learn rules for selecting a subset of these search spaces for further exploration.

Several investigators (Orelup *et al.*, 1988; Tong, 1988; Powell, 1990; Hoeltzel and Chieng, 1987) have developed alternative artificial-intelligence techniques for controlling iterative parameter-design optimization. Gelsey et al. (1995; 1996) describe a Search Space Toolkit which assists in determining properties of the search space that can be used for reformulation. Choy and Agogino (1986) describe a system that automates (Papalambros and Wilde, 1988)'s method of using monotonicity analysis to detect constraint activity.

In Williams and Cagan (1994), Williams and Cagan present *activity analysis*, a technique inspired by monotonicity analysis. Their technique is similar to the technique described in this paper, except that they use qualitative reasoning instead of machine learning to determine which constraints will be active at the optimum. Their technique has the advantages that it does not require training data, and that the reformulation is guaranteed not to lose the global optimum. It has the disadvantage that it requires that the objective function and constraint functions be symbolically differentiable and composed of simple arithmetic operations; it would therefore not be applicable to the complex simulators used in the experiments described in this paper.

7. Future Work

This paper has described on-going work, and there are thus a number of directions for future work. These fall into two groups: extending this work to more difficult design tasks, and improving results by using other learning methods.

7.1. OTHER DESIGN TASKS

The results presented here apply to a constrained class of yacht-design goals, those comprised of a single leg. One question is how this approach can be applied to courses comprised of varying numbers of legs. We believe that we could get reasonable optimization performance by using the trees learned from single-leg courses to perform multi-leg optimization in the following way: If a constraint should be incorporated for every leg of the racecourse, then incorporate it for the full, multi-leg course. We need to test how well optimization performs when handling racecourses in this manner. We could also attempt to learn directly for multi-leg racecourses. Doing so would raise an interesting machine-learning question, since describing a multi-leg racecourse requires a variable number of attributes, and thus traditional learners such as C4.5 do not directly apply.

In the results presented here, we assume that the only change between the previous design sessions and the current design session is the design goal, expressed as a (*wind speed, heading*) pair. An interesting question is what would happen if in addition to changing the goal, we also changed the constraints, or the simulator, or the form of the goal. We would need to find a way to encode as a set of attributes for the learner whatever had changed.

We believe that the results presented here will easily generalize to situations in which there are more than eight reformulations. We used the results from the same set of 100 optimizations to perform three separate learning tasks (for three constraints), and then combined the rules generated by these three learning sessions to select one of the eight reformulations. As the number of reformulations grows, the number of constraints, and therefore the amount of CPU time needed for the learning, will grow logarithmically with the number of reformulations. The CPU time needed for learning is currently insignificant compared with the CPU time needed for the subsequent optimizations. We expect that as the number of reformulations grows, the number of training examples needed will remain constant (since the same training examples are used for each constraint), and the amount of CPU time needed for learning will remain insignificant. We plan to test this hypothesis by using other constraints within the yacht design domain, such as the "boat doesn't sink" constraint.

The learning approach could also be used to decide when to reformulate soft constraints as hard constraints. If it were known with a high degree of confidence that a certain soft constraint will not be violated at the optimum for certain goals, then this soft constraint could be converted into a hard constraint for those goals, which would eliminate a ridge from the search space and thereby make optimization more robust (although it would not reduce the dimensionality of the search space). For example, in the training data that we collected, the *beam constraint* was never violated, so it might be replaced safely with a hard constraint.

Other more-difficult problems might involve a less-smooth search space, a higher-dimensional goal space, or a less reliable optimizer. Such problems may arise when we test this method in other domains. In particular, we plan to test it in the domain of aircraft design.

7.2. OTHER LEARNING METHODS

We found that C4.5 performed nearly as well as a hypothetical "omniscient" learner, for the fairly simple design problem that we used in our experiments. Other learning methods, however, might prove useful when attacking some of the harder problems described in the previous subsection. For example, it would be interesting to see how well neural networks, nearest-neighbor methods, or statistical regression would perform. In particular, C4.5, like most decision-tree learners, uses linear, axis-parallel cuts in its decision trees. However, Figure 5 shows how the activity of the beam constraint varies over the goal space in the training data we used — the space is clearly divided into two regions (except for one point which we believe is noise). The border between these regions does not appear to be axis parallel, and appears to be nonlinear. This suggests that better performance might be achieved using an "oblique" decision tree learner, such as OC1 (Murthy *et al.*, 1993), or by attempting to learn nonlinear region boundaries.

As would be expected, even though our yacht-domain results with C4.5 were nearly optimal for 100 examples, results degrade when given less training data. Although it would be interesting to see if other learning methods would have better small-dataset performance, for any learner we would expect performance to be inferior for small enough datasets. One approach for improving results in such small-dataset cases — as well as in other cases where off-the-shelf learners such as C4.5 may not perform well even if given larger datasets — is to integrate background knowledge into the learning process. One form of background knowledge that is often available, such as in the yacht-design domain, is *modality constraints*. This is knowledge that expresses the modality of the learned class with respect to the attributes. For example, we believe that optimal *beam* is monotonically increasing in wind speed, and monotonically decreasing in heading. We also know that the activity of any constraint of the form $f(x_1, x_2, \ldots, x_n) \leq k$ must be monotonic in k, so, for example, the activity of a cost constraint must be mono-

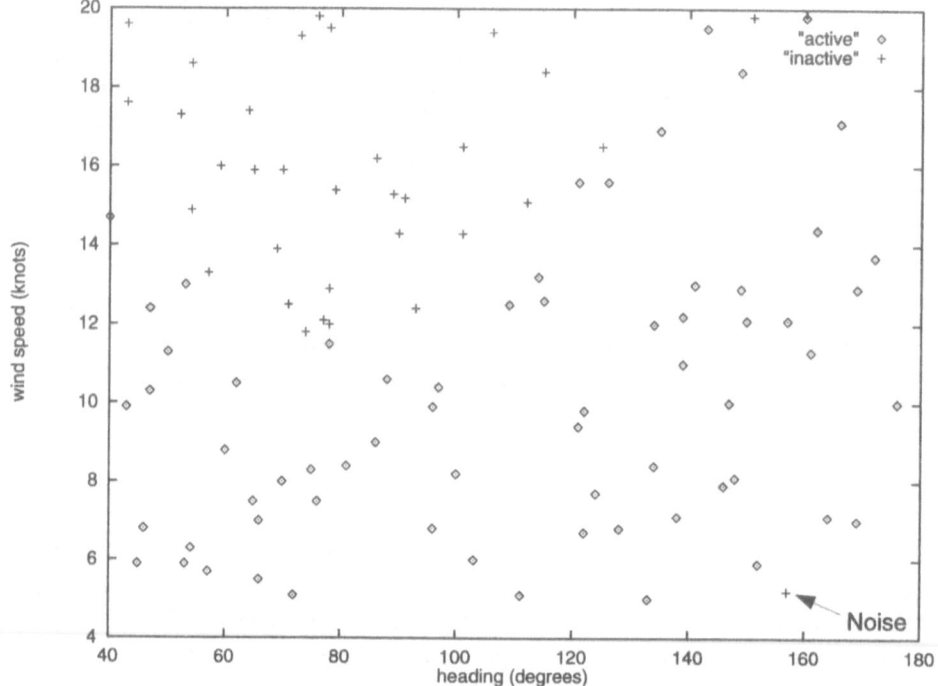

Figure 5. Activity of the beam constraint over the goal space.

tonic in the cost threshold. One open question is how such knowledge could be integrated into learning. One approach would be to use such modality constraints to remove from the training data points that violate the constraints (on the assumption that these points are noise). A second approach is to modify the tree induction algorithm so that it will never construct a tree that violates the constraints.

Finally, even after our learning approach is applied, every additional future optimization can serve as an additional training point for the learning. Thus learning methods that can work in an incremental fashion might also prove useful for this task. In addition, it may prove useful to develop methods that select suitable data prior to learning. For example, when there are not enough existing optimizations to achieve adequate learning results, additional optimizations can be performed to generate further training data. Rather than performing these new optimizations for random goals or for a set of goals that span the goal space, one could allow the learner to choose the goals to be used in the new training data. Background knowledge — such as modality constraints — could prove particularly useful in selecting such goals.

8. Conclusion

We have shown that using the reformulations selected by inductive learning makes design optimization faster, because the reformulation reduces the dimensionality of the search space, and more reliable, because the reformulation makes the search space smoother.

When we started this research, it was not immediately obvious whether inductive learning would be applicable in our chosen domain. In the process of applying inductive learning to this domain, we learned significant lessons about the importance of choosing the right evaluation criteria, the importance of getting good data, and the usefulness of visualization. These lessons are further described in Schwabacher et al. (1995).

Acknowledgments

This research has benefited from numerous discussions with members of the Rutgers HPCD project. In particular, we would like to thank Andrew Gelsey, John Keane, and Brian Davison. This research is part of the Rutgers-based HPCD (Hypercomputing and Design) project supported by the Advanced Research Projects Agency of the Department of Defense through contract ARPA-DABT 63-93-C-0064.

References

Cerbone, G.: 1992, Machine learning in engineering: Techniques to speed up numerical optimization, *Technical Report 92-30-09*, PhD Thesis, Oregon State University, Department of Computer Science.

Char, B., Geddes, K., Gonnet, G., Leong, B., Monagan, M. and Watt, S.: 1992, *First Leaves: A Tutorial Introduction to Maple V*, Springer-Verlag and Waterloo Maple Publishing.

Choy, J. and Agogino, A.: 1986, Symon: Automated symbolic monotonicity analysis system for qualitative design optimization, *Proceedings ASME International Computers in Engineering Conference*.

Ellman,T., Keane, J. and Schwabacher, M.: 1992, The Rutgers CAP Project Design Associate, *Technical Report CAP-TR-7*, Department of Computer Science, Rutgers University, New Brunswick, NJ. ftp://ftp.cs.rutgers.edu/pub/technical-reports/cap-tr-7.ps.Z.

Gelsey, A. and Smith, D.: 1995, A search space toolkit, *Proceedings, 11th IEEE Conference on Artificial Intelligence Applications*, Los Angeles, CA, pp. 117–123.

Gelsey, A., Smith, D., Schwabacher, M., Rasheed, K. and Miyake, K.: 1996, A search space toolkit, *Decision Support Systems, special issue on Unification of Artificial Intelligence with Optimization*, (to appear).

Hoeltzel, D. and Chieng, W.: 1987, Statistical machine learning for the cognitive selection of nonlinear programming algorithms in engineering design optimization, *Advances in Design Automation*, Boston, MA.

IYRU: 1985, *The Rating Rule and Measurement Instructions of the International Twelve Metre Class*, International Yacht Racing Union.

Lawrence, C., Zhou, J. and Tits, A.: 1995, User's guide for CFSQP version 2.3: A C code for solving (large scale) constrained nonlinear (minimax) optimization problems, generating iterates satisfying all inequality constraints, *Technical Report TR-94-16r1*, Institute for Systems Research, University of Maryland.

Letcher, J., Marshall, J., Oliver, J. and Salvesen, N.: 1987, Stars and Stripes, *Scientific American*, **257**(2).

Letcher, J.: 1991, *The Aero/Hydro VPP Manual*. Aero/Hydro, Inc., Southwest Harbor, ME.

Murthy, S., Kasif, S., Salzberg, S. and Beigel, R.: 1993, OC1: Randomized induction of oblique decision trees, *Proceedings of the Eleventh National Conference on Artificial Intelligence*, Washington, DC.

Orelup, M. F., Dixon, J. R., Cohen, P. R. and Simmons, M. K.: 1988, Dominic II: Meta-level control in iterative redesign, *Proceedings of the National Conference on Artificial Intelligence*, St. Paul, MN, pp. 25–30,

Papalambros, P. and Wilde, J.: 1988, *Principles of Optimal Design*, Cambridge University Press, New York, NY.

Powell, D.: 1990, Inter-GEN: A hybrid approach to engineering design optimization. *Technical report*, PhD Thesis, Rensselaer Polytechnic Institute Department of Computer Science.

Quinlan, J. R.: 1993, *C4.5: Programs for Machine Learning*, Morgan Kaufmann, San Mateo, CA.

Schwabacher, M. and Gelsey, A.: 1996, Intelligent gradient-based search of incompletely defined design spaces, *Technical Report HPCD-TR-38*, Department of Computer Science, Rutgers University, New Brunswick, NJ. ftp://ftp.cs.rutgers.edu/pub/technical-reports/hpcd-tr-38.ps.Z.

Schwabacher, M., Hirsh, H. and Ellman, T.: 1994, Learning prototype-selection rules for case-based iterative design, *Proceedings of the Tenth IEEE Conference on Artificial Intelligence for Applications*, San Antonio, Texas.

Schwabacher, M., Hirsh, H. and Ellman, T.: 1995, Inductive learning for engineering design optimization, in D. Aha and P. Riddle (eds), *Working Notes for Applying Machine Learning in Practice: A Workshop at the Twelfth International Machine Learning Conference, (Technical Report AIC-95-023)*, Naval Research Laboratory, Navy Center for Applied Research in Artificial Intelligence, Washington, DC. http://www.aic.nrl.navy.mil/~aha/imlc95-workshop/.

Tong, S. S.: 1988, Coupling symbolic manipulation and numerical simulation for complex engineering designs, *International Association of Mathematics and Computers in Simulation Conference on Expert Systems for Numerical Computing*, Purdue University.

Weiss, S. M. and Kulikowski, C. A.: 1991, *Computer Systems That Learn*, Morgan Kaufmann, San Mateo, CA.

Williams, B. and Cagan, J.: 1994, Activity analysis: The qualitative analysis of stationary points for optimal reasoning, *Proceedings of the Twelfth National Conference on Artificial Intelligence*, Seattle, WA.

9

distributed design

Virtual construction site: Supporting design by multiple methods
in FABEL
*Carl-Helmut Coulon, Wolfgang Gräther, Barbara Schmidt-Belz, Angi
Voβ, Friedrich Gebhardt, Eckehard Groβ and Jörg Walter Schaaf*
A mobile-agent oriented approach to a distributed design
support system
Haruyuki Fujii, Shoichi Nakai, Hiroshi Katukura, Keiichi Hirose
VisionManager: A computer environment for design
evolution capture
Renate Fruchter, Kurt Reiner, Larry Leifer, George Toye

J. S. Gero and F. Sudweeks (eds), Artificial Intelligence in Design '96, 465-483.
© 1996 Kluwer Academic Publishers.

VIRTUAL BUILDING SITE

*Supporting building design by multiple methods in FABEL**

CARL-HELMUT COULON, WOLFGANG GRÄTHER,
BARBARA SCHMIDT-BELZ, ANGI VOβ, FRIEDRICH GEBHARDT,
ECKEHARD GROβ AND JÖRG SCHAAF
GMD
German National Research Center for Information Technology
D-53754 Sankt Augustin, Germany

Abstract. FABEL prototype 3.0 is a research system to support engineers and architects in building design. We introduce and motivate the metaphor of a virtual building site, which has guided the development of this system. In the virtual building site a common data model of the artefact is designed. The user interface is based on a graphic editor which handles design objects of all kinds and at several levels of abstraction and even organizes the application of the tools. There are tools to supply the designer with useful cases, to create pieces of design by adaptation of similar cases or by knowledge-based refinement, and there are tools to assess parts of the design. We describe some of the tools to illustrate the multi-method approach of FABEL.

1. Introduction

FABEL is a German joint research project that aims at supporting architects and civil engineers in designing buildings with a complex infrastructure. When the project was started in 1992, several attempts at building expert systems for this domain had failed, because the knowledge could not be modelled sufficiently and the systems soon became inefficient and untractable.

To cope with such complex domains, we took a twofold approach: Instead of one big expert system, we built small and independent tools. All control was left with the human designers, not with some automatic scheduler. But in order to co-operate on such a complex task, man and machine must be able to communicate what needs to be done, what can be done at some given moment and what is cur-

*This research was supported by the Federal Ministry of Education, Science, Research and Technology (BMBF) within the joint project FABEL under contract no. 01IW104. Project partners in FABEL are GMD – German National Research Center for Information Technology, Sankt Augustin, BSR Consulting GmbH, München, Technical University of Dresden, HTWK Leipzig, University of Freiburg, and University of Karlsruhe.

rently being done. Therefore, a common frame of reference was needed. We call it a virtual building site. The system, FABEL prototype 3.0, is operational since December 95 and will be tested and assessed in the remaining half year of the project. At the previous 'AI in Design' conference, a predecessor, prototype 2.0 for retrieval only, was presented (Voß *et al.*, 1994).

This paper will explain the metaphor of a virtual building site. In short, the whole design process, including the application of tools, is spatially organized and visualized. The design evolves like a building under construction. It can be inspected under different points of view, with different foci. It can be concurrently completed by different engineers with many different (or equal) tools. In FABEL the basic support is retrieval of useful cases, useful under consideration of the current situation. Different concepts of similarity are provided and fully integrated in a powerful shell for storing and retrieving cases. Additional tools can elaborate pieces of designs by adapting cases or generating solutions. Some of them rely on very domain- or task-specific knowledge and therefore have a rather restricted scope of application. Others are more generic, relying on less specific knowledge (or on a wider knowledge base). We will give examples of some tools. All tools can be applied concurrently or in succession, as their scopes of competence allow and as the users like.

2. A Virtual Building Site as a Frame of Reference

Design of physical (or at least visible) artifacts like machines, buildings, cities or plants has a major advantage over the design of abstract artifacts like software. The product has a geometry, sometimes a site, and it can be inspected. The design documents for such products relies on this characteristic. They essentially consist of plans which are usually represented geometrically and annotated by symbolic information. They are developed with CAD systems, and the new generation of these tools supports integrated models of the artifact, rather than collections of independent plans that have to be kept consistent by the designers. The tools for developing integrated models will become more and more comfortable, e.g. by providing 3D visualizations and walk-throughs in virtual reality. Also, in the future they will have to offer support for the early, conceptual phases when composition of design objects, their forms and positions are still approximate and sketchy.

It is a major hypothesis of our approach that *in an adequate design support environment for complex physical artifacts, design information should be organized geometrically in space.* This concerns concrete and abstract design objects, annotations and even the tools to elaborate the design.

This hypothesis leads to the metaphor of a virtual building site, which was proposed by the architect Ludger Hovestadt (1993) from the University of Karlsruhe. Given an integrated design model, a design in progress can be treated much like the real product under construction. The virtual building site can be inspected by projecting it according to the point of view of the observer. This point of view can be changed first of all by "navigating" along the three spatial dimensions and filtering certain types of information. The design can be elaborated by inserting, removing, moving or resizing design objects, and by applying tools at circumscribed places.

Figure 1.   Complex building models contain some thousand design objects. Without filtering and focusing they can not be properly visualized.

While the spatial dimensions are universal, they are not sufficient for focussing. Consider figure 1. It shows a great part of the building model of the Swiss railway company's education center in Murten, which was designed and built by Fritz Haller. Concentrating on some corner of the building would not help, because the plan contains design objects at different levels of abstraction and for different functions (of layout, construction and technical services). Such additional filters allow to focus within the building model on imageable subproblems like designing the columns or the supply-air outlets on one floor. By treating the filters as additional dimensions one obtains a uniform frame of reference that can be treated with a generic set of functions for navigation and positioning[1].

---

[1]In CAD systems additional layers provide a predefined set of projections, in contrast to the unrestricted navigation through projections that can be created on demand by fixing some values in any spatial and other dimensions of the virtual building site.

Figure 2 gives a glance at a virtual building site. It is a layout as produced by the CAD tool of our partner at the University of Karlsruhe. There are design objects at three different levels of abstraction: concrete parts, rectangles and ellipses. The rectangles represent precisely placed design objects that still need to be refined to concrete parts. The ellipses are sketchy elements indicating maximal space reservations, as given by the bounding box of an ellipse[2].

Besides design objects, instances of tools can be placed and applied to parts of the virtual building site (cf. figure 2). The three frames labeled by icons at their upper left corner indicate positions of tools. Such a frame defines the application area of a tool in all dimensions. In order to prevent conflicts, overlapping of application areas is forbidden. For example the tools **AAAO** (figure 2, upper left icon, compare section 4.2.2) and **Topo** (right icon, compare section 4.2.1) work in spatially overlapping locations of the building; nevertheless the application areas are disjoint because **AAAO** places the columns and **Topo** works (in this case) on the air conditioning.

Each activated tool has a user interface. For example the user interface shown in the upper left corner of the figure is used to navigate through the building.

In FABEL, we have about a dozen design support tools. They provide functionality for retrieval, adaptation, assessment (the classic CBR steps (Slade, 1991)), and the generation of new pieces of the plan using other knowledge than cases. They are the subject of section 4. The implementation of the virtual building site and the coordination of all the different tools technically requires a coordination of software processes and user interfaces, which is described in section 5. That section will elaborate the scenario (figure 2) in more detail in order to elucidate the benefits of the described distributed approach.

3. Locating the Support in the PM5-Model of Building Design

The virtual building site is based on a building model (the data) which is part of the PM5 model of the process of building design (cf. (Hovestadt and Schmidt-Belz, 1995)).

PM5 provides us with

- a scheme of several abstraction levels of the building design task,
- a semantic model of *tasks* on these abstraction levels,
- important dependencies and interdependencies between tasks,
- the semantic connection between tasks and the design objects to be produced.

In the context of this paper the most important feature of PM5 is the latter one, that is the connection between tasks and the design objects to be produced.

---

[2]This unusual visualization has two advantages. An ellipse looks more sketchy than a rectangle and one can distinguish even lots of different overlapping spaces, which becomes impossible using rectangles.

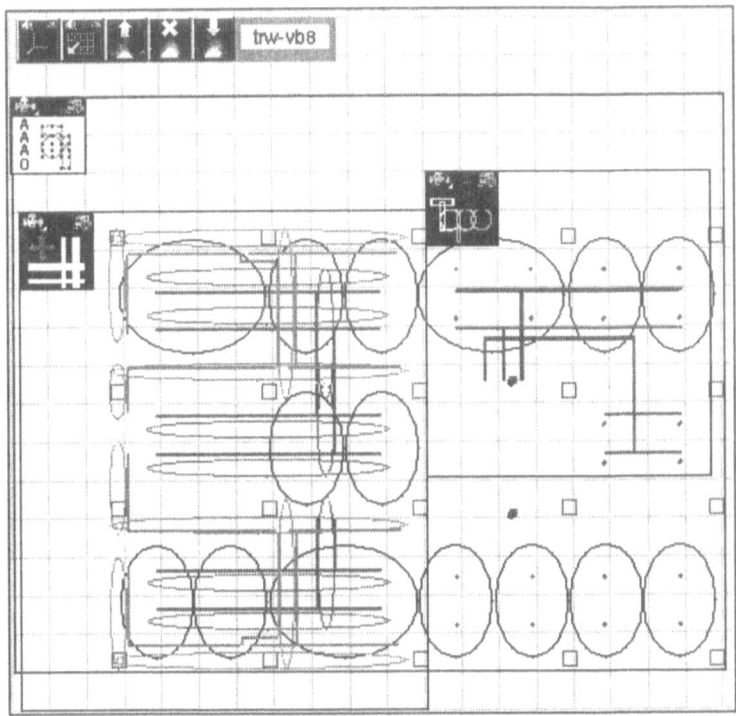

Figure 2. All information stored in the building model is located in the "virtual building site". The whole building model might be even more complex than Figure 1. Most times the user wants to work with part of the building model only. He focuses on it using the navigation panel (top left-hand side) to filter certain types of objects and navigates through the geometric dimensions using common CAD functionality. Frames labeled by icons represent tools not activated. After activating a tool, a minimal user-interface, like the navigation panel, appears presenting the essential functionality of the tool. More functions and detailed information about every tool can be accessed in seperate windows.

In PM5 the tasks are of a benign nature: each task corresponds one-to-one to a certain type of design objects to be created and laid out. For instance, AAAO supports one task (see figure 3) where objects of type "route planning for construction (rp-co)" are created and handled. Therefore a task can be represented by a short term like rp-co. The suggestion is that designers always focus on a certain aspect of the whole building plan, for instance the areas of use in a building (pl-ro), the arrangement of rooms and corridors (pp-ro), the position of furniture (ep-ro), the approximate position of installations (rp-armilla), their exact position (pp-armilla) or the concrete elements that form a ductwork (ep-armilla). When elaborating the design with respect to one type of design objects, the designers have to consider certain other types of design objects already elaborated. Thus, PM5 establishes an idealized structure of tasks together with interdependencies among them.

For some tasks there are models or strategies of how to elaborate this part of the building model. ARMILLA, for instance, is an operational model of how to design the installations for a building (cf. (Haller, 1985)). While retrieval tools are not task-specific, many of the other FABEL tools are, because they rely on additional, specialized knowledge like ARMILLA. Figure 3 shows the scope of FABEL's adaptation and generation tools. For the scope of support by assessment and learning tools we refer to Schmidt-Belz (1995).

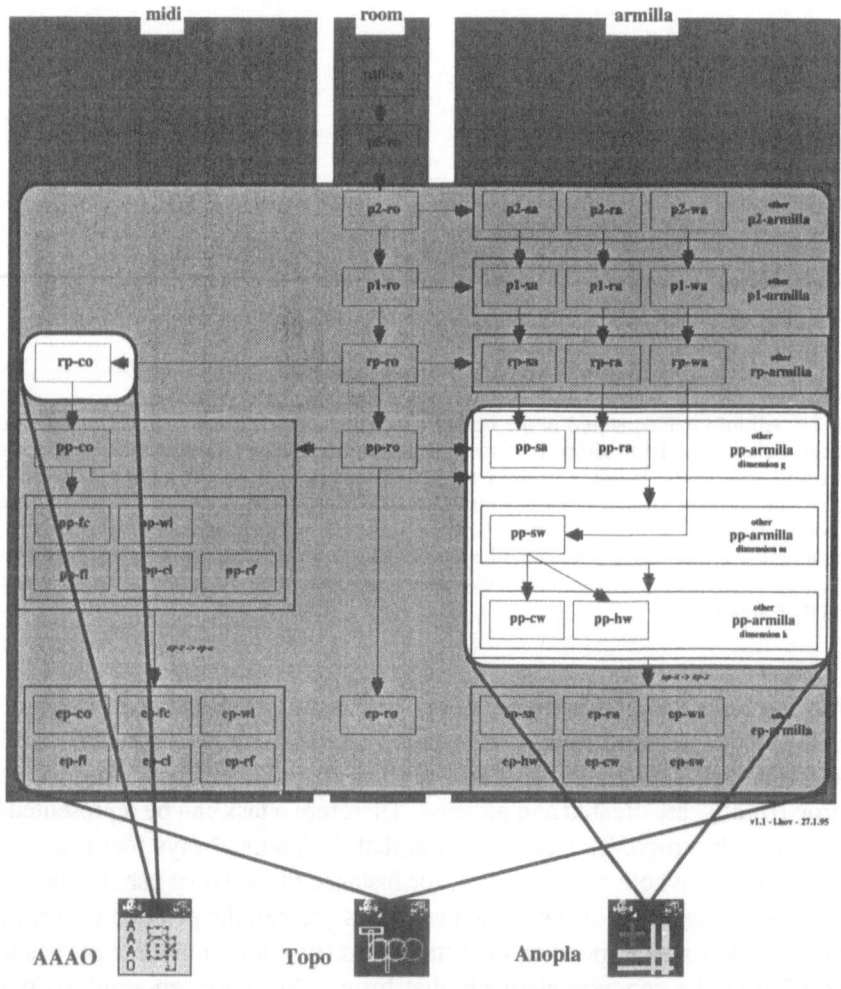

Figure 3. This figure shows which tasks are supported by the three FABEL tools for adaptation and generation presented in this paper. Their scope ranges from a specialist like **AAAO** to a generalist like TOPO.

The scope of generation modules in our FABEL prototype 3.0 varies due to the knowledge available. Anopla relies on ARMILLA knowledge, as we discuss in detail in section 4.2.3.

4. Tools Supporting Complex Design

Tools that are independent must not communicate directly with each other. Instead, they must use a common data space, the virtual building site. Correspondingly, all tools must accept similar input and produce similar output. That means, all tools will accept a piece of plan (that are design objects from the virtual building model) as input and produce design objects (or a piece of plan) as their output (cf. fig. 4[3]).

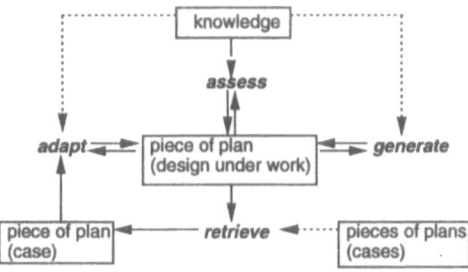

Figure 4. Four kinds of tools: retrieve, adapt, assess and generate.

The retrieval tools hardly use any generic knowledge. Therefore they are applicable to all kinds of plans. In contrast, assessment, generation, and also adaptation require knowledge about the specific task to perform, e.g. about the capacity of pipes, admissible pipe positions or static constraints for the positions of columns. Therefore, their scope is limited to particular steps in the design process (cf. figure 3).

There are two reasons why we offer separate tools for retrieval and adaptation. First, not all cases retrieved can or should be adapted automatically. They serve as inspirations for the user or contain alternatives to consider. On the other hand, the adaptation tools should be able to adapt cases selected by the user himself rather than by a retrieval tool.

Adaptation tools receive two sets of design objects and try to fit one to the other (cf. sections 4.1 and 4.2). Generation tools receive only one plan and try to extend it using generic knowledge (Anopla cf. section 4.3, Roude (Jäschke and Janetzko, 1994) and Syn* (Börner and Faßauer, 1995)). As adaptation usually consists of matching, copying and then fitting the case to the piece of plan under

---

[3] An exception are assessment tools, which produce a commentary to a piece of plan

work, the same kind of knowledge can often be used for case-fitting as well as for case-independent generation. Assessment tools check predicates concerning different aspects like topology, dimensioning or collisions, and attach the result to the checked design object. FABEL's assessment tools DOM (Bartsch-Spörl *et al.*, 1995) and Check-Up (Janetzko and Jäschke, 1995) are not described in this article.

4.1. TOOLS TO FIND USEFUL EXAMPLES

Examples of layouts may serve different purposes in different phases of design. In early design, they are a source of inspiration and alternatives, later they show how to elaborate a partial layout, and finally they can be used for multiplication, e.g. "do all those offices like this one". Thus, the usefulness of an example strongly depends on context. It is a major suggestion of FABEL that *in any context "what is useful" can be captured by a special blend of aspects under which the designs can be analyzed, and that the entire set of aspects can be kept relatively small.* Figure 5 shows the aspects that have been predefined in FABEL for interpreting layouts of buildings (c.f. (Voß *et al.*, 1994)). A layout may be viewed as a set of design objects, as an image (or silhouette), as a structure. It may contain gestalten that indicate certain layout principles, e.g. ring-like, comb-like, or fishbone-like installation patterns, which are recognized by a special tool.

To define an aspect, a representation (formalism and function) and a concept of similarity between concrete representations must be specified. In one of the aspects developed in FABEL, sets of design objects, represented as sets of keywords indicating number and type of design objects, are compared by their overlap. Alternatively, they are represented as a vector of attributes, with an attribute for each type of design objects, and a distance is defined for each attribute. Images, represented as pixel matrices, are the more similar the more elements coincide. Topological structures, represented as graphs, can be compared by their common subgraphs.

Generic retrieval tools prescribe the data structures and schemata for specifying case representations and similarities, and they provide techniques for storing, accessing and comparing cases represented in such data structures. Most commercial tools operate on attribute vectors and distances, text retrieval systems operate on sets of keywords. FABEL offers several generic retrieval tools that can be used for one or more aspects. By far the fastest is ASM (Gräther, 1994), an associative memory that operates on sets of keywords (as node labels). For attributes there is RABIT (Linowski, 1994), a very flexible development tool with an extensible catalogue of distance measures. For the dynamic combination of multiple aspects we have a completely new approach called ASPECT (Schaaf and Voß, 1995).

ASPECT is more general than other retrieval tools as it allows to combine dynamically any kinds of aspects – provided they are defined in terms of a repres-

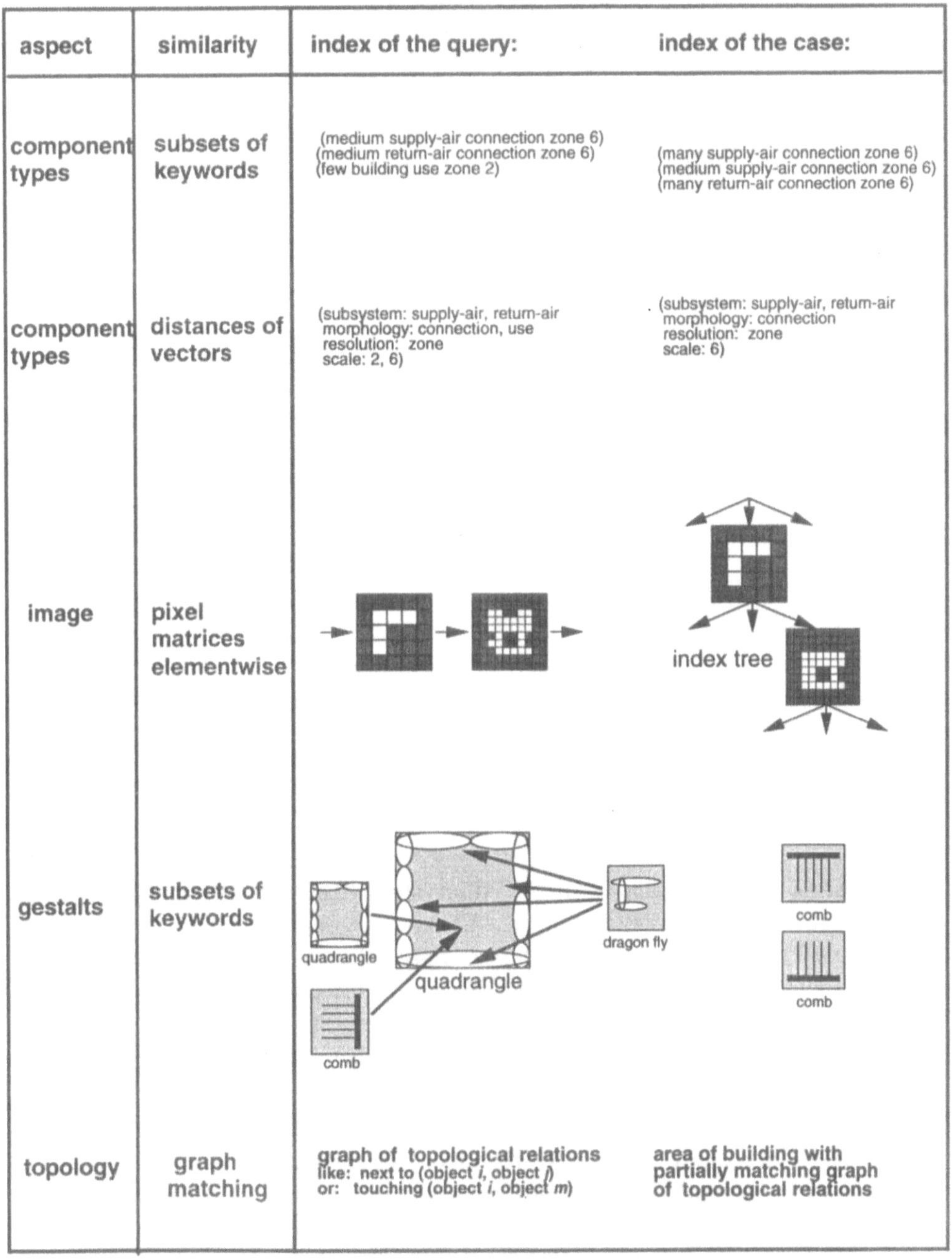
aspect
similarity
index of the query:
index of the case:
component types
subsets of keywords
(medium supply-air connection zone 6)
(medium return-air connection zone 6)
(few building use zone 2)
(many supply-air connection zone 6)
(medium supply-air connection zone 6)
(many return-air connection zone 6)
component types
distances of vectors
(subsystem: supply-air, return-air
morphology: connection, use
resolution: zone
scale: 2, 6)
(subsystem: supply-air, return-air
morphology: connection
resolution: zone
scale: 6)
image
pixel matrices elementwise
index tree
gestalts
subsets of keywords
quadrangle
quadrangle
comb
dragon fly
comb
comb
topology
graph matching
graph of topological relations
like: next to (object i, object j)
or: touching (object i, object m)
area of building with
partially matching graph
of topological relations

Figure 5. Different aspects for retrieving layouts in FABEL.

entation function and a distance function. We use ASPECT to combine structure, images, gestalten and design objects. Together with a query, a set of contextual clues is passed on from which the weights of the aspects are dynamically determined (invoking a backpropagation network). Figure 6 illustrates the metaphors underlying ASPECT. A case is like a polyeder, each side representing one aspect. The similarity (inversely related to distance) between the two cases in the case base under one aspect is attached to a link between their respective sides.

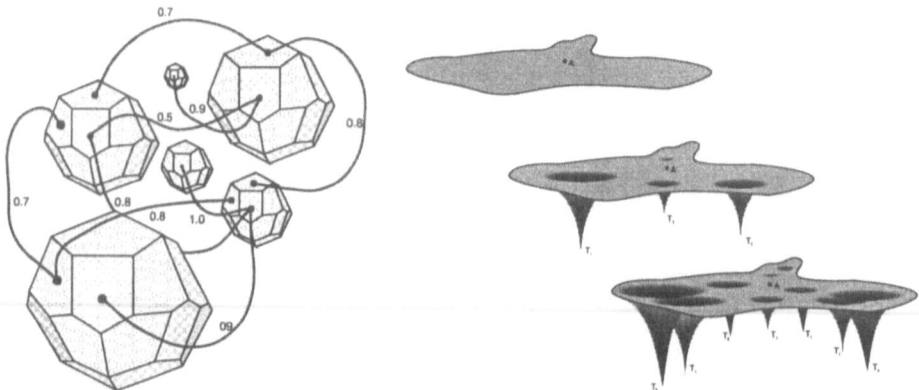

Figure 6. In the case base of ASPECT cases are linked with respect to multiple aspects (sides of polyeders in the picture). Retrieval is like fishing cases from the surface, sinking them with their neighbouring cases to some depth of dissimilarity and retaining those that stay near to the original surface.

ASPECT trades space for time. When a new case is stored, its links to other cases are computed. For efficiency, ASPECT does not store minor similarities, and it will forget aspects seldom used. At retrieval time, the query must be represented in all relevant aspects, and their similarities to certain cases stored must be computed, taking into account the current weights. Response time is reduced by minimizing the set of cases to be used for comparison. To this end, the neighbourhoods pre-computed in the case base are exploited. As figure 6 shows, initially all cases "swim" at the surface. Cases are always "fished" from the surface and sunk as deep as they are dissimilar to the query. Neighbouring cases are torn down, such that the depth of a sunken case always indicates how similar it can be to the query under best circumstances (a triangular inequation is applied here). As time goes by, the surface level decreases monotonically, while surface cases keep to be the most promising candidates for comparison. As soon as the surface level starts sinking, the method adopts an any-time behavior and produces the most similar cases in decreasing order as search goes on.

4.2. TOOLS FOR ADAPTATION AND GENERATION

In this section three tools are presented with respect to the subtasks they support, the knowledge they need, and the methods they employ.

4.2.1. *Topo: transfer and adaptation of topology*

The main focus of the tool Topo is adaptation. It uses extremely little application-specific knowledge and is therefore applicable in a wide range of situations (cf. figure 3). First, it finds common topologies of a given piece of plan and a case. Second, it can transfer and adapt the position of every object of the case which is topologically related to the common topology. An example is shown in figure 7. Additionally Topo provides the tool ASPECT with a representation of topology and a similarity measure for topologies. The third functionality is assessment. Topo builds statistics about usual and unusual topological relations occurring in the case base and can check a given piece of plan with respect to these statistics.

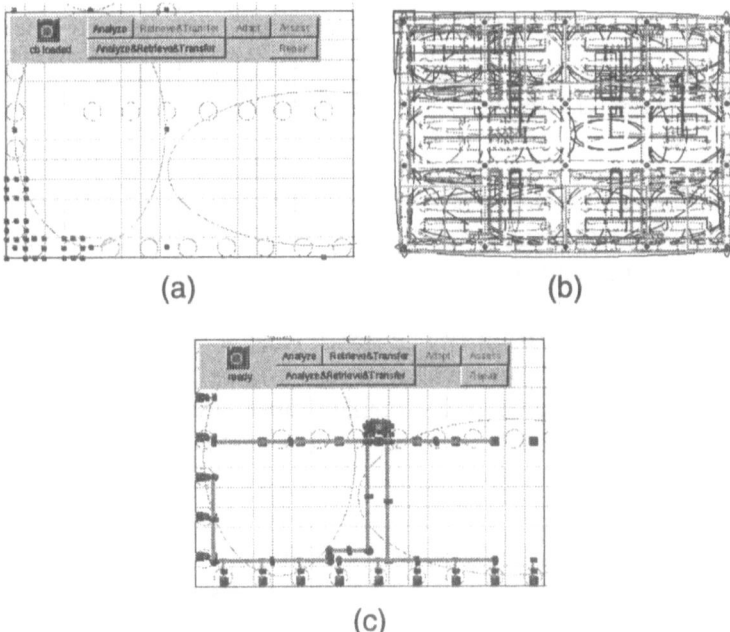

Figure 7. Topo compares the topology of a given piece of plan (a) and a case (b). The selected objects in (a) are zones to be supplied with return-air. The case (b) consists of 3500 objects representing several subsystems of one floor. The result of the transfer of connection lines related to an adequate topology of return-air zones of the case is shown in (c).

The formalism for representing topological relations is the only knowledge used by Topo. The topology is represented by a graph of objects and binary relations of various types. The relations are stored as nodes connected by undirected edges defined by their common objects (Coulon, 1995). The type of a relation is given by the type of the involved objects and their 3-dimensional relation. The type of the instance of a relation in figure 8 is "supply-air-trunk-line$|_x]_y]_z$supply-air-branch-line". It shows a supply-air trunk line touching a supply-air branch line in x-dimension and including it in y- and z-dimension.

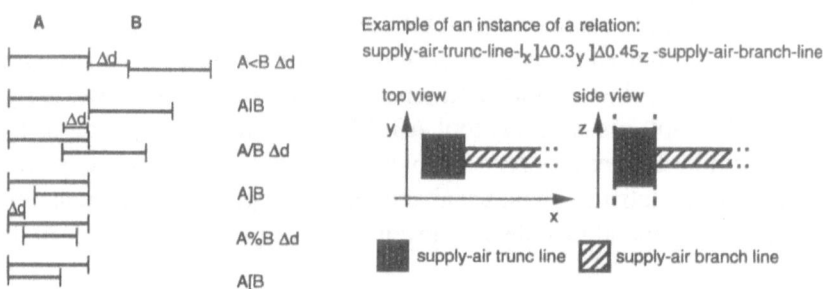

Figure 8. Two objects can be related in 12 different ways for each spatial dimension, the six shown above and their opposites. The symbolical representation is borrowed from the field of pattern recognition (Lee and Hsu, 1992). In order to reconstruct the position of objects some of them are extended by the distance parameter Δd.

The comparison of structure, i.e. topological graphs, is NP-complete (Skiena, 1990; Luks, 1980). Because of this Topo uses an ordinary graph matching algorithm (Barrow and Burstall, 1976) extended by an "intelligent" runtime control (Coulon, 1995). This runtime control provides the user with information about the estimated runtime and actual search progress. Once a common topological substructure of the given piece of plan and the case is found, Topo can transfer every relation of the case connected to the common substructure. In order to select the relations to be transferred Topo uses the process model of section 3 and/or asks the user.

4.2.2. *AAAO – design objects know how to behave*
There is another tool which can adapt cases to the needs of the current situation: AAAO. In our prototype it is specialized in the placement of columns for MIDI buildings[4]. Starting with the layout of rooms or zones of use, AAAO places the columns considering both static demands and architectural and aesthetical rules appropriate for this layout. AAAO can solve a problem by adapting a case as well as by generating a solution starting from any distribution of columns, e.g. a standard distribution.

---

[4]MIDI is a steel framework construction set developed by Fritz Haller (1974) especially for large buildings with high demands on technical services, like schools, offices or laboratories.

Design objects are modeled as active autonomous objects (hence the name AAAO – adaptation by active autonomous objects). They can perceive their local environment, have knowledge about correctness in this environment and have a set of actions to improve it, e.g. to move or to create and destroy objects. The active autonomous objects work in parallel, there is no overall strategy to create the solution. For details of approach and implementation of AAAO see Adami (1995) and Morgenstern (1993).

AAAO is specialized to place columns of a MIDI construction on a floor, where rooms or zones of use are given. The simple example given in figure 9 may illustrate the model. The static and aesthetic demands are modeled as constraints that a position of a column in relation to other columns and zones of use has to satisfy. In the same way many other problems in building design could be modeled where the placement of design objects has to satisfy constraints in a restricted environment of these building elements. Examples are the routes of pipes and other ductwork via the service spaces or the placement of furniture and other equipment in a room.

4.2.3. *ANOPLA: exact and correct placement of pipes*

The starting point for Anopla is a layout with sketched routes (ellipses) of different technical subsystems in the service plenums of a building. Such subsystems are typically supply air, return air, cold water, and rain water systems etc. In a first step Anopla generates the corresponding pipes from the sketches; the pipes are placed on grid lines. In a second step the pipes are moved to valid positions, that is the horizontal co-ordination of the various subsystems takes place. The spatial units for this kind of co-ordination are service plenums of floors or parts of floors. In the following we concentrate on the second step (for more details cf. Gräther (1995)). An example of a problem and a possible solution is shown in figure 10.

Anopla uses knowledge provided by the installation methodology ARMILLA – together with the component based building system MIDI (Haller, 1974) – and heuristics acquired from our expert. The ARMILLA system structures the service plenum vertically into layers and horizontally into lines for branches and lines for pipes. All pipes in layer O1 run in one direction, pipes in layer U1 run in the other direction. Consequently there are no conflicts between pipes in one direction and pipes in the other orthogonal direction.

Pipes can only be "moved" within the adjacent grids. While moving, each pipe segment maintains its connections to the adjoining segments. The constraints on valid positions of pipes can easily be visualized as templates. The templates are derived from ARMILLA (see figure 11).

Thus, the spatial arrangement problem can be formulated as a constraint satisfaction problem. Anopla uses a distributed-agent-based strategy including a decentralized control structure. Our decentralized control structure is comparable to distributed hill-climbing with no backtracking (cf. (Luo *et al.*, 1994)). The imple-

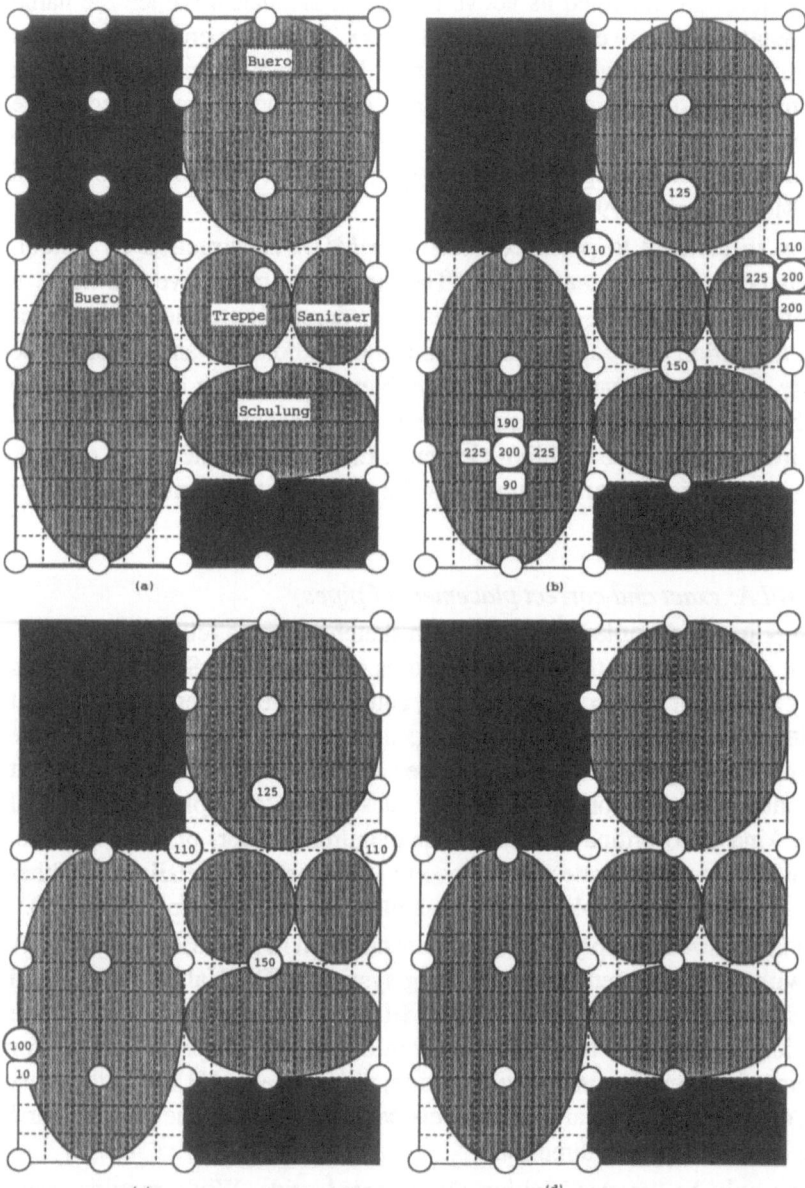

Figure 9. The picture shows four stages of the **AAAO**-model's performance. Picture (a) shows the given floor plan (the actual shape of rooms is the bounding box of the grey ellipses; ellipses can be better distinguished than overlapping or adjacent rectangles) together with an ill-fitting distribution of columns that was taken from a similar case. Picture (b) shows the first evaluation of the positions of columns, the surplus columns outside the building (dark areas) have already been removed. The indicated values are a measure of constraint violation. Picture (c) shows the situation and evaluation after the first action of the columns. While one column can move to a better place, four neighbouring columns cannot improve their position by moving but have to create a new column. Picture (d) shows the solution, which is acceptable.

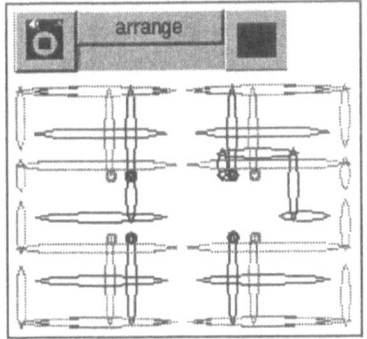 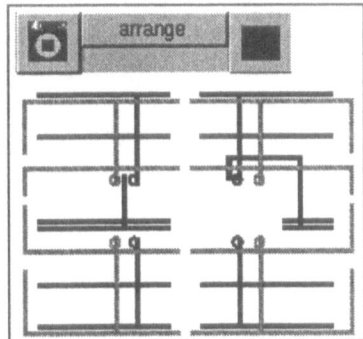

Figure 10. **Anopla** at the virtual building site, placed in the service plenum of a floor. **Left:** Sketched routes for supply air and return air define the problem; some pipes are overlapping. **Right:** Correctly placed pipes. The arrangement process is continually visualized on the screen.

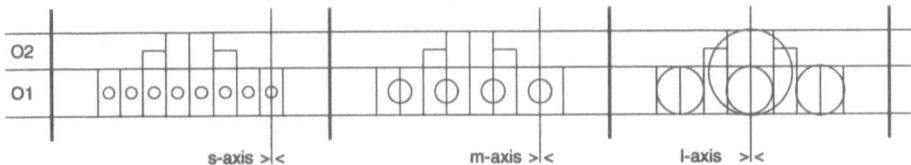

Figure 11. Three generic templates for pipes with small, medium and large diameters in the service plenum. Only templates for the layers O1 and O2 are shown. A horizontal flip generates the corresponding templates for U1 and U2.

mentation is completely object-oriented. Pipes, templates and the service plenum are the main objects. Important methods for pipes are for example: move, move-and-push-away, look-for-position, look-for-better-position, and resize.

5. Putting It All Together

In the first section we argued that a multidimensional frame of reference enables the designers to coordinate different tools operating concurrently on the design. This requires technical coordination of processes and user interfaces. Figure 12 shows the communication network between processes running on different computers and operating systems.

The center of the whole support system is the A5Broker[5] developed at the University of Karlsruhe. It is a database system organizing the building model. All other tools can access the building model by TCP/IP.

---

[5]The prefix "A5" refers to the building model "ARMILLA5"(Hovestadt, 1993). A1 - A3 refer to earlier unsuccesful attempts to build expert systems in this domain.

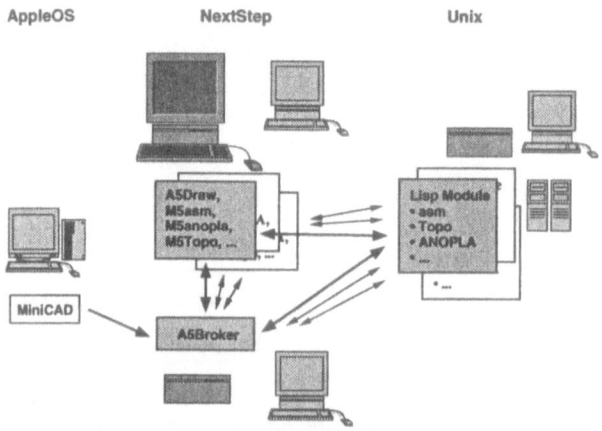

Figure 12. System configuration.

The CAD-tool is called A5Draw, a research prototype which is continuously extended and improved. In order to get access to a large amount of data, the broker was connected to the CAD-tool "MiniCAD", which is the leading CAD tool on the German market for the Apple operating system.

Most of the support tools are implemented in Lisp running on Unix. They communicate with A5Broker and A5Draw by TCP/IP. This distributed approach allows us to apply the same tool to different pieces of design using several processes on different computers.

The interfaces (M5<name of tool>) of the support tools are subprocesses of A5Draw and are positioned in the CAD-plan (cf. figure 2). On the upper left corner of figure 2 the interface of the navigation tool can be seen. It enables the designer to communicate with the database tool. He can specify the types of objects and the area he wants to work on. In order to keep the design consistent, the concerned objects are blocked for all other users.

The black ellipses represent rooms. The tool **AAAO** has already placed the columns represented by tiny squares with respect to the positions of the rooms. On the upper right-hand side **Topo** has matched the topology of the supply-air outlets (grey spots) to part of a case and transferred the topology of a larger supply-air connection line from the case to the layout. On the left-hand side **Anopla** has inserted the connection lines of supply-air and return air (grey and light grey lines) with respect to the position of the connection zones (thin ellipses) and coordinated them.

As a scenario, imagine that we are planning an educational center for the Swiss Railway company ("SBB"). This building should have two floors, a large interior room, and around it offices, class-rooms and laboratories. So far the ground-plan, the MIDI construction and the layout of rooms and service spaces have been designed. This was done according to PM5 beginning with large scales in the pre-planning stages, elaborating route planning and position planning, and finishing with the element planning stage (cf. section 3). The MIDI construction is also completed; the **AAAO** tool has been used to place the columns statically correct. For ARMILLA installations only the pre-planning tasks are finished. All design objects are stored in project "SBB" in the A5Broker. The next tasks are the route planning of subsystems like supply air, return air, cold water, hot water, or sewage.

Assume that two civil engineers at different places will elaborate these tasks. One is a specialist in air-conditioning, the other one is specialized in installations for water subsystems. Both have started the CAD tool A5Draw (probably at different places of the building) and are connected to the same A5Broker; furthermore the FABEL toolbox is on screen and the tools are ready for use ("running" on other computers). The civil engineers open the project "SBB", navigate to the desired position and filter the design objects they want.

The climate engineer starts placing zones for inlets and outlets in the service plenum of the first floor and plans the vertical pipes in the shaft. Then she needs inspiration how to access the outlets. For this reason a retrieval tool is copied from the toolbox and pasted into the actual CAD plan. It is moved to the desired position and resized. Each tool provides a simple remote interface to access its main functionality (e.g. figure 10). She plugs in the tool (connects to the net address of the computer where the tool software is running). She selects a few outlets and one of the vertical pipes to query for similar cases. She browses (clicks) through the retrieved cases and inserts the pipes of one case; afterwards a few adaptations are made by hand. The other engineer uses **Topo** to get support for designing the access pipes for cold water, hot water and sewage. Both save the project "SBB"; the PM5 task route planning for the different installations is finished. Now **Anopla** is activated for the position planning of the installations. When a complex and time-consuming task is being solved by a tool, the designers can meanwhile work on another part of the building with the same or other tools.

6. Conclusions

The proposed metaphor of a virtual building site helps to organize and visualize the very complex process of building design. Users find the whole building model organized and accessible as a kind of hyperplan, where any part – spatially or semantically filtered ad libidum – can be visualized as a plan. When FABEL tools are applied to the building model they are placed in any of these (sub-)plans like tools at a building site. When the users want to inspect the state of the design and see what is currently going on, they may not only inspect data (that is the building model) but will also find the tools, where they were placed ready to be used or where they are currently operating.

Because of the common data space and the tools the metaphor reminds of blackboard architectures (see for instance (Jagannathan *et al.*, 1989)), but in FABEL the control of the whole design process is entirely up to the users. One of the tools (AAAO) could be considered a blackboard architecture in itself where autonomous agents communicate, other tools have a different architecture.

Compared with other CBR systems for architects and engineers, FABEL is unique in the range of tasks supported and the diversified tools supplied. For instance, the building model which underlies the IDIOM system (see Smith et al. (1995), the successor of CADRE) covers only a small part of the A5 building model. Systems like ARCHI II (see Domeshek and Kolodner (1992)) or the system proposed by Oxman (1994) are mainly Hypermedia browsers to supply cases. The indexing by issues, concepts, forms suggested by Oxman (1994) has to be provided by human interpretation of the precedents (cases), while the FABEL system mainly relies on indexes that can be automatically derived from the building model.

The FABEL prototype 3 as described above is operational since December 1995. During the remaining half year of the project, the tools will be further tested and evaluated. Also the architects, that is our project partners at Karlsruhe University, plan to use the system for some realistic building design projects.

References

Adami, P.: 1995, Adaptation by active autonomous objects (AAAO), *in* K. Börner (ed.), *Modules for Design Support*, GMD, Sankt Augustin, pp. 46–50.

Barrow, H. G. and Burstall, R. M.: 1976, Subgraph isomorphism relational structures and maximal cliques, *Information Processing Letters*, **4**, 83–84.

Bartsch-Spörl, B., Bakhtari, S. and Oertel, W.: 1995, Assessment supported by a domain ontology (DOM), *in* K. Börner (ed.), *Modules for Design Support*, GMD, Sankt Augustin, pp. 23–30.

Börner, K. and Faßauer, R.: 1995, Analogical layout design (Syn*), *in* K. Börner (ed.), *Modules for Design Support*, GMD, Sankt Augustin, pp. 59–68.

Coulon, C.-H.: 1995, Automatic indexing, retrieval and reuse of topologies in architectural layouts, *CAAD Futures '95, Proceedings of the Fifth International Conference on Computer-Aided Architectural Design Futures*, Singapore.

Domeshek, E. A. and Kolodner J. L.: 1992, A case-based design aid for architecture, *in* J. S. Gero (ed.), *Artificial Intelligence in Design '92*, Kluwer Academic Publishers, Dordrecht, pp. 497–516.

Gräther, W.: 1994, Computing distances between attribute-value representations in an associative memory, *in* A. Voß (ed.), *Similarity concepts and retrieval methods*, GMD, Sankt Augustin, pp. 12–25.

Gräther, W.: 1995, Exact and correct placement of pipes (ANOPLA), *in* K. Börner (ed.), *Modules for Design Support*, GMD, Sankt Augustin, pp. 69–77.

Haller, F.: 1974, *MIDI - ein offenes system für mehrgeschossige bauten mit integrierter medieninstallation. USM bausysteme haller*, Münsingen.

Haller, F.: 1985, ARMILLA – ein installationsmodell, *IFIB*.

Hovestadt, L. and Schmidt-Belz, B.: 1995, PM5 – a model of building design, *in* B. Schmidt-Belz (ed.), *Scenario of an Intelligent Design Support for Architects by FABEL-IDEA 3*. GMD, Sankt Augustin.

Hovestadt, L.: 1993, Armilla4 – An integrated building model based on visualisation, *Advanced Technologies – architecture – planning – civil engineering, Fourth EuropIA International Con-*

ference on the application of Artificial Intelligence, Robotics and Image Processing to Architecture, Building Engineering, Urban Design and Urban Planning, Delft, The Netherlands, pp. 243–250.

Jagannathan, V., Dodhiawala, R. and Baum, L. S. E.: 1989, *Blackboard Architectures and Applications*, Academic Press, Inc., San Diego.

Janetzko, D. and Jäschke, O.: 1995, Assessing realizations of design tasks (CheckL & CheckUp). in K. Börner (ed.), *Modules for Design Support*, GMD, Sankt Augustin, pp. 15–22.

Janetzko, D. and Jäschke, O.: 1994, Die Verwendung von Operatoren beim Routine-Design, Fabel-Report 20, GMD, Sankt Augustin.

Lee, S.-Y. and Hsu, F.-J.: 1992, Spatial reasoning and similarity retrieval of images using 2D C-string knowledge representation, *Pattern Recognition*, **25**, 305–318.

Linowski, B.: 1994, Computing distances between attribute-value representations in a flat memory, in A. Voß (ed.), *Similarity Concepts and Retrieval Methods*, GMD, Sankt Augustin, pp. 26–35.

Luks, E.: 1980, Isomorphism of bounded valence can be tested in polynomial time, *Proceedings of the 21st Annual Symposium on Foundations of Computing*, IEEE, pp. 42–49.

Luo, Q., Hendry, P. G. and Buchanan, J. T.: 1994, Strategies for distributed constraint satisfaction problems, in M. Klein (ed.), *Distributed Artificial Intelligence: Papers from the Thirteenth International Workshop*, AAAI Press, Menlo Park, CA, pp. 186–200.

Morgenstern, K.: 1993, Anpassung im Bauentwurf mittels aktiver autonomer Objekte. Master's Thesis, Universität Bonn.

Oxman, R.: 1994, Precedents in design: A computational model for the organization of precedent knowledge, *Design Studies*, **15**(3), 141–157.

Schaaf, J. W. and Voß, A.: 1995, Retrieval of similar layouts in FABEL using AspecT. *CAAD Futures '95, Proceedings of the Fifth International Conference on Computer-Aided Architectural Design Futures*, Singapore. School of Architecture, National University of Singapore.

Schmidt-Belz, B.: 1995, Scenario of an intelligent design support for architects by FABEL-IDEA 3. Fabel-Report, GMD, Sankt Augustin.

Skiena, S.: 1990, *Implementing discrete mathematics*, Addision-Wesley Publishing Co.

Slade, S.: 1991, Case-based reasoning: a research paradigm, *AAAI Magazine*, **12**(1).

Smith, I., Lottaz, C. and Faltings, B.: 1995, Spatial composition using cases: IDIOM, in M. Veloso and A. Aamodt (ed.), *Proceedings of the First International Conference on Case-Based Reasoning, ICCBR-95*, Springer, Berlin.

Voß, A., Coulon, C.-H., Gräther, W., Linowski, B., Schaaf, J. W., Bartsch-Spörl, B., Börner, K., Tammer, E.-C., Dürschke, H. and Knauff M.: 1994, Retrieval of similar layouts – about a very hybrid approach in FABEL, in J. Gero and F. Sudweeks (ed.) *Artificial Intelligence in Design '94*, Kluwer Academic Publishers, Dordrecht, pp. 625–640.

J. S. Gero and F. Sudweeks (eds), Artificial Intelligence in Design '96, 485-504.
© 1996 *Kluwer Academic Publishers.*

A MOBILE AGENT-ORIENTED APPROACH TO A DISTRIBUTED DESIGN SUPPORT SYSTEM

HARUYUKI FUJII, SHOICHI NAKAI, HIROSHI KATUKURA AND
KEIICHI HIROSE
Izumi Research Institute, Shimizu Corporation

Abstract. The quality of an architectural design is evaluated from diverse aspects. Application programs that evaluate the quality from a specific aspect have been developed independent of each other. It is one approach to multi-disciplinary design support system to provide a system that incorporates those existing programs without great changes. This paper describes the conception of a *mobile agent-oriented community* (MAOC) to build such a design support system and *agent interchange format* (AIF) for agent exchange in MAOC. In MAOC, agents, represented by AIF, carrying a design problem interact with evaluation programs and problem solvers to solve the problem. This paper also describes hypothetical structure of representation of designs. These proposals are conceptual.

1. Introduction

Researches in architecture have been providing the criteria and methods to predict the quality of a design from diverse aspects such as egress, energy efficiency, structure, etc. Many stand-alone programs that calculate the quality of a building or simulate the performance have been used in design. However, the user has to mediate those programs by hand. A reason for this situation is that the representation of a design in each program is not shared by the others. We haven't made any ontological commitment concerning how to represent a design and design processes. It is one approach to computer-aided multidisciplinary design that those programs are incorporated into a system. The incorporation should be done without affecting each program's external and internal representation of a design.

The authors propose a *translation approach* to the incorporation of the existing programs (and newly implemented programs), and the conception of *a mobile agent-oriented community* (MAOC) and the conception of *an agent interchange format* (AIF) as methods of the translation approach. MAOC is a distributed design support system beyond the platforms and systems. MAOC incorporates existing application programs. AIF is a representation format to exchange *mobile agents* and information between modules in distributed systems like MAOC through a computer network. The long term objective is to provide methods and tools to develop an agent-oriented distributed design support system without affecting the external and internal structure of existing programs. MAOC and AIF are parts of on-going project that explores the development of an environment for multi-disciplinary collaborative architectural design (Fujii, 1989; Nakai et al., 1992; Hirose et al., 1994). MAOC and AIF are similar to ACTORS in the sense that both provide a model of concurrent computation in distributed systems, but differ from ACTORS (Agha, 1986) at the point that an actor in ACTOR passes a task to a new actor in each step in the computation while a mobile agent is alive until the whole problem solving process is done.

2. Design as Problem Solving and its Computer-Aided Support

The authors postulate that a portion of design processes is a collection of problem solving processes. Portable languages to exchange all information conveyed in design and methods for translation among those languages and other design languages, such as drawings, are crucial to the success of practical computer-aided design support systems on this postulate.

Simon et al. (1962) characterized problem solving and classified problem solving processes as follows: Given a set, P, of elements, *problem solving* is to find a member of a subset, S, of P having specified properties. Processes for problem solving are classified into two processes, i.e., *solution generating processes* for finding possible solutions (generating members of P that may belong to S) and *verifying processes* for determining whether a solution proposal is in fact a solution (verifying that an element of P that has been generated does belong to S). Akin (1986) found that architectural design shares some properties with problem solving. He called the design counterpart of solution generating processes *transformations*. Each state in a design problem is transformed into other states by operations. We consider a solution generating and verifying strategy as the following schema: Let Dc be a current state of a design, and Ea,c be the language A expression of Dc that represents the quality of the design. Suppose that a set of solutions ESa expressed in language A is given, and Ea,i is a member of ESa. A verifying process checks if Ea,c is equivalent to Ea,i. If the result is "yes," Dc is a solution from an aspect of the quality that language A expresses. If the result is "no," another proposed solution Dp is generated to decrease the difference between Ea,c and Ea,i, i.e., $Ea,i \backslash Ea,c$. If we know or can infer relations between $Ea,i \backslash Ea,c$ and $Dp \backslash Dc$, operations that transform Dc into Dp are executed. If not, operations are executed arbitrary or based on another inference. When Dp is generated, a verifying process is executed again.

Verifying processes in design are not straightforward. We hardly determine whether elements in a proposed design belong to a solution by merely checking whether each element is in a solution if we don't know the elements in the solution. Instead, we evaluate a proposed design by virtue of the properties that the design ought to have if criteria to evaluate the properties are known. A design is translated into languages that describe the quality of the design to be verified whether all elements in the translated description of the design belong to a solution described in the languages.

Computer-aided solution generating and verifying systems for design problems are made by mediating traditional quality evaluation programs and problem solvers with at least one common language that represents the problem from diverse aspects. Many application programs that evaluate the quality of a design or simulate the performance have been implemented independently from one of specific aspects. DOE-2 predicts the performance from an aspect of comfort and energy efficiency of built environment while NASTRAN analyzes the performance in the structural design context. Some methods that handle more than one aspect in a design have been introduced, too. SHARED, an information model for cooperative product design, represents hierarchical functionalities and abstracted shape of a product to share the information from different aspects and to maintain the coherence among the information (Wong et al., 1992). Gauchel et al. (1992) proposed a multi-disciplinary building modeling system that treats building elements as autonomous objects and relations among the elements as communication among the objects. DOM is intended to endorse the role of modeling a common and shared platform of design knowledge as well as to address the crucial task of representing design decisions and engineering judgments to evaluate design layouts and to support layout construction from scratch (Bakhtari and Oertel, 1995). Multi-layered logic is applied to a framework for representing architectural design knowledge by Clibbon et al. (1995). Each of the systems requires that all modules are written in the same programming language, but we have not reached any consensus on a common language or a representation format to be used in the programs.

We want to take advantage of existing application programs, even though they are written in different programming languages and have their own i/o convention. Since it is not realistic or desirable to require that those programs are (revised to be) implemented in one programming language or system (Gruber, 1992), to mediate existing programs, we need languages independent of machine, programming language, and environment.

3. Language for Domain Knowledge and Task Knowledge

Languages to represent a design and to design and translations among the languages are the keys to the implementation of distributed design support systems. A *language*, from now on, refers to a pair of a set of well-formed strings (or patterns of symbols) over an alphabet in the formal language sense and a way to determine what well-formed strings mean. What a well-formed string means is determined by a semantics and an *ontology*, where an *ontology* is a definition of classes, relations, functions, and other objects (Gruber, 1992) to which an element of an alphabet in a language refers. We call a pair of an alphabet and an ontology is called a *vocabulary*.

3.1. STRUCTURE OF LANGUAGE TO REPRESENT DOMAIN KNOWLEDGE

The authors hypothesized structure of languages to represent knowledge about a design, *domain knowledge,* based on achievements in design science and philosophy of language. The structure consists of three classes, namely, *surface language* class, *universal language* class, and *model* class (Fujii et al., 1996). The distinction between the *surface language* class and the *universal language* class is inspired by the distinction between *S-structure* and *D-structure* of natural languages hypothesized in linguistics.

The authors have adopted one of *machine translation* approaches. Brazier et al. (1994), MacKeller and Peckham (1994), Clayton et al. (1994), and Chaplin et al. (1994) proposed methods to relate heterogeneous representations of the same design with each other. They employed rules to map an element(s) of a representation to an element(s) of another or demon-like rules to update representations if one of them is updated. The approach proposed by the authors differs from those methods. A representation is regarded as a *language* that has syntactic structure. A representation is translated into an intermediate language(s) that conveys all information concerning representations to be translated and its target, then translated into another representation(s). The translation is done by transforming syntactic structure of a representation into that of another representation. Coyne et al. (1990) addressed problems in knowledge representation for design, i.e., representation of information about (1) a design and its vocabulary, (2) the goal of a design such as intended interpretation and functionalities of a design, (3) the knowledge concerning design, manipulation of a design. In their idea, information about a design should be translated into information about intended interpretation and functionalities of the design and vice versa. Behind their conception, they possess the claim that a language for design ought to be rich enough to allow one to represent all information concerning design. This claim is plausible. The world that we talk about is limited by what we can express in our language and theory (Putnam, 1983).

Figure 1 depicts hypothetical structure of knowledge representation languages for an architectural design. A *universal language* is characterized by the combination of a domain independent formalism, knowledge representation format, and a domain specific *ontology* of terms formulated in the formalism. We refer to the definition of the formalism *syntax*[1], and that of the *ontology semantics*[2]. An *ontology* defines a set of the concepts composed of entities, relations among the entities, and relations among the relations and the entities, recursively. A *universal language* must contain the information that is the union of information represented by each *surface language*. The *surface language* class consists of intuitive representations such as drawings, three dimensional graphic descriptions, circuit diagrams, natural language texts, etc. A portion of the information represented by a *universal language*, which is relevant to the focus of a *surface language*, is translated into the *surface language*. For instance, a drawing is a geometric representation of the spatial organization of a building while a circulation diagram is a topological representation of the adjacencies among spaces in a building. Each *surface language* also has syntax and *vocabulary*. The syntax of a *surface language* is defined corresponding to the conventional use of symbols used in the language. Each symbol in the *vocabulary* of a *surface language* must have the counterpart in an *ontology*. The meanings of *surface language* expressions are determined by defining the semantics of the *surface language* with correspondence to the *universal language* or by defining a translator between both languages.

The structure of knowledge representation concerning design has another class that the authors call *model* class. The *model* class contains, in a logical sense, the model(s) of *universal language* expressions of a design. An *architectural world model*, which is expressed by a *universal language*, is regarded as the reference of *surface language* expressions of the same design. The *model* contains a set of instances of the concepts in an *ontology* concerning the design. Whether a *surface language* expression is true or false is determined with respect to an *architectural world model*. Each *model*

---

[1] Glymour (1992) explains a syntax and a semantics of a formal language as follows: a syntax is a set of explicit rules that make it possible to determine in a mechanical way whether or not a string of symbols is well-formed. A semantics specifies precisely the conditions under which a set of values for variables in the language satisfies a formula in the language.

[2] The semantics explained here is different from a term *meaning* often used in the architectural research community. A meaning of the *meaning* is considered as the correspondence among surface languages. We define m-entail in Definition 3.2 to avoid this confusion. For example, we say, "the 'square' *m--entails* a 'table'" instead of "the 'square' *means* a 'table'" if both 'square' and 'table' are terms of *surface languages*.

Definition 3.1
A set of sentences A *entails* a sentence B if and only if B is true in every relation structure for which all sentences in A are true (Glymour 1992).

Definition 3.2
Let A and B be surface languages different from each other, and let C be a universal language which the semantics of A and that of B are specified with respect to. A collection of formulas Fa expressed in A *m-entails* a collection of formulas Fb expressed in B if and only if the representation of Fa in C entails every sentence in the representation of Fb in C.

in the same architectural world renders every expression in all *surface languages* true.

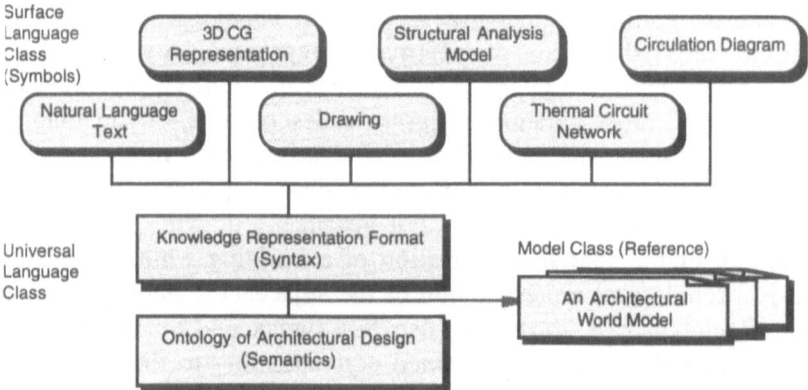

Figure 1. Conceptual structure of knowledge representation in architectural design.

3.2. LANGUAGE TO REPRESENT TASK KNOWLEDGE

Process language class composed of languages to represent *task knowledge,* knowledge about to design, is required since *domain languages,* languages for *domain knowledge,* are not informative enough to represent and execute operations to make changes in a design. Syntax of the *process language* is defined independent of program languages and systems like syntax of *domain languages. Process languages* are differentiated from *domain languages* by virtue of the semantics. While semantics of a *domain language* is reference semantics, that of a *process language* is denotational semantics[3]. Semantics of a *process language* is partially dependent on program languages and systems in the sense that some terms in the language must correspond with operations or actions in the world with which the semantics is defined. For instance, if the semantics of a language is defined with respect to a world in a computer network, terms in the language must have computer operation counterparts. A lexicon is the union of a collection of the reserved atomic symbols that directly correspond to the operations and a collection of symbols that doesn't. An *ontology* that defines the relations among terms includes the reserved atomic symbols as its primitive elements.

3.3. AN AGENT INTERCHANGE FORMAT

Sharing of *domain knowledge* and *task knowledge* is a key to a computer-aided distributed design support system where design support systems and traditional quality assessment applications work together.

---

[3] We are following a classification of semantics by Stefik (1995).

Existing AI methods are not enough to represent and share task knowledge while some strategies are helpful for an ontological commitment concerning *domain knowledge*. The pair of Interlingua and FrameKit is proposed upon a machine translation context (Mitamura and Nyberg III, 1992), while the combination of KIF (Geneserth and Fikes, 1992) and Ontolingua is proposed from an aspect of the portability of knowledge (Gruber, 1992). They provide formal languages to represent the common knowledge shared in a system. To define a *universal language*, we can take advantage of the abstract relation between Interlingua or KIF and FrameKit or Ontolingua. Terms defined by the combinations can tell and ask about information in the shared knowledge. However, they might not be able to be interpreted as operations upon the knowledge. As far as *task knowledge* is concerned, MACE (Gasser et al., 1987) was implemented as a testbed for distributed reasoning systems where modules called agents collaborate with each other to solve problems that can not be solved by a single agent. JAVA and Telescript provide the practical means of developing distributed systems, but agents aren't portable beyond the environment. Safety and security for each system participating in the exchange of agents are of great concern. It is not desirable that agents are directly accessible into a module where the agents are stranger. We can't be sure that there is no chance for transported agents to destroy a system where they visit.

The authors propose a concept of *agent interchange format* (AIF) to send an agent carrying *domain knowledge* and *task knowledge* back and forth between modules in MAOC. AIF is a domain, language, and system independent format to represent the information characterizing certain agents in MAOC. A module that receives an AIF formula makes an instance of a certain class from the information in the formula. The instance behaves as an agent in the target module.

Figure 2 shows a diagram describing how AIF works. Suppose that there are mainly two modules implemented in different languages or different machines. One of the modules that we call lang-b world is written in language lang-b, while the other system lang-c world is written in lang-c. A mobile agent Mb in lang-b world, which is interacting with agent HCI-01 (1), is required to interact with agent TrA-01 in lang-c world to solve a problem. Mb is translated into MA that is an AIF formula (2), then MA is sent out to lang-c world, outside lang-b world (3). When MA arrives, lang-c world changes MA into MA' by changing the information about where the AIF formula currently is (4), then agent Mc in lang-c world is generated from the information carried by MA' (5). After interacting with agent TrA-01 (6), Mc is translated into an AIF formula MA" (7) that is to be sent to another module from lang-c world (8). Note that an AIF formula is not an agent but a packet of information from which an agent is instantiated in a module.

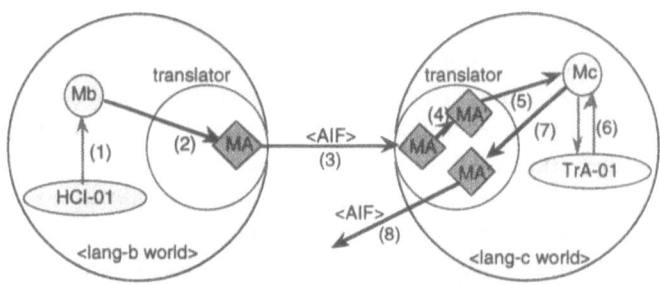

Figure 2. Agent transfer between different agent communities.

A demon method triggered when an AIF formula arrives at each module that is referred to as lang-b world or lang-c world in Figure 2. Figure 3 shows the demon method. We are following a convention of object-oriented programming. An AIF formula is translated into a local expression with feature-value pairs to be instantiated as an agent of class *traveler* explained later. The agent is pushed into a queue, and a trigger for *ifArrivedDemon* is sent to let the agent interact with other agents in the module.

$$AIF_{local} := translate(AIF)$$
$$currentTraveler := traveler \leftarrow makeInstance(AIF_{local})$$
$$myself_{static-agent} :: iQueue \leftarrow push(currentTraveler)$$
$$myself_{static-agent} \leftarrow ifArrivedDemon$$

Figure 3. A demon method triggered if an AIF formula arrives.

4. Mobile Agent-Oriented Community

There are two strategies of agent-oriented approach: (1) Some *software agents*, who play their particular role in a larger scale problem solving, organize a collaborative team each time when a problem is given. This strategy requires that agents organize and control the team in the whole process. The communication between the agents is mediated by a message written in a query language. Khedro (WWW) have accepted this strategy and introduced *Federation* architecture. (2) An agent, which the authors call a *mobile agent,* handles a problem. A *mobile agent* is instantiated when a new problem is given. A *mobile agent* assigned a problem creates a plan to solve the problem and executes each step of the plan by interacting with agents, *static agents,* that solves a part of the problem. The communication between *static agents* is mediated by *mobile agents*. There is no *static agent* that controls a problem solving team. Instead, a *mobile agent* takes responsibility for its own problem solving. The authors have adopted the latter strategy and introduce MAOC. The following sections give an explanation of MAOC with its components and the functionalities.

4.1. A SCENARIO

MAOC is expected to support design problem solving that consists of the situation described in the following scenario: You are an architect who is in charge of the design of a building. You might want to collaborate with a civil engineer and an HVAC (Heating, Ventilating, and Air Conditioning) engineer or problem solving systems representing the engineers. If you don't know whom to collaborate with, someone who knows introduce engineers to you. Your project team uses an intelligent CAAD system with which each member of the team manipulates the data of the building from its own perspective. An HVAC engineer retrieves the information about the thermal comfort of the building such as the heat capacity and thermal resistance of the building elements in use as well as the space organization in order to estimate the heating and cooling load. He translates the appropriate information into the thermal circuit network and constructs the input data for an application program that estimates the heating and cooling load. He gives you a comment whether the balance between thermal comfort and energy efficiency is kept and how the design ought to be from an aspect of his expertise. A civil engineer executes a similar schema of tasks from an aspect of durability. The interaction among you, the civil engineer, and the HVAC engineer helps the team to find a solution for the building design.

If we see the interaction from a point of view of a participant, each participant does is to retrieve a particular information from the CAAD system, translate it into the input form for an application program with a particular aspect, let the program evaluate the input and get the result, translate some portion of the result into the form that the CAAD system accepts, and feedback the translated result to the other participants.

4.2. OUTLINE OF A MOBILE AGENT-ORIENTED COMMUNITY

We define *a mobile agent-oriented distributed design support system* to be one that is composed of a set of autonomous modules that we call *static agents*, *mobile agents*, and a set of paths, a computer network, for *mobile agents* to be sent back and forth among *static agents*. We call such a system *a mobile agent-oriented community* (MAOC) for short. A *mobile agent* carries a design problem and a plan to solve it. A *mobile agent* is an object in a certain class when it is in a module, while it is a packet of information when it is outside a module. A *static agent* is a module that *mobile agents* interact with through a speech act to solve a problem carried by the *mobile agent*. A *static agent* responds to a *mobile agent*'s speech act with respect to the type of the speech act and helps a *mobile agent* to solve a design problem. Three types of speech acts should be taken into the account, namely, declarative, interrogative, and imperative (Levinson, 1983). In an analogical sense, a *mobile agent* is a traveler seeking a solution of a problem while a *static agent* is a place where a traveler can visit to tell, ask, or request someone something related to the problem solving. With respect to the communication structure, MAOC falls into a fusion of blackboard systems and message-passing systems. Each problem is shared by *mobile agents* and *static agents* through a blackboard, while a *static agent* sends *mobile agents* enclosing messages to other static agents whose names are explicitly known by *static agents* called *travel agents*, but either *mobile agents* or *static agents* (except agents called travel agents) don't have knowledge about other agents.

Each mobile agent takes responsibility for its own problem solving. It is not assumed that there is a single static agent that maintains control of the whole community. In the initial stage of problem solving, each *traveler* begins with a design problem and a vague and incomplete plan to solve the problem. The plan is then decomposed into a sequence of well-organized sub-plans that are relatively concrete and complete through the interaction with *static agents*. Ideally, the entire community finds a complete plan. To provide the functionalities, an architecture for MAOC must have: (a) a convention for ensuring that the activities of the *static agents* in the community are organized so that the *static agents* interact with *mobile agents* to solve problems that are beyond the power of each *static agent*, (b) an infrastructure for communication on which *mobile agents* are sent among *static agents*, (c) a formalism for domain knowledge sharing, (d) a formalism for task knowledge sharing, and (e) a mechanism that maintains the consistency of information in the community.

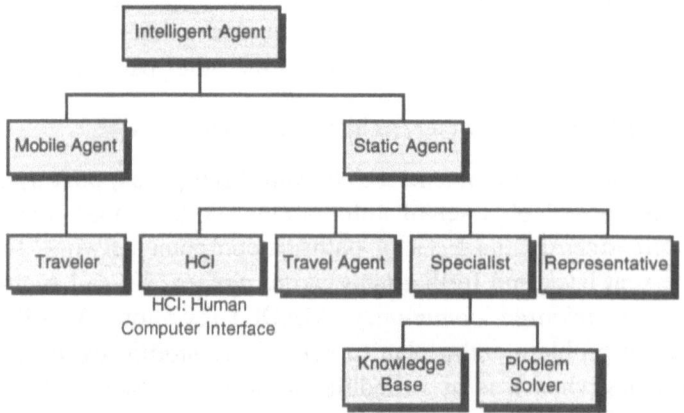

Figure 4. Hierarchy of intelligent-agent classes in the system.

Figure 4 shows the class hierarchy of *intelligent agents* in MAOC. The *intelligent agent* class has features id and status. An identification symbol, a value of id, is unique to each instance of intelligent agent class. The status of an instance is indicated by the value of status that is either busy or idle.

MAOC consists of the instances of *traveler*, HCI (human computer interface), *travel agent, specialist,* and *representative* classes and traditional application programs. Figure 5 shows a conceptual architecture of MAOC. The solid lines indicate the passes on which instances of *mobile agent* class can traverse from one agent to another.

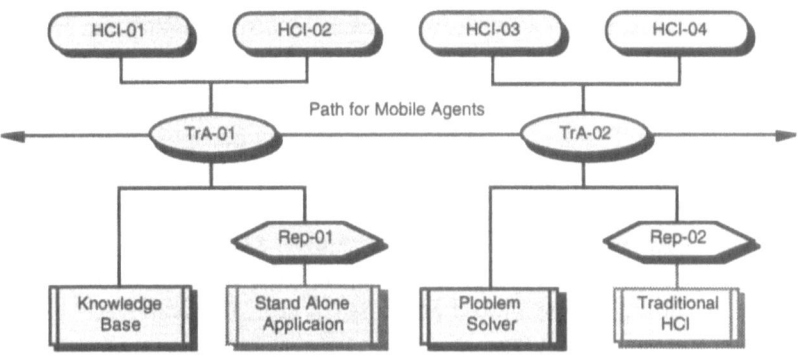

Figure 5. Conceptual model of an agent system.

A *static agent* becomes active if a *mobile agent* visits the agent. The demon method described in Figure 6 is fired when a *static agent* receives either a *mobile agent* or a clock tick. If a *mobile agent* is sent as an AIF expression, a *mobile agent* corresponding to the expression is instantiated.

$$\textbf{if } \neg\lceil iQueue :: contents = \varnothing \rceil$$
$$\textbf{then } currentAgent := iQueue \leftarrow pop$$
$$myself_{Agency} \leftarrow agencyMethod(currentAgent)$$
$$oQueue \leftarrow push(currentAgent)$$

Figure 6. A demon method for <u>Static Agent</u> Class.

4.3. TRAVELER CLASS

Traveler class is a sub-class of *mobile agent* class. A *traveler* is an instance of the *traveler* class. AIF is a portable information packet that conveys the features and values required to make a *traveler*. A *traveler* is characterized by values of the following features:

<id, status, home, ontology, intention, BelSoa, Context, Plan, Message>

The value of home indicates where a *mobile agent* was generated. The value of ontology declares the ontology with which *domain knowledge* and *task knowledge* are defined. An instance's belief about the state of affairs around itself is stored in BelSoa. The history composed of what a *mobile agent* did and how the state of affairs changed is stored in its Context. A *mobile agent* executes the top priority instruction in Plan, renews Plan, and changes BelSoa and Context until intention of the *mobile agent* is satisfied, where intention represents the goal of *mobile agent*'s problem. The content of Message represents a speech act and *domain knowledge* related to the speech act. A *mobile agent* walks through a computer network whose nodes are *static agents*, i.e., *travel agents, representatives, specialists,* and *HCI's.*

4.4. HCI CLASS

An instance of HCI class, an HCI, is a human computer interface, that facilitates the interaction between the user and MAOC. An HCI displays the *surface language* expressions of a design in question and the interaction between the use and MAOC. The user can assert, query, or request something concerning a design problem to an HCI, and the HCI responds to the input. Each input is interpreted as a speech act. A *traveler* encapsulating the speech act is generated. The *traveler* is sent to a designated *static agent*. In addition, an HCI receives a *traveler* carrying a speech act. The HCI, then, retrieves the speech act from the *traveler* translates and displays the information that is expected to be given to the user. For example, if a user inputs a sentence that requests a finite element model of *building1*, an HCI generates a *traveler* described below. We use an AIF formula to give concrete idea about the *traveler*.

```
(AIF  (id traveler1)
      (status idle)
      (home hci1)
      (ontology *design-process*)
      (intention (get-response-to Message))
      (BelSoa ((at hci1)))
      (Context ((at hci1)))
      (Plan  ((traverse traModel hci1) (get-response-to Message)))
      (message-language KQML+)
      (Message (KQML+  (id message-01)
                       (sender hci1)
                       (receiver traModel)
                       (message-type imperative)
                       (ontology *architectural-design*)
                       (content-format KIF)
                       (Content (KIF  (id kif1)
                                      (expression (femModelOf *building1*)))))))
```

The *traveler* has the identification symbol traveler1. The status idle means that the *traveler* is ready to receive a message. An s-expression (home hci1) shows that the *traveler* is generated by hci1. An s-expression (BelSoa ((at hci1))) represents that the *traveler* believes that it is at hci1 now. The Context slot contains the information about actions that the *traveler* executed and the history of the belief related to the actions. The Plan slot contains the sequence of actions that the *traveler* and other *static agents* execute to respond to the speech act given by the user. The *traveler* is instructed, by hci1, to traverse from hci1 to traModel and process the enclosed message at traModel. We assume that traModel is a designated *travel agent* to hci1. Since hci1 doesn't know either how to get a finite element model of *building1* or who can solve the problem, hci1 decides to send the *traveler* to traModel. The Message slot contains the information about the speech act given by the user. The message-type slot shows that the type of the speech act is imperative. The Content slot contains a LISP-like representation, (femModelOf *building1*), of a finite element model of *building1*.

4.5. TRAVEL AGENT CLASS

Each instance of *travel agent* class, a *travel agent*, maintains explicit information about other agents, i.e., id, class, address, functionalities, and can retrieve an in-coming *traveler*'s information, i.e., intention, plan, message, etc. A *travel agent* is composed of an in-coming queue, an interpreter, a processor, a target language generator, and an out-going queue. A *traveler* sent to a *travel agent* is stored in the in-coming queue. When its status is idle, the *travel agent* gets the *traveler* on the top of the in-coming queue and excerpts the speech act and the plan of the *traveler*. A *travel agent* performs as follows: (1) gets the intention and the plan of a *traveler*, (2) decomposes the plan into sub-plans, (3) assigns the sub-plans to various other *static agents*, (4) makes a plan composed of the sub-plans and the *static agents* to which the sub-plans are assigned, and (5) replaces the old plan of the *traveler* with a new plan. Let us take a look at the transition of a *traveler* in AIF. When *traveler* traveler1, which is described in the section 4.4, has arrived at a *travel agent* named traModel. It is identical to the AIF formula generated by hci1 except the fact that BelSoa and Context have changed.

```
(AIF  (id traveler1)
      (status busy)
      (home hci1)
      (ontology *design-process*)
      (intention (get-response-to Message))
      (BelSoa ((at traModel)))
      (Context ((at traModel)(traverse traModel hci1)(at hci1))
      (Plan ((get-response-to Message)))
      (Message ... ))
```

An instance, traveler1, equivalent to the content of the AIF formula is made in the environment where traModel works. Suppose that traModel believes repModel will satisfy the traveler1's intention. TraModel gives traveler1 a new plan contained in the following AIF expression. If repModel is out of the scope of the environment or the machine where traModel works, traveler1 is translated into an AIF formula described below and sent to the

environment or the machine where repModel works. If they are in the same
environment or machine, traveler1 will be sent as an instance to repModel

```
(AIF  (id traveler1)
      (status busy)
      (home hci1)
      (intention (get-response-to Message))
      (BelSoa ((at traModel)))
      (Context (...))
      (Plan ((traverse repModel traModel)(get-response-to Message)))
      (Message ... ))
```

4.6. SPECIALIST CLASS AND REPRESENTATIVE CLASS

Specialist class is a sub-class of the static agent class whose responsibility is
to execute a task within a particular aspect. The specialist class consists of
problem solver class and knowledge base class. Instances of problem solver
class are either programs for numerical analysis or problem solvers such as
programs for finite element modeling, earthquake response analysis
programs, programs analyzing energy efficiency, search engines, planners,
etc. while instances of knowledge base are information retrieval systems or
temporal data storage such as blackboards, etc. The representative class
serves to incorporate stand-alone application programs into MAOC Most of
the stand-alone applications are numerical analysis programs that assess the
quality of a building from a particular aspect. A representative translates a
part of message carried by a traveler into a local convention used as the i/o
data for the stand-alone program for which the representative serves. When a
traveler visits a representative and the representative is not busy, the
representative excerpts a sequence of instructions assigned to itself from the
plan carried by the traveler, generates a set of instructions and data, and
submits the set to a stand-alone program. A traveler waits for a message
from a representative in the out-going queue. If the program returns any
result after a certain time interval, the representative translates the result into
a message and gives the translated message to the traveler. If a representative
is busy when a traveler arrives, it lets the traveler wait for its turn in the in-
coming queue until the representative becomes idle. A specialist executes the
same process with no help from a representative.

The AIF formula of traveler1 arrives at repModel as follows.

```
(AIF  (id traveler1)
      (status busy)
      (home hci1)
      (ontology *design-process*)
      (intention (get-response-to Message))
      (BelSoa ((at repModel)))
      (Context (...))
      (Plan ((get-response-to Message)))
      (Message ... ))
```

The AIF formula is interpreted and translated into an instance of the environment where repModel works. repModel looks at traveler1's intention, plan, and message. Since the intention of traveler1 in this example is to get a finite element model of *building1*, repModel returns *femModel1* as the result. RepModel encloses the result in a message and gives it to traveler1 with a plan for traveler1 to go back to its home, hci1, and report the result. RepModel generates the following AIF formula and sends it to hci1. Hci1 displays the solution, *femModel1*, for this small problem.

```
(AIF  (id traveler1)
      (status busy)
      (home hci1)
      (ontology *design-process*)
      (intention (get-response-to Message))
      (BelSoa ((at repModel)))
      (Context (...)
      (Plan ((report Message)))
      (Message (KQML+  (id message-02)
                       (sender repModel)
                       (reciever hci1)
                       (message-type declarative)
                       (language AIL)
                       (format KIF)
                       (Content (KIF  (id kif2)
                                      (expression *femModel1*))))))
```

5. Technical Issues underlying MAOC

This chapter describes a few reasoning algorithms that contribute to MAOC. The first section explains two algorithms to make AIF rich and easy to handle. The second section explains how an articulated plan is generated from a vague plan by an example.

5.1. REASONING ALGORITHMS

The authors use natural language processing (NLP) techniques to take advantage of expressive power of natural language (NL) expressions that are not cumbersome to manipulate. NL represents multiple levels of abstraction. NL does not have to describe everything explicitly.

NL-like expressions such as definite noun phrases to refer to conceptions in a model and incremental model-based reasoning to complement knowledge from incomplete or partial information are being used. Algorithms for these techniques were originally implemented for an NLP system (Fujii, 1994, 1995). An abductive reasoning algorithm to instantiate a definite noun phrase by a concept is shown in Figure 7. A definite noun phrase is indicated by np that is an object. We take it for granted that we have a relation $isa(x,y)$ that is true if y is a sub-class or instance of x. C_{global} is a set of all concepts in a model and $C_{current}$ is a subset of C_{global} that is currently focused on.

DefiniteNPInstantiation(np)
$\mathsf{Inst} := \left\{ x \middle| x \in C_{current} \ \& \ isa(np, x) \right\}$
if $\mathsf{Inst} = \varnothing$
   then $\mathsf{Inst} := \left\{ x \middle| x \in C_{global} \ \& \ isa(np, x) \right\}$
     if $\mathsf{Inst} = \varnothing$
       then $\mathsf{Inst} := \{y\}$ *s.t.* $y \notin C_{global} \ \& \ y :: class = np :: class$
return Inst

Figure 7. An algorithm for instantiation of a definite noun phrase.

SenseModelsPairs$\left(S_{new}, M_{given}, \mathsf{IR} \right)$
   S_{new}; the set of the interpretation(s) of the new sentence
   M_{given}; the set of the possible model(s) w.r.t. the given context
   IR; the set of the inference rules for (model - based) reasoning
$\mathsf{SR} := \varnothing$
for each $s \in S_{new}$
   do $\mathsf{M} := \varnothing$
     for each $\Gamma \in M_{given}$
       do if $\Gamma \nabla_{\mathsf{IR}} \neg s$
         then $\mathsf{M} := \mathsf{M} \cup \left\{ \Phi \middle| \lceil \Gamma, s \to \Phi \rceil \in \mathsf{IR} \right\}$
     if $\mathsf{M} \neq \varnothing$
       then $\mathsf{SR} := \langle \Gamma, \mathsf{M} \rangle \cup \mathsf{SR}$
Return SR

Figure 8. An algorithm for model-based incremental knowledge acquisition.

Figure 8 shows an algorithm for incremental knowledge acquisition. Knowledge acquisition in design is often considered abductive. A person to whom new information about a design is given revises a model(s) that the person previously has for the design so that the revised model(s) renders the new information true, where $\Gamma \nabla_{\mathsf{IR}} \neg s$ means $\neg s$ is not proved from Γ with a set of inference rules IR, and $\langle \Gamma, \mathsf{M} \rangle$ means an ordered pair of Γ and M.

5.2. PLANS AND ACTIONS

Planning that articulates a vague and incomplete plan of a *traveler* is done by two ways. A generative grammar is one way to articulate a plan when the pattern of the articulation is previously known. The other way is to use means-ends-analysis. We show an example of planning using a generative grammar. A context free grammar that generates plans is shown in Figure 9, where VP is a set of vague and incomplete plans that are considered as non-terminal symbols in the grammar. CA is a set of concrete actions that is considered as terminal symbols. Rule is a set of rewriting rules of the symbols. An element(s) *g* of CA indicates a vague and incomplete plan(s) from which a concrete plan(s) is generated.

TypicalPlan = L(G) s.t. G=<VP, CA, Rule, g>

Figure 9. A schema of a context free grammar generating plans.

Suppose that there are instances of travel agent class, i.e., traGen, traModel, instances of representative class, i.e., repHvac, repModel, and instances of traveler class, i.e., traveler1, traveler2 in MAOC. Application programs, progHvacLoad and progHvacModel, are mediated by repHvac and repModel, respectively, to collaborate with agents in MAOC. TraGen and traModel have rewriting rules to articulate vague and incomplete plans, heatLoad, coolLoad, hvacLoad, and hvacModel, into sequences of concrete actions, pHeatLoad, pCoolLoad, and pModel. The pseudo-codes below show how the instances are defined in an experimental MAOC. We omit traverse between an agent to another for simplicity.

```
instance <- new(travelAgent,traGen)
   ::rule <- push(<traGen,heatLoad> --> <traModel,hvacModel><repHvac,pHeatLoad>)
   ::rule<- push(<traGen,coolLoad> --> <traModel,hvacModel><repHvac,pCoolLoad>)
   ::rule<- push(<traGen,hvacLoad> --> <traGen,heatLoad><repHvac,pCoolLoad>)

instance <- new(travelAgent,traModel)
   ::rule<- push(<traGen,hvacModel> --> <repModel,pModel>)

instance <- new(representative,repHvac)
   ::the_trad_app := progHvacLoad

instance <- new(representative,repModel)
   :the_trad_app := progHvacModel
```

Let traveler1 and traveler2 be instances of traveler class. Each has a vague plan. Instance traveler1 plans to let traGen execute coolLoad and go back to its home. Similarly, traveler2 plans to let traModel execute heatLoad and go back to its origin. These instances are defined as follows.

```
instance <- new(traveler, traveler1)
   ::plan <- push(<traGen,coolLoad><home,eoa>)

instance <- new(traveler, traveler2)
   ::plan <- push(<traGen,hvacLoad><home,eoa>)
```

The following is a log from the experimental MAOC describing how *travelers* and *static agents* act to execute each *travelers*'s plan. As mentioned in section 3.3, demon methods are triggered when a *mobile agent* arrives at a

static agent or a unit time passes. More than one agent can execute their
process almost at once in MAOC.

```
agent <- temporal_trigger
Time:87788; traveler1 lets traGen execute coolLoad
Time:87791; traveler1 comes to traGen
Time:87794; traGen decomposes <traGen,coolLoad>
                        into <traModel,hvacModel><repHvac,pCoolLoad>
Time:87797; traveler2 lets traGen execute hvacLoad
Time:87799; traveler2 comes to traGen
Time:87802; traGen decomposes <traGen,hvacLoad>
                        into <traGen,heatLoad><repHvac,pCoolLoad>

agent <- temporal_trigger
Time:88810; traveler1 lets traModel execute hvacModel
Time:88812; traveler1 comes to traModel
Time:88815; traModel decomposes <traModel,hvacModel> into <repModel,pModel>
Time:88819; traveler2 lets traGen execute heatLoad
Time:88822; traveler2 comes to traGen
Time:88825; traGen decomposes <traGen,heatLoad>
                        into <traModel,hvacModel><repHvac,pHeatLoad>

agent <- temporal_trigger
Time:90452; traveler1 lets repModel execute model
Time:90455; traveler1 comes to repModel
Time:90457; repModel sends model to progHvacModel being done at 90517
Time:90461; traveler2 lets traModel execute hvacModel
Time:90464; traveler2 comes to traModel
Time:90466; traModel decomposes <traModel,hvacModel> into <repModel,pModel>

agent <- temporal_trigger
Time:91976; traveler1 lets repHvac execute pCoolLoad
Time:91979; traveler1 comes to repHvac
Time:91981; repHvac sends pCoolLoad to progHvacLoad being done at 92041
Time:91985; traveler2 lets repModel execute pModel
Time:91988; traveler2 comes to repModel
Time:91990; progHvacModel is idle
Time:91992; repModel sends pModel to progHvacModel being done at 92052

agent <- temporal_trigger
Time:94499; traveler1 comes back home
Time:94501; traveler2 lets repHvac execute pHeatLoad
Time:94504; traveler2 comes to repHvac
Time:94506; progHvacLoad is idle
Time:94508; repHvac sends pHeatLoad to progHvacLoad being done at 94568

agent <- temporal_trigger
Time:96932; traveler2 lets repHvac execute pCoolLoad
Time:96935; traveler2 comes to repHvac
Time:96937; progHvacLoad is idle
Time:96939; repHvac sends pCoolLoad to progHvacLoad being done at 96999

agent <- temporal_trigger
Time:97885; traveler2 comes back home
```

The log traces actions. Traveler1 visits traGen to let it articulate coolLoad, and traGen decomposes it into <traModel,hvacModel> <repHvac,pCoolLoad>. Since hvacModel is vague, traveler1 visits traModel to articulate hvacModel. TraModel decomposes hvacModel into <repModel,pModel>. At this point, traveler1 has a complete plan, <repModel,pModel> <repHvac,pCoolLoad>. Traveler1 visits repModel and repHvac as in the plan, lets each agent execute the action assigned to the agent. RepHvac and repModel let the traditional applications, which each agent represents, execute the actions. Traveler2 acts in a similar way concurrently.

6. Continuing Research

Since MAOC and AIF are an on-going project, only experimental systems have been implemented two of which are shown in this paper. It is hard to give the empirical evaluation fort the project at this point. One thing that is clear is that we don't have to be anxious about the security of each module in MAOC since AIF is sent in a form of s-expression and no program code is sent in MAOC. The conceptions are expected to be explained from a formal aspect to evaluate MAOC and AIF from a theoretical aspect.

7. Conclusion

This paper described the conception of a *mobile agent-oriented community* (MAOC) that incorporates stand-alone programs that predict the building performance and the conception of *agent interchange format* (AIF), a format for *mobile agent* exchange in MAOC. The combination of MAOC and AIF provide a model of concurrent computation in distributed design support system without much anxiety about security. This paper described hypothetical structure of languages to represent knowledge concerning design, too. The structure is the assumption on which diverse aspects in design are treated. MAOC and AIF. They are independent of platforms, operation systems, programming languages, etc. The authors convinced themselves that it is on the right track to use MAOC and AIF to construct a practical design support system incorporating existing stand-alone building performance analysis programs.

References

Agha, G.: 1986, *ACTORS: A Model of Concurrent Computation in Distributed Systems*, The MIT Press, Cambridge.

Akin, Ö.: 1986, *Psychology of Architectural Design*, Pion, London.

Bakhtari, S. and Oertel, W.: 1995, DOM: An active assistance system for architectural and engineering design, *Proceedings CAAD Futures'95*, Vol. 1.

Brazier, F., Van Langen, P., Ruttkay, Zs. and Treur, J.: 1994, On formal specification of design tasks, *in* J. S. Gero and F. Sudweeks (eds), *Artificial Inteligence in Design '94*, Kluwer, Dordrecht, pp. 535-552.

Chaplin, R., Li, M., Oh, V., Sharpe, J. and Yan, X.: 1994, Integrated computer support for interdisciplinary system design, *in* J. S. Gero and F. Sudweeks (eds), *Artificial Inteligence in Design '94*, Kluwer, Dordrecht, pp. 591-608.

Clayton, M., Fruchter, R., Krawinkler, H. and Teicholz, P.: 1994, Interpretation objects for multi-disciplinary design, *in* J. S. Gero and F. Sudweeks (eds), *Artificial Inteligence in Design '94*, Kluwer, Dordrecht, pp. 573-590.

Clibbon, K., Candy, L. and Edmonds, E.: 1995, A logic-based framework for representing architectural design knowledge, *CAAD Futures'95*, Vol. 2.

Coyne, R., Rosenman, M., Radford, A., Balachandran, M. and Gero, J. S.: 1990, *Knowledge-Based Design Systems*, Addison-Wesley.

Fujii, H.: 1989, Dynamic simulation of the interaction between interior environment and individual behavior, *Proceedings 12th Symposium on Computer Technology of Information, Systems and Applocations*, Architectural Institute of Japan, Tokyo.(in Japanese)

Fujii, H.: 1994, Understanding spatial descriptions, *Master Project Report*, Department of Philosophy, Carnegie Mellon University.

Fujii, H.: 1995, Incorporation of natural language processing and a generative system, *Proceedings CAAD Futures '95*, Vol. 1.

Fujii, H., Katukura, H. and Nakai, S.: 1996, Formal representation of an agent who plans and acts. Journal of Architecture, *Planning and Environmental Engineering*, No.482, Architectural Institute of Japan, Tokyo. (in Japanese)

Gasser, L., Braganza, C. and Herman, N.: 1988, Implementing distributed artificial intelligence system using MACE, *in* A. Bond and L. Gasser (eds), *Readings in Distributed Artificial Intelligence*, Morgan Kaufmann, San Mateo.

Gauchel, J., Wyk, S., Bhat, R. and Hovestadt, L.: 1992, Building modeling based on concepts of autonomy, *in* J. S. Gero (ed.), *Artificial Intelligence in Design '92*, Kluwer, Dordrecht, pp.181-197.

Genesereth, M. and Fikes, R.: 1992, Knowledge interchange format version 3.0 reference manual, *Report Logic-92-1*, Computer Science Department, Stanford University.

Glymour, C.: 1992, *Thinking Things Through*, The MIT Press, Cambridge.

Gruber, T.: 1992, A translation approach to portable ontology specifications, *Technical Report KSL 92-71*, Knowledge Systems Laboratory, Stanford University, Stanford.

Khedro, T., Genesereth, M. and Teicholz, P.: *Concurrent Engineering Through Interoperative Software Agents*, FCDA, Stanford University (in World Wide Web).

Hirose, K., Katukura, H. and Nakai, S.: 1994, Knowledge sharing and reuse in structural analysis: An application for distributed environment, *Proceedings of The 17th Symposium on Computer Technology of Information, Systems and Applocations*, Architectural Institute of Japan, Tokyo. (in Japanese)

Levinson, S.: 1983, *Pragmatics*, Cambridge University Press, Cambridge.

MacKeller, B. and Peckham, J.: 1994, Specifying multiple representations of design objects in SORAC, *in* J. S. Gero and F. Sudweeks (eds), *Artificial Inteligence in Design '94*, Kluwer, Dordrecht, pp. 555-572.

Mitamura, T. and Nyberg III, E.: 1992, Hierarchical lexical structure and interpretive mapping in machine translation, *Proceedings of COLING-92*.

Nakai, S., Katukura, H., Ebihara, M. and Hirose, K.: 1992, Agent-oriented problem solving for structural analysis, *Japan-US Workshop on Expert Systems and AI Applications in Civil and Structural Engineering*.

Putnam, H. 1983: *Realism and Reason - Philosophical Papers, Volume. 3*, Cambridge University Press, Cambridge.

Simon, H., Newell, A. and Shaw, J.: 1979, The process of creative thinking, *Models of Thought, Volume. I*, Yale University Press, New Haven.

Stefik, M.: 1995, *Introduction to Knowledge Systems*, Morgan & Kaufmann, San Francisco.

Wong, A., Sriram, D. and Logcher, R.: 1992, SHARED: an information model for cooperative product development, *IEEE Computer*, **March**.

J. S. Gero and F. Sudweeks (eds), Artificial Intelligence in Design '96, 505-524.
© *1996 Kluwer Academic Publishers.*

VISIONMANAGER: A COMPUTER ENVIRONMENT FOR DESIGN EVOLUTION CAPTURE

RENATE FRUCHTER
Center for Integrated Facility Engineering
Stanford University
550 Panama Street
Stanford CA 94305 USA

AND

KURT REINER, LARRY LEIFER, AND GEORGE TOYE
Center for Design Research
Stanford University
560 Panama Street
Stanford CA 94305 USA

Abstract. Computer-based design evolution capture in a multi-disciplinary project environment remains a difficult problem. This paper describes **VisionManager**, a prototype for design evolution capture, visualization, and reuse in support of multi-disciplinary collaborative team work. Based on our research experience, our hypothesis is that one of the key factors in reducing life-cycle cost is improved communication, coordination and cooperation among team members. VisionManager accommodates and integrates many perspectives within a design and manufacturing enterprise and allows team members to: (1) augment shared graphic design models with the team members' design intents, interests, and responsibilities, (2) capture versions at different levels of granularity, such as, feature, discipline perspective, and project level, (3) create private, public, and consensus versions in a hierarchical archive, (4) infer shared interests and route change notifications with regard to a modified feature or perspective, (5) visualize the design evolution of features, discipline perspectives, and the overall project based on captured semantics, and (6) reuse previous alternatives. VisionManager is distinguished from the state-of-the-art file-based document management systems and proposes a model-based and content-based approach for design evolution capture, visualization, and reuse.

1. Introduction

In many design and manufacturing enterprises, product development and process management is done by large geographically distributed multi-disciplinary teams (e.g., engineering, marketing, manufacturing, financing, etc.). Past decades have seen revolutionary increases in the complexity of products and the power of tools used to describe and analyze them. However, the fundamental tools of collaboration and project evolution management have remained unchanged. The limitations inherent in these tools actually limit the quality and performance of the products they describe, while adding to their cost and time to market.

This paper describes the model-based and content-based approach of **VisionManager** to capture, visualize and reuse both the design product (i.e., the graphic models of the artifact) and the design process (i.e., the evolution of explored designs and the corresponding reasons for the decisions). Our hypothesis is that one of the key factors in reducing life-cycle cost is improved communication, coordination and cooperation among members in a multi-disciplinary team. We conjecture that design and knowledge capture, representation, sharing, reasoning and re-use is far less costly than the re-invention of comparable design and/or knowledge. Our objectives in developing the VisionManager prototype are to:

- improve the quality of the product and the process, and
- reduce time consuming and error prone efforts to capture, access, share and visualize product model alternatives, information, knowledge, design intents and decisions throughout the project evolution.

Other important issues in computer support for multi-disciplinary design, such as constraint management, multiple graphic representations (Coyne 90), and design synthesis, are not addressed in this research.

1.1. OBSERVATIONS

Our observations of traditional teamwork indicate that:

- Team members develop their solutions independently as well as collaboratively.
- Each team member develops multiple alternatives. Evolution of discipline solutions and interactions among professionals are hard to document and track.
- Unsatisfactory changes prompt team members to backtrack to earlier solutions, which many times have to be recreated.
- Different discipline solutions interact with each other. The process of identifying shared interests is ad-hoc and based on participants' imperfect memories. This error-prone and time consuming process rapidly leads to inconsistencies and conflicts.
- Meetings are usually the forum in which inconsistencies are detected and resolved before the project can progress.

In addition, the current conventional project information development and management process is based on:

- Individual notebooks, recording background information and results of reasoning and calculations. Notebooks are private documents and are not shared among team members.
- Memos, generated by computers but handled as paper documents, distributed to selective team members, and filed. Paper memos can not be easily updated and are hard to retrieve.
- Graphics and other data, indexed by drawing number and date are generally hard to recover and in their paper form laborious to annotate and update.
- Documentation, in the form of successive approved versions under configuration control often are filed as signed off paper documents.

1.2 CYCLES OF DESIGN EVOLUTION CAPTURE

Based on our observation of multi-disciplinary design teams at work and on evidence presented in design theory literature (Asimov, 1962) (Schoen, 1983) we view design as:

- a *social activity* in which professionals in a multi-disciplinary team propose alternatives, and interact to negotiate a consensus solution.
- an *exploratory activity* in which ambiguity is maintained in alternative solutions until individual professionals and the team commit to one or more particular solution(s). Throughout this iterative process, design modifications are necessary to address industry specifications and change orders. A modification may be considered too minor to be worthy of documentation. Consequently only the person who performed this modification may be aware of it. Furthermore, the consequences of modifications may not have been thoroughly considered. One way to avoid ill-advised modifications is to capture rationale in change notifications, linked with the product model, identify shared interest and inform team members in a timely fashion.
- a *re-design activity* in which designers recreate or revisit previous proposals or ideas. A deeper understanding of the original design will reduce the chances of dangerous modifications when it is redesigned.

We have formalized the design evolution capture in collaborative teamwork as an iterative transition among two communication cycles:
1. An *asynchronous collaboration cycle*, in which team members work independently at concurrent or different times on discipline subsystems. This cycle consists of a *design loop* and a *version loop* (Figure 1). In the design loop, the team member proposes a graphic model of a discipline subsystem, interprets the graphic model by annotating its features with semantic meaning and creating semantic models, critiques the proposed alternative, and saves the model as a private version. In the version loop,

the designer detects shared cross-disciplinary interests in parts of the subassembly captured in the private version, sends notifications regarding these subassembly parts to the interested team members, and saves the private version as a public version. The public version can be accessed by the rest of the team.

2. A *synchronous collaboration cycle* occurs in face-to-face meetings. In this cycle team members consider the public versions they created, define the overall design of the future device, negotiate design modifications and cross-disciplinary conflicts to achieve consensus, and archive a consensus version. As the team members try to resolve conflicts, they re-enter the asynchronous cycle and propose new public versions (Figure 2).

These private, public and consensus versions are stored in a hierarchical archive that defines the explored design space.

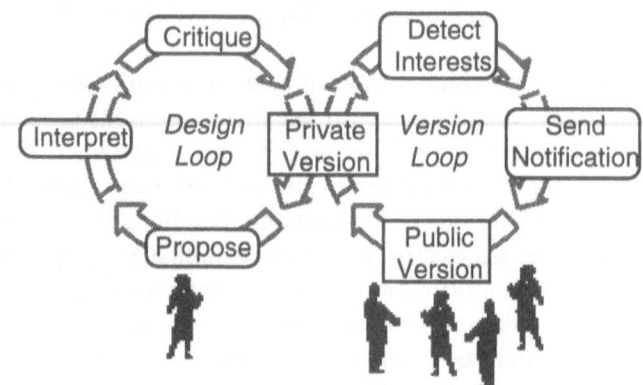

Figure 1. VisionManager supports asynchronous collaborative design cycles.

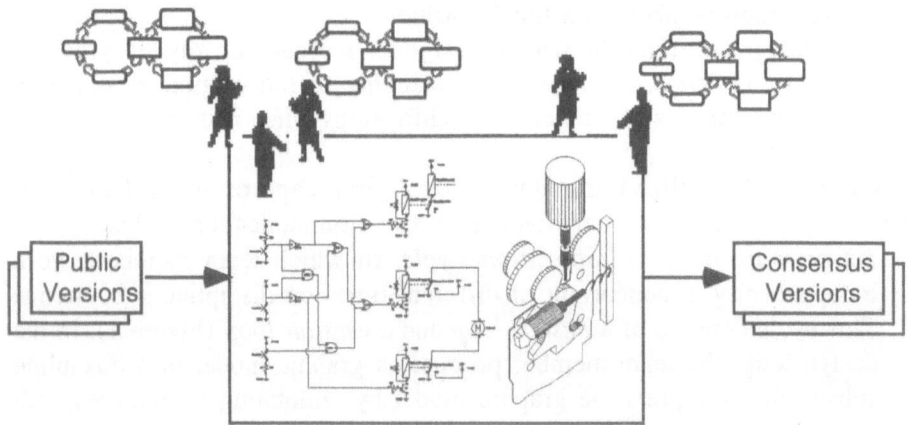

Figure 2. VisionManager supports synchronous collaborative design cycles.

1.3. DESIGN CAPTURE TOOLS

To date, product documentation tools offer a file-based approach to archive, access, and retrieve design evolution. Referencing a specific design version in such documentation systems means referencing a specific file or group of files. Time-stamps, filenames, and directory hierarchies provide document organization, but provide limited information about the file contents. Commercial version management systems such as Autodesk's WorkCenter and Control Data Corp.'s Design Vault, off-load organizational and tracking duties from the designer, but their focus is still on the files, not on the design and its semantic content. VisionManager allows the designer to view and access different portions of the design based on semantic and contextual content without having to worry about which file contained which design version at what time.

2. Vision Manager Prototype

We developed the VisionManager prototype to test our formalization and explore VisionManager's use in asynchronous and synchronous collaboration cycles. VisionManager is implemented using: AutoCAD[1] as the geometric modeling environment, the Illustra Server[2] (an object-relational DBMS) for storing the product models, and Internet email for routing notifications. VisionManager builds on the modeling capabilities provided by our Semantic Modeling Extension (SME) to AutoCAD (Clayton, 1994). SME is accessed from within AutoCAD using an additional pulldown menu. VisionManager is implemented using AutoLISP, DCL, C, and SQL statements. VisionManager is developed on SUN[3] workstations and SME currently runs on UNIX, PC, and Macintosh[4] computers.

2.1. SCENARIO

The following mechatronic system design scenario illustrates the concepts behind VisionManager. The mechatronic device chosen for our test case is a real project performed by a team of mechanical engineering students in the Design and Manufacturing Program of the Mechanical Engineering Department at Stanford. The goal of the project was to design and build an automobile door latch system that combines the function of latch, power lock and cinching into a single assembly. The project was proposed and sponsored by General Motors Corp.

---

[1] AutoCAD is a trademark of Autodesk, Inc.
[2] The Illustra Server is a trademark of Illustra Information Technologies, Inc.
[3] SUN is a trademark of SUN Microsystems, Inc.
[4] Machintosh is a trademark of Macintosh, Inc.

The scenario begins with the team proposing a schematic solution which includes the different subsystems of the door latch system and their interactions. The solution consists of four assemblies which must be integrated with the existing automobile door and forkbolt. (Figure 3) The *Sensors* subassembly will detect the door position and state of the locking system. The *Logic Circuits* subassembly will receive the sensor data, process it to determine the desired actions, and control the actuators. The *Actuators* subassembly provides power to the drive mechanisms. The *Mechanisms* subassembly then moves the *Forkbolt* to either lock or cinch the *Door*.

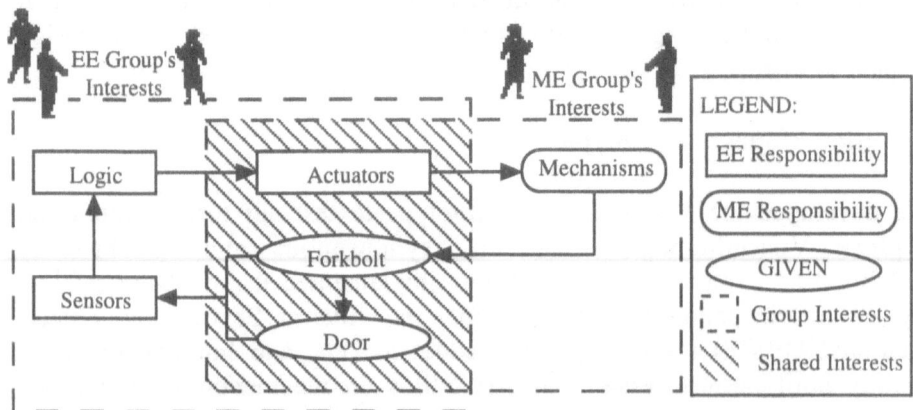

Figure 3. Design schematic showing responsibilities and areas of shared interest.

The design progresses in an iterative mode through:
- *propose* design alternatives in a shared graphic model,
- *interpret* the shared product model into semantic discipline models,
- *gather networked information* by using the discipline models to customize their search for additional discipline information,
- *analyze and evaluate* the discipline models to derive behavior and compare it to function,
- *explain* the results to other members of the team,
- *route change notifications* for proposed changes,
- *capture* and *visualize design evolution* that integrates the many perspectives within a design and manufacturing enterprise.

Gathering network information, analyzing and evaluating designs using networked services and explaining evaluation results are beyond the scope of this discussion. These tasks and the tools supporting them have been presented in a previous paper (Fruchter and Reiner, 1995).

2.1.1. The Design Team

The project team is comprised of individuals or sub-teams, referred to as *Groups*, which are responsible for different aspects of the design. In this scenario, the team has two groups: electrical engineering (EE) and mechanical engineering (ME). It is the responsibility of the groups, or individual designers within the groups, to specify the appropriate technologies that will complete their discipline design solutions. Each group will define their area of:

- *responsibility*, the subsystems for which they propose design alternatives
- *interest*, the subsystems whose changes may affect their design proposals

Figure 3 illustrates the ME and EE groups' responsibilities and interests, as well as shared interests between the two groups.

The electrical group's device detects the state of the automobile door and control the actuators to drive the cinching mechanism. The electrical group is *responsible* for the *Sensors*, *Logic Circuit*, and *Actuators* subassemblies (Figure 3). In addition, they are *interested* in the *Forkbolt* and *Door*. For example, they are interested in the *Forkbolt* specifications because the *Sensor* system must be integrated with the forkbolt. If the forkbolt design changes, it will affect the sensor configuration.

The mechanical group's device must transfer the actuator output to the forkbolt. The mechanical group is *responsible* for the *Mechanisms* subassembly which links the actuators to the forkbolt (Figure 3). In addition to this subassembly, they are *interested* in the *Actuators*, *Forkbolt* and *Door*. For example, they are interested in the *Door* because it imposes spatial constraints on the mechanism design.

2.1.2. The Design Models

VisionManager promotes an object-oriented approach to modeling and annotating the device models. The VisionManager modeling tools are flexible and allow the designers to organize their models along any line: departmental, area of specialty, or product systems. In this scenario, the team has divided the design into three different model types: an overall *product model*, two *assembly models,* and several *subassembly models*.

The product model is comprised of the electrical and mechanical groups' assembly models (Figure 4). An assembly model is composed of the systems or subassemblies for which the group is responsible. The electrical assembly model consists of the *Sensors*, *Logic*, and *Actuators* subassembly models (Figure 5). Similarly, the mechanical assembly model consists of the *Mechanisms* subassembly model.

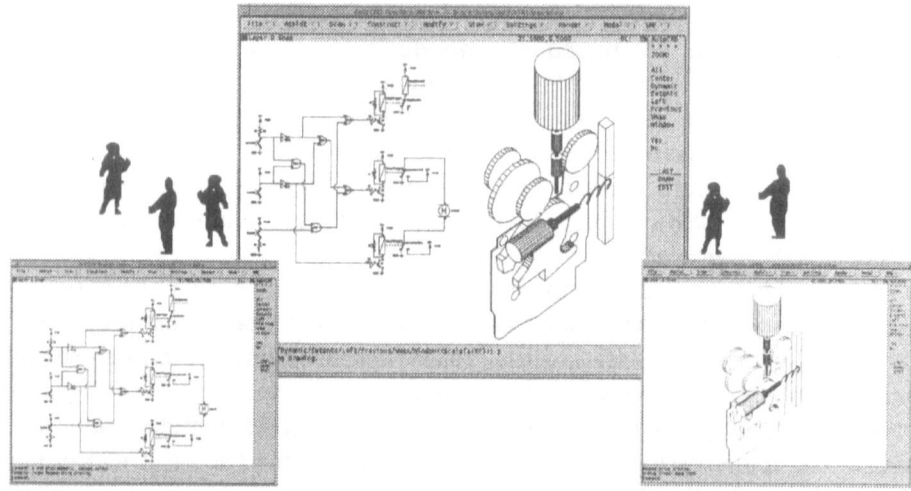

Figure 4. The Product Model is comprised of the electrical and mechanical assemblies.

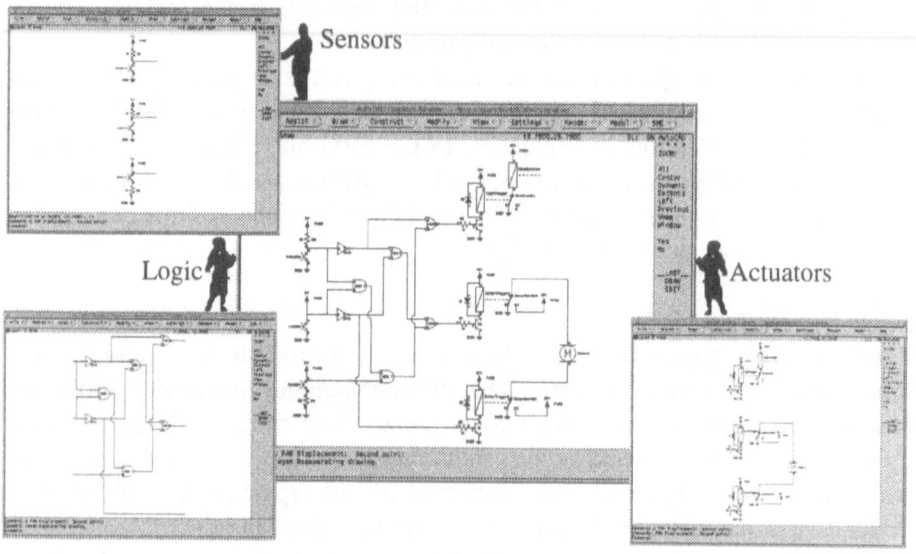

Figure 5. The electrical assembly model is composed of three subassembly models:
Sensors, Logic, and Actuators.

2.2. MEMORY ORGANIZATION

The product model in VisionManager consists of graphics, semantic annotations, design notes, and person-to-person notifications. The Semantic Modeling Extension (SME) provides interactive mechanisms which enable designers to map shared graphic entities to multiple symbolic representations

(Figure 6). The three primary SME object types *Manager Objects*, *Interpretation Objects*, and *Feature Objects,* provide a flexible structure for indexing, storing, and retrieving knowledge and data. (Figure 7) The designers use these objects in conjunction with *Person Objects*, *Graphics Objects*, *Note Objects*, *HyperLink Objects*, and *Notification Objects* to capture and express their intents behind the graphic entities.

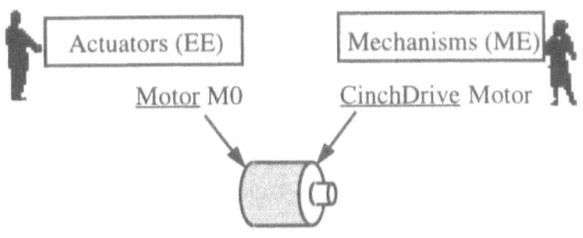

Figure 6. Multiple interpretations of a shared graphic entity.

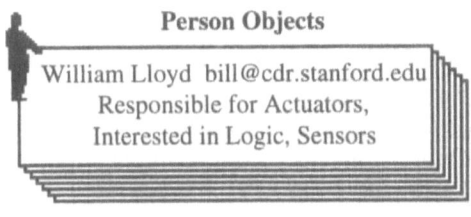

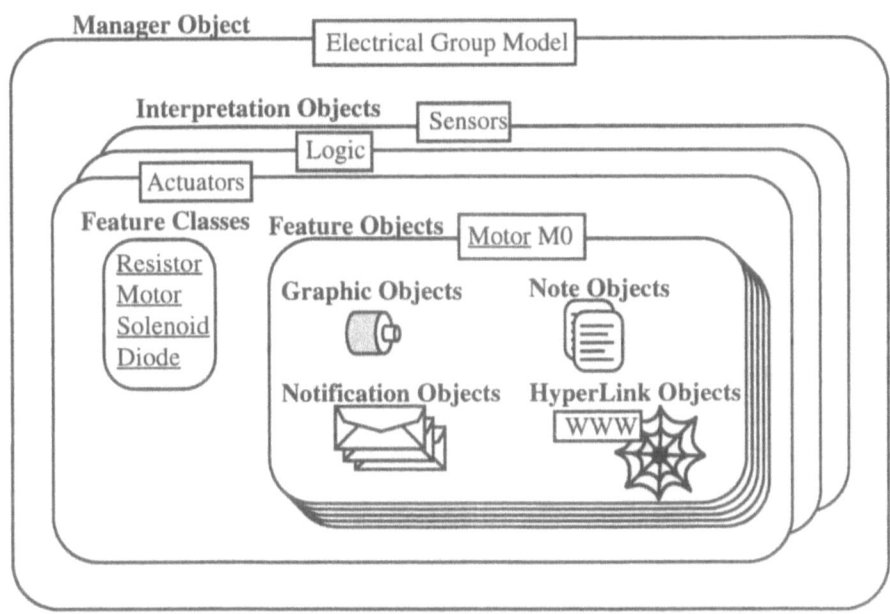

Figure 7. VisionManager modeling objects.

Manager Objects provide a means by which a person or team can group multiple interpretations of the design. It encapsulates a list of *Interpretation Objects* and a list of *Person Objects* associated with the project. In Figure 7 a Manager Object is used to encapsulate the Electrical Group's interpretations of the design.

Interpretation Objects encapsulate features for a particular perspective. An *Interpretation Object* has two primary attributes: a list of *Feature Classes* and a list of *Feature Objects* (Figure 7). Feature Classes provide an ontology to describe the semantic meaning of the graphics within a context. This ontology can be defined or augmented by the user at run-time. The list of *Feature Objects* is edited by the user to contain the instances from a particular graphic model which are relevant to an interpretation.

Feature Objects capture the link between graphic entities and symbolic entities (Figure 6). We define a feature to be a constituent element of a design which has meaning to a designer within a particular context. The basic components of a *Feature Object* are a *Feature Class*, an identifier or *Feature Name*, and a list of *Graphics Objects*. Other information objects can be linked to *Feature Objects* such as *Note Objects*, *HyperLink Objects*, and *Notification Objects* (Figure 7). *Feature Objects* allow graphic entities to have multiple meanings within different interpretations.

Person Objects serve as a record of the project participants and their declared roles and interests. A *Person Object* consists of the designer's name, a user-name, a user-password, an email address, a list of responsibilities, and a list of interests, (Figure 7). *Person Objects* can be added, updated, and deleted by the users. The lists of interests and responsibilities are used by VisionManager to infer which team members should be sent email notifications about changes to a portion of the design. VisionManager relies on the list of responsibilities to verify that a particular designer is allowed to store an updated *Interpretation Object* in the database.

Graphics Objects contain Drawing Interchange File (DXF) representations of the graphic model entities. A graphic entity may be shared among many *Feature Objects*.

Note Objects contain text written by the project members. *Note Objects* are used to capture the design rationale or other design related information that a designer traditionally records in notebooks, memos, etc. Notes are encapsulated in *Feature Objects* to describe design requirements or intents. VisionManager's *Note Browser* allows the user to browse and search *Note Objects* in order to locate specific *Feature Objects* or *Interpretation Objects*.

HyperLink Objects provide a mechanism to link a *Feature Object* to sources of information. VisionManager currently handles references to World Wide Web (WWW) pages and electronic images. A feature in the graphic model could be linked to component specification sheets available on the WWW or a photo of a prototype.

Notification Objects record the communications among the designers and are routed in asynchronous mode. These notifications can be used to solicit feedback, to give approval, to broadcast change notifications, or to initiate negotiations. A *Notification Object* consists of:

- *Feature Objects*, the focus of the notification
- affected *Interpretation Objects*, share an interest in the *Feature Objects*
- *Person Objects,* the mailing list, and
- a *Note Object,* describes the rationale or situation.

Notification Objects are stored as a part of *Feature Objects* in the shared product model.

2.3. DESIGN EVOLUTION CAPTURE

VisionManager indexes versions at different levels of granularity, i.e., *Feature Object, Interpretation Object,* and *Manager Object,* and allows the user to capture different types of versions such as private, public and consensus. The following sections elaborate on how VisionManager supports design evolution capture.

2.3.1. Version Composition

The three primary object types in SME are the basis for the three different version types in the VisionManager.

- *Feature Versions* represent the most primitive versions which the user can access. These capture the evolution of a single design feature.
- *Interpretation Versions* are the second type of version in VisionManager. A series of Interpretation Versions illustrate the evolution of the *Feature Classes* and the *Feature Objects* belonging to a particular *Interpretation Object*. This allows the user to view the design as it develops from a particular perspective.
- *Manager Versions* allow the user to trace the evolution of a group of *Interpretation Objects*. With Manager Versions the change in an assembly model, (e.g., Electrical) can be traced over the project lifespan.

The content of Feature, Interpretation and Manager versions is the same as their SME modeling counterparts, with the addition of several flags. These flags are set by designer to indicate the version *type,* (e.g., public, private, consensus). At present, the version type is set by the user at the Interpretation Version level and inherited by all of its *Feature Objects*.

2.3.2. Public and Private Versions

A public version is one which can be accessed by anyone with access privilege to the project database. Private versions allow the designer or team to retain a milestone of the work without sharing it with the rest of the team. There are different levels of privacy: private to the group, the sub-group, and individual. Access to the version is granted and confirmed via *Person Objects*. The labels and number of levels of privacy can be defined by the user at run-time to match the project organization.

2.3.3. Consensus Versions

A consensus version of an interpretation is one which has been accepted by the entire team. This version may not be complete in terms of its design status, but the team-members agree that it does not conflict with other interpretations. There are different levels of consensus: project-wide consensus, group consensus, sub-group consensus. The consensus types can be defined by the user at run-time to match the team organization.

2.4. CONTENT-BASED DESIGN ACCESS AND RETRIEVAL

In the following we describe how VisionManager enables designers or the team to check-in, visualize and check-out version from the archive.

2.4.1. Version Check-In

VisionManager relies on a centralized database to record the version evolution of the product model. Since the design is not being stored in files or directory hierarchies, all access to current or historic versions is accomplished through VisionManager's check-in and check-out facilities.

Explored designs are indexed and stored through VisionManager's check-in process. Check-in consists of the following steps:

- *Designer specifies the check-in granularity level.* VisionManager requires designers to check-in their design at either the *Manager Object* or the *Interpretation Object* level.
- *VisionManager verifies responsibility and access permission.* VisionManager references the project's *Person Objects* to verify that the check-in is being performed by a designer who is responsible for the *Interpretation Object*.
- *VisionManager archives the design version.* VisionManager compares the *Interpretation Object's Feature Classes* and each of its *Feature Objects* to the previously stored versions. Each component of the *Feature Object* is checked : *Graphic Objects, Note Objects, Notification Objects*, and *HyperLink Objects*. To reduce the storage of redundant information, the system records only those objects which have changed.

Interpretation Objects are retained by the VisionManager as alternatives.

For instance, in our scenario, the sensor designer is considering 3 different sensor technologies: infrared emitter/detector pairs, contact switches, and fiber optic sensors (Figure 8). Each option is captured in an *Interpretation Object* and may undergo several revisions. Infrared sensors are the preferred solution at this stage and the designers make their solution publicly accessible as sensor version S_1. The rationale for selecting this technology over the other two alternatives is recorded as *Note Objects* linked to *Feature Objects*.

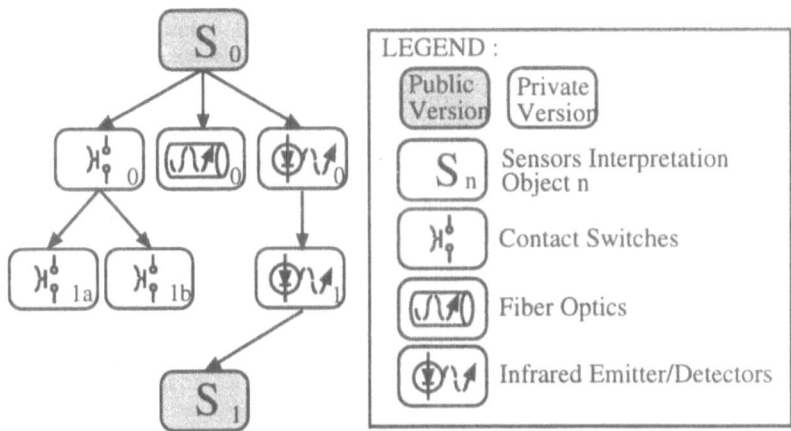

Figure 8. A *public-to-public* design evolution generated by the Sensors sub-assembly designer while exploring three alternatives.

2.4.2. Shared Interest Detection and Notifications

When an interpretation is checked in, VisionManager assists the designer by inferring which features have been changed and who shares an interest in those changes. A feature is changed if any part of the *Feature Object's* information, (e.g., *Graphic Objects, Notes Objects*) has been altered since the previous version. Once a changed feature is found, VisionManager searches all of the *Interpretation Objects* for *Feature Objects* which share graphics with the changed feature. This process yields a list of changed *Feature Objects* and affected *Interpretation Objects*. VisionManager identifies which team members are interested in the changes by comparing the declared interests in the *Person Objects* to the list of affected *Interpretations Objects*. With this information and a text note written by the user, the system builds the *Notification Object*. This facility is invoked:

- by the user to check consistency with other public versions,
- optionally when a private version is stored in the database, or
- automatically when a public or consensus version is stored.

It is always left to the users discretion to modify any part of the *Notification Object*, or discard it. For instance, even though VisionManager has identified several persons who may be interested in a change, the user may want to solicit feedback from only one of them. The user removes the other persons from the mailing list before sending the email. Notification

Objects not only serve as a means of communication among the designers, they are also a record of the rationale behind feature changes.

In our scenario, the mechanism designer generated three potential solutions: screw/nut, linkage, gears (Figure 9). Since there are two required functions, cinching and locking, the designer initially specifies two separate gear trains for driving the forkbolt. The mechanism designer sends a message to the actuator designer indicating the need for two motors that fall within certain speed & torque ranges. The Actuators designer specifies the motors and sends a notification back to the Mechanism designer. After incorporation of the motors in a private version, the designer determines that there is no room for two drive trains and two motors within the confines of the car door. The redesign uses only one motor and a solenoid. Each time a notification is sent, the originator's design must be versioned for public access. This insures accuracy when referencing graphics.

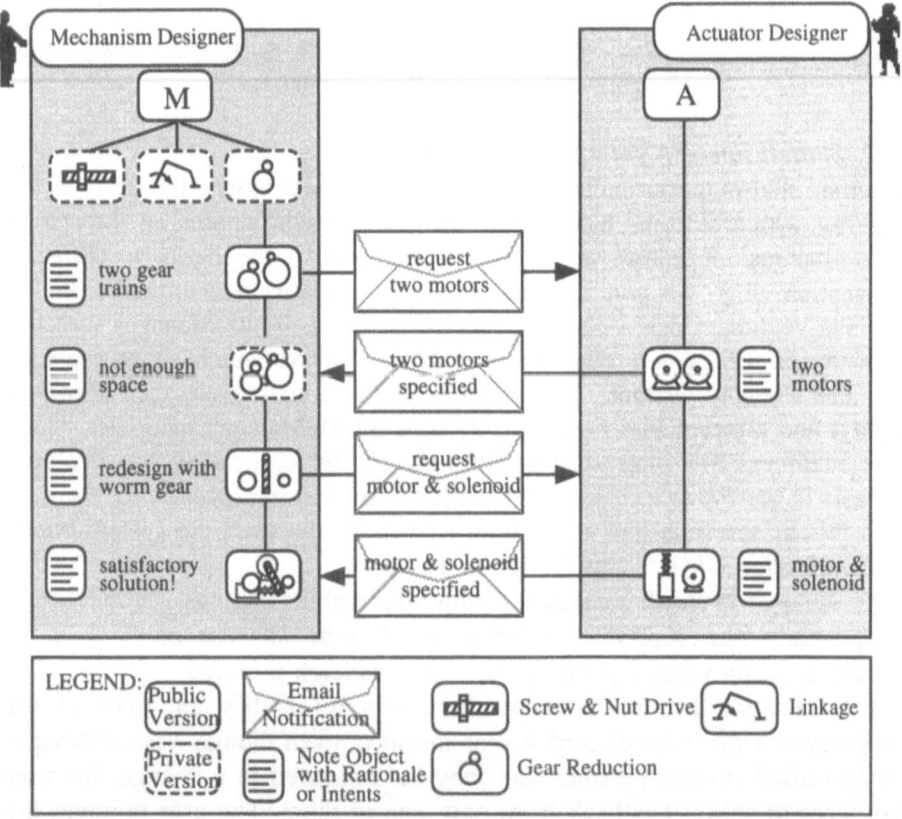

Figure 9. Email notifications sent between the Mechanism and Actuator designers.

2.5. VISUALIZATION AND REUSE OF DESIGN EVOLUTION

The design evolution is only as useful as the visualization and retrieval mechanisms which allow users to revisit design alternatives. VisionManager supports visualization by enabling the user to browse the project evolution based on its content. The browsing and filtering mechanisms leverage model graphics and semantics rather than filenames and time-stamps.

The user first specifies the level of detail at which he/she wants to view the evolution: Feature Versions, Interpretation Versions, or Manager Versions. Specification of the subject of the evolution can be provided by either entering a text string or selecting a graphic. For example, to specify an *Interpretation Object*, the user could either type : "Actuators" or click on one of the graphic entities in the CAD model that is a member of the Actuators *Interpretation Object*. If the entity is shared among several *Interpretation Objects*, the user must specify from which point-of-view the evolution should be viewed (e.g., Actuators, Mechanisms, Figure 10).

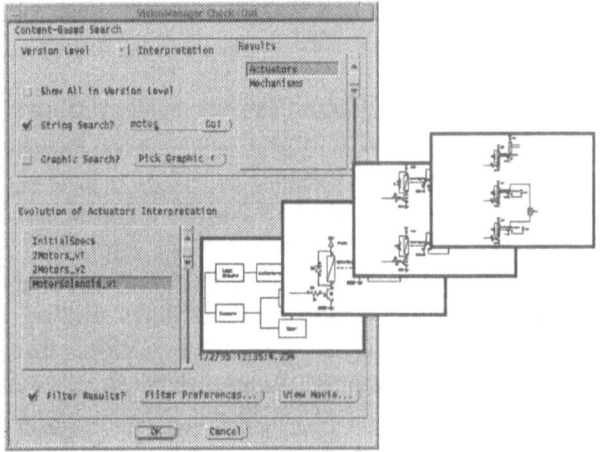

Figure 10. Check-Out of an Actuators version after searching all Interpretation Objects for the string "motor".

2.5.1. Evolution Filtering

In a project with many *Interpretations Objects*, the version space will be large and difficult to navigate without computer assistance. Evolutions can be played back as movies. This may not be the most efficient means to locate a particular version. In addition to referencing components or assemblies by their semantic model information (i.e., *Feature Class* and *Name*), VisionManager provides mechanisms for content-based evolution filtering.

The most basic filters utilize the flags, (e.g., Hot-Item, Daily-Backup), set by the user when checking in an *Interpretation Object*. For instance, the user may wish to view all "Hot-Item" EE versions at the Manager Version level.

In a similar fashion, the user can also specify which type of versions to view, (e.g., public, consensus).

VisionManager also allows the user to specify a filter based on an "increment of change" in the version evolution. In order to gauge the change between two given versions, VisionManager must first measure the *volume* of *Feature Objects, Interpretation Objects* and *Manager Objects*. The user can set preferences to base *Feature Object* volume on the number of : *Graphic Objects, Note Objects, HyperLink Objects*, and/or *Notification Objects*. An *Interpretation Object's* volume can either be the number of *Feature Objects* it contains or the sum of the *Feature Objects'* volumes. Similarly, a *Manager Object's* volume can either be the number of *Interpretation Objects* it contains or the sum of the *Interpretation Objects'* volumes. For example, the user may request to view the Sensor *Interpretation Object* at increments of five *Feature Object* changes. The system will produce a subset of the full Interpretation Version evolution in which five features have been changed between subsequent versions.

2.5.2. Design Reuse

The check-out function in the VisionManager system allows designers to backtrack to previously versioned interpretations, or to incorporate portions of earlier versions in their current design.

When an *Interpretation Object* is checked-out, its contents, (i.e., *Feature Classes* and *Feature Objects*) are merged into the AutoCAD design space. The graphic entities contained by the *Graphic Objects* are added to the model only if they are not already present. The relative position and orientation of graphic entities is preserved within, but not across, Interpretation Versions. Consequently, VisionManager relies on the user to place the graphic entities in the model. When a *Feature Object* is checked-out, its contents (i.e., *Feature Class, Feature Name, Graphic Objects*, etc.) can be added to a current *Interpretation Object* in the AutoCAD space. Alternatively, the "shell" of the *Interpretation Object* (i.e., just its *Feature Classes*) can be merged as well. The description of the mechanisms which are used to maintain consistency between the AutoCAD design space and the Illustra DBMS is beyond the scope of the paper.

3. Related Research

The points of departure for this research are the following domains: collaboration technologies, database version and configuration management, and design rationale and documentation.

3.1. COLLABORATION TECHNOLOGIES.

Communication of intents, problems, and decisions is critical in achieving better cooperation among professionals across organizations. In recent years, commercial efforts, (e.g., CORBA) and research work addressing issues in concurrent engineering, such as information sharing and exchange, multi-criteria representation, analysis and evaluation, has been of growing interest. Two recent projects propose integrated approaches for multi-criteria and multi-disciplinary semantic representation and reasoning (Clayton, 1994; Fruchter and Reiner, 1995; Saad, 1995). Other research directions have proposed different integration frameworks to address the information exchange needs for collaborative work, such as, blackboard architectures to link CAD with expert systems (Fenves, 1990; Phol, 1990), blackboard object-oriented database framework (Sriram, 1991), and federated agent architecture (Cutkosky, 1993; Khedro, 1995). The prototypes developed in these studies do not include capabilities for versioning and documentation of design rationale.

3.2. VERSION AND CONFIGURATION MANAGEMENT

Previous research on integration has focused on the use of database management systems as the primary integration scheme to enable design software applications to share data. In this scheme, software applications are able to store, access and update data through a central data base or distributed database. Different systems have been developed for project data management to support collaborative work. A recent study explores a relation database approach to change management in a CAD environment to support collaborative engineering (Krishnamurthy, 1995). The proposed change management consists of three layers: versions, assemblies, and configurations. A version is a specific design description of a primitive entity, an assembly integrates individual versions to describe the state of a composite entity, as well as a design in a discipline. Configurations integrate discipline designs to describe the overall project as a collection of the different discipline drawings. These approaches track only the evolution of graphic entities without tracking evolution of design rationale or cross-disciplinary communication.

3.3. DESIGN RATIONALE

Three major approaches to design rationale have been proposed in the past decade: history-based rationale, argumentation-based, and device model based rationale.

History-based rationale approaches, such as the electronic notebook document design by recording the sequence of events that happen during design (Lakin, 1989). The approach requires a low overhead for recording design activities, however, does not consider the specific use of these documents, detection of interests, or the needs of document users. This approach has the same pitfalls as the documents produced in typical practice, i.e., there is no link between the product model and the created document. The access to relevant information is even harder than with the current design documents due to the increased amount of information recorded.

The argumentation-based approach is derived from hypertext research, where the goal is to provide uniform structure to a potentially diverse medium (Kunz,, 1970; McCall,, 1987; Chung, 1994). The document is recorded as non-interpreted text. This approach requires designers to learn the documentation and access methods, and adds a high documentation overhead to the design process.

Device-based approach takes a model-based approach used in diagnosis expert systems (Gruber, 1991; Baudin, 1989). The key in this approach is to develop reusable devise models. This approach does not address the requirements posed by team work, such as shared interests, notifications, negotiation. This approach requires the designer to formalize the device model. Consequently increases the designer's overhead in documenting the design and is unrealistic during conceptual design. More recent research explored mechanisms for active design documentation that uses an initial domain specific design model able to generate and explain standard design decisions. In this approach designers can adjust the initial design model. (Bicharra Garcia, 1993).

4. Conclusions

This paper presents an initial effort in the development of a computer environment to support capture, visualization and reuse of design in multi-disciplinary teamwork. VisionManager proposes a model-based and content-based paradigm for design evolution capture. The system provides capabilities to:

- augment a shared graphic design model design with the team members' ontology, design intents, interests, and responsibilities,
- capture versions at different levels of granularity, such as, feature level, discipline perspective level, and project level,
- create private, public, and consensus versions in a hierarchical archive,
- infer the perspectives and corresponding ontologies of a shared feature,
- infer shared interests and route change notifications with regard to a modified feature or perspective,

- visualize the design evolution of features, discipline perspectives, and the overall project based on captured semantics, and
- reuse previous alternatives captured in archived versions.

VisionManager is particularly promising since it provides a rich knowledge archive of explored perspectives, shared interests, alternatives, and the design intent behind them, that designers will not be able to remember on their own. This archive is built up during the design process, rather than as an independent process performed by an expert assistant who is familiar with the technicalities of indexing and retrieval.

VisionManager provides a mechanism for creating the "team memory" or the "corporate memory." The "team memory" can serve a number of roles:

- archive of design evolution for different users, such as, current team members, project managers, clients, or new team members who join the team later and need to become acquainted with previous decisions,
- learning resource for apprentices, who need to learn the organization's design practice, and
- case-study resource for future projects.

We plan to test VisionManager in two learning environments, one is mechatronic system design (Toye, 1993), and the other is computer integrated architecture/engineering/construction (Fruchter and Krawinkler, 1995).

References

Asimov, W.: 1962, *Introduction to Design*, Prentice-Hall, Englewood Cliffs, NJ.

Baudin, C., Uderwood, J., Baya, V., Mabogunje, A.: 1993, Using device models to facilitate the retrieval of multimedia design information, *Proceedings 13th IJCAI*, pp. 1237-1243.

Chung, P. W. H. and Goodwin, R.: 1994, Representing design history, *in* J. S. Gero and F. Sudweeks (eds), *Artificial Intelligence in Design '94*, Kluwer, Dordrecht, pp. 735-751.

Clayton, J. M., Fruchter, R., Krawinkler, H. and Teicholz, P.: 1994, Interpretation objects for multi-disciplinary design, *in* J. S. Gero and F. Sudweeks (eds), *Artificial Intelligence in Design '94*, Kluwer, Dordrecht, pp. 573-590.

Coyne, R. D., Rosenman, M. A., Radford, A. D., Balachandran, M., and Gero, J. S.: 1990, *Knowledge-Based Design Systems*, Addison-Wesley Publishing Company.

Cutkosky, M., Engelmore, R., Fikes, R., Gruber, T. R., Genesereth, M., Mark, W., Tenenbaum, J. M., Weber, J. C: 1993, PACT: An experiment in integrating concurrent engineering systems, *IEEE Computer*, special issue on Computer Support for Concurrent Engineering, pp. 28-37.

Fenves, S. J., Fleming, U., Hendrickson, C., Maher, M. L. and Schmitt, G.: 1990, Integrated software environment for building design and construction, *Computer-Aided Design*, **22**(1), 27-36.

Fruchter, R. and Krawinkler, H.: 1995, A/E/C teamwork, *Proceedings Second ASCE Congress of Computing in Civil Engineering*, Atlanta, pp. 441-448.

Fruchter, R., Reiner, K., Toye, G. and Leifer, L.: 1995, Collaborative mechatronic system design, *Proceedings of CERA95 Conference*, Washington D.C., pp. 231-242

Gruber, T.R.: 1989, *The Acquisition of Strategic Knowledge*, Boston Academic Press,

Bicharra Garcia, A. C., Howard, C. and Stefik, M. J.: 1993, Active design documentats: A new approach for supporting documentation in preliminary routine design, *CIFE Tech. Report Nr. 82*, CIFE, Stanford University, Stanford.

Khedro, T.: 1995, AgentCAD for cooperative design, to be published in *Proccedings CAAD Futures 95*, Singapore.

Krishnamurthy, K. and Law, K. H.: 1995, Configuration management in a CAD paradigm, to appear in *Proceedings 1995 International Mechanical Engineering Congress*, San Francisco.

Kunz, W. and Rittel, H.: 1970, *Issues of Elements of Information Systems*, Center for Planning an Development Research, UC. Berkeley.

Lakin, F., Wambaug, H., Leifer, L., Cannon, D. and Sivard, C.: 1989, The Electronic Design Notebook: Performing medium and processing medium, *Visual Computer: International Journal of Computer Graphics*, 5(4), 214-226.

McCall, R., 1987, PHIBIS: Procedurally hierarchical issue-based information systems, *Planning and Design Acquisition*, Boston, MA.

Pohl, J. and Chapman, A.: 1990, Expert system for architectural design, *Journal of Real Estate Construction*, 1, 29-45.

Saad, M.: 1995, *Shared Understanding in Synchronous Collaobrative Building Design*, PhD Thesis, University of Sydney.

Schon, D.: 1983, *The Reflective Practitioner*, Basic Books, Inc., New York.

Sriram, D., Logcher, R., Wong, A. and Ahmed, S.: 1991, An object-oriented framework for collaborative engineering design, *in* D. Sriram, R. Logcher and S. Fukada (eds), *Computer-Aided Cooperative Product Development*, Springer-Verlag, Berlin.

Toye, G., Cutkosky, M. R., Leifer, L., Tennenbaum, M., Glicksman, M. J.: 1993, SHARE: A methodology and environment for collaborative product development, *2nd IEEE Workshop on Enabling Technologies Infrastructure for Collaborative Enterprises*, pp. 33-47.

10

rules, models and theories in design

J. S. Gero and F. Sudweeks (eds), Artificial Intelligence in Design '96, 527-540.

ELICITATION OF RULES FOR GRAPHIC DESIGN EVALUATION

GEORGE GLAZE
Faculty of Visual Communications
West Hertfordshire College
Watford WD1 3EZ UK

JEFF JOHNSON AND NIGEL CROSS
Design Discipline
Faculty of Technology
The Open University
Milton Keynes MK7 6AA UK

Abstract. This paper reports a set of experiments in which graphic design rules were elicited from two graphic design experts. In these experiments no attempt is made to judge a design to be 'good', but the rules are sufficient to diagnose a design as 'bad'. The experiments show that graphic design rules can be elicited from expert graphic designers in a sufficiently explicit way for them to become operational in a computerised expert system. The experiments show that the graphic design experts do not apply their rules consistently, but when the rules are made explicit they agree entirely with the diagnosis of the rule based system. These experiments therefore support the thesis that there are rules governing the aesthetics of graphic design, as perceived by practising graphic designers, which can be implemented on computers. These rules may not guarantee 'good' design but may assist the novice to produce designs which are 'not bad'.

1. Introduction

This paper deals with the issue of making explicit the implicit aesthetic judgement of graphic design professionals. It addresses the issues of the expression of intuitive, emotive and configurational reporting and the difficulties associated with the interpretation of the professional's diverse and specialised comments. It reports a procedure of rule elicitation through the diagnosis and reporting (in a critique) by two professionals of bad design symptoms in a sample of single-sheet (A4 size) poster desktop publications. A new procedure for the diagnosis of bad symptoms in a population of such desktop publications is presented.

Also examined are the effectiveness and reliability of the rules formulated from the professional's intuitive design critique. This focuses on the inconsistency of the professionals' diagnoses of bad symptoms and on factors which might inhibit accurate diagnosis and consistent reporting.

Sets of formalized rules were elicited from two expert graphic designers, based on their critiques of a sample of A4-size amateur desktop poster publications. The expressions and bad symptoms reported by the experts were interpreted into rules. The formalized rules were used for the diagnosis of bad symptoms in a second sample of similar desktop publications. The experts subsequently diagnosed bad symptoms in this sample also. The results of the experts' diagnoses of bad symptoms in the second sample are compared with the application of the formalised rules used systematically. The performance of the two procedures is analysed to determine the correspondence of the systematic, rule-based diagnoses with those of the experts.

We do not attempt to predict 'good' design; only to identify symptoms of 'bad' design. Application of a rule based system that identified 'bad' design symptoms could at least result in the elimination of such symptoms, and therefore enable a user to produce designs that are 'not bad'. By definition, a design that is not diagnosed to be bad is 'not bad'; all 'good' designs will be 'not bad', but some 'not bad' designs may not be good.

2. Knowledge Elicitation for Graphic Design

Knowledge elicitation in graphic design is difficult particularly because there is little consensus on the subtle features which contribute to the aesthetics of a visual representation and ambiguity in the application of a design prescription.

Although graphic designers, like some other experts, find it difficult to articulate their implicit strategies, they do at times communicate their expertise verbally and by demonstration: for example the apprentice approach to design practice and some forms of design education confirm the success of these methods of knowledge transmission.

Tunnicliffe and Scrivener (1991) offer a procedure devised for the elicitation of knowledge in the graphic design domain. In their opinion traditional methods of knowledge elicitation are inappropriate to graphic design because they offer a 'reductionist' view more suited to scientific problems. They contend that to reduce the overall problem of designing to discrete parts for analysis distorts the knowledge elicited and may miss essential design knowledge. They emphasise that graphic designing relies on narrative, metaphors, analogies and gestalts. In design knowledge elicitation, they suggest, it is necessary to arrive at information relating to the interdependence of elements in a design, e.g. components such as headline text, body text and pictures, and 'weight, balance, colour, impact and emphasis'. They used a procedure based on a combination of *protocol analysis* and *teachback* techniques. They report that these techniques allow the graphic designers to perform in a way natural to them. These procedures provide a holistic view of the graphic designer's processes and facilitate the subsequent analysis of the verbal and visual data generated and recorded.

Of the available techniques of knowledge acquisition and elicitation, *critiquing* was considered to be the most appropriate for our own study. Designers routinely criticise their own work using explicit language and therefore should be happy to criticise the work of others. Critiquing is a relatively simple and inexpensive procedure to arrange and provides a rich source of data for analysis. In our procedure, expert graphic designers were asked to comment critically on graphic design work produced by (anonymous) others.

3. Rules Elicited from Graphic Design Experts

Normally, only one expert is used as the source for eliciting rules for rule based systems. Moore and Miles (1991) have discussed the reasons for this, and suggested some advantages of using more than one expert, particularly in a design domain context. We used two graphic design experts for this study, but did not attempt to amalgamate their separate sets of rules into one set. The inconsistencies, and even occasional contradictions, between experts in any given domain can make impossible the amalgamation of rules elicited from different experts.

Our chosen experts worked independently of each other, and did not know that more than one expert was being consulted, nor what the purpose of the experiment was. Not giving the experts a full explanation of the purpose of the task and their contribution is seen in retrospect to have contributed to inconsistency and perhaps poor motivation. The experts were unaware that their comments eventually would be formulated into rules.

3.1. PREPARATION OF THE GRAPHIC DESIGN SAMPLES

We were particularly interested in providing an evaluation system which might be used to assist inexperienced, novice 'designers' in the production of well designed, simple desktop-published documents. For example, many people use wordprocessors to produce simple 'posters' of single sheets to make announcements. These are often poorly designed and laid out from a graphic design point of view. Our aim was to see if we could develop a set of evaluation rules that could be applied to such documents. Application of the rules in a rule based system could then provide a design evaluation feedback to the 'designer', who could then adjust their design to produce a more aesthetically satisfactory design.

We collected 40 random examples of single-sheet, A4-size posters from the notice boards of our faculty building. These were drawn from the many examples of informal notices that proliferate on such notice boards, e.g. advertizing articles for sale, clubs to join, forthcoming meetings, etc. The posters were photocopied sequentially during the same print run. The specimens were duplicated to be the same size as the originals on A4 sheets.

Forty questionnaire sheets were given to each of two graphic design experts, together with the forty copied posters and a covering letter explaining what the experts were required to do. The experts were requested to inspect the sample of posters identifying any 'bad symptoms' of graphic design in the specimens. Comments on each 'bad symptom' were to be reported on the accompanying questionnaire. Their comments were expected to be their personal view on the symptoms reported.

The experts were asked to assess the relative importance of individual symptoms by assigning a weighting on a scale 1, 2 or 3 corresponding to their considered importance; 1 representing not very important, 2 important and 3 very important. Written comments on each of their findings was also encouraged to expose the constraints on the expert's diagnosis. The level of detail was to be to the expert's discretion but would clearly be a factor in the classification and discrimination of reported symptoms.

Both experts employed a similar approach to identifying design defects in the specimens. Bad symptoms were identified on the specimens by a freely drawn line encircling the feature which was also referred to in the questionnaire. Where there was more than one symptom, each was identified by an alphanumeric code. The comments on the associated questionnaire were similarly coded.

3.2. DESCRIPTION BY THE EXPERTS OF THE 'BAD SYMPTOMS' DETECTED IN EACH SPECIMEN OF THE SAMPLE

In their descriptions and comments on the bad symptoms identified, the experts, as expected, used a mixture of ambiguous expressions, specialist terminology, and emotive statements. These comments were used to establish the implicit or explicit 'rules' that were being applied by each of the experts.

Data analysis began with the transcription and interpretation of the experts' responses. Where necessary a response was recomposed into logical English statements embracing the essence of the expert's individual comment. Subsequently each statement was formalised to simulate (for each expert) a robust set of rules appropriate for a simple rule-based application.

3.3. INTERPRETATION OF FREE FORMAT SPEECH TO FORM TECHNICALLY DEFINED AND RIGOROUS RULES

The first author's experience and familiarity with the conventions, terminology and phraseology of the graphic design domain informed the interpretation of the experts' comments. The rules were initially formulated by this researcher who attempted to interpret the expert comments.

However it was recognised that the interpretation of idiosyncratic words and phrases is often variable between individuals. Specialist idiom can also be inconsistent at different times and in different situations. Accurate interpretation relies on the precise and logical definitions of specialist expressions. Consequently the interpretation was verified by asking the experts who originated the comments to confirm the meaning of many of the terms received in their responses. The information received was used in formulating an experts' glossary.

3.4. FORMULATION OF RULES

Rules were formulated from the symptoms of bad design identified by the experts. The process of formulating a rule was iterative, involving an initial formulation, testing and perhaps one or more attempts at reformulation before eventual acceptance.

Follow-up interviews with the experts were also used to ensure that an identified bad symptom was being correctly interpreted.

The following are examples of some of the bad symptoms reported by the experts and the rules elicited from their comments.

3.4.1 Examples of expert (A) symptoms and rules
SYMPTOM(S): "Bad gap (word spacing)."
REASON(S): "Such unfortunate alignment, causing rivers, in which the words on the line below are closer than the words on the same line, distract from their intended sense."
RULE FORMULATED:
Line space should be greater than word space

SYMPTOM(S): "The easy use of space at the top of the page is not echoed in the graceless foot of the page.
"Not visually centred."
REASON(S): "One is left with a feeling that the elements are out of phase - this must be counter productive."
RULES FORMULATED:
i. *If there are top and bottom margins then the top margin should be less than or equal to the bottom margin*

ii. Where two or more vertically adjacent blocks of type are centred then the horizontal separation of their centres should be less than 1mm

3.4.2 Examples of expert (B) symptoms and rules
SYMPTOM(S): "Far too close to the top of the page."
REASON(S): "Top heavy, otherwise OK."
RULE FORMULATED:
Not all blocks should be above the centre of the page

SYMPTOM(S): "More space at top than bottom.
"Thrown together, some centred type, some ranged left, some ranged left and then set right.
"Crooked pasting up."
REASON(S): "Just a mess."
RULES FORMULATED:
i. If there are top and bottom margins then the top margin should be less than or equal to the bottom margin
ii. When more than 70% of blocks are centred then all blocks should be centred
iii. Where more than 70% of blocks are centred then for any non-centred block on one side of the centre line there should be a block on the opposite side such that a reflection of either block about the centre line intersects the other to produce a block with area greater than 90% of both blocks.
iv. Blocks should be parallel to the horizontal edges of the page or at an angle greater than 10 degrees

Although there was overlap of similar, and sometimes identical, rules used by the two experts, there were also substantial differences between the two rule sets.

4. Testing the Rules Elicited from the Experts

The objective now was to compare the performance of a simulated rule-based system with that of the expert on a new set of publications from the same general population. The purpose was to determine the effectiveness of the rules and whether the expert's application of the rules could be predicted on a new sample of posters.

4.1. EXPERIMENTAL DESIGN AND METHOD OF ANALYSIS

4.1.1. Procedure
(a) Apply the rule-based diagnostic system to a new sample of posters unseen by the design experts.

The lists of rules previously elicited from the experts were applied to a new sample of posters. The results were recorded as a prediction of the expert's evaluation.

(b) Obtain the expert's diagnoses of bad design symptoms on the new sample of poster layouts.

The unmarked samples were then sent to the experts for their diagnosis, as for the first sample of posters.

4.1.2. Rule sets

Crucial to the experiment was the need to determine beforehand the most appropriate statistical design for the data characteristics. A study of the rule sets revealed that several applied insufficiently to be useful for the purpose of statistical analysis. Pilot studies of the statistical method had shown that rules which applied to fewer than 15 posters would generate unreliable results. Nothing at all could be concluded where a rule applied to fewer than 5 posters. Consequently several rules were eliminated because they failed to satisfy these pre-conditions, leaving 17 A-rules and 13 B-rules.

4.1.3. Sample size

From a study of the statistical requirements it appeared that 30 was the minimum number of specimens required to ensure reliable results from the implementation. It was also essential to determine the minimum sample size to ensure that the experts' time was not used unnecessarily. Consequently a new set of 30 posters was collected in the same way as before.

4.1.4. The experts' evaluation of the posters

Following the same procedure as described earlier, the new samples were distributed to experts (A) and (B) for their independent observations and comments on the bad symptoms identified in the specimen layouts in the sample.

Separate lists of the experts' comments were made and later compared. It was noted that the degree of detail varied between the experts and from one specimen to the other as expected from the previous responses. Often different symptoms in the same specimens were identified by the experts. Moreover designers describe the same symptom using personal expressions in a personal blend of specialist and emotive terms. They also attach different levels of importance to the same symptom. This suggests that the criteria designers employ are different and their values are variable. In the experts' responses to the new sample, subtle variations to previous comments were apparent and some new comments were reported.

We examined the returned samples and each of the experts' comments to determine which of the formalised rules applied. This procedure relied on the accurate translation and matching of rules formulated from the comments reported in the old with those reported in the diagnosis of the new samples. A list of the bad features abstracted from the comments reported by each expert and the number of times the comments applied was recorded, see Tables 1 and 2.

TABLE 1. Comparison of a rule based diagnosis using (A)'s rules, with expert (A)'s own diagnosis of bad symptoms in the same sample.

Rule		Number of posters to which the rule could be applied	Bad symptoms diagnosed by rule based system	Bad symptoms diagnosed by expert A	Bad symptoms diagnosed by expert A in follow-up
A1	If the are left and right margins, then they should be equal; + or - 2mm, excepting when a picture bleeds off the page	28	17	0	17
A2	If there are top and bottom margins, then the top margin should be less than or equal to bottom margin	28	8	0	8
A3	Blocks should not be in contact - vertically	23	6	2	6
A4	Blocks should not be in contact - horizontally	23	5	0	5
A5	Word blocks should not touch - horizontally	30	0	0	0
A6	Vertically adjacent blocks should be vertically left and right aligned	17	3	1	3
A7	When more than 70% of blocks are centred then all blocks should be centred	23	9	4	9
A8	Blocks of capital letters should be equal to or less than 4 lines	25	4	1	4
A9	For three or more vertically adjacent text blocks of the same fount, the space between the tops and bottoms should be equal	17	5	0	5
A10	The longest line in a non-list block of text greater than four lines should be a minimum of twenty characters	15	0	0	0
A11	There should be less than four typefaces	30	3	3	3
A12	There should be less than five type sizes; excepting lists, tables and diagrams	30	8	2	8
A13	The words in a line should be separated by less than or equal width to a lowercase 'n' in the fount used; excepting lists and tables	30	12	9	12
A14	Characters in a word should not touch; excepting ligatures, script or italic text and imported artwork	30	2	0	2
A15	Ascenders and descenders in consecutive lines of text should not touch	26	0	0	0
A16	The largest block must be above the horizontal centre of the page. If more than one large blocks of equal size, then at least one must obey this rule	30	11	2	11
A17	The distance between centre lines of vertically adjacent blocks of centred strings of words must be less than 1 mm	30	23	1	23

While there were forty specimens in the old sample there were just thirty specimens in the new sample. The general standard of design of the new sample of poster layouts was reported by the experts to be 'much better' than the design of layouts in the old sample. As a result fewer bad symptoms were reported by the experts in the new sample.

4.1.5. Application of the rule sets to the new posters

The specimens were also evaluated using a systematic application of the previously formulated rules. The results are shown in Tables 1 and 2. The figures show (i) the number of posters in the sample to which the rule was applicable; (ii) the number of times 'bad' symptoms were observed in those posters by systematic application of the rule; (iii) the number of times 'bad' symptoms comparable to the rule were reported by the expert; (iv) the number of times 'bad' symptoms were identified by the expert when subsequently talked-through the rules by the experimenter in a follow-up interview.

TABLE 2. Comparison of a rule based diagnosis using (B)'s rules, with expert (B)'s own diagnosis of bad symptoms in the same sample.

Rule		Number of posters to which the rule could be applied	Bad symptoms diagnosed by rule based system	Bad symptoms diagnosed by expert B	Bad symptoms diagnosed by expert B in follow-up
B1	If there are left and right margins, then they should be equal; + or - 2 mm	28	17	1	17
B2	If there are top and bottom margins, then the top margin should be less than or equal to bottom margin	28	8	8	8
B3	Characters should be recognisable	30	1	0	1
B4	Characters should not touch	30	12	0	12
B5	The largest block must be above the horizontal centre of the page. If more than one large block of equal size then at least one must obey this rule	30	11	1	11
B6	There should be less than four typefaces; excepting any imported artworks	30	2	2	2
B7	If there are three or more vertically adjacent blocks of the same fount then the distance between the adjacent edges of any two of these	15	8	3	8
B8	Vertically adjacent characters should not form a word	30	0	0	0
B9	The width of blocks below the centre of the page should be less than at least one block above the centre of the page	30	12	1	14
B10	The type size of text below the centre of the page should be less than the largest type size above the centre of the page	30	2	0	2
B11	The bottom margin should be less than 1/3 of the page height	30	0	0	0
B12	Within a box rule the space at the top, the top should be less than or equal to the space at the bottom	15	10	1	10
B13	Blocks should be parallel to the horizontal edges of the page or at an angle greater than 10 degrees from the horizontal	30	2	0	2

4.1.6. Analysis

The analysis is based on the association between our identification of bad symptoms by the systematic application of the expert's prior rules and the expert's own reported bad symptoms. Our assumption was that the experts would use the same rules for the new sample as for the old. Our aim was to determine if the experts tended to agree with the application of their rules.

Fisher's exact test (Siegel, 1956) was used in the analysis of the data. The Fisher exact test is appropriate where figures smaller than 5 occur in the data. On this basis of this analysis Expert (A) used rules 11 and 13 consistently, and Expert (B) used rules 2 and 6 consistently. These were the only cases in which a statistically significant relationship was found. In all other cases we have no reason to reject the null hypothesis that the designers did *not* use their rules to judge the designs.

This was an unexpected negative result which could have seriously undermined the thesis that elements of graphic design aesthetics are rule-based, and can be built into computer systems. However, follow-up interviews then produced an equally unexpected positive result. In these the researcher pointed out to the experts that their previous 'rules' had been violated, and the experts then agreed 100% that these 'rules' had been correctly applied in the simulated rule based system.

5. Follow-Up Interviews

Analysis of the data revealed discrepancies perhaps due to ambiguities in some of the definitions derived from both experts' comments. To facilitate verification and clarification of these issues follow-up interviews with the experts were convened. On these occasions, the experts were informed of the rules that had been elicited from their earlier diagnoses of bad symptoms.

5.1. FOLLOW-UP INTERVIEW WITH (A)

The elicited rules were discussed with (A) after the implementation tests were completed. The discussion revealed inconsistencies in the expert's reporting of bad symptoms and conditions which affected the application of rules and their interdependence. In some instances the expert had not applied his 'rules' because there were additional factors which conditioned the rule's application in the second sample.

After the follow-up interview, several of the rules were revised, adding conditions of their application. The definitions were also refined; e.g.

(A13) Original: The words in a line should be separated by less than or equal width to a lowercase 'n' in the font used; excepting lists and tables

Revised: *The words in a line should be separated by less than or equal width to a lowercase 'i' in the font used; excepting lists and tables.*

Expert (A) acknowledged that there were anomalies in his reporting of some bad symptoms; e.g. although it was agreed that

(A8) Blocks of capital letters should be equal to or less than 4 lines

this was reported on only one occasion when the expert gave the following reason for the bad symptom as "The bottom 7 lines are unreadable".

Although reported by Expert (A) to be important this symptom was not reported in three other instances where we observed that the rule applied. In our systematic application of the rule these bad symptoms were routinely identified, reported and shown as mismatches in the result of the implementation. Some possible other factors in the identification of this bad symptom are:

(i) The lines of type are centred; so are the three other instances in the other specimens.

(ii) The type was set solid, i.e. with no additional spacing between consecutive lines of type.

(iii) The type size is 24 points.

(iv) The block of capital letters reported is positioned at the bottom of the page.

(v) The type style is a serif typeface. Would there have been a difference if a sans serif face had been used?

The association of these factors and perhaps others which contribute to the symptom being reported as bad was not explicitly addressed at the follow-up interview. However it was observed that the other three instances had some leading (additional line spacing); it is evident that the addition of leading would improve the readability of this block of type.

However a block of three lines of type in the same layout and with the same type parameter as above was not judged by the expert to be bad, so *the number* of lines (between 4 and 7) is the crucial factor in this instance. Enhancement of the rules and their re-application to the sample of posters produced complete matches in the observed instances.

5.2. FOLLOW-UP INTERVIEW WITH (B)

The same procedure was employed as applied to the meeting with (A). The researcher showed the list of rules to (B) and both referred to them during the discussion. The discussion with (B) revealed conditions which affected the application of rules and their interdependence as had the interview with (A).

The new information obtained at the follow-up interview was used to enhance the formulation and subsequent application of the (B) rules. There were two notable instances of rule enhancement:

(B13) Original: Blocks should be parallel to the horizontal edges of the page or at an angle greater than 10 degrees from the horizontal was modified to:

> *The horizontal edges of blocks should be parallel to the horizontal edges of the page.*

(B2) If there are top and bottom margins, then the top margin should be less than or equal to the bottom margin
was modified to:

> *If there are top and bottom margins, then the top margin should be less than the bottom margin.*

Enhancement of the rules produced complete matches in the observed instances in the re-application of the re-formulated rules to the sample.

5.3. COMMENTARY ON THE FOLLOW-UP INTERVIEWS

The results show that for the two designers studied, these designers' intuitive judgement of bad design in a population of A4 desktop publications can be elicited and formulated into a coherent and rigorous set of rules which are generalizable to disparate specimens within the general population.

Analysis of the results shows that these designers (perhaps unknowingly) do use rules to diagnose bad design symptoms. These designers' intuitive knowledge can be interpreted as rules. A graphic designer consistently reuses some of the same 'rules' in the evaluation of different specimens.

From their comments, it seems that the experts attempt mentally to recompose a layout when assessing its defects before offering their diagnosis. Indeed alternative solutions were sometimes offered. Also the experts evidently attempt to determine the prevailing constraints in an attempt to evolve a more appropriate alternative solution. This can be deduced from the recommendations and suggestions they offer in their reasons why a symptom is bad.

A graphic designer's perception of layout design also appears to be conditioned by the number of bad symptoms present in a layout, i.e. the fewer the number of defects the more forgiving the assessor. Consequently a major defect is sometimes overlooked where there are few or no other bad symptoms. Where there are many bad symptoms the expert's diagnosis is more stringent.

Graphic designers may apply their rules inconsistently to different specimens. The evidence suggests that the performance of the rules benefits from feed-back to their source, i.e. it was observed that rule application was highly consistent with the expert after the follow-up interview.

It has been demonstrated that the systematic rule based procedure is more consistent than the expert in the application of rules formulated from the expert's comments.

6. Some Possible Causes of the Experts' Inconsistency

What were the reasons for the experts' inconsistency in the application of their rules? The experiment shows that the expert's judgement is not absolute, it is provisional and inconsistent. Inconsistency suggests that the expert is perhaps uncertain of the application of the rule and/or the factors affecting its application. It may also reveal the addition of new rules. But this is difficult to verify. It may show the withdrawal of a previously used rule. The addition or withdrawal of a rule in regrading marks the point when, where and in what circumstances a rule is used.

The experts agreed with the rules formulated from their critiques of the specimens in the samples when their rules were discussed at the follow-up interview. In some instances the experts had not applied their 'rules' because there were additional factors which conditioned the rule's application in the second sample. One reason for this perhaps was due to the higher standard of design reported by the experts in the second sample of specimens. As reported earlier, the result was that the experts were more forgiving of the bad symptoms observed in the second sample.

Inconsistency is perhaps due in part to the monotony of the repetitious nature of following a diagnostic procedure. The variability within the sample may be another factor, e.g. some contained imagery while others were text only.

Inconsistency may result from visual illusions and perceptual constancy. The expert's cognitive style may be field-dependent or field-independent, i.e. in the perception of a publication some people may be more affected by the content and others by the page. Inconsistency may be due to interest in or dislike for the subject of the narrative. The expert's attitude may be affected by low motivation resulting from the lack of interest and lack of reward.

7. Conclusions

Pye (1978) suggests that 'people do not unanimously agree about what is beautiful and what is not, for they do not unanimously agree about anything whatever'. Here it has been demonstrated that complete agreement between the successive critiques of the same expert is unlikely and suggests a reason that agreement in different individuals' opinions should not be expected. This suggests why a rule based system performs aesthetic evaluation with greater consistency and reliability than experts.

We have described a novel method for the elicitation of aesthetic rules of graphic design evaluation from two graphic design experts. The procedure devised relies on the graphic design experts' independent diagnosis of symptoms of bad design in a sample of graphic layouts.

The diagnosis of 'bad' design symptoms in a graphic layout is shown to be an effective method of making explicit the graphic designer's implicit intuitive, aesthetic judgement. It is argued that the presence of a bad design symptom in a graphic layout indicates that the design is not good and that the absence of bad design symptoms suggests that the design is not bad.

The procedure devised required the graphic design experts to provide their rationale for their diagnosis and this was used to inform the application of their rules to the bad symptoms identified in the graphic layouts.

The subtle expressions used by the graphic design experts provided interpretational difficulties which were resolved by the formulation of the rules into representational statements suitable for use in a computer system using the methodology of 'The Expert System Test'. This methodology supports the unambiguous application of a rule in the diagnosis of a bad symptom.

The results of our tests show that the systematic application of graphic design rules performs more consistently in design evaluation than the experts from whom they were elicited.

It can be concluded that:

• Rules for aesthetic evaluation of graphic design can be elicited from an expert and consistently applied in a rule based system.
• Expert designers do not apply their own implicit evaluation rules consistently.
• When an expert designer has their rules made explicit to them, they will apply them consistently, in agreement with the rule based system.

The rules of aesthetic evaluation identified in this study are acceptable to expert graphic designers, perhaps representing "canons" which they believe, i.e. they are real aesthetic (bad, not bad) evaluation rules. With the rules made explicit, they could be built into a computerised graphics system to support more naive users.

References

Moore, C. J. and Miles, J. C.: 1991, Knowledge elicitation using more than one expert to coverthe same domain, *Artificial Intelligence Review*, **5**, 255-271.
Pye, D.: 1978, *The Nature andAesthetics of Design*, The Herbert Press Ltd., London.
Siegel, S.: 1956, *Nonparametric Statistics for the Behavioral Sciences*, McGraw-Hill Book Company, London.
Tunnicliffe, A. J. and Scrivener, S. A. R.: 1991, Knowledge elicitation in design, *Design Studies*, **12**(2), 73-80.

J. S. Gero and F. Sudweeks (eds), Artificial Intelligence in Design '96, 541-559.
© 1996 *Kluwer Academic Publishers.*

A MODEL-BASED TOOL FOR FINDING FAULTS IN HARDWARE DESIGNS

MARKUS STUMPTNER AND FRANZ WOTAWA
Christian Doppler Laboratory for Expert Systems
Institut für Informationssysteme
Technische Universität Wien

Abstract. The state of the art in integrated circuit design is the use of special hardware description languages such as VHDL. The designs programmed in VHDL are refined up to the point where the physical realization of the new circuit or board can be created automatically. Before that stage is reached, the designs are tested by simulating them and comparing their output to that prescribed by the specification. A significant part of the design effort is taken up by detection of unacceptable deviations from this specification and the correction of such faults. This paper deals with the development of VHDLDIAG, a knowledge-based design aid for VHDL programs, with the goal of reducing time spent in fault detection and localization in very large designs (hundreds of thousands of lines of code). Size and variability of these programs makes it infeasible in practice to use techniques based on a detailed representation of program semantics. Instead, the functional topology of the program is derived from the source code. Model-based Diagnosis is then applied to find or at least focus in on the component(s) in the program that caused the behavioral divergence. The support given to the developer is sufficiently detailed to yield substantial reductions in the fault localization costs when compared to the current manpower-intensive approach. A prototype is currently being tested as an integral part of the standard computer-aided design environment. Discrimination between diagnoses can be improved by use of multiple test cases (as well as interactive input by the developer).

1. Introduction

The current state of the art in the design of integrated circuits is based on heavy use of hardware specification languages. Starting from a specification, a new ASIC or circuit board is designed by developing a description in such a language, which can then be executed to simulate the functionality of the circuit. This significantly increases the chance that errors in the design can be found and corrected before the physical circuit is produced, thus reducing the costs of the overall design process (throwing away and redoing the masks and tooling for a circuit design that was found to be defective is an extremely expensive proposition). As the design process continues, the design is continually refined, until a level of detail has been achieved where it can be transformed automatically into a representation at the logic gate level. The gate level description is used as the basis for layouting and the production of the physical circuit, a process which also requires little human interference.

As a result, IC design is increasingly getting similar to the software design process, and the search for faults in the programs that describe the designs tends to absorb a significant part of the design effort in the earlier stages of the design process, all the more so since large hardware designs (comprising multiple ASIC's and microprocessors) can reach dimensions of several 100.000 lines of VHDL code and thousands of components and signals at the top level. For such designs, typically written by large design teams (or multiple teams at different physical locations), fault detection and localization becomes a very time-consuming activity.

This paper describes the principles behind the VHDLDIAG tool that is used as a design aid in the development of hardware descriptions in VHDL (Very High Speed Integrated Circuit Hardware Description Language), which is probably the most widely used of these languages. We use techniques of model-based diagnosis for creating a simple internal representation of the design, checking test runs for errors, and locating the source of the errors. If unique identification is not possible, the tool helps at least in focusing the attention of the user (i.e., the hardware designer) on those parts of the system where the problem originates, proposes signals whose observation will reduce the set of diagnoses, and can also continue analysis on the basis of observations entered interactively by the user. The system is used in conjunction with existing commercial design support tools (e.g., simulators and graphical design tools) and is intended for use in all design phases where VHDL is used.

The paper is structured as follows: Section 2 takes a closer look at the hardware design cycle using VHDL and the requirements for a knowledge-base debugging tool. Sections 3 and 4 describe the representation and reasoning techniques on the basis of a simple example. Finally, we discuss experiences with implementation and test use of the system and compare our approach with other work related to knowledge-based software design support.

2. The Hardware Design Cycle

Figure 1 gives an overview over the hardware design cycle using VHDL. Typically, the design of an integrated circuit (either an ASIC or a circuit board with multiple ASICs mounted on it) starts with a specification delineating the functional requirements.

This specification document can be given as either a pure textual document or as a VHDL program. It usually consists of a coarse description of the inten-

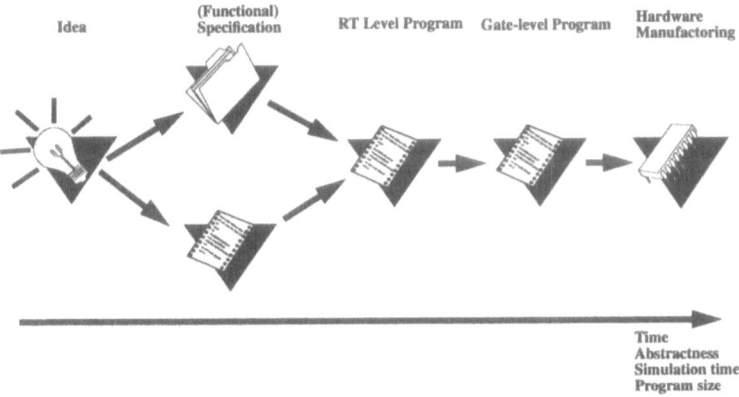

Figure 1. The hardware design cycle.

ded architecture of the design, i.e., a partitioning into components, the functional description of all parts and their interactions, and other design information such as expected fault probabilities, used technology, timing constraints and physical constraints (i.e., working temperature range). Finally, a testing guide will be supplied. If the basic partitioning and the functional descriptions are already written in VHDL, one gains the big advantage of being able to test or verify the specification using the mental model of the expected hardware device. Otherwise, a very simple VHDL design will be created initially to reproduce as closely as possible the specified behavior without paying attention to architectural or structural notions.

From here on, design progresses by the addition of more detailed substructure to the design, i.e., the description of the design in terms of components and their subcomponents. The functional behavior of the design is distributed across the components so their interaction will reproduce the originally specified behavior. Testing of the new versions of the design is achieved through simulating its behavior and having the designer observe the data values occurring on the signal lines between components. The graphs of signals over time are called waveforms (see Figure 2). The waveform traces of the program to be tested, usually comprising several 10.000 signal changes, are compared to the traces generated by the specification on the same input values. From here on, we will refer to the former as the *implementation waveform* and the latter as the *specification waveform*. A discrepancy between the two may indicate the existence of a fault in the design.

The next abstraction level reached after the functional specification is the so-called register transfer level (RTL). The RTL design will be much larger than the original specification, increasing simulation and therefore testing time. A typical RT programs could have about 6 MB source code and produce simulation runs lasting from one hour to several days, depending on the simulated amount of real time. After finishing the RT program implementation, a synthesis tool con-

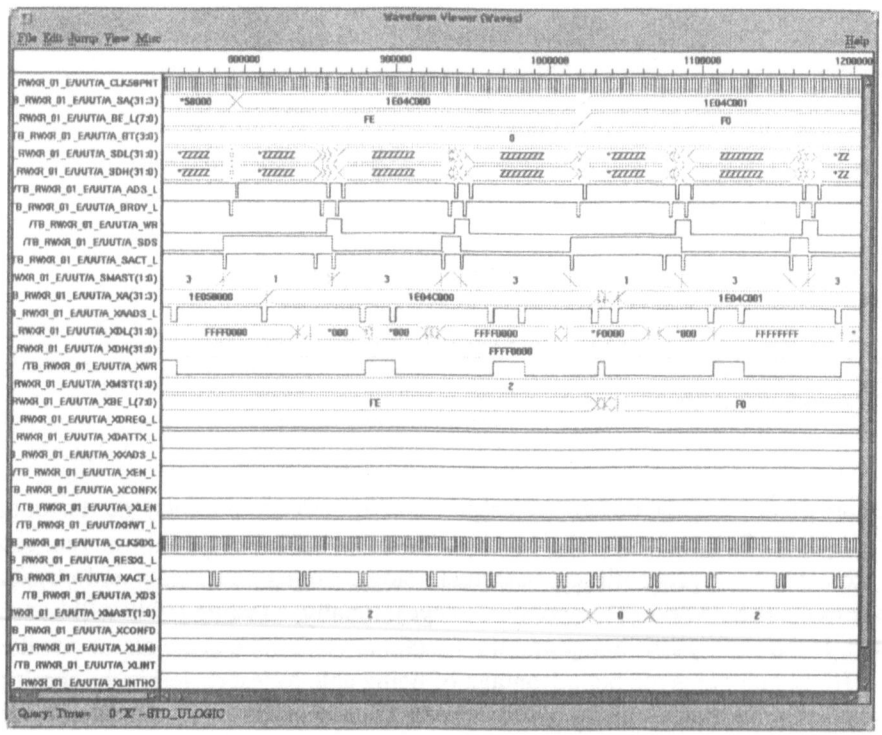

Figure 2. A typical waveform trace.

verts it to a gate-level program with a limited amount of user interaction (which deals with specification of certain parameters such as, say, the width of a bus). Gate-level programs are then simulated as well. The resulting waveform traces are checked for correctness. Because of the automatic conversion, a detected misbehavior of a gate-level program (that performed correctly at RTL level) can only be caused by timing problems and virtually never by a functional fault. Timing faults are mainly corrected using the synthesis tool and trying different parameter values. The gate-level VHDL code is not modified by hand. In contrast, faults in RT programs or specifications are directly corrected. While the assumption usually holds that the more abstract (already tested) program is the correct one, the designer also has to check for the possibility that this is not the case.

Figure 3 shows the difference between the abstraction levels using a small part of a VHDL behavior and a gate-level program.

Next, we take a brief look at the testing considerations for the various design phases.

Functional Specification In addition to preliminary information about structure and behavior, the functional specification als contains information about the

```
...                              ...
signal S,A,B: integer := 0 ;     signal S,A,B: bit_vector( 1 to 4 ) := "0000" ;
...                              ...
S <= A + B ;                     c1: carryadder(A(1),B(1),'0',S(1),QC1);
...                              c2: carryadder(A(2),B(2),QC1,S(2),QC2);
                                 c3: carryadder(A(3),B(3),QC2,S(3),QC3);
                                 c4: carryadder(A(4),B(4),QC3,S(4),Q);
                                 ...
```

(1) Behavior   (2) Gate-level

Figure 3. VHDL behavior vs. gate-level: Adding two numbers.

intended gate technology, environmental considerations, etc. It can only be verified by comparison to the underlying mental model or older versions of itself, possibly also written in VHDL (in the case of regression tests).

Register Transfer Program Multiple versions of the RTL program will usually be created until the program meets the requirements (i.e., matches the behavior of the specification). Every version must be tested using most of the test cases. Usually up to 20 versions are shipped out for testing. Comparing test results and finding the faults is a major task while programming in register transfer level. A study has shown that up to 60 percent of the effort in this design stage are used for these tasks.

Gate-level Because simulation time and compare time are very high compared to RTL programs, automatic detection of discrepancies is also recommended for the gate-level phase. A simulation run is typically compared with the simulation run of the associated RT program. Subsequent gate-level versions are not directly compared to each other.

3. Knowledge-Based VHDL Design Support

The original goal of the project was to develop a tool that would reduce the overall design effort, without engendering significant changes in the overall structure of the design process (which is codified and fixed by the funding institution). The main interest lay in reducing the amount of time for each individual simulation/fault detection/fault correction cycle, as well as (by improving the quality of the detection and correction stages) possibly reducing the number of cycles. Figure 4 shows the design subcycle. The time involved in a single iteration depends strongly on the complexity and size of the program (i.e., the time increases as the design progresses).

Of the three stages, simulation is already being handled by highly optimized commercial tools. Detection and correction are largely manual tasks open to improvement.

Of major importance is the requirement that any tools generated should not result in the need to alter or embellish the design cycle or the designs themselves.

The effort of using the tools should be kept to a minimum, i.e., while entering numerical parameters would be considered adequate, for example, developing a separate representation for every design was out of the question. This is important since designs often involve integration of VHDL code coming from different sources which may or may not be using the same tools (e.g., circuits from subcontractors or from the extensive standard libraries supplied by the simulator companies).

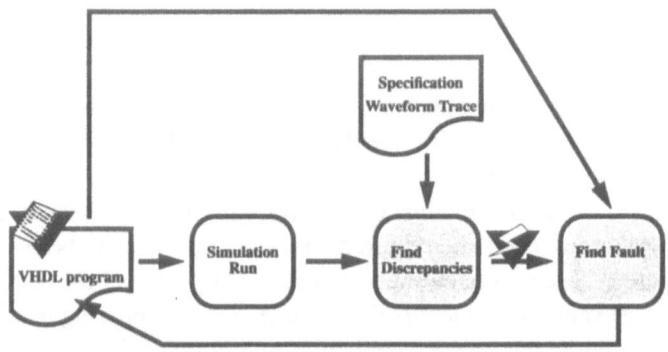

Figure 4. The VHDL design subcycle.

What is required therefore is a generic approach that will allow the mapping of the semantics of a VHDL program to the somewhat abstracted internal representation. Given a discrepancy, the resulting model of the program should then be analyzed to find or at least limit the area of the program where the fault can have originated. Therefore, the model must represent the structure and the functional and causal relationships between the signals in the VHDL program. This can be achieved with an adaptation of the representation and reasoning mechanisms commonly used in model-based diagnosis. However, apart from the model, we also need to derive the observations that describe the actual fault which occurred. Since virtually all the information available about a simulation run is contained in the resulting waveforms, the comparison of specification and implementation waveforms is a crucial prerequisite to checking the correctness of the program and an integrated part of the tool. Therefore we deal with this issue first. Section 4 deals with the diagnosis process.

3.1. FINDING FAULTS IN THE DESIGN

Currently, comparison is done manually by the user (hardware developer) with the help of test benches which can be used to format the output in an appropriate manner. However, the test benches themselves require significant development effort and are by their nature both specific to a particular circuit and abstraction level. Ultimately, a major part of the detection effort is spent by the designer in scrolling along an implementation waveform such as in figure 2, trying to spot timing or signal value differences with regard to the matching signal from the other trace.

The VHDLDIAG diagnosis tool eliminates the need to write testbenches by executing an automatic comparison between implementation and specification trace. Internally, a waveform trace is represented as a set of *events*, each of which is a tuple (s, v, t) representing the fact that a certain signal s changed to value v at time t. The value of each signal at an arbitrary point in time is the value produced by the last event up to that time. Instead of checking the waveforms manually, the designer now merely needs to provide

— the specification of which signals are to be simulated.
— the choice of the exact comparison operator.

All the traced signals can be compared in a single comparison run after the simulation is finished. [1] Note that programs have to be syntactically correct to be simulated. Also, the approach obviously cannot deal with errors that do not produce divergent signal values, i.e., a nonterminating loop inside a user-defined function, which will simply cause the simulation to hang.

The need to provide a choice of compare operators arises from the fact that every program manifestation may have a (slightly) modified behavior that is expressed by subtle waveform trace differences without impeding the correct function of the design. These differences can occur, for example, because abstract hardware descriptions do not take delay times into account while gate-level descriptions do, or because a simulation run contains minor errors (i.e., double initialization cycles) that do not influence the overall outcome. Testing that both traces are identical is therefore not always an useful way to detect errors, despite the deterministic nature of the simulation. Another (real world) example would be that of two otherwise identical simulation runs using different clock cycles. In such a case an identity comparison of signal value behaviors would deliver literally thousands of errors. The different compare operators can be used to introduce tolerances into the comparison process.

Identity (I) Comparing for identity means that an error occurs if at some time t a signal s has different values in both traces.

Identity plus Tolerance (IT) The introduction of a tolerance limit allows a small temporal difference δ between two waveforms. An event that occurs at time t in one waveform trace must occur in the interval $[t - \delta, t + \delta]$ within the second trace.

---

[1] The number of signals to be traced and compared is currently not restricted by the diagnosis tool but by the simulators which do not have the capacity to trace a large number of signals, as the ability to consider more than about a hundred signals per run (out of, say, 10,000 in a large ASIC) was beyond the capability of any human observer.

Functional (F) A functional comparison ignores the absolute time values in both traces. Instead, the events in both waveform traces are grouped according to the partial order defined by their time values. All events occuring at the same time are summarized in a so-called state. Two waveforms are said to be equal if their sequences of states are equal.

Functional plus Tolerance (FT) This mode allows an event a to change its position with regard to an event b in the functional ordering, as long as the relative position of a to b is not displaced by more than δ along the original time axis.

The following is an estimate of the utility of the different compare modes based on use with actual trace files (0 – never used, 1 – always used). Several modes can be used for comparing the same pair of traces. Interactive tuning of the compare parameters is then used to improve the hit rate beyond the values in the table.

Compare Mode	Degree of Utilization	Fault detected but no fault	No fault detected but real fault
I	0.7	0.5	0
IT	0.5	0.4	0.1
F	0.5	0.2	0.1
FT	0.2	0.2	0.2

From a more theoretical viewpoint we can order the compare modes by their relative strictness. In a two dimensional space, each dimension consisting of possible waveform traces, we denote by R_m the number discrepancies reported by each mode $m \in \{I, IT, F, FT\}$. The following relations hold for all possible traces:

$$R_I \subset R_F \subset R_{FT}$$
$$R_I \subset R_{IT} \subset R_{FT}$$

The comparison process also allows for sampling, i.e., the restriction of the comparison to specific, possibly periodic time points or intervals during the simulation.

The need to provide tolerance intervals for values instead of event times did not arise as floating-point signals occur rarely in the designs. It is, however, possible for different data types to be used for a pair of matching signals in the two traces. Gate-level programs, for example, typically use a four-valued logic and bit vectors where behavior and register transfer programs use nine-valued logics and integers, respectively. A simple value comparison would result in many spurious errors. Therefore, the designer can specify equivalence classes for data values to be used during the comparison.

In summary, the compare functionality implemented in the diagnosis tool is generic and not tied to a particular commercial simulation environment.

It makes test bench generation unnecessary and the required parameters can usually be chosen without problems by the designer. The result is a set of observations that are delivered to the actual model-based diagnosis engine, of the form $(Signal, \{(Value, Time)|\ldots\})$ for correct signal events and $(Signal, Value, Expected\_Value, Time)$ for discrepancies, i.e. deviations from the correct behavior. These observations are the basis for deducing which (probably unobserved) parts of the system may contain the fault.

4. Locating Design Defects

After finding the symptoms (signals showing a discrepancy), we have to find their source (as mentioned, typically only a few percent of the signals in a system are observed).

4.1. ADAPTING MODEL-BASED DIAGNOSIS TO DESIGN PROBLEMS

The model-based approach is based on the notion of providing a representation of the interactions underlying the correct behavior of a device. By describing the structure of a system and the function of its components, it is in principle possible both to reason about the way to achieve desired behavior (Stroulia et al., 1992), i.e., to synthesize a design (although this requires a very detailed model), as well as to ask for the possible reasons why the desired behavior was not achieved. In the diagnosis community, the model-based approach has achieved wide recognition due to the advantages already mentioned: once an adequate model has been developed for a particular domain, it can be used to diagnose different actual systems from that domain. In addition, the model can be used to search for single or multiple faults in the system without alteration.

The usual model-based system representation in diagnosis can be adapted to the design of VHDL programs without much trouble. A system is assumed to consist of a set of components $COMP$, whose correct behavior is described by a logical theory called *system description* (SD). The assumption that a component C behaves correctly is expressed by the fact $ok(C)$. The set of observations OBS contains statements about the actual, observed behavior of the system.

Using the standard consistency-based view as defined by Reiter (1987), a diagnosis Δ for a VHDL program is a subset of $COMPS$ such that the assumption of incorrectness for exactly the components in Δ is consistent with the observations:

$$SD \cup OBS \cup \{ok(c)|c \notin \Delta\} \cup \{\neg ok(c)|c \in \Delta\} \not\models \bot$$

The basis for this is that an incorrect output value (where the incorrectness could be observed directly or derived from observations of other signals) cannot be produced by a correctly functioning component with correct inputs. Therefore, to make the system consistent and avoid a contradiction, the component must be assumed to work incorrectly. In practical terms, one is interested in finding minimal diagnoses, i.e., a minimal set of components whose malfunction explains the misbehavior of the system (otherwise, one could explain every error by simply assume every component to be malfunctioning). We will return to the special features involved in diagnosing designs instead of finished artifacts, after discussing the representation.

Developing a model-based representation for VHDL programs faced two problems. First, the definition of formal semantics for VHDL is an open research topic (Kloos and Treuer, 1986), although the definition of the VHDL languages as an IEEE standard means that the existing commercial VHDL environments are reasonably compatible. Second, the size of the programs involved precludes the use of a more intricate representation. Merely executing (i.e., simulating) a VHDL program using a highly optimized commercial simulator takes from hours to days of real time on a high-end workstation. Therefore, diagnosing a complete logical representation of the full VHDL program and its semantics is not feasible. A strongly abstracted view of the design must be used for diagnosis. Accordingly, our representation abstracts over values and time points (Hamscher, 1991), but retains the capability to distinguish between the initialization phase and operating mode of a circuit, a requirement for handling feedback loops.

Additional criteria for the choice of representation were:

- No diagnoses may be excluded due to abstractions. In other words, misleading the designer is worse than offering him an answer that does not uniquely identify the component involved.
- Integration with available commercial simulation packages.
- Computational costs must be minimized by requiring very few additional simulation runs.

The principal idea is to abstract as much as possible over time and values on the one hand, while preserving the capability to discriminate between substantial parts of the VHDL-code on the other hand. Stronger discrimination between diagnoses can be achieved by applying multiple test cases and measurement selection (i.e., specifying signals that offer good chances of discriminating between diagnoses when included in a trace). Further discrimination can be achieved by requesting the user to evaluate the correctness of particular signals.

4.2. A VHDL EXAMPLE

To give an example of how to design models and find faults in VHDL, we introduce a simple VHDL program which represents a device we have called D75. The D75 is a sequential device with five input ports a, b, c, d, e and three outputs f, g, and h.

A possible initial design of the D75, consisting of four components, is shown in Figure 5. A designer who uses VHDL must now implement a program for every component to get the same behavior as the specification. The behavior can

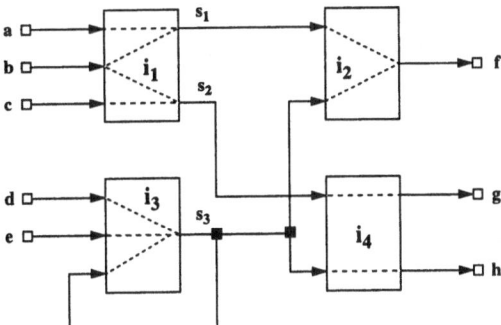

Figure 5. The D75 - Diagnosis components.

be tested by checking the waveform traces for every test case specified by the designer. If this check leads to discrepancies we want to find the component that causes the misbehavior. The dotted lines in Figure 5 show the functional dependencies between inputs and outputs of the components.

The following VHDL code fragment describes the internal topology of the D75 as shown in figure 5. (A full VHDL program would include additional declarations; they are not required for understanding the example and hence omitted.)

```
entity d75_e is
    port( signal a,b,c,d,e: in integer;
          signal f,g,h: out integer);
end d75_e;

architecture a2 of d75_e is
    ...
    signal s1,s2,s3: integer := 0;
    ...
begin
    i1: c1 port map (a,b,c,s1,s2);
    i2: c2 port map (s1,s3,f);
    i3: c3 port map (d,e,s3,s3);
    i4: c4 port map (s2,s3,g,h);
end a2;
```

This VHDL top level implementation of the D75 shows the most important VHDL components, *entities* and *architectures*. Entities are interface declarations, whereas architectures are behavior definitions, i.e., they associate a behavior with an entity. An architecture consists of a set of *concurrent statements*, so called because they are executed in parallel (i1 ... i4 in the example above). Communication between concurrent statements is achieved by *signals*. Signals are defined in the declaration part of the architecture or entity, or in the interface declaration (port ...) of the entity. VHDL is strongly typed. Each entity has a type associated with it (this means that multiple entities of the same type can exist). For brevity, we have omitted the type definitions c1, c2, c3, and c4. For details on VHDL syntax and semantics see, e.g., IEEE (1988).

The behavior of the concurrent statements in the D75 is described by signal assignments for every concurrent statement. We write in$i$ for the i-th input and out$i$ for the i-th output of the concurrent statement. We order the signals of the port maps so that input signals come first and outputs last.

```
i1 port map(in1,in2,in3,out1,out2)
      out1 <= in1 * in2;
      out2 <= (-1)*(in2 * in3);
i2 port map(in1,in2,out1)
      out1 <= in1 + in2;
i3 port map(in1,in2,in3,out1)
      out1 <= in1 + in2 + in3 after 5 ns;
i4 port map(in1,in2,out1,out2)
      out1 <= in1;
      out2 <= in2;
```

A concurrent statement represents a parallel process associated with a particular component in the system. Due to the hierarchical structure of entities (each of which has a set of concurrent statements associated with it), VHDL code is partitioned into hierarchies of concurrent statements which usually bottom out in a sequence of sequential statements, since the computation of a new signal value (which in our case only uses standard arithmetic operators and a simple delay) can be arbitrarily complex, using library functions and user-defined functions that use the full expressiveness of the imperative language parts of VHDL (these bear a strong resemblance to the language Ada). The current version of the VHDLDIAG system does not analyze the semantics of sequential statements, its granularity is at the level of concurrent statements.

4.3. SYSTEM DESCRIPTION

Since the simulation is executed by commercial simulation tools, the internal state of the system as it is simulated is not visible, apart from the signal values that are contained in the trace. Therefore, it is not possible to determine by observation in which manner exactly a particular (e.g., incorrect) data value was computed. VHDLDIAG analyzes the VHDL code statically and computes the *functional dependencies* between signals (i.e., which signals are possibly involved in the computation of which other signals) regardless of whether the actual branch of the program that contains the signal assignment in question is actually executed at runtime. The system model based on functional dependencies is then used to search for the components which are responsible for the discrepancies. For example, in the D75 circuit, s_1 depends functionally on signals a and b.

The main idea of our abstraction is to ignore time and values of the signals. In our system description we only talk about correct or incorrect signals which means that we do not care when the discrepancy occurred and what the expected value would be. A simple ordinal time axis is retained for the handling of feed-back loops. The length of the time axis is equal to the number of components (since we are ignoring actual time values, n time steps are sufficient to propagate any observation through all components in the system).

On one hand this abstraction has the advantage that we do not need to reason about thousands of time points for several hundreds of signals. On the other hand it has the obvious drawback that we can exclude fewer diagnoses than with a detailed model. However, programmers use this abstraction during debugging very effectively to identify the faulty concurrent statement.

4.3.1. *Components*

The system description modeling the component behavior of a VHDL program is a set of sentences (one for each value of T) on the time axis) for every concurrent statement $C \in COMPS$ and $out_i \in out(C)$ of the form

$$ok(C) \wedge ok(in_1)_T \wedge \ldots \wedge ok(in_n)_T \rightarrow ok(C, out_i)_{T+1}$$

and for the initialization

$$ok(C) \rightarrow ok(C, out_i)_0$$

where out_i depends functionally on the input signals $in_1 \ldots in_n$. In effect these rules state that the output of a component is correct if the component is correct and the input signals on which the output depends are correct. For example, the following describes the concurrent statement $i4$ in the D75:

$$ok(i4) \wedge ok(s2)_T \rightarrow ok(i4, g)_{T+1}$$
$$ok(i4) \wedge ok(s3)_T \rightarrow ok(i4, h)_{T+1}$$

where $T \in \{1, \ldots, 4\}$

4.3.2. *Connections*

Let $Driver(s) = \{c | c \in COMPS \wedge s \in out(c)\}$ where s is a signal. The system description modeling the connections of the VHDL program given by E and A is the set of sentences which includes for each signal $s \in Signals(A, E) \wedge s \in out(c) \wedge c \in COMPS$ the sentence

$$\bigwedge_{c_j \in Driver(s)} ok(c_j, s)_T \rightarrow ok(s)_T$$

i.e., the output of a signal is correct if all signal values being placed on it are correct. For example, given component $i3$ from our example the rule

$$ok(i3, s3)_T \rightarrow ok(s3)_T$$

must be element of the system description.

For a more detailed discussion of the representational issues, see Friedrich et al. (1995).

4.4. COMPUTING DIAGNOSES

Given the system description, Diagnoses are computed by applying the standard hitting set DAG method as described in Reiter (1987), Greiner et al. (1989)

and de Kleer (1995). In principle, the method is based on computing the so-called *conflict set*. In MBD terminology, a conflict is a disjunction of abnormality assumptions for individual components that is implied by $SD \cup OBS$. In other words, at least one of the components in C must be abnormal for SD to be consistent with OBS. Computing a minimal hitting set for the set of minimal conflicts yields a diagnosis. The system description basically has the effect of propagating correctness assumption claims along signals as long as components are assumed correct. If an inconsistency is derived because the correctness is propagated to a signal that was observed to be incorrect, the components involved in the propagation are part of the conflict. E.g., in Figure 5, if signals $s1$ and $s3$ are known to be ok, but $\neg ok(f) \in OBS$ (f is known to be not ok), and $ok(i_2)$ is assumed, then $ok(f)$ will be derived, yielding a contradiction that results in the conflict $\{i_2\}$ (i.e., i_2 must be part of any diagnosis given these particular observations). Various heuristics are used to restrict the size of the hitting set DAG and improve performance. For more detail, see Friedrich et al. (1995).

The diagnosis tool (called VHDLDIAG) has been implemented in Smalltalk (Visualworks) and is currently used in actual production work by selected designers. Below are typical performance figures, produced with an actual ASIC:

Components at top level	379
Number of interface signals at top level	210
Number of signals at top level	412
Maximum number of levels	5
Average number of levels	~ 2
Size of source code	6 MB
Number of gates in produced circuit	$> 100,000$
Simulation and Compare time	30 minutes
Actual Diagnosis time	2 minutes

4.4.1. *The assumption of model correctness*

In a certain respect the problem of diagnosing a design is unique in the realm of model-based reasoning. In conventional model-based diagnosis, the system description is an exact specification not only of the overall behavior of the system, but of its individual parts. For example, when diagnosing the hardware implementation of a 16-bit adder, the adder's system description will describe the behavior of the logical gates from which the adder is composed. A fault is assumed to occur because one of the components does not act according to its specification.

In diagnosing a design, however, the assumption that the specification will be a complete representation of the structure of the artifact is obviously invalid. In our case, the internal structure and the way in which the behavior is described will differ widely between a functional specification and its RTL implementation

– the implementation will usually contain many internal components and signals which have no counterpart at all in the functional specification. The only part of the specification that is directly usable is the waveform trace generated by the specification. We are therefore forced to base our model of the VHDL implementation on analysis of the code of the implementation itself. That implies, however, that it is the model that reflects the incorrectness of the design and whose output (the implementation trace) is confronted with observations that are correct (the specification trace), whereas in traditional diagnosis problems, the model is correct and it is the observations, made from the behavior of the actual system, that reflect on the incorrect behavior. In addition, the question of how a design defect may manifest itself in the model leads us to the related issue of so-called structural faults.

4.4.2. *The assumption of structural correctness*

Structural faults are faults that do not occur because a component is functioning incorrectly, but because there is a missing or *additional* connection between two components, as in a bridge fault in electrical engineering. The bridge fault problem was first noted by Davis (1984) and mainly excluded from consideration in subsequent work on diagnosis. However, it is very relevant when diagnosis is applied to designs, and in particular, software designs. The use of an incorrect argument in an expression (e.g., by using a different variable name, switching the ordering of arguments), or the omission of part of a complex expression constitute typical examples of such faults.

The usual way for dealing with structural faults is to assume the existence of a different, complementary model that allows to reason about the likelihood of such faults (i.e., modelling of spatial neighbourhood in the case of bridge faults). In software, such models could take the shape of considering name misspellings, variable switchings, or attempts to repair expressions (i.e., synthesize missing parts) to provide correct functionality. This is an open research issue.

5. Discussion

The notion of using knowledge-based systems as a support for software and related design tasks is not new. Previous research efforts usually aimed at working with languages that had strictly limited expressiveness, such as telecommunications protocols (Riese, 1993) or logic programming (Console et al., 1993). In addition they usually dealt with relatively small problems so that employing a representation that expressed the complete semantics of the problem domain was feasible. In our case, an adequate subset of the language had to be identified so that the representation would be abstract enough to deal with large programs, but would work even if the full expressiveness of VHDL were used in the programs. The result was a representation that focuses on the functional structure of the program. Even if full discrimination (i.e., finding a unique minimal diagnosis) is not achieved in a particular case, the tool still aids the designer in understanding the functional dependencies in the circuit. Information flow is described and important areas are identified so the designer can start debugging with a more detailed focus. Other salient points of the problem domain include:

- the existence of hidden structural differences between specification and implementation – signals may not match or have no counterpart, components may have no counterpart, the internals of components may have been changed.
- functional differences between specification and implementation – e.g., program control structures may be replaced by interacting subcomponents
- parts of the system may be unrepresentable – for example, the simulator may be coupled with actual existing hardware that is too complex to be integrated into the design (e.g., a commercial microprocessor to be mounted on a circuit board that is being simulated).
- No changes to the design
- Requirements for external axiomatizations are inacceptable, as their preparation is time-consuming and requires additional training for designers who already have to cope with the complexity of VHDL. Finally, in case of designs coming from different teams, (sometimes subcontractors in different companies using different development methodologies), or in case of designs that incorporate complex existing hardware (i.e., the abovementioned board designs which include a microprocessor and therefore require interfacing to an actual, physical processor for simulation) external axiomatizations are often impossible to obtain.

The last point bears some elaboration. First, it effectively disqualifies the verification-based systems which are available on the market and often applied successfully in similar circumstances. These systems do require a separate axiomatization of the properties to be verified (for each circuit), and even the best available are very limited in the size ($\sim$ 1 K lines of actual VHDL code) as well as complexity of the systems they can be applied to (e.g. Barrow (1983) and Burch (1994)). In addition, while they can be used to verify individual properties of a program, they do not try to pinpoint the location of the error. Thus they complement the simulation/debugging approach rather than avoiding it.

In design, model-based techniques have been used to guide the redesign of a component or system (Goel and Chandrasekaran, 1989; Stroulia et al., 1992). Like the verification approach, this concept requires the existence of an independent functional specification which describes (qualitatively or quantitatively) the variables and causal relations in the system. In the VHDL design domain, such a specification unfortunately does not exist. Instead, waveform traces provide specifications about particular aspects of the system behavior (test cases) on the basis of limited observations of the system. The only description about the composition and detailed workings of the design is the possibly incorrect program itself. A description of the functionality of larger program components is therefore not available. What is available as the basis of a system model are the semantics of individual VHDL statements.

With our current abstract representation, the guidance for redesign apart from pointing out the violating statement is limited. Consider the statement S<= f(E1,...,En). If it is assumed to behave incorrectly, this might be due to a number of reasons which would lead to different redesign approaches, such as changing one of the expressions Ei, changing the function f, changing the target S or eliminating the whole statement. Providing capabilities for distinguishing between these possibilities requires detailed reasoning about sequential statement semantics and variable/signal values. The local integration of such mechanisms in a manner that does not significantly degrade overall performance (and again limit practical usability) is an open topic. One possible approach would be the use of fault modes, which are commonly used in hardware diagnosis to discern between different types of faults and prune the search space. Software diagnosis must in principle deal with an unlimited number of fault models including structural faults. The possibility of expressing structural faults through fault modes has been recently examined in Böttcher (1994).

Finally, much work has been done in knowledge-based debugging support for software. We cite two examples. The PROUST system (Johnson, 1986) requires a separate description of the program to be manually developed and entered so that assumptions about the intentions of the programmer can be gleaned from comparison with the description. Similarly, in Allemang and Chandrasekaran (1991), it is necessary to provide an axiomatization of the intended functionality. The **Talus** system (Murray, 1988) attempts to match student solutions to example programs to find the the the source of divergent outputs. While structural variations between the example and student programs are allowed, the system still requires a direct match between the subresults produced inside a function and also uses the full semantics of the language used (a functional subset of LISP) to reason about the program, an approach that is not acceptable as a general method in our case, where the computational effort would be too high. Both for PROUST and **Talus**, the goal is to aid students, i.e., novice programmers in finding as many bugs as possible in small programs. In contrast, in our case, the goal is to help experts or at least experienced users orient themselves in very large programs of different structure and functional architecture.

6. Conclusion

In this paper, we have described the VHDLDIAG tool which provides design support by using model-based reasoning for determining the source of errors in hardware designs that are written in the VHDL specification language. One of the basic requirements was that the tool should fit into the standard design process used. The tool parses the standard VHDL source code written by the designers, and derives observations about execution correctness by automatically comparing the waveform traces produced by specification and more detailed implementation versions of the VHDL design.

The system uses a model of the functional structure of the design to identify components that are responsible for incorrect behavior. If a test case does not allow complete discrimination of the components involved, multiple test cases, automatically generated proposals for measurement selection, and finally interactive input from the designer can be used for restricting search further.

The tool has been implemented in Visualworks/Smalltalk and has been successfully used for finding faults in full-scale, actual ASIC designs. It is currently being tested in its future production environment. Results so far indicate savings of up to 10 % of the whole design cycle. Possible future improvements include a more complete representation of VHDL semantics. In particular, we will investigate the representational issues of the sequential parts of the language along the lines of design for imperative languages as described in Allemang and Chandrasekaran (1991) and Liver (1994). While computationally more expensive, this representation could be used (strictly locally) to increase the discriminatory power if the standard representation produces too many diagnosis candidates. In the vein of the tutoring environments discussed in the previous section, we also intend to utilize this for providing limited repair capability for designs.

Acknowledgement

This work was supported by Siemens Austria under project grant DDV GR 21/96106/4).

References

Allemang, D. and Chandrasekaran, B.: 1991, Maintaining knowledge about temporal intervals, *Proceedings 6th Knowledge-Based Software Engineering Conference*, IEEE, pp. 136–143.

Barrow, B.: 1983, Proving the correctness of digital hardware designs, *Proceedings AAAI*, pp. 17–21.

Böttcher, C.: 1994, No faults in structure? How to diagnose hidden interaction, *Proceedings IJCAI*, Montreal.

Burch, J. R.: 1994, Symbolic model checking for sequential circuit verification, *IEEE Transactions on Computer-Aided Design of Circuits and Systems*, 13(4), 401–423.

Console, L., Friedrich, G. and Theseider Dupré, D.: 1993, Model-based diagnosis meets error diagnosis in logic programs, *Proceedings IJCAI*, Morgan Kaufmann, pp. 1494–1499.

Davis, R.: 1984, Diagnostic reasoning based on structure and behavior, *Artificial Intelligence*, 24:347–410, 1984.

de Kleer, J.: 1995, Focusing on probable diagnoses, *Proceedings AAAI*, Morgan Kaufmann, pp.842–848.

Friedrich, G., Stumptner, M. and Wotawa, F.: 1995, Model-based diagnosis of hardware designs, In *Proceedings on the Sixth International Workshop on Principles of Diagnosis*, Goslar.

Goel, A. and Chandrasekaran, B.: 1989, Functional representation of designs and redesign problem solving, *Proceedings IJCAI*, Morgan Kaufmann, pp.1388–1394.

Greiner, R., Smith, B. A. and Wilkerson, R. W.: 1989, A correction to the algorithm in Reiter's theory of diagnosis, *Artificial Intelligence*, **41**(1), 79–88.

Hamscher, W. C.: 1991, Modeling digital circuits for troubleshooting, *Artificial Intelligence*, **51**(1-3), 223–271.

Johnson, W. L.: 1986, *Intention-Based Diagnosis of Novice Programming Errors*, Pitman Publishing.

Kloos, C. D. and Breuer, P. T. (eds): 1995, *Formal Semantics for VHDL* , Kluwer, Dordrecht.

Liver, B.: 1994, Modeling software systems for diagnosis, *Proceedings Fifth International Workshop on Principles of Diagnosis*, New Paltz, NY, pp. 179–184.

Murray, W. R.: 1988, *Automatic Program Debugging for Intelligent Tutoring Systems*, Pitman Publishing.

Reiter, R.: 1987, A theory of diagnosis from first principles, *Artificial Intelligence*, **32**, 57–95.

Riese, M.: 1993, *Model-based diagnosis of communication protocols*, PhD thesis, EPFL, Lausanne.

Stroulia, E., Shankar, M., Goel, A. and Penberthy, L.: 1992, A model-based approach to blame-assignment in design, *in* J. S. Gero (ed.), *Artificial Intelligence in Design '92*, Kluwer, Dordrecht, pp. 519–537.

IEEE: 1988, Standard VHDL Language Reference Manual LRM Std 1076-1987.

J. S. Gero and F. Sudweeks (eds), Artificial Intelligence in Design '96, 561-579.

ON KNOWLEDGE LEVEL THEORIES OF DESIGN PROCESS

TIM SMITHERS

Euskal Herriko Unibertsitatea, Informatika Fakultatea
649 Postakutxa, Euskal Herria, 20080 Donostia, Espaina

Abstract. AI in Design research is a mixture of scientific and engineering activities, with the emphasis on the latter. So far, almost all of this work has been carried out in the absence of usable theories of design process. Yet, as in other areas of technological development and application, such theoretical understanding could make a big difference to the effectiveness and applicability of the techniques and systems developed. This paper first briefly reviews the nature of theories in science, and the nature of the design process with respect to this. It then reviews some important problems with attempts to understand design in terms of cognition in AI in Design research. It suggests that what we need instead are Knowledge Level theories of design process. It also identifies the need for greater care and precision in the use of the terms theory, model, method, and description in the field. It ends by identifying some important benefits the development, testing, and use of Knowledge Level theories of design process could have on the field of AI in Design as a whole.

1. Introduction

A review of the proceedings of the previous Artificial Intelligence in Design (AID) conferences (Gero, 1991; Gero, 1992; Gero and Sudweeks, 1994), shows AID research to be a mixture of scientific and engineering activities. The (nearly) overwhelming emphasis is, however, on the engineering activities concerned with the development, testing, and application of AI techniques and systems for use in design of various kinds in various domains. (A review of AID work published in journals yields a similar conclusion.)

So far, all of this engineering activity has been carried out in the absence of any usable theory or theories of design process. Yet, as in other areas of technological development and application, such theoretical understanding could make a big difference to the effectiveness and applicability of the computational techniques and systems developed and applied by AID research.

The history of science and technology shows that theoretical understanding is *not* a necessary prerequisite for the development and effective application of technology. This is true for AID technology too. Good empirical understanding is sufficient to build some useful systems, but theoretical understanding can and does result in better, more effective, more efficient, and more acceptable applications and products. Theory also makes new ideas, techniques, and applications possible since we cannot discover everything from empirical practice and investigation alone. If this were true, we would not need any science!

For example, the steam engines of Boulton and Watt, used to power the industrial revolution in England during the late 18th and early 19th century (Dickinson and Jenkins, 1981), were all designed and built before any theory of heat engines was developed. Thermodynamics came later, developed first (nearly) by Carnot in 1824, then, after work by Joule and Kelvin, (more completely) by Clausius, who published his two laws on the relationship between the flow of heat and mechanical work in 1850 (Hills, 1989). Boulton and Watt successfully designed and built all their steam engines on the basis of an empirical understanding they, and others, developed by building and testing a large number of steam engines. Today, however, thermodynamics, together with the theory of fluid dynamics, forms an essential input to the design of efficient and cleaner internal combustion engines. The design and manufacture of the modern car engine would *not* be possible without this theoretical understanding.

Another example of useful technology coming before theoretical understanding was the Tea Clippers (fast square rigged sailing ships) of the 19th Century, which were designed, constructed, and sailed to considerable (financial) success, with no theoretical understanding of the fluid dynamics that governs the behaviour of both sails and hulls. Today, in contrast, this theoretical understanding forms an essential part of the design of ocean racing yachts. A rare example of theory coming before the technology is the computer: Turing and others had largely worked out the basics of the theory of computation before the first computers were designed and built during and just after the Second World War.

Some theory is, of course, used and applied in AID research, but this is either mathematical theory or theory from other fields and domains. What we don't have is any usable theoretical understanding of design process, the process that AID is fundamentally concerned with: AID does not have the equivalents of theories of thermodynamics or fluid dynamics.

A survey of AID research shows not just a lack of development of a usable theory (or theories) of design process, it also demonstrates widespread ignorance and neglect of related and relevant work on the fundamental nature of design process by researchers in the Design Research community. See Akin (1978; 1986; 1988), Alexander (1971) Archea (1987), Bazjanac (1974) Cross et al. (1981), Cross (1989), Darke (1979), Hillier and Leaman (1974; 1976), Hubka (1982), Jones (1991), Lawson (1990), Mitchell (1990), Rowe (1987), Schön (1983; 1985; 1987; 1992a; 1992b), for some particularly relevant work in this area. As a consequence, in much AID research, designing gets characterised by what we can get computer programs to do, rather than what really goes on when professional designers design. A completely inadequate characterisation of the full richness and complexity of real designing, and of what AI-based designing systems or design support systems must effectively deal with in real applications.

This is not the only isolation that AID research demonstrates. There is also very little evidence of a general awareness and understanding of some significant and again relevant developments in Knowledge Engineering. In the last ten years or so there has been a steady and effective development of understanding, techniques, and general methods for the systematic design and construction of knowledge systems. See Akkermans et al. (1993), Breuker and Wielinga (1988), Breuker and Velde (1994), Bylander and Chandrasekaran (1988), Schreiber et al. (1993; 1994), Steels (1990), Wielinga and Breuker (1987), and references therein, for example. As a consequence, in AID, we still see systems built in ways reminiscent of the construction of the first generation Expert Systems of the 1970s and early 1980s. We see very little of the kind of modelling of expertise that modern Knowledge Engineering methods explicitly advocate and actively support. The adoption and use of these modern Knowledge Engineering methods and techniques could significantly improve the system building activities of AID research. AID should not have to repeat the lessons of the field of knowledge systems to develop such methods for itself.

This rather critical view of current AID research is not intended to suggest that there has been no attempts to develop theories of design, see Akman et al. (1990), Arciszewski and Michalski (1994), Brazier et al. (1994; 1995), Dasgupta (1991), Gero and Maher (1990), Gero (1994), Tomiyama (1994), Treur (1991), Smithers (1992; 1995), Smithers and Troxell (1990), Smithers et al. (1994), Yoshikawa (1981) and references therein. However, none of these are explicit attempts to develop Knowledge Level theories of design process, nor can they properly be understood as such. It is also not the case that there has been no work reported in AID on modern knowledge acquisition methods and relate topics, see Alberts (1994), Alberts and Dikker (1994), Bernaras (1994), for example. Nor is it the case that there are no examples of a good appreciation and understanding of the relevant Design Research literature, see Blessing (1994), and McDonnell (1994), for example. However, all of these works represent isolated exceptions rather than a connected body of work within AID research, and hardly any of the theories have been or are being used in practice, and thus effectively tested.

If AID research is to move on from its present largely empirical and somewhat isolated practice, it needs to become more actively concerned with the development of usable theories of design process. It must develop theories which acknowledge and reflect our understanding of designing as one of the most remarkable and rich kinds of knowledge intensive intelligent behaviour we can observe and study: remarkable in its achievements and rich in its diversity and generality. The study and investigation of design in AID research should not, deliberately or inadvertently, diminish or dismiss this remarkableness and richness. It should lead to a better understanding and appreciation of them.

In this paper I propose that the appropriate kind of theory or theories of design process for AID to be trying to develop, test, and use, are *Knowledge Level theories of design process*. Knowledge Level theories could both properly connect up with the important body of work in Knowledge Engineering on Knowledge Level modelling of expertise and the design and construction of knowledge systems, and they could embody, in an appropriately abstract way, the insights and understanding we have of the nature of design process developed in Design Research. First, I briefly review the nature of scientific theories, and argue for a more disciplined use of some important terms. Some problems with cognitive theories of design process in the context of AID research are then discussed. I then review Newell's concept of the Knowledge Level and the extensions developed and used in modern Knowledge Engineering methods. Following this I present a set of Knowledge Level principles to be used in the development of Knowledge Level theories of design process. I end with a discussion of the role of Knowledge Level theories of design process in AID research.

2. On the Nature of Scientific Theories

The aim of scientific theories, in a broad sense, is to offer general understanding and explanations of the phenomena we observe. The role of explanation is to "remove puzzlement," (Wilkes, 1989), to demystify, and to "increase intelligibility," (Boden, 1962). Theories in science are the vehicles for delivering this general understanding and explanation. They are abstract statements about the general properties, characteristics, and underlying processes and mechanisms of *all* (possible) instances of a real phenomenon. What makes it a theory is that it is expressed in a way that makes no reference to, and in no way depends upon any particular instances (examples) of the phenomenon.

To be a good theory does not require it to support predictions, but it must be able to support the construction of effective explanations of particular instances or classes of the phenomenon covered. Evolution by Natural Selection is an example of a theory, widely regarded in biology as providing the best explanation of the origin of species, but which does not support the making of predictions of what new species *will* evolve, for example.

To construct effective explanations, all the terms and concepts used to form a theory must be operationalised. These operationalisations must make it possible, in practice, to identify and classify, unambiguously, particular examples and states of the phenomenon covered by the theory. This operationalisation of the terms and concepts used in a theory depend, in turn, on the successful establishment of appropriate *scientific observables*, see Jackson (1995). An observable, in this sense, is a symbolic representation of some aspect of the phenomenon being theorised about. It is the result of a projection process which must have two properties: (1) it must be possible for other interested persons to make the same observation, to make the same relation between the aspect of the phenomenon and its symbolic characterisation; and (2) it must be possible to record and preserve its value or values for future reference and comparison. It is this communally established projection process and recordability of observables that operationalises the concepts and terms of a scientific theory.

Thus the process of developing and testing theories in science involves, in large part, the development and use of terms and concepts that can be operationalised well enough to make them effective in forming explanations. It is an ongoing and often difficult process, with no notion of absolute truth or correctness involved. The doing of science, and, in particular, the development of scientific theories, is thus an essentially social process (Hull, 1988).

The development and testing of theories is the concern of science. It is *not* the concern of engineering, though much engineering is often required in the process of developing and testing scientific theories. Engineering is concerned with building things, and with the materials, techniques, and methods required to do this, including the effective use of relevant scientific theories. The practices of science and engineering are thus often intertwined, but, nonetheless, to be distinguished.

We also need to be clear that mathematical theories (including logic theories) are *not* scientific theories. They are formal constructions in the formal domain. Mathematics provides formal languages with which to express scientific theories and models. The use of mathematical theory for presentation and development does not, however, mean that what is presented and developed is necessarily a scientific theory: no amount of use of formalism necessarily makes anything a scientific theory of anything.

2.1. ON THEORIES, MODELS, DESCRIPTIONS, AND METHODS

There is a widespread habit in the AID literature (and elsewhere) to mix up the use of the terms *theory*, *model*, *description*, and *method*. For example, the "model of design process" presented in Smithers (1992) and Smithers and Troxell (1990) should be called a theory of design process: it is presented as a general statement about designing in general, and is not a model of some particular kind of designing in some particular domain. (In Smithers et al. (1994), this misuse is corrected.) Another example is to be found in Gero (1994). This describes steps towards a theory of exploration in designing, not a model of exploration, as the author describe it. There are many more examples where theory and model are used interchangeably, but theories and models are quite different things! So too are descriptions!

Theories, as we have seen, are general statements which make no reference to and do not depend upon particular instances of the phenomenon they are supposed to be theories about. Models, on the other hand, *do* refer to particular instances or classes of instances of a phenomenon, they *must* in order to be models since models must be models of something. Models are best constructed from a theory. In this case the theory provides a `kit of parts' from which to build them, but to complete a model we have to introduce boundary conditions and initial conditions and any particular constraints that make it a sufficient model. Models are thus particularisations and specialisations of theory.

Models can also be built using empirical understanding and knowledge. This is what we have to do when we no theoretical understanding. In this case they are particularisations of this empirical knowledge. Such models may be poor models because we cannot be sure of their consistency, whereas we can if we start from a good theory. It is also difficult and sometimes impossible to know when an empirical model is no longer safe to use, to know when the implicit modelling assumptions it depends upon no longer hold. This is why engineering does not require theory, but works better when it can use theoretical understanding.

The empirical understanding that forms the basis for all theory construction and model building (directly or indirectly) comes from attempts to *describe* what we observe. The terms, concepts, analogies, and metaphors we use in forming such descriptions are the breeding ground for the theoretical constructs and observables that we seek to operationalise, on the way to forming good theories. These descriptions are, however, *not* theories or models, and cannot be. Before we have a theory or a model of what is described we must add further structure that supports the explanation (and perhaps, prediction) forming that we expect form theories. This structure is not to be found in the description: the description is of what needs explaining, it *cannot* therefore, also be the explanation! We therefore need to guard against taking descriptions to be models or theories: in particular, descriptions of what designers do when they design are not straightforwardly models or theories of design process and should not be presented as such.

Design methods (normative statements about ways of designing) though particular to certain types of designing, are also *not* models or theories of design process. Design methods (for humans or machines) specify actions, and an organization to be adopted. The aim being to ensure a good and consistent quality of designing. In principle, design methods should be derived from a model of the particular kind of the design process involved. And, from the preceding, the model itself is best derived from a theory of design process. If design methods are presented in the absence of a model and theory it is difficult to know that the method itself is consistent. In other words, to know if, when it is applied, we will get a stable design process under all conditions: one that reliably results in a convergence on good designs. It is basically impossible to know if it is truly normative, as it is intended to be.

Thus, it can be seen that important differences exist between theories, models, descriptions, and methods of designing. Furthermore, if the usage of these terms does not respect these differences important distinctions are lost.

3. On the Nature of Theories of Design Process

For there to be a scientific theory or theories of the design process, designing must exist as a uniquely identifiable phenomenon. That is, it must have the status of being a *natural kind*. This is a real question! Not all things that were once thought to be real phenomenon in the history of science have turned out to be so. We can be mistaken in what we try to theories about: Phlogiston, the Ether, and Cold Fusion, for example. It could be that there is no fundamental difference between designing and doing other kinds of knowledge intensive activities, such as planning, scheduling, diagnosing, tutoring, playing chess, telling and understanding stories, etc., and some people do indeed claim this (Newell *et al.*, 1958; Newell and Simon, 1972; Newell, 1990). If this is true we should not be expecting or attempting to develop theories of design process, rather, some more general theory of problem solving, as Newell and Simon have long advocated. However, for many in the field of AID research, designing is a fundamentally distinct kind of process, and so they adopt the working hypothesis that we can form theories of design process. The same working hypothesis underlies all the attempts to understand designing in Design Research, though it is seldom, if ever, acknowledged, and is perhaps not even recognised as being one of the starting assumptions in this field.

3.1. PROBLEMS WITH THEORIES OF DESIGN AS COGNITION

A very common further assumption in AID, and in Design Research in general, is that designing is a cognitive phenomenon, a cognitive process, where cognition here refers to the whole range of individual and collective behaviour that people engage when solving problems, when designing, for example. For theories of Design as Cognition, designing is thus defined by what people do when they design. Consequently, we see attempts to develop cognitive theories of design: theories of *Design as Cognition*. (They are often referred to as cognitive models, but see section 2.1 above. Worse still, descriptions of human design behaviour are sometimes presented as models of design.) This assumption, that any theory of design process must be a cognitive theory, is so widespread that often it is not even made explicit. It is as if it is not possible to have any other kind of theory of design process.

There are, however, some serious methodological problems with any attempt to develop a theory of Design as Cognition as a theory of design process for AID Problems that go largely unnoticed and unacknowledged by those engaged in this kind of work.

Designing forms only a part of the full range of human (cognitive) behaviour that people are in general capable of, and typically engage in. Designing is a particularly sophisticated kind of behaviour, drawing on numerous human cognitive capacities in complicated ways. It is, nonetheless, embedded in the full range of human cognitive behaviour. This is what Simon argues at some length in Simon (1981).

The problem is that Cognitive Science does not yet have any well established theoretical understanding of the cognitive capacities used during design, and during other kinds of sophisticated human behaviour (Newell, 1990) notwithstanding. As a consequence, the terms and concepts used to present theories of Design as Cognition cannot be operationalised well enough to support the construction of effective explanations of human design behaviour: why designers do what they do, when they do it, and how they do it—we are not asking for predictions here, just good explanations! Instead, they have a more descriptive folk-theoretic status: they can be effective in describing what happens, but not in explaining why and how it does. This is a problem currently shared by all cognitive theories of human behaviour which build upon, or otherwise take as a starting point, the concepts of folk psychology. Folk theoretic concepts like beliefs, desires, goals, ideas, mental images, etc., do not yet have any effective operationalisations. They have yet to be established as scientific observables, and perhaps may never be so, see Stich (1986), and Churchland (1988; 1989), for some arguments as to why not.

A simple way of seeing what consequences this problem has for trying to understand the design process, is to ask how you tell, from a theory of Design as Cognition, when a particular human designer (or group of designers) is (or are) doing design and when not—just asking them provides no basis for an explanation! If a theory of design cannot be used to establish when a designer is doing design and when not, it is not a theory of design! But, without a much more complete general theory of human cognition, any attempt at developing and testing a theory of Design as Cognition is going to have a very difficult time making precisely this kind of necessary distinction. A more complete general theory of human cognition is a necessary prerequisite to the development of any theory of Design as Cognition, which would thus be a refinement of the general theory. AID research can neither wait for such a general theory (or theories) of cognition, nor can it develop one by itself, nor, as we will see, does it need to. Though it can and should contribute to this more general project.

Another problem faced by any theory of Design as Cognition is that it needs to deal with, and ultimately explain, all the variations and differences we see in all human designing. The practices of experienced professional designers is full of contingencies: it is deeply influenced by the particular education, training, and previous experiences of the people involved Human design behaviour does not follow straightforwardly from a set of identifiable process laws. The actual practices and behaviour of human designers seldom, at least on the surface, demonstrate how things *must* be done. We seldom see the necessities of designing, or know when we do. We see much more of how things *can* be done, and not how they *must* be done. Going from the observation and description of all this contingent human design behaviour to theories embodying necessary and sufficient properties, characteristics, and laws of design process is not a simple and direct possibility, as much of the history of Design Research shows, see Cross (1984) and McDonnell (1994). Indeed, the history of Design Research shows that it is only recently that people in the field have recognised the need for more reflection on the fundamental nature of design process, and less on prescriptions of how it should be done. There is also growing concern for the wider aspects and implications of this basic contingent nature of human problem solving behaviour in the field of computer science and computer systems applications. See, for example, the special issues of the Communications of the Association for Computing Machinery on *Participatory Design* (Kuhn and Muller, 1993), *Social Computing* (Schuler, 1994), *Requirements Gathering* (Holtzblatt and Beyer, 1995), and *Representations of Work*, (Suchman, 1995a).

Any explanations or models derived from theories of Design as Cognition must be presented in terms of human behaviour, albeit abstract terms. Thus, a further methodological problem with such theories is that any attempt to develop methods of design from them must specify design behaviour: they must say what designers *must* do. Specifying what people must do is not in general an easy thing to do, and telling designers how they must act in order to design well has never been very successful in practice. Rather than feeling they are well guided to design well, human designers typically feel unnecessarily constrained or otherwise prevented from doing things in the way they want or need to. This dissatisfaction derives, as we can see, from the fact that any attempt to specify design behaviour will tend to remove or limit the natural contingent nature of human design behaviour. They may also have wider implications within design teams an organisations related to ownership, power, and who knows how the work is actually done, see Suchman (1995b) and Bannon (1995). When design methods are presented and adopted as design management methods or strategies, they can, however, be more successful, see Taguchi (1986; 1989), for example.

4. Designing as a Knowledge Process

If the scientific aim of AID is to understand and explain human design practice, we have to develop a theory or theories of human designing, theories of Design as Cognition, no matter how hard this proves to be. Taking design process to be defined by what people do when they design is, however, not the only way of attempting to understand design process. In particular, theories of Design as Cognition are not the only kind of theories of design process we can seek to develop in AID. Just as AI research has demonstrated that it is possible to realise certain kinds of expert behaviour in computation-based knowledge systems, AID research has also demonstrated that certain aspects and activities of designing can be understood and realised in the same way, albeit in rather limited demonstrations, so far. An important conclusion that can be drawn from the largely empirical efforts and results of AID research up to now, is that human design practice is not the only way we can seek to implement design process. Computation, and knowledge systems, in particular, offer an alternative, and, of more practical importance, an additional form of implementation that we can seek to effectively combine with human design competence and practice.

4.1. IMPLEMENTATION THEORIES OF DESIGN

Within AID, theories of Design as Cognition can thus be understood as *implementation* theories of design—theories about how design processes are implemented by human designers. However, it is not just an alternative theory of implementation that AID needs. It needs to develop an implementation independent theory of design process. This is because we need a theory of design process which covers both kinds of implementation, human designing and knowledge system-based designing. This is especially true if we are to have the much needed theoretical basis for developing effective AI-based design support systems — which is where we can reasonably expect most to be gained by the widespread application and use of knowledge systems technology in design.

4.2. NEWELL'S KNOWLEDGE LEVEL

This need for an implementation independent level of understanding of intelligent problem solving behaviour is not particular to AID. It is a need shared by all attempts in AI research to understand knowledge intensive problem solving behaviour, and was recognised a long time ago by Allen Newell. In his Knowledge Level paper, Newell (1981) argued for a separation of the (computational) Symbol Level into two levels. One he continued to call the Symbol Level, at which the symbol processing needed to realise some particular kind of problem solving behaviour should be described and specified. The other level, he called the *Knowledge Level*. He presented this fundamental idea as a hypothesis, his Knowledge Level Hypothesis, which states that:

> *There exists a distinct computer system level lying immediately above the symbol level, which is characterised by knowledge as the medium and the principle of rationality as the law of behaviour.*

Knowledge, the medium at the Knowledge Level, is understood as a competence notion, "being a potential for generating action." The principle of rationality, the law of behaviour at the Knowledge Level, says that actions are selected to attain the agent's goals.

The key insight that this hypothesis embodies is that the concept of knowledge can be used to effectively abstract away from the details and particularities of human or machine problem solving behaviour, yet still be used to capture the essential nature and character of the problem solving process itself. By presenting knowledge as a competence notion, as a potential to act in a way that solves problems, Newell established the basis for effectively operationalising the concept of knowledge, and thus for making it a practical scientific observable in the understanding and explaining of human and computer-based problem solving behaviour. People and knowledge systems solve certain problems because they know certain things, and they know how and when they can use what they know to solve those problems. At Newell's Knowledge Level, problem solving becomes a knowledge process. Designing too can be understood as a kind of knowledge process. We can therefore seek to develop a theoretical understanding of this knowledge process: we can attempt to develop a *Knowledge Level* theory (or theories) of design process.

4.3. THE KNOWLEDGE ENGINEERING KNOWLEDGE LEVEL

Starting from Newell's concept of the *Knowledge Level*, the field of knowledge engineering (KE) has developed techniques for Knowledge Level modelling of expertise. This has been one of the most significant developments in AI in the past ten years (van de Velde, 1993; Akkermans *et al.*, 1993). Here knowledge is, again, taken as a competence notion. It is this "competence-like" notion of knowledge that supports its effective operationalisation into a practical observable for the modelling of human problem solving behaviour. It supports effective abstraction over the particular activities of people, and it provides the basis for forming models of different kinds of human problem solving behaviour. The abstracting away of the particularities and contingencies of the human problem solving behaviour makes it possible to say what knowledge is necessary and sufficient for different kinds of expert behaviour, and how it needs to be organised, generated, and used to realise the kind of problem solving behaviour being modeled.

There are, however, two important differences between Newell's Knowledge Level, KL_N, and the concept of the Knowledge Level subsequently developed and used in modern knowledge engineering methods, KL_{KE}. KL_{KE} models of expert behaviour make *no* cognitive claims, that is, they do not attempt to say what the human problem solving behaviour must be. They cannot do this since KL_{KE} models of expert behaviour are not intended as models of human implementation. They are intended to support, in a uniform and systematic way, the formulation of implementation specifications for computer-based systems. It is the implementation independent Knowledge Level modeling of problem solving behaviour that forms the core of knowledge engineering methods such as CommonKADS (Breuker and Wielinga, 1989; Schreiber *et al.*, 1994). Newell, and others, in contrast, make the claim that a proper KL_N description and explanation of a problem solving computation-based system can be used to say how the same problem solving must be done by humans. They explicitly seek to establish cognitive claims via KL_N theories and models by equating the symbol processing involved with cognition.

A second important difference between KL_N and the modern knowledge engineering concept of Knowledge Level, KL_{KE}, is that at the KL_{KE} knowledge is differentiated and structured, whereas, for Newell, the knowledge an agent is identified as having at the KL_N is one undifferentiated body. This extension of Newell's original idea is fundamental to the success of KL_{KE} modelling of human problem solving behaviour and to supporting the subsequent development of knowledge system applications to replicate it. The role of models of expertise in modern knowledge system development methods, such as CommonKADS, is to effectively model the necessary and sufficient elements of expert problem solving behaviour in an implementation independent way, in terms of a knowledge process. This is so that they can both capture the essential nature and characteristics of the problem solving process, as displayed in the human expert, without attempting to deal with all the detailed variations, differences, and contingencies normally seen, and still provide the basis for developing implementation specifications of computer-based knowledge systems that replicate, often only in part, the modeled expert behaviour. It is the lack of any necessary cognitive commitment in KL_{KE} models and the recognition of the role differentiation and structure of knowledge that provides this implementation independence, yet supports the effective transfer from human implementation to computer-based implementation (again, at least in part).

This ability to capture the fundamental nature of knowledge intensive problem solving behaviour, as seen in human expert behaviour, in a way that abstracts away from all the cognitive complications, is what is needed in a theory or theories of design process in AID research. The principles underlying KL_{KE} modelling of expertise could thus provide the basis for developing appropriate KL_{KE} theories of design process.

KL_{KE} theories of design process thus offer a way of understanding and explaining designing in a way that does not depend upon a (more) complete cognitive science, and in a way that is more easily related to the concepts and techniques of computation, as developed in AID research. They would identify the necessary and sufficient types of knowledge, the roles they play, the interactions between these types, and their organisation for the knowledge process to be a design process. As KL_{KE} theories they would not depend upon, and so would not attempt to explain, what people do when they do design. As general theories of design process they could, however, be used to build KL_{KE} models of particular kinds of designing, in particular domains, which could then be used to understand how people and organisations implement this design process, and how computation might be effectively (and acceptably) introduced into this human practice. In particular, any KL_{KE} theory of design process should be able to be used to establish operationalised distinctions between such widely used *folk*-categories as *routine design, original design*, and *creative design*, or be able to be used to show why these are *not*, after all, proper kinds of designing.

5. Knowledge Level Principles for Knowledge Level Theories

The major activity in the development of effective modern knowledge engineering methods, has been towards the modelling of expertise. There has so far been little attempt to develop KL_{KE} theories of problem solving expertise, but see Benjamins and Jansweijer (1994) for an exception. The models of expertise that have been constructed according to these KE methods are therefore empirical models, models based on observational data of human expert behaviour. They are not theory-based models. For this, we would need KL_{KE} theories. It is, however, recognised that theory-based models are not just desirable but important for strengthening the KE methods that depend upon such modelling activities (Akkermans, 1995; van Harmelen, 1995).

In an important attempt to make explicit the principles and underlying assumptions of the practice of expertise modelling, as supported by the Common-KADS method, Akkermans et al. (1994), presents a set of knowledge differentiating and structuring principles. These KL_{KE} modelling principles are used here as a basis for a set of principles for developing KL_{KE} theories of design process. They are intended to identify the assumptions needed for the development of such KL_{KE} theories, and they have been derived quite directly from the Akkermans et al KL_{KE} modelling principles. The five principles are:

- **1. The Knowledge Application Principle**: All instances of effective real-world designing can be viewed as the rationalisable application of appropriate domain and task knowledge by a designing agent, or a team of designing agents.

- **2. The KL Principle**: The Knowledge Level is the appropriate level for developing a general theory of design process that captures the necessary and sufficient properties and organisation of those knowledge processes that are design processes. It calls for the theoretical description of design problem solving at a conceptual level that is independent of representation and implementation media, mechanisms and practices.

- **3. The Role Limiting Principle**: All knowledge used and developed by a design process is structured with each part having identifiable, stable, and restricted roles within the totality of the design process.

- **4. The Differentiated Rationality Principle**: Within the limitations of a particular structuring and role assignment of knowledge, the principle of rationality must be further specialised to account for the way in which designs are produced by applying the appropriate type of knowledge in the appropriate way. This specialisation of the principle of rationality we can call the *rationality of exploration*, since it is exploration that is at the heart of all effective designing.

- **5. The Knowledge Typing Principle**: A Knowledge Level theory of design process specifies three different categories of knowledge: domain knowledge; task knowledge; and inference knowledge. Within these categories further generic types of knowledge can be distinguished.

Any KL_{KE} theory of design process developed according to these principles should thus explicitly present the necessary and sufficient categories of knowledge involved, how they must be and can be organised according to their particular and distinct roles within in the design process as a whole, and what form the rationality of exploration must have in operating over this knowledge and its organisation to realise effective designing.

6. KL_{KE} Theories of Design Process in AI in Design

As I established at the beginning of this paper, the large majority of the activities within the field of AID research are concerned with the engineering of AI-based design systems, and, more typically, AI-based design support systems. Often, this engineering research is focused on particular techniques and on particular aspects of designing in particular design domains. This strong emphasis on the problems of building effective AI-based design systems and design support system is, of course, both natural and appropriate for AID research. We should not expect the field to change significantly in this respect as more theoretical understanding of design process is developed and used within it. We can, however, ask in what way should the development of KL_{KE} theories of design process impact on AID research in general?

The development, testing, and use of KL_{KE} theories of design process should result in a closer and more effective contact with Design Research in general and with the field of Knowledge Engineering. By attempting to develop theories of design process that abstract away from the details and complications of human professional design practice, it should be possible to build into such theories the insights into the fundamental nature of designing and the abstractions used to present them, that have been developed by workers Design Research. In other words, KL_{KE} theories of design process provide a way of expressing, in a theoretical form, the results of the work in Design Research that AI in Design is currently largely unaware of. This would both significantly strengthen work in AID research, and offer a new and important direction for work in Design Research.

By developing KL_{KE} theories of design process based upon essential similar principles now embodied in modern KE methods for the design and implementation of application knowledge systems, it will be possible for design theory to be used directly in the construction and development of models of particular kinds of designing in particular domains, models of design expertise. This would both strengthen the models used in this crucial KE activity, and contribute to the the development of KL_{KE} theories of expert behaviour in general. It would therefore lead directly to the better engineering of AI-based design and design support systems for application. It should also make possible a much clearer and sounder understanding of how computation can be effectively and acceptably applied in human design practice.

A further development could be in the development and application of KL_{KE} methods of designing explicitly based upon established theoretical understanding and modelling of design processes. These would form an essential part of the enterprise models now being developed to support organisation restructuring and more flexible and responsive management methods and strategies, without suffering from the effects of having to be prescriptive of human behaviour. The responsibility of how a KL_{KE} design method is to be applied in professional design practice would be and would remain the responsibility of the professional designers involved: they could continue to own and protect the secrete of how they design.

7. Conclusions

Theories are the vehicles for delivering scientific understanding and explanations of phenomenon—puzzlement reducing, demystifying, and intelligibility increasing descriptions of things. To do this, theories must employ terms and concepts that can be practically operationalised.

Human design is but one kind of human behaviour. Theories of design as cognition must therefore be properly connected up to and embedded in a general theory of human cognition. But we currently have very little of such a general theory. As a consequence theories of Design as Cognition in AI in Design become dangerously free floating and of little effective use in explaining design process: they can't even be used reliably to say what designing is and is not.

Knowledge Engineering Knowledge Level theories of design process are a more appropriate type of theory for AID research to be developing. By abstracting away from the particularities of human behaviour, and the as yet largely mysterious cognitive goings on, KL_{KE} theories and models of design process escape the problem of there not yet being a sufficient general theory of human cognition. As for other kinds of expert behaviour like planning, scheduling, diagnosing, etc., they also allow us to better understand how other technologies, in particular computational technologies, can be effectively and acceptably introduced into human design practice.

The use of the terms theories, models, descriptions, and methods, is currently very undisciplined in AID research. This means it is very difficult to identify what are proper theoretical contributions to the field. Without good theoretical developments, the field cannot expect to progress towards good explanations of design process, nor towards more effective technological developments and applications. AID research thus needs to devote more effort to the development, testing, and use of KL_{KE} theories of design process, and, by so doing, establish proper and more effective connections with the important work being done in Design Research and Knowledge Engineering. The principles presented above provide a sound basis for developing such Knowledge Level theories of design process.

Acknowledgments

I am grateful to Hans Akkermans, Amaia Bernaras, Francis Brazier, Brian Logan, Janet McDonnell, Rolf Pfeifer, Luc Steels, Jan Treur, Wade Troxell, Frank van Harmelen, Pieter van Langen, and Walter van de Velde, for numerous discussions that have been influential in the development of the work presented here. An earlier version parts of this paper were presented in Smithers (1995). I am also grateful to Amaia Bernaras and four anonymous referees for helpful comments on an earlier version of this paper. Finally, I am happy to acknowledge the financial support of the University of the Basque Country for my current position, and the Faculty of Informatics, in particular, for providing a very supportive academic home.

References

Akman, V., ten Hagen, P. J. W., and Tomiyama, T.: 1990, A fundamental and theoretical framework for an intelligent CAD System, *Computer-Aided Design*, **22**(6), 368–376.

Akin, O.: 1978, How do architects design, *in* J-C. Latombe (ed.), *Artificial Intelligence and Pattern Recognition in Computer Aided Design*, North-Holland, pp. 65–98.

Akin, O.: 1986, *Psychology of Architectural Design*, Pion Limited, London.

Akin, O.: 1988, Expertise of the architect, *in* M. D. Rychener (ed.), *Expert Systems for Engineering Design*, Academic Press.

Akkermans, J. M., van Harmelen, F., Guss Schreiber, A. T., Wielinga, B. J.: 1993, A formalization of knowledge-kevel models for knowledge acquisition, *International Journal of Intelligent Systems*, **8**, 169–208.

Akkermans, J. M., van de Velde, W., Wielinga, B. J., and Schreiber, A. T.: 1994, Rational: principles underlying expertise modelling, *in* B. J. Wielinga (ed.), *Expertise Model Definition Document*, Chapter 1, KAD-II project document KADS-II/M2/UvA/026/5.0, pp. 5–9.

Akkermans, J. M.: 1995, In personal discussion, Amsterdam, January.

Alberts, L. K.: 1994, YMIR: a sharable ontology for formal representation of engineering-design knowledge, *in* J. S. Gero and E. Tyugu (eds), *Formal Design Methods for CAD*, Elsevier, pp. 3–32.

Alberts, L. K. and Dikker, F.: 1994. Integrating standards and synthesis knowledge using the YMIR ontology, *in* J. S. Gero and F. Sudweeks (eds), *Artificial Intelligence in Design '94*, Kluwer, Dordrecht, pp. 517–534.

Alexander, C.: 1971. *Notes on the Synthesis of Form*, Harvard University Press, paperback edition, first published in 1964.

Archea, J.: 1987. Puzzle-making: what architects do when no one is looking, *in* Kalay, Y. E. (ed.), *Principle of Computer-Aided Design: Computability of Design*, John Wiley and Sons, pp. 37–52.

Arciszewski, T. and Michalski, R. S.: 1994. Inferential design theory: a conceptual outline, *in* J. S. Gero and F. Sudweeks (eds), *Artificial Intelligence in Design '94*, Kluwer, Dordrecht, pp. 295–308.

Bannon, L. J.: 1995. The politics of design: representing work, *in* (Suchman, 1995a), pp. 66–68.

Bazjanac, V.: 1974. Architectural design theory: models of the design process, *in* Spillers, W. R. (ed), *Basic Questions of Design Theory*, North-Holland, pp. 2–19.

Benjammins, R. and Jansweijer, W.: 1994. Towards a competence theory of diagnosis, *IEEE Expert*, **9**(5), pp. 43–52.

Bernaras,.: 1994. Problem-oriented and task-oriented models of design in the CommonKADS framework, *in* J. S. Gero and F. Sudweeks (eds), *Artificial Intelligence in Design '94*, Kluwer, Dordrecht, pp. 499-516.

Blessing, L. T. M.: 1994. *A Process-Based approach to Computer-Supported Engineering Design*, PhD. Thesis, University of Twente, Enschede, The Netherlands.

Boden, M. A.: 1962. The paradox of explanation, Proc. Aristotelian Soc., n.s., pp. 159–178. Reprinted *in* Boden M. A., 1981, *Minds and Mechanisms*, Ithaca, N.Y., Cornell University Press.

Brazier, F. M. T., van Langen, P. H. G., Ruttkay, Zs., and Treur, J.: 1994. On formal specification of design tasks, *in* J. S. Gero and F. Sudweeks (eds), *Artificial Intelligence in Design '94*, Kluwer, Dordrecht, pp. 535–552.

Brazier, F. M. T., van Langen, P. H. G., and Treur, J.: 1995, A logical theory of design, *in* J. S. Gero and F. Sudweeks (eds), *Advances in Formal Design Methods for CAD*, Preprints of the IFIP WG 5.2 Workshop on Formal Design Methods for Computer-Aided Design, Key Centre of Design Computing, University of Sydney, pp. 247–271.

Breuker, J. A. and Wielinga, B. J.: 1988, Models of expertise in knowledge acquisition, *in* P. Guida and G. Tasso (eds), *Topics in Expert Systems Design: Methodologies and Tools*, North-Holland, Amsterdam.

Breuker, J. A. and Wielinga, B. J: 1989, Model-driven knowledge acquisition, *in* P. Guida and G. Tasso (eds), *Topics in the Design of Expert Systems*, North-Holland, Amsterdam, pp. 265–296.

Breuker, J. A. and van de Velde, W.: 1994, *The CommonKADS Library of Expertise Modeling*, IOS Press.

Bylander, T and Chandrasekaran, B.: 1988, Generic tasks in knowledge-based reasoning: The right level of abstraction for knowledge acquisition, *in* B. Gaines and J. Boose (eds), *Knowledge Acquisition for Knowledge Based Systems*, Academic Press.

Churchland, P. M.: 1988, On the ontological status of intentional states: Nailing folk psychology to its perch, *Behavioural and Brain Sciences*, **11**(3), 507–508.

Churchland, P. M.: 1989, *A Neurocomputational Perspective: The Nature of Mind and the Structure of Science*, The MIT Press.

Cross, N., Naughton, J., and Walker, D.: 1981, Design method and scientific method, *Design Studies*, **2**(4), 195–201.

Cross, N. (ed.): 1984, *Developments in Design Methodology*, John Wiley.

Cross, N.: 1989, *Engineering Design Methods*, John Wiley.

Darke, J: 1979, The primary generator and the design process, *Design Studies*, **1**(1), 36–44.

Dasgupta, S.: 1991, *Design Theory and Computer Science, Processes and Methodology of Computer Systems Design*, Cambridge University Press.

Dickinson, H. W. and Jenkins, R: 1981, *James Watt and the Steam Engine*, Encore Editions. First published in 1927.

Gero, J. S. (ed.): 1991, *Artificial Intelligence in Design '91*, Butterworth-Heinemann, Oxford.

Gero, J. S. (ed.): 1992. *Artificial Intelligence in Design '92*, Kluwer, Dordrecht.

Gero, J. S.: 1994, Towards a model of exploration in computer-aided design, *in* J. S. Gero and E. Tyugu (eds), *Formal Design Methods for CAD*, Elsevier, pp. 315–336.

Gero, J. S. and Maher, M. L.: 1990, Theoretical requirements for creative design by analogy, *in* P. A. Fitzhorn (ed.), Proceedings First International Workshop on *Formal Methods in Engineering Design, Manufacturing, and Assembly*, The Broadmoor Hotel, Colorado Springs, Colorado, USA, January 15–17, 1990. Available from Department of Mechanical Engineering, Colorado State University.

Gero, J. S. and Sudweeks, F. (eds): 1994, *Artificial Intelligence in Design '94*, Kluwer, Dordrecht.

Hillier, W. and Leaman, A.: 1974, How is design possible, *Journal of Architectural Research*, **3**, 4–11.

Hillier, W. and Leaman, A.: 1976, Architecture as a discipline, *Journal of Architectural Research*, **5**, 8–32.

Hills, R. L.: 1989, *Power from Steam, A History of the Stationary Steam Engine*, Cambridge University Press.

Holtzblatt, K. and Beyer, H. R. (eds): 1995, Requirements gathering, the human factor, *Communications of the ACM*, **38**(5), 31–88.

Hubka, V.: 1982. *Principles of Engineering Design*, Butterworth Scientific, translation by W. E. Eder.

Hull, D. L.: 1988, *Science as Process: An Evolutionary Account of the Social and Conceptual De-*

velopment of Science, The University of Chicago Press, Chicago.

Atlee Jackson, E.: 1995, No provable limits to "scientific knowledge", *Complexity*, **1**(2), 14–17.

Jones, J. C.: 1991, *Designing Designing*, Architecture Design and Technology Press.

Kyhn, S. and Muller, M. J. (eds): 1993, Participatory design, *Communications of the ACM*, **36**(4), 25–103.

Lawson, B: 1990, *How Designers Think*, Academic Press.

McDonnell, J. T.: 1994, *Supporting Engineering Design Using Knowledge Based Systems Technology with a Case Study in Electricity Distribution Network Design*, PhD Thesis, Department of Computer Science, Brunel University, England.

Mitchell, W. J.: 1990, *The Logic of Architecture*, The MIT Press.

Newell, A.: 1981, The knowledge level, *AI Magazine*, **1**(2), 1–20. Also in *Artificial Intelligence*, **18**(1), 87–127, 1982.

Newell, A.: 1990, *Unified Theories of Cognition*, Harvard University Press.

Newell, A., Shaw, J. C. and Simon, H. A.: 1958, Elements of a theory of human problem solving, *Psychological Review*, **65**, 151–166.

Newell, A. and Simon, H. A.: 1972, *Human Problem Solving*, Prentice-Hall, Englewood Cliffs, N. J.

Rowe, P. G.: 1987, *Design Thinking*, The MIT Press.

Schön, D. A.: 1983, *The Reflective Practitioner, How Professionals Think in Action*, Basic Books.

Schön, D. A.: 1985, *The Design Studio, An Exploration of its Traditions and Potential*, RIBA Publications.

Schön, D. A.: 1987, *Educating the Reflective Practitioner*, Jossey-Bass.

Schön, D. A.: 1992, Designing as a reflective conversation with the materials of a design situation, *Knowledge Based Systems*, **5**(1), 3–14.

Schön, D. A.: 1992, Kinds of seeing and their functions in Ddesigning, *Design Studies*, **13**(2), 135–156.

Schreiber, A. T., Wielinga, B. J., and Breuker, J. A. (eds): 1993, KADS: a principled approach to knowledge-based systems development, *Knowledge-Based Systems Book Series*, **11**, Academic Press.

Schreiber, A. T., Wielinga, B. J., de Hoog, R., Akkermans, J. M. and van de Velde, W.: 1994, CommonKADS: A comprehensive methodology for KBS development, *IEEE Expert*, **9**(6), 28–37.

Schuler, d. (ed.): 1994, Social computing, *Communications of the ACM*, **37**(1), 29–80. 126–127.

Simon, H. A.: 1981, *The Sciences of the Artificial*, Second Edition, MIT Press. (First edition, 1969.)

Smithers, T. and Troxell, W. O.: 1990, Design is intelligent behaviour, but what's the formalism?, *AI EDAM*, **4**(2), 89–98.

Smithers, T.: 1992, Design as exploration: Puzzle-making and puzzle solving, *Exploration-Based Models of Design and Search-Based Models of Design*, Workshop Notes, AID' 92, CMU, Pittsburgh, June.

Smithers, T. (ed.): 1994, *The Nature and Role of Theory in AI in Design Research*, Workshop Notes, AID' 94, Swiss Federal Institute of Technology, Lasuanne, Switzerland, August. Available from the Faculty of Informatics, University of the Basque Country, San Sebastián, Spain.

Smithers, T., Corne, D., and Ross, P.: 1994, On computing exploration and solving design Pproblems, *in* J. S. Gero and E. Tyugu (eds.), *Formal Design Methods for CAD*, Elsevier, pp. 293–313.

Smithers, T.: 1995. AI in design needs knowledge level theories, *in* J. S. Gero and F. Sudweeks (eds), *Proceedings Fourth Workshop on Research Directions for Artificial Intelligence in Design*, Department of Architecture and Design Science, University of Sydney, pp. 73–79.

Steels, L.: 1990, Components of expertise, *AI Magazine*, **11**(2), 30–61.

Stich, S.: 1986, *From Folk Psychology to Cognitive Science, The Case Against Belief*, A Bradford Book, The MIT Press.

Suchman, L. A. (ed.): 1995, Representations of Work, Special Issue of *Communications of the ACM*, **38**(9), 33–68.

Suchman, L. A.: 1995, Making work visible, *in* L. A. Suchman (ed.), *Communications of the ACM*, **38**(9), 56–64.

Taguchi, G.: 1986, *Introduction to Quality Engineering*, Asian Productivity Organization, First Edition.

Taguchi, G., Elsayed, E. A., and Hsiang, T. C.: 1989, *Quality Engineering in Production Systems*, McGraw-Hill Book Company, First Edition.

Tomiyama, T.: 1994, From general design theory to knowledge-intensive engineering, *AIEDAM*, 8(4), 319–333.

Treur, J.: 1991, A logical framework for design processes, *in* P. J. W. ten Hagen and P. J. Veerkamp (eds), *Intelligent CAD Systems III*, Proceedings of the Third Eurographics Workshop on Intelligent CAD Systems, Springer-Verlag, pp. 3–20.

van de Velde, W.: 1993, Issues in knowledge level modelling, *in* D. J-M. David, J-P. Krivine and R. Simmons (eds), *Second Generation in Expert Systems*, Springer-Verlag, Berlin, pp. 211-231.

van Harmelen, F.: 1995, In personal discussion, Amsterdam, January.

Wielinga, B. J. and Breuker, J. A.: 1987, Models of expertise, *in* B. du Boulay, D. Hogg and L. Steels (eds), *Advances in Artificial Intelligence II*, Elsevier Science, pp. 497–509.

Wilkes, K.: 1989, Explanation–how not to miss the point, *in* A. Montefiore and D. Noble (eds), *Goals, No Goals, and Own Goals*, Unwin Hyman, London, pp. 194–210.

Yoshikawa, H.: 1981, General design theory and a CAD system, *in* T. Sata and E. A. Warman (eds), *Man-Machine Communication in CAD/CAM*, North-Holland, pp. 35–58.

11

conceptual design

A representation scheme to support conceptual design of
mechatronic systems
Martin K. Stacey, Helen C. Sharp, Marian Petre, George Rzevski,
Rodney A. Buckland
Generating conceptual solutions on FuncSION: Evolution of a
functional synthesiser
Amaresh Chakrabarti, Ming Xi Tang
Adopting a minimum commitment principle for computer aided
geometric design systems
Xiaohong Guan, Ken J. MacCallum

J. S. Gero and F. Sudweeks (eds), Artificial Intelligence in Design '96, 583-602.
© 1996 *Kluwer Academic Publishers.*

A REPRESENTATION SCHEME TO SUPPORT CONCEPTUAL DESIGN OF MECHATRONIC SYSTEMS

MARTIN STACEY, HELEN SHARP, MARIAN PETRE
Computing Department, The Open University,
Milton Keynes, UK

AND

GEORGE RZEVSKI, RODNEY BUCKLAND
Design Discipline, The Open University,
Milton Keynes, UK

Abstract. This paper outlines DROOL, a novel knowledge representation scheme for supporting early stages of the design of multi-technology systems. The key idea behind DROOL is that mechatronic systems should be considered as interlocking flows of matter, energy and information. This provides a powerful framework for thinking about important aspects of conceptual design, and an equally powerful structure for organising a design representation scheme. The paper presents an object model for DROOL, and shows how the scheme describes design concepts at various levels of abstraction, in terms of component hierarchies and interlocking flows. The development of DROOL is part of the FACADE Project, whose aim is to develop intelligent computer tools to facilitate communication among designers with different specialties in concurrent engineering projects. The FACADE System comprises a suite of interfaces with different visual representations to support different conceptual design activities, with a single underlying product representation: DROOL. Development is concentrating on two interfaces with alternative visual representations in which designers can describe the flows of matter, energy and information in mechatronic systems: blob diagrams and concept arrays.

1. Introduction

The FACADE Project (Rzevski, 1995) aims to develop computer based support for the conceptual design of mechatronic systems in a concurrent engineering context. Its primary goal is to facilitate the communication of conceptual design ideas and their implications between the different members of a design team.

The design of mechatronic systems is characterised by engineers from different disciplines collaborating to design a complex artefact, all of whom have particular perspectives on the artefact, and work with different views of it. The design of any complex product is a combination of many different design activities, in which designers create, modify and evaluate different elements of the future product. But these different design activities are linked: designers collaborate, and individual designers switch quickly between views. They need to see how their own and others' design decisions affect other aspects of the design, which are described using other notations and visual representations. This requires an intelligent design support system that can display the different aspects of a design in different visual representations, and that can propagate changes, implications and constraints between them.

Each design activity requires different computer support: it needs a design environment making particular aspects of the design explicit in its visual representation of one view of the design, and providing a set of operations that are natural and easy to use with that view. Conceptual design requires tools differing fundamentally from conventional CAD systems. CAD packages require precise details which designers do not want to specify or guess in conceptual design; indeed having to do this may inhibit creative thinking during the early stages of design (Faltings, 1991).

Our intelligent design support tool comprises a suite of design environments supporting different views of the artefact, which communicate with an underlying product model. This paper describes the DROOL knowledge representation scheme used to generate product models in the FACADE System.

In the next section we present a view of conceptual design, and discuss two contrasting visual representations of conceptual designs that are the focus of the development of the FACADE System: Concept Arrays and Blob Diagrams. Section 3 introduces the notion of orthogonal object lattices describing different groupings of objects in DROOL product models, and presents an object model for DROOL. In section 4 we describe how DROOL objects represent different aspects of engineering product models. In section 5 we illustrate the use of DROOL with an example of a conceptual design of a mechatronic system: FireSat, a satellite intended to monitor bush fires in Australia. We conclude in section 6 by summarising the present development of DROOL and outlining potential ways of increasing its scope.

2. Conceptual Design of Mechatronic Systems

Conceptual design is concerned with making major decisions about what the artefact does and how it works, what its major components are and how they interact. This stage is followed by embodiment design, working out the details of how the chosen principles and mechanisms are implemented in an artefact constructed from available and manufacturable parts. Examples of mechatronic products which we consider include coffee makers, satellites, knitting machines, photocopiers and industrial robots, and the conceptual design of each one will involve many different design activities.

It is common for engineers doing conceptual design to think about the functions or physical entities that they need in the design, and think about how they have to be connected, without thinking immediately about their physical shapes or how they should be laid out in three dimensional space. Within the time available we can only build interfaces to support a few design activities, and we are concentrating on those activities we regard as central to the design of a wide range of mechatronic systems. So we have chosen to exclude computer support for spatial conceptual design, spatial layout design or the design of appearance, to focus on design activities that are usually prior to spatial design (Rzevski et al, 1995).

2.1. PROVISIONALITY AND ABSTRACTION

Conceptual design is on the whole characterised by provisionality, uncertainty and imprecision. At the same time, there may be very precise constraints that need to be accommodated. Any tool to support conceptual design must acknowledge this (cf Hewson's work on sketching in design, 1994), and provide the ability to work with any mixture of precise decisions and constraints with uncertainty and imprecision; it must allow designers to suspend decisions and activities at any point and return to them later. Similarly, conceptual design involves reasoning about design elements and the relationships between them at a wide range of levels of abstraction. These design elements are typically types of mechanisms and physical principles, fleshed out with approximate sizes or capacities. But they might include components conceived as abstract functions, or specific major components around which the rest of the system is designed. While the elements of conceptual designs are almost always implicitly physical objects or combinations of objects, they are thought about in functional terms. A design tool must provide the ability to work with concepts at different levels of abstraction, to switch between abstraction levels, and include elements at very different abstraction levels in the same product model.

Many computer design support systems have been developed to support structured design methodologies that expect designers to work systematically from abstract and general requirements and concepts to more concrete and detailed designs, for example Modessa (Kersten, 1995), MAX (De Vries, 1994), Schemebuilder (for instance, Bracewell et al, 1993, 1995) and the Norwegian Mechatronics Design Methodology programme (Hildre and Aaslund, 1995). These methodologies are intended to encourage designers to explore the space of alternative approaches to achieving their goals without being locked into a concrete solution too soon; they have the disadvantage that designers often find them too rigid, and often want to work by modifying relatively concrete designs, for good reasons (Rzevski et al, 1995). Other systems are less ideologically committed, but focus on later stages of the design process to support constraint-based design and consistency maintenance, for example the Edinburgh Designer System (Smithers et al, 1990), which evolved in parallel with a sophisticated artificial intelligence oriented view of engineering design.

The FACADE System is being designed to avoid imposing a discipline on designers by requiring them to follow a top-down methodology. Instead it is intended to allow designers to make use of whatever discipline they find useful, without restricting their freedom to work concretely or abstractly, or mix abstraction levels.

2.2. CONCEPT ARRAYS

Designers of complex mechatronic products need to think about how matter, energy and information flow through the system. For example, the flows in a photocopier include the transmission of paper from paper tray to collection rack, the transmission of electric current through the paper transport mechanisms, the flash, and the focusing mechanism, and the transmission of information from the control pad to the focusing mechanism. We argue elsewhere that thinking in terms of the flows in a mechatronic system and the types of operations different components of the system perform on the flows is a valuable conceptual tool in design (Rzevski, 1995b). Reasoning explicitly about flows enables designers to think about how different parts of a flow fit together, and how primary flows require subsidiary flows (for example of electricity or information), so designers can evaluate the integrity of the flow processes and check the completeness of the conceptual design.

These flows comprise the *process system* in Andreasen's (1980) Theory of Domains, which states that the synthesis of machines consists in successively establishing four systems: the process system; the *functional system*, a structure of functions or effects needed in the machine to create the transformations; the *organic system*, a structure of organs, each realising one or more functions through physical effects; and the *parts system*, a structure of single machine parts making up the embodiment of the machine. Our view is that this is too complex: while processes are primary, the concepts that engineers use in conceptual design are typically a combination of process, function and organ. We propose the use of a set of *Concept Arrays* to facilitate thinking about flows (Rzevski et al, 1995). A Concept Array shows the elements of a design tabulated in an array with columns for each type of flow, and rows for Input, Storage, Processing and Conversion, Transfer and Transmission, Use, and Disposal. Membership of an individual flow can be shown by marking or colour coding its participant elements. (Concept Arrays are different from the evaluation charts described by Pugh

(1990), and Quality Function Deployment arrays (see for instance Day, 1993), which chart concepts against issues or evaluation criteria to facilitate the evaluation and selection of concepts.)

Designers use the stack of arrays to develop the design by listing the user's requirements in the first requirements array, and listing more concrete versions of the requirements and concepts in subsequent arrays. For example, the array shown in Figure 1 contains the major flows of matter (fuel), energy (electric current, heat and kinetic energy) and information (camera images) for a satellite. When concepts have been selected for particular flows, this process can be repeated recursively for the subsidiary flows needed to make the primary flow work. In this approach to conceptual design, describing the ordering of the design elements is secondary to naming them. The links between them are conceptually implicit. In the early stages of the design process, naming the flows and what they transmit, and the design elements that participate in them, is sufficient to record the information designers wish to convey to themselves and each other.

	Information	Energy	Matter
Input	CCD Camera with thin-film filters for IR and visible	500W Solar arrays	(Propellant loaded at launch site)
Storage	32 MByte RAM Tape recorder	Ni-Cd battery Momentum wheels S/c orbit	Spherical propellant tanks
Processing/ Conversion	JPEG Image compression	Electrical heaters Transmitter SSAs Articulation drives 20N thrusters 200N thrusters	20N thrusters for attitude control 200N thrusters for orbit control
Transfer/ Transmission	S/c transmitter in Ka band	Regulated 28V power bus	Pressurised propellant transfer system
Use		S/c power bus Payload	
Disposal		S/c heat pipes	(S/c to parking orbit at EOL)

Figure 1. A concept array for FireSat.

The development of the FACADE System includes a design environment to support designing using Concept Arrays. The primary design action is naming or renaming design elements participating in flows by writing in cells in arrays, which are displayed as tables. Additional information can be supplied through dialogues and form filling, when the designer needs to define the design elements, or requires more complete record keeping or further reasoning from the design support system, for example about costs or the integrity and completeness of the flows. The design system tracks the relationships between the increasingly concrete requirements and concepts named in corresponding cells in the hierarchy of arrays, and infers connections between the design elements to construct representations of the flow, which can be checked for completeness and integrity.

2.3. BLOB DIAGRAMS

Schematic diagrams can provide an alternative graphical representation of a product in the early stages of design, in which rough sketches, symbols or icons represent elements of the artefact which it is known will be needed, although details (precise or otherwise) are not yet known. We call them blobs to avoid making any commitment about what they are; their shapes are frequently unimportant or are symbols rather than pictures. Lines between the blobs represent connections of some description. These may be physical conduits for petrol or electricity, or a functional relationship such as 'powers' or 'stabilises'. As the design progresses, design choices are made and the provisionality lessens. The informal graphical representations designers use in blob design afford both provisionality and the explicit representation of structure, especially logical structure.

We are developing a design environment for supporting blob design in the FACADE System. This includes a graphical interface with which the designer produces blobs and the connections between them by drawing them. The designer can add detailed information about the design elements denoted by the blobs by filling in forms and answering questions in managed dialogues. This interface should show blobs, that appear vague and uncertain (ideally in ways that symbolise particular types of uncertainty), until particular embodiments of the concepts are chosen.

Blob design includes two very different design situations. In one, the structure of a design element is known, so we know what inputs and outputs it can have. The choice of input or output channel determines both what is transmitted and the direction of flow. Explicit labelling of links is required only to disambiguate the choice of channel. In the other situation, the task involves constructing the objects themselves, as well as networks of objects, so that an underspecified object may have an indeterminate set of inputs and outputs. So links between objects need to be actively labelled with what they transmit; the required input and output channels and internal structures are added to the design elements denoted by the blobs.

The design environment communicates the identities of the concepts represented by the blobs and the nature of the connections between them to the product model when they are drawn, as well as additional information specified using forms and dialogues. The knowledge representation formalism should support both design processes, and so both directions of flow of information: from component to connection, and from connection to component.

2.4. A COMMON THEME: CONNECTIVITY DESIGN

Both the design activities we have described here are fundamentally concerned with deciding what the important elements in the design are and what they do, and deciding how they are connected. The two types of visual representation both show individual components (at some level of abstraction); the blob representation shows the connections between them; the array representation leaves the connections implicit but highlights the role the components play in the flows of matter, energy and information.

So blob design and designing with Concept Arrays are both ways of designing one aspect of an artefact, which we call connectivity design. We take the view that connectivity design is the central part of the conceptual design of a wide range of mechatronic systems. Not all: in many cases the critical part of conceptual design is thinking about the shape of the artefact or about how its parts interact in three dimensional space (Stacey, in preparation). Even when connectivity design is not central to the conceptual design, thinking in terms of connectivity is essential to developing a conceptual design into an embodiment, as is spatial design.

2.5. CONSIDERATIONS FOR A KNOWLEDGE REPRESENTATION SCHEME

A design support system to support different design activities contributing to connectivity design requires a representation scheme that can represent design elements and their properties at different levels of abstraction; and that can represent the connections between the elements. In the simplest outlines of designs, the representation needs to label the connections with their types and identities; more sophisticated descriptions must include what the connections carry, and how the different design elements influence the content of these transmissions. The representation requirements for design activities involving spatial thinking and spatial visual representations are very different; we are excluding these concerns from the design of DROOL, except to consider how the scheme might be extended to include them at a later stage. A product representation for conceptual design needs to include different types of information without forcing artificial distinctions between concept categories on the designer, although these categories may be significant for the structure of the product representation. Theoretical studies of the design of mechatronic systems have highlighted the range of different types of product models that have complementary roles in mechatronic design (reviewed by Buur, 1990). Various taxonomies have been proposed besides Andreasen's (1980). Hildre and Aaslund (1995) argue that no single product model can capture all aspects of a mechatronic system; our view is that a single representation scheme can cover many different aspects of a design by linking objects describing different aspects of a design to their physical manifestations. (We discuss this further in sections 4.3 and 6.5.)

The development of the DROOL knowledge representation scheme has been motivated by the following important guiding principles for the design support system the FACADE Project is developing.

- *Separation of Configuration and Function.* A component of a machine may take part in several different functional systems and processes, and what they are may only be determined in the course of designing the machine. So the scheme should distinguish between configurational elements of designs (physical things described at some level of abstraction: what components *are*) and functional elements (what components *do* when they participate in flows), so that the mappings between the configurational and functional elements can be created and changed in the course of the development of the design. This principle is a major difference between DROOL and the many alternative product representation schemes.
- *Provisionality and Abstraction.* The scheme should represent design elements at different levels of abstraction, and represent the uncertain or provisional status of aspects of the design. It should enable the designer to change the design by replacing design elements with more concrete versions, more abstract generalisations, or alternative instantiations of the same abstraction. It should enable the free combination of design elements described at different levels of abstraction in the same product model.
- *Multiple Views.* The scheme should integrate representations of different aspects of the products, to track the relationships between them, and to enable designers to combine concepts formulated in different terms. The scheme's primary component should be representations of flows of matter, energy and information.
- *Incremental Development.* The scheme should permit the incremental development of conceptual designs from the specification of requirements to embodiment design, by the addition of information to concept descriptions, and the replacement of abstract concepts with concrete ones. But it should not constrain the designer to follow any particular methodology or order of design actions.

3. A Representation Scheme for Mechatronic Devices

In this section we outline the DROOL knowledge representation scheme: Design Representations in Orthogonal Object Lattices. This is an object oriented scheme implemented in VisualWorks, a descendant of Smalltalk-80 produced by ParcPlace Systems, Inc. The scheme takes its name from the use of alternative groupings of design elements, as lattices of objects, to define different views of the product model and its place in a hierarchy of abstractions.

3.1. ALTERNATIVE MODELS OF MECHATRONIC SYSTEMS

DROOL is one of a number of object-oriented formalisms for product models, several of which have features in common with it. Gui (1991) presents a simpler component-connector model for configuration design. A number of projects employ the idea of grouping device components differently for different system views on a design. We took inspiration from the EDC Product Model developed by Tim Murdoch and Nigel Ball at Cambridge University, a more elaborate scheme than DROOL intended more for detailed configuration design (Murdoch and Ball, 1994; Murdoch, 1995; Ball and Murdoch, 1995). It uses the concept of system to group assemblies relevant to particular system views of designs. DROOL does this, but configurational elements are secondary to functional elements in the system views. Other object oriented knowledge representation schemes developed to support different views on designs, to meet different requirements, include SHARED (Wong and Sriram, 1993) in engineering and ICON (Brown et al, 1995) in building design.

Of course, component-connector models of devices have been around a long time in the field of qualitative reasoning. In their seminal work on the qualitative physics of flows and pressures, De Kleer and Brown (1984) employ *valves, pressure sensors, pressure references, conduits* and *terminals* (that is, outside connections) as objects; they point out the limitations of component-connector models (p21). Barrow (1984) uses *modules* which have *ports* as well as *states* and state and output equations, and *connected* assertions, in his VERIFY system for verifying the correctness of digital circuits.

3.2. ORTHOGONAL OBJECT LATTICES

Different types of relationship group subsets of the objects in a product model into independent but interlocking graph structures. These different lattices of objects provide alternative views of the design which are independent but interlock. We term these structures lattices because they are ordered, but need not be trees. DROOL includes three types of lattice; the inclusion of qualitative spatial relationships and geometry would require others.

Configuration Views: Composition Trees. Almost all machines are hierarchical in structure: they comprise aggregations of components, which themselves have components. DROOL supports the hierarchical description of assemblies as trees of components, with the relationships 'Is-Component-Of' and 'Includes'.

- *Process Views: Flow Networks.* DROOL represents flows of matter, energy and information explicitly as objects, as well as the design components and connections that make up those flows. It also uses explicit objects to represent the causal interactions that link the different flows into networks.

- *Alternative Versions: Abstraction Lattice.* DROOL can link representations of components at different levels of abstraction with the relationships 'Is-Generalisation-Of' and 'Is-Specialisation-Of'. This enables designers to change designs by following these links to substitute more abstract or more concrete design components. As assemblies can be generalised in different ways they form an abstraction lattice rather than a tree.

3.3. AN OBJECT MODEL

The major conceptually significant objects in the DROOL knowledge representation scheme are shown in Figure 2. This is an ERMIA diagram (Green and Benyon, 1995, in press) of the object model, which shows the major conceptual relationships between the objects. Each box denotes a class of entities, and the lines show essential and optional relationships between the entities that belong to each class.

Section 4.1 introduces the objects representing configurational elements of the design (physical structures described at some level of abstraction, and how they are physically connected). These are *Assemblies*, *Ports* and *Conduits*. The objects representing process elements in the design are introduced in section 4.2; these are *Flows*, *Transforms*, *Roles*, *Transmissions*, *Transform-Interactions*, and *Flowing-Entities*. In DROOL the process elements of the design are functional aspects of the configurational elements. There is an incomplete mapping between the configurational objects, shown on the left of Figure 2, and the process objects shown on the right: assemblies implement transforms, ports take roles, and conduits convey transmissions. We discuss the motivation for separating form and function in this way in section 4.3.

Configuration objects are used to represent the attributes and variables the other objects in the product model can have. *Interdependency* objects represent the relationships between configurations, including design constraints. DROOL permits the constraints and dependencies between aspects of the system to be specified freely, and at differing degrees of precision and detail, because the designer may find it useful to highlight all sorts of relationships.

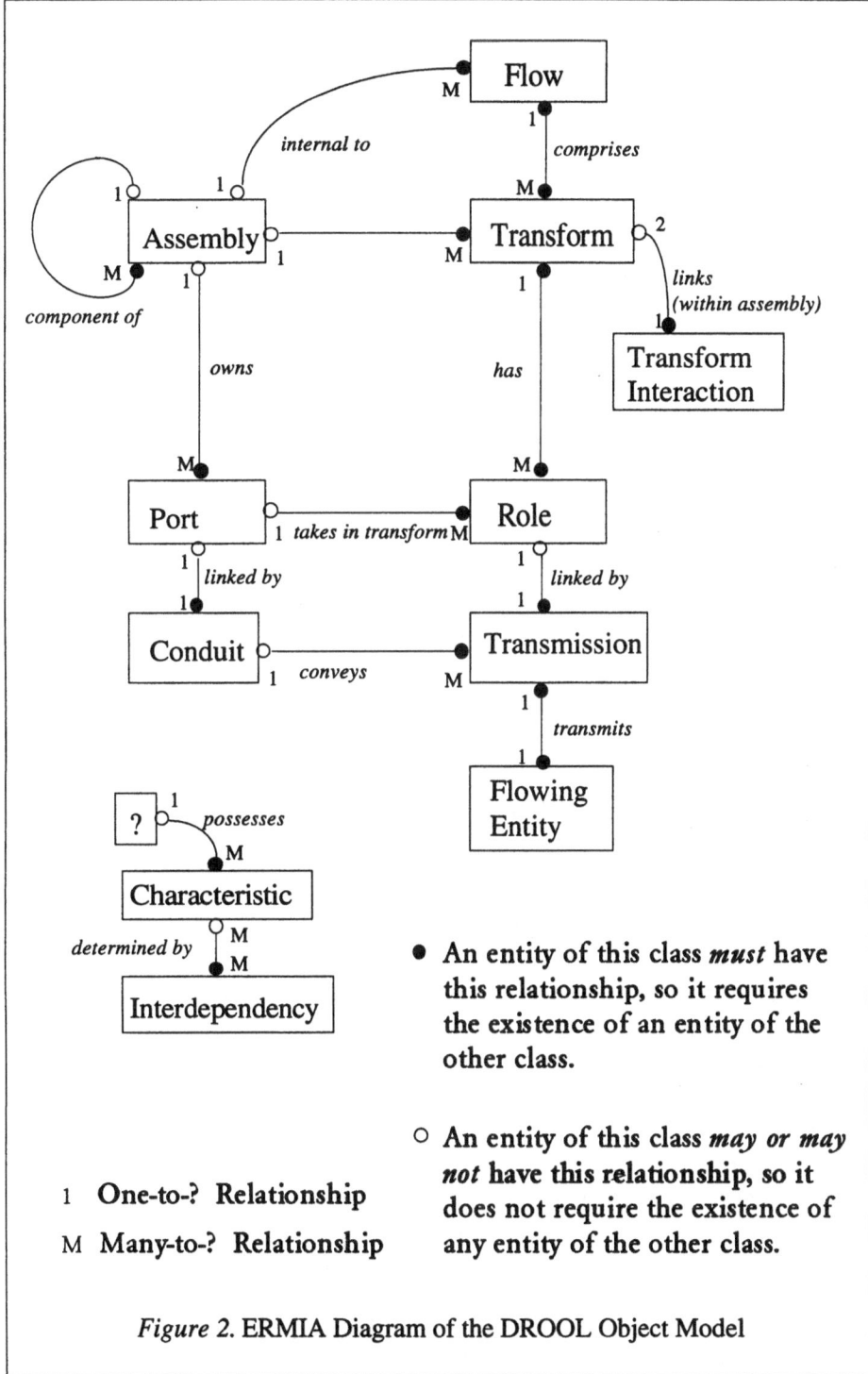

Figure 2. ERMIA Diagram of the DROOL Object Model

4. Representing Mechatronic Systems in DROOL

In this section we describe how the objects listed in section 3.3 are used to model aspects of the conceptual designs of mechatronic systems. DROOL embodies a fundamental design decision to separate the objects representing aspects of machine function from the objects representing the identity and configuration of components of the machine. Although DROOL is designed primarily to support functional descriptions in terms of flows of matter, energy and information, these functional descriptions are attached to physical descriptions, so we begin with how DROOL models the configurational structure of mechatronic systems in section 4.1, and discuss how DROOL models processes in section 4.2, and the relation between function and configuration in section 4.3. Section 4.4 discusses how models of transforms can be implemented in DROOL. Section 4.5 considers networks of flows as maps of causal influence. Section 4.6 deals with how abstraction is handled.

4.1. REPRESENTING THE CONFIGURATIONAL STRUCTURE OF MACHINES

We assume that the elements that make up the conceptual design of a mechatronic system are, in some sense, physical objects. So if an element in a conceptual design is specified as a requirement *'Do X'*, it is modelled in DROOL as a component *'Thing to do X'*. All such physical things are modelled with Assembly objects. The physical composition of an assembly is modelled by listing its components, which are also assemblies; these components may be functionally and structurally related in all the ways DROOL supports.

Configuration design consists in creating or modifying physical objects, and composing them to create new composite objects. New assemblies are created in DROOL in three ways: (1) by defining them ab initio; (2) by adding or removing information from the descriptions of previous assemblies, to make them more or less abstract; and (3) by grouping or merging previously existing assemblies. Assemblies are modified by changing their set of characteristics, and the values of the characteristics, and by adding or removing other objects.

The present version of DROOL supports one type of physical relationship between assemblies other than composition. This is connection through conduits that can convey flows of matter, energy and information. They are made by using Conduit objects representing physical connections to connect special parts of assemblies called ports, which we represent with Port objects. Port objects belong to the lowest level Assembly object for which they are defined; they include restrictions on the types of flow they can convey. DROOL permits the inclusion in product models of assemblies that do not play a direct part in any flows of matter, energy or information, but serve some other function like housing or anchoring other components. But we have not yet incorporated any mechanisms for describing spatial relationships or other aspects of configuration, to model the way assemblies are composed of components.

4.2. REPRESENTING PROCESSES

The DROOL view of functional systems is that they comprise a network of functional units that communicate by passing matter, energy or information between them; these functional systems are processes defined in terms of single flows of one flowing-entity. The flows are represented by Flow objects, which point to chains or networks of Transform objects linked through Role objects by Transmission objects. The Transform objects represent functional units that act on the flow of the flowing-entity in some way. The Transmission objects represent the state of the flow between Transform objects; they do not influence the flow. Transmission objects point to Flowing-Entity objects that represent the state of the flowing-entity conveyed by the flow between each pair of transforms. The Role objects define the different inputs and outputs of each transform, so that these are represented independently of physical embodiment or connections to other transforms. Transform-Interaction objects represent explicitly the interactions between transforms in different flows.

4.3. RELATING FUNCTION TO CONFIGURATION IN DROOL

Transform objects belong to Assembly objects: they represent *how* the assembly acts *on* the flowing-entity. Similarly transmissions depend on conduits to convey them, and ports fulfil roles in transforms. Why then does DROOL separate function from physical existence by using parallel sets of objects? The first reason is that we find it useful to think of functional units, transforms, as primary elements in conceptual designs of mechatronic systems. The second reason is that a single physical conduit can convey more than one flow (for example a pipe carrying a flowing liquid and heat), and we require the ability to express this relation explicitly. The third reason is pragmatic: assemblies may have several functional aspects, which may be added or removed in the course of design, so representing them as first-class objects is both conceptually and computationally easier.

The DROOL approach to relating function to configuration rests on a number of assumptions, that have guided our choice of objects and the relationships between them. Each action on a flow is performed by a physical thing, so each functional unit (a transform) is implemented by a physical component of the machine (an assembly). A transform passes matter, energy or information (the flowing-entity) to other transforms in the *same flow* located at *other assemblies*. (Flows and flowing-entities are not changed at or by transmissions.) Flows interact, notably when flows of information guide the behaviour of flows of matter or energy. In order to model the interactions between flows, we employ the axiom that interactions happen at physical locations, and that these physical locations are the assemblies implementing the transforms: transforms can influence other transforms in *other flows* located at the *same assembly*.

4.4. MODELLING FLOWS OF ENERGY, MATTER AND INFORMATION

The simplest DROOL product models are produced simply by naming and linking components of the design participating in flows of matter, energy and information. DROOL is also designed to support more detailed representations of designs. This requires adding information to the transform objects about how transforms make changes to the flowing-entity and to the properties of the flow at its inputs and outputs. This is the rationale for using objects to represent the state of the flow between transforms, and the state of the flowing-entity. In the terms of Andreasen's (1980) Theory of Domains, this information is part of the functional system and organic system levels of the product model.

DROOL is designed to support the development of representations of transforms that enable a full range of inferences about its behaviour. This involves defining a transform function. This states how the characteristics of the transform and its assembly, and of the transmissions and flowing-entities at its inputs and outputs, determine the behaviour of the transform and are determined by that behaviour.

According to this view, the transforms are parameterised finite state machines: their behaviour is determined both by discrete changes of state (corresponding to the choice of function with which the outputs are calculated from the inputs in an envisionment of the system's behaviour) and continuous changes of parameters (corresponding to the inputs to a particular function for computing outputs).

4.5. MAPPING CAUSALITY IN DROOL

We take the metaphysical view that the behaviour of each part of a mechatronic system is *caused* by flows of matter, energy and information; so causal influences are conveyed between assemblies by flows, and are conveyed between flows at assemblies. The flow of the flowing-entity (for example a liquid, or an object being manufactured) is not the same as the flow of causal influence around the system; in many systems the flowing-entity travels in one direction only, but an emergent property of the flow (such as the pressure of a flowing liquid) can travel in the other direction, so that events downstream can affect what happens upstream. Thus a transform function can influence and be influenced by transmission characteristics at both inputs and outputs, but can only be influenced by the characteristics of the flowing-entity at its inputs, and can only determine the characteristics of the flowing-entity at its outputs. Modelling the organisation of a mechatronic device in DROOL as a network of flows of matter, energy and information thus creates a map of the causal influences in the system.

4.6. REPRESENTING ABSTRACTION IN DROOL

DROOL handles changes in abstraction and generality by adding and removing information from assemblies. This information includes the characteristics of the assemblies and their ranges and values; the attributes of the assemblies' ports and their values; the sets of subassemblies belonging to the assemblies, and the existence of transforms associated with the assemblies, as well as their transform methods.

DROOL represents functional decompositions by expanding functional units into sequences of more concrete functional units: it represents simple, powerful transforms as associated with highly abstract assemblies; alternative decompositions are represented as specialisations of these assemblies that have subassemblies whose transforms in that system are a sequence constituting the functional decomposition.

We envisage the development of a permanent assembly library describing useful high level abstractions, some components of actual completed designs, and a few useful middle level descriptions. However any assembly library will contain only a small fraction of the potentially useful assembly representations, so most assemblies will be constructed by modifying others, especially by composing high level abstract assemblies and adding detail to them.

5. FireSat: An Example from Outer Space

We illustrate the Concept Array approach and the use of DROOL to represent conceptual designs with a design for FireSat (Rzevski and Buckland, 1995), a proposed satellite system to monitor bush fires in Australia. The array shown in Figure 1 is a partial concept array; it follows the arrays developed to describe the customer's requirements and more precise operational specifications of those requirements. It only includes the parts of the satellite system that go into orbit; the characteristics of the ground segment are an integral part of the conceptual design of a satellite, but we have deleted the information flow in the ground segment from the array for simplicity and clarity.

The array shows parts of several flows, which include: The flow of information from the on-board camera to the microprocessor for data compression to the data storage system to the satellite's transmitter. The flow of electrical energy from a solar array to a battery to a regulated power bus (with power conditioning circuitry) to the transmitter, camera, articulation drives and heaters. The flow of matter (propellant) from the tanks to the thrusters and out into space. The conversion of chemical energy (in the propellant) to kinetic and potential energy (of the satellite as a whole; an orbit is a dynamic configuration of potential and kinetic energy). DROOL represents each of the entries in the Concept Array with a transform, each belonging to an assembly, which represents an action performed by a component of the satellite. A new assembly is created or retrieved by DROOL and added to the product model when a label for a new component is added to a Concept Array; if the component is mentioned in another cell in the array, another transform is added to the assembly if an appropriate one does not exist.

In Figure 1, the concept listed for the transmission of propellant is 'pressurised propellant transfer system'. This can be specialised in two ways, by choosing a blowdown system, suitable for monopropellant thrusters, in which compressed helium stored in the propellant tank pushes out the propellant, or by choosing a regulated system, more suitable for bipropellant thrusters, in which compressed helium in separate tanks is allowed to flow into the fuel and oxidiser tanks to push out the fuel and oxidiser. In a DROOL representation of the design of a satellite, this choice is encoded by substituting an assembly representing a blowdown system or regulated system for the assembly representing the more general concept.

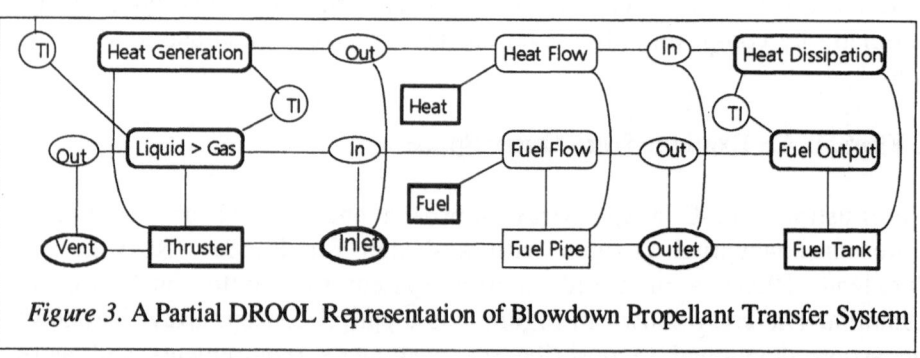

Figure 3. A Partial DROOL Representation of Blowdown Propellant Transfer System

A partial DROOL representation of a blowdown propellant transfer system is shown in Figure 3. The assembly representing the propellant tank is shown in the lower right hand corner; it is linked through a port representing the outlet to a conduit representing the propellant pipe; this is linked to the thruster assembly through its inlet port. The propellant tank assembly has a transform representing its output of propellant, which is part of the propellant flow (the flow object is not shown). This transform is linked through its out role to a propellant flow transmission associated with the propellant pipe conduit, which has a flowing-entity object representing the state of the propellant in the pipe. The propellant flow transmission is linked through an in role to the transform representing the burning of the liquid propellant to produce gas. The out role of this transform is associated with the vent port of the thruster. It is linked by a transform-interaction to a transform representing the conversion of chemical energy into linear and rotational kinetic energy (not shown).

The flow of heat from the thrusters to the propellant tanks is a significant influence on the design of a satellite, and a factor limiting the use of the thrusters. To model this flow, we add another set of functional objects representing the flow of heat to the same set of configurational objects. We include a heat generation transform in the representation of the thruster, which is linked through a heat flow transmission (associated with the propellant pipe conduit) to a heat dissipation transform in the representation of the propellant tank. The generation of heat is caused by the burning of fuel; this influence is represented by a transform-interaction. Similarly, the thermal properties of the propellant tank influence its emission of propellant; this influence is represented by a transform-interaction.

If we decide to choose bipropellant thrusters, we replace the propellant pipe conduit with separate conduits for fuel and oxidiser, and a pair of regulated pressurised propellant transfer systems. Each has an assembly representing the helium tank and an assembly representing the fuel or oxidiser tank, connected by a helium pipe conduit. The helium tank representation includes a transform representing the output of pressurised helium, which is linked to a helium flow transmission. This is linked to the helium input transform of the fuel tank assembly, which is linked through a transform-interaction (describing the effect of the helium pressure on the fuel pressure) to the fuel output transform.

6. Present and Future Developments

The DROOL view of conceptual designs comes from engineering experience, observations and rational reconstructions of design processes; we have not yet tested it systematically. DROOL and the first prototype design environments are being implemented. Using the prototypes of our intelligent support system to test the validity of our view of conceptual design is a central part of the future development of the FACADE Project.

We argue in this paper that DROOL has the power to represent the essential features of the conceptual designs of a wide range of mechatronic systems. But its set of relationship types is insufficient to support some important design activities, notably spatial design at both the conceptual and embodiment stages. DROOL has been designed to be open-ended, so that additional features can be added without forcing a redesign of its core elements. We conclude by noting some of its limitations and potential extensions.

6.1. SPATIAL RELATIONSHIPS

DROOL does not support spatial relationships. This limits its ability to model the configuration of mechanical systems and support configuration design activities. While including labelled binary relations between assemblies is computationally easy, supporting inferences about spatial relationships requires more sophistication. To do this, SHARED (Wong and Sriram 1993) employs qualitative spatial relationships based on a point interval algebraic formulation (Mukerjee, 1991).

6.2. GEOMETRY

Supporting the transition from conceptual design to detailed embodiment design requires the addition of geometric descriptions to a representation scheme designed for conceptual designs, such as DROOL. The approach to this that best fits the structure of DROOL models is that taken in the EDC Product Model (Murdoch and Ball, 1994), which employs feature objects to represent geometry-owning regions of assemblies; the division of assemblies into features need not correspond to its decomposition into subassemblies.

6.3. FLOWS WITHOUT CONDUITS

Some flows of matter, energy and information do not flow through defined conduits constructed to convey them. A vitally important example is heat produced as a side effect by the operation of devices for doing other jobs. Thinking about these flows can be important for making major design decisions, for example in designing a satellite to ensure thermal control. They can be described at various levels of detail, which would make different demands for extensions to DROOL. The simplest approach is simply to define a flow with a transform asserting that, say, heat is lost by the assembly. This requires no extensions to DROOL, but does not allow us to say what happens to the heat or other flowing-entity after it is lost by the assembly. The next simplest approach is defining null ports without conduits for emitting and absorbing the flow, and defining a flow of, say, heat as a network with transmissions linking all the major emitters with all the major absorbers. This could be used with characteristics describing the limits on the capacities of each transform, to guide qualitative design decisions about where barriers or specially designed disposal mechanisms are required. Accurate modelling of flows without conduits requires geometric modelling of spatial layouts.

6.4. DESIGN ELEMENTS THAT ARE NOT COMPONENTS

Some types of conceptual design involve reasoning with design elements that are not components of the machine in any sense, so are excluded from the DROOL representation. We exclude these design activities from the scope of the FACADE System. We note that not-component concepts are usually spatial, for example the trajectories of machine parts, objects being processed and radiation, so including mechanisms for spatial reasoning is a prerequisite for including not-component concepts in the scope of DROOL.

6.5 OTHER FUNCTIONAL VIEWS

The interlocking flow view is of course not the only way to think about the functioning of mechatronic systems. Other types of functional system view could be represented by adding slots and objects to the present version of DROOL, provided they share its basic assumptions: (1) Functional units have physical locations at physical components of the device (assemblies), and communicate with other functional units having physical locations (not necessarily different ones). (2) Actions on physical or functional units can be described by changes to parameter values. (3) Actions on an object being processed can be described in terms of its states before and after the action.

Acknowledgements

The research described in this paper has been supported by EPSRC Grant GR/ J48689 to George Rzevski, Helen Sharp and Marian Petre. Claudia Eckert made helpful comments on an earlier draft of this paper, as did the reviewers.

References

Andreasen, M. M.: 1980, *Syntesemetoder på systemgrundlag*, PhD Thesis, Lunds Tekniska Högskola.

Ball, N. R. and Murdoch, T.N.S.: 1995, A layered framework for sharing design data. *Proceedings of the 10th International Conference on Engineering Design*, Heurista, Prague, Czech Republic, pp. 1471-1476.

Barrow, H. G.: 1984, VERIFY: A program for proving correctness of digital hardware designs, *Artificial Intelligence*, **24**, 437-491.

Bracewell, R. H., Bradley, D. A., Chaplin, R. V., Langdon, P. M. and Sharpe, J. E. E.: 1993, Schemebuilder: A design aid for the conceptual design stages of product design, *Proceedings of the 9th International Conference on Engineering Design*, Heurista, Den Haag, pp 1311-1318.

Bracewell, R. H., Chaplin, R. V., Langdon, P. M., Li, M., Oh, V. K., Sharpe, J. E. E. and Yan, X. T.: 1995, Integrated platform for ai support of complex design (part i): rapid development of schemes from first principles, *in* J. E. E. Sharpe and V. K. Oh (eds) *AI System Support for Conceptual Design*, Springer-Verlag, Berlin.

Brown, F. E., Cooper, G. S., Ford, S., Aouad, G., Brandon, P., Child, T., Kirkham, J. A., Oxman, R. and Young, B.: 1995, An integrated approach to CAD: modelling concepts in building design and construction, *Design Studies*, 16(3), 327-347.

Buur, J.: 1990, *A Theoretical Approach to Mechatronics Design*, PhD Thesis, Institute for Engineering Design, Technical University of Denmark, Lyngby.

Day, R. G.: 1993, *Quality Function Deployment*, ASQC Quality Press, Milwaukee.

De Kleer, J. and Brown, J. S.: 1984, A qualitative physics based on confluences, *Artificial Intelligence*, 24, 7-83.

Faltings, B.: 1991, Qualitative models in conceptual design, *in* J. S. Gero (ed.) *Artificial Intelligence in Design '91*, Butterworth-Heinemann, London, pp. 645-663.

Green, T. R. G. and Benyon, D. R.: 1995, Displays as data structures: Entity-relationship models of information artefacts, *in* K. Nordby, P. H. Helmersen, D. J. Gilmore and S. A. Arnesen (eds) *Human Computer Interaction: Interact '95*, Chapman & Hall, London, pp. 55-60.

Green, T. R. G. and Benyon, D. R.: in press, The skull beneath the skin: entity-relationship modelling of information artefacts, *International Journal of Human Computer Studies*.

Gui, J. K.: 1991, Object-oriented assembly and assembly design process modeling, *Journal of Engineering Design*, 2(2), 141-149.

Hewson, R.: 1994, *Marking and making: a characterisation of sketching for typographic design*, PhD Thesis, Institute of Educational Technology, The Open University, Milton Keynes.

Hildre, H. P. and Aaslund, K.: 1995, Conceptual design for mechatronics, *in* J. E. E. Sharpe and V. K. Oh (eds.) *AI System Support for Conceptual Design*, Springer-Verlag, Berlin.

Kersten, T.: 1995, 'Modessa' A computer based conceptual design support system, *in* J. E. E. Sharpe and V. K. Oh (eds.) *AI System Support for Conceptual Design*, Springer-Verlag, Berlin.

Mukerjee, A.: 1991, Qualitative geometric design, *in* J. Rossignac and J. Turner (eds) *Proceedings of the First ACM/SIGGRAPH Symposium on Solid Modeling Foundations and CAD/CAM Applications*, ACM Press.

Murdoch, T. N. S.: 1995, Sharing design data, *Technical Report CUED/C-EDC/TR-28*, Cambridge University Engineering Department, Cambridge.

Murdoch, T. N. S. and Ball, N. R.: 1994, The development of an edc product data model, *Technical Report CUED/C-EDC/TR-21*, Cambridge University Engineering Department, Cambridge.

Pugh, S.: 1990, *Total Design*, Addison-Wesley, Wokingham.

Rzevski, G.: 1995a, FACADE: Concurrent engineering applied to multi-technology products, *Proceedings of the International Workshop on Concurrent/Simultaneous Engineering Frameworks and Applications*, Lisbon, Portugal.

Rzevski, G.: 1995b, Intelligent systems: Issues and trends, *Proceedings of the International Conference on Intelligent Manufacturing*, Wuhan, China.

Rzevski, G. and Buckland, R. A.: 1995, FireSat: A satellite designed using concept arrays, *Technical Report 9502*, The Open University, Centre for the Design of Intelligent Systems, Milton Keynes.

Rzevski, G., Buckland, R. A., Petre, M., Stacey, M. K., and Sharp, H. S.: 1995, Conceptual design of mechatronic systems, *Technical Report 9503*, The Open University, Centre for the Design of Intelligent Systems, Milton Keynes.

Smithers, T., Conkie, A., Doheny, J., Logan, B., Millington, K. and Tang, M. X.: 1990, Design as intelligent behaviour: An AI in design research programme, *Artificial Intelligence in Engineering*, 5(2), 78-109.

Stacey, M. K.: in preparation, Mental Processes in Engineering Design.

Vries, T. J. A. de: 1994, *Conceptual Design of Controlled Electro-mechanical Systems, a Modeling Perspective*, PhD Thesis, Universiteit Twente, Enschede.

Wong, A. and Sriram, D.: 1993, SHARED: An information model for cooperative product development, *Research in Engineering Design*, 5(1), 21-39.

J. S. Gero and F. Sudweeks (eds), Artificial Intelligence in Design '96, 603-622.
© 1996 *Kluwer Academic Publishers.*

GENERATING CONCEPTUAL SOLUTIONS ON FUNCSION: EVOLUTION OF A FUNCTIONAL SYNTHESISER

AMARESH CHAKRABARTI AND MING XI TANG
Engineering Design Centre, Department of Engineering
University of Cambridge
Trumpington Street, Cambridge CB1 1PZ, UK

Abstract. FuncSION is a software that can synthesise, using a database of functional elements, an exhaustive set of solution concepts to satisfy functional requirements of a design problem. It is intended to stimulate designers' thinking by providing a framework where these solutions are offered to the designers for exploration in the conceptual design stage. Reported in this paper are some of the testing results using FuncSION in two case studies and three hands on experiments, in terms of its ability to (i) offer a wide range of new, interesting and useful ideas, and (ii) facilitate exploration of these ideas in an effective way. The main results are: it does provide useful ideas and interesting insights to the designers, but does this at the cost of having to deal with a potentially huge list of candidate solutions which are hard to explore sufficiently. Based on these results, a scheme for coping with a large number of solutions without losing explorability is proposed, whereby designs could be generated and explored at multiple levels of abstraction, using pre-defined as well as customised clustering strategies at any of these levels. An implementation of the scheme, in terms of a user editable hierarchical database of elements and solutions, and a general algorithm for synthesis at multiple levels are proposed. A set of clustering strategies for identifying and grouping the solutions considered by experienced designers to be redundant and wasteful is also discussed, with some initial testing results.

1. Introduction

In transmission design, the major functional requirement of a design is to transmit and transform forces and motions. This can be expressed as a transformation from a set of input characteristics to a set of output characteristics. Each of these characteristics may be required to change with time. The transformation at an instant between the input and the output characteristics is an *instantaneous transformation*. An ordered set of such transformations can be used to express the overall functional requirement of a problem. In this approach, a *solution concept* is an abstract description of a system, of identifiable *functional elements,* that can satisfy given functional requirements. For instance, a solution concept, for transmitting a force on the same plane but into a different direction and position, could be a system in which an input rack takes the input force to rotate a pinion, which moves an output rack in the required direction to provide the required output force.

The instantaneous transformation of a given system can be deduced using the information about its constituent elements, connections, and their rules of combination. FuncSION (acronym for **Func**tional Synthesiser for **Input Output Networks**) is a system developed at the EDC in Cambridge University that synthesises solution concepts, using functional elements from a database as illustrated in Figure 1, to fulfil a given functional requirement of a design in terms of its required instantaneous transformation, so that designs so synthesised fulfils the required function at one instant of time.

In the representation that FuncSION uses, a design problem is defined as a transformation between a set of instantaneous *input* (and *output vectors,* each of which has a set of characteristics such as *kind* (the type of I/O, such as force, rotation etc.), *orientation* (the spatial axis along which an I/O is oriented) *sense* (the sense of the I/O along the spatial axis) a *position* (spatial co-ordinates) and *magnitude,* to represent the required I/O characteristics. FuncSION allows the user-definition of a database of functional elements, where each element is expressed as a transformer which transforms an input vector of given characteristics into an output vector of specified characteristics (for example, a screw element can transform an input vector of the rotation kind into an output vector of the translation kind so that they are co-axial to each other).

In FuncSION, the synthesis of a solution concept is supported in a three step process. The first step involves Kind Synthesis, where an exhaustive search algorithm is used to synthesise a set of topological networks of causally connected functional elements, each of which is structurally feasible and can transform the give input kind into the output kind required of a design problem. Concepts generated by this procedure is exhaustive, i.e., all possible combinations of elements, from a given database of elements and using a specified maximum number of the transformations allowed in any solution concept, which fulfils the given functional requirement, are generated. In the second and third steps, possible alternative spatial configurations (i.e., spatial layouts) of each such topological candidate solution concept can be generated, using two further procedures called orientation and sense synthesis, so that the concepts can also satisfy the orientation and sense constraints imposed by the design problem. The representations for the design problem, design solution concepts and synthesis procedures used in FuncSION are reported in Chakrabarti and Bligh (1994).

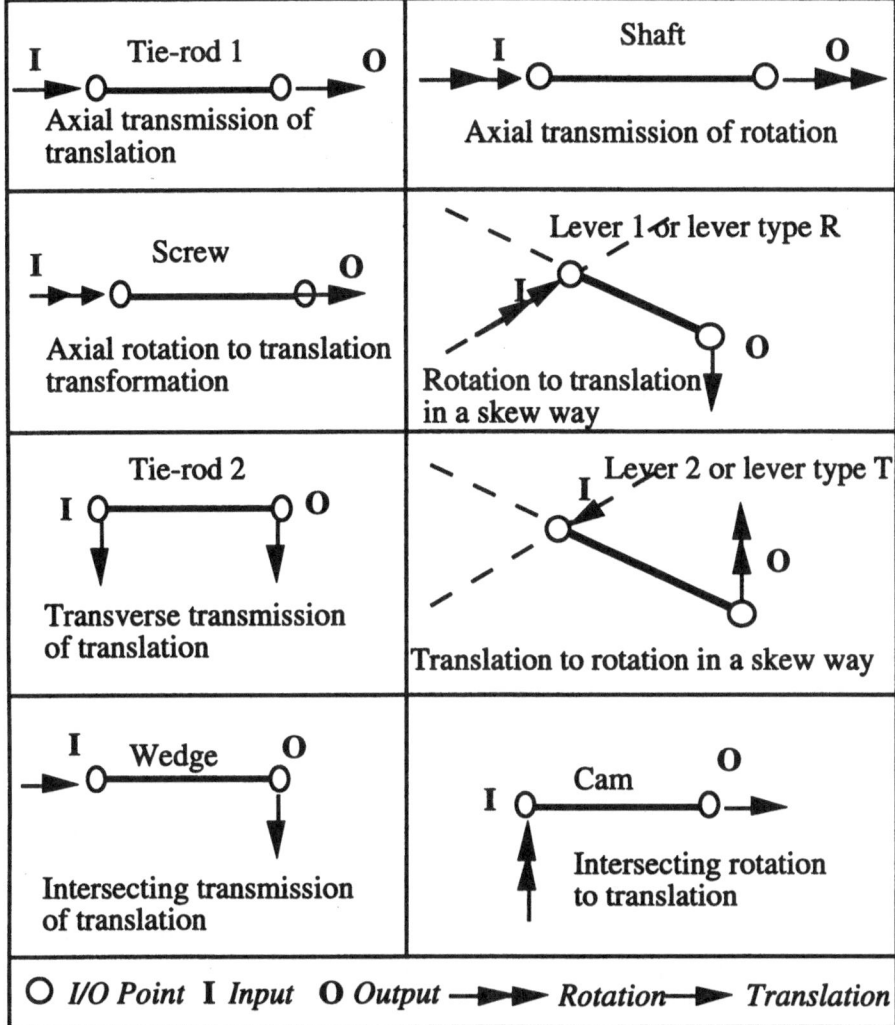

Figure 1. The elements used by FuncSION in the synthesis of conceptual solutions.

The objective of this paper is to present the development of FuncSION based on its application in real design cases and its evaluation by experienced designers. In Section 2 the results of testing FuncSION are discussed. From these tests, the main research problems to be tackled before utilising the potential of FuncSION are identified in Section 3 and Section 4. A scheme for solving these problems and its implementation is presented in Section 5, with some of initial test results.

2. Testing

2.1. MAS PROJECT CASE STUDIES

The Mobile Arm Support (MAS) project was intended to design a means for enhancing the mobility of Muscular Dystrophy (MD) sufferers. People having this disorder have little or no lifting strength in their arms, although they do not lose any of the finer controls of their fingers. In the task clarification phase of the project, it was found that the MD sufferers are capable of using their inertia to move their arms in horizontal plane in absence of significant surface resistance. It was decided that an arm support would be designed as a means of enhancing mobility, which would be able to provide powered vertical motion of the arm. There should be enough freedom in the horizontal direction for the users to use their own strength to move their arms in the horizontal plane. ₁ne project ran in two phases spanning a total of over three years, which led to the development of two prototypes.

2.1.1. Comparison with Designs generated in MAS I Project

The two designers who worked in MAS I project were assisted by a brainstorming session, which gave them an initial pool of ideas. They explored these ideas, and eventually came up with three variants, one of which was selected for embodiment.

   As a retrospective study, an input-output requirement, which describes the intended instantaneous function of the arm support, was given independently to FuncSION, for it to generate ideas and their spatial configurations. As vertical mobility was the main requirement, the input was either a translation or a rotation, which could be in any of the three reference directions, and the output was specified as a vertical translation. With the two specifications given to it (one is a torque to force transformation, and the other, a force to force transformation) a total of 162 ideas were generated. These ideas were compared with those that designers generated.

   There were a total of 73 ideas generated by the designers. Most of these ideas are either physical effect-like solutions, or incomplete and incomprehensible, or from a different domain of knowledge, and thus are not within the realm of FuncSION, leaving a total of 27, not necessarily distinct ideas which could be compared with the solutions FuncSION generated. FuncSION managed to generate 22 of these.

   Of the 162 solutions that FuncSION suggested, 13 (the number of distinct ideas of the 22) were generated by the designers. However, given that FuncSION allows the same element to be used more than once in a solution, it often generates a number of solutions which might be considered as variants of other ideas (possibly giving an inflated impression of its originality). Also, some solutions generated might be too expensive to be considered by the designers at all. One method to compare designers' solutions with solutions generated by FuncSION, in the above context, might be to group designs generated by FuncSION into a number of clusters and to eliminate those clusters which contain "expensive" solutions. If an idea exists in the designers' documents which can be abstracted as one of the solutions in a cluster from FuncSION, then to assume that this cluster has been considered by the designers. There are two problems with this method. One is the issues of what criteria should be used to group designs as similar, and to classify designs as wasteful / expensive. The second is that the designs generated by the designers are often at a different level of abstraction than those generated by FuncSION. If these solutions are at a higher level of abstraction, these cannot be discussed within the realm of FuncSION (e.g., those ideas that are physical-effect-like). If these are at a lower level of abstraction, we need to abstract them to the right level before these can be compared with solutions generated by FuncSION; this has two consequent difficulties.

Take the instance of the "shaft rack and pinion" solution (Bauert, 1993) as an example, this could be interpreted as a "shaft lever tie rod" solution before it is compared with FuncSION (because a pinion and a rack could be abstracted as a lever and an axial tie-rod respectively).The difficulties are that if we interpret, by having spotted an instance of the "shaft rack and pinion" design, that the designers have considered the whole cluster that represents "shaft levers tie rod", then we might have made two layers of mis-conception: whether or not the designers considered "shaft lever tie rod" solution class as a whole (various possible embodiments of this class at the level of abstraction at which the designers considered their design); whether or not they considered the whole cluster of "shaft lever tie rod" type solutions (the variants of this solution at the same level of abstraction such as "lever tie rod", "shaft lever" etc.).

We have tried to deal with the first problem, of finding criteria for clustering designs, by carrying out a set of further experiments with experienced designers, and identifying their common notions of wasteful/expensive and similar as clustering heuristics. The second problem, about comparison of designers' idea-instances with FuncSION's solution clusters, was dealt with by this assumption that if designers' instances can be abstracted into more than one solution in a cluster, then they must have considered the solution types represented by that whole cluster. However, if there is just one single or no idea-instance that could be abstracted as a solution in a cluster from FuncSION, then the designers did not consider this solution cluster.

There were interesting and inexpensive solutions that were suggested by the computer, which designers did not conceive (one example of which is a single link lever connecting an input rotation to a tie-rod via a four bar linkage to provide an output hand motion). It was interesting to note that some solutions which were regarded by the designers as distinct solutions were regarded by the computer as topologically the same (e.g., the final two solutions in MAS I, see Figure 2). This signifies the importance of considering spatial configurations as distinct solutions.

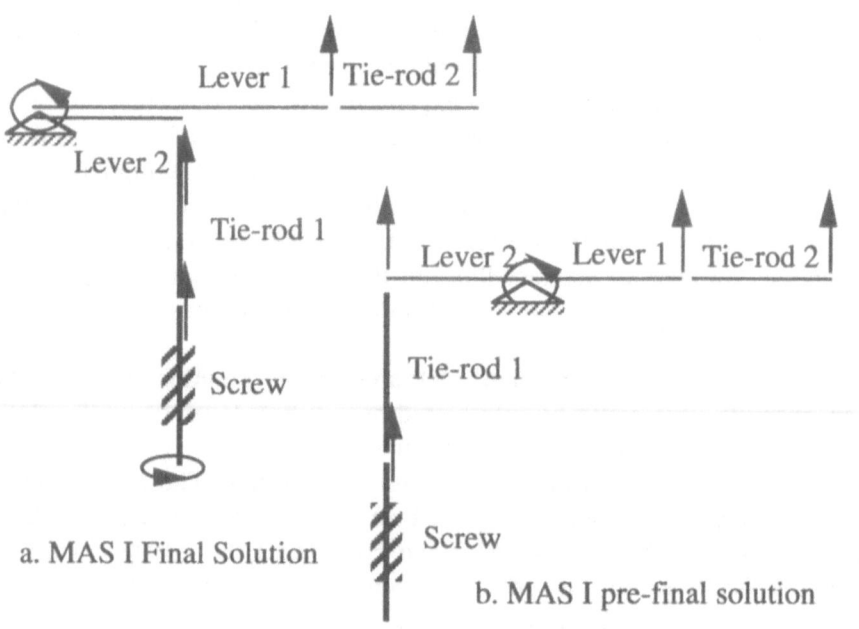

Figure 2. The final solutions in MAS I project.

2.1.2. *Comparisons with Designs Generated in MAS II Project*

In phase two of MAS project, designers were given the designs generated by FuncSION in Phase I, along with the other existing ones, for consideration. They went through these as an exercise, hardly taking note of them as serious solutions, and got on with designing as they otherwise would (and again did not come up with those feasible designs as in phase I). Possible reasons might have been that (i) right from the beginning this was taken as a redesign exercise, with the intention of modifying the previous designs to alleviate the existing problems; (ii) concepts generated by FuncSION were not easy to understand due to their user-non-friendly abstract representation, and lack of visualisation of how they worked; (iii) there were too many solutions to browse; (iv) there were a large number of infeasible, expensive, or similar solutions which discouraged the designers to explore further. However, these are only guesses, and needed validating before they can be given serious consideration. We thus did some further testing for an evaluation, which is discussed below.

2.2. HANDS ON EXPERIMENTS BY EXPERIENCED DESIGNERS

Three experienced designers were asked to use the system to evaluate the solutions generated by the computer for aspects of their originality, feasibility, redundancy and wastefulness, and to comment on whether and how they would modify the ideas they find unacceptable to make them acceptable. They were also asked to make comments on the ease of use of the package, and to make any other observations or suggestions.

2.2.1. Experience of Designer A

Designer A went through the MAS design exercise twice using a different database of basic elements each time. In the first experiment he used 5 elements: two lever types, two types of transitional elements, and a screw type element. He wanted to check whether or not FuncSION produced the solution he had in mind, which it did. He then went through the second experiment, where he used another database of five elements. He could not think of any sensible solution using these elements, and wanted to check whether FuncSION could surprise him. There were three solutions which he found useful and interesting.

However, there were a large number of solutions which he thought were redundant (with repetitive transitional elements, e.g., three shafts in series as a distinct solution to two shafts) or wasteful (e.g., having two cams in a single I/O design). For instance, of the 20 solutions in this second experiment, there were 4 redundant and 3 wasteful designs, which totalled 33% of the total number of solutions.

He found that he could not cope with more than 20 solutions at a time, and suggested that browsing the solutions using user-defined categories (such as all solutions with a screw, or all solutions with lever only) would make handling large number of solutions easier. The solutions were difficult to visualise or interpret as we expected, and he thought having iconic representations coupled with simulation facilities would make visualisation easier.

2.2.2. Experience of Designer B

Designer B went through the MAS exercise thrice, each time with a further reduced database of elements and less number of elements to be used in a solution concept, so as to bring down a large number of solutions to explore (from 700 to 50 to 6). In each of these cases, he found that a solution concept having additional tie-rods or shafts were just variations on the original theme.

In the above cases, designs with a lever preceding a screw were considered wasteful, as long as levers were being interpreted as links and not as temporal elements such as gears. Once this bias was removed, however, some of the solutions having levers preceding gears were now considered geared variations of the rest of the solution. Similarly, a solution having three levers in series followed by a tie rod was originally considered as a feasible but not exciting solution when the levers were interpreted as links. But when the levers were interpreted as gears, the designer found the same solution a clever new idea as this became a rack and pinion solution. This means that being able to see the solution at levels of greater detail reveals more insights as to how useful it might be. He felt that the solutions in these exercises gave him six distinct ideas. The first one consisted of two cams, which he felt, unlike Designer A, was not a wasteful idea, but a new idea he did not think of. The other clusters were screws with tie-rods and levers, levers and tie-rods, two cams connected by a lever, a cam driving two levers, and a cam connected with tie-rods, which included both MAS I and MAS II final solutions! Having a vertical tie rod on a screw gave him the idea of using a sleeve to isolate the transitional component of the screw (an insight).

Regarding visualisation issues, levers were not understood as being capable of being abstractions of gears, in the beginning. Also, a lever, as used in FuncSION, was not a conventional see-saw type but a more fundamental element which could be combined in various ways to produce bell crank levers as well as see-saws. Cams, in the present representation, were hard to visualise, and it was hard to visualise tie rods as axial links.

Regarding the procedural bits, Designer B felt that when he found an interesting Cam based design, he wanted to explore all the Cam based designs. So a user-defined clustering facility such as find all designs that have a shaft in the middle would be useful. He suggested that it would be useful to do synthesis with only output specified.

2.2.3. Experience of Designer C

Designer C went through the MAS exercise twice. In the first run, he chose three elements from the database, and asked for solutions having at most three elements. There were two solutions: one with two levers and the other with two levers and a cam in between. He expected both the solutions and there were no surprises, although both were perfectly reasonable solutions. He then chose a 4 element-database, and solved again for the same requirements. This time there were twenty solutions, several redundant (the heuristic was that one or many translators in series, or at the beginning or end), though he thought these variations might be useful for optimisation purposes. Here also, the solutions did not surprise him. This is not surprising because the database chosen was very limited and there was not much scope for innovation. He felt that visualisation would be improved if the symbols were more self-explanatory, and if two solutions could be seen alongside one another.

3. Observations and Discussion

Designers in the above experiments found that FuncSION in general generates a range of interesting solutions, and often comes up with surprisingly clever ideas and insight. However, it also generates a large number of redundant and expensive solutions, and this makes it difficult or frustrating to evaluate and explore the ideas to any depth. They had some difficulty in visualising the designs at the present representation, and could visualise only when these designs were shown at a lower level of abstraction.

On the whole the above experiments suggested a common pattern of when solutions were considered similar: if two solutions are different only by a transitional element (e.g., a tie rod, or a shaft), then they are similar. The consideration for wastefulness was not as straightforward, however.

If this criterion of what redundant solutions mean were applied to cluster the designs FuncSION suggested in the MAS I and II cases described before, and the wasteful designs it suggested were clustered using the wastefulness criterion, the rest of its designs could fall into 12 different clusters of solutions of an average size of about 12. Of these, only 4 clusters were considered at any length at all by the designers in MAS I and MAS II, while just a single low-level instance was found for 2 of the 8 other clusters. This indicated the potential of FuncSION for suggesting different ideas and idea types. It is important to note that the above clusters were the result of solving the MAS problem using a database of 5 elements only. If this database were increased to 7 for instance, the number of clusters, after eliminating the expensive solutions would be as high as 29, of which only 6 would then have been considered by the designers, and only 4 of these at any length!

Based on the experience gained from these case studies and experiments, it was felt that FuncSION needed further attention in three different areas:

3.1. TOO MANY SOLUTIONS

One of the problems associated with the synthesis approach adopted by FuncSION is that the system may generate so many solutions that it is difficult for the designer to even browse through them. Some of these solutions were considered by the designers to be redundant (which is a variation another which uses additional, non-essential elements), or wasteful. However, the exploration of 'redundant' solutions often might be useful if the non-redundant ones cannot provide some additional functions which are not within the realm of the main function.

Take the solution that was generated by FuncSION for Phase II of the MAS project (see Figure 3) as an example. There are two consecutive tie-rods of the same type, which might appear to be redundant unless one were trying to provide two extra degrees of freedom for the movement of the output point in the horizontal plane.

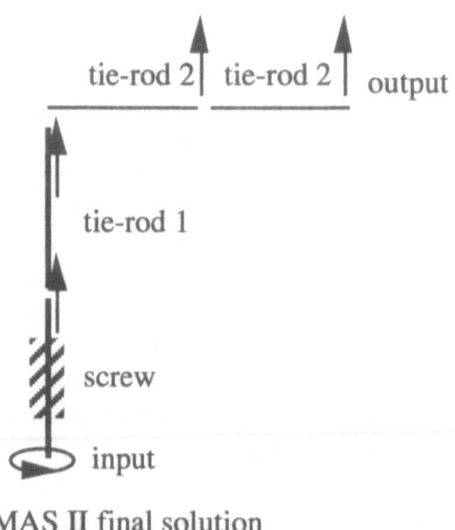

MAS II final solution

Figure 3. Redundancy can be useful.

Take the MAS I project as another example in which the final two solutions are considered by FuncSION to be topologically the same (Figure 2). The two solutions are only different in terms of the sense configuration. However, for the designers, one was considered to be a substantial modification of the another as it made the design more compact. So, whether a design is to be considered redundant or not, depends largely on other requirements that the design might have. Also, the exploration of redundant solutions might be useful if these non-redundant ones can provide some additional functions which were not originally thought of.

However, it was clear that far too many solutions were typically produced by FuncSION for the designer to meaningfully explore. For example, take a typical case of synthesis where only 32 topologically distinct solutions are generated from a database of 5 elements and a single I/O function, each of these can have at least 4 spatial configurations, each of which can have at least 3 physical concepts, giving a total of 384 solutions. The conclusion is that a strategy is needed to generate or present these solutions whereby they could be browsed through without being overwhelmed by them.

3.2. TOO DIFFICULT TO INTERPRET AND TO VISUALISE

Two main issues concerning the interpretation and the visualisation of the synthesis results were considered vital by the designers who evaluated the FuncSION system. The first is that the representation of elements in the database is too abstract. For example a shaft looks similar to a screw in terms of input/output function. The second is that the static representation for functional elements and conceptual solutions makes it hard for the designer to image the likely behaviour of an element or a conceptual solution, thus contributing little towards supporting designers' creative thinking. That is, the expected behaviour of each element or a solution needs to be visualised in order to give the designers more information. Thus a means of visualising solutions and their elements should be developed.

3.3. SOME DESIGNS DO NOT FUNCTION TEMPORALLY

So far, all the solutions that FuncSION generated work at one instant of time. For instance, a lever type element represents a transformation from an instantaneous input to an instantaneous output. This could be an abstraction of a gear or belt type element, which can provide translation at that point for an extended length of time, or it could be a link type lever whose position and direction of output change with time. The conclusion is that a temporal reasoning facility is required to evaluate the potential of each such solution to function temporally.

4. Objectives Revisited

The central objective of FuncSION is to provide an environment which would stimulate designers thinking by supporting them to explore a wide range of ideas and, if they wish, variants of these ideas, so as to increase their chances of developing new, interesting and useful designs. Two factors contribute to developing such ideas: there must be a wide range of computer generated ideas for them to explore, and these solutions must be explored and evaluated sufficiently by the designers.

In order to generate a wide range of ideas, FuncSION needs a wide variety of elements in its database, which in turn produces a large number of solutions, many of which are similar to each other. If a means could be developed to cluster these solutions into groups of similar solutions, and also weed out solutions that are considered "wasteful" by the designers for a specified requirement, then this number could be more manageable.

The evaluation on FuncSION indicated that it is easier to explore solutions if they are not too many, and if they can be visualised easily. It is easier to visualise a solution if it, and its component elements, can be seen at a sufficient degree of detail (in terms of their behaviour as well as their spatial relations), i.e., the less abstract it is. On the other hand, the more abstract the database used by FuncSION is (i.e., where the an element can represent a large number of less abstract elements), the less the number of solutions generated will be. Therefore, there is a conflict about the right level of abstraction, see Figure 4. If it is too high, the solutions will be more difficult to visualise, and consequently to explore, whereas, if it is too low, there will be too many detailed solutions to explore. What further complicates this issue is that the "right" level of abstraction varies according to the experience of the designer. Experienced designers might only need to look at designs at a higher level of abstraction than an inexperienced designer, and still be able to imagine their details and evaluate them, while inexperienced designers might need more visualisation support, i.e., further degrees of possible detail of the solutions before they could evaluate them.

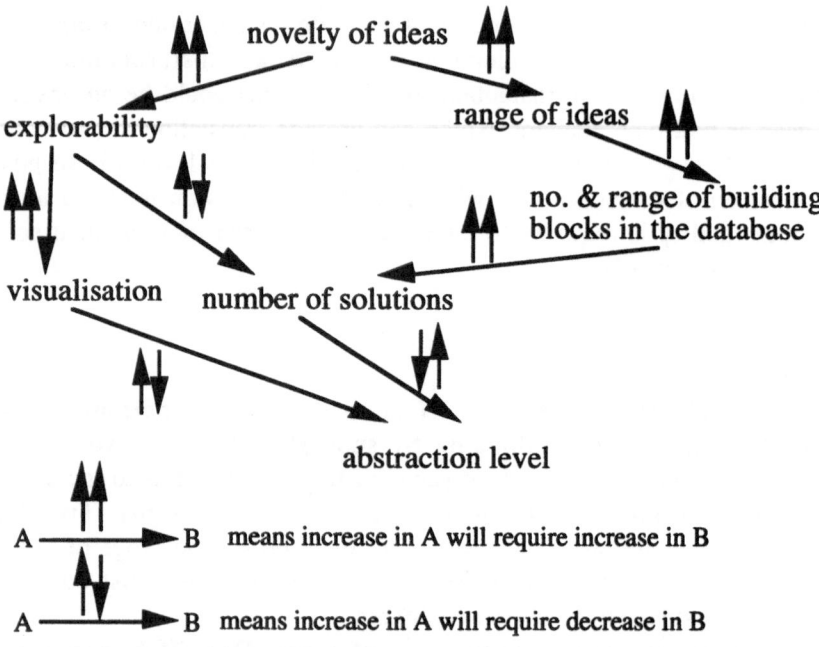

Figure 4. The factors that affect the novelty of ideas in a design.

Based on the results of evaluation, we conclude that FuncSION must be able to offer designers a wide variety of solutions to explore; the number of these solutions need to be small without compromising their range. support Designers should be visualise and browse through these designs at various degrees of detail, and should be able to see the variants a given design could have for he evaluates them. They should be able to put their own clusters on the solution space, based on criteria such as what they consider, or is generally considered wasteful, and invoke any particular cluster at their will.

5. Further Development, Implementation and Evaluation

Four strategies have been initially identified to avoid over-generalisation of the synthesis solutions in a computer-based system. These strategies are
- to use a hierarchical functional element structure,
- to provide synthesis programs which operate at different levels of abstraction,
- to use design heuristics to cluster design solutions, and
- to provide alternative control strategies for visualisation and browsing.

A new version of FuncSION has been implemented using a knowledge-based system development tool called GoldWorksIII^{TM} on a SparcStation. It consists of a database of functional elements and their transformation rules, a functional synthesiser with synthesis algorithms that operate at different levels of abstraction, and a graphical user interface for browsing through and visualising the solution concept generated.

5.1. DATABASE

A way of eliminating unnecessary combinations of synthesis solutions is to allow the designers to choose the types of functional elements and their interfaces from a hierarchical structure. For instance, a lever element, at a lower level of abstraction, can be split into link type, gear type, pulley type, etc., while an axial tie rod can be split into axial links, chains, belts, ropes, racks etc.

Once a solution is generated at a given level of abstraction, it should be possible for the designers to navigate through the solution, or parts of it, at other levels of abstraction before making any change. The functional database should be hierarchically structured to enable the designers to edit or modify the elements at various levels of the hierarchy.

An object-oriented product data model is used to build a database of functional elements which can be selected to synthesise solutions based on a user defined input/output requirement specification. Each functional element in the database has a *type* which is associated with a set of rules that determines how it responds to different orientation or sense inputs. In the current implementation there are 72 such rules.

5.2. ALGORITHMS FOR CLUSTERING AND BROWSING SOLUTIONS

The implementation of the original version of FuncSION used specific features of the functional elements to solve the design problems at a specific level. It is therefore unable to deal with a hierarchical structure of functional elements and therefore is abstraction-level-specific. However, it is possible to extend the algorithm so that it can solve multiple input/output synthesis problems using a set of black boxes with inputs and outputs having attributes, the exact values of which would depend on the level of the functional element hierarchy. The key idea is to separate data from the synthesis procedures, and wrap them both with a common interface. In the new version of FuncSION, a three-steps strategy is used to synthesise solutions. The first step generates an exhaustive set of candidate solution concepts for a selected set of functional elements. The second step tests the feasibility and functionality of the solution concepts to eliminate infeasible ones. The third step clusters the feasible solution concepts based on user selected heuristics.

For example, suppose we have three elements (1 2 3) (in this list each number represents a functional element and the actual elements can be filled in later). In the generate stage, all the combinations of these three elements are generated, resulting in a list of candidate solution concept structures (1, 2, 3, (1 2), (2 1), (1 3), (3 1), (2 3), (3 2), (1 2 3), (3 2 1), (1 3 2), (3 1 2), (2 1 3), (2 3 1)). In this list, (1 2), for example, means that the connection form element 1 to element 2 forms a possible solution concept structure.

In the test stage, each candidate solution concept structure is mapped to a chosen level of the functional element hierarchy in the database, retrieving the real element attributes. The compatibility of functional elements within each candidate solution concept can then be tested. This is done by removing those which are incompatible in terms of input/output transformation. For example, if the output of element 1 does not match the input of element 2, then the solution concept structure (1 2) is incompatible. All the compatible solution concepts must also be tested using the input/output requirement specification. The results of this process form the solutions of the kind synthesis step (Chakrabarti and Bligh, 1994).

Each kind synthesis solution can then be selected by the designers for orientation and sense synthesis. The orientation synthesis is done by propagating an input orientation from the input point to the output point of a kind synthesis solution concept using the orientation transformation rules, the result of orientation is a list of orientation synthesis solution concepts The sense synthesis is done by propagating an input sense from the input to the output point of an orientation synthesis solution concept. Both orientation and sense synthesis generate multiple solution concepts because one element typically responds to the same orientation and sense input in more than one way and can have alternative spatial configurations.

The outcome of kind, orientation and sense synthesis may still be a large set of solution concepts with alternative spatial configurations. The clustering heuristics discussed above can then be selected by the designers and applied to these solutions to group solution concepts with distinct features.

A solution concept generated at one level of the functional element hierarchy can be specialised in a number of different ways. We have so far implemented the following:

1. any solution concept generated by the system at one level of the functional element hierarchy can be mapped to a lower level by substituting the elements in the solution concept with those at a lower level. This may produce a list of combinations. For example, if a solution (lever -> tie-rod) is mapped to a lower level, then for a hierarchy with two possible variants of a lever (a gear and a link-liver) and a tie-rod (a rack and a link-type-tie-rod), there will be 4 low level combinations, i.e., (gear -> rack), (gear -> link-type-tie-rod), (link-lever -> rack) and (link-lever-> link-type-tie-rod). The lower level elements may introduce interface constraints that would render some combinations invalid (in this example the second and the third solutions are invalid). A program has been designed to work out only the valid mappings.

2. a solution concept can be modified by a designer by replacing any part of it with an element or an interface at a lower level of abstraction. This allows the designers to specialise or further constrain a solution concept in a *depth-first* manner. For example, if a solution concept contains an a lever, then it is possible for the designer to modify this element by looking at its sub-class or super-class elements. Any modification made by a designer is checked by the system to ensure the consistency of the solution.

3. solution concepts generated at a low level can be clustered into a higher level by merging low level elements or interfaces into higher level ones.

It is necessary to integrate design heuristics into the synthesis process in order to offer the designers a wide range of solutions and their variants in a controlled manner. We define a *variant solution* in the following ways:

• Designer's preference, i.e., the solutions that a designer would consider as the variants of another design,

• Generalised solution concepts based on experiments, i.e., what we have found universally as variants from the hands-on experiments (this can grow as one does more experiments with the designers), or

• Variations of past design examples even though they may have been noted by the designers as wasteful. Here a wasteful solution is the one considered by the designers as inefficient or too expensive.

All these could form part of a library of heuristics or filters that could be integrated with a systematic synthesis program to weed out the solutions which may be generally regarded as being "bad ideas". This results in an organised concept solution tree instead of a huge number of solutions at the same level of abstraction.

A number of heuristics have been found useful in the experiments and can be selectively (by the designers) applied to the synthesis program to cluster the solutions generated by the computer. These heuristics include: fixing the number of transformations; each element is used more than once; each element is used no more than once; each element is used at least once; each element is used exactly once; no element is used repeatedly more than a specified times; no element is repeatedly used consecutively; same tie-rods are not directly connected; no translators such as shaft, tie-rod etc. are used; fixing the input/output elements; and only input is specified while the output is left open.

5.3. GRAPHICAL USER INTERFACE

The synthesis algorithm described in Chakrabarti and Bligh (1994) used a *breadth-first* search strategy to generate general to specific synthesis solutions. While this remains a useful control strategy in the new development, a number of alternative control strategies must also be used for the designers to explore the whole solution concept tree. Some of these control strategies are:
- to allow the designers to path through all the levels of the solution concept tree;
- to pick up of a few solution concepts from a user-defined level on a random basis before generating all the possible solution concepts at that level;
- to set default values for the numbers of solution concepts to be generated at each level of the functional element hierarchy.

A new graphical user interface is designed to allow the designers to control the functional synthesis process with visualisation support by
- ionising each functional element for an easy selection;
- simulating the behaviour of individual components as well as the solution concepts generated from individual elements;
- helping the designers to browse through the solution concept tree.

Simulation is an important way of supporting the understanding of a synthesis solution so as to help with its selection and modification. While a 3D modelling tool can be used to visualise the final solution after embodiment design, it is only necessary at the functional synthesis stage to use a two dimensional graphical display scheme. Within this scheme, each functional element has an iconic image that can be actively manipulated within a graphical window.

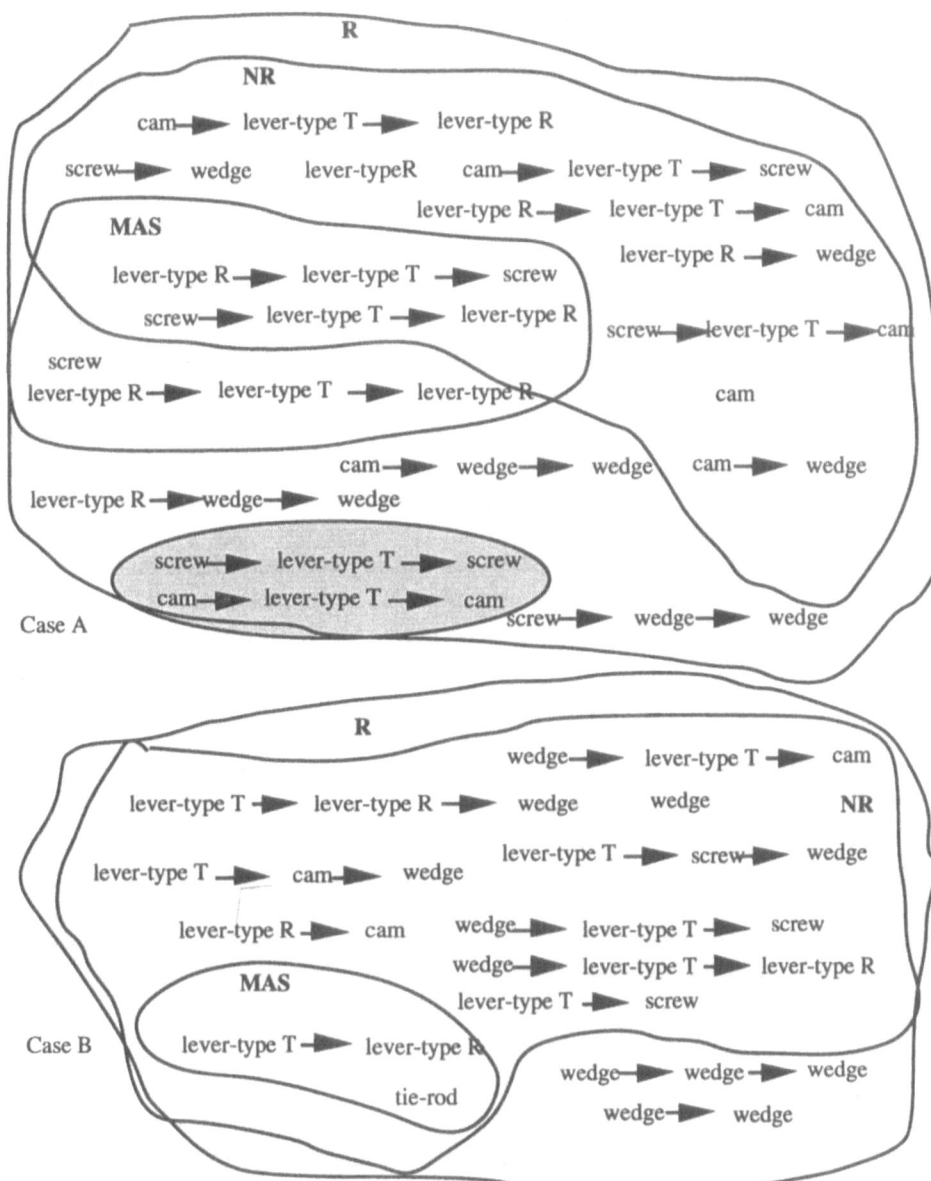

R: idea clusters generated by the algorithm that allows repitition of elements

NR: idea clusters generated by the algorithm that does not allow repitition of elements

idea clusters that would be considered wasteful by the designers

MAS: idea clusters generated by the designers in MAS projects

Figure 5. A comparison between a Repeat and a Non-repeat algorithm.

5.4. EVALUATION OF THE NEW VERSION

In order to evaluate the newly implemented system, we have produced some test cases using some of the heuristics discussed above. Figure 5 shows how the solution clusters, produced by a repeated (each element is used more than once) and a non-repeated (each element is used no more than once) algorithm in two of the test cases, relate to the ideas generated by the designers in the MAS project.

In Case A (in the case of a rotation to translation input/output requirement using 7 elements with a maximum of 3 allowed transformers per solution), the repeated algorithm produced 18 clusters, 2 of which would have been considered by the designers as wasteful, and 5 of the remaining ones were independently generated by the designers. For the same case, the non-repeated algorithm generated 12 solution clusters with no wasteful clusters, but failed to produce 2 of the 5 clusters which were independently generated by the designers.

In Case B, the number of clusters for the repeated algorithm is 13 (in the case of a translation to translation input/output requirement using 7 elements with a maximum of 3 allowed transformers per solution), none of which would have been considered wasteful by the designers. The number of clusters produced by the non-repeat algorithm in this case is 11, which included the two that were independently touched upon by the designers.

The indication is that the non-repeated algorithm generates less number of variants or redundant solutions and thus less number of wasteful solutions, but at the cost of omitting some of the solution clusters which would still be regarded useful and important by the designers. This simply pontificates the point that it is a heuristic and not a general principle. It should therefore only be used in situations where the designers are given a prior warning about its possible consequences.

6. Related Work, Conclusions and Further Work

There are three main areas which relate to this piece of work. One is computational synthesis approaches and approaches that they take to cope with complexity, one is design methodology and how generation aspects could be supported, and the third is the systems and user interface issues.

There have been evidences in design theory and methodology that it is important to generate a range of designs and explore them sufficiently before homing in on specific designs for further development. In fact in some of the protocol studies done in the recent past, it has been found that the best approach in conceptual phase has been a consecutive expansion and narrowing down of ideas (Dylla, 1989; Fricke, 1992). It has been a major problem however, in synthesis support systems as well as in manual methods suggested in design methodology (Pahl and Beitz, 1984) as to how to explore designs without compromising their range.

As mentioned in Lee *et al* (1992), granularity of building blocks is particularly important for managing complexity, and they felt complexity could be tackled using a few important parameters at a time. However, this is only part of the problem. Even if the problem is solved using few parameters at a time, there would still be a large number of feasible alternatives to compare, evaluate and modify. We feel that the major part of complexity arises from the conflict about level of abstraction right for getting high explorability as well as wide range of solutions. Our approach tackles this in three new ways. One is to clustering designs based on designers' heuristics of similar designs; the second is to provide range by generating solutions at a high level of abstraction, while allowing visualisation at a low level for each of these solutions, and the third is by bringing designer in the navigation process which is essential for design support systems.

In conclusion, this new version of FuncSION provides a database of hierarchical functional components and their interfaces for the user to select. The system generates synthesis solutions using an algorithm at a level of abstraction selected by the designer. The solutions generated by the system can be clustered using the heuristics chosen by the designer, allowing the designer to switch between multiple solutions and to concentrate on the interesting ones. Visualisation and simulation techniques are provided for the designer to explore and browse the hierarchical structure of functional components and the tree of synthesis solutions.

Initial testing results have shown that the integration of a hierarchical functional component database with systematic synthesis techniques, the heuristics for clustering, visualisation and simulation contributed to stimulate the designers' think. The newly developed version of FuncSION provided a good basis for utilising AI techniques in functional modelling of mechanical engineering design.

Work is being carried out to fully incorporate the control strategies and clustering heuristics discussed in this paper, and to enhance the visualisation facilities further with a fully animated graphical user interface. This new version of FuncSION is being integrated with an embodiment generator and a kinematic analysis system to form an integrated functional modelling system.

Acknowledgements

The work presented in this paper is currently being funded by the EPSRC. We would like to acknowledge the support from Dr Stuart Burgess, Dr Thomas Bligh, Mark Nowack and Doug Isgrove who acted as the designers in the experiments reported in this paper. We would like also to acknowledge the support from Dr Nigel Ball, Dr Lucienne Blessing and Dr Tim Murdoch for their support in the development of the past and current version of FuncSION.

References

Ball, N. R. and Bauert, F.: 1992, The integrated design framework: Supporting the design process using a blackboard system, *in* J. S. Gero (ed.), *Artificial Intelligence in Design '92*, Kluwer, Dordrecht, pp. 21-38.

Bauert, F.: 1993, The mobile arm support phase in design, manufacture, testing, software tools, *Technical Report CUED/C-EDC/TR 13*, Cambridge University.

Chakrabarti, A. and Bligh, T. P.: 1994, A two-step approach to conceptual design of mechanical device, *in* J. S. Gero and F. Sudweeks (eds), *Artificial Intelligence in Design '94*, Kluwer, Dordrecht, pp. 21-38.

Ehrlenspiel, K. and Dylla, N. D.: 1989, Experimental Investigation of the design process, *in*: V. Hubka (ed.), *Proceeding of ICED89, International Conference on Engineering Design*, Mechanical Engineering Publication, Bury St Edmunds, Vol. 1, pp. 77-95.

Fricke, G.: 1992, Experimental investigation of individual processes in engineering design, *in* N. Cross, K. Doorst and N. Roozenburg (eds), *Research in Design Thinking*, Delft University Press, Delft, pp.105-109.

Johnson, A. L et al: 1993, Modelling functionality in CAD: Implications for product representation, *Proceedings of the 9th International Conference on Engineering Design*.

Lee, C-L., Iyenger, G. and Kota, S.: 1992, Automated configuration design of hydraulic systems, *in* J. S. Gero (ed.), *Artificial Intelligence in Design '92*, Kluwer, Dordrecht, pp. 61-82.

Pahl, G. and Beitz, W.: 1984, *Engineering Design*, Design Council, London.

Thornton, A.: 1993, *Constraint Specification and Satisfaction in Embodiment Design*, PhD Thesis, University of Cambridge, Department of Engineering.

J. S. Gero and F. Sudweeks (eds), Artificial Intelligence in Design '96, 623-639.
© 1996 *Kluwer Academic Publishers.*

ADOPTING A MINIMUM COMMITMENT PRINCIPLE FOR COMPUTER AIDED GEOMETRIC DESIGN SYSTEMS

XIAOHONG GUAN AND KEN J. MACCALLUM
CAD Centre
Department of Design, Manufacture and Engineering Management
University of Strathclyde
75 Montrose Street, Glasgow G1 1XJ, UK

Abstract. In spite of the ever increasing capacity, existing computer aided geometric design systems are normally only used in practice to model and analyse what has already been designed. They do not seem to contribute to design as much as they might be able to, mainly due to their inability to support designers in the conceptualisation and synthesis stages. In this paper, we propose a minimum commitment modelling principle for developing computer aided geometric design systems with the objective of preserving in the systems the maximum design solution space conceived by a designer in the process of geometric design. We assert that, by adopting this principle, these systems have the potential of extending their support to the early concept exploration stages and the gradual refinement of the established concepts, and therefore have a better capability to assist the entire geometric design process. Aspects of minimum commitment have been used in a limited way in a few design related systems and in building computer based planning systems, but there seems no clear formulation of the principle and its implications for CAD systems. An approach to applying the principle in building a geometric design system is discussed based on our ongoing research into a computational geometric design approach using vague geometry.

1. Introduction

For the past three decades, we have seen significant evolution of computer aided design (CAD) systems from simple, *ad hoc*, special-purpose systems, such as numerical control tools, 2D interactive drawing and draughting tools, to more sophisticated, integrated, general-purpose systems based on 3D geometric modelling, parametric, variational and design with feature techniques. These systems have been positively transforming or influencing many aspects of the traditional engineering design process in various ways.

However, it is commonly recognised that these systems only offer support in

modelling and analysing what has already been designed and do not contribute
to design as much as they might be able to. This is mainly because of their inability to support designers in the conceptualisation and synthesis stages. This, we
believe, results from an *activity gap* and an *information gap* between the systems
and practical design. The activity gap refers to the mismatch between the practical
design activities and the way in which they are carried out and those which are
supported by these systems. The information gap refers to the mismatch between
the type and level of information which is available and handled by designers in
practice and that which is required when using the systems.

To reduce, minimise or eliminate these gaps, we propose a minimum commitment modelling principle for such systems which aims at maintaining the geometric design solution space conceived by the designer. In the following section,
an informal model of the geometric design process adopted in this research is
summarised. In Section 3, we establish the principal goal for a computer aided
geometric design (CAGD) system and the minimum commitment principle for
achieving it. Here, a CAGD system refers to that which can be used by a designer
to develop or model the geometry of a product. Existing applications of the principle to related areas are reviewed briefly in Section 4. In Section 5, our approach
to applying the principle in building a geometric design system using vague geometry is discussed.

2. A Geometric Design Model

In our research, we have established and adopted a model of geometric design
based on the various views of general engineering design, such as those described
in Ervin and Gross (1987), Suh (1990), Tjalve (1979), and Tomiyama and Yoshikawa (1985). In this model, geometric design is viewed as a process of establishing the set of geometric properties of a product to the extent that is required for
physically manufacturing the product. The relevant product component structure,
complete or partial, should already have been developed before starting the geometric design, but may be modified during the process. The term `component' is
used throughout this paper in a wider sense than that which is used in domains
such as manufacturing. Dependent on the context, it may correspond to `subsys-
tem', `subassembly', `component', or `part'.

The geometric properties are defined by a set of geometric parameters which
characterise or specify the overall shape and size of the product as well as the
shape, size, location, and orientation of the constituent components. These properties are completely and uniquely defined if values of all the geometric parameters are fully and uniquely specified.

Values of the geometric parameters may be specified by a designer through a
set of geometric constraints. These constraints define an n-dimensional solution
space for the geometry of the corresponding components or product. This solu-

tion space is termed here as the *conceived solution space*, that is, the geometric design solution space conceived by the designer mentally. It may be large at the early stages of design due to, for example, the vagueness of information, and is iteratively modified or refined as new or more precise information arrives. Geometric design can therefore be regarded as a process of incrementally modifying and refining the conceived solution space until it degenerates into a point at which all the parameters are uniquely defined.

Whenever an activity or action of constraint manipulation is carried out which changes the possible values of those geometric parameters (thus the conceived solution space) defined by the manipulated constraints, it is said that a decision is made towards the geometric design (or properties) of the product. Examples of such constraint manipulation activities include specifying new as well as deleting and modifying existing constraints. Such a decision is considered to be a commitment since it requires the dedication to the consequences or effect, e.g. modification, reduction or expansion of the solution space in a particular direction, resulting from its implementation.

For those geometric parameters that have a domain of real numbers (i.e. with numeric values), one decision on a geometric parameter is said to show greater commitment of the designer than another one on the same parameter if the total sum of the widths of all the closed value ranges it defines for the parameter is less than that defined by the other decision. The *width* of a closed real interval/range is the difference of its upper bound and lower bound. In this case, the corresponding solution space is reduced. Otherwise, if the total sum of the corresponding widths is greater than that of the other, it is regarded as showing less commitment and results in a larger solution space.

3. The Minimum Commitment Principle for CAGD Systems

We believe that a CAGD system should play the role of an Intelligent Design Assistant (IDA) (Duffy *et al.*, 1985) which, although playing a role secondary to the designer, actively participates in the design process. Here, we note the key abilities of such a CAGD system to model the geometric design solutions conceived by the designer and to take care of the management, analysis and presentation of the modelled solutions. Following the design model described in the previous section, the geometric models established in the system based on the constraints on geometric parameters specified by the designer define a solution space which corresponds to the solution space conceived by the designer (i.e. the conceived solution space). This geometric solution space captured inside the system is termed as the *modelled solution space* (Figure 1).

Traditionally, instead of capturing and maintaining a solution space, CAGD systems tend to support the modelling of one or a small number of concretely and precisely defined distinct solutions (points) within the solution space. Thus,

Figure 1. Faithful interpretation in a computer-based design support system of design information supplied by the designer and preservation of the solution space defined.

single and specific choices have to be made on the geometry of the object being considered for it to be modelled in such a system. This reflects a trial-and-error approach to design. Here, we assert that an IDA-based CAGD system should aim at supporting the capture and maintaining of the design solution space. Further, the modelled space should be as close as possible to that intended by the designer, i.e. the conceived solution space (Figure 1). With respect to the geometric design model presented earlier, this means that the computational scheme used in such a system should be as faithful as possible in interpreting the geometric constraints to facilitate the modelling and maintaining of the solution space defined by these constraints, rather than a specific solution that satisfies these constraints. A system capable of achieving this may therefore have an improved capability to support the conceptualisation and synthesis stages of the design process.

To achieve the above goal, we propose the following principle for IDA-based CAGD systems:

Minimum Commitment Modelling Principle   A commitment that is modelled in a CAGD system should not be greater than that desired and requested by the user.

This principle states that the level of commitment which the system models is the minimum needed to describe the range of possible solutions intended by the designer. With respect to the geometric design model described in Section 2, this means that the system, when used, should neither force the designer to constrain prematurely or unnecessarily the geometric parameters that characterise the geometric properties of a product, nor make unnecessary assumptions or reductions of the intended solution space by opting for representing or recording only a specific value for each of the parameters that satisfies the given constraints.

The term `minimum' commitment is chosen to emphasise the minimum necessary for the design to progress in accordance with the designer's decisions. Minimum commitment therefore implies a space of possible designs, within which alternative solution points may be investigated without committing to any solutions. The principle serves to support two arguments:

- a designer should not be forced to define a specific (even default) object or value for representation in the system, when only vague (e.g. approximate or incomplete) information is known;
- a designer should be able to use the computer's representation of a minimum commitment model to help explore a space which is not easily to visualise cognitively except as solution points.

However, the principle *does not prevent commitments being made by a designer at any stage if they properly reflect the designer's conception.* It does not impose any restrictions or make any judgement on how designers should pursue design. We do not intend to claim in any way that the designers must follow a minimum commitment route in their design practice, although the principle has been proposed for engineering design as the review in the next section will explicate. Here, we emphasise that the CAGD system supports such a principle in relation to its user, the designer.

A corollary of the above principle is that changes to a modelled solution space proceed incrementally to advance the status of a design while maintaining minimum commitment. Consequently, we propose that the minimum commitment modelling principle is complemented by the Principle of Incremental Refinement:

Principle of Incremental Refinement   A CAGD system should support the incremental refinement of a design solution space, i.e. the continuous evolution of vague concepts into a complete and precise solution, in steps which are sufficiently small to maintain commitment at a minimum.

4. Existing Applications of Minimum Commitment

Minimum commitment can be considered as a common sense principle or strategy for dealing with uncertain, complex situations typical in design. It suggests concentrating on what are important or known, rather than those that are minor or require unknown or unavailable information (thus requiring guesses or assumptions based on insufficient data, information and knowledge). A simple application of this principle can be seen in writing, e.g., technical papers or thesis. In this case, the author normally avoids committing to detailed writing before an appropriate outline or plan (probably selected from a set of alternatives) is laid out.

In this section, we briefly examine the use of the principle in such areas as engineering design, computer based planning and design systems.

4.1. ENGINEERING DESIGN

In his text on engineering design, Asimow put forward a minimum commitment principle for design as follows:

> *In the solution of a design problem at any stage of the process, commitments which will fix future design decisions must not be made beyond what is necessary to execute the immediate solution. This will allow the maximum freedom in finding solutions to subproblems at the lower level of design.* (Asimow, 1962)

Minimum commitment here does not suggest that the designer avoids decisions which are necessary and important for advancing the design. Instead, the designer should ignore those unnecessary aspects or unimportant details to give themselves an as large as possible amount of play in the subsequent exploration. This is therefore different from what Janis and Mann (1979) described as the `procrastination in defensive avoidance' which refers to the situation where the decision maker, when facing a difficult decision, avoids necessary decisions.

Dym has also considered minimum commitment (he and others have used the term *least commitment* instead) as a general design approach or strategy which requires the designers to:

> *make as few commitments as possible to any particular configuration because the data available are perhaps too abstract or very uncertain at this point in the design process.* (Dym, 1994)

Least commitment is regarded as `a (good) habit of thought' which `militates against making decisions before there is reason to make them'. It is considered as of particularly importance for the early stages of design `where consequences of any one design decision are likely to be propagated far down the line'.

French also discourages premature detailed studies in conceptual design stage since they `may prove to have been a complete waste of time when some hitherto unconsidered facet of the problem is studied, and shows that a radical revision of ideas is necessary' (French, 1992). Using an example (designing a way of centering discs on shafts in high-speed rotors), he further explained the importance of not making (even unconsciously) unnecessary design decisions without some sort of review at early stages to avoid imposing unnecessary difficulties in downstream process ((French, 1992), p.7). This is consistent with one of the principles for solution search proposed by Hubka (1987) generalised from experience. It states that `concretisation too early can confine the considerations in a particular direction'. It is considered as imposing design `prejudice', `mental set' and `fixation'.

On the other hand, if some commitments have to be made, then the designer should carry out proper exploration and evaluation of alternative decisions. Cognitive research shows that, although `obsessional mulling over the uncertainties of a major decision and preoccupation with the search for an ideal choice often lead nowhere and may even be detrimental', a decision maker has a greater chance of making better and sound decisions if, among other suggested actions, a thorough search of alternatives and relevant information is carried out before evaluating and making an actual choice (Janis and Mann, 1979). If we consider designing as a process of decision making which advances the design from its initial state (the problem definition) to the final state (the finished design model), then the designer should make a good exploration of alternatives before sticking to a specific solution if he/she is to make better design decisions. However, as revealed by Goel and Pirolli's study (1989), designers adopt a limited commitment mode control strategy in design evaluation to resolve/negotiate the `tension between keeping options open for as long as possible and making commitments'.

To summarise, it is well known that design is an iterative and incremental refinement process, and experiences the transition from uncertainty to certainty. At the early stages of this process, designers concentrate on concept exploration and synthesis and may still be defining the design problem more clearly. Under this situation, it is better that they delay premature and unnecessary detailed decisions but concentrate on design conceptualisation and synthesis, since unnecessary early commitments may have the following negative effect:

— a reduced solution space;
— small details committed at an earlier stage restrict the freedom in subsequent design;
— possible conflict as new information arrives;
— wasted effort on the details if an incorrect major decision is made.

If, however, some commitments have to be made, then proper exploration and evaluation should be carried out.

4.2. PLANNING SYSTEMS

Minimum commitment, again known as least commitment, has been used in building computer based planning systems in artificial intelligence research. Originally introduced by Sacerdoti (1977) in developing a planning system called NOAH, the idea behind the least commitment principle is to delay decisions until one has as much useful information as possible for making them. This application in the planning systems is best summarised by Weld:

Instead of committing prematurely to a complete, total ordered sequence of actions, plans are represented as a partially ordered sequence, and the planning algorithm practices least commitment planning - only the essential ordering decisions are recorded. (Weld, 1994)

The key method used in a planner to achieve the delay of decisions on the orders of actions is to specify only the necessary ordering constraints on the actions that are satisfied through relevant constraint satisfaction algorithms, and to refine the constraints gradually. Another embodiment of this principle is the use of abstract planning operators that use variables to avoid premature commitment to specific planning choices. As demonstrated by Weld (1994), adoption of least commitment can increase the computational efficiency of the planning system.

4.3. COMPUTER-BASED DESIGN SYSTEMS

While traditionally CAD systems are only capable of modelling concrete and precise solutions which requires commitments of the users to precise and single choices, some existing design related systems adopt a least commitment approach in some aspects to permit ranges of values or choices. For example, in a constraint based 2D layout system, WRIGHT (Baykan *et al.*, 1992), Baykan and Fox regard spatial layout as generating configurations of design units that satisfy given spatial relations and limits on dimensions. These spatial relations and dimensions are handled through a constraint propagation method. A least commitment based approach is adopted which only removes from variable ranges those values that violate a constraint and which, instead of choosing specific locations, selects constraints to be satisfied by design units.

Another such example is the ALADIN system reported in (Farinacci *et al.*, 1992) for aiding the design of aluminum alloys for aerospace application. In the system, alloy design is treated as a planning problem where `the final alloy design is a sequence of steps to be taken in a product plant in order to produce the alloy'. This has enabled the system to use, in the targeted alloy design domain, a least commitment approach similar to that which has been widely used in computer based planning systems (as reviewed in the previous subsection). Thus, design hypotheses are described as ranges of values which `are kept as broad as possible until more data is present to force them to be restricted, which allows the system to avoid trial-and-error in selecting values'.

Based on an approach of `design by least commitment', Mäntyla et al proposed the use of *relaxed feature models* in a generative process planning system to avoid the problem of over-specification of geometric models that persists in existing process planning systems which restricts the subsequent manufacturing options (Mäntyla *et al.*, 1989). This design by least commitment approach encourages the designer to `systematically avoid making design decisions that unnecessarily limit the freedom of later process planning in the search for good manufacturing solutions'. In the feature relaxation approach, a choice of including **round-**

ing at the end of a **slot** feature or leaving it out, for example, is considered as a commitment to certain applicable manufacturing operations. Such a commitment (e.g. either to include the **rounding** or not to), if not functionally significant or essential, is thought to be harmful and better be relaxed so as to generate a better manufacturing plan. Feature relaxation is achieved by examining the variations of a feature model of a part that can be generated by varying the types of the features of the model or geometric attributes of the features. The notion of *relaxation groups* is introduced which consists of features that can be treated as variations of each other. Features contained in a relaxation group are used for relaxation. For instance, **internal slot** and **break slot** are classified in the same relaxation group, thus the shape of the **rounding** at the end of an **internal slot** can be relaxed to that of a **break slot** which gives some additional process alternatives. By introducing certain ʾvaguenessʾ into a feature model, feature relaxation is thought to facilitate a systematic search of the space of similar parts which reveals manufacturing alternatives that were not present in the original model. The process planner, therefore, has ʾmore freedom to create a better process planʾ as well as provide more meaningful feedback during manufacturability analysis.

Hei-Or et al described a *relaxed parametric design* paradigm where ʾdecisions which needlessly limit the freedom of design in later stages are avoidedʾ (Hel-Or *et al.*, 1994). Existing parametric design systems are considered to cause over-specification and overwork since the process of correcting under-constrained and over-constrained models (that are easily produced in the process of defining a full and exact specification of the constraint models required by these systems) is time-consuming and error-prone. To overcome these problems, the relaxed parametric design paradigm uses ʾsoft constraintsʾ, i.e. constraints that need not to be satisfied exactly. A *probabilistic-constraints* scheme is developed to implement the relaxed parametric design paradigm. Instead of specifying and solving rigid constraints as in conventional parametric design systems, this scheme uses soft constraints which are associated with certain softness functions. A softness function specifies the amount of rigidity with which the constraint is to be satisfied. This scheme treats the relaxed parametric model as a static stochastic system. The softness functions of the constraints are expressed as covariance matrices. Kalman filter is used to solve the corresponding parametric system. A simple 2D parametric modeller has been implemented to test the algorithm developed.

Encouraging least commitment design practice is also one of the precepts that have driven the development of the feature-based thin-walled component design system reported in Nielsen (1991). In this system, feature-forms are represented by a set of virtual boundariesʾ which are geometric abstractions such as mid-planes, centrelines, and locating points. In using the system, one needs not to supply information required for completely defining the feature-forms in 3D. Minimum commitment design is encouraged by supporting the use of such abstract feature-forms which can be modified incrementally.

4.4. SUMMARY

So far, we have briefly reviewed the existing applications of the minimum commitment principle in a number of areas including design and computational support systems. We have seen the benefit of adopting minimum or least commitment principle in engineering design. The use of this principle in developing computational systems has so far been directed mainly to computer-based planning systems. A few computer-based design or manufacturing process planning systems have been developed in which we have seen the use of the minimum commitment principle in some ways and to certain extent.

We believe that the minimum commitment principle can be further exploited and extended to develop systems for supporting the early geometric design and ultimately to the entire process (and probably to other aspects of design as well). Early stages of design are the period in which a designer has least information and most uncertainty. Therefore, early stages of design are the phases where the designer, ultimately the design, can benefit most effectively from adopting minimum commitment principle. However, cognitive limitations of visualising and evolving a solution space may make designers commit too early to concrete solution points in spite of the uncertainty. Thus, there exists the need to support this visualisation and evolution without making unnecessary commitment. In the next section, we will briefly introduce our ongoing research effort directed towards this aspect.

5. An Approach to Minimum Commitment for CAGD

We have been developing an approach to geometric design guided by the minimum commitment modelling and incremental refinement principles and the geometric design model described earlier in this paper. Following the design model, the geometry of an object is characterised in this approach by a hierarchy of component arrangements, each component having various geometric parameters. The value space of these geometric parameters are derived, using the techniques of constraint reasoning, from high-level geometric relationships or constraints given by designers and are represented uniformly by real interval numbers. For a more detailed description of the approach supporting the system, see Guan (1993) and Guan and MacCallum (1995).

The prototype system embedding this approach is being developed using Common Lisp and Common Lisp Object System (CLOS) (Guan *et al.*, 1995). It currently is capable of modelling a set of primitive shapes (such as `cuboid`, `cylinder`, `sphere`), independent size constraints of inequlity or equality types (such as `width` $\approx$`20.4`, `depth`$\leq$`34`, `radius=6`), and basic spatial relationships (such as `above`, `right`, `front`). The minimum commitment modelling and incremental refinement principles have been embodied in the system in two main aspects:

– supporting the modelling of `vague geometry', and

— supporting the progression of vague modelling.

Vague geometry here refers to the vague expression of form used by a designer, most often at the early design stage. Vagueness is, typically, an inherent part of a process of evolving ideas from abstract to concrete. It reflects a designer's desire to communicate overall appearance or overall concept, to illustrate abstract concepts, or to illustrate concepts in ways which given economy of effort. It may also reflect lack of knowledge or certainty of some aspects of the geometry at certain stage of design.

The system aims to support the designer in modelling vague geometry. For instance, suppose we know that a component, G, of a product being designed can be modelled roughly as a cuboid. We are not sure of the exact size of the cuboid, but know that its width is definitely within the range of 10 to 15 units, its depth is exactly 6 units, and its height is about 6 units. Further, we know so far that G is going to be placed above another component, geometry1. The conceived solution space described by these fragments of information is visually illustrated in Figure 2(a). It is the solution space for the geometric properties of the component G which we have conceived so far.

Using the system, the above vaguely specified information can be used to build the geometric model for the component G using the following operation[1]:

```
(create-geometry :shape 'cuboid
                 :size-constraints '((width = 10 -> 15)
                                     (depth = 6)
                                     (height ^= 6))
                 :spatial-relationships '((above geometry1)))
```

Figure 2(b) illustrates the solution space corresponding to the geometric model or representation of G established in the system. The width of G being within the range of 10 to 15 is captured directly by an interval whose lower bound and upper bound are 10 and 15, respectively. The height of G being about 6 is also represented by an interval. The lower bound and the upper bound of this interval are determined by the given 6 and a user-defined *degree of approximation* which is used as the *width* of the interval. Here, this degree of approximation is assumed to be 2 which yields the value range for the depth of G to be 5 to 7. The depth of G being exactly 6 is also represented by an interval whose lower and upper bounds are both 6.

In representing the location information G above geometry1, the notion of *geometric configuration space* is introduced in the system to provide a geometric bound to all objects considered in a specific level of configuration. It is a 3D cuboid space that is associated with a 3D right-handed Cartesian co-ordinate system (e.g. *OXYZ* in the figure, note that only the *OXZ* projection is illustrated). The location of a component in a specific configuration is characterised by a *datum*

---

[1]Note that, although the system is used here by directly calling the corresponding Lisp functions or CLOS methods, a graphic user interface is available for the system.

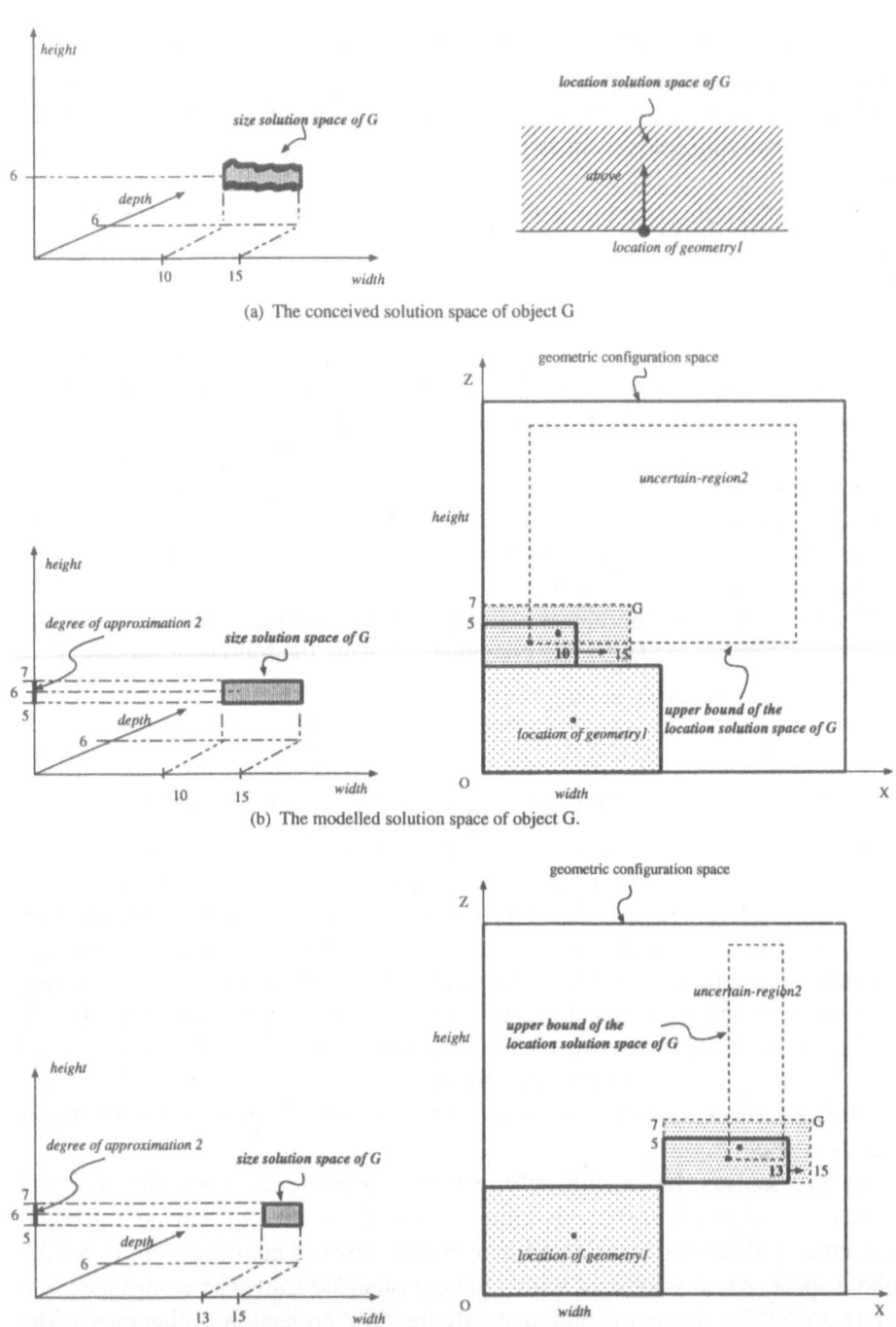

(a) The conceived solution space of object G

(b) The modelled solution space of object G.

(c) The modelled solution space of object G after refinement.

Figure 2.   Conceived solution space and modelled solution space in the vague geometry based system.

point on the component such as the geometric centre. This datum point is situated in a 3D *uncertain region*, represented by three intervals, which captures the approximation associated with the location of the component and is the minimum space that includes all the possible solutions of the location (within the corresponding geometric configuration space). Instead of choosing one specific point position for G that conforms to the given spatial relation, G above geometry1 is modelled in such a way that it defines the boundary of the location uncertain region (uncertain-region2) of component G along the *OZ* co-ordinate axis direction (Figure 2(b)).

This way of interpreting or representing geometric information reduces the level of commitment and thus preserves the solution space in two senses:

- From the user's point of view, early stage vague information can be used directly by the system. The user is neither forced to specify precisely at which point the component should be in the configuration space, nor forced to choose for its width one specific value between 10 and 15 and for its height one value close to 6. In other words, the user is not forced to specify and work on one specific solution prematurely.
- The system, on the other hand, is built with the goal of making as few unnecessary or unessential assumptions as possible in interpreting such vague information internally. Thus unlike conventional systems that model one or a few specific solution points in the conceived solution space, it aims to model the space by representing and managing the corresponding boundaries or bounds.

Figure 3(a) shows a graphic display of the geometric model of G generated by the system. In the current implementation, the solution spaces associated with the size and location of a component, here G, resulting from a (vague) geometric specification is not presented in the display. Thus, only an instance (corresponding to the minimum size and the lowerest-leftest-frontest position in the location uncertain region) of the defined solutions is displayed, although the corresponding size ranges and location uncertain region are represented in the system. The issue of suitable visualisation of size ranges and location uncertain regions calls for further research.

Naturally, we will want to make progress on the vague model towards a complete and precise solution. In the above example, we should be able to refine the size and location of the component G as new information arrives or when we feel necessary or suitable. For example, if we now know that the width of component G should be no less than 13 units and that it should also be located to the right of geometry1, we can then issue the following commands to the system:

```
(add-size-constraint '((width geometry2) >= 13))
(specify-location geometry2
                  :spatial-relationships '((right geometry1)))
```

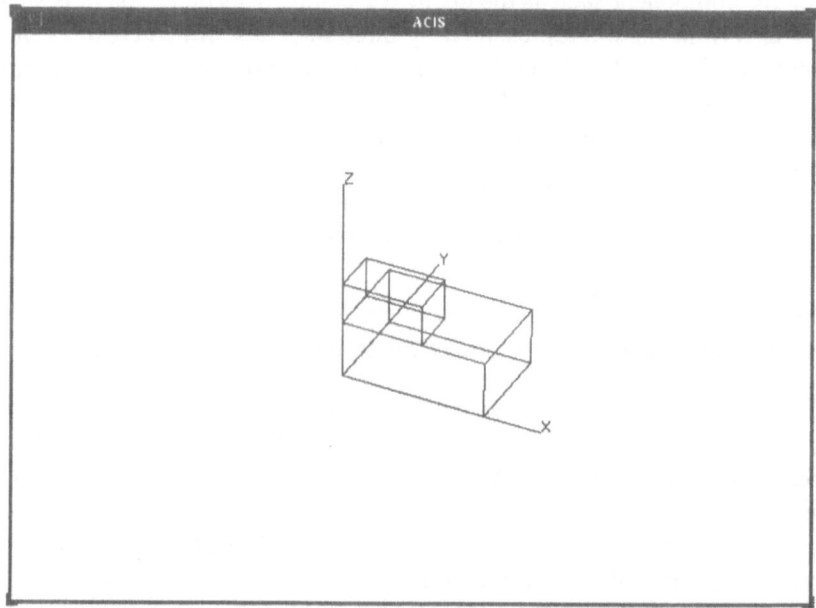

(a) The geometric configuration that contains an instance of the geometric model of G.

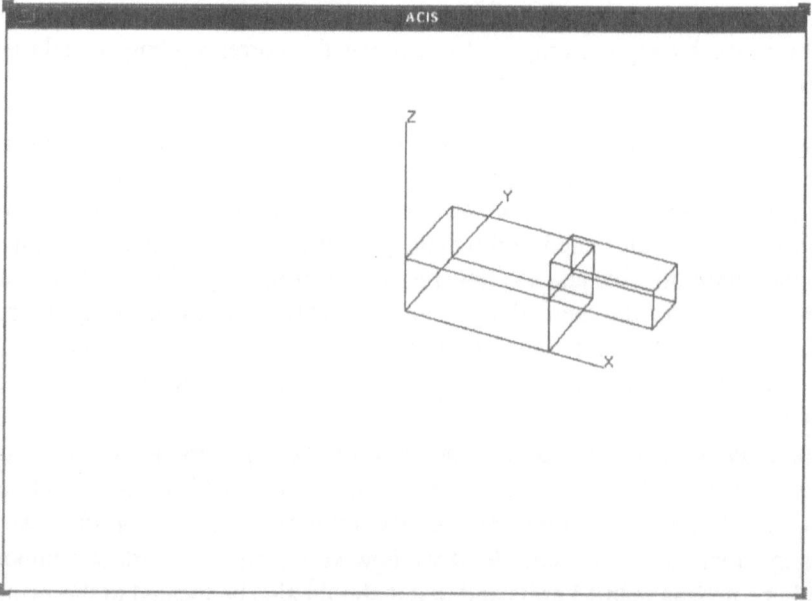

(b) G refined.

Figure 3.   Graphical presentation of geometric models in the system.

As a result, the modelled solution space of component G is refined from that illustrated in Figure 2(b) to that in Figure 2(c). Note that `geometry2` in the above commands is the unique name (identifier) generated by the system for the component G. In the current implementation, this unique name is used when the component is referred to. Again, a graphical display of the refined model is shown in Figure 3(b).

It should be pointed out that:

— Besides refining the solution space by reducing the vagueness of the existing model, a piece of information in an existing model can be removed when a piece of new information is added in and is in conflict with the existing one. This means that the evolution of the (shape and size of) solution space include not only an incremental reduction, but also an expansion. For instance, the `above geometry1` relation in the previous example can be retracted if a `below geometry1` relation is introduced for G and we decide to substitute it for the `above` relation.
— When using the system, the user makes decisions in terms of what shape, size and location a component should have. The system models and maintains the consequences of these decisions. Since the decisions are made incrementally and interactively, the modelling process in the system is also incremental and interactive. The system does not automate the decision making process involved in geometric design.

6. Conclusions and Future Work

By introducing the notions of conceived solution space and modelled solution space, we have formulated a major goal of effective CAGD systems. That is to capture and maintain a solution space and to minimise the difference between the modelled solution space and the conceived solution space. A minimum commitment modelling principle, togther with that of incremental refinement, is proposed subsequently for guiding the development of such a system. Supporting this principle means that a CAGD system should model, record or interpret design information as faithfully or precisely as possible by not making or requiring its user to make unnecessary assumptions that would lead to a modelled solution space distorted from the conceived solution space.

We distinguish minimum commitment modelling principle for CAGD systems from that for engineering design itself and focused on the former. While adopting a minimum commitment strategy in engineering design reduces backtracking, the proposed minimum commitment modelling principle for CAGD systems helps the systems to preserve a design solution space thus encourages the early modelling of geometric concepts and their exploration as well as the adoption of the principle in design itself.

We have, in this paper, described our initial effort in developing a computer-based geometric design system that adopts the minimum commitment modelling and incremental refinement principles through supporting the modelling of vague geometry and the evolution of such vague concepts into complete geometric models of design. Through this, we seek to explore the implications of adopting the principles in the area of geometric design, and also to obtain useful insight into the possible application of them to other aspects of design.

The planned research programme for developing the system further include, among others, the following areas.

- *Extension of the approach to modelling various types of vague geometry.* We have established a taxonomy of vague geometry by classifying *vagueness* into approximation, abstraction and incompleteness and *geometry* into shape, size, location and orientation. Our research so far has been concentrated on some of these aspects. A natural continuation is to study those aspects of vague geometry that have not been considered, such as vague shapes, and to develop the corresponding modelling methods.
- *Development and enhancement of the ability of the system to enable the gradual evolution of vague models into complete and precise design.* Such evolution can be the refinement of the rough size or location of an object as demonstrated in this paper. It can also be the refinement of an outline geometric model of a product into (i) those of its constituent subsystems, and (ii) a finer, more precise or complex shape of the product.
- *Visualisation or presentation of the modelled solution space.* Although it captures the geometric design solution space, the system currently does not yet have an effective visual means of presenting or conveying the captured space to the user. Research with respect to this would increase the usability of the system and encourage the user to engage in exploration of alternative solutions.

Finally, research could be carried out with respect to alternative or better ways of interpreting or modelling the conceived solution space which enable or maximise, in a computer-based system, the benefit of minimum commitment modelling principle.

7. Acknowledgement

The authors wish to acknowledge the support received from EPSRC, UK for the research described in this paper.

References

Asimow, M.: 1962, *Introduction to Design*, Prentice-Hall, Inc., Englewood Cliffs.
Baykan, C. A. and Fox, M. S.: 1992, WRIGHT: a constraint based spatial layout system, *in* C. Tong and D. Sriram (eds), *AI in Engineering Design*, Academic Press, Inc.

Dym, C. L.: 1994, *Engineering Design: A Synthesis of Views*, Cambridge University Press.

Ervin, S. M. and Gross, M. D.: 1987, Roadlab - a constraint based laboratory for road design, *Artificial Intelligence in Engineering*, **2**(4), 224–234.

French, M.: 1992, *Form, Structure and Mechanism*, MacMillan Education Ltd.

Goel, V. and Pirolli, P.: 1989, Motivating the notion of generic design within information processing theory: the design problem space, *AI Magazine*, **Spring**, 19–36.

Guan, X.: 1993, *Computational Support for Early Geometric Design*, PhD Thesis, University of Strathclyde, Glasgow, September.

Guan, X. and MacCallum, K. J.: 1995, Modelling of vague and precise geometric information for supporting the entire design process, *in* M. Mäntyla, T. Tomiyama, and S. Finger (eds), *Preprints of the IFIP WG5.2 First Workshop on Knowledge Intensive CAD*, IFIP.

Guan, X., Stevenson, D. A. and MacCallum, K. J.: 1995, A prototype system for early geometric configuration, *Proceedings of the Third International Conference on Computer Integrated Manufacturing (ICCIM)*, Singapore, July 11-14.

Hel-Or, Y., Rappoport, A. and Werman, M.: 1994, Relaxed parametric design with probabilistic constraints, *Computer-Aided Design*, **26**(6), 426–434.

Hubka, V.: 1987, *Principles of Engineering Design*, Springer-Verlag, heurista edition.

Janis, I. L. and Mann, L.: 1979, *Decision Making: A Psychological Analysis of Conflict, Choice, and Commitment*, Free Press, New York.

Duffy, A. H. B., MacCallum, K. J. and Green, S.: 1985, An intelligent concept design assistant, *in* H. Yoshikawa and E. A. Warman (eds), *Design Theory for CAD: Proceedings of the IFIP WG 5.2 Working Conference on Design Theory for CAD 1985 (Tokyo)*, North-Holland, pp. 301–317.

Mäntyla, M., Opas, J. and Puhakka, J.: 1989, Generative process planning of prismatic parts by feature relaxation, *Proceedings of ASME Design Automation Conference*, **1**, pp. 49–60.

Nielsen, E. H.: 1991, *Designing Mechanical Components with Features: Representing the Form and Intent of In-progress Design for Automated Modification and Evaluation*, PhD Dissertation, Department of Mechanical Engineering, University of Massachusetts.

Farinacci, M., Hulthage, I., Rychener, M. and Fox, M.: 1992, ALADIN: an innovative materials design system, *in* C. Tong and D. Sriram (eds), *Artificial Intelligence in Engineering Design*, **II**, Academic Press, Inc, pp. 215–262.

Sacerdoti, E. D.: 1977, *A Structure for Plans and Behavior.* Elsevier North-Holland, Inc.

Suh, N. P.: 1990, *The Principles of Design*, Oxford University Press.

Tjalve, E.: 1979, *A Short Course in Industrial Design*, Newnes-Butterworths.

Tomiyama, T. and Yoshikawa, H.: 1985, Extended general design theory, *in* H. Yoshikawa and E. A. Warman (eds.), *Design Theory for CAD, Proceedings of the IFIP WG5.2 Working Conference 1985 (Tokyo)*, North-Holland, pp. 95–124.

Weld, D. S.: 1994, An introduction to least commitment planning, *AI Magazine*, **15**(4), 27–61.

12

spatial and layout planning in design

The generation of form using an evolutionary approach
Michael A. Rosenman
Evolutionary layout design
Walter Hower, Manfred Rosendahl, Derrick Köstner
DOM-ARCADE: Assistance services for construction, evaluation,
and adaptation of design layouts
Shirin Bakhtari, Brigitte Bartsch-Spörl, Wolfgang Oertel

J. S. Gero and F. Sudweeks (eds), Artificial Intelligence in Design '96, 643-662.
© 1996 *Kluwer Academic Publishers.*

THE GENERATION OF FORM USING AN EVOLUTIONARY APPROACH

MICHAEL A. ROSENMAN
Key Centre of Design Computing
Department of Architectural and Design Science
University of Sydney NSW 2006 Australia

Abstract. This paper presents an evolutionary approach to design using a hierarchical growth model. It argues that the evolutionary approach fits well to the well-known generate-and-test approach in design and is especially suited to design situations where the (inter)relationships between complex arrangements of elements and their behaviour are not known. The evolutionary approach is used as the computational method for the synthesis and evaluation stage of the design process. A hierarchical model of design is used to avoid the combinatorial problems involved in linear models. The concepts are exemplified in the context of the design of house plans.

1. Introduction

Design is a purposeful knowledge-based human activity whose aim is to create structural descriptions, i.e. form, which when realized satisfy the particular given intended purposes (Rosenman and Gero, 1994). Design may be categorized as routine or non-routine with the latter further categorized as innovative or creative. The lesser the knowledge about existing relationships between the requirements and the form to satisfy those requirements, the more the design problem tends towards creative design. Thus, for non-routine design, a knowledge-lean methodology is necessary. Natural evolution has produced a large variety of forms well-suited to their environment, this process being capable of acceleration by selective breeding. This suggests that, even though resources of time and technology are limited, the use of an evolutionary approach may lead to a mechanism useful for providing meaningful design solutions in a non-routine design environment . An example of the capabilities of an evolutionary approach in a creative domain is the generation of creative Art forms through the use of simple geometric primitives and rules (Todd and Latham, 1992). This paper is an initial investigation into the possibilities of the approach .

2. An Evolutionary Model Of Design

2.1. GENETIC ALGORITHMS

Genetic algorithms (GAs) are a class of algorithms, based on the adaptive process of natural evolution, which employ a general uniform knowledge-lean methodology without preconceived prejudices as to the solution (Beasley et al., 1993; Holland, 1975). While GAs have been used to solve a variety of problems mainly optimization, learning and control problems, (Goldberg,, 1989; Grefenstette et al., 1989), there has been very some limited research and applications in design (Gero et al.,1994; Jo, 1993, Jo and Gero, 1995; Schnier and Gero, 1995; Woodbury, 1993). The trend which has developed in GA applications is to encode the genotype as a string of characters, usually binary. Each such character represents a one-to-one correspondence to a property in the phenotype, as distinct from nature where the genotype encodes instructions for the generation of living form.

2.2 EVOLUTION AND DESIGN

The evolutionary approach is basically a generate and test approach which corresponds well to the procedures for design synthesis and evaluation in the design process. The specific characteristics of the approach are:

- a large pool or population of members (e.g. design solutions);
- members are selected for 'survival' using a biased random selection mechanism based on their 'fitness', i.e. relation to a fitness function;
- new members are generated from the existing ones using evolutionary mechanisms as crossover, inversion, 'gene splicing' and mutation.

The advantages of the evolutionary approach are:

- more diverse sections of the state space can be investigated than with other methods, thus tending to discover a variety of potential solutions with less fear of immediate convergence to single local points;
- a probabilistic selection method directs the random generative process to produce meaningful and satisfactory solutions.

2.2.1. Growth of Form

Whereas the process of generation of form in living systems involves the placement of different kinds of protein in particular locations, the process of generation of form in design involves the placement of units of different kinds of material in particular locations. Normally in the design domain, we represent objects in terms of the shape of their envelope. However, we may describe an object, at a lower level of abstraction, as a composition of material units (cells or building blocks), where the type and scale of such units are chosen depending on the context and suitability for the level of abstraction required. An object can then be described by the location of each such unit and can be 'grown' by locating a required number of such units, one at a time in sequence, Figure 1.

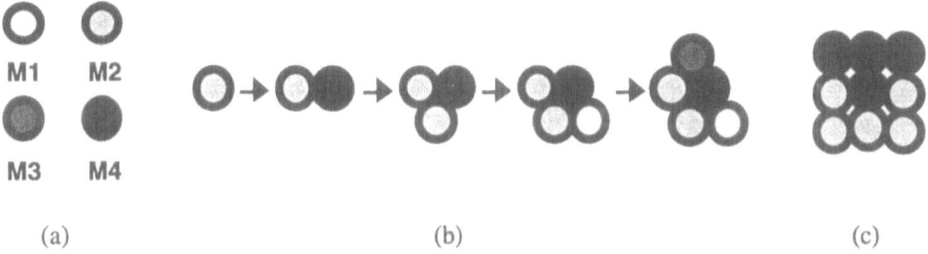

M1   M2

M3   M4

(a)   (b)   (c)

Figure 1. A growth model of form; (a)units of materials, (b) growth through location of units of material, (c) distinct elements.

The form produced will depend on the form of the unit material and the location procedures, i.e. rules of growth, used. Different location procedures will create different geometries, as in crystal growth. In general, random aggregation of different units of materials will be generated, Figure 1(b). Distinct elements may emerge as a case of aggregations of the same material units, Figure 1(c).

The genotype for a homogeneous object is thus the sequence of coded instructions for selecting and locating material units, analogous to the DNA string in natural evolution. When this code is interpreted and executed, the phenotype, i.e. the object (or rather its representation), will be generated. A general model of form growth can be proposed as:

> For given total units of material required
>   **SELECT**   a unit of material, Mm
>   **LOCATE**   unit of material, Mm (using locating procedures) relative to
>   other units

A gene in such a model becomes (Ot, Mm, L(Mm)), where L(Mm) is the instruction for locating the unit of material Mm relative to the generated object at each step, Ot. Initially Ot is a single unit. The genotype is a sequence of such genes. Where a homogenous object is considered, the material identification is constant and the gene is basically a sequence of location operators. Obviously, such a model is computationally infeasible in general and a more computationally feasible approach is required.

2.2.2. Elements, Components and Assemblies

An object may be simple or complex. A simple object is termed an *element* and by definition is homogeneous otherwise it can be decomposed into separate elements each being homogeneous. Note that composite materials like reinforced concrete, are treated as conceptually homogeneous. An element is thus a composition of material units. A complex object is termed an *assembly* and is composed of objects termed *components*. Recursively, components may be assemblies or elements. A formal description is as follows:

O :-	$A \mid E$	*An object is an assembly or an element*
A :-	$(C, R(C))$	*An assembly is a set of components and a*
C :-	$\{\cup(C_i\}$	*set of relationships among the components*
$R(C)$:-	$\{\cup(R_k(C_i, C_j))\}$	
C_i :-	O	*ith component is an object*
E :-	$(M_m, R(M_m))$	*An element of material M_m is a set of*
M_m :-	$\{\cup(M_{mp}\}$	*material units of type M_m and a set of*
$R(M_m)$:-	$\{\cup(R_l(M_{mp}, M_{mq}))\}$	*relationships among the material units*
M_{mi} :-	ith unit of material M_m	

2.2.3. Levels of Abstraction and Parametric Design

The 'size' of the unit of material will vary according to the problem and the level of abstraction required. Obviously, the smaller the unit, the larger the genotype required to generate an object with the associated problems of an exponential increase in computation time.

2.2.4. Evaluation of Design Solutions

The evaluation of designs is carried out by interpreting the generated design solution, the phenotype, and determining its behaviours according to a set of behavioural requirements formulated from the design requirements. The actual levels of performance of the object's behaviours may be determined using causal knowledge in the form of formulae, rules, etc. or by users exercising judgment in the case of qualitative behaviours such as aesthetic quality. Such judgments are subjective and personal and they may be made as a complex evaluation of many factors without rationalization of the separate factors. The designer takes full responsibility for such evaluations. Such subjective evaluation coupled with user interaction is the approach taken in the generation of *Biomorphs* (Dawkins, 1986) and creative Art forms (Todd and Latham).

Since an object exhibits more than one behaviour, the evaluation of the fitness of the object is a multiobjective problem and hence will involve evaluation using concepts similar to Pareto optimization (Horn et al., 1994; Jo, 1993; Jo and Gero, 1995). Constraints can be implemented through the use of penalty functions.

The issue of 'emergent' behaviour and functionality, i.e. the evolution of the fitness function itself, while of importance, will not be treated in this paper.

2.2.5. Design Through Hierarchical Decomposition / Aggregation

Simon (1969) points out, that even though organisms are very complex, it is only possible for them to evolve if their structure is organised hierachically. The above formulation allows for the generation of objects through the recursive generation of its components until a level is reached where the generation becomes one of generating an element. Such an approach assumes a knowledge of the decomposition structure of an object.

The advantages of a hierarchical approach are that only those factors relevant to the design of that component are considered and factors relevant to the relationships between components are treated at their assembly level. Instead of one long genotype consisting of a large number of low-level genes, the genotype is composed of a set of chromosomes relevant to their particular level. In addition to reducing the combinatorial problem substantially, parallelism is supported since all the different chromosomes (components) at a particular level can be generated in parallel. If the set of possible alternatives of component types is sufficiently rich, i.e. large and varied enough, then many different combinations of members of different such sets are possible, at the next level, with a good chance of satisfying the criteria and constraints at that level. Only when no such possible combination satisfies such criteria is there a need for some generation of new alternatives at the lower level.

There are basically two approaches. The first is a top-down approach, used by Cramer (1985) and also in Genetic Programming (Koza,1992; Rosca and Ballard, 1994). This approach considers the entire object tree at the one time, with crossover occurring between corresponding subtrees at any level. The second is a bottom-up multi-level approach where, although the overall decomposition structure is known, the composition of the various levels is not. At each level, a component is generated from a combination of components from the level immediately below. At each level, an initial population is generated and then evolved over a number of generations until a satisfactory population of objects at that level is obtained. Members of that population are then selected as suitable components for generating the initial population at the next level. The process is repeated for all levels, Figure 2.

In a flat model of form generation, a genotype will consist of a string of a very large number of basic genes. In a hierarchical model, there are a number of component chromosomes, at different levels, consisting of much shorter strings of genes which are the chromosomes at the next lower level. All in all, the total number of basic genes willthe same in the flat and hierarchical models.

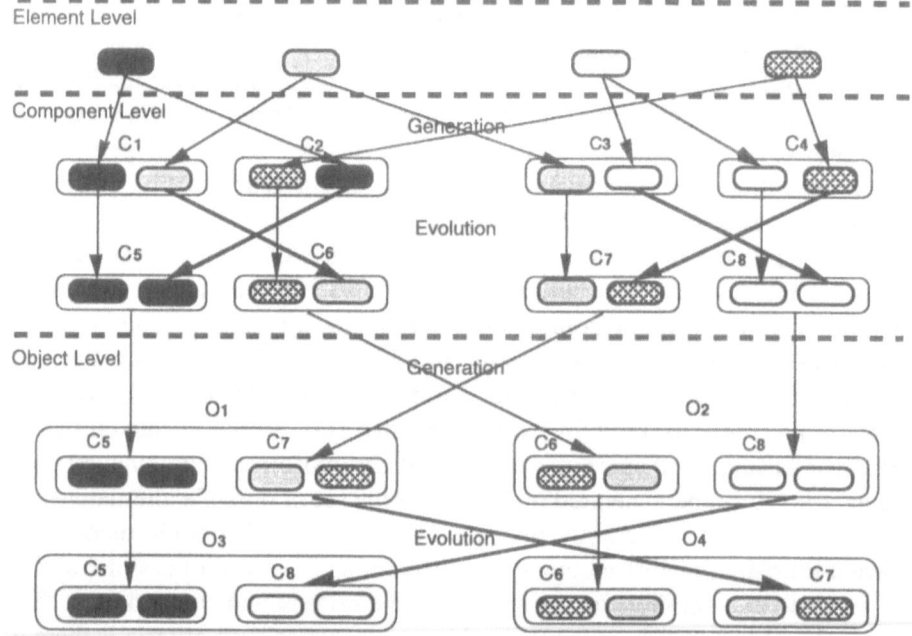

Figure 2. Multi-level combination and propagation.

2.3. DESIGN GRAMMARS

In order to generate an object (a design solution) a generative method, such as a design grammar, is required. A design grammar deals with a vocabulary of design elements and transformation on these elements and hence defines a design space (Woodbury,1993). In shape grammars, the vocabulary consists of shapes and the grammar rules define transformations on these shapes (Stiny, 1980). Successive application of shape grammar rules generates shape compositions which may be related to designs such as buildings (Stiny and Mitchell, 1978). While design grammars provide a generative capability, they are syntactic mechanisms without the evaluative mechanisms for directing the generation towards meaningful solutions.

2.3.1. Recipes and Blueprints—Genotypes and Phenotypes
A recipe (or plan) is a set of instructions or operations, whereas a blueprint is a representation of the solution (Dawkins, 1986; Woodbury, 1993).

The aim of the design process, in an evolutionary approach, is the attainment of a set of instructions, a genotype, that when executed, yields a design description of a product, a phenotype, whose interpreted behaviours satisfy a set of required behaviours, the fitness function. In this approach, a grammar rule is a gene, the plan (sequence of rules) is the genotype and the design solution is the phenotype.

The advantage of the use of grammar rules as the genetic information is in the simplicity of the information needed to be kept, since the generative rules are fewer in number than solution parameters and generally less complex. Moreover, small changes in such rules or their combinations can lead to large and unexpected changes in the design solutions, a desirable property for creative design (Woodbury, 1993).

2.3.2. The Evolution of New Rules and Plans

There are basically two approaches in the generation of genotypes of design grammar rules, analogous to the two approaches in classifier systems, the Michigan and Pitt approaches (Goldberg, 1989; Wilson and Goldberg, 1989). The first approach, as taken by Gero et al. (1994), attempts to 'learn' new grammar rules. The second approach, taken in this paper, is based on the premise that the grammar rules are fundamental operators, which cannot be decomposed or recomposed, that the particular grammar contains all required rules and that the aim of the design process is to find satisfactory sequences of such rules.

2.3.3. A General Model for An Evolutionary Approach to Design

The general model of design using an evolutionary approach may be stated as follows:

```
for all levels in the object hierarchy
    for all components at that level
        GENERATE    initial population of members (by synthesizing lower level
                    units)
        EVOLVE      population until satisfactory
```

3. A House Design Example - Space Generation

The above concepts can be exemplified through the generation of 2-D plans for single-storey houses. The work of Jo (1993) and Jo and Gero (1995) demonstrated that a single-level approach was not able to converge towards satisfactory solutions mainly due to the interactions of the various factors of the fitness function required for the various elements.

3.1. A HOUSE SPATIAL HIERARCHY

A house can be considered to be composed of a number of zones, such as living zone, entertainment zone, bed zone, utility zone, etc. Each zone is composed of a number of rooms (or spaces), such as living room, dining room, bedroom, hall, bathroom, etc. Different houses are composed of different zones where each zone may be composed of different rooms. Each room is composed of a

number of space units. Generally, in a design such as a house, the space unit
will be constant. The scale (level of abstraction) of the space unit depends on
the precision required in differences between various possible room sizes. The
smaller the unit, the longer the genotype for a given size of room but the greater
the shape alternatives.

3.2. GENERATION - THE DESIGN GRAMMAR

In the above formulation, the generation of spaces, basically comes down to
locating spatial component units for that level. At the room level, the
component unit is a fundamental unit of space. At the zone level, the component
unit is a room and at the house level the component unit is a zone.

The design grammar used here is based on the method for constructing
polygonal shapes represented as closed loops of edge vectors (Rosenman,
1995). The grammar is based on a single fundamental rule which states that any
two polygons, Pi and Pj, may be joined through the conjunction of negative
edge vectors, V_1 and V_2, that are equal in magnitude and opposite in direction.
The conjoining of these vectors results in an internal edge (neutralized vector)
and a renaming of the new polygon, Pk. The edge conjoining rule, R1is:

$$Vm + Vn \quad = \quad Vmn \quad = \quad 0 \quad .... \quad \textbf{R1}$$

Rule R1 is commutative and applies for all values of vector direction. This rule
ensure that new cells are always added at the perimeter of the new resultant
shape. This rule is shown diagrammatically in Figure 3.

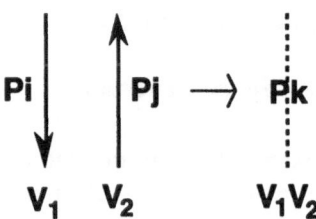

Figure 3. Edge vector rule for the construction of 2-D shapes.

The fundamental conjoining rule can be specialized for different types of
geometries. Orthogonal geometries are based on the following four vectors of
unit length: W = (1, 90), N = (1, 0), E = (1, 270), S= (1, 180).

so that Rule1 becomes:

$$N + S \quad = \quad NS \quad = \quad 0 \quad .... \quad \textbf{R1a}$$
$$E + W \quad = \quad EW \quad = \quad 0 \quad .... \quad \textbf{R1b}$$

These two (sub)rules allow for the generation of all polyminoes. Orthogonal geometries will be used in this paper without loss of generality. Other (sub)rules may be formed for other geometries (Rosenman, 1995).

3.2.1. Genotype and Phenotype

A polygon is described by its sequence of edge vectors. A suffix is used to identify individual edges of the same vector type. Thus the square cell of Figure 4 is described as (W1, N1, E1, S1). The sequence of edge vectors describing the (polymino) shape is the phenotype in a geometric sense. This provides the description of the shape's structure from which its behaviour may be derived.

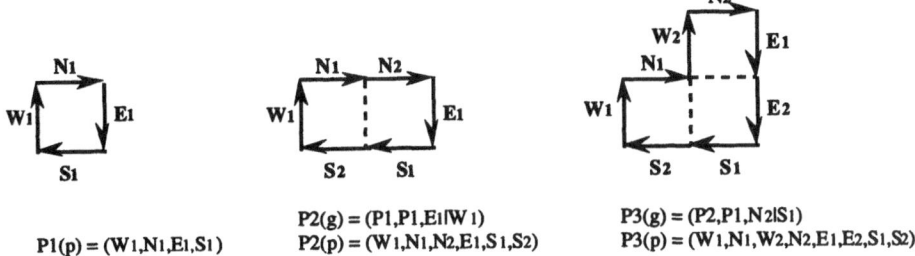

P1(p) = (W1,N1,E1,S1)

P2(g) = (P1,P1,E1|W1)
P2(p) = (W1,N1,N2,E1,S1,S2)

P3(g) = (P2,P1,N2|S1)
P3(p) = (W1,N1,W2,N2,E1,E2,S1,S2)

Figure 4. Generation of a trimino.

The genotype for any generated polymino is the sequence of the two subshapes (polyminoes) used and the two edges joined. An example of the generation of a trimino is shown in Figure 4. Figure 4 shows a basic unit or cell, P1, which provides a starting point for the generation of polyminoes. Each generated shape is accompanied by its genotype and phenotype. The generation of these polyminoes occurs from a random selection of edges in the first shape conjoined with a random selection from equal and opposite edges in the second shape. At each step in the generation, the phenotype is reinterpreted to generate a new edge vector description and the conjoining (sub)rules applied. The genotype for the generated trimino is given as (P2, P1, N2|S1). This can be expanded as ((P1, P1, E1|W1), P1, N2|S1). When the same units are used for generation, the unit can be omitted and the genotype represented as the sequence of edge vector conjoinings. That is P3(g) = (E1|W1, N2|S1).

The length of the genotype (and phenotype) depends on the size of the polymino to be generated, that is on the area of the polymino. This corresponds to required room sizes. For different room types, minimum and maximum area constraints can be given so that polyminoes will be generated (randomly) with areas within those constraints.

Once a population of different rooms is generated for each room type in a given zone, the zone can be generated through the conjoining of rooms in a progressive fashion. Because of the cell-type structure of the polygons, the conjoining may occur at any appropriate pair of of cell edges. Therefore, a large number of possible zone forms can be generated from two rooms. An example of some possibilities arising from the conjoining of two polyminoes is given in Figure 5.

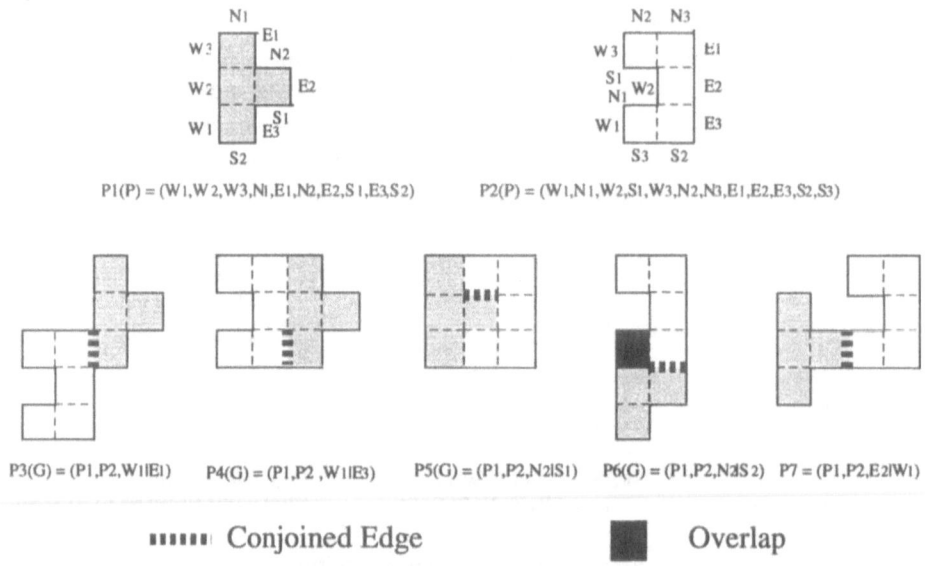

P3(G) = (P1,P2,W1IE1)   P4(G) = (P1,P2 ,W1IE3)   P5(G) = (P1,P2,N2IS1)   P6(G) = (P1,P2,N3IS 2)   P7 = (P1,P2,E2IW1)

▪▪▪▪▪▪ Conjoined Edge   ■ Overlap

Figure 5. Some examples of conjoining two polyminoes.

The two polyminoes, P1 and P2, represent instances of two different room types and the polyminoes resulting from the joining of the two rooms represent instances of a particular zone type. When one pair of edges are conjoined other edges may also be conjoined, e.g. P4, P5 and P6. In the case of overlap, as in P6, the resultant shape is discarded.

The same process used for generating zones is used to generate houses. The joining of different instances of different zone types generates different instances of houses.

3.2.2. Order of Selection

At the zone and house level, the order of selection of the units to be joined may influence the solution and its performance. For example, if living rooms and dining rooms are chosen as the first two rooms in the living zone, they will always be adjacent. However, there are problems with choosing a random order of room selections for every zone instance generation for the same population as future crossovers may lose some room types and include more than one of the same type. Thus for any population, the same order of room types must be kept. A given order may be chosen randomly or from an algorithm based on the number (and/or strength) of interconnections required in an interconnection matrix as used, for example, by Coyne (1988) in his plan of actions.

3.3. THE EVOLUTION OF HOUSE DESIGNS

The above grammar can be used to generate initial populations for each level in the spatial hierarchy. Each such initial population is then evolved, as necessary, so that solutions are 'adapted' to design requirements.

3.3.1. The Evaluation Criteria—Fitness Functions

At each level, different fitness functions apply according to the requirements for that level. While the requirements for designs of houses involve many factors, many of which cannot be quantified or adequately formulated in a fitness function, some simple factors will be used initially to test the feasibility of the approach. For this example, the fitness function for rooms will consist of minimizing the perimeter to area ratio and the number of angles. This requirement tends to produce useful compact forms. For zones, the fitness function will consist of minimizing a sum of adjacency requirements between rooms reflecting functional requirements. At the house level, the fitness function will consist of minimizing a sum of adjacency requirements between rooms in one zone and rooms in other zones. This has the tendency to select those arrangement of zones where adjacency interrelations are required between rooms of different zones. In addition to these quantative assessments, qualitative assessments will be made subjectively and interactively by a user/designer.

Although the above criteria have been described in terms of optimizing functions, the aim is not to produce global optimum solutions but rather to direct the evolutionary process to produce populations of good solutions either as components for higher levels or at the final level itself. So that, even though the global optimum solution for the shape of a room using the above ctiteria, may be known, this may not be the optimum solution at the zone and and house levels. By selecting other non-optimal but good solutions, according to the given criteria, good unexpected results may be achieved for the overall design.

Other factors are required in more realistic design contexts. For example, a house needs to meet site requirements, both in terms of size and orientation for view or climate. Such factors can be formulated as constraints at various levels. These constraints can be handled explicitly as survival factors or as penalty functions. That is, a solution which does not meet a constraint is eliminated or, alternatively, a penalty can be added to the fitness of the solution according to the degree of violation. It is argued that with a sufficiently large population of room and zone alternatives such constraints can be met. If not, then redesign, i.e. new zone and/or room forms, must be produced.

3.3.2. Propagation—Crossover

Simple crossover is used for the production of 'child' members during the evolution process. Looking first at the room level to see the effect of such a crossover process, crossover can occur at any of the four sites as shown in

Figure 6(a) with two results as shown in Figure 6(b). Since we are always dealing with cells of the same space unit, the cell identification in the genotype representation has been omitted for simplicity.

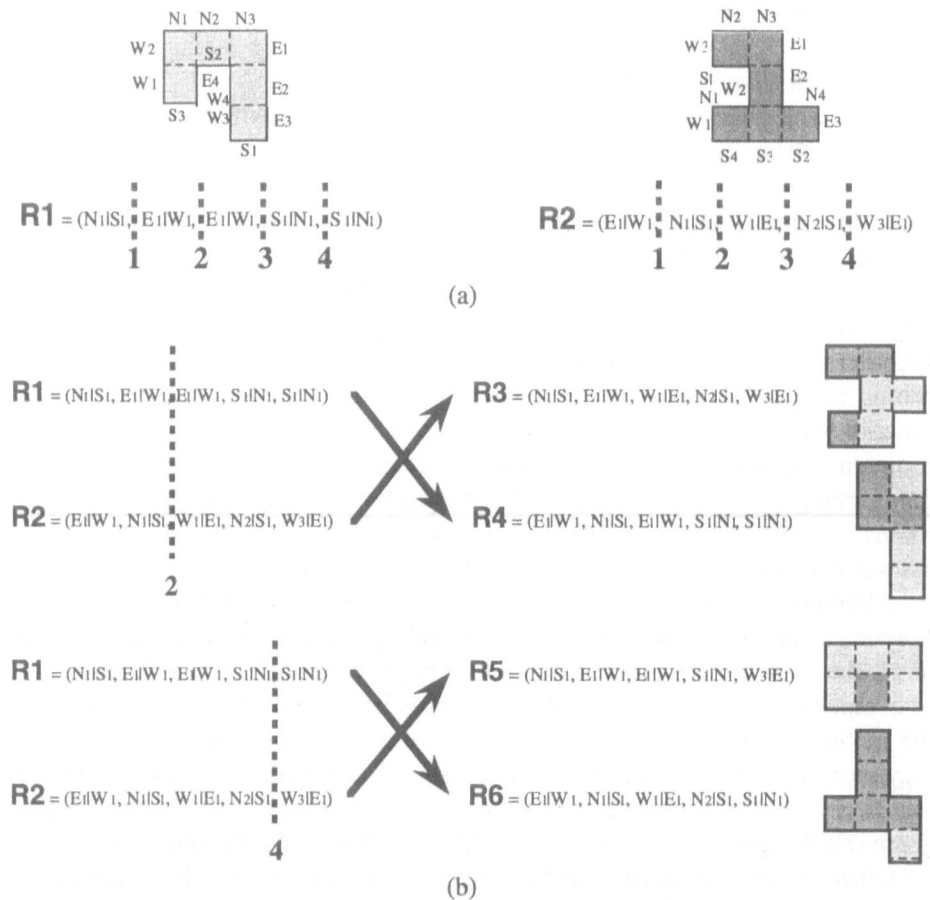

Figure 6. Crossover at room level; (a) initial rooms R1 and R2 generated from unit square cell U1, (b) crossover at sites 2 and 4.

At the zone level, crossover occurs as shown in Figure 7. Two initial instances of living zones, Z1 and Z2 are shown in Figure 7(a). Each zone has one instance of each of living room, dining room and entrance. Figure 7(b) shows crossover for one of the four possible sites. A similar process is followed at the house level.

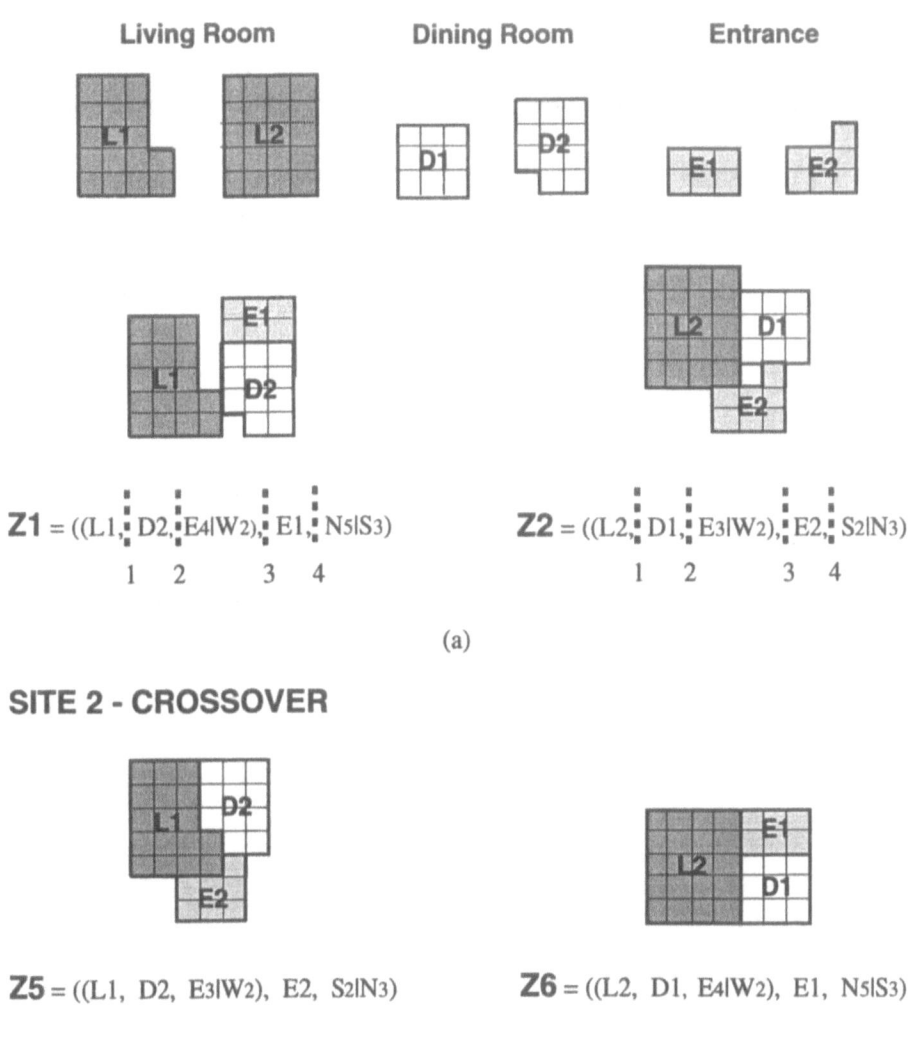

$Z1$ = ((L1, D2, E4|W2), E1, N5|S3)

1 2   3   4

$Z2$ = ((L2, D1, E3|W2), E2, S2|N3)

1 2   3   4

(a)

SITE 2 - CROSSOVER

$Z5$ = ((L1, D2, E3|W2), E2, S2|N3)

$Z6$ = ((L2, D1, E4|W2), E1, N5|S3)

(b)

Figure 7. Examples of zone crossover; (a) rooms and initial zones, Z1 and Z2, (b) crossover at Site 2.

4. Implementation And Results

A computer program written in C++ and Tcl-Tk under the Sun Solaris OpenWindows environment is under implementation. Currently, only the simple criteria described previously have been used. Each evolution run, for all levels, tends to converge fairly quickly to some solution which may not necessarily be the best or, in some limited cases, even acceptable. The usual method to break out of such convergence is to introduce some level of mutation.

Rather than use a mutation operator, it was found that a more efficient strategy was to generate multiple runs with different initial randomly generated populations. This produces a variety of gene pools thus covering a more diverse area of the possible design space. Moreover, such runs can be generated in parallel. Users can nominate the population size, number of generations for each run and select rooms, zones and houses from any generation in any run as suitable for final room, zone or house populations.

Results are shown in the following figures, Figures 8 to 11 for room, zone and house solutions.

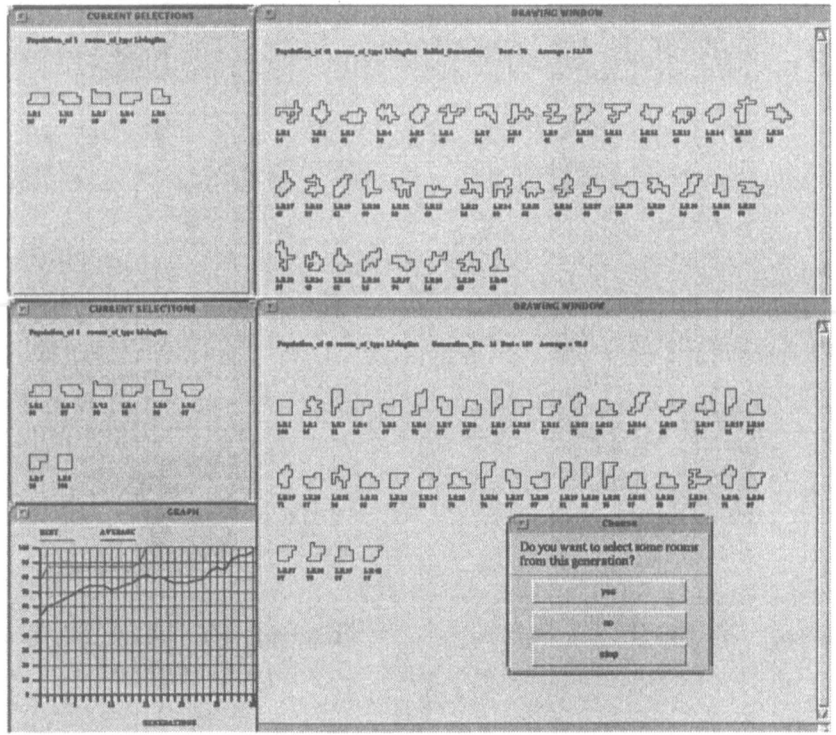

Figure 8. Results of living room generation.

The upper part of Figure 8 shows 5 Living Room shapes selected from previous runs. with a randomly generated initial population of 40 Living Rooms in a new run. The lower part of Figure 8 shows the evolution of this population through 16 generations. Three new shapes are selected, during this evolution. The graph shows the evolution through 30 generations where the population converged on the square shape (evolved at the 16th generation). The upper line in the graph shows the evolution of the best solution while the lower line shows the evolution of the population average. The Dining Room, Kitchen and Entrance were generated in a similar way.

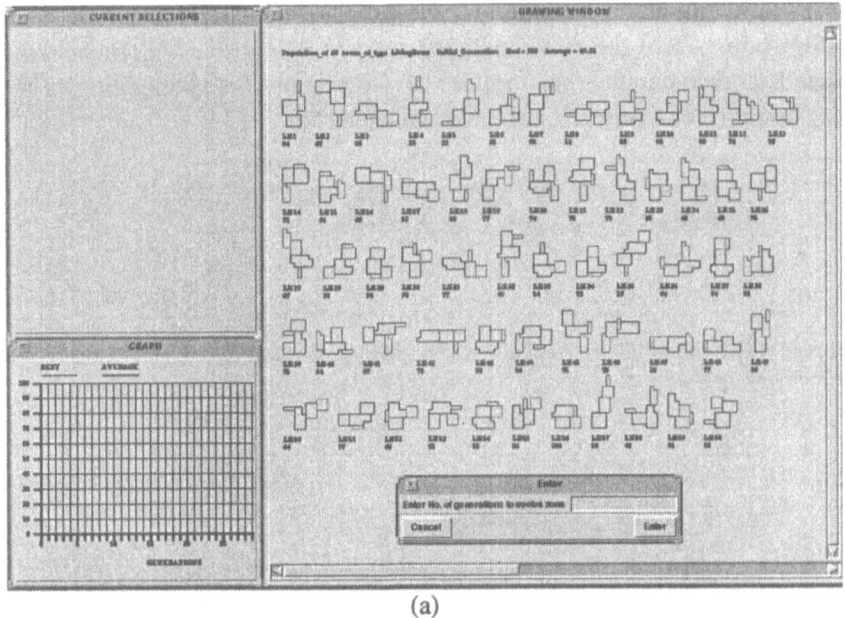

(a)

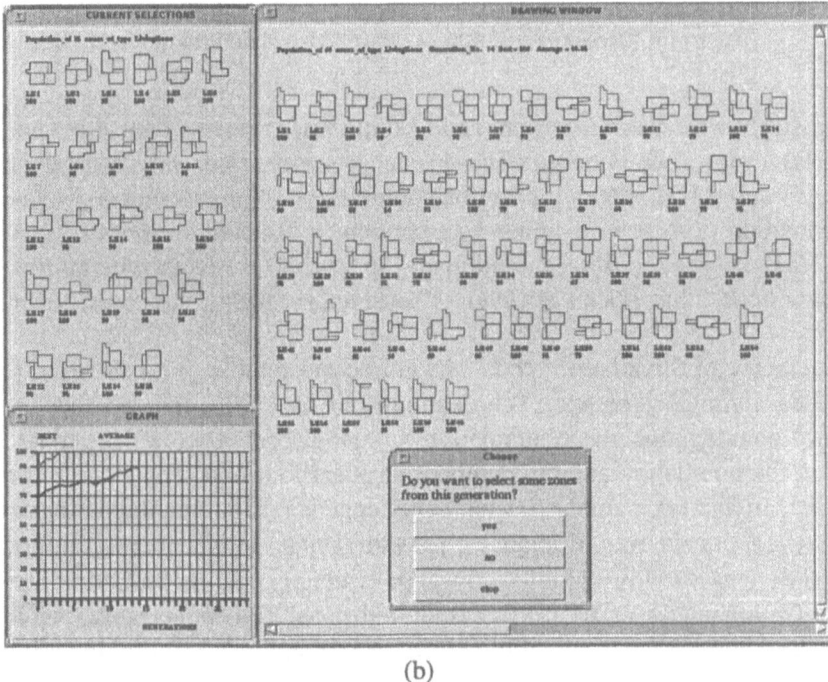

(b)

Figure 9. Results of Living Zone generation; (a) initial Living Zone population, (b) evolved population.

Figure 9(a) shows an initial population for the Living Zone, randomly generated by selecting rooms from the final selctions for Living Room, Dining Room, Kitchen and Entrance populations. Figure 9(b) shows the 14th generation of the final run. Twenty five Living Zones have been selected.

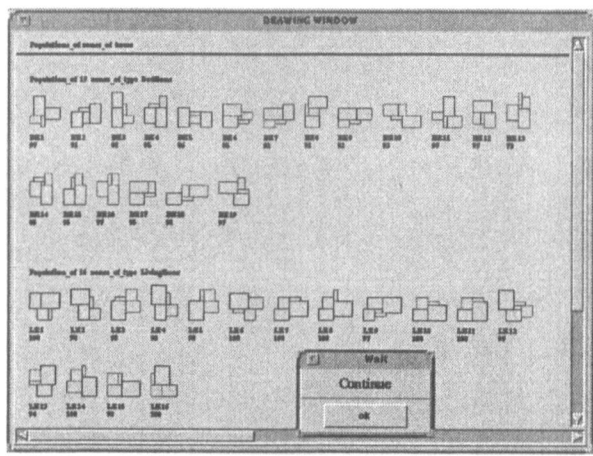

Figure 10. Results of Bed and Living Zones generation.

Figure 10 shows a set of Bedroom and Living Zones selected during a later run. The Bedroom Zone consists of a Master Bedroom, Bedroom, Bathroom and Hall. Figure 11(a) shows the initial population of the second run in the house generation process (one house was previously selected in the first run). Figure 11(b) shows the 10th generation in the evolution of this population. Three new houses have been selected as satisfactory during the course of the evolution.

The total area of this house type is 80 sqm, corresponding to a genotype of length 79 in a single genotype. The example of Jo (1993), showed that no satisfactory convergence was obtained with a single genotype of this length. The size of the population and the number of generations and runs required to generate a satisfactory number of members depends on the genotype length. The longest genotype was of length 19, for the Living Room (area 20 units). Convergence was usually achieved by approximately 25 generations. The addition of more zones and rooms presents no problem for the hierarchical approach used here although it would present extra combinatorial problems for the single genotype approach.

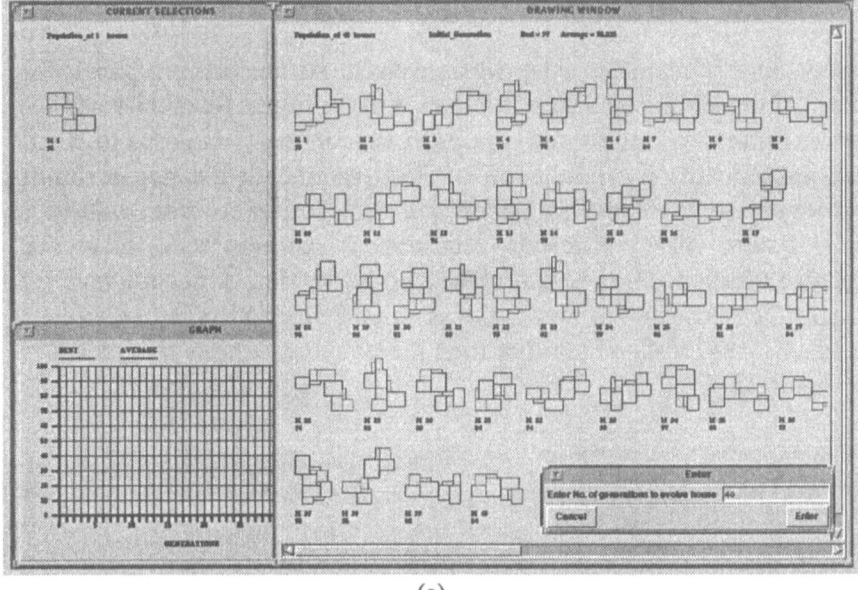

(a)

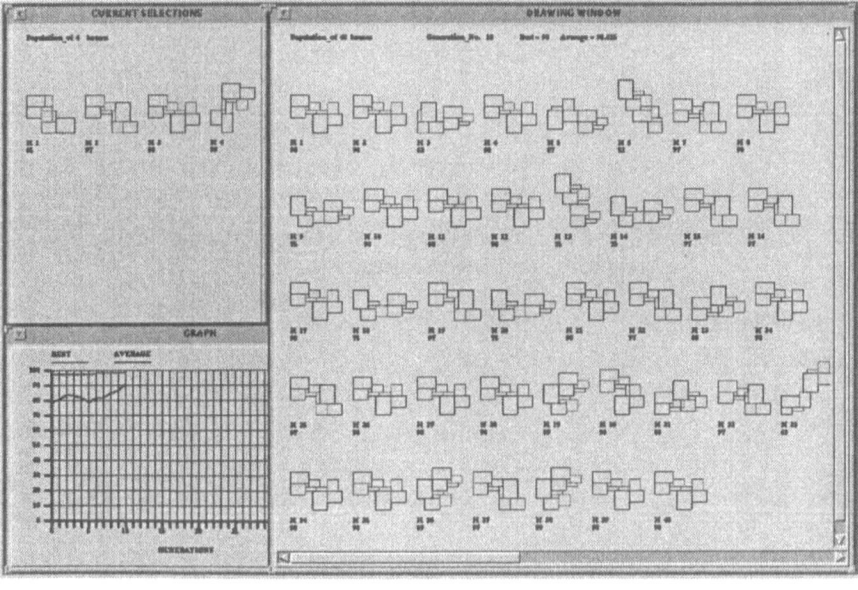

(b)

Figure 11. Results of House Generation; (a) initial house population, (b) evolved
population of houses.

5. Changing The Fitness Function

Once a population of members exists for a given fitness function this population can be used as the initial population if a change in the fitness function occurs. In this way, existing designs adapted to a previous context can evolve to create new designs adapted to a new context. For example, if the house fitness function involves the minimization of house perimeter to area ratio as a reflection of some energy efficiency requirement, compact house plans will result, as for example in Figure 12. If subsequently, the fitness function involves maximizing the perimeter to area ratio as some reflection of a requirement desiring maximum cross-ventilation, then the existing designs may adapt to those in Figure 13.

$LZ1 = (((L1, D1, S_3|N_2), E1, S_4|N_1), K1, E_9|W_2)$

$BZ1 = (((MB1, HA1, S_1|N_1), BA1, W_6|E_2), B1, S_2|N_2)$

$LZ2 = (((L1, D1, S_1|N_4), E2, W_4|E_2), K2, S_2|N_3)$

$BZ2 = (((MB2, HA2, S_1|N_2), BA2, S_2|N_2), B1, W_5|E_2)$

$H1 = (LZ1, BZ1, W_1|E_8)$

Area	= 70
Perimeter	= 38
P/A Ratio	= 0.543

$H2 = (LZ2, BZ2, W_1|E_3)$

Area	= 77
Perimeter	= 42
P/A Ratio	= 0.545

Figure 12. Two compact house plans resulting from a fitness function involving minimum perimeter to area ratio.

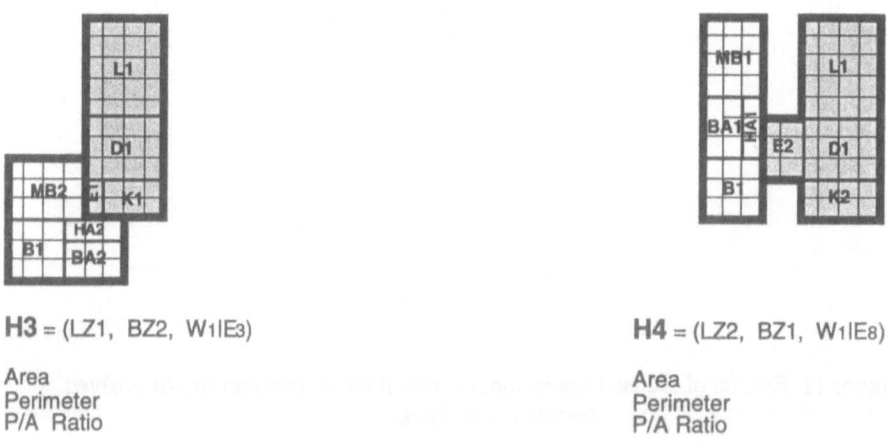

$H3 = (LZ1, BZ2, W_1|E_3)$

Area
Perimeter
P/A Ratio

$H4 = (LZ2, BZ1, W_1|E_8)$

Area
Perimeter
P/A Ratio

Figure 13. Evolution of house plans from rxisting house plans resulting from a new fitness function involving maximum perimeter to area ratio.

Note that a change in the fitness function at one level does not affect the fitness function at lower levels. Thus a change in the fitness function of the house design, does not necessarily involve a change in the fitness function of the zones and the rooms. If a change in the fitness function of zones is required, such as also maximizing perimeter to area ratio, the new populations of zones must be produced from the existing room populations before new house designs can be generated anew.

6. Summary

This paper has presented concepts for a general evolutionary approach to the generation of design solutions based on the growth of cells in a hierarchical organization. While the example presented is based on the generation of 2-D plans through the synthesis of a fundamental 2-D space unit, the approach can be generalized to the synthesis of any material cells. Although the example was based on orthogonal geometry, the method for growth is general for any polygonal geometry and may be extended to polyhedral geometry.

The main advantage of a hierarchical approach is that at each higher level the number of components making up the assembly at that level is reduced and genotypes are shorter. The fitness function relates only to the requirements for that component. It is argued that because a number of possibilities is considered, the effect of suboptimization is mitigated. In addition to reducing combinatorial problems, parallelism is supported.

As an alternative to mutation in the evolution of populations, the use of multiple runs with new randomly generated initial populations was used. This also has the advantage of allowing parallel processing.

While the various fitness functions at the different levels of the component hierarchy have involved optimization criteria, the goal of the approach is not optimization of these factors per se but rather their use as a driving force in the generation of satisfactory form. Thus the method produces form by 'growing' a set of fundamental units (cells) of material according to 'growing' rules whose sequence is directed by adapting to a given environment.

Further work will involve the inclusion of more realistic criteria and constraints as well as investigating the need for the recursive generation of new lower level components when no satisfactory assembly can be generated. Efficiency issues need to be investigated.

An area of further research is that of incorporating schema-based representations with an evolutionary approach for generation (Jo, 1993; Jo and Gero, 1995). A functional decomposition approach would be required in which distinct components to satisfy those functions are at least identified.

Acknowledgements

This work is partially supported by the Australian Research Council.

References

Beasley, D., Bull, D. R. and Martin R. R.: 1993, An overview of genetic algorithms: Part 1, fundamentals, *University Computing*, **15**(2), 58-69.

Coyne, R. D.: 1988, *Logic Models of Design*, Pitman, London.

Cramer, N. L.: 1985, A representation for the adaptive generation of simple sequential programs, *Proceedings of an International Conference on Genetic Algorithms and their Application*, pp. 183-187.

Dawkins, R.: 1986, *The Blind Watchmaker*, Penguin Books.

Gero, J. S., Louis, S. J. and Kundu, S.: 1994, Evolutionary learning of novel grammars for design improvement, *AIEDAM*, **8**(3), 83-94.

Goldberg, D. E.: 1989, *Genetic Algorithms in Search, Optimization, and Machine Learning*, Addison-Wesley, Reading, Mass.

Grefenstette, J. J. and Baker, J. E.: 1989, How genetic algorithms work; a critical look at implicit parallelism, *in* J. D. Schaffer (ed.), *Proceedings of the Third International Conference on Genetic Algorithms*, Morgan Kaufmann, San Mateo, CA, pp. 20-27.

Holland, J. H.: 1975, *Adaptation in Natural and Artificial Systems*, The University of Michigan Press, Ann Arbor.

Horn, J., Nafpliotis, N. and Goldberg, D. E.: 1994, A niched Pareto genetic algorithm for multiobjective optimization, *Proceedings of the First IEEE Conference on Evolutionary Computation (ICEC '94), Vol1*, IEEE World Congress on Computational Intelligence, Pistcataway, NJ: IEEE Service Center, pp. 82-87.

Jo, J. H.: 1993, *A Computational Design Process Model using a Genetic Evolution Approach*, PhD Thesis, Department of Architecural and Design Science, University of Sydney (unpublished).

Jo, J. H. and Gero, J. S.: 1995, A genetic search approach to space layout planning, *Architectural Science Review*, **38**(1), 37-46.

Koza, J. R.: 1992, *Genetic Programming: On the Programming of Computers by Means of Natural Selection*, MIT Press, Cambridge, Mass.

Rosca, J. P. and Ballard, D. H.: 1994, Hierarchical self-organization in genetic programming, *Proceedings of the Eleventh International Conference on Machine Learning*, Morgan Kaufmann, San Mateo, CA, pp. 252-258.

Rosenman, M. A.: 1995, An edge vector representation for the construction of 2-dimensional shapes, *Environment and Planning B: Planning and Design*, **22**, 191-212.

Rosenman, M. A. and Gero, J. S.: 1994, The what, the how, and the why in design, *Applied Artificial Intelligence*, **8**(2), 199-218.

Schnier, T. and Gero, J. S.: 1995, Learning representations for evolutionary computation, *in* X. Yao (ed.), *AI'95 Eighth Australian Joint Conference on Artificial Intelligence*, World Scientific, Singapore, pp. 387-394.

Simon, H. A.: 1969, *The Sciences of the Artificial*, MIT Press, Cambridge, Mass.

Stiny, G. and Mitchell, W.: 1978, The Palladian Grammar, *Environment and Planning B*, **5**, 5-18.

Stiny, G.: 1980, Introduction to shape and shape grammars, *Environment and Planning B*, **7**, 343-351.

Todd, S. and Latham, W.: 1992, *Evolutionary Art and Computers*, Academic Press, London.

Wilson, S. W. and Goldberg, D. E.: 1989, A critical review of classifier systems, *in* J. D. Schaffer (ed.), *Proceedings of the Third International Conference on Genetic Algorithms*, Morgan Kaufmann, San Mateo, CA, pp. 244-255.

Woodbury, R. F.: 1993, A genetic approach to creative design, *in* J. S. Gero and M. L. Maher (eds), *Modelling Creativity and Knowledge-Based Creative Design*, Lawrence Erlbaum, Hillsdale, NJ, pp. 211-232.

J. S. Gero and F. Sudweeks (eds), Artificial Intelligence in Design '96, 663-680.

EVOLUTIONARY LAYOUT DESIGN

WALTER HOWER
Department of Computer Science, University College Cork, NUI,
College Rd, Cork, Republic of Ireland

AND

MANFRED ROSENDAHL AND DERRICK KÖSTNER
Institut für Informatik, Fachbereich 4, Universität Koblenz-Landau,
Rheinau 1, D-56075 Koblenz, Federal Republic of Germany

Abstract. The present work treats the computation of heterogeneous layout configurations; distinct shapes as rectangles and triangles have to get placed into a target frame. In our design application the main restriction is the requirement that the objects must not overlap. Here, we further constrain the problem to obey the following requirement: The user shall be able to interact with the (semi-) automatic layout system in a way such that s/he may pick an object to place it in a subarea, offered by the system, in an arbitrary manner without the need to think about the placement of the other objects. (After an algorithm's termination a globally consistent layout should be guaranteed.) Thereby, the user still has degrees of freedom to finally arrange the objects. Such a realization shall also enhance the acceptance of the system by the user because entire solution classes, obtained by topological layout relations (instead of maintaining single co-ordinate points), are offered. The current work employs *evolutionary* computing techniques in order to get timely computations. This paper shows the use of interesting artificial intelligence techniques in the design area with close connections to a class of combinatorial problems in operational research with a wide range of applications in business and industry.

1. Introduction

The present work treats the computation of heterogeneous layout configurations; distinct shapes as rectangles and triangles have to get placed into a target frame. In the areas of artificial intelligence (AI) and computer-aided design (CAD) this problem can be formulated as a so-called constraint satisfaction problem (CSP). Hower (1995)[1] may serve as an initial entry point to the CSP realm.[2] In our design

---

[1] Hower and Jacobi (1994) presents a distributed realization. (Abel et al. (1995) illustrates the processing of spatial joins in distributed spatial database systems.)

[2] The complexity of backtracking-like algorithms are discussed in Zahn and Hower (1996).

application the main restriction is the requirement that the objects (CSP variables) must not overlap (which forms the constraints).[3] Here, we further constrain the problem to obey the following requirement:

> The user shall be able to interact with the (semi-) automatic layout system in a way such that s/he may pick an object to place it in a subarea, offered by the system, in an arbitrary manner without the need to think about the placement of the other objects. (After an algorithm's termination a globally consistent layout should be guaranteed.)

Due to the combinatorial search space we are facing because of the exponential number of combinations of the various possibilities of layout configurations we supersede our exhaustive *topological CSP* layout solver TOPCSP (Hower, 1996) incorporating a complete constraint satisfaction algorithm for a while and approach the problem in an incomplete heuristic manner.[4] The current work therefore employs *evolutionary* computing techniques in order to hope for timely computations. (Also in Dowsland and Dowsland (1992), where a review of packing problems[5] is given, such heuristics are identified to have promising potential.) We subsume by the term evolutionary techniques not only "genetic algorithms" and "evolution strategies"; we also include "simulated annealing" in our experiments to broaden the comparison.

The present paper is organized as follows: Section 2 introduces us to the terminology of evolutionary computing. Section 3 appreciates other relevant work. Section 4 describes our approach dealing with the layout problem. Section 5 concludes with final remarks.

2. Terminology

In a *genetic algorithm* (GA) the terms *chromosome, gene,* and *pool* got their inspiration by their relation to natural evolution. In real life often the best or mostly adapted individuals of a species will survive in the course of evolution. To transfer this to an algorithm a number of chromosomes are grouped into a pool. Each chromosome consists of the same definite number of genes often realized by a bit string. The quality of each chromosome, its *fitness,* is determined by a function of its bit string called the *fitness function.* A new set of chromosomes is generated from the pool by the application of mainly two operators: *crossover* and *mutation.* The crossover takes two chromosomes (called parents) and creates two new ones (called children) by swapping one or more of randomly chosen genes. The mutation is only applied to one chromosome and generates a child by tipping over one

---

[3] You may consult Hower and Graf (1995) which surveys the literature on constraint-based design.

[4] Thanks to Anna Thornton who has encouraged us in lively discussions after the CoPiCAD-94 workshop (Hower *et al.,* 1994) to try "adaptive" techniques.

[5] see also Hadjiconstantinou and Christofides (1995)

gene (or more genes, as used here), also randomly chosen. The fitness function then evaluates the fitness of the newly created chromosomes. The pool of the next generation is formed by only selecting the chromosomes with the best fitness of all individuals of the old generation in conjunction with the new generation.

The *simulated annealing* (SA) algorithm has been inspired by the physical annealing of solids. Nevertheless, there is a strong analogy to the GA approach; therefore, we subsume the SA approach under the label of evolutionary computing (EC). In its pure form one single bit string instead of a pool of chromosomes is maintained. Thus, mutation is the only operator. Similar to the GA a function that specifies a measure of goodness (sometimes called cost or objective function) of the bit string is employed. However, the most important difference to the GA is the fact that in one specific case the SA allows the adoption of a child with a worse cost. This decision is determined by the following "probability" $P := e^{-\frac{cost(child)-cost(parent)}{T}}$, where T is a normalization factor (traditionally called "temperature") that decreases in the course of the various generations. (Generally, a minimization is performed; i.e., a small cost value is better than a big one.) A generated random number R in the range from 0 to 1 is compared to P and only if $R < P$ then the probabilistic acceptance of adopting the worse child chromosome is achieved. Otherwise (if $R \geq P$) the parent is kept. In the following we will use the notions fitness (function), chromosome, and gene for the SA, too, in order to avoid the change of terminology in the text.

The *evolution strategy* (ES) can in some sense be described by a GA with (additional) specific features concerning the parameters.[6] For instance, each chromosome may contain so-called strategy variables which have an influence on the evaluation of chromosomes of further generations. This strategy variables will be changed by mutation and crossover as well as the other genes. This mechanism is called *self-adaptation*. Moreover, in the ES there are two different procedures of selecting the chromosomes for the next generation. In general, by the (μ, λ)-selection ("comma strategy") the best μ of the λ children become the parents of the new generation. The $(\mu + \lambda)$-variant ("plus strategy") also takes the parents into consideration as the GA does as well.

3. Relevant Work

3.1. BASIC MATERIAL

The article by Bremermann (1962) represents one of the earlier papers of illustrating some evolution ideas in the area of computer science. Davis (1987) collects various papers in the GA (mainly) and SA fields. Goldberg (1989) is one of the mostly cited GA books. Forrest (1993) also introduces into this computation philosophy. Beasley *et al.* (1993a; 1993b) treat the area in an even broader

---

[6]Normally no bit string representation is chosen; real values are preferred.

way. The book by Michalewica (1994) highlights GAs, too. The book by Aarts and Korst (1989) is one of the mostly cited books on the SA topic (mainly) and "Boltzmann Machines". Rutenbar (1989) presents a brief overview of the SA philosophy. Eglese (1990) represents a more thorough SA discussion. Nakakuki and Sadeh (1994) points to keep track of some information gained in the SA computation process (and thereby follows the idea of *tabu search* (TS)[7] a little bit). Dockx and Lutsko (1994) proposes a hybrid framework comprising both a genetic algorithm and a simulated annealing component. Leung *et al.* (1994) investigates both a GA and SA as well as a TS approach and favours the latter one. Kido *et al.* (1994) combines all these three EC techniques. Reeves (1993) collects material concerning (at least) these heuristics in a comfortable book format. Schwefel (1995) illustrates the ES approach of computation; Schwefel and Rudolph (1995) introduces further parameters where we are able to indicate separately both the number of the parent individuals and the number of reproduction iterations. De Jong and Spears (1993) briefly reviews EC procedures in general. Rechenberg (1994) and Fogel (1995) present the EC world in an exhaustive manner.

3.2. RELATED ARTICLES

In Stube (1993) and Hower (1993; 1996) only rectangles are treated while obeying "non-overlap" constraints.

In Tsang and Warwick (1990), Paredis (1993), Eiben *et al.* (1994), and Bowen and Dozier (1995) GA techniques are employed in CSPs. Minton *et al.* (1994) heuristically repairs inconsistent assignments (dealing with over-constrained CSPs) in a hill climbing manner—in some sense similar to the philosophy of local changes produced by the EC operations.

In Davis (1985) a GA application to bin packing is briefly sketched. Smith (1985) and Falkenauer and Delchambre (1992) use a GA in bin packing, too. Kröger (1995) also pursues a (parallel) GA approach to pack rectangles. Reeves (1995) is a comprehensive study of this topic, too. Chan *et al* (1991) instantiates such a procedure to perform module placement in VLSI. Tam (1992) works with a GA to deal with rectangular shapes in floor planning. Kado *et al.* (1995) proposes a hybrid architecture comprising a GA and a slicing tree structure component. Thornton (1994) compares GA and SA realizations to deal with mechanical design constraints and prefers the latter one. Kämpke (1988) uses SA in bin packing. Bolz and Wittur (1990) employs SA in order to position a line between two rectangles as well as to center a short text sentence inside a box. Cagan (1994) combines the idea of shape grammars with the SA technique to treat a geometrical knapsack problem (such that the pieces do not overlap). Lüders and Ernest (1995) approaches the problem how to automatically produce screen layouts via SA (regarding the task as a combinatorial optimization problem). Souilah (1995)

---

[7] see, for instance, Løkketangen (1995)

also uses SA; there, a non-overlapping design is called "free layout". Bland and Dawson (1991) uses a TS strategy in configuration design. Hower *et al.* (1995) mainly focusses on the EC treatment of rectangles and just briefly points to triangles.

4. The Approach

4.1. REPRESENTATION

Here, we focus on the treatment of rectangles and triangles with two edges being parallel to the co-ordinate axes. Therefore, we are able to classify the various possibilities of non-overlapping layout configurations into at most six cases[8] when relating two objects to each other. Every layout relation concerning two specific objects gets a (topological relation) number ($\in \{1, 2, 3, 4, 5, 6\}$). Furthermore, when we deal with n objects we have $n \cdot (n - 1)/2$ combinations to relate each of the n objects with the other ones in a pairwise manner; this number is the number of genes of each chromosome (and therefore its length). Thus, in our *topo*logical *EC* implementation ToPEc each of our chromosomes consists of strings of topological layout relations; every gene is provided with a relation number for each binary combination of objects. Thereby, the *fitness*, which may range from 1 to $n \cdot (n - 1)/2$, indicates how many object combinations can successfully be arranged.

4.2. ARCHITECTURE

In our realization there are three EC possibilities (ES, GA, SA) to solve a given problem. After choosing one of them there are several specifications that can be made.

4.2.1. *General Parameters*

One thing to declare is the *quality of mutation/crossover*. This means that we can specify a probability that indicates whether we change a chosen relation of the chromosome in work randomly, or whether we take advantage of the knowledge at which point in the string (proceeding from the front/top of a string) an inconsistency occurs in the intended layout configuration. At this point it is worthwhile to mention that only those layout relations are potential candidates to get selected by the EC operators which really allow a non-overlapping placement when just considering the two corresponding objects which are currently in the focus. (For instance, if two objects are too wide to get placed next to each other the relations *is left of / is right of* are already excluded from the set of possible layout relations by the program prior to the evolutionary run.) The genes of a chromosome are vis-

---

[8]four cases with respect to two rectangles, four, five, or six cases with respect to a rectangle/triangle in conjunction with a triangle (depending on the specific case)

ited from the top where the binary relation concerning object 1 and object 2 is located. The forthcoming genes with their layout configurations are then considered in addition to the possibilities we still have so far, and so on, until the final gene is processed which relates the objects with the numbers $n - 1$ and n. In our TOPEC realizations it is also possible to specify the number of interventions to a single chromosome or a pair of chromosomes per generation. A parameter *changes per generation* declares up to how many changes are made, in relation to the number of genes in the chromosome. After this, the *number of generations* of all chromosomes for one run can be edited. The next parameters handle more global issues. The *pool quality* determines a part of the pool creation from one run to another. If a chromosome achieved a good but not perfect fitness ($< n \cdot (n - 1)/2$) it is useful to partially integrate it into the pool of the following run, to give it a chance to become a solution after a few more generations. Furthermore, if one or more solutions are found in one run, the odds are good that there are further solutions nearby, i.e. they can perhaps be found with a few changes to the chromosome. So these chromosomes are undertaking a one-point mutation to join the pool of the following run. Now, the pool quality indicates which percentage of chromosomes ("near-miss" strings and modified solutions) should be adopted from one run to the next. (A heuristic chooses the most promising chromosome(s) of the last run.) The rest of the new pool is built in the same way as during the initialization phase described in 4.2.2. The *number of runs* can be stated as well. Additionally, the parameter *distinct solutions* can be activated (to save space) or not (to save time).

4.2.2. *Specific Features*

If we select the ES or GA a pool is created where we may choose whether we want to take a fixed pool size (of 20 for instance), or whether we favour the usage of a button to indicate some functional dependence (of the pool size) related to n, the number of objects. To avoid redundancies the pool creation is organized as follows: The first third of the genes of each chromosome is chosen systematically, the last two thirds randomly. This really cannot prevent that there are two identical chromosomes in the pool but in most cases it will do so. One parameter is the *relation of mutation/crossover*. This is specified by a probability whether during the run of the GA (or ES) the mutation or the crossover operator is used (to change the layout relations in one single chromosome or in a pair of chromosomes, respectively).

In our ES we mainly work with the same parameters as the GA does. Furthermore, at the moment we make no use of self-adaptation; however, it is planned to introduce it in the future. The important difference to the GA is the fact that in our present realization of the ES we support the comma strategy. Thereby, every time the new generation of children is preferred to the parents.

The SA only runs with one single chromosome; so, the *pool quality* indicates the probability of adopting the one chromosome or initiating a new one. (We also modified the SA to run with a pool and introduced the possibility of a crossover. However, tests showed that this is not advantageous; the results got even worse.)

4.3. ILLUSTRATION

4.3.1. *Rectangles*

Let us illustrate the approach by a small example. The target rectangle has the dimension 10×10, and the four rectangles to be placed into the target frame have the following measures of width (w) and height (h):

$$w_1 := 8, h_1 := 3; w_2 := 5, h_2 := 5; w_3 := 3, h_3 := 4; w_4 := 2, h_4 := 6.$$

Figure 1 shows the starting situation.

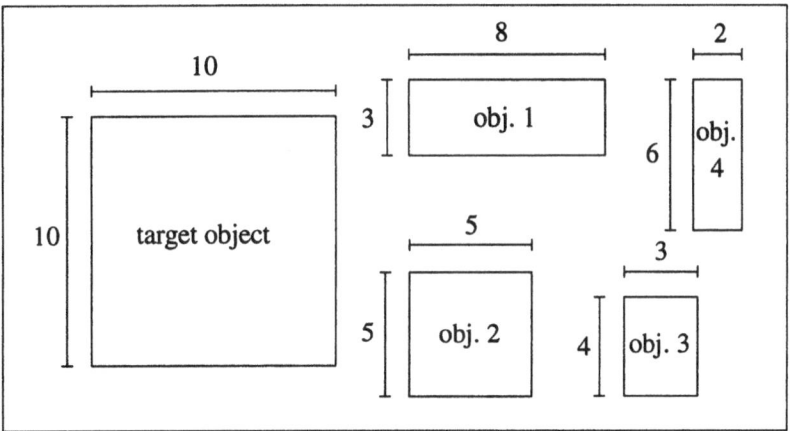

Figure 1. Initial situation of a placement of rectangles.

Here, we deal with four objects; so, we have six genes (layout relations) in each chromosome. Restricting this small example to rectangles there are only four possible relations for each of the six pairs of rectangles related to each other (*is above, is below, is right of,* and *is left of*) to avoid that they overlap. We now look at an arbitrarily chosen chromosome shown in Figure 2a.

The algorithm starts its work from the top of the string and first reaches the relation "object 1 *is above* object 2". (In our implementation this information will be projected to intervals of co-ordinates as presented elsewhere.) This relation and also the next one "object 1 *is above* object 3" can get fulfilled.

However, trying to obey the following relation "object 2 *is above* object 3" is leading to an inconsistency because the sum of the heights of the first three objects ($3 + 4 + 5 = 12$) is bigger than the target height (10). The fitness function of the EC will now provide this chromosome with a fitness of 2 because the first two

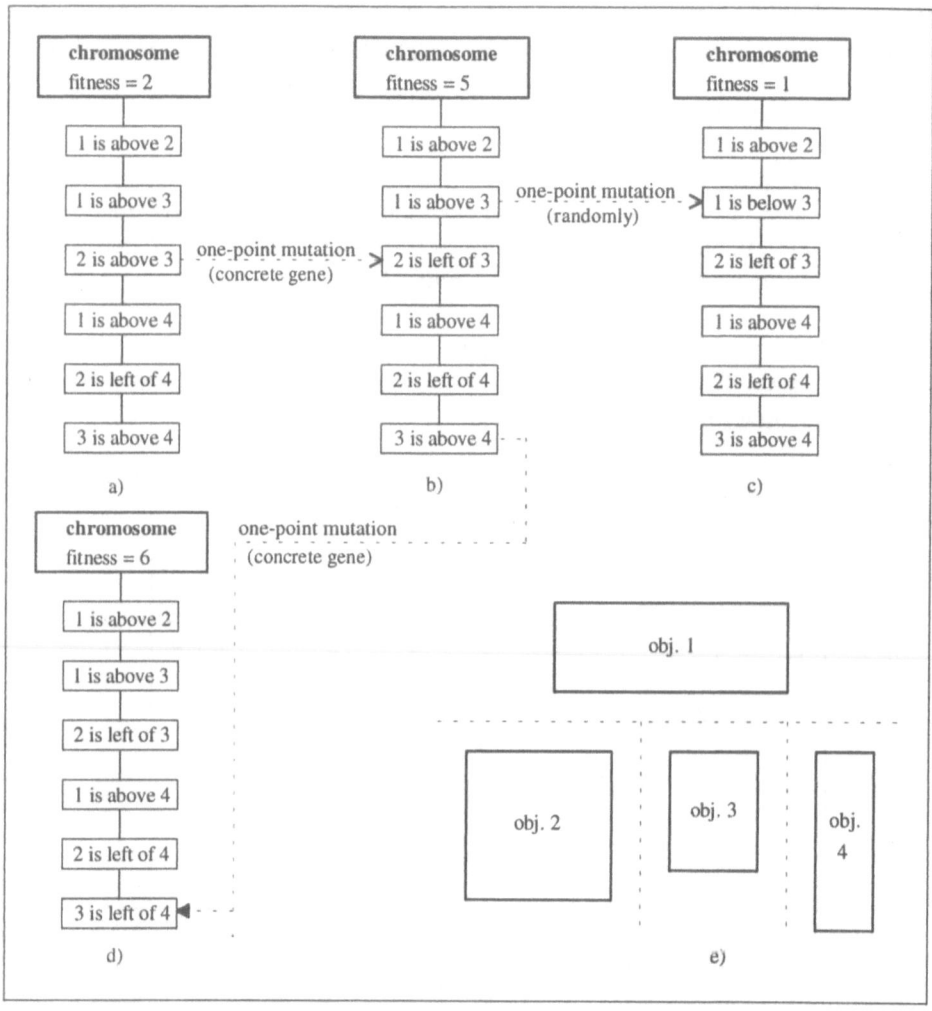

Figure 2. Snapshot of potential EC operations and a solution layout.

relations of it could be passed consistently. (Here, the order of the binary combinations of the objects in consideration affects the fitness value. However, due to the
fact that the numbering of the objects is in some sense arbitrary, and furthermore,
the layout relations get chosen in a random way, there is actually no real effect on
the approach.) Now a child of this chromosome (a new generation) will be created
(see Figure 2b). Here, let the EC operator be the mutation and let it work on one
concrete gene. Then in the child chromosome the third relation "object 2 *is above*
object 3" is changed to another possible binary relation because the evaluation
of the parent chromosome has stopped there (as mentioned above in the passage
w.r.t. *quality of mutation/crossover*). The new chromosome gets a fitness of 5 as

its evaluation passes the first five relations without complications. This time the sixth relation causes an inconsistency because the relations "object 1 *is above* object 3", "object 1 *is above* object 4", and "object 3 *is above* object 4" are leading to a height of 13 $(3 + 4 + 6 = 13)$ which exceeds the height of the target frame (10). Anyway, the parent is replaced by the child chromosome (in the GA and SA because of the better fitness and in the ES in any case because of the comma strategy in our present implementation), and the child becomes the new parent. To show the individual working method of the three strategies the next generation is created with a mutation at a random position in the new parent chromosome (see Figure 2c). The second relation is changed from "object 1 *is above* object 3" to "object 1 *is below* object 3" which causes an inconsistency for a similar reason as for the first chromosome. (It is not possible to put the first three objects on top of each other.) This leads to a fitness of 1 for the new child chromosome. This value is worse than the fitness of the parent. Now we take a separate look at the treatment of this situation by each of the three different EC techniques.

The GA always chooses the best fitness, i.e. in the example it keeps the parent chromosome for further processing. However, for the SA we have the following probability $P := e^{-\frac{cost(parent)-cost(child)}{T}}$. (In contrast to the original SA we work with a maximization of fitness instead of minimization. Thus, in this term the positions of *cost(child)* and *cost(parent)* are exchanged.) The comparison of P with the random number R decides whether the parent or the child will survive. Due to the comma strategy in our realization the ES always proceeds with the child chromosome no matter whether it is better or not. In the example here we go on as the GA would do and create a new generation based on the last parent chromosome (see Figure 2b). As in the first case let the EC operator be the mutation and let it work on one concrete gene; here the relation will be "object 3 *is above* object 4". In the new child chromosome this gene is mutated to "object 3 *is left of* object 4" (see Figure 2d) which leads to a fitness of 6; thereby, one solution class of the problem is obtained.[9] Figure 2e shows the corresponding final layout class.

Experimental Results.

Here we compare TOPEC with our previous implementation TOPCSP. Let us take the example prepared in Figure 3.

TOPCSP, which computes *all* layouts, works with intervals and relations between objects. In our example it computes 2,888 classes of globally consistent solutions for placing the objects.[10]

---

[9] We work on topological relations; hence, the solution consists of intervals of co-ordinates for each object.

[10] At first glance this number perhaps sounds very large because of the small target object with a dimension of 8 × 8 that leads to maximal 64 pairs of co-ordinates of the corresponding reference points to place one of the six objects (when considering just integer values). However, please remember, when we enlarge the edges of the target and placing objects with a factor of 1,000 (for

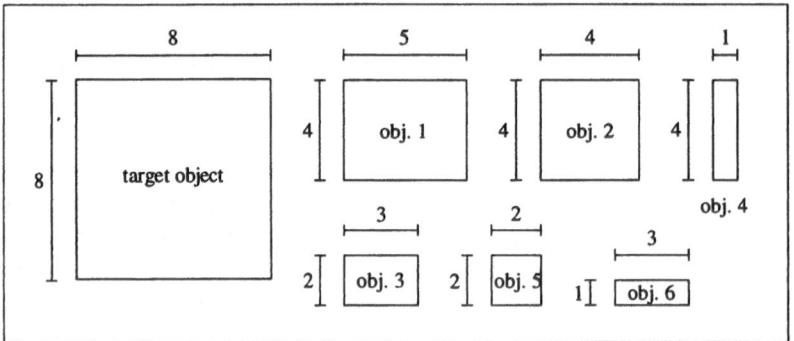

Figure 3. Rectangle set for the comparison of TopCsp and TopEc.

For our comparison we run the several EC methods with varying parameters on the example. The parameter *quality of mutation/crossover* is fixed for all runs by the relation 80%/20% which means that in 80% of the cases there is a definite place of change, and in 20% of the cases the place of change is chosen randomly; furthermore, also the *changes per generation* may be predefined by "up to 20%". Table 1 shows the results of our test runs. It has to be read as follows: For instance, row 1 means that, in order to get 10% of all solutions, a GA with the indicated parameters takes 4% of the time TopCsp has used to derive all solutions. One important point that does not appear in the table is the computation of the first solution. In every tested configuration the first solution was computed so quickly that it was not possible to determine the run time—a welcome aspect of the current implementation.

Explanation of the header of Tables 1 and 2:

example: numbering of the example
EC: choice of the specific evolutionary computing technique
mu/cr: relation of *mu*tation/*cr*ossover
pq: *p*ool *q*uality
ps: *p*ool *s*ize
ng: *n*umber of *g*enerations
10%, 20%, 30%: 10%, 20%, or 30% (resp.) of all (TopCsp) solutions
(which are caught by the corresponding EC technique in the time fragment indicated in the numbered row related to the TopCsp time).

Now, let us go into the details of Table 1. The first point to mention is that it seems as if the crossover operator is of no real use to the layout problem. When

instance) there are 64 million pairs of co-ordinates but TopCsp still needs the same space and time to compute still 2,888 solution classes really covering all layout possibilities.

example	EC	mu/cr	pq	ps	ng	10%	20%	30%
1	GA	50/50	75%	20	20	4%	17%	84%
2	GA	95/5	75%	20	20	2%	10%	52%
3	GA	95/5	35%	20	20	2%	9%	41%
4	SA	100/0	75%	1	20	3%	9%	40%
5	SA	100/0	35%	1	20	4%	11%	49%
6	SA	100/0	75%	1	40	3%	8%	34%
7	ES	50/50	75%	20	20	4%	19%	102%
8	ES	95/5	75%	20	20	3%	13%	62%

TABLE 1. EC runs on rectangles based on topological relations.

we look at the rows 1 and 2 where only the parameter *relation of mutation/crossover* is changed it becomes obvious that using the crossover is a waste of time. The same can be observed in rows 7 and 8. To get a further feeling of the various parameters let us have a look upon row 3 where in the GA the parameter *pool quality* is reduced to 35% leading to better results (compared to row 2). (Please remember that a pool quality of 75% indicates that three-quarters of the chromosomes of the previous run will be adopted into the pool of the next run. Therefore, during the course of the runs several chromosomes may exist in a multiplicity of generations, compared to one single run.)[11] So when in row 3 the reduction of the pool quality leads to better results then we may derive that the chromosomes take no advantage of more than 20 generations. By way of contrast, in the SA shown in the rows 4, 5, and 6, the reduction of the *pool quality* (transition from row 4 to row 5) yields worse results. Thus, in row 6 we again take the parameters of row 4 with the exception of the number of the generations; now, we allow 40 (instead of 20) generations per run which leads to even better results. The reason for this behaviour could be the fact that the SA can escape from local optima on its way to find further solutions. Thereby the SA is able to compute more different solutions in the course of time and outperforms the GA in this respect. As mentioned above we still do not use the power of self-adaptation of the ES which probably causes the corresponding results to be the worst of the three techniques. (For instance, in the example row 7 the ES takes longer to get only 30% of all solutions than the constraint solver needs to compute all the layout possibilities in its entirety.)

---

[11] By a pool quality of 100% every chromosome would be adopted into a new run being equivalent to an enlargement of the number of generations during one run.

On the average we observe the following results: In 3% of the time which is needed by the exhaustive constraint solver TOPCSP to catch all solutions we already get 10% of the solutions via TOPEC; in 12% of the time we obtain 20% of the solutions. However, to reach 30% of the solutions we need 58% of the time required by constraint satisfaction.

We therefore may conclude that the evolutionary techniques presented here represent a promising framework for the task to catch one solution or a few of numerous solutions quickly. However, when we are really interested in the complete computation of the entire solution space of all layout possibilities the (incomplete) heuristics put up a bad performance; as we know, these techniques cannot guarantee to catch the various layout solutions at all.

Remarks.

Our experiments show that the choice of the appropriate approach to the packing and layout configuration problem of placing the objects actually depends on the demand on the degree of the completeness of the solution space. A further aspect which should be taken into consideration is the specific use of such an automatic layout system: In the case of an inconsistency the complete constraint solver may detect the non-possibility of a layout very early whereas the incomplete heuristics do not provide the user with the information that there is no layout possibility at all. (For instance, when we enlarge the problem introduced in Figure 3 by a working rectangle with the dimension 4×3 TOPCSP quickly reports the inconsistency whereas TOPEC still test numerous layouts—without giving the definite answer that it is not possible to get just a single layout.) Thus, a parallel architecture would be conceivable where a complete approach (in regard to the inconsistency detection) is used as well as an incomplete one in order to get the answer whether (or not) we have an inconsistent problem as quickly as possible.

4.3.2. *Triangles*

Now it is time to mention that all the techniques also work on heterogeneous layout problems comprising rectangles and triangles where two edges are parallel to the co-ordinate axes. Figure 4 presents a screenshot after an exemplary program run to show the actual status of the work.

In order to compare TOPEC to TOPCSP in regard to a problem containing triangles let us consider the example problem prepared in Figure 5.

Table 2 shows the result concerning the GA. Table 3 shows the result concerning the SA.
Explanation of the header of Table 3:

example: numbering of the example
EC: choice of the specific evolutionary computing technique
mu_t: number of *mutations* per *temperature*

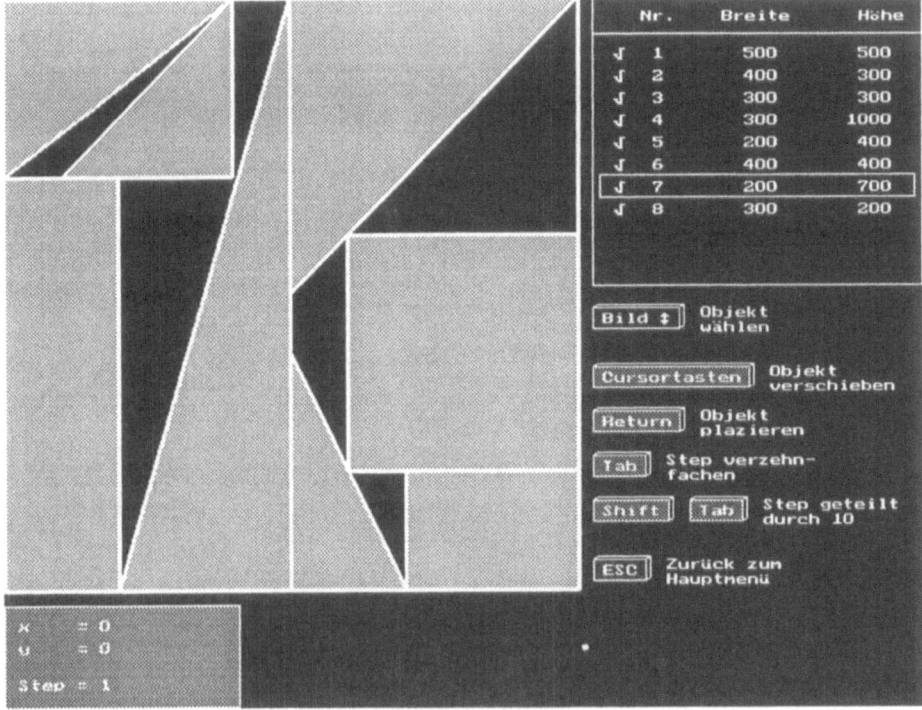

Figure 4. Screenshot of a heterogeneous layout.

example	EC	mu/cr	pq	ps	ng	10%	20%	30%
1	GA	50/50	75%	20	20	14%	30%	87%
2	GA	95/5	75%	20	20	12%	29%	69%
3	GA	95/5	35%	20	20	9%	29%	61%

TABLE 2. TOPEC(GA) compared to TOPCSP on triangles.

add/mul: *add*itive or *mul*tiplicative temperature reduction parameter
$reduce_t$: degree to *reduce* the *t*emperature
        (additive: $t := t + reduce_t$, multiplicative: $t := t * reduce_t$)
10%, 20%, 30%: 10%, 20%, or 30% (resp.) of all (TOPCSP) solutions
(which are caught by the corresponding EC technique in the time fragment indicated in
the numbered row related to the TOPCSP time).

Incidentally, we obtain even worse results than the ones presented in 4.3.1

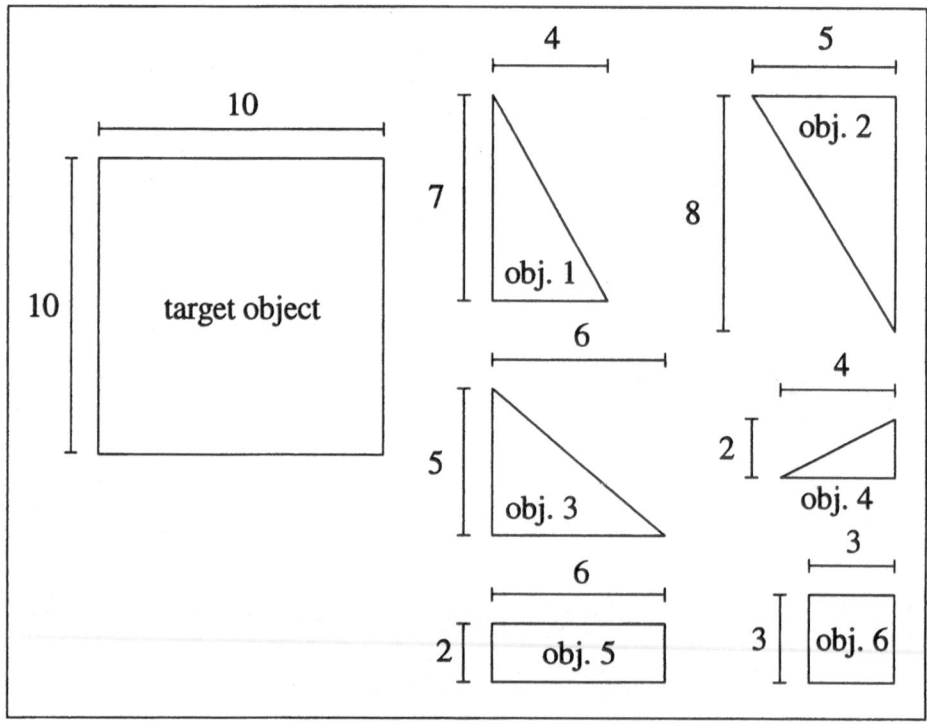

Figure 5. Placement problem comprising rectangles and triangles.

example	EC	mu_t	add/mul	$reduce_t$	10%	20%	30%
1	SA	30	add	-0.01	10%	27%	58%
2	SA	50	add	-0.01	11%	40%	94%
3	SA	30	add	-0.03	12%	38%	139%
4	SA	30	mul	0.99	10%	19%	55%
5	SA	50	mul	0.99	11%	24%	77%
6	SA	30	mul	0.97	11%	32%	82%

TABLE 3. TOPEC(SA) compared to TOPCSP on triangles.

which you may consult for a further discussion. (For instance, in the example row 3 of Table 3 SA takes longer to get only 30% of all solutions than the constraint solver needs to compute all the layout possibilities in its entirety.)

Figure 6 shows the screenshot of a solution layout.

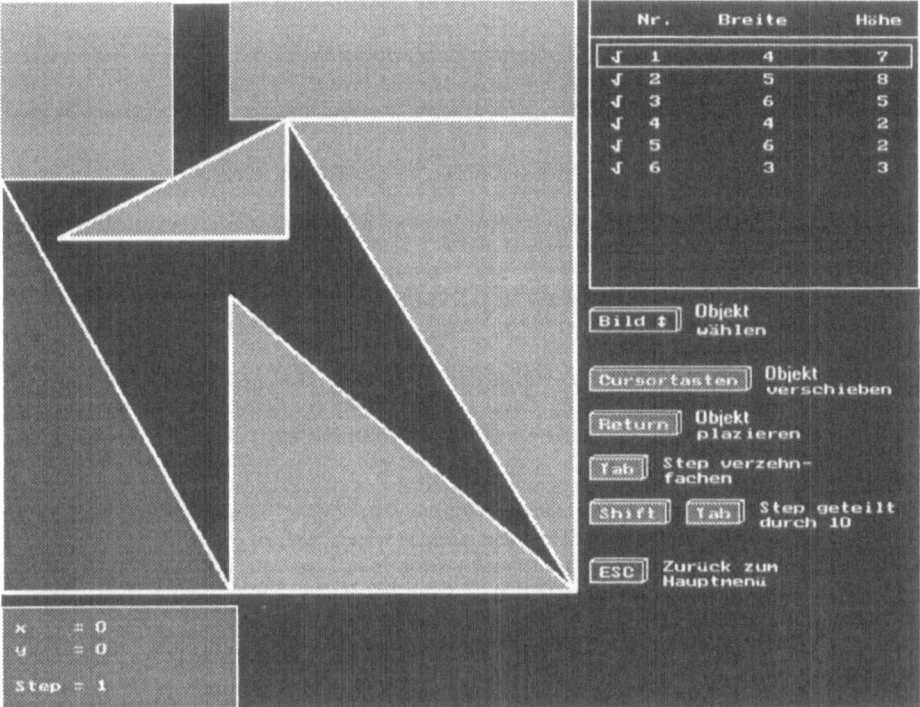

Figure 6. Screenshot belonging to Figure 5.

5. Final Remarks

In our computer-aided layout design system the user is able to prefer placing certain objects in a desired area. This feature is obtained by the fact that after the algorithms' termination a globally consistent layout is guaranteed. Thereby, our system reflects a semi-automatic layout engine where for each object to be placed a subarea is offered to the user in which s/he still has degrees of freedom to finally arrange the objects. Such a realization also enhances the acceptance of the system by the user because entire solution classes, obtained by topological layout relations (instead of maintaining single co-ordinate points), are offered.[12] We have seen that the use of an exhaustive layout solver is not necessary when only a few placements suffice, and it is out of the question when a large number of objects (along with the inherent combinatorial explosion of the various layout possibilities) have to be placed. Then, heuristics like evolutionary procedures as described here shall be preferred. To conclude, this paper shows the use of interesting techniques of AI in Design with close connections to a class of combinatorial problems in operational research with a wide range of applications in business and industry.

---

[12]Grigni *et al.* (1995) favours topological inference, too.

References

Aarts, E. and Korst, J.: 1989, *Simulated Annealing and Boltzmann Machines*. Wiley—Interscience Series in Discrete Mathematics and Optimization. John Wiley, Chichester, England.

Abel, D. J., Ooi, B. C., Tan, X-L., Power, R. and Yu, J. X.: 1995, Spatial join strategies in distributed spatial DBMS, *in* M. J. Egenhofer and J. R. Herring (eds), *Advances in Spatial Databases*, Vol. 951, *Lecture Notes in Computer Science*, Proceedings, Springer-Verlag, Berlin/Heidelberg, pp. 348–367.

Beasley, D., Bull, D. R. and Martin, R. R.: 1993, An overview of genetic algorithms: Part 1, Fundamentals, *University Computing*, 15(2), 58–69.

Beasley, D., Bull, D. R. and Martin, R. R.: 1993, An overview of genetic algorithms: Part 2, Research Topics. *University Computing*, 15(4), 170–181.

Bland, J. A. and Dawson, G. P.: 1991, Tabu search and design optimization, *Computer-Aided Design*, 23(3), 195–201.

Bolz, D. and Wittur, K.: 1990, Die Umsetzung deklarativer Beschreibungen von Graphiken durch Simulated Annealing, *in* K. Kansy and P. Wißkirchen (eds), *Graphik und KI*, Vol. 239 of *Informatik-Fachberichte*, pp. 68–77. Proceedings, Springer-Verlag, GI-Fachgespräch, Königswinter, 3./4. April.

Bowen, J. and Dozier, G.: 1995, Solving constraint satisfaction problems using a genetic/systematic search hybrid that realizes when to quit, *The 6th International Conference on Genetic Algorithms (ICGA-95)*, University of Pittsburgh, Pennsylvania, USA, July 15–19.

Bremermann, H. J.: 1962, Optimization through evolution and recombination, *in* M. C. Yovits, G. T. Jacobi and G. D. Goldstein (eds), *Self-organizing systems 1962*, Spartan Books, Washington, D.C., pp. 93–106.

Cagan, J.: 1994, Shape annealing solution to the constrained geometric knapsack problem, *Computer-Aided Design*, 26(10), 763–770.

Chan, H., Mazumder, P. and Shahookar, K.: 1991, Macro-cell and module placement by genetic adaptive search with bitmap-represented chromosome, *INTEGRATION, the VLSI journal*, 12, 49–77.

Davis, L.: 1985, Applying adaptive algorithms to epistatic domains, *in* A. Joshi (ed.), *IJCAI 85*, Proceedings of the Ninth International Joint Conference on Artificial Intelligence, Vol. 1, Distributed by Morgan Kaufmann Publishers, Inc., Los Altos, California, pp. 162–164.

Davis, L. (ed.): 1987, *Genetic Algorithms and Simulated Annealing*, Research Notes in Artificial Intelligence, Pitman, London/Morgan Kaufmann, Los Altos, California.

De Jong, K. and Spears, W.: 1993, On the state of evolutionary computation, *in* S. Forrest (ed.), *ICGA-93*, Proceedings of the Fifth International Conference on Genetic Algorithms, Morgan Kaufmann, San Mateo, California, pp 618–623.

Dockx, K. and Lutsko, J. F.: 1994, SA/GA: Survival of the fittest in Alaska. *in* P. Cheeseman and R. W. Oldford (eds), *Selecting Models from Data: AI and Statistics IV*, Springer-Verlag, pp. 463–469. Fourth International Workshop on Artificial Intelligence and Statistics, Ft. Lauderdale, Florida, USA, 1993.

Dowsland, K. A. and Dowsland, W. B.: 1992, Packing problems, *European Journal of Operational Research*, 56, 2–14.

Eglese, R. W.: 1990, Simulated annealing: A tool for operational research, *European Journal of Operational Research*, 46(3), 271–281.

Eiben, A. E., Raué, P-E. and Ruttkay, Zs.: 1994, Heuristic genetic algorithms for constrained problems. Part II: Empirical Results, Rapportnr. IR-351, Artificial Intelligence Group, Faculteit der Wiskunde en Informatica, Vrije Universiteit Amsterdam, The Netherlands.

Falkenauer, E. and Delchambre, A.: 1992, A genetic algorithm for bin packing and line balancing, *Proceedings of the 1992 IEEE International Conference on Robotics and Automation*, Nice, France, pp. 1186-1192.

Fogel, D. B.: 1995, *Evolutionary Computation: Toward a New Philosophy of Machine Intelligence*. IEEE Press, Piscataway, NJ.

Forrest, S., 1993. Genetic algorithms: Principles of natural selection applied to computation, *Science*, 261, 872–878.

Goldberg, D. E.: 1989, *Genetic Algorithms in Search, Optimization, and Machine Learning*, Addison-Wesley.

Grigni, M., Papadias, D. and Papadimitriou, C.: 1995, Topological inference, *in* C. S. Mellish (ed.), *IJCAI-95, Proceedings of the Fourteenth International Joint Conference on Artificial Intelligence*, Vol. 1, distributed by Morgan Kaufmann, San Mateo, California, pp. 901–906.

Hadjiconstantinou, E. and Christofides, N.: 1995, An exact algorithm for general, orthogonal, two-dimensional knapsack problems, *European Journal of Operational Research*, **83**(1), 39–56.

Hower, W. and Graf, W. H.: 1995, Research in constraint-based layout, visualization, CAD, and related topics: A bibliographical survey, *International Workshop on Constraints for Graphics and Visualization (CGV '95)*, Cassis, France, September 18, DFKI Research Report RR-95-12, Deutsches Forschungszentrum für Künstliche Intelligenz GmbH, Saarbrücken.

Hower, W. and Jacobi, S.: 1994, A distributed realization for constraint satisfaction, *in* H. Kitano et al. (eds), *Parallel Processing for Artificial Intelligence 2*, Vol. 15, *Machine Intelligence and Pattern Recognition series*, Elsevier Science, Amsterdam, The Netherlands, pp. 107–116.

Hower, W., Rosendahl, M. and Berling, R.: 1993, Constraint processing in human-computer interaction with an emphasis on intelligent CAD, *in* M. J. Smith and G. Salvendy (eds), *Human-Computer Interaction: Applications and Case Studies*, Vol. 19A, *Advances in Human Factors / Ergonomics*, Elsevier Science, Amsterdam, The Netherlands, pp. 243–248. Proceedings of the Fifth International Conference on Human-Computer Interaction (HCI International '93, Orlando, Florida, USA), Volume 1.

Hower, W., Haroud, D. and Ruttkay, Z. (eds): 1994, *Constraint Processing in Computer-Aided Design (CoPiCAD-94)*, Workshop Notes, Third International Conference on Artificial Intelligence in Design (AID'94), Swiss Federal Institute of Technology, Lausanne, Switzerland, 15–18 August 1994.

Hower, W., Köstner, D. and Rosendahl, M.: 1995, Computer-aided layout by evolutionary computing, *in* Veltkamp, R. C. and Blake, E. H. (eds), *Proceedings of the Fifth Eurographics workshop on Programming Paradigms in Graphics, EUROGRAPHICS '95*, CWI, Amsterdam, The Netherlands, pp. 251–269.

Hower, W.: 1995, Constraint satisfaction — Algorithms and complexity analysis, *Information Processing Letters*, **55**(3), 171–178.

Hower, W.: 1996, Bottom-up layout generation, *Informatica*, **20**(1).

Kado, K., Ross, P. and Corne, D.: 1995, A study of genetic algorithm hybrids for facility layout problems, *The 6th International Conference on Genetic Algorithms (ICGA-95)*, University of Pittsburgh, Pennsylvania, July 15-19.

Kämpke, T.: 1988, Simulated annealing: Use of a new tool in bin packing, *Annals of Operations Research*, **16**, 327–332.

Kido, T., Takagi, K. and Nakanishi, M.: 1994, Analysis and comparisons of genetic algorithm, simulated annealing, TABU search, and evolutionary combination algorithm, *Informatica*, Special Issue on Artificial Life, **18**(4), 399–410.

Kröger, B.: 1995, Guillotineable bin packing: A genetic approach, *European Journal of Operational Research*, Special Issue: Cutting and Packing, **84**(3), 645–661.

Leung, W., Sheung, J., Chan, H. W. and Chan, M.: 1994, An empirical study on search algorithms that utilize randomness, *PRICAI'94, Proceedings of the Third Pacific Rim International Conference on Artificial Intelligence*, International Academic Publishers, Beijing, PR China, pp. 64–69.

Løkketangen, A.: 1995, Tabu search—Using the search experience to guide the search process. An introduction with examples, *AI Communications*, **8**(2), 78–85.

Lüders, P. and Ernst, R.: 1995, Das Automatisierte Bildschirmlayout — Ein Kombinatorisches Optimierungsproblem? *Informatik Forschung und Entwicklung*, **10**(1), 1–13.

Michalewicz, Z.: 1994, *Genetic Algorithms + Data Structures = Evolution Programs*, Artificial Intelligence, Springer-Verlag, Berlin/Heidelberg, second, extended edition.

Minton, S., Johnston, M. D., Philips, A. B. and Laird, P.: 1994, Minimizing conflicts: A heuristic repair method for constraint satisfaction and scheduling problems, *in* E. C. Freuder and A. K. Mackworth (eds), *Constraint-Based Reasoning*, Special Issues of *Artificial Intelligence*, A Bradford Book, The MIT Press, Cambridge, Massachusetts, pp.161–205.

Nakakuki, Y. and Sadeh, N.: 1994, Increasing the efficiency of simulated annealing search by learning to recognize (un)promising runs, *Proceedings AAAI-94, Twelfth National Conference on Artificial Intelligence*, AAAI Press, Menlo Park, CA.

Paredis, J.: 1993, Genetic state-space search for constrained optimization problems, *in* R. Bajcsy (ed.), *IJCAI-93, Proceedings of the Thirteenth International Joint Conference on Artificial Intelligence*, Chambéry, Savoie, France, August 28 – September 3, 1993. IJCAII. Vol. 2, distributed by Morgan Kaufmann, San Mateo, California, pp 967–972.

Rechenberg, I.: 1994, *Evolutionsstrategie '94*, volume 1 of *Werkstatt Bionik und Evolutionstechnik*, frommann-holzboog, Stuttgart.

Reeves, C. R. (ed.): 1993, *Modern Heuristic Techniques for Combinatorial Problems*, Advanced topics in computer science, Blackwell Scientific, Oxford.

Reeves, C. R.; 1995, Hybrid genetic algorithms for bin-packing and related problems, *Annals of Operations Research*, J.C. Baltzer A.G. Scientific Publishing Company.

Rutenbar, R. A.: 1989, Simulated annealing algorithms: An overview, *IEEE Circuits and Devices Magazine*, pp. 19–26.

Schwefel, H-P. and Rudolph, G.: 1995, Contemporary evolution strategies, *in* F. Morán, A. Morena, J. J. Merelo and Chacón, P. (eds), *Advances in Artificial Life*, Vol. 929, *Lecture Notes in Artificial Intelligence, Subseries of Lecture Notes in Computer Science*, Proceedings of the Third European Conference on Artificial Life, Springer-Verlag, Berlin/Heidelberg, pp. 893–907.

Schwefel, H.-P.: 1995, *Evolution and Optimum Seeking*, Sixth-Generation Computer Technology Series, John Wiley, New York, NY.

Smith, D.: 1985, Bin packing with adaptive search, *in* J. J. Grefenstette (ed.), *Proceedings of an International Conference on Genetic Algorithms and their Applications*, Lawrence Erlbaum Associates, Hillsdale, NJ, pp. 202–207.

Souilah, A.: 1995, Simulated annealing for manufacturing systems layout design, *European Journal of Operational Research*, **82**(3), 592–614.

Stube, B.: 1993, Ein konstruktives Verfahren zur Plazierung allgemeiner Strukturen unter Berücksichtigung von Constraints, *in* A. Iwainsky (ed.), *Computergrafik und automatisierte Layoutsynthese*, Wartburg, 11.–13. Oktoberr, GI 3. Workshop, pp. 71–81.

Tam, K. Y.: 1992, Genetic algorithms, function optimization, and facility layout design, *European Journal of Operational Research*, **63**(2), 322–346.

Thornton, A. C.: 1994, Genetic algorithms versus simulated annealing: satisfaction of large sets of algebraic mechanical design constraints, *in* J. S. Gero and F. Sudweeks (eds), *Artificial Intelligence in Design '94*, Kluwer Academic Publishers, Dordrecht, The Netherlands, pp. 381–398.

Tsang, E. P. K. and Warwick, T.: 1990, Applying genetic algorithms to constraint satisfaction optimization problems, *in* L. C. Aiello (ed.), *ECAI 90, Proceedings of the 9th European Conference on Artificial Intelligence*, Pitman Publishing, London, England, pp. 649–654.

Zahn, M. and Hower, W.: 1996, Backtracking along with constraint processing and their time complexities, *Journal of Experimental and Theoretical Artificial Intelligence*, **8**(1).

J. S. Gero and F. Sudweeks (eds), Artificial Intelligence in Design '96, 681-699.
© 1996 *Kluwer Academic Publishers.*

DOM-ARCADE: ASSISTANCE SERVICES FOR CONSTRUCTION, EVALUATION, AND ADAPTATION OF DESIGN LAYOUTS

SHIRIN BAKHTARI AND BRIGITTE BARTSCH-SPÖRL

BSR Consulting GmbH
Wirtstrasse 38,
D-81539 Munich, Germany

AND

WOLFGANG OERTEL

Technical University of Dresden
Department of Artificial Intelligence,
D-01062 Dresden, Germany

Abstract. This paper summarizes some aspects of the process and the result of our research and development activities[1] aimed at building an active assistance system for designing the technical installation systems for highly complex buildings. The emphasis in the development of DOM-ARCADE is set on the conceptual identification and computational realization of the considered domain ontology. The key issue in modelling the domain ontology is to make a symbiosis of two kinds of knowledge: the deep domain knowledge and the decision making knowledge. The deep knowledge provides the ability of a semantic interpretation of the syntactically specified design layouts. The design decision making knowledge is represented as a set of design specialists (critics) that are grouped under the ARCADE part of the system.

The DOM-ARCADE makes extensive use of its deep design knowledge and supports designers by suggestions how to assess the quality and reliability of their designed layouts, how to ensure the integrity of their proposed layouts within the current project of interest, how to overcome deficiencies in the layout, and how to construct a layout from scratch that fits the frame of reference.

---

[1]This research was supported by the Federal Ministry of Education, Science, Research and Technology (BMBF) within the joint project FABEL under contract no. 01IW104. Project partners in FABEL are GMD – German National Research Center for Information Technology, Sankt Augustin, BSR Consulting GmbH, München, Technical University of Dresden, HTWK Leipzig, University of Freiburg, and University of Karlsruhe.

1. Introduction

This paper summarizes some aspects of the process and the result of our research and development activities aimed at building an active assistance system for designing the technical installation systems like e.g. air conditioning, sewage, etc. considering the designed shape for highly complex buildings, e.g. office buildings.

The DOM-ARCADE system is designed to undertake the role of an active design assistant and makes extensive use of its deep design knowledge and decision making knowledge in order to support designers by suggestions how to assess the quality and reliability of their designed layouts, how to ensure the integrity of their proposed layouts within the current project of interest, how to overcome deficiencies in the layout, and how to construct a layout from scratch that fits the frame of reference.

Our research and development program aimed at building an active design assistance system was two-pronged: As computer scientists we studied first the current design practice with the goal to spell out what are the specific characteristics of the design domain and what functionalities should comprise a design assistance system in order to make a qualitative leap forward with regard to the actual design practice. Secondly, in response to the domain specific requirements we were considered with the development of a new integration concept that can meet the articulated requirements.

An essential characteristic of most complex real-world applications, such as design, is that the activities are almost carried out in a multiple discipline collaborative framework using a common and shared knowledge platform. Within such a collaborative framework the domain knowledge shared among different designers and engineers – having different foci of interest on the domain knowledge – is referred to as an ontology. There are notable research and development investigations in this field, such as the ARPA funded Knowledge Sharing and Reuse Effort at Stanford (Neches et al., 1991), the YMIR ontology modelling (Alberts et al., 1994) that is proposed by the group at Twente University for design applications, and the ROOMMOVE system proposed in Branki et al. (1994).

The emphasis in the development of DOM-ARCADE is set on the identification and computational representation of the domain ontology. It is notable that for our purpose, the key issue in modelling the design ontology is to make a symbiosis of two kinds of design knowledge: the deep design knowledge and the decision making knowledge.

The representation of the deep design knowledge provides the ability of a semantic interpretation of the syntactically specified design layouts. It enables to review a proposed layout and determine the meaning and permissible use of the involved design objects and their geometric-topological relations. The decision making knowledge, on the other side, is represented as a set of different design specialists that operate on the basis of the deep design knowledge. We group the design specialists by function and their scope of application in different service units. These service units build an ARCADE along the edge of the DOM which is viewed as the heart of the system. The following design specialists comprise the support functionality of our system:

- DOM-A assesses the quality and the reliability of the proposed layouts as well as their integrity within a certain project of interest.
- DOM-R rectifies the proposed layout in terms of shortcomings and errors of omission. In case there are shortcomings that can not get rectified, the design specialist may also remote and reuse similar layouts from a layout-management-base that can cover them.
- DOM-C supports the completion of a partially specified layout. It also assists problem solving from first principles.
- DOM-AD effects adaptation. It provides suggestions how to make a not yet satisfactory layout suitable to meet the requirements of the target task.
- DOM-E undertakes the role of an authority that monitors and evaluates the non-linear process of adaptation.

The objectives of this paper are the following: In section 2 we discuss some significant issues central to our approach to building active design assistance systems. Then we answer the following questions: How is the knowledge in DOM represented and how are the design specialists specified and realized. A scenario explains further how the DOM-ARCADE is used. The technical underpinnings are explained in section 6. We conclude by a summary of what is gained.

2. Recounting the DOM-ARCADE Development Cascade

The purpose of recounting the incremental development of a set of functionalities is to share the idea behind and the history of our stepwise refinement of the conceptualization and of the cascade of system developments that resulted in DOM-ARCADE[2].

In the beginning of the FABEL project (Fabel , 1993), we studied our application domain which is designing the technical installation for highly complex buildings, e.g. office buildings. The first task that had been under our consideration was to figure out what are the specific characteristics of the considered domain and what are the requirements in terms of needed, demanded, and desired support services. The main interests of the architects and engineers were in support for tedious and error-prone tasks as well as for problems concerning the integrity of the rather insular designed layouts that have to be integrated into a workable final state of the design (Bakhtari *et al.*, 1994). A summary of the results of our knowledge engineering process together with suggestions how to meet these requirements is given in Bartsch-Spörl *et al.* (1996).

---

[2]The technical underpinnings are described in section 6.

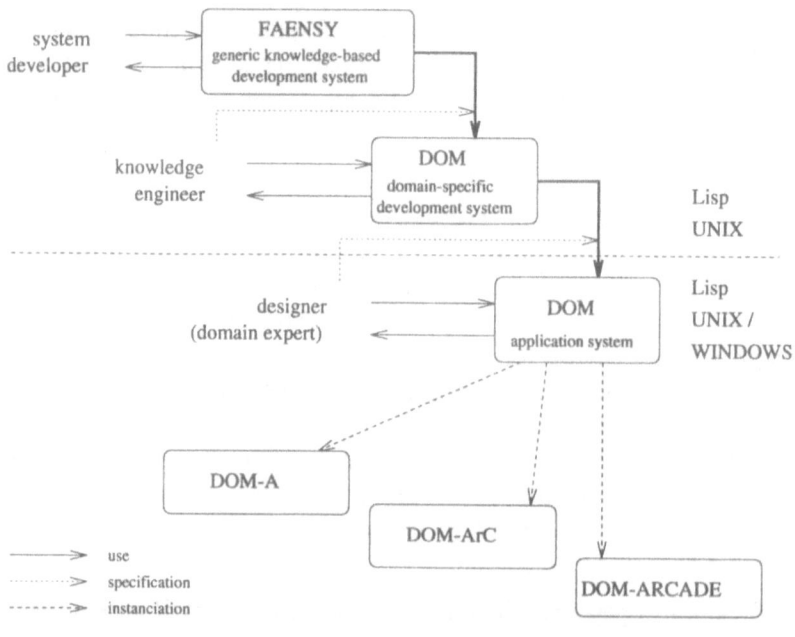

Figure 1. The DOM-ARCADE system development path.

Subsequently, we turned to studying the scientific literature about design principles and processes and applying AI methods for design tasks and more specific for building design. We investigated the topic of problem classes and adequate AI methods for design. The essential insight is that we have to deal mainly with both routine and innovative design tasks (Bakhtari *et al.*, 1994) and to start with a rather new integration approach that bundles together a mixture of non-AI and AI-methods with the main goal to offer design support functions that come in small units and can be used on a voluntary basis (Bartsch-Spörl, 1995). Such an approach has the advantage that it does not force the designers to follow a rigid flow of work. It also does not put pressure on the system developers to produce a functionally complete system from the very beginning. Instead, it allows to start with e.g. a useful set of basic functions and to deliver sophisticated further functionalities at a later stage.

The DOM-ARCADE application system – as shown in Figure 1 – is the result of a set of prototypes and system developments, starting with the generic knowledge-based development system, FAENSY (Oertel, 1994) which has been implemented in Allegro Common Lisp and runs on UNIX workstations with a user interface based on Tcl/Tk (Ousterhout, 1993). The specification of the deep design knowledge was the first step towards the domain ontology modelling, abbreviated to DOM (Bakhtari *et al.*, 1995a).

The DOM development system has especially been designed for development issues in the course of ontological engineering. In favour of more efficient development, we decided to carry on with our development framework that is to develop and test on UNIX workstations and run the application system under WINDOWS. The specification and realization of the design specialists followed in terms of quality assessment (A) (Bakhtari *et al.*, 1995b), rectification (R), construction (C) (Bakhtari *et al.*, 1995c), adaptation (AD), and the evaluation of the non-linear process of adaptation (E).

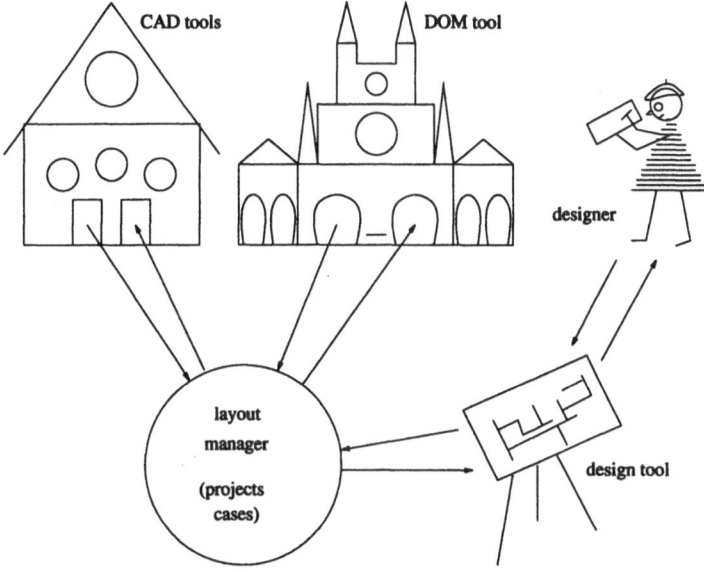

Figure 2. The DOM-ARCADE in its CAD-Environment.

The provided functionalities in the DOM-ARCADE are accessible in a CAD environment[3]. Our system is accessible through AUTOCAD[4], through DANCER (Hovestadt, 1993) for specification of conceptual design layouts, and also through the FABEL case-base manager (Walther *et al.*, 1994) as shown in Figure 2.

In fact, most of the current CAD systems provide a platform for compositions of syntax primitives of the design vocabulary, e.g. lines, points, etc. This is also valid for the design platform DANCER that we use in the FABEL project.

Hence, the first step towards the realization of the DOM system is the formulation of identification rules that enable the access to the semantics of the syntactically specified design objects and their topological arrangement in a proposed layout.

---

[3]DOM in the german language means cathedral. The translation is given to make the figure comprehensible.

[4]AUTOCAD is a trademark of AutoDesk.

3. How is the Deep Design Knowledge Represented?

The DOM system incorporates and maintains a rich core of deep design knowledge that enables the system to append semantic knowledge to the syntax primitives, e.g. a line could be the visualization primitive for a supply-air duct, a sewage-
vertical-pipeline, etc. The represented deep knowledge in the DOM establishes
the meaning and permissible use of the design objects and the topological relations between these objects (Bakhtari *et al.*, 1995a). The concerning knowledge
includes a set of appropriate models for the involved subsystems as well as rules
of interest, e.g. for admissible topological relations.

The deep design knowledge is built up as a network over a set of generic semantic structures that we call *concepts*. The relations between the concepts are
identified by a set of identification rules and are specified as associations. The
most important associations are specialization (taxonomy) and partialization (par-
tonomy). The knowledge how to classify design objects relies on taxonomic knowledge and uses identification and construction rules that classify the syntactic design
entities into classes of concepts.

Due to the subsystem for which the design concepts are used – e.g. supply-
air subsystem or heating subsystem – we have also formalized some subsystem
characteristic features that are also integrated within the whole network. Since the
concepts are allocated in a concept network, their subsystem characteristic features are also determinable. The relations between entities are identified in terms
of their corresponding *topological relations*, e.g. contact, overlap, etc. Further, the
design objects which have to be viewed as a whole will get aggregated and made
explicit as instances of the *aggregate concepts*. The knowledge about aggregations allows to compose higher order objects like e.g. a pipeline-system.

The process of semantic interpretation enables to establish useful information
about the meaning and permissible use of the involved concepts and topological
arrangements in the proposed layout. As soon as a design layout has been specified – partially or completely – through a design platform and transferred to
the DOM-ARCADE system, the process of semantic interpretation can be invoked.
Each proposed design layout goes first through a qualification and aptitude test
due to a syntax examination. The knowledge needed for the aptitude test comprises a set of design specific constraints – e.g. examination rules for acceptable
values for the attributes, consistency checks, etc. – that are represented for all subsystems that may be involved in the proposed layout. Each design object is then
represented in terms of its *geometrical data*, *descriptive data*, and *visualization
data*.

Figure 3 illustrates an example of a layout that has been drawn on a design
platform and transferred to the DOM-ARCADE system. The proposed layout includes the design of four subsystems that can be identified in accordance with
their specified visualization data. By agreement, the design objects for the sewage

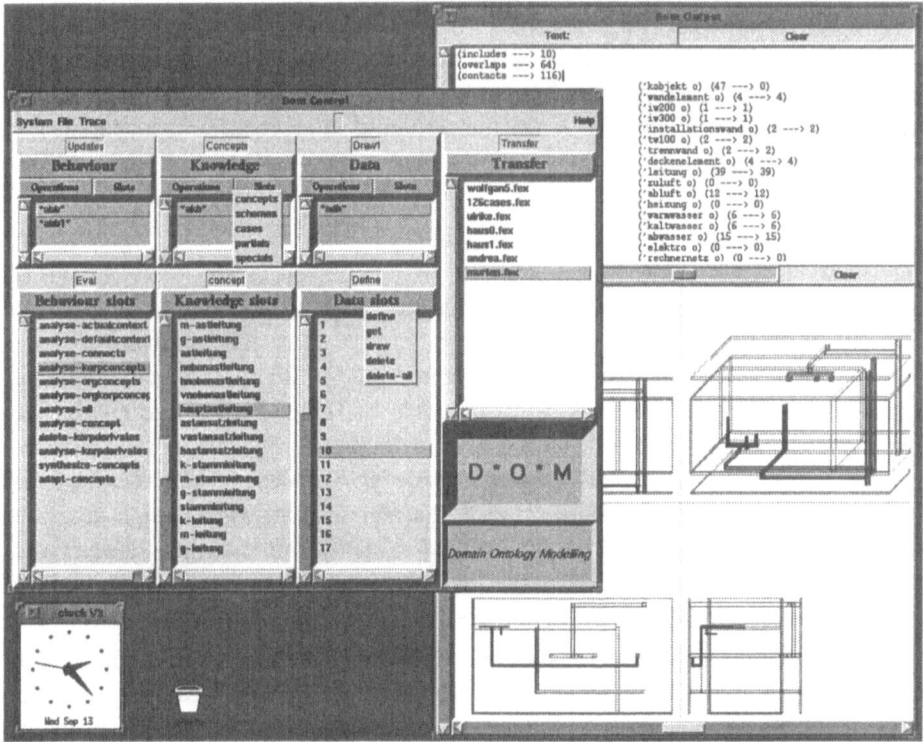

Figure 3. Four views of a proposed layout including four subsystems on the interface of the DOM-ARCADE development system.

subsystem are in dark-blue, the cold-water-supply subsystem in green, etc.[5].

The DOM provides a semantic view on the syntactically qualified design objects and identifies these as instances of generic semantic structures – concepts. All aggregate concepts are further built upon the meanwhile determined semantics. The syntactically specified relations between the concepts go through a test for their admissible design due to the specific subsystem for which they have been drawn. These relations get first identified in terms of their corresponding semantic topological relations e.g. connect, contact, overlap, etc. The question that has to be answered at this stage is whether the designed relations may get qualified through the semantic interpretation. Each designed topological relation, e.g. the connecting duct for the in- and outlets of the sewage subsystem, has to be positioned in a certain manner within the predefined technical service shaft. The predefined space of the shaft is defined by boarders – in the ceilings and walls – that surround the actual room space – shown as a block in Figure 3.

---

[5]Because of publishing reasons we reproduced the layout in black and white, thus, the distinction between the subsystems is unfortunately not really replicable.

The determination of the semantics, permissible use of the design objects, and admissible topological relations between the objects due to each subsystem proceeds on the basis of an analysis function that is invoked through the interface of the DOM-ARCADE system. We make the semantic interpretation transparent for the user through a textual window upon the layout.

4. The Design Specialists in the ARCADE

The pursuit of the representation of the decision making knowledge in terms of a variety of design specialists is to develop tangible tools for architects and engineers which enable the system to get adapted within the continuous process of synthetic and analytic activities of designing a building. Hence, the qualitative leap forward that can make a meaningful contribution to the current design practice is to make extensive use of the deep knowledge – as described in the previous section – and let design specialists review the layout in terms of design decision making knowledge to ensure the quality and reliability of each proposed layout at multiple stages of elaboration as well as their integrity in the final state of the design of the building (Bakhtari *et al.*, 1995c).

The representation of deep design knowledge enables the system to determine the semantics and the permissible use of design objects and their topological relations, while the design specialists incorporate the knowledge needed for problem solving. Each of the design specialists has its territory of specific knowledge and the corresponding problem solving ability. DOM-A for instance, reviews a layout in order to assess the quality of it by applying a rule-based problem solving method. DOM-R on the other side, uses mainly case-based techniques (Kolodner, 1993) to remote and reuse similar former layouts in order to cover the target requirements. Each design layout may be reviewed by one or more of the design specialists that are introduced in the following sections. A scenario of the their deployment for a proposed layout is given in section 5.

4.1. DOM-A

DOM-A addresses a crucial task in building design assistance systems which is the quality assessment. The knowledge for the quality assessment specialist is formalized on the basis of two categories of functions: engineering judgements and specific design decisions.

Engineering judgements comprise functionalities which are of quite general necessity and usefulness for architectural and engineering design. Design decisions are specific agreements while designing a new building. These agreements are subject of negotiations between the designers who are going to be involved in designing a new building at the beginning of the project. We call these project specific agreements *maxims* and group them within a frame of reference.

With the distinction of the mentioned two categories we reckon on the following merit. Since the frame of reference includes maxims and agreements for a particular project, the designer may want to switch between different frames of reference for different purposes. Thus the part concerning the frame of reference is being held interchangeable.

The criterion of success in quality assessment is that of being conform referring to the engineering judgements and to the design decisions. Beside some valid standards for the design of buildings, one of the engineering judgement rules is the coherence assessment.

- Coherence assessment examines all involved geometrical and topological arrangements in a proposed layout for their coherent connections. In general, the examination can be viewed as a check for a closed loop or chain of well arranged duct-systems that connect all well situated in- and outlets together in order to satisfy a subsystem specific functionality.

The frame of reference in the current implementation of our system includes design agreements, principles, and rationals that are documented as the ARMILLA methodology (Haller, 1985). ARMILLA is a rational methodology for spatial ordering and organization for designing the technical installation systems for highly complex buildings. Our concerning frame of reference includes e.g. the following functionalities:

- Spatial ordering and organization maxims that check the position of the involved concepts and their topological relations for their conformance relative to the frame of reference.
- Coordination maxims that review a layout including more than one subsystem for allowed and not allowed collisions. The coordination maxims give regulations and priorities for the spatial organization of different subsystems within the layer structure of the technical service space.
- Configuration maxims for the spatial ordering of in- and outlets.

4.2. DOM-R

The DOM-R specialist operates as a tolerant and cooperative critic. The problem solving of this specialist is two-fold: Rectification and remote and reuse of similar layouts that have turned out to be successfully used for problem solving in similar cases.

For manners of rectification, it incorporates knowledge about syntactically well-formed layouts. It has problem solving knowledge about

- how to rectify the shortcomings and errors of omissions,
- how to adjust the geometrical dimensions in order to get the design objects well situated, and

— how to use the scope of validity in order to get some necessary modifications on the proposed layout done.

Actually, the rectification was – in our primary program – planned as a first step towards the development of an adaptation specialist. As the amount of knowledge grew rapidly while specifying the knowledge needed and its functionality got more and more useful and also necessary for general purposes, we decided to group the functionality and put it with the knowledge about how to remote former cases together.

In case the rectification process can not cover an identified shortcoming or a mistake DOM-R applies a case-based technique (Aamodt *et al.*, 1994) and retrieves a similar layout that can cover the requirements. Case-based methods can offer comprehensive support in knowledge-intensive and structurally complex domains. Whenever a decision has to be made, we try to access the cases which we have tackled successfully and reuse them in order to get the current case of interest appropriately solved. If the process of retrieval fails then we ask the designer to solve the raised problem.

With the help of the retrieval functionality, it is possible to find cases that are related syntactically or semantically to the layout actually handled on the design board.

4.3. DOM-C

Over the period of our ontological engineering, we found out that the knowledge for quality assessment has much in common with the knowledge for the construction of new layouts from scratch.

The layout construction functionalities are applicable for generating a solution from scratch as well as for the completion of a partially specified case. The design of subsystems from scratch proceeds on the basis of the deep design knowledge, the knowledge that is specified for the applicability of engineering rules, and the knowledge included in the concerning frame of reference. The construction specialist can be asked for assistance in several steps and at multiple stages of elaboration. It may serve as inspiration or as a coach for the designer while designing a specific layout for a certain subsystem by indicating alternatives. This specialist serves as an assistant during the following design stages:

— the completion of a partially specified case,
— the indication of possible alternatives, and
— assistance for case construction from scratch.

4.4. DOM-AD

Whereas DOM-R is used for small and mostly local rectification steps, the more complicated adaptation problems (Blumenthal *et al.*, 1994) are tackled by DOM-

AD. We say "tackled" and not "solved" because DOM-AD operates in an open world domain where we can neither guarantee that the target problem is solvable nor that our domain knowledge and problem solving capabilities are sufficient to solve the adaptation problem at hand. Though in practice, the specialist has shown that in very many cases it came up with solutions that are acceptable both for the system's background knowledge and for our users as well.

When we speak of adaptation then we mean a modification in the layout of current interest that is in conflict with at least one part of the deep design knowledge, or with the decision making knowledge, or with the articulated requirements of the target problem. We say that DOM-AD was successful if a sequence of modification actions has led to a new constellation in the considered layout with the following characteristics:

- all involved design objects are identified as instances of the generic semantic structures (concepts) and their usage is permissible – in regard to the represented deep design knowledge,
- all relations between the objects have a corresponding topological relation within the knowledge network of the system – in regard to the represented deep design knowledge,
- the designed subsystems fulfil their functionality – concerning the design specialist, and
- the designed layout fits the frame of reference.

When we look at our approach to adaptation from a more problem solving methodological point of view then it can be characterized as an ontology-driven process. Compared to other approaches that proceed in the context of case-based adaptation (Voss *et al.*, 1996) we tackle a complex open world problem by a remarkably well-structured knowledge-based approach. In particular, we rather re-use ontological knowledge which is the rational behind the designed layouts and do not use former layouts (cases) to carry out the adaptation.

4.5. DOM-E

The knowledge of the specialist that monitors the non-linear process of adaptation – DOM-E – relies in general on the evaluation knowledge. During each iteration of the adaptation process a goal is identified as one of the following alternatives:

- a result of the quality assessment process which has turned out to be non-conform with the rationals of the underlying ontology, or,
- a user requirement for modifying some involved concepts or their topological arrangements.

In case the adaptation actions at the first level were successful, we examine the obtained solution according to the frame of reference and assess the quality of the obtained result. If new discrepancies occur, then we first classify and specify the new discrepancies and view at the new list of discrepancies as our new list of goals.

If the evaluation monitoring specialist identifies that there is at least one discrepancy that is non-resolvable it stops the process. A discrepancy is viewed as non-resolvable if it can not get embedded within the whole knowledge network in the DOM.

5. A Scenario

The central issue of this section is to give a general idea of how the DOM-ARCADE is used and how the design specialists operate on a proposed layout. As we have outlined previously, the DOM-ARCADE is accessible as soon as a layout has been specified – partially or completely – through a design platform and transferred to our system. An example of a proposed layout is shown in Figure 4 with the following criteria[6]:

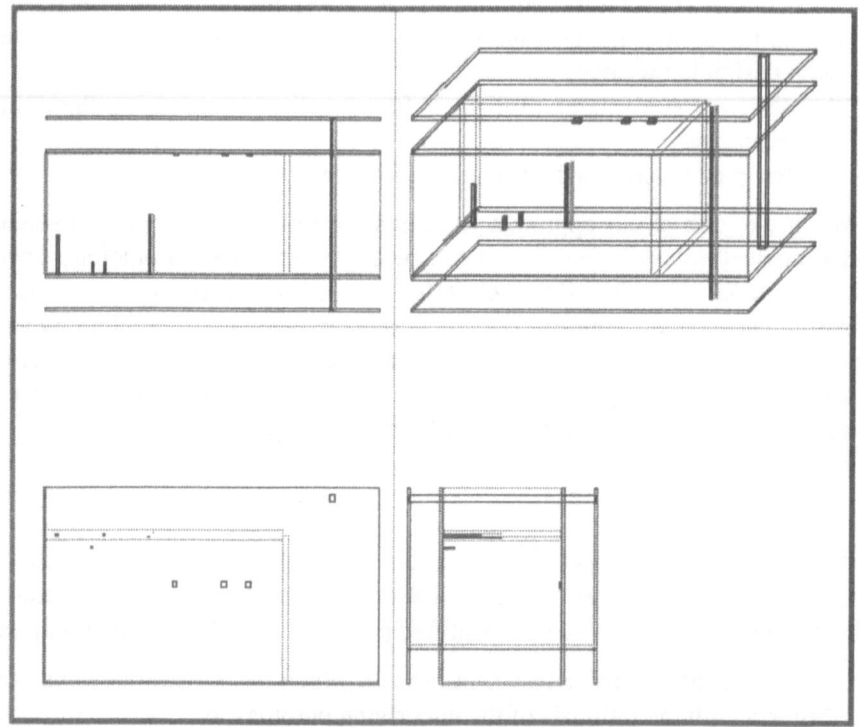

Figure 4. A proposed layout.

---

[6] As mentioned before, because of publishing reasons we had to reproduce all layouts in black and white, thus, the distinction between the subsystems is really hard.

- The layout includes the design of four subsystems: The return-air subsystem (in blue), the sewage subsystem (in dark-blue), the cold-water-supply subsystem (in green), and the warm-water-supply subsystem (in brown).
- The layout has been designed for a certain project of interest which is called the murten-project[7].
- The frame of reference is specified as the one which includes the ARMILLA methodology of designing the technical installation systems.
- The call for assistance is articulated as follows: Assess the quality and reliability of each subsystem and the integrity of all involved subsystems within the design of the murten-project, modify and complete the layout if necessary.

In the first stage, all syntactic design objects and relations between the objects involved in the proposed case will be specified in terms of a system internal representation due to each subsystem. The DOM constitutes a semantic view on the syntactically qualified design objects and identifies these as instances of generic semantic structures – concepts – e.g. as an outlet, a main-connecting-duct, etc. The concepts are – as mentioned above – allocated in a concept network where their subsystem characteristic features get determined. The relations between entities get identified in terms of their corresponding topological relations e.g. contact, overlap, etc. Further, the design entities which have to be viewed as a whole will get aggregated and made explicit as instances of the aggregate concepts.

The rectification specialist DOM-R gets active and operates in terms of a syntax qualification and aptitude examination. The quality assessment specialist DOM-A gets invoked and reviews the layout on the basis of its semantic interpretation. The DOM-A applies first the engineering judgement rules. Its result is the following:

- There is no vertical-main-duct (that goes through all stories of the building and lets water run through) for the sewage subsystem designed, which means that the designed sewage-inlets are not connected. This leads to a water-jam at that story and violates the engineering rules. Since this is an essential and decisive mistake, it has to get mended before the examination can proceed.

Now the cue is given to the DOM-C specialist. This specialist is told to generate a vertical-main-duct for the sewage subsystem as well as to generate the ducts missing that are needed for the connection of the designed sewage-inlets to the vertical-main-duct. Since each duct has to be designed and placed in accordance with regard to the frame of reference, the first step is to take the considered organization and configuration principles into account.

Figure 5 shows the generated organization, e.g. logical partitions and layer structures as well as the configuration concepts, e.g. in- and outlet positions for

---

[7]Murten is the name of the city in Switzerland where this building is built.

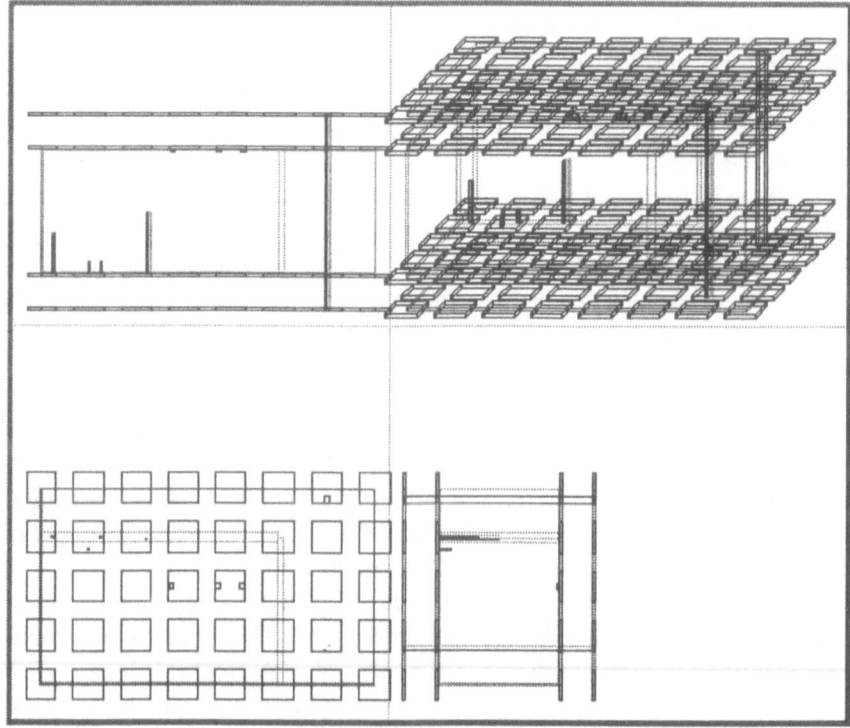

Figure 5. The proposed layout with the generated relevant organization and configuration principles.

the proposed layout. In accordance with the frame of reference, the completion rules get active and generate the ducts missing. The result is shown in Figure 6.

The assessment specialist DOM-A reviews the new constellation. As the generated sewage system passes the examination, the DOM-A continues its process and applies all the rules for the integrity test which considers first the workability of the designed subsystems regarding to the collision maxims, specific design rules, and algorithms that e.g. deal with priorities and arrangement maxims. The result is formulated as the following:

- The in- and outlet connecting ducts for the subsystems cold-water-supply (green) and warm-water-supply (brown) are designed too close to each other. Their positions violate the minimum distance between these two subsystems. The layout design has to ensure that one really gets cold-water if it is wanted. So, adapt the position of the two mentioned ducts to the laying structures and constraints that are specified in the frame of reference.

Now, the adaptation specialist DOM-AD is told to carry on with the process. The specialist recalls all the organization and configuration concepts that have been generated for the construction and completion specialist. Further, some of

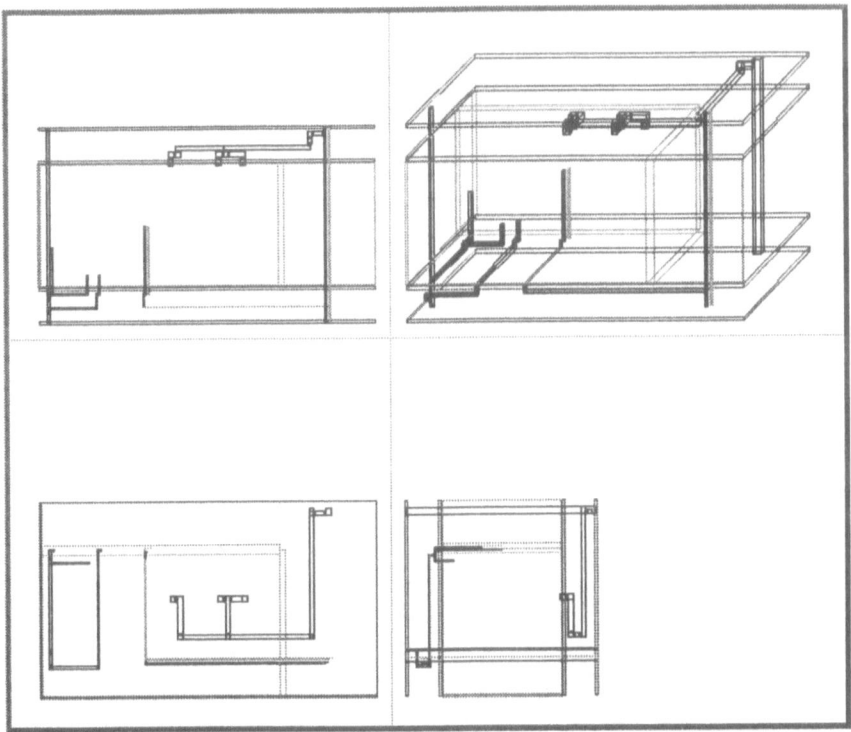

Figure 6. The layout after construction of the missing sewage ducts.

the relevant adaptation rules and algorithms are invoked. The question of which adaptation rules are to be viewed as relevant and which algorithm is to be applied and whether the adaptation process may proceed or is to be stopped is answered by DOM-E. In the case of our example the adaptation is successfully done in one step. Figure 7 shows the result of the whole process that is:

- A coherent layout after the application of the quality assessment functions, construction assistance and its adaptation in order to fit the frame of reference and which can be successfully integrated within the whole design of the murten-project.

6. Technical Underpinning

One major characteristic of the architectural and engineering design is that the requisite knowledge is accumulated experimentally. The important implication is that we have to deal with incomplete knowledge and to take precautions for a stepwise extension as well as for a goal-oriented modification of the knowledge without incurring the full cost of re-representation and re-organization of the whole system.

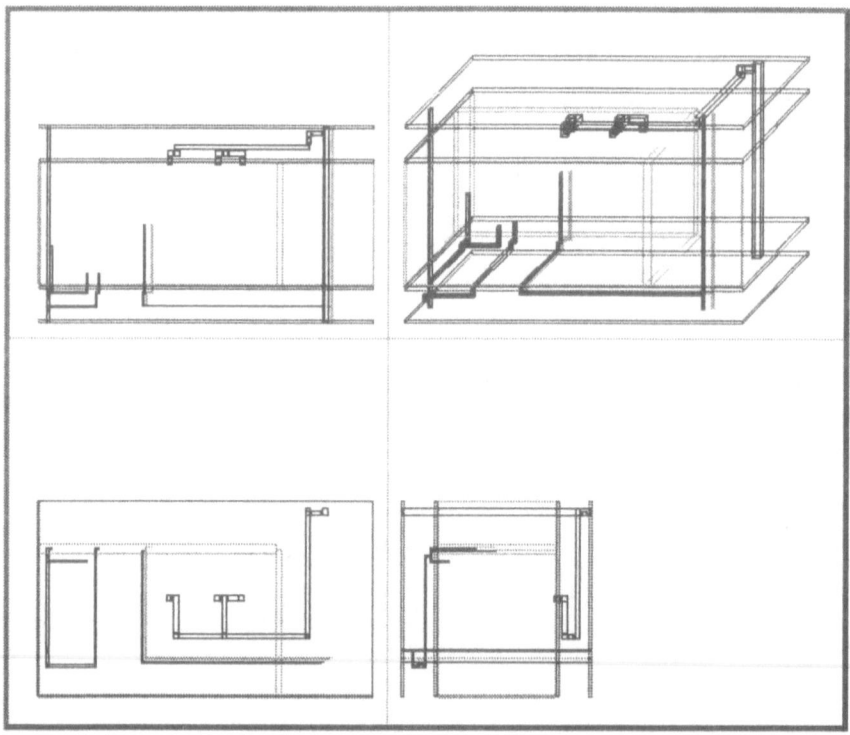

Figure 7.   The result: A coherent layout after the application of the quality assessment functions, construction assistance and its adaptation to fit the frame of reference.

Therefore, each step of knowledge extension has to proceed in a guided way and the scope of completion or modification of some knowledge chunks should be kept local. Thus, to put it concisely, the DOM-ARCADE has a throughout object-oriented realization which allows software developers to create units of generic functionality and shareable units of information within a collaborative framework for that particular domain. Following the object-oriented knowledge modelling, the deep design knowledge is represented as a hierarchy of knowledge elements. Knowledge elements can be concepts, schemes, cases, specializations, or partial-izations.

(defclass knowledge ()

((concepts) (schemes) (cases) (specials) (partials)))

The extension of the scope of knowledge is easily done through addition of single knowledge elements within this framework. Concepts are the generic know-ledge elements. They can be regarded as classes of layout objects. But, the scope of knowledge in the DOM system includes also cases – layouts that can be re-mote and reused – and schemes. Schemes are a set of design patterns for cer-tain subsystems, e.g. patterns for an orthogonal and well-ordered supply-air system design. The current implementation of our system includes patterns that are defined within the ARMILLA frame of reference.

The knowledge model is represented as a network with various kinds of associations, e.g. specialization, partialization. The specialization part represents the current state of the involved knowledge elements in terms of a taxonomy, the part-of-relation between the elements is represented by the partialization. Concepts, for instance, include the following features:

(defstruct (concept (:type list))

name crspace crtime analyse synthesize adapt)

Each concept name has a corresponding classification term in the real world, e.g. trunk, duct, in- or outlets, etc. The creation-space (crspace) and the creation-time (crtime) determine where and when the concept was defined. Corresponding to the methods of a class, the procedural knowledge – the problem solving knowledge of each design specialist – is represented in the parts analyse, synthesize, and adapt.

Since a proposed layout includes arrangements of physical and abstract domain objects, the main issue is to determine and establish links between the objects that comprise a proposed layout and the specified classes in the formal representation. In other words, the objects that are involved in the proposed layout are subjects of classification due to a set of identification and/or construction rules. The specific arrangement of the objects is also kept in terms of aggregate relations and topological relations. Examples for topological relations are e.g. contact, overlap, etc.

(defclass data ()

((objects) (conceptnames) (aggregates) (contact) ... (includes)))

The organization and interaction between the system components is also organized in the object-oriented paradigm. The main system classes are: Transfer, Data, Knowledge, and Behavior, where each of which includes static components (slots) and dynamic components (methods). A class specification together with its set of instances build a so-called base. So we get a transfer base, a data base, a knowledge base, and a behavior base. Each of these classes is specified by a set of slots and a set of methods. The slots are containers for sets, for instance sets of files, building objects, connections, concepts, or assistance operations. The methods are defined to operate on the whole instances or on single objects of the sets that are stored in the slots of instances. Examples for this kind of methods are create, delete, get, eval, draw, load, save, open, close.

We reckon with the merit that it is possible to work simultaneously with a set of transfer bases, data bases, knowledge bases, and behavior bases and to connect them with each other in several not pre-determined ways. Our goal is to make an isomorphic mapping between the structure of the system kernel and the structure of the user interface. Having achieved this goal it guarantees a clear, transparent organization of the whole system and allows the user to influence internal processes consciously and goal-driven.

As mentioned previously, the DOM-ARCADE development system is implemented in Allegro Common Lisp and runs on UNIX workstations with a user interface based on Tcl/Tk (Ousterhout, 1993) using the generic knowledge-based development tool FAENSY. The DOM development system is especially designed for development issues in the course of ontological engineering. The DOM-ARCADE application runs also under WINDOWS.

7. What is gained?

Since we followed an application-driven research and development approach, a significant feature of the DOM-ARCADE systems is that it does not require a change of the designers working practice. It undertakes the role of a competent and cooperative design assistant and is adapted within the continuous loops of construction, evaluation, and adaptation issues of designing an artefact.

Our research and development program is accentuated by the domain ontology modelling. We make extensive use of the ontology for the realization of a variety of building design support functionalities. We followed the goal to develop support functions that come in small units and can therefore be used as tangible tools by architects and engineers on a voluntary basis. The support functions are offered to the designers as design specialists that assist whenever they are asked, thus, they do not intrude into a design process, and may be launched either standalone or in combination with other specialists.

The functionality of the system can be extended easily to either more design specialists or to a variety of further project specific agreements that will be specified as different frames of reference and the user may let the layouts get reviewed due to different frames of reference.

One example for the limitations of the system deployment is the following: With our adaptation approach we cannot overcome the risk of seeing from time to time an adaptation attempt fail. This situation could be improved to a certain extent by a new adaptation-oriented retrieval approach (Smyth et al., 1995) delivering as its result either nothing or a solution that will be adaptable without risk. But even then we will never be able to guarantee a success rate of 100 percent because of the open world domain we live and operate in.

References

Aamodt, A. and Plaza, E.: 1994, Case-based reasoning: Foundational issues. Methodological variations, and system approaches, *AI Communication*, 7(1), 39-59.

Alberts, L. K. and Dikker, F.: 1994, Integrating standards and synthesis knowledge using the YMIR ontology, *in* J. S. Gero and F. Sudweeks (eds), *Artificial Intelligence in Design '94*, Kluwer, Dordrecht, pp. 517-534.

Bakhtari, S. and Bartsch-Spörl, B.: 1994, Bridging the gap between AI technology and design requirements, in J. S. Gero and F. Sudweeks (eds), *Artificial Intelligence in Design '94*, Kluwer, Dordrecht, pp. 753–768.

Bakhtari, S., Bartsch-Spörl, B., Oertel, W. and Eltz, U.: 1995a, *DOM: Domain Ontology Modelling for Architectural and Engineering Design*, Fabel-Report (33), GMD, Sankt Augustin.

Bakhtari, S. and Oertel, W.: 1995b, DOM-ArC: An active decision support system for quality assessment of cases, in A. Aamodt and M. Veloso (eds), *ICCBR-95 Case-Based Reasoning. Research and Developments*, Lecture Notes in Computer Science, Springer-Verlag, Berlin, pp. 381–390.

Bakhtari, S. and Oertel, W.: 1995c, DOM: An active assistance system for architectural and engineering design, *Proceedings of the 6th International Conference on Computer-Aided Architectural Design, CAAD Futures-95*.

Bartsch-Spörl, B.: 1995, Towards the integration of case-based, schema-based and model-based reasoning for supporting complex design tasks, in A. Aamodt and M. Veloso (eds), *ICCBR-95 Case-Based Reasoning. Research and Developments*, Lecture Notes in Computer Science, Springer-Verlag, Berlin, pp. 145-156.

Bartsch-Spörl, B. and Bakhtari, S.: 1996, A support system for building design—Experiences and convictions from the FABEL Project, in J. Sharpe (ed.), *AI System Support for Conceptual Design*, Springer-Verlag, London, pp. 279-297.

Branki, C., Douglas, J., Bailey, D.: 1994, Agent communications server for shared ontologies in planning and design, *Workshop Notes: A Semantic Basis for Sharing Knowledge and Data in Design (AID-94)*, Lausanne, Switzerland, pp. 20-27.

Blumenthal, B. F. and Porter, B. W.: 1994, Analysis and empirical studies of derivational analogy, *Artificial Intelligence*, **67**, 287–327.

Fabel-Consortium: 1993, *A Survey of FABEL*, Fabel-Report Nr. 2, GMD, Sankt Augustin.

Haller, F.: 1985, *ARMILLA - ein Installationsmodell. Institut für Baugestaltung, Baukonstruktion und Entwerfen*, Universität Karlsruhe.

Hovestadt, L.: 1993, A4 Digital Building: Extensive computer support for the design, construction, and management of buildings, in U. Flemming and S. van Wyk (eds), *Proceedings of the Fifth International Conference on Computer-Aided Architectural Design Futures: CAAD Futures*, North-Holland, Amsterdam, pp. 405–422.

Kolodner, J.: 1993, *Case-based Reasoning*, Morgan Kaufmann, San Mateo, CA.

Neches, R., Fikes, R., Finin, T., Gruber, T., Patil, R., Senator, T. and Swartout, W.: 1991, Enabling technology for knowledge sharing, *AI Magazine*.

Oertel, W.: 1994, *FAENSY: Fabel Development System*, FABEL Report Nr. 27, GMD, Sankt Augustin.

Ousterhout, J. K.: 1993, *An Introduction to Tcl and Tk*, Addison-Wesley.

Smyth, B. and Keane, M. T.: 1995, Experiments on adaptation-guided retrieval in case-based design, in A. Aamodt and M. Veloso (eds), *ICCBR-95 Case-Based Reasoning. Research and Developments*, Lecture Notes in Computer Science, Springer-Verlag, Berlin, pp. 313-324.

Voss, A., Bartsch-Spörl, B. and Oxman, R.: 1996, A Study of Case Adaptation Systems, in J. S. Gero and F. Sudweeks (eds), *Artificial Intelligence in Design '96*, Kluwer, Dordrecht (this volume).

Walther, J., Graether, W., Oertel, W., Schmidt-Belz, B. and Voss, A.: 1994, An open architecture for multiple case retrieval methods, in M. Keane, J. P. Haton, and M. Manago (eds), *Proceedings of the Second European Workshop on Case-Based Reasoning EWCBR-94*, AcknoSoft Press, pp. 373–380.

13

creativity and innovation in design

J. S. Gero and F. Sudweeks (eds), Artificial Intelligence in Design '96, 703-722.
© *1996 Kluwer Academic Publishers.*

EMERGENT BEHAVIOUR IN CO-EVOLUTIONARY DESIGN

JOSIAH POON AND MARY LOU MAHER
Key Centre of Design Computing
Department of Architectural and Design Science
University of Sydney NSW 2006 Australia

Abstract. An important aspect of creative design is the concept of emergence. Though emergence is important, its mechanism is either not well understood or it is limited to the domain of shapes. This deficiency can be compensated by considering definitions of emergent behaviour from the *Artificial Life (ALife)* research community. With these new insights, it is proposed that a computational technique, called evolving representations of design genes, can be extended to emergent behaviour. We demonstrate emergent behaviour in a co-evolutionary model of design. This co-evolutionary approach to design allows a solution space *(structure space)* to evolve in response to a problem space *(behaviour space)*. Since the behaviour space is now an active participant, behaviour may emerge with new structures at the end of the design process. This paper hypothesizes that emergent behaviour can be identified using the same technique. The floor plan example of Gero and Schnier (1995) is extended to demonstrate how behaviour can emerge in a co-evolutionary design process.

1. Introduction

Emergence has recently drawn the attention of research workers in the *design community* (Gero and Yan 1994, Gero, Damski and Han 1995, Edmonds and Soufi 1992, Liu 1995). This is an important research agenda because emergence contributes to creative design (Gero 1992). However, the definitions offered from the *design community* are usually applied to shape only. Hence, this paper attempts to borrow characteristics of emergence identified from other research communities, i.e. *ALife community*, to enhance our understanding of emergent behaviour. This cross-breed of explanation is expected to help us to develop a general theory of emergence for *design*. The theory should be general enough to cover not only shape emergence or emergent structure, but also emergent behaviour.

Section 2 carries out a small survey on definitions offered by the two research communities and to identify key concepts for emergence. It is then followed by a section on the discussion of evolving representation of design genes and how it satisfies our refined understanding of emergence. Section 4 introduces co-evolutionary design. A discussion of how the evolving representation can model emergent behaviour and how complex behaviour can emerge from the process are presented in section 5. The floor plan example used by Gero and Schnier (1995) is extended in section 6 to illustrate how the process works.

2. Emergence

Emergence is, very often, used and discussed by the *design community* without a proper definition, e.g. Liu (1995). The process which leads to emergence is largely unknown. The omission of a definition makes the development of a computational model for emergence difficult. *ALife community* is another research group which frequently makes reference to the terminology of *emergence*. Hence, it is beneficial to visit definitions offered by these two communities to elicit the similarity and differences when this term is used. The expectation from this compare and contrast exercise aims to help us generalize emergence in design beyond emergent shape.

2.1. HOW IS EMERGENCE BEING DEFINED?

The most notable definition on emergence offered by the *design community* comes from Gero (1992), which says

"*A property that is only implicit, i.e. not represented explicitly, is said to be an emergent property if it can be made explicit.*"

The context that Gero addresses is shape emergence. A process to discover emergent shapes is reported by Gero and Yan (1994). The process comprises de-construction followed by re-construction, which in their own words, is shape hiding and shape emergence (Figure 1).

A few more definitions from the *design community* are presented here which relate primarily to shape emergence:

"*... drawing might be thought of as a visual image together with an associated description that imposes structure upon it. Thus a drawing may be thought of as a structured entity. From this perspective, an emergent shape occurs when a revised description, or structure, is discovered ...*" (Edmonds and Soufi 1992)

"*... emergent subshapes are... emergent entities and relationships – ones that they never explicitly input ...*" (Mitchell 1993)

There are some research works which are loosely related to *design community*. For example, the research focus of Finke, Ward and Smith (1992) is to explore the cognitive processes and structures that contribute to creative thinking and discovery. They regard emergence as one of the properties for preinventive structures.

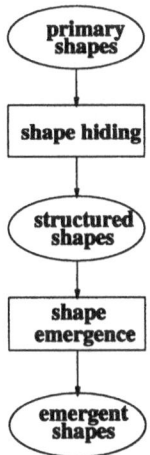

Figure 1.   A process model for shape emergence (Gero and Yan, 1994).

"... emergence, which refers to the extent to which unexpected features and relations appear in the preinventive structure. By definition, these features and relations are not anticipated in advance and become apparent only after the preinventive structure is completely formed..."

In the work of Hofstadter and McGraw (1993) to create alphabetic style, they did not give a definition for emergence but offered an implementation

"... implemented as a large number of small codelets that are run in simulated parallel ..."

Research workers in *ALife community* (or non-design community) offer more definitions and explanation/elaboration to *emergence*. For example, to name a few:

"... emergent properties ... collections of units at a lower level of organization, through their interaction, often give rise to properties that are not the mere superposition of their individual contributions ..." (Taylor 1990)

"The key idea is that functionality is made to emerge as a global side-effect of some intensive, local interactions among components that make up the system ..." (Maes 1990)

"Emergent functionality means that a function is not achieved directly by a component or a hierarchical system of components, but indirectly by the inter-action of more primitive components among themselves and with the world." (Steels 1991)

Forrest (1990) echoed Hofstadter's idea that *"..information which is absent at lower levels can exist at the level of collective activities.."* and considered this is the essence of the following constituents of emergent computation:

1. A collection of agents, each following explicit instructions;
2. Interactions among the agents (according to the instructions), which form implicit global patterns at the macroscopic level, i.e. epiphenomena;
3. A natural interpretation of the epiphenomena as computations.

Mataric (1993) characterizes emergent behaviour in the field of swarm intelligence as below:

1. it is manifested by global states or time-extended patterns which are not explicitly programmed in, but result from local interactions among a system's components;
2. it is considered interesting based on some observer-established metric.

One common theme connects these definitions/explanations in the ALife community: global behaviour occurs as a result of local interactions among low level units.

2.2. SIMILARITIES, DISSIMILARITIES AND INSPIRATIONS

The obvious similarities between the two research communities in defining this terminology is *to make implicit to explicit*. That is to say, a property which is not intentionally or explicitly described or input or programmed in the original representation becomes an explicit, known pattern. At the surface, the two communities seem to have different approaches to explain how this process occurs. The *design community* uses de-construction and re-construction to explain emergence, while *ALife community* emphasizes emergence as a collective behaviour of local interactions among lower level units. However, similarities can be found on detailed studies. According to the data-driven approach by Gero and Damski (1994), new shapes are derived by

1. *de-construction*: hiding the original shapes; removing the existing relationships between existing data points
2. *re-construction*: establishing new relationships between data points where the fitness of these new relationships are assessed by available schema or the aesthetic judgment of human designers

The discovery of new relationships between data points can be argued to be an equivalent process of local interaction between low level units (as proposed in the *ALife community*). It is also observed that emergence in the *design community* has the premise that new behaviours are identified on top of existing behaviour or the de-construction of existing behaviour. Having said these similarities, additional ideas which are shared by the *ALife community* are characterized as follows:

– **Lower level building units**. This is not explicitly mentioned in Gero's definition, though it is implicitly referred to in the data-driven approach in Gero and Damski (1994), i.e. half planes.

- **Bottom-up approach.** *ALife community* tends to elaborate the definition more by stating the emergent function/computation is a result of the interaction of lower level units. The data-driven approach proposed by Gero and Damski (1994) can be regarded as a bottom-up approach.
- **Local interaction but global effect.** The effect of local interactions between low level units (bottom-up approach) is manifested at a higher level. In GA terms, the resulting behaviour is displayed at the phenotype [1] and not at the genotype [2] level.
- **Self-organization, collective phenomena and collective behaviour.** Forrest (1990) regards these are the three overlapping themes essential to emergence.
- **The whole is greater than the sum of its parts.** Using the example of Gero and Schnier (1995), if we sequence four pen movement commands one after another:

  1. draw one unit length to the right;
  2. draw one unit length down;
  3. draw one unit length to the left;
  4. draw one unit length up;

  the end result is not just two horizontal lines and two vertical lines, we have an emerged behaviour of a square.
- **External observer.** Forrest (1990) highlights that the emergent pattern is an *interpretation* to the epiphenomena. In other words, there is an external observer to interpret the pattern where this emergent pattern makes sense to the observer.

These extra items for emergence are hidden in the work of Gero and Damski (1994). That is to say, the data-driven approach (in shape emergence) fits well into the framework of emergence described by *ALife community* because the process to create a new shape starts from the lower level units (the half planes).

The *best* way to summarize the concept of emergence is to combine definitions and explanations from these two communities. Hence, emergence can be defined as *"global pattern as a result of local interactions of low level units"* and includes the characteristics: de-construction and reconstruction, low level units, local interaction, global effect, self-organization, collective phenomena and behaviour.

---

[1] It can be a living organism for biological sytems or a design solution for design systems;
[2] It is a way of representing or encoding the information which is used to produce the phenotype.

3. Evolving New Representations of Design Genes

In the work of Gero and Schnier (1995), new evolved complex representations from each genetic cycle are deposited to the pool of basic genes [3] for building genotypes, i.e. the size of alphabets for coding increases. The solution space is incrementally restructured by the insertion of evolving structure elements into the pool of building blocks.

To further clarify this idea, there is a pool which serves as the source to provide materials to build up genotypes (see Figure 2). In each genetic cycle, a process is executed to combine basic genes randomly from this pool to form genotypes for fitness evaluation. At the start, the pool only contains 7 basic genes of {s1, s2, .. s6, s7}. Good and bad phenotypes and their corresponding genotypes are separated after the evaluation process at each cycle. A process is invoked to identify common useful representations (the building blocks) among these genotypes which contribute to their success. The quality of each pair of genes is evaluated to determine the correlation of the pair and the goodness of the phenotype. Hence, as shown in Figure 2, all adjacent genes in the genotype are grouped together (ie. s1 with s3, s3 and s6, s6 and s7) and they are measured for their contribution to the fitness value of the corresponding design. After the pair-wise performance evaluation, each pair of genes that achieves a threshold value or above will become a single entity. So, for example, s1 and s3 evolved to become *es1* and s3 and s6 become *es2*. These evolved structures will join the existing building-block genes in the pool to develop design solutions in the next generation.

In other words, the number of alphabets has increased from 7 to 9 and the set of basic genes becomes {s1, s2, .. s6, s7, es1, es2}. These 9 basic genes are used to generate solution genotypes at the next generation. An evolved representation, say, es1 (a basic gene by then), may interact with another basic gene. This will give rise to another evolved representation which is more complex. It has to be noted that the adjacency of two genetic units do not imply the adjacency of characteristics in the phenotype[4]. The phenotype of an evolved representation may be a disjoint manifestation. Two observations can be made if we proceed with the genetic cycles using the evolving representation approach: (1) the number of basic genes (alphabets) increases and (2) the evolved representation is getting more and more complex.

Though their work primarily concentrates on genetic engineering, it does not exclude an implicit end result of emergence. To qualify their work on emergence, we recall the definition offered by Maes (1990) where it states that emergence occurs as a global side effect (which is the result of "good fitness") of some in-

---

[3]Gene is the biological unit of heredity which occupies its own place on a genotype. This can be viewed as the design variable for design systems.

[4]see evolved gene 340 and 349 in Figure 12 of Gero and Schnier (1995)

tensive, local interactions among components (the "adjacency" of genes in the genotype).

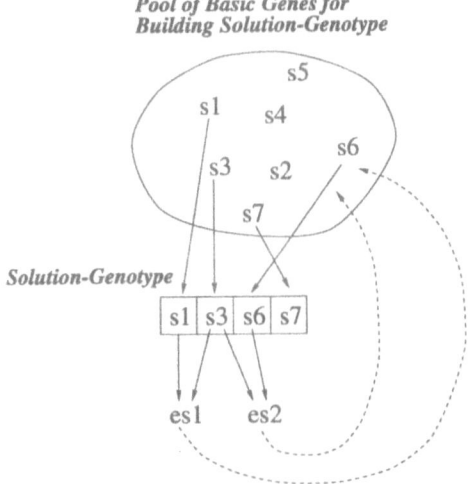

Figure 2. Evolving representation according to Gero and Schnier (1995).

4. Co-Evolutionary Design

Design as a sequential process which moves from the formulation of the problem to the synthesis of solutions faces a lot of challenges (Corne, Smithers and Ross 1993, Gero 1993, Jonas 1993, Maher and Poon 1996). The central theme behind these views is that design should be considered as an iterative process where there is interplay between problem reformulation and solution generation. According to the evolutionary design process model offered by Hybs and Gero (1992), the formulation of functional requirements is to define expected behaviour, B_e, which is represented as the problem space. The solution space can be considered to contain structure elements where the design process is to search the right combination of structure elements to satisfy the requirements, B_e. The behaviour (B_s) exhibited by the current structural combination is compared against B_e in the evaluation process. Reformulation, which is defined as $S \rightarrow B_e$, is conducted if necessary.

This idea of the co-evolutionary nature of design is graphically illustrated in Figure 3 as the interaction of problem space (the required behaviour) and solution space (the potential structural combinations). The diagram highlights the co-evolution of the behaviour-space with the structure-space over time and has the following characteristics:

- There are two distinct search spaces: behaviour-space and structure-space.
- These state spaces interact over a time spectrum.

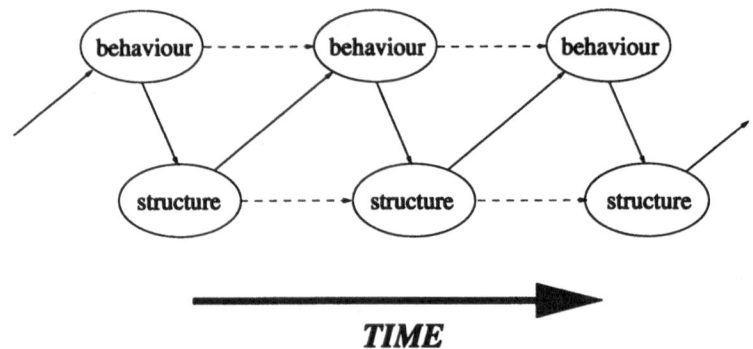

Figure 3. Co-evolution of behaviour-space and structure-space.

– Horizontal movement is an evolutionary process.
– Diagonal movement is a search process where goals lead to solution. This can be the

    • Downward arrow: *"Problem leads to Solution"* or *synthesis* where $B_e \rightarrow S(B_s)$.

    • Upward arrow: *"Solution refocuses the Problem"* or *reformulation* where $S \rightarrow B_e$.

The behaviour-space(t) is the design goal (the required behaviour) at time t and structure-space(t) is the solution space which defines the current search space for design solutions. The structure-space(t) provides not only a state space where a design solution can be found, but it also prompts new requirements for behaviour-space(t+1) which were not in the original behaviour-space at time t. This is represented by the upward arrow from structure-space at t to behaviour-space at time $t+1$. The upward arrow is an inverse operation where structure-space(t) becomes the goal and a "search" is carried out in the behaviour-space, the space at time $t+1$, for a "solution". This iterative relationship between behaviour-space and structure-space evolves over time.

This model of exploration depicts an evolutionary system, or in fact, two evolutionary systems. The evolutionary systems are the behaviour-space and the structure-space. The evolution of each space is guided by the most recent population in the other space. This model is called co-evolution and provides the basis for a computational model of design exploration. The basis for co-evolution is a simple genetic algorithm (Goldberg 1989) where special consideration is given to the representation and application of the fitness function so that the problem definition can change in response to the current solution space. A computational implementation of co-evolutionary design is reported in Maher and Poon (1995) and Maher, Poon and Boulanger (1995).

5. Emergent Behaviour in Co-Evolutionary Design

The work of Maher and Poon (1996) and Gero and Schnier (1995) leads to two
research issues. They are highlighted here and will be further discussed in the
following two sections:

1. A mechanism for evolving more complex representations of behaviour.
2. Emergent behaviour in co-evolutionary design.

5.1. EVOLVING MORE COMPLEX REPRESENTATIONS OF BEHAVIOUR

Emergence is defined as a recognition of collective phenomena resulting from
local interactions of low level units. A complex evolving representation can thus
be classified as an emergent representation. To follow this line of logic, a complex
evolving behaviour is an emergent behaviour. In general, the methods of Gero
and Schnier (1995) do not specify whether the evolving representation is structure or behaviour. In this example, the initial basic genes represent simple pen
movements that we assume represent structure or shape. The example extends a
basic GA (Goldberg 1989) by the incorporation of evolving genes. The solutions
are represented as genotypes and the performances are measured by a constant
fitness function. The evolution in Gero and Schnier (1995) example brings forth
evolved structures. The simplest way to emerge complex behaviour is a direct application of the algorithm of evolving representation to behaviour variables, as
illustrated in Figure 4.

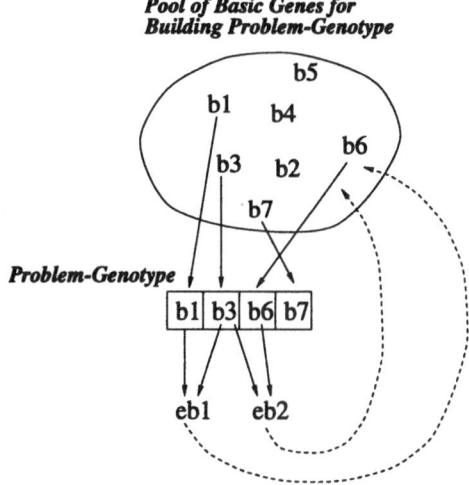

Figure 4. Direct application of evolving representation to behaviour space.

5.2. INTERACTIONS BETWEEN BEHAVIOUR VARIABLES

Adjacency is considered to be the interaction between genes in the example of
Gero and Schnier (1995). Their rationale of using "adjacency" is the consecutive
drawing commands used to draw the floor plan. At the level of behaviour space,
if the idea of Gero and Schnier (1995) is followed closely, we can apply the same
mechanism for behaviour variables to interact, i.e. only neighboring genes are allowed to interact locally. However, there is an implicit assumption behind their
approach: the evolved structure is an ordered pair where $(a, b) \neq (b, a)$. For example, in Figure 5, using a movement of one unit left and one unit up, a different
order of these two movements produces different results.

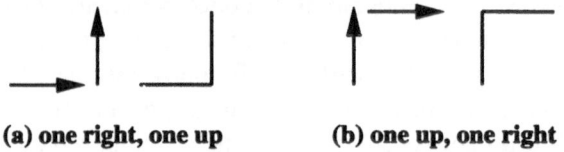

(a) one right, one up   **(b) one up, one right**

Figure 5. Example to demonstrate different phenotypes result from arranging same genes in different orders.

The ordered pair adjacency interaction is a good approach if a chronological
sequence is required, otherwise, this can be relaxed to keep the pair in a set, i.e.
$\{a, b\}$. To further relax the original approach is to break another implicit assumption. Their consideration of "adjacency" assumes the genes to be arranged in a
linear manner. The linear presentation of genes is a human notation. The chemical
notation for water, H_2O, does not imply the *actual* configuration of the molecule.
In fact, a water molecule is not arranged as H-H-O, but H-O-H. Hence, we can
treat the behaviour variables to float around in a space and can interact with other
variables, either strongly or weakly. Since there is no fixed *a priori* arrangement
of genes, local interaction to neighboring genes means much more interactions
than a linear arrangement.

The consideration of how two behaviour variables interact depends upon the
necessity of chronological control and the basic assumption about genes arrangement. In summary, the interaction scheme between two behaviour variables can
be classified as the following patterns with decreasing constraints (Figure 6):

  (a) linear arrangement of genes and an evolved gene is an ordered pair
  (b) linear arrangement of genes and an evolved gene is a set
  (c) no predefined sequence of genes and an evolved gene is an ordered pair
  (d) no predefined sequence of genes and an evolved gene is a set

5.3. EMERGENT BEHAVIOURS CORRELATE WITH GOALS

It has been pointed out in section 4 that the formulation of functional requirements is to define expected behaviour, B_e (Hybs and Gero 1992), i.e. expected be-

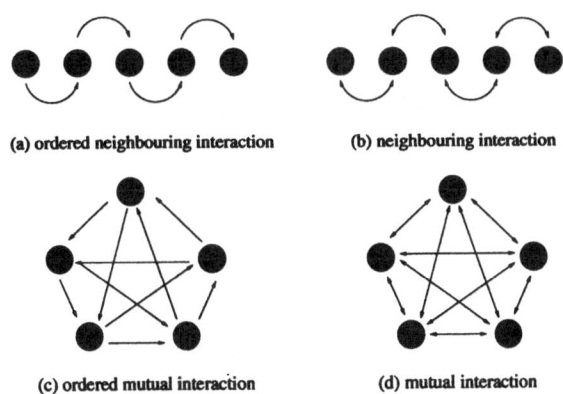

(a) ordered neighbouring interaction

(b) neighbouring interaction

(c) ordered mutual interaction

(d) mutual interaction

Figure 6. Different interaction schemes for "adjacent" genes.

haviour is derived from functions. Emergent behaviour must then be goal-oriented as well where this complex behaviour cannot emerge outside the context of a fitness function. For example, if the design goal is to maximize the area of a floor plan, taking no explicit consideration of the appropriateness of room arrangement, the fitness evaluation to a design solution only derives a numerical result about the area of a floor plan. Since there is no way to assess the quality of a floor plan, there will not be any complex behaviour of room arrangement emerging from the design process. If emergent behaviours of room arrangement are sought, the fitness function must be modified to cater for room arrangement as part of its evaluation.

Though emergent behaviour is goal related, this does not necessarily represent a causal relationship. When Fogel (1995) discusses the identification of "good" building blocks in the context of genetic algorithms, he suggests that there are no viable credit assignment algorithms for isolated genetic structures or behavioural traits because these elements are highly integrated. According to his arguments, credit assignment does not exist in evolution, it is a human construction. Hence, emerged behaviour patterns should be carefully interpreted as a statistical correlation between the complex evolved behaviour and the fitness of the phenotype.

5.4. EMERGENT BEHAVIOUR BASED ON FITNESS MEASUREMENT IN CO-EVOLUTIONARY DESIGN

In Gero and Schnier (1995), complex behaviour is learnt from a target case. In other words, emergent behaviour is identified outside the design process. However, in co-evolutionary design, learning of emergent behaviour is part of the design process. Learning is interwoven with the synthesis of solutions throughout the design cycles. The behaviour emerged by learning from a design is a static behaviour where emergent behaviour is dynamic and responsive to the changing solution space in co-evolutionary design.

Complex behaviour that emerges from co-evolutionary design case can be learnt using the evolving representation algorithm. The end result is a collection of many useful behaviour patterns. Emergent behaviour is derived from the evaluation of behaviour-phenotypes by the current best structure. A close match indicates a "good" phenotype. These evolved behaviours can be used to help find design solutions in a new design context.

The original idea of evolving representation is further extended. The evolved structure remains in the algorithm. The emergent behaviours, which are equally important, are added to the pool for building the behaviour-genotype.

In the original thesis (Gero and Schnier 1995), the insertion of these evolved structures to the pool reduces the search space and transforms the search space. Here, an emergent behaviour is critical because it can reformulate the behaviour space. The evolved behaviour serves as constraints and re-draws the boundary of the structure-space.

The method used to co-evolve behaviours and structures is shown in figure 7. These are the core steps to be found in every cycle of an iterative design. There are four main parts in the method.

The first part is to synthesize design solutions (**Generate Structure-Phenotypes**) and the second part is to learn useful structures from the "good" structure-phenotypes (**Identify Useful Structure-Gene-Pairs**). Figure 8 is a generic diagram to describe these two main parts. In fact, this diagram can be applied to structure dimension (part 1 and 2 in Figure 7) and behaviour dimension (part 3 and 4 in Figure 7). This generic diagram shows a pool of basic genes $\{r_1, r_2, .., r_6, r_7\}$ which is the source to construct genotypes. Each genotype is combined from the basic genes in a stochastic manner and therefore the length of each genotype may vary. After the mapping process, phenotypes are evaluated by the current best phenotype from the other population. When we are at part one of the method, this will be current best behaviour (CBB) from the behaviour dimension. The fitnesses of phenotypes help us in two ways:

- To determine the current best (CB) which, at this stage, is current best structure (CBS). The CBS becomes the target case for the fitness function to assess the behaviour-phenotypes in the next evaluation.
- To divide the population into "good" and "bad" individuals; the degree of "goodness" indicates one's closeness to the requirements.

Genotypes of the "good" phenotypes are selected and the gene-pairs are analysed. For all these "good" genotypes, adjacent pairs of genes are loaded to a table. Whenever a genotype carries the current adjacent pair, the fitness of the phenotype is included to the pair's contribution. Those pairs, for example, $\{er_1, er_2\}$, which contribute above a threshold value, are inserted to the pool of basic genes. In other words, the genes that build up genotypes are evolving.

---

1. Generate Structure-Phenotypes

- Build structure-genotypes from structure-genes in base pool
- Crossover structure-genotypes
- Map structure-genotypes to structure-phenotypes
- Evaluate structure-phenotypes against current best behaviour (CBB)
- Classify structure-phenotypes into "good" and "bad"
- Select best performing structure-phenotype as current best structure (CBS)

2. Identify Useful Structure-Gene-Pairs

- Identify all structure-gene-pairs in "good" structure-phenotypes
- Compute contribution of structure-gene-pairs
- Select structure-gene-pairs above threshold to become {es1, .. esn}

3. Generate Behaviour-Phenotypes

- Build behaviour-genotypes from behaviour-genes in base pool
- Crossover behaviour-genotypes
- Map behaviour-genotypes to behaviour-phenotypes
- Evaluate behaviour-phenotypes against CBS
- Classify behaviour-phenotypes into "good" and "bad"
- Select best performing behaviour-phenotype as CBB

4. Identify Useful Behaviour-Gene-Pairs

- Identify all behaviour-gene-pairs in "good" behaviour-phenotypes
- Compute contribution of behaviour-gene-pairs
- Select behaviour-gene-pairs above threshold to become {eb1, .. ebn}

Figure 7. Method to co-evolve behaviours and structures.

The third part is to reformulate behavioural requirements (**Generate Behaviour-Phenotypes**). The first three steps again follow the generic mechanism as displayed in Figure 8. The performance of phenotypes are evaluated by the CBS and the best performing phenotype becomes the CBB. CBB is used to evaluate structure-phenotypes in the next cycle.

The final part is another learning process where emergent behaviour occurs (**Identify Useful Behaviour-Gene-Pairs**). The complex emerged behavioural patterns are added to the pool of basic genes for the generation of behaviour-genotype in the next cycle. The algorithm to calculate the contribution of a pair of behaviour genes to the fitness of a "good" phenotype is:

1. Put all adjacent behaviour pairs to a table.
2. Set the initial contribution of each pair of these behaviour genes to zero.
3. Identify genotypes which exhibit the behaviour described in the pair from the pool of solutions.
4. Add the fitness of the phenotype which bears the current adjacent pairs to the appropriate entry in the table.

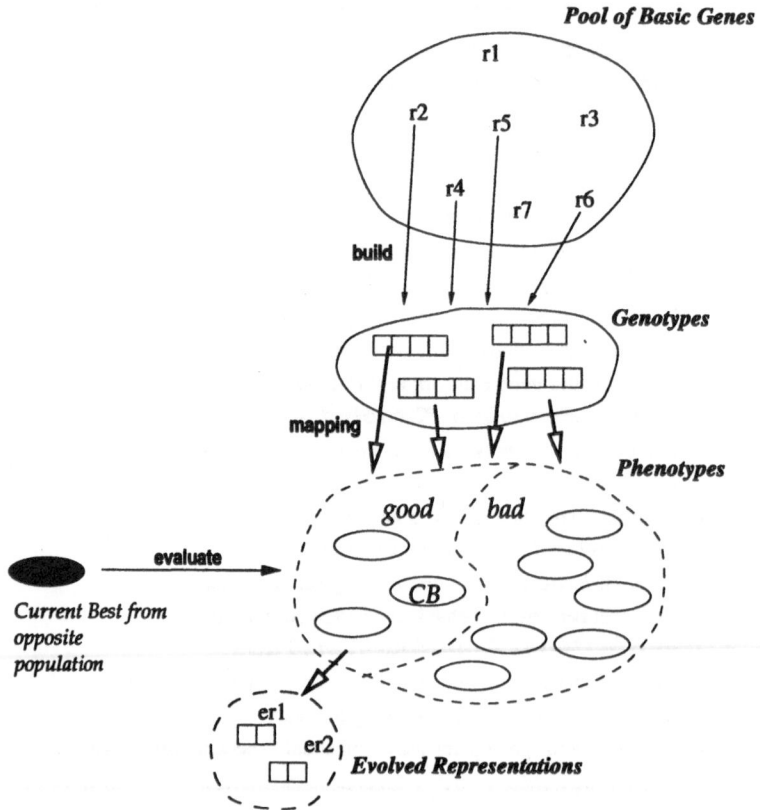

Figure 8. Generic mechanism to generate new representations.

This general algorithm can pick up behaviour patterns which are commonly exhibited by the "good" population in the pool of solutions. To include a preliminary selection such that only solutions from the structure space which are evolved-structure-carriers, this algorithm can pick up behaviours which are generally displayed among these evolved-structure-carrier solutions.

The fitness function varies at each cycle. A different CBB is used to measure the performance of generated structures, and a different CBS is used to measure performance of generated behaviour. This approach to forming a fitness function is like having a different target case each time. However, both the CBB and CBS encapsulate new complex knowledge that emerge with the design process.

6. An Example: Emergent Behaviour in Floor Plans

We extend the example of Gero and Schnier (1995) to include a representation of behaviour. Figure 9 shows the genotype and phenotype representations of structure and behaviour. In the original example of floor plan layout in Gero and

Schnier (1995), genotypes are built from "basic" genes (the pen movements of
←, ↓, → and ↑) and evolved complex genes (which are learned from a target
design case). The phenotype is represented as a floor plan. In their example, the
basic genes and a phenotype are considered to be along the dimension of the
structure space in Figure 9.

In the work of Gero and Schnier (1995), two processes are carried out: *learning* and *design*. The learning process is first executed and completed before the
commencement of the design process. These two processes have different criteria
to measure the fitness of a phenotype. During the *learning* process, the fitness of a
phenotype is classified as "good" or "bad" according to its closeness to the target
design case. "Good" phenotypes are chosen where statistical correlations between
the recurrence of gene-pairs and fitness of phenotypes are identified. Gene-pairs
which achieve better than a user-defined threshold value will be included as a
complex representation to be remembered. While the goal at *design* stage is to
design a floor plan which has minimal overall wall length under three design constraints. One of the generated designs can be found in Figure 10.

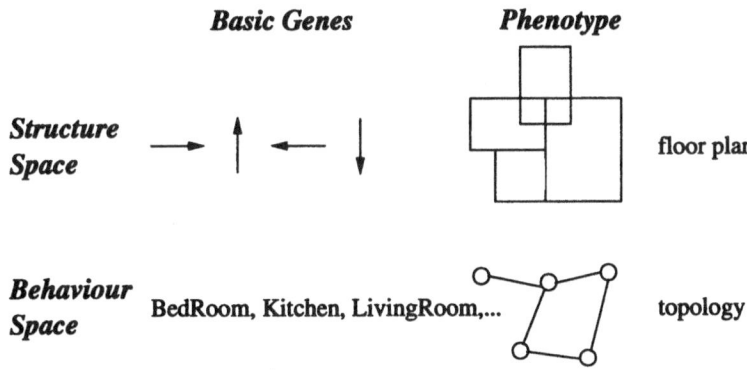

Figure 9. Basic genes and phenotype in the structure space and behaviour space of a floor plan
design.

Here, we extend the example in two ways such that (1) pen movements include a jump (for example, → →) to indicate an entrance; (2) appropriate behaviours are required to be exhibited by the room arrangements. The objective
is to re-use the implicit behaviour between areas on the floor plan. The "basic"
genes for the behaviour space is a collection of room-types and the phenotype is
a topology to describe relationships between these rooms (along the dimension of
behaviour space in Figure 9). The fitness of a behaviour-phenotype is measured
in terms of its closeness to the topology of the given design. The quality of each
gene-pair is again a numerical value to correlate occurrence and performance of
phenotypes. A threshold value is defined to select complex behaviours as emergence.

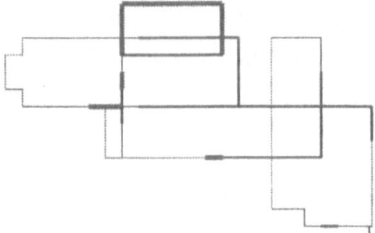

Figure 10. Generated floor plans.

6.1. ROOM ADJACENCY

We define the adjacency of two rooms A and B, adjacent(A, B), where there is
an entrance or pathway to lead from room A to B, and vice versa. However, to
resolve relationships such as **adjacent(adjacent(adjacent(A, B), C), D)**, this becomes much harder with our primitive definition of adjacency. Hence, *adjacency*
is defined as

- there is an entrance or pathway to lead from room A to B, and vice versa, as
  shown in Figure 11(a) where a solid line indicates a wall and a dashed line
  to represent an entrance; or
- A, B and C are adjacent *iff* A is adjacent to B and B is adjacent to C and C is
  adjacent to A, e.g. Figure 11(b).

Figure 11. Examples of adjacency.

The floor plan of a house in Figure 12 provides an illustration of the representation. This floor plan has ten room-types: {Lounge, DiningRoom, Kitchen, Bath,
Bedroom1, Bedroom2, MasterBed, Ensuite, Entry, Corridor} which makes up 10
edges on the adjacency graph. An adjacency graph is a graph where the nodes
represent the room-types and the edges denote the two room-types are adjacent to
each other.

Adjacency pairs of the enclosed areas are stored in a list, L_{adj}. These adjacency pairs will be compared against the required adjacency. Redundant genes are
allowed in the L_{adj} where the length of L_{adj} varies from floor plan to floor plan.
The relationship between genotypes and phenotypes, structure space and behaviour space can be found in Figure 13.

Structure-Phenotype:
Floor Plan

Behaviour-Phenotype:
Adjacency Graph

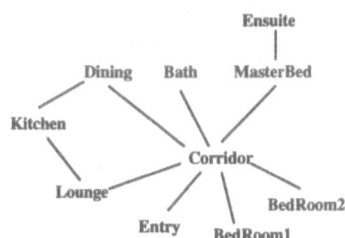

Figure 12. Representation of behaviour.

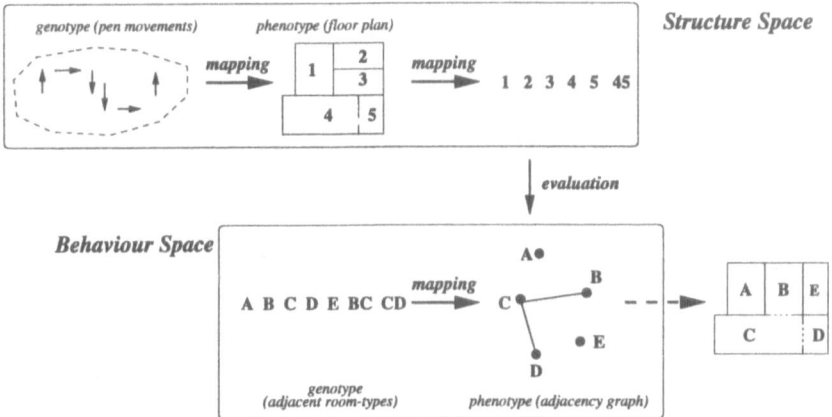

Figure 13. Relationships between genotype and phenotype, structure space and behaviour space of a floor plan design.

6.2. EVALUATION AND EMERGENCE

The characteristics of a co-evolutionary design is that it has a changing fitness function. The fitness function evolves with time while it maintains to be relevant to the initial expectation. Hence, we define a fitness function in a co-evolutionary paradigm as a concatenation of initial requirements R_{init} and current best behaviour (or structure), i.e. Fitness = R_{init} + CB (B or S).

During the conceptual design stage, user requirement is usually a vague one which may be a list of room-types, with a possible specification of some room-type adjacency. Any initial room-type adjacency will be defined as the current best behaviour (CBB).

During the synthesis of solution, $B_e \rightarrow S(B_s)$, genotypes in the structure space are first generated to describe the turtle-graphic drawing for the phenotypes (floor plans). The phenotypes are then matched against the R_{init} and CBB. The

CBB is empty or very simple at this initial stage, however, this will be getting more and more complex along with the time spectrum. There are two matchings to complete. The first matches the number of enclosed areas in a floor plan solution (which represent rooms) and the number of room-types in R_{init}. The second match compares the adjacencies on the floor plan solution to the adjacencies in CBB. Using the adjacency graph in Figure 13 for illustration, the pairs BC and CD require the floor plan to have a room (if it is assigned to be Room-C) to have openings (entrance) to Room-B and Room-D. This CBB does not specify the connections between other rooms. This CBB is satisfied as soon as there exists a pathway to lead from Room-C to Room-B and another pathway to Room-D. The best performing floor plan after the evaluation is to become the current best structure (CBS). The phenotypes are then classified into two categories, the "good" and the "bad". Adjacent pairs of genotypes in the "good" subpopulation are analysed to identify how these pairs contribution to the "goodness" in the phenotypes. Pairs which demonstrate high statistical correlation to fitness of structure-phenotypes are added to the pool of basic pen movements. Each of these evolved structures will be labelled as a single entity and can be re-used in subsequent cycles.

During the reformulation of problem specifications, $S \rightarrow B_e$, current best floor plan (CBS) from the structure-space is used as the goal to assess performance of relationships in behaviour-space. Many adjacency graphs are generated to satisfy the CBS on hand. Each graph is first tested if it has the same number of rooms as on the floor plan of CBS. Then the adjacency of the CBS is matched against the adjacency graph. Since there is no-function-in-structure, the CBS does not have an implicit assumption behind each enclosed areas. Phenotypes in the behaviour dimension are submitting ideas to propose how these areas can be best utilised. The best adjacency graph will be kept as a CBB for the evaluation of structures in the next cycle. The population of behaviour-phenotypes are also classified into "good" and "bad". Genes in the "good" population are analysed pairwise. Pairs are added to the pool of behaviour-genes if they display strong evidence of correlation with "fit" phenotypes.

Let's say the pair {Corridor, Lounge} is an emergent behaviour from a cycle. This pair will become a single entity. The recurrence of this pair of adjacent rooms in subsequent cycles will be replaced by the single entity. The end result after the process is a list of emergent (and complex) representation of useful behaviours to be applied immediately within the design process. The emergent behaviours begin with simple patterns, but they become more complex when a pair of adjacent genes is emergent behaviours from previous cycles.

The behaviour-space and structure-space represent a cooperative type of co-evolution. The fitness function in the co-evolutionary paradigm is an emergent and complex one. The function imposes more sophisticated requirements while proposed solutions are getting more complex as time proceeds.

7. Conclusion

Emergence has been applied primarily to shapes in the *design community*. This limitation can be complemented by the various explanations from the *ALife community* where it is usually applied to behaviour. "Local interactions of low level units" is the key to emergence which can be achieved by a data-driven approach (Gero and Damski 1994). However, this approach still awaits to be tested outside the domain of shapes. For example, in a 2D visual representation in Figure 14, the triangle (denoted by the solid line) evolved outside the data points is found by the extrapolation of line segments (relationships) between data points. This can be resolved using the half plane concept proposed by Gero and Damski (1994). However, for the symbolic representation outside the *non*-shape domain, the problem lies on the "search" for a generalized relationships between low level units, before these relationships start to interact which leads to emergence.

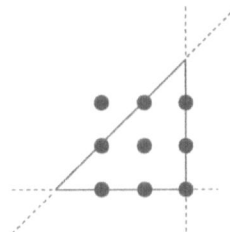

Figure 14. Triangle emerged from extrapolation of line segments.

This paper further draws on the work by Gero and Schnier (1995) and argues its bottom-up approach to evolving representation qualifies this algorithm to be a computational tool for emergence. It is discussed that an evolving complex behaviour is an emergent behaviour. Emergence can take place in co-evolutionary design where the learning of emergent pattern is part of the design process and is in response to alternative solutions for structure or behaviour.

Acknowledgements

This work is supported by an Australian Postgraduate Award.

References

Corne, D., Smithers, T. and Ross, P.: 1993, Solving design problems by computational exploration, *in* J. S. Gero and F. Sudweeks (eds), *Preprints of the IFIP WG5.2 Workshop on Formal Design Methods for Computer-Aided Design*, pp. 249–270.

Edmonds, E. and Soufi, B.: 1992, The computational modelling of emergent shapes in design, *in* J. S. Gero and F. Sudweeks (eds), *Preprints of the Second International Conference on Computational Models of Creative Design*, Dept of Architectural and Design Science, Unveristy of Sydney, pp. 173–189.

Finke, R. A., Ward, T. B. and Smith, S. M.: 1992, *Creative cognition: Theory, research, and applications*, MIT Press.

Fogel, D. B.: 1995, *Evolutionary computation: Toward a new philosophy of machine intelligence*, IEEE Press.

Forrest, S.: 1990, Emergent computation: Self-organizing, collective and cooperative phenomena in natural and artificial computing networks, *Physica* **42**, 1–11.

Gero, J.: 1992, Creativity, emergence and evolution in design, *in* J. S. Gero and F. Sudweeks (eds), *Preprints of the Second International Round-Table Conference on Computational Models of Creative Design*, Key Centre of Design Computing.

Gero, J.: 1993, Towards a model of exploration in computer-aided design, *in* J. S. Gero and F. Sudweeks (eds), *Preprints of the IFIP WG5.2 Workshop on Formal Design Methods for Computer-Aided Design*, Key Centre of Design Computing, University of Sydney, pp. 271–279.

Gero, J., Damski, J. and Han, J.: 1995, Emergence in caad systems, *in* M. Tan (ed.), *Proceedings of CAAD Futures'95*.

Gero, J. S. and Damski, J.: 1994, Object emergence in 3d using a data driven approach, *in* J. S. Gero and F. Sudweeks (eds), *Artificial Intelligence in Design'94*, Kluwer, Dordrecht, pp. 419–436.

Gero, J. and Schnier, T.: 1995, Evolving representation of design cases and their use in creative design, *in* J. Gero and F. Sudweeks (eds), *Preprints of the Third International Conference on Computational Models of Creative Design*, Key Centre of Design Computing, University of Sydney, Sydney.

Gero, J. and Yan, M.: 1994, Shape emergence by symbolic reasoning, *Environment and Planning B: Planning and Design* **21**, 191–218.

Goldberg, D. E.: 1989, *Genetic algorithms: In search of optimization and machine learning*, Addison-Wesley.

Hofstadter, D. and McGraw, G.: 1993, Letter spirit: An emergent model of the perception and creation of alphabetic style, *Technical Report CRCC-68*, Center for Research on Concepts and Cognition, Indiana University.

Hybs, I. and Gero, J. S.: 1992, An evolutionary process model of design, *Design Studies* **13**(3), 273–290.

Jonas, W.: 1993, Design as problem-solving? or: Here is the solution - what was the problem, *Design Studies* **14**(2), 157–170.

Liu, Y.-T.: 1995, Some phenomena of seeing shapes in design, *Design Studies* **16**, 367–385.

Maes, P.: 1990, Situated agents can have goals, *in* P. Maes (ed.), *Designing Autonomous Agents*, MIT/Elsevier, pp. 49–70.

Maher, M. L. and Poon, J.: 1995, Co-evolution of the fitness function and design solution for design exploration, *Proceedings of IEEE International Conference on Evolutionary Computing*, IEEE.

Maher, M. L. and Poon, J.: 1996, Modelling design exploration as co-evolution, *Microcomputers in Civil Engineering (Special Issues on Evolutionary Systems in Design)*, to appear.

Maher, M. L., Poon, J. and Boulanger, S.: 1995, Formalising design exploration as co-evolution: A combined gene approach, *in* J. S. Gero and F. Sudweeks (eds), *Preprints of the Second IFIP WG5.2 Workshop on Advances in Formal Design Methods for CAD*, Key Centre of Design Computing, University of Sydney, Sydney, pp. 1–28.

Mataric, M. J.: 1993, Designing emergent behaviors: From local interactions to collective intelligence, *in* H. L. R. Jean-Arcady Meyer and S. W. Wilson (eds), *From Animals to Animats 2: Proceedings of the Second International Conference on Simulation of Adaptive Behavior*, MIT Press, pp. 432–441.

Mitchell, W, J.: 1993, A computational view of design creativity, *in* J. S. Gero and M. L. Maher (eds), *Modelling Creativity and Knowledge-Based Creative Design*, Lawrence Erlbaum Associates.

Steels, L.: 1991, Towards a theory of emergent functionality, *in* J.-A. Meyer and S. W. Wilson (eds), *From Animals to Animats: Proceedings of the First International Conference on Simulation of Adaptive Behavior*, MIT Press, pp. 451–461.

Taylor, C. E.: 1990, "Fleshing out" artificial life II, *in* J. D. F. Christopher G. Langton, Charles Taylor and S. Rasmussen (eds), *Artificial Life II: Proceedings of the Workshop on Artificial Life*, Addison-Wesley, pp. 25–38.

J. S. Gero and F. Sudweeks (eds), Artificial Intelligence in Design '96, 723-742.

INNOVATIVE DESIGN BASED ON SHARABLE PHYSICAL KNOWLEDGE

VALERI V. SUSHKOV AND LAMMERT K. ALBERTS

AND

NICHOLAAS J. I. MARS
Department of Computer Science
University of Twente
PO Box 217, 7500 AE Enschede, The Netherlands

1. Introduction

Building a computational model of the early engineering design phases is a necessary condition to develop automated support for the entire process of computeraided design. Over the last few years, many research efforts have been concentrated in this area but mostly limited to modeling reasoning process about conceptual design within a single engineering domain. However, as follows from a nature of the innovative design one of its main characteristics is the use of knowledge unknown in the domain a priori.

According to a systematic design methodology (Pahl *et al.*, 1984), the entire cycle of a design process can be presented in a top-down manner:

1. Formulation of initial design specifications (IDS). In the innovative design phase, IDS may only indicate certain key features of the required design and neglect its detailed aspects.
2. Generation of design concepts that would meet the IDS.
3. If several alternative concepts have been generated, further revision and refinement of the IDS based on specific aspects of alternatives. Evaluation of the alternatives to select a concept that would meet all the requirements.
4. Formulation of the exact functional specifications with respect to the concept selected.
5. Generation of a new design description through instantiation of conceptual knowledge into specific design.

The most difficult step in this methodology is formulation of the IDS in such a way that it would be possible to systematically find alternative concepts that would be capable of meeting them. This means that explicit relationships between all possible requirements that may arise and available natural and design knowledge have to be established.

To develop a framework for modeling domain-independent innovative design one has to find answers to the following questions:

 — what is needed to overcome the ill-structuredness of initial problem formulations;
 — what knowledge is needed to generate new design concepts according to specifications given;
 — in what form should this knowledge be represented;
 — what reasoning method should be used to manipulate this knowledge effectively.

The paper discusses intermediate results obtained during developing the Invention Designer project aimed at automation of the early design phases. A product-based model of innovative design presented below is based on a systematic approach to structuring conceptual knowledge for innovative design.

To overcame ill-structuredness of problem formulations, an ontology of design specifications is proposed.

2. Types of Innovative Design

The essence of engineering design process is the mapping between known physical knowledge given a required function into a description of a realisable design product. Every engineering domain contains a set of fundamental physical laws (or "first principles") thus providing a basis for designing new products through combining and instantiating these principles. The use of first principles governs tasks of routine design or redesign rather well. The problem arises when a designer experiences the lack of domain-dependent fundamental knowledge on the basis of which a new product can be designed.

Even a quick look at the evolution of modern technology reveals that one of its trends is to create products embedding diverse physical principles. Due to complex technical, ergonomic and ecological constraints even simple products tend to integrate various technologies. In many cases, a complex and unreliable design based on the principles of one domain can be replaced by a simple and reliable design built upon the principles of another domain. Essential there is that knowledge required for innovative design can be drawn from different domain and a new product can utilize physical principles from different domains.

Transferring physical knowledge from one domain to another can significantly simplify a design. Suppose, a problem is formulated as to prevent an electrical motor from overheating. One of the known solutions within the electrical domain uses a temperature sensor which reads the current temperature value and an electronic system switching the power supply off when the threshold value of the temperature is reached. This problem can be solved much easier and the overall design can be made more reliable if the poles of the motor are made of an alloy with a Curie point equal to the required threshold value of the temperature. When the temperature reaches the threshold value, the magnetic properties of the poles change and the motor stops (Petrovich and Tsourikov, 1986). The necessity of introducing a complex and unreliable additional design has disappeared.

As a consequence, it is possible to distinguish between two types of innovative designs with respect to engineering domains where a new design was generated:

1. **Domain-dependent** innovative design, when a new product is entirely based on previous domain knowledge and physical principles.
2. **Domain-independent** innovative design, when a new product is based on a new physical principle that has not been used in the domain before.

To organize the cross-domain search for appropriate design knowledge at the level of specific design descriptions is nearly impossible: all available specific design knowledge of each engineering domain would have to be stored and properly indexed. Another problem is that it is unclear how to adapt previous knowledge to a new situation - no proven technique for conceptual adaptation of specific knowledge is available yet. The task can be simplified by organizing the search at the level of fundamental physical principles and using a procedure based on synthesis rather than on adaptation.

There are several known approaches exploring ways of using modeled physical knowledge in the innovative stages of engineering design, such as: design from first principles (Williams, 1990); synthesis of new design concepts based on Bond Graph approach (Zaripov, 1988; Malmqvist, 1993; Schmekel, 1992); systematic design on the basis of Design Catalogues (Pahl *et al.*, 1984); and modeling domain-dependent engineering principles in form of primitive generic design components and synthesis of design concepts from these primitives (Alberts, 1993).

All of these approaches deal with instantiating physical principles into specific design descriptions. The common idea behind the methods is the direct mapping between *physical* functions and relevant physical principles. This may produce the desired results in a case when it is known what physical function is required. On the other hand, in the innovative design phase, a problem might be formulated in terms that have nothing to do with physical entities. To overcome this situation, one should understand how to translate ambiguous formulations inherent to conceptual design phases into functions expressed in physical or design terms.

Using physical handbooks and encyclopedia does not help much because physics as described in those sources is not adequately structured and organized properly with respect to engineering tasks. A systematic approach to organize this knowledge is required. Conceptual physical knowledge can be drawn from classes of specific designs and should describe general functional and design aspects of groups of designs based on the same physical principle.

In order to develop a knowledge-based support for the domain-independent innovative design a unifying approach to structuring and representing various types of knowledge is required. This means breaking barriers created by the historical division of engineering sciences into different domains. Therefore, fundamental principles that would have to be used to provide the domain-independent innovative design are to be represented in a form that would enable reasoning about conceptual aspects of existing artificial systems essential to design new products.

3. TIPS Pointer to Physical Effects

An approach to tackle the problems mentioned in the previous section is proposed by the Theory of Inventive Problem Solving (TIPS) (Altshuller, 1988; Sushkov *et al.*, 1995). Apart studying evolutionary trends of artificial systems and discovering general principles of innovative design, Altshuller describes the framework for organizing the use of fundamental physical knowledge in creative and innovative designs.

The physical knowledge is represented in a form of generally described physical phenomena and can be retrieved and used in a new situation according to the *generalized technical function* performed by a group of different designs based on the same phenomenon. Two designs that do not necessarily belong to the same engineering domain and based on the same phenomenon which is used to produce the same function are regarded as analogous designs. For instance, a phenomenon of thermal expansion may be used to perform a function "to displace an object" in various contexts.

In TIPS, all the physical phenomena are collected in *TIPS Pointer of Physical effects (PPE)* (Selutski, 1987). The physical effects in PPE are structured according to generalized technical functions. The functions are labeled generalized since they indicate some non-precise technical result that can be achieved by a whole group of different designs, for instance, *to control an electromagnetic field.* The current set of functions which is used in the software system Invention Machine [1] counts hundreds of functions covering most of the possible IDS.

Apart from the database of physical effects and lists of functions, the applicability of each effect is illustrated by analogous design cases drawn from various engineering domains. The analogous cases are used to demonstrate how the same effect is used to achieve the same function in different design implementations. A fragment of the organization of the PPE is shown in Figure 1.

As shown in the picture, PPE defines functions in a way different from what we can see, for instance, in Design Catalogues (Pahl *et al.*, 1984). The functions

---

[1] Invention Machine is a trademark of Invention Machine Lab, Minsk, Belarus

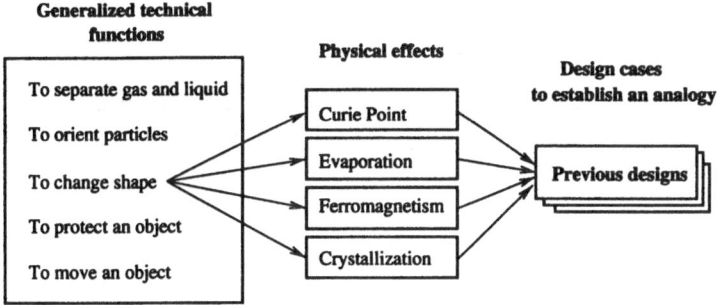

Figure 1. A fragment of the pointer to physical effects.

formulated in PPE cover the whole classes of manipulations with abstract physical quantities, without exact identification what engineering domain they belong to. Examples of technical functions in PPE are *to protect substance, to separate gas and liquid, to increase speed, to control displacement.*

Summarizing, PPE has three basic features:

1. Technical functions are used as indices of physical effects. They were collected and generalized after studying a large number of patents.
2. A technical function can be mapped onto a single effect directly performing the necessary physical function as well as onto an effect which does not produce the required final result itself but is a necessary auxiliary effect for obtaining the function.
3. PPE also includes a collection of design cases, illustrating design applications of each effect. The collection is designed to help in establishing an explicit analogy between the previous case and a new problem.

A drawback of PPE is that no formal representation of the physical effects is suggested. As a consequence, a solution to an inventive problem is only suggested as a set of physical effects without an indication of what a design implementation based on these effects will look like. This leaves the evaluation of the applicability of the retrieved effects to the responsibility of a human designer. Also, specific design implementations used in PPE as analogous design cases are domain-specific and do not always help in establishing an analogy with a new situation.

4. Integration of Experiential and Exact Knowledge

The purpose of product-based modeling of design is to enable reasoning about the relationships between design products and its functions (Tong *et al.*, 1992). It is possible to distinguish between two strategies for mapping between previously used designs and functions (Figure 2):

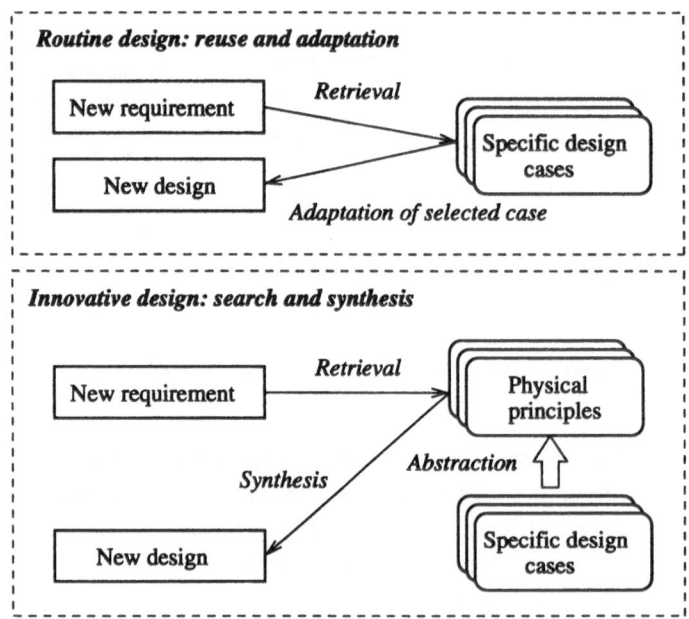

Figure 2. Two ways of using of previous knowledge in innovative design.

1. Reuse of design knowledge incorporated in previous designs and adaptation of the knowledge to a new situation by using AI techniques for reusing past experience, like case-based reasoning (Sycara *et al.*, 1989; Kolodner *et al.*, 1993). This makes it possible to index previous design cases and use the indices to retrieve analogous cases. However, the probability of finding a previous design case similar to a given one is strongly limited by 'missing' knowledge. This point is crucial for the innovative phase of design. It does not seem possible to index and store all aspects of a design solution: the set of possible initial requirements and all physical and design aspects of a final artifact is infinite.

2. Use of previous design knowledge represented in terms of primitive design components - generic building blocks. A function given is decomposed into a set of primitive functions and a new design description is synthesized by combining corresponding components fulfilling each primitive function (Zaripov, 1988; Alberts, 1993).

Using the physical principles in innovative design can lead to two types of innovative design solutions obtained by the use of physical principles: i) the principle was previously used as a basis of designing some artifact in some engineering domain and ii) the knowledge has not been instantiated into any design yet. The model of the innovative design introduced in the paper is limited to the first time of innovative design solutions.

Analysis of more than 400.000 patents made by Altshuller indicated that the second situation is only met in two percent of all inventions (Altshuller, 1988), virtually all of which are based on recent scientific discoveries that had not been used before. Brown and Chandrasekaran 1985 also subdivide the innovative design category into *inventions* and *innovations*; the latter are based on innovative use of existing knowledge or designs. This means that for most new design problems requiring an inventive solution existing knowledge can be used and no generation of new fundamental knowledge is required. We regard this as a strong argument why a systematic approach to innovative design can be based on using previous experience.

We argue that the best way to model innovative design based on fundamental knowledge is to integrate both experiential knowledge that deal with managing IDS and exact knowledge represented in form of generic building blocks. The role of exact knowledge in our model is twofold: first it enables one to represent various types of conceptual knowledge in a uniform way, second it is used for automated synthesis of new design concepts by combining and instantiating fundamental physical knowledge. We argue that a knowledge base incorporating all available physical principles can be effectively used to organize the use of deep physical knowledge to provide knowledge-based support for computer-aided innovative design.

We distinguish between two major directions in which our current research is being carried out:

1. Modeling physical principles to be sharable between different engineering domains. We regard this as a task of generic knowledge modeling since *deep physical knowledge*
   can be used to generate a number of various types of specific design knowledge. Structuring and organizing modeled generic physical knowledge into a knowledge base.
2. Translation of initial key requirements into physical functions. In this activity, one has to formalize the collection of experientially defined *technical* functions of PPE and relate it to a predefined set of *physical* functions.

In the next sections, we will discuss a framework for modeling physical principles on the basis of YMIR – an ontological approach to modeling generic design knowledge, examples of modeled principles, and present fragments of the ontology for IDS.

5. YMIR: a Sharable Ontology for Modeling Design Knowledge

To model generic physical knowledge for innovative design we use YMIR, a sharable ontology for modeling design knowledge in a uniform way. YMIR was developed at the Knowledge-Based Systems Group of the University of Twente (Alberts, 1993) and defines a taxonomy of concepts for the formal description of design knowledge in different domains. The concepts in YMIR for the elements from which to synthesize technical system descriptions are called *generic system models (GSM)*. These generalized concepts have been defined in terms of network models in System Theory (Shearer *et al.*, 1969). Generic system models explicitly incorporate the relation between such features of an engineering system as behavior and form. The advantage of YMIR is that the modeling framework is applicable to all domains in which technical systems can be described as system-theoretical network models.

In contrast to Bond Graphs, YMIR is not limited to only model energy aspects of a design. Using material and geometrical aspects of designs to model their behaviors makes it possible to incorporate necessary information on material properties of designs. This claim is important since, in many situations, initial key requirements are formulated in terms of changing parameters of material components. As a consequence, the requirements can not be directly expressed in terms of energy transformations without knowing a particular effect which can meet those requirements.

6. Generic Physical Principle (GPP)

In YMIR, the GSM is used as a basic concept to represent generic design knowledge. The same framework can be used to model fundamental physical knowledge from the engineering point of view. In that case, a physical principle is represented as a system that has input and output ports and its behavior comprises different physical functions. A GSM representing a physical principle is more abstract in nature than a GSM representing specific design knowledge because it has no influence of design constraints and other specific limitations.

Any part of a system performing some specific function can be modeled as a set of fundamental physical phenomena occuring in the design. This makes it possible to apply the same physical model to represent different groups of designs. From a knowledge-based point of view, any physical phenomenon might be represented as a tuple:

$$E = <E, M, C, t>,$$

where E is a set of energy parameters, M is a set of material parameters, C is a set of context variables (conditions making the effect occur), and t is the time variable. On the basis of this definition the concept of a *Generic Physical Principle (GPP)* is introduced. A GPP is a model of an abstract physical system based on a specific physical phenomenon. The basic
model of a GPP has three parts:

1. *Behavior*: a set of relations between input and output energy flows.
2. *Form*: material and geometrical parameters which determine the behavior of GPP.
3. *Structure*: lists of ports for input and output energy flows.

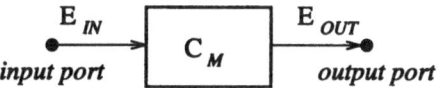

Figure 3. Graphical form of GPP.

The basic idea behind the GPP concept is that the behavior of a system based on any physical principle can be expressed as a set of related physical functions, each of which, in turn, can be instantiated into a multitude of technical functions by adding specific contexts. Therefore, a GPP can be regarded as a high-level model of generic design knowledge which can be instantiated into a primitive design concept by adding constraints and domain-specific information.

A general form of GPP is depicted in Figure 3. Here, C_m is a specific physical property of a material which provides the transformation of an input energy flow E_{in} into an output energy flow E_{out}. Two types of GPPs are distinguished:

1. *Homogeneous GPP*: input and output flows are of the same type of energy.
2. *Heterogeneous GPP*: input and output energy flows are of different types of energy. The elements transforming one type of energy into another are known as *transducers* in System Theory.

Summarizing, GPP incorporates both high-level *energy transformation* knowledge as a capability of some physical phenomenon to produce physical functions and *material* aspects that make this transformation possible. Availability of material-related information in GPP enables one to translate between a GPP and more specific design description.

A uniform way of representing GPPs and using the law of energy conservation make it possible to link several primitive GPPs into a more complex structure labeled a *Generic Design Concept (GDC)* (one of the possible definitions of a Design Concept notion in general is given in and). GDC is a synthesized concept of a new design which does not contradict physical realizability.

7. Modeling at Macro- and Microscopic Levels

Each GPP is modeled as a system model at the appropriate level of abstraction. The choice of the level of abstraction depends on what energy and form-related information on a physical effect is crucial to translate between physical and engineering levels and to check the applicability of translated knowledge against functional requirements.

Let us consider, for instance, the effect of Joule heat. The Joule heat arises in a conductive material when electrical current passes through the material. The heat is generated as a consequence of increasing internal energy of the material through collisions of migrating electrons which lose own energy. This effect is used in many technical systems, like electric stoves, toasters and hair dryers.

There are, at least, two levels of abstraction at which the modeling might be performed: *microscopic* and *macroscopic*. Modeling physical behavior at the microscopic level requires establishing explicit relationships between energy characteristics of migrating electrons and the general increase of internal energy of the material. To model the effect under macroscopic observations means to investigate what information is crucial to achieve an externally observed result: arising of heat in a resistive material as a consequence of the current passing through the material. More precisely, we have to establish relations between input and output energy flows, and to specify how particular form-related aspects of a material are involved in the relations.

The latter gives no clear understanding what internal physical processes cause the effect but exactly shows what is needed to cause the resulting effect and how one or another property of a material subjected to an action of external energy flow can be used in a practical way. In that case, the internal physical behavior does not need to be modeled unless we are interested in in comprehensive modeling of the processes behind the effect. However, to make a final decision what level of abstraction would be more appropriate for modeling, first we need to analyze what information on the effect is necessary and enough to translate between the physical model and its design implementation.

This method of modeling an external behavior of a system as a black-box might be used for all the types of transducers. An external energy flow amplifies some internal physical process inside the material which is very weak to be observed or can not be applied within the required range of parameters. In other words, the source of energy already presents in a system and may produce much more required energy by converting flow of some other type of energy. It turn, existence of the weak energy source is provided by some other external energy flow acting on the material (like gravitational forces or thermal radiation of environment). If this external flow has no influence on the system workability within the required interval of parameters, it might be omitted while modeling.

8. Modeling Physical Principles as GPPs

YMIR introduces two types of variables that are used to describe a behavior of a system:

1. *System variables* which describe energy flows in a system. Two
   types of system variables are distinguished: *implicit* variables which describe
   the energy flow through the system (e.g., electrical current and force flow),
   and *explicit* variables, which describe the potential differences across the
   system (e.g., voltage and displacement respectively).
2. *Form-related variables* which describe principal material and

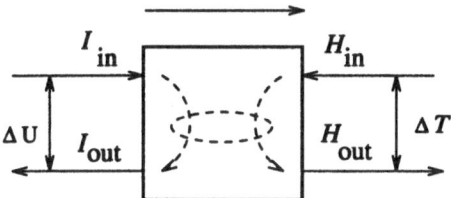

Figure 4. General model of the effect of Joule heat.

geometrical properties of a design. They serve as parameters in the equations relating the system variables (e.g., length, modulus of elasticity).

Besides, derived forms of both types of variables can be used as arguments of functions. For instance, power and kinetic energy are derived system variables and area and volume are derived form-related variables.

The behavior of a generic system model is expressed as a set of relations between input and output system variables, where form-related variables serve as relating parameters. The physical behavior of the effect of Joule heat can be defined by the tuple:

$$\Re =<< U_{in}, I_{in} >, < T_{out}, H_{out} >, F >$$

where U_{in} and I_{in} are input voltage and current, T_{out}, H_{out} are output temperature difference and the heat flow, and F is a set of form-related variables which are used as parameters in the equations of energy transformations.

The model of the effect of Joule heat discussed above is depicted in Figure 4. The arrows indicate directions of specific energy flows. The arrow above the box indicates the direction of the energy transformation.

A relation between the pairs of input or output system variables is established in a correct way easily whereas to establish the relation between input and output system variables at the same level of abstraction is not possible. The problem arises because modeling of the same physical property of a material in terms of both domains is not allowed. In terms of electrical domain, the relation between the voltage U and current I is established by the Ohm law $I = U/R$ whereas an electrical resistance R may not be used as a form-related variable to establish the relation between system variables characterizing the domain of heat and mass transfer. To establish this relation one has to use appropriate form-related variables of the thermal domain, like thermal capacity or thermal resistivity. To establish a relation between input and output energy flows one should investigate how form-related variables from both domains are interrelated.

As said in previous section, a solution to this problem consists in modeling the external behavior of the effect and to hide microscopic-level information inside a black box. Therefore, the behavior of the effect can be defined as a relation between both input system variables and relevant output system variables. In the effect of Joule heat, such variable is the difference between values of the temperature of the material before and after heating. Thus, the observed temperature difference is a time-dependent function of the internal process of heat and mass transfer occuring inside the material.

The model of the effect of Joule heat is a system in which an output is defined as a thermal capacitance. A particular behavior of the system depicted in Figure 4, where a subject of interest (or essence of the effect, in other words) is the relation between the temperature difference and input voltage and time required to achieve the difference. According to the law of energy conservation, we assume that all electrical energy is transformed into a thermal energy generated by the heat source, and the thermal capacitance is created by the material from which the material component is made of.

As known from System theory, thermodynamic power is defined as a product of entropy flow and temperature. Therefore, a relation of electrical-thermal transformation can be written as $E\Delta T = I\Delta U$, where E is entropy flow through the system. However, entropy was introduced as an associated parameter with the flow of heat through a thermal resistance, since the thermal resistance dissipates no energy and the net heat flow is always zero. Hence, it is assumed that $E\Delta T = H$, where H is the heat flow.

As a result of the heat flow through a given material, this material stores internal energy by virtue of temperature rise. For a real thermal system including non-dot material component the relation between the temperature rise and amount of heat flowing through the component is:

$$\Delta T_{out} = \int_{t_1}^{t_2} \frac{H}{M_m C_m} \, dt;$$

where the system variable ΔT_{out} is the temperature difference for the time period from t_1 to t_2, and form-related-variables are: M_m and C_m are the mass and the specific heat of the resistive material respectively.

To define the external behavior of the effect of Joule heat we relate both pairs of system variables as

$$\frac{dT}{dt} = f(\frac{dU}{dt});$$

and substituting the heat flow variable with a relation for electrical power in terms of input voltage and electrical resistance, we obtain the equation for a particular behavior of the effect of Joule heat:

$$\Delta T_{out} = \int_{t_1}^{t_2} \frac{U_{in}^2}{R_m M_m C_m} \, dt.$$

Although the effect of Joule heat is non-reversible, this relation is correct unless the rise of temperature influences the relation between input system variables. At into account another physical effect which always accompanies the effect of Joule heat: electrical resistivity of a conductor is changed as a consequence of heating the conductor. This influence can be regarded as a feedback, and this relation must be included in the model of the system forming the effect as a part of its behavior.

To model this part of the effect behavior, first the electrical resistance is defined as a form-derived variable in terms of form variables:

$$R = \rho_M \frac{L_m}{A_m};$$

where: ρ_M is electrical resistivity of the material, L_M is the length of the conductor and A_M is its cross-sectional area. For temperature intervals of as much as few hundred degrees, electrical resistivity is related to temperature by a linear expression of the form:

$$\rho = \rho_0(1 + \alpha_M \Delta T)|_{\Delta T = T - T_0},$$

where α_M is the temperature coefficient of the resistivity, and ρ_0 is the electrical resistivity at a temperature T_0.

Substituting $R = \rho_0 L/A$, we can define the behavior of the system based on the Joule effect as:

$$\Re = \Delta T_{out} = \int_{t_1}^{t_2} \frac{U_{in}^2}{\rho L_m A_m M_m C_m} \, dt;$$

with the feedback constraint equation is $\rho = \rho_0(1 + \alpha_M \Delta T_{out})$.

Summarizing, an overall model of the effect of Joule heat includes three parts: form, structure specified by input and output ports, and behavior as a set of relations between system variables (Tables 1, 2, and 3).

TABLE 1. GPP behavior

Type of relation	Physical model	Mathematical expression
Input system variables	$R_I = \langle I_{in}, U_{in}, R \rangle$	$I = \frac{A_m U}{\rho_M L_m}$
Output system variables	$R_O = \langle H_{out}, T_{out}, C \rangle$	$\Delta T_{out} = \int_{t_1}^{t_2} \frac{H}{M_M C_M} \, dt$
Transformation	$R_T = \langle T_{out}, I_{in}, U_{in}, H_{out} \rangle$	$\Delta T_{out} = \int_{t_1}^{t_2} \frac{U_{in}^2}{\rho L_M A_M M_M C_M} \, dt$
Feedback relation	$R_F = \langle I_{in}, T_{out} \rangle$	$I_{in} = U/\rho_0(1 + \alpha_M \Delta T_{out})$

The same approach is used to model the effect of thermal expansion. The effect of thermal expansion states that all substances change their shapes as a consequence of undergoing changes in temperature. The effect can be widely applied

TABLE 2. GPP form

Form parameter	Form variable
Mass of an object	M_m
Specific heat of a material	C_m
Resistivity of the material	ρ
Initial resistivity of the material	ρ_0
Length of the object	L
Cross-sectional area of the object	A_M
Temperature coefficient of resistivity of the material	α_M

TABLE 3. GPP structure

Input ports	Output ports
I_{in}, U_{in}	H_{out}, T_{out}

in various technical systems where the precise change of mechanical parameters
is needed. For instance, it is used to precisely control the displacement of a table
in a microscope.

The black-box model of the effect of thermal expansion for changing a linear
size of a material is depicted in Figure 5. In this model, an input heat flow H_{in}
causes elementary mechanical deformations of a crystal grid of the material. Under macroscopic observations, these elementary deformations result in changing
a linear size of a whole material.

In terms which relate thermal and mechanical domains, the behavior of the
effect can be expressed as:

$$\Re = << H_{in}, T_{in} >, < D_{out}, F_{out} >, F >;$$

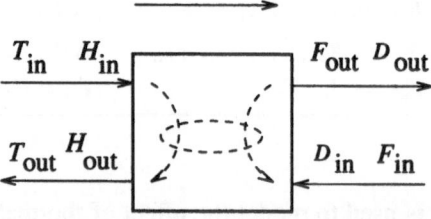

Figure 5. Black-box model of the effect of thermal expansion.

where H_{in} and T_{in} are input heat flow and temperature, D_{out} and F_{out} are output displacement and mechanical force correspondingly, and F is a set of form-related variables which provide transformation of energy.

The basic relation between linear size of a material and temperature is:

$$\Delta L = \int_{T_1}^{T_2} \alpha_M L_0 dT;$$

where ΔL is the increase in length, L_0 is the initial length of the material element, T is the temperature. T_1 and T_2 are the initial and final values of the temperature, and α_M is the coefficient of thermal expansion for the material M.

In terms of system theory, and following the law of energy conservation we can write the following relation for the behavior of the effect of thermal expansion:

$$\frac{dD}{dt} = f\left(\frac{dT}{dt}, F\right);$$

or in an exact form:

$$\Delta D = \int_{T_1}^{T_2} \alpha_M L_0 dT;$$

where ΔD is the linear displacement of the element subjected to heating.

9. Generic Design Concept

A generic design concept is synthesized on the basis of causal chaining several GPPs into a physical system whose overall behavior achieves the function required on the basis of energy resources given. Two GPPs may be connected if two conditions are satisfied:

1. Output energy flow of the first GPP and input energy flow of the second GPP are of the same type,
2. Values of parameters of both input and output energy flows belong to the same interval.

The GDC includes at least, two components: an energy source and a GPP that provides the required transformation of energy or material parameter. Summarizing the contents of previous section, a topology for Generic Design Concept can be synthesized which would be capable of performing the function "controlled displacement" on the basis of two effect described above: Joule heat and thermal expansion (Figure 6).

10. Mapping Between IDS and GPP

The behavior of any GPP can be observed from two points of view which may be of a particular interest to a designer:

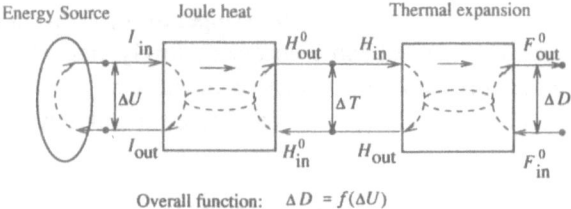

Figure 6. Example of generic design concept.

- A material property of some component of the effect provides the transformation of an input energy flow under the set of specific conditions.
- The transformation of the input energy flow into the output energy flow leads to the change of the value of a material property of some system's component.

These two viewpoints are used to distinguish between two classes of operations that can be produced on the basis of the same effect: modification of an energy flow and modification of a material parameter. To retrieve a specific GPP that would be capable of performing one of these operations, an initial requirement must be translated into a physical function which represents one of those operations. Then, a GPP that has a behavior including the required function is retrieved from the collection of GPPs.

To provide this translation, a new research direction has been recently initiated within the Invention Designer project. It is argued that to develop a formal method that would provide the adequate translation an ontological approach to represent IDS is needed. An ontology of IDS defines a taxonomy of possible concepts of IDS and distinguishes what levels of abstractions are needed to translate from one IDS concept into another.

Analysis of the TIPS Pointer to Physical Effects leads to the conclusion that generalized technical functions can be identified with a limited set of fundamental physical functions which, in turn, can be expressed in terms of operations over physical parameters. If to express the physical entities in terms of system or form-related variables, one can come up with the classification of *physical functions* which may be then mapped to different GPPs.

On the basis of this classification, we can represent physical functions performed by a GPPs in a qualitative form. Using these qualitative notations we bridge a gap between the generalized technical functions and physical functions. Since any system or form-related variable can be directly associated with its derivations, the basic function has to be defined once and then its instances can be generated.

Currently, all possible IDS fall into four categories: commonsense, constrained, behavioral and functional. Due to the lack of space we omit a detailed description

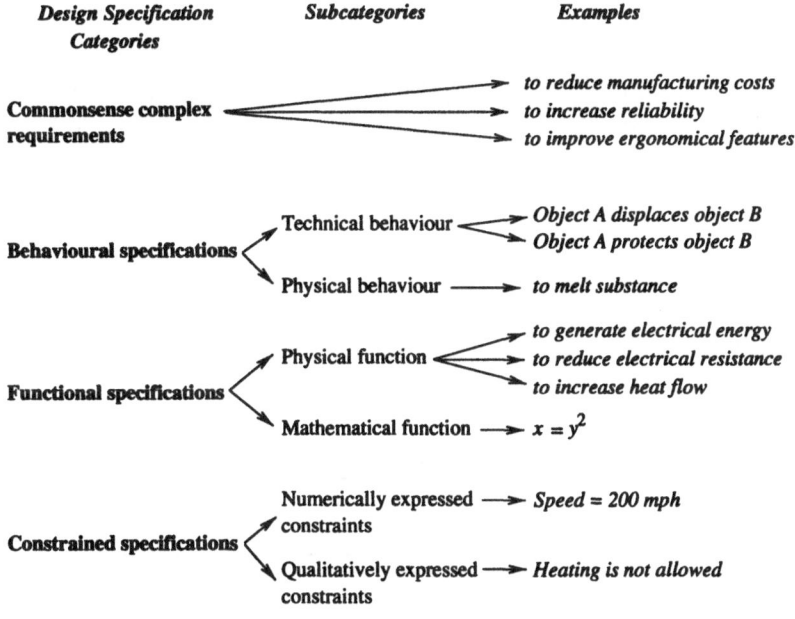

Figure 7. IDS ontology.

of the ontology. The IDS taxonomy with examples is shown in Figure 7.

Translation between the first and other categories is not clear how to achieve yet by automated means. For this reason a current research concentrates on establishing an adequate translation between behavioral IDS, constrained IDS and physical functions.

11. Invention Designer

Invention Designer developed as an implementation of the concepts introduced in previous sections illustrates applicability of sharable physical knowledge to innovative design. Current knowledge base of Invention Designer consists of 25 GPPs which are retrieved according to given specifications of a physical function and energy sources that are allowed to use. An example of working with Invention Designer is show in Figure 8.

An inference mechanism consists in search for a GPP whose behavior matches the function given. Then, other GPPs that can link the GPP selected to the energy source are searched.

Currently, the alternatives can not be evaluated against specific constraints and requirements. Therefore, the Invention Designer supports the step "Generation of design concepts that would meet the IDS" of the schema proposed in Section 1.

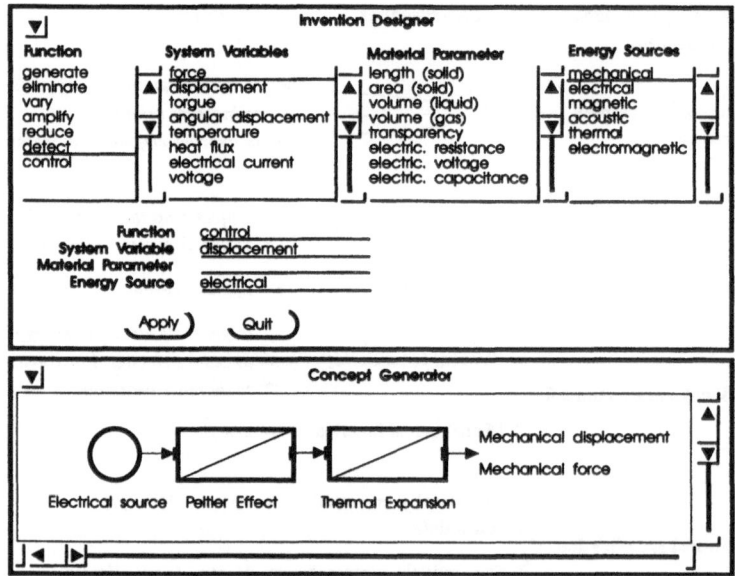

Figure 8. An example of working with Invention Designer. A causal network consisting of two effects - Peltier effect and the effect of thermal expansion was generated to perform the function "to control mechanical displacement. The search was limited by specifying available source of electrical energy.

12. Related Research

The use of fundamental laws of natural sciences to model and represent knowledge for innovative engineering design is studied by a number of research groups worldwide. Among them are the University of Tokyo (Ishi *et al.*, 1993; Taura *et al.*, 1995), investigating how physical phenomena can be modeled to build knowledge-intensive CAD systems, European groups systematizing physical knowledge to be used in creative design (Linde, 1994), and innovative design (Taleb-Bendiab, 1993; Schmekel, 1992) as well as groups in the USA (Tsourikov, 1993; Williams, 1990) and Russia (Zaripov, 1988). To model design knowledge in a uniform and sharable way, a growing interest in using ontological approaches should be pointed out (Alberts, 1993).

13. Conclusions

We discussed a possibility to develop a model of innovative design based on the use of sharable physical knowledge in a systematic way. The concept of Generic Physical Principle makes it possible to develop a collection of sharable physical phenomena the use of which will be independent of specific engineering domain.

Among the principal features of the approach discussed are:

1. A basic knowledge structure is introduced for modeling sharable physical

knowledge – Generic Physical Principle. It incorporates both energy and material aspects of physical principles that are needed to make instantiations of physical knowledge into design concepts.

2. An ontological approach to modeling design knowledge is applied to model different types of physical principles in a uniform way and to evaluate future designs against physical realizability at the earliest stages.
3. To organize and structure a collection of GPPs and to complete a knowledge base of GPPs the available TIPS Pointer of Physical Effects is used as a primary knowledge source.
4. A uniform way of modeling and representing GPPs makes it possible to combine GPPs into more complex sharable structures labeled Generic Design Concepts.
5. Modeling of GPPs is based on system-theoretical approach which helps to make abstractions of physical functions performed by GPPs and to develop an ontology for design specifications.

Currently, a knowledge base of GPPs on the basis of the approach discussed above is being developed. It will serve as a testbed for verifying the concepts introduced. In the short-run we plan to develop a knowledge base of 200 GPPs and to implement a mechanism for constraint management.

One of the problems still to be solved is what to do when IDS can not be translated into any of predefined functions. This means that an initial problem statement requires reformulation. The reformulation of the problem can be done by using the Theory of Inventive Problem Solving techniques (Altshuller, 1988) which help to obtain a formulation as a kind of particular function recognizable by PPE. However these techniques are not ready to be modeled formally yet.

One of the research directions which requires thorough study is the development of an ontology for IDS. A taxonomy of abstract design functions will help to translate initial key requirements into exact physical functions.

References

Alberts, L. K.: 1993, YMIR: a domain-independent ontology for the formal representation of engineering-design knowledge, *IFIP Workshop on Formal Design Methods for CAD*, Tallinn, Estonia, pp. 139–152.

Andersson, K.: 1993, A vocabulary for conceptual design – Part of a design grammar, *in* J. S. Gero and E. Tyugu (eds), *Formal Design Methods for CAD*, Elsever Science, Amsterdam, pp. 157–171.

Altshuller, G. S.: 1988, *Creativity as an Exact Science*, Gordon and Breach Scientific Publishers, New York.

Brown, D. and Chandrasekaran, B.: 1985, Expert systems for a class of mechanical design activity, *in* J. S. Gero (ed), *Knowledge Engineering in Computer-Aided Design*, North-Holland, Amsterdam.

Ishi, M., Tomiyama, T. and Youshikava, H.: 1993. A synthetic reasoning method for conceptual design, *in* M. Wozny and G. Olling (eds), *Towards Worlds Class Manufacturing*, Elsiever Science, North–Holland.

Kolodner, L. K. and Wills, L. M.: 1993, Case-based creative design, *Artificial Intelligence and Creativity*, **Autumn**, 50–57.

Linde, H.: 1994, *COIS – A Contradiction-Oriented Innovation Strategy*, Fachhochschule Coburg, Germany.

Malmqvist, J.: 1993, Computer-aided conceptual design of energy-transforming technical systems, *Proceedings of International Conference on Engineering Design ICED'93*, The Hague, August 17-19, pp. 1541–1550.

Pahl, G. and Beitz, W.: 1984, *Engineering Design: a Systematic Approach*. Springer Verlag.

Petrovich, N. T. and Tsourikov, V.M: 1986, *A Way to Invention*, Evrika, Molodaya Gvardia, Moscow (in Russian).

Schmekel, H.: 1992, *A System for Conceptual Design Based on General and Systematic Principles of Design*, Doctoral Thesis, The Royal Institute of Technology, Stockholm.

Selutsky, A. B. (ed.): 1987, *Daring Formulae of Creativity*, Karelia, Petrozavodsk (In Russian).

Shearer, J. L., Murphy A. T. and Richardson, H. H.: 1969, *Introduction To System Dynamics*, Addison-Wesley.

Sushkov, V. V, Mars, N. J. I. and Wognum, P. M.: 1995, Introduction to TIPS: a Theory For Creative Design, *Artificial Intelligence in Engineering*, **9**.

Sycara, K. and Navinchandra, D.: 1989, Integrating case-based reasoning and qualitative reasoning in design, *in* J. S. Gero (ed.), *AI in Design*, Computational Mechanics, UK.

Taleb-Bendiab.: 1993, CONCEPTDESIGNER: a Knowledge-Based System for Conceptual Engineering Design, ICED'93, The Hague, The Netherlands, pp. 1303–1310.

Tong, C. and Sriram, D. (eds): 1992, *Artificial Intelligence in Engineering Design, Volume II: Models of innovative design, reasoning about physical systems, and reasoning about geometry*, Academic Press Inc.

Taura, T., Koyama, T. and Kawaguchi, T.: 1995, Research on natural law database, *Design Symposium*, Tokyo.

Tsourikov, V. M.: 1993, Inventive machine: Second generation, *AI and Society*, **7**(1), 62–78.

Williams, B. C.: 1990, Interaction-based invention: Designing novel devices from first principles, *Proceedings 2nd AAAI Workshop on Model Based Reasoning*, AAAI, Boston, MA, pp. 168–175.

Zaripov, M. F.: 1988, *Energy-Informational Method of Scientific and Engineering Creativity*, VNIIPI, Moscow (In Russian).

J. S. Gero and F. Sudweeks (eds), Artificial Intelligence in Design '96, 743-759.
© *1996 Kluwer Academic Publishers.*

ASSISTING CREATIVITY BY COMPOSITE REPRESENTATION

EWA GRABSKA
Institute of Computer Science, Jagiellonian University,
ul. Nawojki 11, 30-072 Cracow, Poland

AND

ADAM BORKOWSKI
Institute of Fundamental Technological Research
Polish Academy of Sciences
ul. Swietokrzyska 21, 00-049 Warsaw, Poland

Abstract. The paper presents a composite syntactic-semantic representation of objects that is particularly suited for creative design in engineering. This representation is to a great extent oriented towards visual evaluation. The role of emergence both in art and in engineering design is discussed firstly. It is shown on examples taken from Escher's prints, Civil Engineering and Mechanical Engineering that the emergence can occur with respect to the shape of the object or to its internal structure or topology. The composite representation supports the search for alternative solutions in both domains. It has been implemented as a design tool including the editor of graph grammars, the generator of composition graphs, the library of primitives and the visualisation module.

1. Introduction

Design is associated with patterning, shaping and form giving at many levels. In particular, creative design is closely associated with the action of drawing and with the visual perception of the result. Computers certainly offer substantial support for visual thinking and imagery since they allow the designer to generate and manipulate drawings easily. Moreover, due to the possibility to generate patterns iteratively computer allows the user to analyze and evaluate new classes of pictures, e.g., fractals.

In this paper we propose an approach to creative design, in which *visual evaluation* is an important mode of work of a designer. Our proposal is based on the *composite representation* allowing to integrate the product modelling and the process modelling in the conceptual phase. We discuss design within the framework of graph transformations which are described by rules transforming subgraphs of limited size (Schneider and Ehrig, 1994; Borkowska and Grabska, 1995). The result of product modelling is the artifact represented as a union of transformed primitives being basic geometrical shapes. However, graphical design is not reduced to the problem of selection and arrangement of such basic shapes. If we choose primitives that are to be components of the graphical solution, transform and arrange them according to a certain conception, then we often obtain a drawing which exhibits attributes quite different from the mere summation of the properties of all its components. The components interact with one another (Coyne *et al.*, 1990). Therefore visual evaluation gives an opportunity for designers to see shapes in their own drawings that they had not conciously constructed. Such a phenomenon, called *emergence*, attracted large scientific interest recently (Edmonds, 1992; Gero and Yan, 1994; Weld, 1994).

In the following sections we discuss the features of emergence in art and in engineering design. In the case of art the emergent attributes of shape, color or texture influence the visual impact of the work of art on the spectator. In engineering a discovery can be triggered by an emergence, provided the visualisation of different aspects of the considered object is available.

It is commonly agreed that a computer tool assisting design should enable the user to browse comfortably through alternative solutions. Several interesting paradigms were reported recently in the literature, among them a maze concept by Boulanger and Smith (1994). We believe that the graphical environment described in this paper serves the same purpose.

2. Emergence in art

Our attempt to bring design and computer systems together depends on joining the computer generation with the human perception of the result in the form of visual representations. Although this paper deals with computational aspects of creative design, we do not agree with Chomsky's opinion that rewrite rules can be identified with the creativity governed by rules (Coyne *et al.*, 1990). Our approach to design is closer to M. C. Escher's approach because his work as the mathematician and the designer was only preparatory to his work as the graphic artist (Schattschneider, 1990).

We adapted Escher's strict separation of structural (syntactic) and semantic aspects of the artwork in the process of creative design. Our approach to design is *atomistic*. Artifacts are formed from simple building blocks called *primitives*. The primitives are our atoms, the indivisibles, but just as physical atoms have internal structure and can be split into elementary particles, in the same way the primitives may sometimes be further divided. Thus, we deal with the hierarchical organization of an artifact. Given the primitives, we define rules limiting the way they may be combined together. These rules determine the *syntax* of an artifact and allow us to define its *structure*.

Conventionally, we describe an artifact by means of a graphical model with primitives defined as simple geometric shapes appropriate to the artifact's type. Each occurance of a primitive within a model is called an *instance*. Instances are defined by specifying their transformations such as sizes, position, etc. In other words, they are transformed geometric primitives forming *components* of the graphical model.

In this section we consider the notion of emergence in its basic meaning ap-

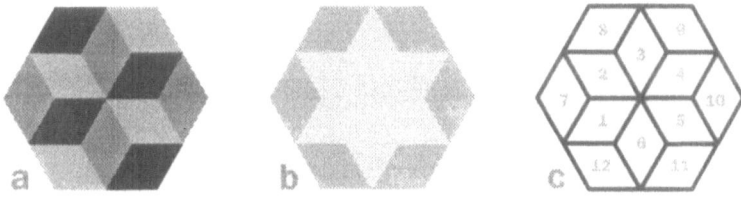

Figure 1.   Emergent shapes: a) Escher's visual trick; b) new coloring; c) syntax.

plicable to art. The well known artworks of Escher will serve us as an illustration. The pattern in Figure 1.a shows a well known visual trick used by Escher for many prints, e.g., for "Cycle", "Metamorphose", and "Concave and Convex" (MacGillavry, 1986). A set of three adjoining parallelograms, repeated in two directions, can be interpreted as a stack of blocks in two different ways. Seen from below, the black face is the bottom of each block. On the contrary, seen from above, the black faces are the tops of blocks. This double interpretation of Escher's pattern was intended. The pattern was designed to be ambiguos. For both interpretations the pattern is composed of twelve instances of the parallelogram (Figurere 1.c). On the other hand, in the structure of the pattern two primitives: a cube and a parallelogram can be distinguished. Moreover, there exist also additional two-dimensional interpretations of the pattern, e.g., this figure can be seen as the star surrounded with six parallelograms. Reflection upon coloring the pattern and changing colors of its components may lead to the new pattern in which the central star is seen as a structural element, i.e., a primitive (Figure 1.a). Color belongs to the attributes which are essential both in extracting new properties of the pattern and in its modification. It is possible that the star was not conciously constructed by Escher, i.e., it might be an emergent shape. Escher strictly separated the syntactic and semantic aspects of his work of art in the process of creative design. We illustrate this problem using his woodcut "Square Limit". The woodcut presents fish-shaped elements which are regularly and continuosly halved in the direction from within outward (see: p.315 in Schattschneider (1990)). Escher used recursion for gradual reduction of the shapes because he wanted to obtain an infinite number of component elements in the bounded area of the plane. The structure of the "Square Limit" proposed by him is shown in Figure 2.b. Each triangle corresponds to one component of the woodcut. In the next step of creat-

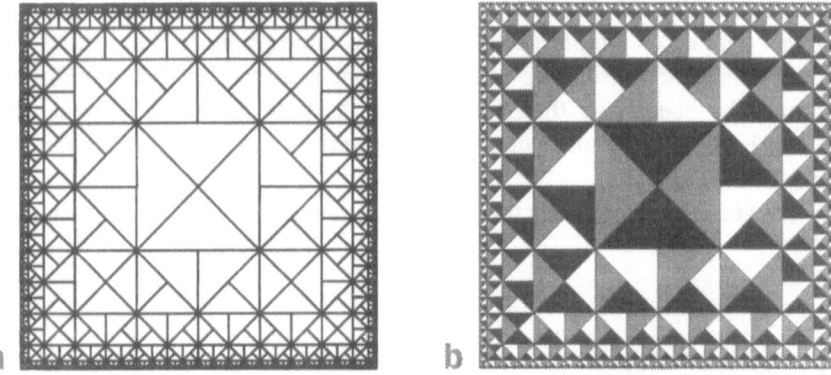

Figure 2. "Square Limit": a) structure; b) coloring.

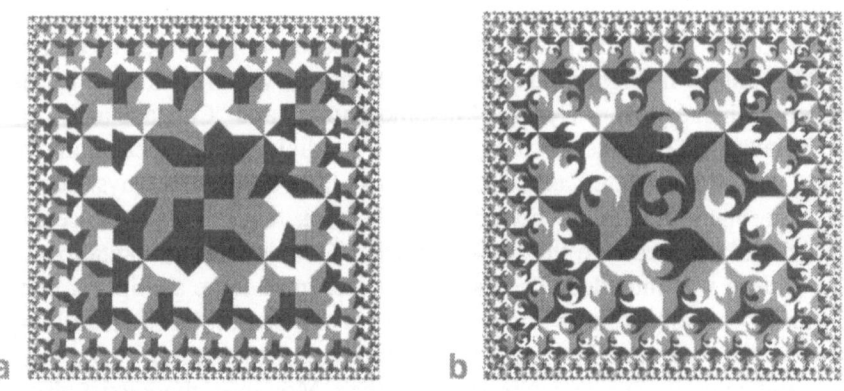

Figure 3. Two visual variations of Escher's Square Limit.

ive work the structure was equipped with the color attribute. Figure 2.b presents Escher's proposal of coloring of the "Square Limit". It is worth to notice that different coloring the same structure gives the observer an opportunity to see shapes that had not been conciously constructed by the designer (Weld, 1994). Then Escher could look for shapes replacing the triangles. He has choosen the fish shaped one. Two different realizations proposed by us are shown in Figure 3. In Sec. 4 we shall generalize Escher's approach to design. One of the key ideas of our proposal is to specify a single graph defining the internal structure of the designed object and to let the designer play with visual variations by modifying the realization scheme.

3. Emergence and discovery in engineering design

Contrary to pure art, the shape of an engineering object is subordinate to its function and to the desired qualitative attributes of the designed artifact. Hence, the notion of emergence should be understood here in a broader sense. Acquiring a new idea by an alternative interpretation of a picture remains essential. However, this time the designer may look not only at the object itself but also at the visual representation of such important data as, e.g., the stress field inside the designed

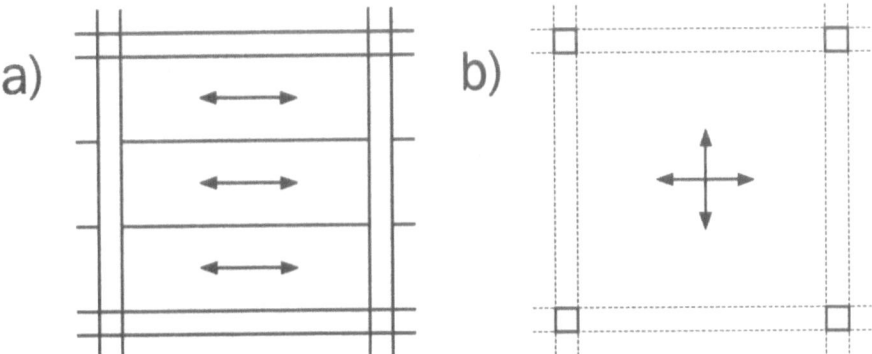

Figure 4.   Alternative solutions of a deck: a) unidirectional slabs; b) bidirectional plate.

object or the shape of its natural vibration.

Since in engineering we prefer to deal with unambiguous descriptions, it happens rarely that the visualisation of a single attribute of an object can be interpreted in many ways. More freedom in interpretation occurs when visual counterparts of several attributes or alternative solutions are perceived simultaneously.

Figure 5.   Examples of a reinforced concrete deck: a) discrete solution; b) smooth solution.

It is a conscious or subconscious process of merging, combining and modifying

Figure 6. Exhibition hall in Torino (Italy) (the authors – R. Biscaretti di Ruffia and P. L. Nervi.)

such pictures in our mind that leads to discoveries in design. Some of such processes can be classified as emergent.

Let us consider several examples from Civil and Mechanical Engineering. In
spite of their apparent variety, load carrying structures in Civil Engineering are
composed of a small number of primitives. One can consider a column, a beam
and an arch as such primary blocks, since a plate can be regarded as the bidirectional beam, a shell - as the bidirectional arch and a cable structure is the reversed
arch. Figure 4.a shows the simplest solution of a deck over a given rectangular
area: the rows of columns are connected by beams and the spans between them
are covered by slabs. Such a deck would have an appearance similar to the cast-
in-situ one shown in Figure 5.a, except that the junctions between slabs could be
aesthetically nonacceptable. The designer would probably take into account not
only this circumstance. He or she would also consider the load carrying capacity:
a plate working in 2 directions (Figure 4.b) is obviously stronger for the same
thickness than a collection of uni-directional slabs. Pursuing further the idea of
smooth flow of stresses in the structure, one might prefer to remove the beams

completely and to choose the structure shown in Figure 5.b. A concept of plate
emerges from the initial scheme of Figure 4.a when the flow of stresses is re-
interpreted in the bidirectional sense. Similar situations arrise for more complicated structures. Consider covering a huge area of an exhibition hall or a hangar for
aeroplanes. Since arch is reasonable choice when dealing with long spans, one
might compose the roof of a large number of adjacent arches (Figure 6). Nervi
managed to make this simple structure aesthetically appealing by gradually leading forces to the foundation via intermediate pylons. However, we conjecture that
the hangar build by the same designer in early 1940's (Figure 7) impresses the observer even more. It is a principally bidirectional structure: the arches arranged

Figure 7.   Hangar for aeroplanes in Italy (the author – P. L. Nervi).

into a grid inclined at 45 degrees to the boundary of the building interact at each
node and transmit the forces to the truss at the bottom of the roof. At the same
time this roof can be viewed as a cylindrical shell with the diaphragms at the ends
and the stiffeners a long the edges. Such a duality in the interpretation of Figure 7
resembles the visual effects discussed in Sec. 2.

   The diversity of objects in Machine Engineering is much broader. As an example of discovery that can be interpreted in terms of emergency let us consider
the design of the motorcycle. For a long time the scheme taken from the bicycle
prevailed: the rear wheel and the fender were attached separately to the load carrying frame. Looking at the stress flow diagrams and considering the functional

scheme of the motorcycle one eventually came upon the idea that the fender could
bear the load as well. Such a solution was adopted indeed some years ago and

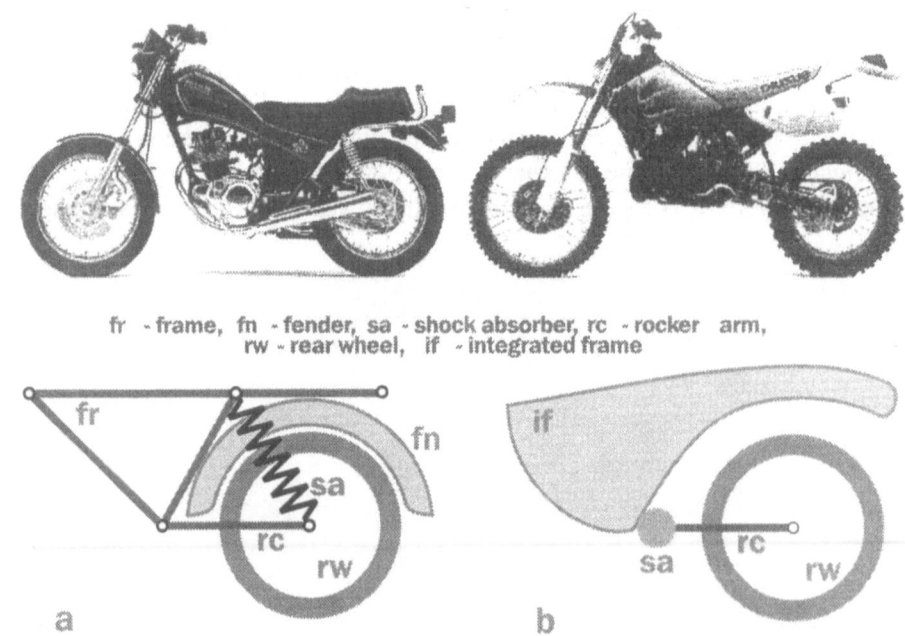

fr - frame, fn - fender, sa - shock absorber, rc - rocker arm,
rw - rear wheel, if - integrated frame

Figure 8. Conceptual solutions of motorcycles: a) conventional scheme with the shock-absorber
attached to the upper part of the load-carrying structure; b) solution with shock-absorber integrated
into the hinge.

most modern motorcycles have an integrated load carrying structure. Note that
several aspects influenced the design simultaneosly: the forces must be transmitted from the body of the motorcycle to the wheel, the driver must be protected
against the mud and the form of the bike must be aesthetically acceptable. Recently a more radical change in the solution of the rear part of the motorcycle
has been introduced. Figure 8.a shows the previous arrangement: the rear wheel
is attached to the main block by means of 2 elements - a horizontal rocker-arm
and a vertical or sligthly skew shock-absorber. The rocker-arm is pin-jointed to
the frame of the bike in order to allow the wheel to move vertically. Hence, its
only function is to keep the wheel at a constant horizontal distance from the main
block. Vertical forces are taken over of by the shock-absorber.

Once upon a time some ingenious designer noticed that an elastic-damping
element can be incorporated into the hinge that connects the rocker-arm to the
main block of the motorcycle. As a result the radically new solution depicted in
Figure 8.b was found. Due to its simplicity and the advantages it shows in heavy
terrain, such a solution prevails nowadays in cross-country machines.

The discovery of the integrated load-carrying block can be viewed as the *shape emergency*: looking at the old-fashioned motorcycle the designer could envision the unification of certain parts. On the contrary, the transition between the schemes shown in Figure. 8.a and 8.b involves the change of structure or topology. Therefore, we are confronted here with the *topological emergency*. It plays crucial role in any kind of creative design.

4. Composite syntactic-semantic representation

Creativity can be discussed in relation to both artifact and process (Coyne *et al.*, 1990). In this section we propose a composite representation which integrates the two aspects of creative design. Our representation includes the following elements: the composition graphs (CP-graphs) generated by a graph grammar, the realization schemes and the control diagrams.

The syntax of an artifact is described by a CP-graph. Such a graph has two types of node labels. For a given node, a label of the first type contains the name of a primitive and a label of the second type lists the *bonds*. The number of node bonds expresses the maximal number of the connections between the primitive corresponding to the node and other primitives. Each CP-graph has two kinds of bonds: *engaged* bonds which are connected by means of edges and the remaining bonds called *free* bonds.

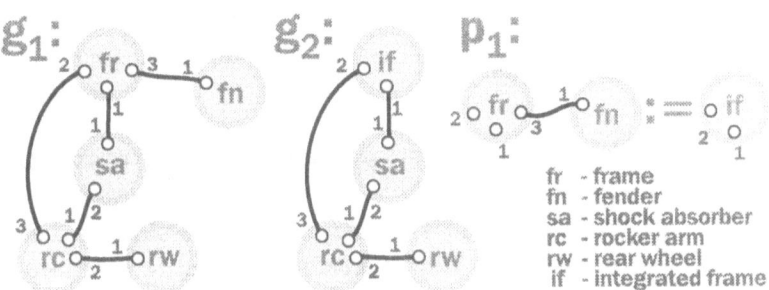

Figure 9. CP-graph representation of shape emergence for conceptual solutions of motorcycles.

The CP-graph g_1 shown in Figure 9 represents the conventional solution of a motorcycle depicted in Figure 8.a. The primitives are the following: a frame (fr), a fender (fn), a shock absorber (sa), a rocker-arm (rc), and a rear whell (rw) . The abbreviations of the primitives fr, fn, sa, rc, and rw are the node labels of the CP-graph g_1. We draw a picture of a CP-graph representing bonds as small circles placed in the nodes and showing edges as the lines connecting the pairs of bonds. The CP-graph g_2 shown in Figure 9 is the result of shape emergence, in which integrated load carrying block was proposed (see: Figure 8). On the other hand, the CP-graph g_2 can be obtained by applying the rule p_1 to the graph g_1.

In general, a rule p is composed of two CP-graphs - left-hand side and right-hand side, which are denoted by $l(p)$ and $r(p)$, respectively. The replacement operator := means that the CP-graph $l(p)$ is to be replaced by the CP-graph $r(p)$. We assume that for each rule p, $l(p)$ and $r(p)$ have the same number of free bonds with the same numeration. For example, the number of free-bonds of $l(p_1)$ and $r(p_1)$ of the rule p_1 in Figure 9 equals 2, and the order numbers of the bonds are 1 and 2.

To apply the rule p_1 to the CP-graph g_1 we remove the $l(p_1)$ from g_1 and insert $r(p_1)$. Then we replace two edges between the node fr and two nodes sa and rc of graph g_1 by the edges connecting free bonds of the node constituting $r(p_1)$ with the bonds of the same two nodes in the same bond order.

The topological emergence considered in Sec. 3 and related to the new solution of a motorcycle (Figure 8.b) can be described by the rule p_2 shown in Figure 10. The graph g_3 in this figure is the result of application of p_2 to the graph

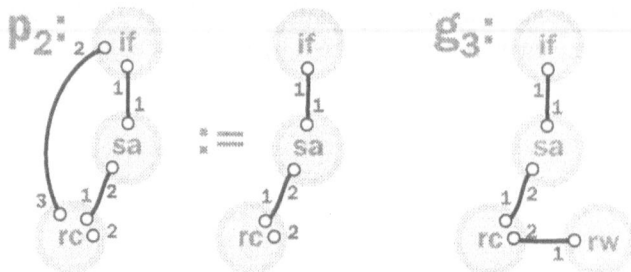

Figure 10.   CP-graph representation of topological emergence in designing motorcycles.

g_2. The new conceptual solution of a motorcycle is the result of two modification of the conventional solution. This creative process can be described by two CP-graph rules constituting the *graph grammar*. The process of transforming the CP-graph g_1 by the rule p_1 and then the CP-graph g_2 by the rule p_2 is a *derivation* of the graph grammar.

Having a set of rules we want to generate such CP-graphs which are syntactic descriptions of plausible solutions. In other words, some means of representing and using control knowledge for CP-graph grammmars is needed. In our approach we represent such a knowledge by the *control diagram* which is a connected directed labelled graph. With exception of the initial and the final node labelled I and F, respectively, all other nodes of a control diagram are labelled by production names. Applying a derivation process according to the order stated in the control diagram, we start with a production which corresponds to the label of a direct

Figure 11. Control diagram of the creative process for motorcycles.

succesor of the initial node. The derivation proces stops when the final node is reached. Figure 11 presents a very simple control diagram of the creative process for motocycles described by the two rules p_1 and p_2 and presented above. We assume that CP-graph g_1 is an initial graph. The control diagram has 2 paths leading to 2 solutions. The first path corresponds to the derivation from the CP-graph g_1 to the CP-graph g_3 by means of the rules p_1 and p_2. The second path allows us to apply the rule p_1 to the CP-graph g_1 and generate the CP-graph g_2.

Usually, control diagrams are more complex. We shall have an opportunity of seeing it in the next example.

Let us consider Escher's visual trick discussed in Sec. 2. The CP-graph g_1 in Figure 12 shows the syntax of the pattern composed of 12 primitives, which

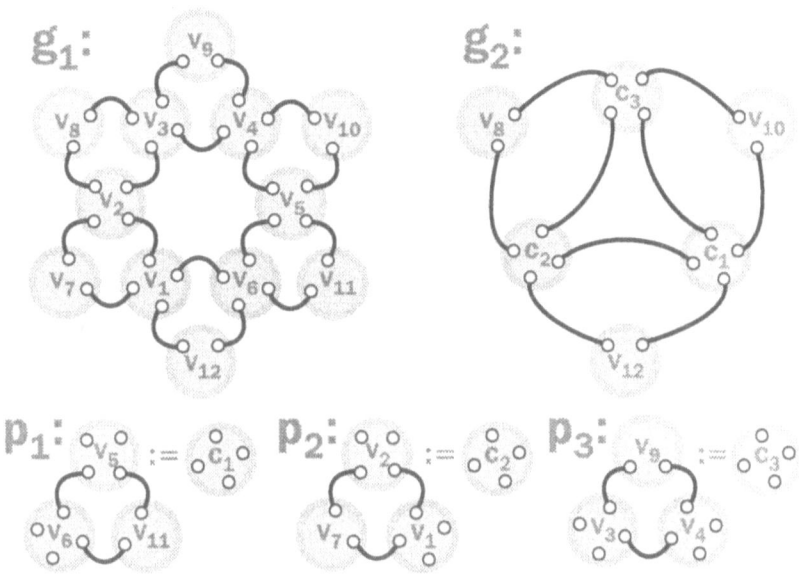

Figure 12. CP-graph representation of a cube interpretation of Escher's visual trick.

was presented in Figure 1. Each node corresponds to a single parallelogram. Node bonds corresponding to the fragments of the parallelogram assigned to the nodes, namely to its sides. The edges connecting bonds represent the coincidence relation that says that there exists coincidence of the appropriate pairs of sides of the neighbouring parallelograms.

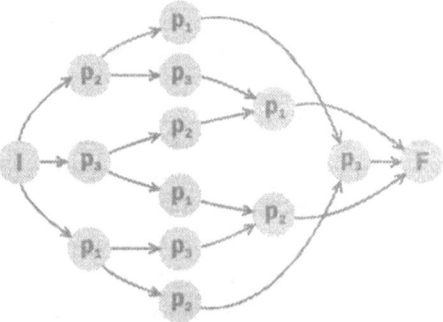

Figure 13. Control diagram of the set of rules shown in Figure 12.

The CP-graph g_2 in Figure 12 corresponds to the interpretation of the pattern seen from below. In this case we distinguish three cubes with black bottoms, and three parallelograms. The sequence p_1, p_2, p_3 of the graph rules in this figure al-

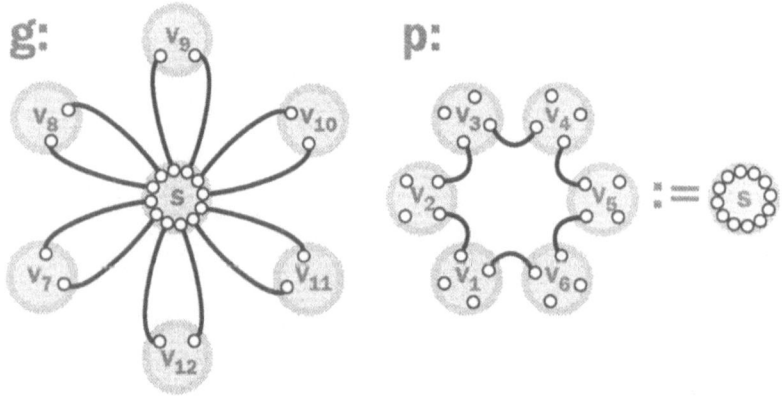

Figure 14. CP-graph representation of star interpretation of Escher's visual trick.

lows us to derive the graph g_2 from the graph g_1.

It is worth to noting that these 3 rules are independent. Each permutation of the sequence p_1, p_2, p_3 of rules leads us from the initial graph g_1 to the graph g_2. The control diagram shown in Figure 13 describes the derivations. The number of paths from the node I to the node F is equal to 6, i.e., to the number of permutation of the sequence. Different paths in the control diagram are associated with different order of perception of the succesive cubes in Figure 1.

The CP-graph g in Figure 14 is the result of applying of the graph rule p shown in this figure to the CP-graph g_1 in Figure 13. The rule p describes joining six parallelograms together in one star. In the structure g the edges also represent the coincidence relation of appropriate sides.

As it has been considered for Escher's "Square Limit" a CP-graph enables us to play with shapes of primitives for different graphical models defined for the structure. Figure 15 presents our propositions of two graphical models defined

Figure 15. Two realization of the syntax shown in Figure 14.

for the syntax shown in Figure 14. In our approach shapes are seperated from the syntax. They are defined in a *realization scheme* being the third element of our composite representation.

A realization scheme is composed of 3 units. For the first of them the user defines a set of primitives together with a set of admissible transformations in the Euclidean space. We assume that the set of admissible transformations contains at least similarities. A question arises in what way a set of admissible transformations and a set of primitives have to be chosen in order to make the most of computer in designing. Obviously, the optimal choice depends on the purposes that the designer is to gain. It seems that the most universal choice would be if affine transformations were accepted as the set of addmissible transformations. We must emphasize that sometimes it is better to have more prototypical parts and less admissible transformations. For instance, if we want to built a graphical model of the fragments of conic curves it is better to accept all conic curves as primitives (they will be parametrized) and to take isometries as a set of admissible transformations (Grabska, 1993). Non-geometrical attributes, like color, texture, etc, that are to be associated with primitives are also defined within the first unit of the realization scheme.

Figure 16. Coloring relation between bonds.

The second unit of a realization scheme contains CP-graph nodes with labels corresponding to the names of primitives. In other words, the designer assigns the primitives to the CP-graph nodes and relates the fragments of primitives to the bonds of nodes corresponding to these primitives. Additionally, the attributes are transfered from primitives to CP-graph nodes.

The third unit contains the descriptions of conditions related to an artifact and its component. Moreover, the transformations that match the components of graphical models are determined. First of all, these are the topological relations between primitives of an artifact. But we also consider non-geometrical relations between components. Names of relations are defined in the syntax (a CP-graph) of an artifact as edge labels and then in the third unit of the realization scheme such relations are represented by means of the appropriate predicates.

Let us consider non-topological relation between bonds. Figure 16 presents the next realization of the syntax shown in Figure 14 and the CP-graph being the syntax enriched with 12 bonds representing component fragments of the realization (2 small circles for each of 6 non-central components) and 6 new edges labelled by c for connecting these bonds. We experiment with coloring small circles by means of the relation between colors which is taken to be equality, i. e., an edge labelled by c connects two bonds if the circles corresponding to the bonds are unicolored. This relation is satisfied for the realization shown in Figure 16.

5. Remarks on implementation

The composite representation was implemented in a prototype program developed by the first author and her students. It is written in C++ under Microsoft Windows Operating System for PC computers and is source-code portable to other hardware platforms for which MS Windows is available.

The prototype program is a bridge design tool allowing the user to acomplish an efficient search of alternative conceptual solutions. A CP-graph grammar with 33 rules generating different types of bridges has been outlined in Borkowski and Grabska (1995). The main modules of the code are: an editor of CP-graph grammars and control diagrams, a generator of CP-graphs, and a realization block with a browser of primitives. Each of them is accessible through a separate window of the Windows 3.1 environment. The editor allows the user to define rules of a CP-graph grammar and a control diagrams for them. In the generator, CP-graph structures of designed bridges are derived using sequences of rules supplied by the editor. The realization block enables the user to assign primitives to nodes of CP-graph rules and change technical parameters of bridges.

Every designing session must consist of several steps. The first step in designing a new bridge with this application is to specify a terrain. It may be done by dragging a mouse. All information concerning the shape of terrain must be entered at this point of designing process. It may be done by dragging a mouse or reading the profile of the valley from a disk file. Next the type of bridge, the number of pylons and the length of bridge have to be specified. It is done by using a dialog window into which all needed data is to be entered. Also the choice whether the bridge is to be generated automatically or by hand is to be made here by clicking an appropriate box in this dialog window. Pressing the OK button results either in generating and drawing on the screen a bridge based on the entered data (if automatic generation was choosen) or in drawing a skeleton of a bridge (i.e. showing the positions of pylons). In the later case the bridge has to be generated by adding subsequent elements by hand. After the bridge has been drawn the pylons are equally spaced. If it is necessary to modify their position, this can be done by dragging with a mouse their position lines. A removal of any individual pylon is also possible. Now, when a general structure of a bridge has been fixed, the upper or lower arcs may be added to the beam in order to improve the aesthetics. At every point of the design process the user can display the CP-graph, corresponding to current stage of a bridge. A transition from, e.g., continuous beam to cable bridge is accomplished by mere replacement of the primitives. The rules that exclude certain sizes or combinations of elements are included into the predicates of the realization scheme implemented in the realization block. Thus, the program keeps the designer in the domain of admissible solutions.

The most radical and, therefore, most interesting changes occur when the CP-graph is modified. Unfortunately a bridge is not the best object for such search: its topology is too stiff. Therefore, we are experimenting at present with a more advanced version of the program. This version is thought as a general purpose knowledge representation scheme preserving the main feature of the composite model – the clear separation of syntactic and semantic elements. Implemented as an expert system shell such a program can be easily ported to a certain domain of engineering design by means of the exchangeable libraries of primitives and the domain specific realization schemes.

6. Conclusions

Up to now the best means of considering conceptual solutions has been sketching. A thorough study of traditional sketching allows us to share opinion that contemporary CAD tools still appear to be too cumbersome for genuine sketching (Goldschmidt, 1991). On the other hand, computers offer some new possibilities and ideas for this creative activity, which are not available by traditional means, for instance visualization of recursive and iterative methods. One of reasons to use these possibilities can be intention of employing the theory of fractals as a new paradigm of architecture (Schmitt, 1987; Grabska *et al.*, 1994).

In order to persuade designers to use computers in the process of creative design we must equip CAD tools with "an intelligence" that traditional tools of sketching, a pencil and a paper, do not possess. It is commonly agreed that an intelligent drawing pad should incorporate the knowledge about constraints imposed upon the designed artifact. It seems, however, to be not sufficient. The composite representation we outlined above forces the designer to think about objects at 2 levels: the higher level of structural properties and the lower level of geometry, primitives etc. Defining syntax and semantics in an explicit way is the real benefit of CAD. We believe that such a mode might be advantageous for creativity.

One of the main topics discussed in our paper is emergence. We demonstrated on real life examples that the emergence can occur on both levels of our representation scheme. By visualizing the internal structure of the designed artifact through its CP-graph the designer has a chance to grasp the essence of the current solution. Whether he or she comes upon a sparking new idea looking at that graph depends entirely upon the "human component" – our role as the developers of new tools is giving the user a chance for creative work.

References

Boulanger, S. and Smith, I.: 1994, Models of design processes, *in* I. Smith (ed.), *Proceedings of the EGSEAI Workshop*, Lausanne, pp. 132–145.

Borkowski, A. and Grabska, E.: 1995, Representing designs by composition graphs, *in* IABSE Colloquium Bergamo 1995, *Knowledge Support Systems in Civil Engineering*, IABSE Reports, Vol. 72, Zürich, pp. 27-36.

Coyne, R. D., Rosenman, M. A., Radford, A. D., Balachandran, M. and Gero, J. S.: 1990, *Knowledged-Based Design System*, Addison-Wesley Publishing Company, Inc.

Edmonds, E. A.: 1992, Knowledge-based systems and new paradigm for creativity, *in* J. S. Gero and M. L. Maher (eds), *Modelling Creativity and Knowledged-Based Creative Design*, Lawrence Erlbaum, NJ, pp. 269-282.

Gero, J. S. and Yan, M.: 1994, Shape emergence using symbolic reasoning, *Environment and Planning B: Planning and Design*, **21**, pp. 191-218.

Goldschmidt, G: 1991, The dialectics of sketching, *Creativity Research Journal*, **4**(2), 123-143.

Grabska, E.: 1993, Theoretical concepts of graphical modeling: Realization of CP-graphs, *Machine GRAPHICS and VISION*, **2**(1), 3-38.

Grabska, E.: 1995, Visual evaluation in design space, *in* T. Oksala (ed.), *Proceedings of the Decon 95 - Design: Emergence, Content*, Baden-Baden, pp. 1-7.

Grabska, E., Oksala, T. and Seppanen, J: 1994, Fractal dimensions in architecture design: Between artificial and natural beauty, *Acta Polytechnica Scandynavica*, Civil Engineering and Building Construction Series No. 19, Helsinki, pp. 29-35.

MacGillavry, C. H.: 1986, The symmetry of M. C. Escher's "Impossible" images, *Comp. and Math. with Appls*, **12B**(1/2), 123–138.

Schattschneider, D.: 1990, *Vision of Symmetry: Notebook, Periodic Drawings, and Related Work of M. C. Escher*, W. H. Freeman and Company.

Schmitt, G.: 1987, Expert systems and iterative fractal generators in design and evaluation, *Proceedings of CAAD Futures'87*, Eindhoven.

Schneider, H. J. and Ehrig H.: 1994, *Graph Transformations in Computer Science*, Lecture Notes in Computer Science, 776, Springer-Verlag, Berlin.

J. S. Gero and F. Sudweeks (eds), Artificial Intelligence in Design '96, 761-780.

SKELETON-BASED TECHNIQUES FOR THE CREATIVE SYNTHESIS OF STRUCTURAL SHAPES

DEREK M. STAL AND GEORGE M. TURKIYYAH
University of Washington, Box 352700
Seattle, WA 98195 USA

Abstract. A dominant problem in engineering design is that very often an optimal design lies outside the search space defined at the start point of the design process. To access new state-space that may contain the optimal design, it is necessary to change the problem formulation during design by changing the design variables or grammar of the problem. This generation and exploration of new search spaces is referred to as creative design. We describe a creative shape generation methodology that combines ideas from computational geometry and numerical optimization and can synthesize and optimize shapes from high level functional specifications of performance requirements and design objectives. The methodology uses a geometric abstraction of shape—the skeleton—as the basis for shape representation and parameterization. This parameterization supports creative design as it is adaptive and changes throughout the design process to accommodate the shape changes that occur. The design process iterates on the construction of skeleton-based design spaces and the formulation of search problems that explore these spaces. This methodology is capable of generating new shapes with geometries and topologies significantly different from those of the starting shape and which lie outside the initial design space. Two examples illustrate the implemented synthesis process.

1. Introduction

The generation and manipulation of three dimensional geometric shapes to achieve certain engineering objectives is a problem generic to a large number of design applications where "form follows function", i.e., where the designed physical artifacts derive their ability to fulfill their function to a large extent from their shape. The design of such artifacts is often determined by the need to optimize criteria such as weight, volume, fit, etc. as well as the need to satisfy appropriate performance objectives. Depending on the particular design context, various functional and behavioral objectives such as controlling fluid motion, heat dissipation, buckling, structural vibrations, peak stresses or displacements may be used.

The problem of generating shapes that optimally satisfy required behavioral objectives and manufacturing constraints has recently become of significant practical importance in a number of technological applications. The increased popularity and broader use of composites and high performance materials to which the structural shape of both individual components and complete systems is critically important, have brought to the forefront the need for effective shape design methodologies. Similarly, with the growing acceptance of layered manufacturing processes that allow effective production of general geometries and topologies, the need for generative design methodologies that synthesize general geometric shapes from functional specifications is being recognized as a necessary tool for increased design productivity.

Unfortunately, in the current state of the art, there are no adequate *performance-based* design tools—tools that allow designers to describe requirements and objectives, and directly obtain design descriptions that meet design goals. Available computer-aided design tools provide only very low level abstractions that do not allow designers to express functional specifications as the means of generating geometric shapes. Similarly, state-of-the-art shape optimization techniques, while very effective in fine-tuning already conceived and parameterized shapes, do not have generative design capabilities: they cannot handle the generation of design parameterizations, large shape changes or topological transitions adequately.

Consider, for example, the task of designing a minimal material thermal fin that fits within a given volume. The goal of such a design would be to produce a shape that can dissipate enough heat to keep the maximum temperature below an appropriate threshold without compromising the structural integrity of the part. Using today's tools, not only is a designer required to generate parameterizations for the shape but available methods limit, due to the static nature of the representations they use, the kind of designs they can be generated. Starting from a three-spoke fin design, the current generation of design systems could, at best, generate a "better" three-spoke fin; a more efficient design, however, could be a four-fin design that existing design processes are unable to produce.

Thus, optimal shapes very often possess features that are incompletely defined in the original object structure. In order to access these optimal forms, the space being searched must also evolve during the design process. This is referred to as *creative* design and can be characterized as the exploration of space that is only partially defined in the initial design domain (Brown, 1992). Therefore, creative processes produce new variables and types during design that move or extend the state space of potential designs and admit solutions, potentially optimal ones, that were not accessible from the starting problem formulation (Gero, 1990).

In this paper we propose and demonstrate a technique, combining ideas from computational geometry and numerical optimization to produce a creative design methodology that promises to deliver performance-based capabilities to design and optimization systems. Geometric reasoning allows us to modify and update the design state space during the design process while numerical optimization allows us to search these subspaces efficiently. As a result, the methodology is capable of generating new geometric shapes from a description of design goals and requirements.

2. Requirements for a Creative Shape Design Process

Engineering design, in general, involves three major types of variables: functional, behavioral, and structural. Function describes the performance criteria for which object or shape is being designed; behavior is a measure of to what level the functional requirements are satisfied; and structure represents what will be manifested in the final artifact (Gero, 1994). There corresponds to each of these variables a space which represents their respective domains and ranges and is referred to as the *design state space*. The state space incorporates all the characteristics of the design variables present in the initial object and represents the locus of all possible designs with those variables and their bounds.

Traditionally, design is based upon *a priori* knowledge of the variables which define structure and function, and to provide a starting point for a *search* through possible combinations of constrained values to find an optimal solution (Gero, 1994). The range of the values of variables is not only fixed *a priori*, but also typically very small. For problems in which the values are real numbers, the search becomes a numerical optimization and all accessible solutions lie within the domain of the state-space defined by the initial variables. This design process, whose search space is substantially smaller than the space of all possible designs, is referred to as *routine* design.

"Innovative" design, on the other hand, seeks to expand the state-space by manipulating the ranges of design variables. Thus, it can be characterized as the *exploration* of a well-defined space (Coyne *et al.*, 1987) which spans the domain of all possible designs defined by variable set, or object formulation. Since the same variables are used to describe the object, however, the underlying structure will remain unchanged regardless of how much it is altered. For example, in the design of a three-dimensional form the geometry may be significantly changed during the search process but the topology will remain the same.

Optimal designs, however, often lie outside the space spanned by the initial design variables. These solutions are therefore unattainable through design processes that are routine or innovative in nature. Thus, in order to find optimal designs, there is a need to not only search and explore well-defined spaces, but also implement models of design that allow expansion/modification of the state-space. Such *creative* methodologies not only change the ranges of design variables but produce new variables during the design. The exploration process no longer consists of a single search through a well-defined space but rather a series of searches through spaces generated at each step in an iterative design process. A diagram of the design state-spaces inspired by Gero (1990) is shown in Figure 1.

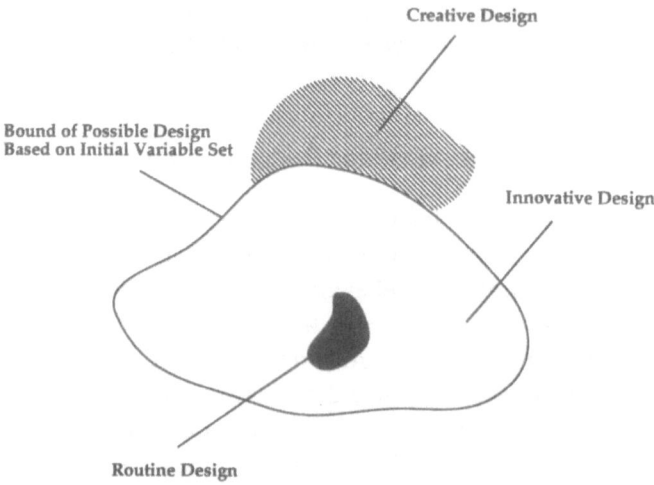

Figure 1. State-space of routine, innovative, and creative designs.

Several strategies have been proposed to generate, expand, search, and explore design spaces. Schmidt *et al.* (1994) and Cagan *et al.* (1993) have developed simulated shape annealing algorithms for shape design, the latter also incorporating interactive shape grammar generators into the design process. Various genetic and evolutionary algorithms, such as those of Carlson (1994), Goldberg (1989), Maher (1994a), and Zhao *et al.* (1994) have also been described. Genetic algorithms operate analogously to evolutionary systems: various features in the design process are combined through genetic crossover. The "offspring" are evaluated and the most successful survive the selection process. These designs are again combined and the process continues until a single optimal design is found. Chapman *et al.* (1994) have added a conceptual interpolation feature to this process which allows new designs to be generated from existing designs with similar functions. Case-based algorithms (Maher, 1994b) and design processes that incorporate homogenization principles (Papalambros, 1990) have also been proposed.

The difficulties in generating optimal shapes in a Euclidean space are those inherently involved in creative design: given a set of design goals, objectives, and possible constraints, the final design space is not defined *a priori* in a computationally usable form. As a result, the objective and constraints are not specified in terms of specific design variables but rather in terms of properties and characteristics of the final unknown shape. For example, geometric constraints (i.e, maximum thickness, maximum radius of curvature, etc.) are defined in terms of shape features that are generated only as the design evolves. Moreover, some constraints may not even be activated during the design process if the features they are predicated on do not appear in the shape.

In order for a design process to be able to produce novel shapes, it must possess several features. First, it must be capable of changing parameterizations during the design process. This is an essential characteristic of creative design and, since a good shape parameterization of the optimal shape is not generally known a-priori, it is unlikely that the initial parameterization is going to be adequate throughout the design process. Therefore, there is a need for mechanisms to recognize when a given parameterization is no longer adequate as well as mechanisms for modifying these parameterizations during the exploration of new state-spaces. At each stage of design, the design process must also be able to assign appropriate values to these parameters in order to guide the synthesis towards the optimal solution.

Secondly, the synthesis must constrain the design space to valid shapes. During design, it is generally necessary to constrain the values that design variables may assume to those values that produce valid Euclidean shapes only—no self intersections. Without this restriction, physically impossible geometries may be generated, rendering the final design unusable. Enforcing these requirements is difficult when using, for example, control points of splines or other similar standard boundary representations as design parameters, because of the large number of constraints generally needed to prevent all pairwise boundary segment or boundary facet intersections.

Third, a spatially adaptive description of the evolving shape must be provided. Different regions of a shape are likely to require different resolutions during the design process. For computational purposes, an effective design procedure should be able to take advantage of the fact that regions of a shape that do not have interesting features or where features are not evolving during the design process can be parameterized using a coarse resolution. Such spatial adaptivity is particularly important in three dimensions because of the large number of design parameters that would be needed for a non-adaptive description. Additionally, topological modifications to the design must be permitted.

Finally, the design process must be able to express and manipulate general constraints. In order to be applicable over a broad range of problems, a design methodology must handle geometric and behavioral design requirements and constraints. These include constraints on shape characteristics such as area, volume, thickness, curvature, etc. as well as behavior quantities such as stresses, displacements, temperatures, etc.

3. Overview of the Design Process

We describe a shape generation methodology that satisfies the above requirements and can synthesize and optimize shapes from high level functional specifications of performance requirements and design objectives. This methodology uses a geometric abstraction of shape—the skeleton—as the basis for shape representation and parameterization. This parameterization is adaptive and changes throughout the design process to accommodate the shape changes that occur. The design process iterates on the construction of skeleton-based design spaces, formulation of search problems in these spaces, and solution of these problems using numerical optimization procedures. This methodology is capable of generating new shapes with geometries and topologies significantly different from those of a starting shape.

The design strategy consists of a sequence of motions through geometrically valid design subspaces. Each step consists of the formulation and solution of a search problem in a specific subspace. At each design step, the current shape is parameterized and the problem requirements are expressed in terms of the design variables. This parameterization constrains the search to a specific portion of Euclidean space. The resulting numerical optimization problem is then solved using standard numerical techniques until convergence, i.e., until no further progress can be made with the current design parameterization. A new parameterization is then generated which allows the exploration of a different region of the design space and allows the design to proceed further towards the optimal solution. The process is repeated until convergence to a final shape.

The advantages of adopting such a strategy, the repeated formulation and solution of a sequence of search problems of moderate size, are twofold. First, each design subspace constructed in the sequence is computationally tractable, as opposed to the intractable space that contains all possible shape geometries and topologies. Second, we are able to capitalize on known and efficient numerical optimization strategies for searching multidimensional spaces.

The design subspaces generated, i.e., the shape parameterizations used during the design, are based on the skeleton (or medial axis) of a shape—a lower-dimensional geometric abstraction that captures significant intrinsic shape characteristics as will be described in the next section. At every design iteration, the set of parameters used to define the design is based on sampling the skeleton of the current shape. The lengths of rays emanating from these selected points are the design parameters during this design step. The numerical optimization process determines the values of these parameters that produce an optimal shape. During optimization, the lengths of the two rays emanating from a given skeletal point are not constrained to have the same value. The object may shrink on one side of the skeleton and expand on the other. This implies that the final shape will have a different skeleton that will be used as the basis of the parameterization for the next design step.

Starting from an initial shape, the overall design process can be summarized as follows:

1. *Generate the skeleton and construct a skeleton-based parameterization of the current shape.* The skeleton is an abstract description of the shape consisting of the lines/surfaces of symmetry of the shape. Skeleton-based shape

parameterizations may be defined to consist of a set of design variables (ray lengths) whose values, together with the skeleton, can reconstruct a valid shape. Thus, this parameterization defines the search space which includes all possible combinations of values of the design variables that do not violate shape integrity constraints.

2. *Formulate a search problem in the newly constructed design space.* By expressing the objectives and constraints in terms of the design variables, a search problem to find the optimal combination of the values of the design variables can be formulated. Since the design variables consist of a vector of n scalars, the search problem can be formulated as a constrained n-dimensional non-linear optimization problem. Initial values for the design variables are assigned so that the starting configuration represents the geometry and topology of the current shape.

3. *Solve the optimization problem and repeat steps 1,2,3 until convergence.* The solution of the optimization problem results in new values of the design variables that define a new shape. The shape geometry and topology can change at iteration. Topology changes occur when ray pairs collapse on the interior skeleton or extend to the exterior skeleton, as described in the next section. Geometric changes will arise from the solution of the problem optimization which changes the values of the design variables to satisfy supplied goals and constraints. When the geometry and topology do not change between iterations, convergence has been reached.

The resulting framework achieves powerful capabilities for the design of general shapes which overcomes the problems associated with current shape design procedures. In particular, it does not preimpose a fixed parameterization on the design space nor does it preimpose a specific structure such as a Cartesian grid on the design domain, both of which are key elements of a creative design process. Therefore, the class of optimal shapes that can be obtained is not restricted and the process always generates valid Euclidean shapes. This method handles a large class of design constraints and is able to adequately support large geometric changes and topological modifications.

4. The Skeleton: A universal shape (re) parameterizer

In this section, we describe the skeleton and the central role it plays in the design process. The skeleton, or *medial axis* (MA), of a two-dimensional object may be described as the set of centers of maximal inscribed discs, i.e., which contain no other discs and touch the boundary at two points. Every point on the skeleton is therefore equidistant from at least two points on the boundary. The skeleton was first devised by Blum (1973) who used it for the representation and classification of biological shapes and it has since been used in a number of applications, particularly in computer vision (Faugeras, 1993). On the skeleton, points that are equidistant from more than two boundary points are called *junction points* and those equidistant from only one point are *end points*. Figure 2 shows interior and exterior skeletons of various shapes as dotted lines.

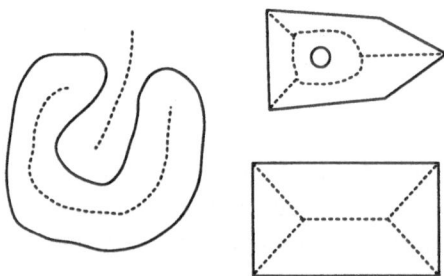

Figure 2. Examples of shape skeletons (skeletons are shown as dotted lines).

We use the skeleton of a shape as the basis for shape representation. A shape can have an interior as well as an exterior skeleton. The exterior skeleton can be defined as the topological closure of the locus of maximal inscribed disks included in the complement of the shape. In three dimensions, the skeleton can be similarly defined as the set of points that are minimally equidistant from two boundary points and is composed of a set of skeletal patches (surfaces). Surface patches intersect at junction curves consisting of points minimally equidistant from three boundary points. We use the interior skeleton to parameterize a shape during optimization. Numerous automatic methods for generating skeletons have been proposed, based on both discrete and geometric approaches. A review and discussion of the extensive variety of generation techniques is presented in Turkiyyah *et al.* (1996) and will not be repeated here.

4.1. SHAPE PARAMETERIZATION

The skeleton is an intuitive and appealing representation. It captures all significant aspects of intrinsic shape characteristics. For example, elongated shapes have a skeleton arc that follows their middle axis, pointed sub-shapes have a skeleton arc that follows the bisector of the angle made by the two lines, and rounded shapes have skeletons ending at distances equal to the minimum radius of curvature of the rounded boundary. Similarly, in 3D thin elongated skeletal regions correspond to long thin protrusions. The topology of the skeleton graph is also directly related to the topology of the shape. The number of cycles in the 2D skeleton graph and the number of closed surfaces in the 3D skeleton hypergraph correspond to the number of internal holes in the shape. The skeleton thus characterizes the basic shape characteristics. The geometry of the skeleton does not depend on a specific coordinate system but only on the object shape and boundary curvature, and

changees continuously as the shape changes. As a result, a skeleton based representation is appropriate for defining and supporting shape evolution during shape
synthesis. Further, the skeleton represents a shape in terms of a graph which is
more amenable to manipulations inside design systems.

Line segments that join a skeleton point to its corresponding boundary point
will be termed *rays*. Each regular point of the skeleton is the base of two rays
emanating from it (Figure 3). Rays are perpendicular to the boundary, provided
that the corresponding boundary (i.e., at the tip of the ray) is locally smooth. If a
ray tip is located at a position that has a discontinuous derivative, then the direction of the ray will fall between the right-side and left-side perpendiculars to the
boundary. Three or more skeleton arcs meet at a junction in 2D and three or more
surface patches meet at junctions in 3D. Junction and end points of the skeleton
have three and one associated rays, respectively. It is important to note here that
a shape can be recovered from the skeleton and the lengths of the rays by joining
the ray tips.

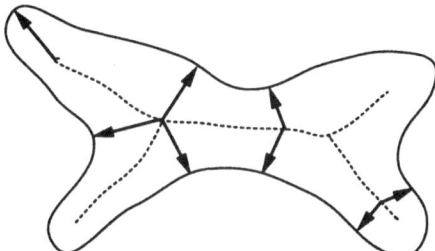

Figure 3. Skeleton-based rays: Design variables.

Perturbations in the lengths of the rays emanating from the skeleton, define
shape changes. As a result, the skeleton of a shape allows the automatic generation of shape parameterizations (a set of ray lengths) and, therefore, of a design
space that can be searched during each iteration of the design process. We use a
finite set of skeleton rays as design variables (Figure 3): they form the vector of
parameters that parameterize the shape. By joining the tips of rays we can reproduce a shape from its skeleton and the values of these parameters. Changes in the
magnitudes of rays cause shape changes and may even alter shape topology as
will be shown below. The number of rays to be used in a given design iteration
depends on the nature of the problem, the total number of desired design variables, the required shape resolution, the stage of the design, etc. and may depend
on the curvature of the skeleton arcs/patches. Thus, the state-space is continually
changing and more possible designs become accessible.

There are two principal properties that make a skeleton-based parameterization an ideal candidate for shape design parameterization:

– Shape integrity requirements can be formulated as simple bounds on design variables. The skeleton-based strategy proposed can impose integrity constraints by simply placing skeleton-based lower and upper bounds on design variables $\underline{d_i} \le d_i \le \overline{d_i}$ $(i = 1, 2, \cdots, n)$ where $\underline{d_i}$ and $\overline{d_i}$ are respectively the lower and the upper bounds on the design variables, derived from the interior and exterior skeletons respectively. As shown in Figure 4, the interior skeleton is the lower bound for the rays emanating from it, while the exterior skeleton is the upper bound for these rays. Qualitatively, the interior skeleton limit guarantees that the shape does not produce "negative regions" by collapsing upon itself, while the exterior skeleton limit guarantees that the shape will not overlap, i.e. produce "double regions".

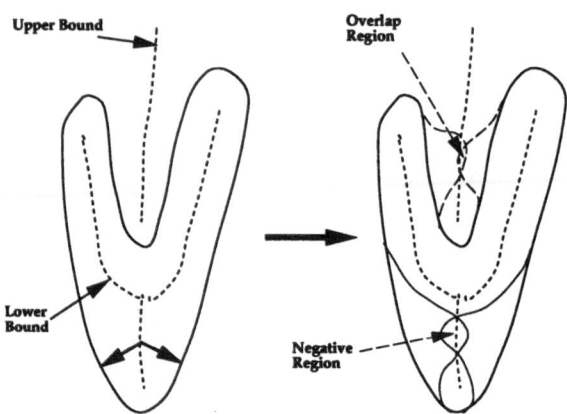

Figure 4.   Interior and exterior skeletons guarantee shape integrity.

– Topological transitions can occur naturally in the design space defined by the skeleton design variables. Changes in the values of design parameters (produced by, for example, a numerical optimization process) result in improved shapes that may be qualitatively different from the initial one: indentations, protrusions and holes can appear or be removed. Large shape changes as well as topological changes may occur. For example, when pairs of rays, emanating from a same location, simultaneously reach their lower bounds on the interior skeleton, portions of zero thickness appear and are removed. When rays from different branches meet one another at their respective upper bounds, defined by the exterior skeleton, they may be interpreted as defining continuity of the shape at these locations and hence make the current shape continuous.

4.2. RESKELETONIZATION AS A MEANS OF EXPANDING THE DESIGN SPACE

As discussed, creative design algorithms must be able modify the search space during design. This is done by changing the design variables which are used to define an object and, hence, the search formulation. The skeleton provides us with powerful tool for this task. For any object, a skeleton can be found and the object can subsequently be reparameterized with rays. These rays are then used as design variables during an optimization process that modifies the values of these variables (ray lengths) and generates a new shape. The new object is then skeletonized and another parameterization is found with the new skeleton. Now, the description of the object has changed and a new design space may be searched that was inaccessible from the original parameterization. This use of the skeleton as a tool for exploration of new search space is illustrated next.

Consider the object shown in Figure 5. The skeleton for the shape is shown as a dotted curve and the rays which define the parameterization are depicted as solid arrows. Two rays extend from each node on the skeleton, one on each side. During optimization the rays become the design variables: the length of each ray may be varied from zero to an exterior skeleton or other design bounds. The orientation of the rays, however, will *not* change during the optimization. For this illustration, the design criteria are hypothetical goals that produce the shapes in the subsequent figures.

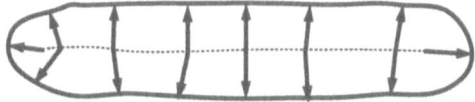

Figure 5. Original object parameterization defining the initial search space.

Two intermediate shapes obtained during the search (optimization) process are shown in Figures 6(a) and 6(b). The final shape at the end of the first optimization is shown in Figure 6(c). Note that during the optimization, the orientation and origin of each ray has not changed—only the lengths of the rays have been modified. For this reason, the accessible shapes from the optimization are restricted to those that can be generated with the current rays and their orientations.

In figure 6(d), the shape from the end of the first optimization has been reparameterized. The new skeleton is shown as the dotted curves and the new rays are shown as arrows emanating from it. The the shapes in Figure 6(c) and Figure 6(c) are substantially similar. However, during the reparameterization process, some of the original geometry may be lost due to the use of a finite number of rays. If an infinite number of rays were used to define the object, then the reparameterization would indeed be an exact replication of the original. However, since computation and time limitations restrict the design process, minimal ray densities that provide

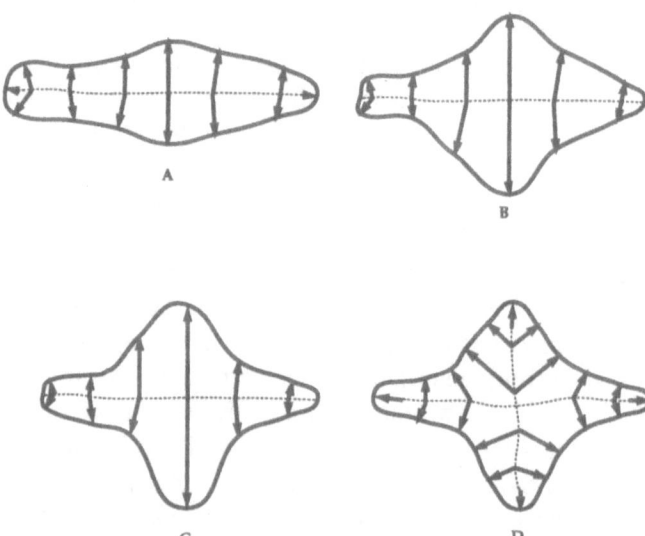

Figure 6. Shape evolution within a fixed design space is shown in (A), (B), and (C). (D) shows a reskeletonization that generates a new design space.

an accurate approximation of the original shape must be used. In this example, the new reparameterization in 6(d) (16 rays) provides a very close approximation of the original shape in 6(c).

Subsequent iterations of the design process (optimizations and reparameterizations) are shown in Figures 7, 8, 9, and 10. Intermediate shape evolutions from the optimization process within a design space are not shown. In each figure, the shape is generated by changing the lengths of the rays of the previous parameterization and then, when no further progress can be made with this parameterization, the shape is reparameterized with a new skeleton and corresponding rays, thus redefining the design variables for the next iteration of the design process. Therefore, novel shapes may evolve, since the search space is redefined at each iteration and the exploration of state-space is determined during the design process.

During each design iteration, the solution progresses as far towards the optimal shape as is possible within the current search space (using a numerical optimization procedure). As the parameterization, and hence, the search space changes, however, the design moves incrementally closer to the optimal shape. This is illustrated in Figures 11 and 12. The letters that label each shape in Figure 11 correspond to the search spaces shown in Figure 12. The dotted curve represents the evolution of the object during each optimization and includes the intermediate forms generated during the traversal of a individual search space, such as those shown in Figures 6(a) and 6(b). When the optimization reaches the boundary of

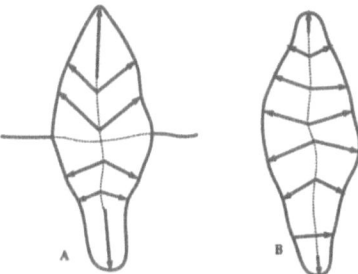

Figure 7. Second iteration: Evolution and reparameterization.

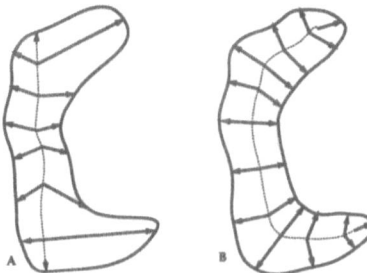

Figure 8. Third iteration: Evolution and reparameterization.

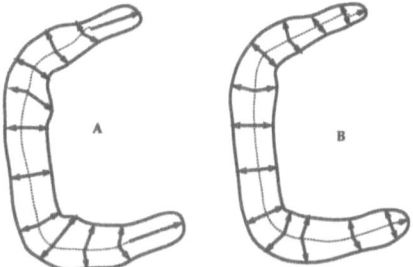

Figure 9. Fourth iteration: Evolution and reparameterization.

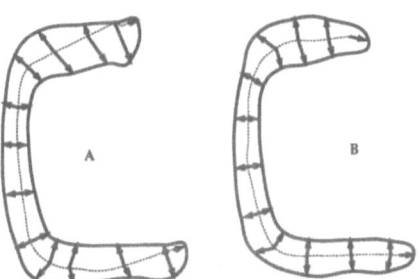

Figure 10. Fifth iteration: Evolution and reparameterization.

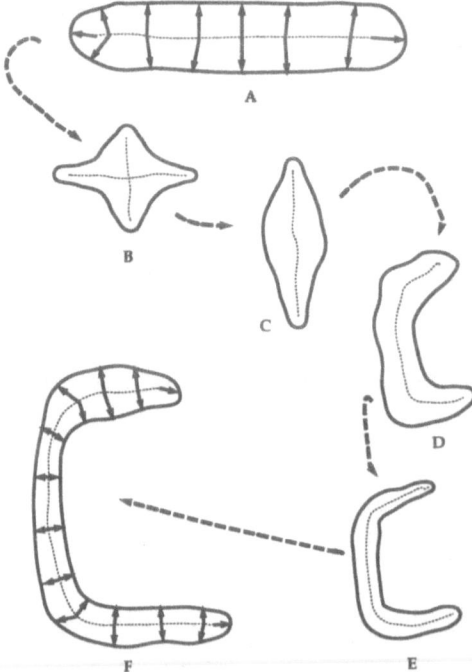

Figure 11. Summary of example design.

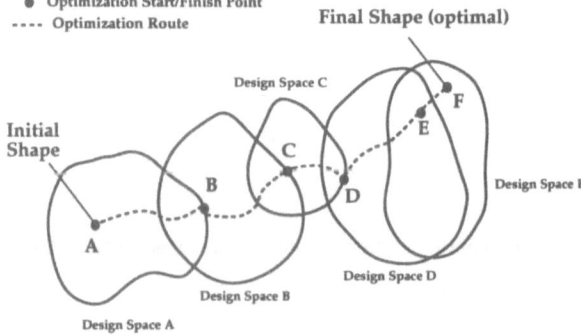

Figure 12. Search space generation and exploration.

the search space (Figure 6(c)), the object is reparameterized (Figure 6(d)) and a new search space is defined. The end of each optimization and reparameterization is shown as a solid dot. Thus, the design progresses toward the optimal solution efficiently by searching a series of smaller spaces. Even if there was a search space that included both the initial and final points of the design (A and F), it would be much less efficient to search it than to search multiple smaller spaces.

Thus, the reparameterization through skeletonization of successive shapes supports not only a creative design process, but a more efficient one as well.

5. Examples

We illustrate the methodology through two examples. The first one shows how large changes may occur in the context of a problem with relatively simple requirements and constraints. The second example demonstrates the design process in the presence of behavioral requirements expressed as constraints on stresses.

5.1. GENERATION OF AN I-SECTION

Consider the problem of finding the shape of a solid of fixed area which maximizes the moment of inertia about a horizontal axis. The design is constrained to fit in an exterior rectangular box of given dimensions with no thickness of any portion of the solid being smaller than a specified threshold.

Figure 13 shows the evolution of the design (only the upper half of the shape is shown—the lower half is symmetric). Starting from a rectangular initial shape (upper left), the skeleton is computed (dotted lines) and design variables (arrows) are selected at junctions and at midpoints of skeleton arcs. We have chosen only the interior skeleton to define design variables. Upper and lower bounds are imposed on the design variables so that the shape stays within the outer box (thin dotted line) and variables emanating from the main stem of the skeleton (variables where the the notion of the "thickness" of an "elongated subshape feature" makes sense) are constrained to be smaller than half the minimum allowable thickness. Algebraic expressions defining the moment of inertia and the area of the solid can be easily written in terms of the design variables (domain integrals) and the resulting optimization problem is solved.

The first row in Figure 13 shows the iterates produced by a sequential quadratic optimization (SQP) process. The shape evolves as the values of design variables (lengths of skeleton rays) change. The upper right diagram shows the resulting shape (thick line) defined by the values of the design variables at the end of the numerical optimization process. The skeleton of the initial shape is also shown in the diagram for reference.

The shape produced at the end of the first design iteration is used as the starting shape for another design optimization problem since the current parameterization cannot make any further progress towards the optimal shape. The new design iteration is initiated by computing the skeleton of the current shape. Again, a discrete sampling of the skeleton is used to define a set of rays parameterizing the shape and defining a new search space. As in the first iteration, junctions and mid-points of skeleton arcs are used but naturally different densities may be used in different regions (and may indeed result in a more efficient process). The second row in the diagram illustrates the evolution of the shape under the con-

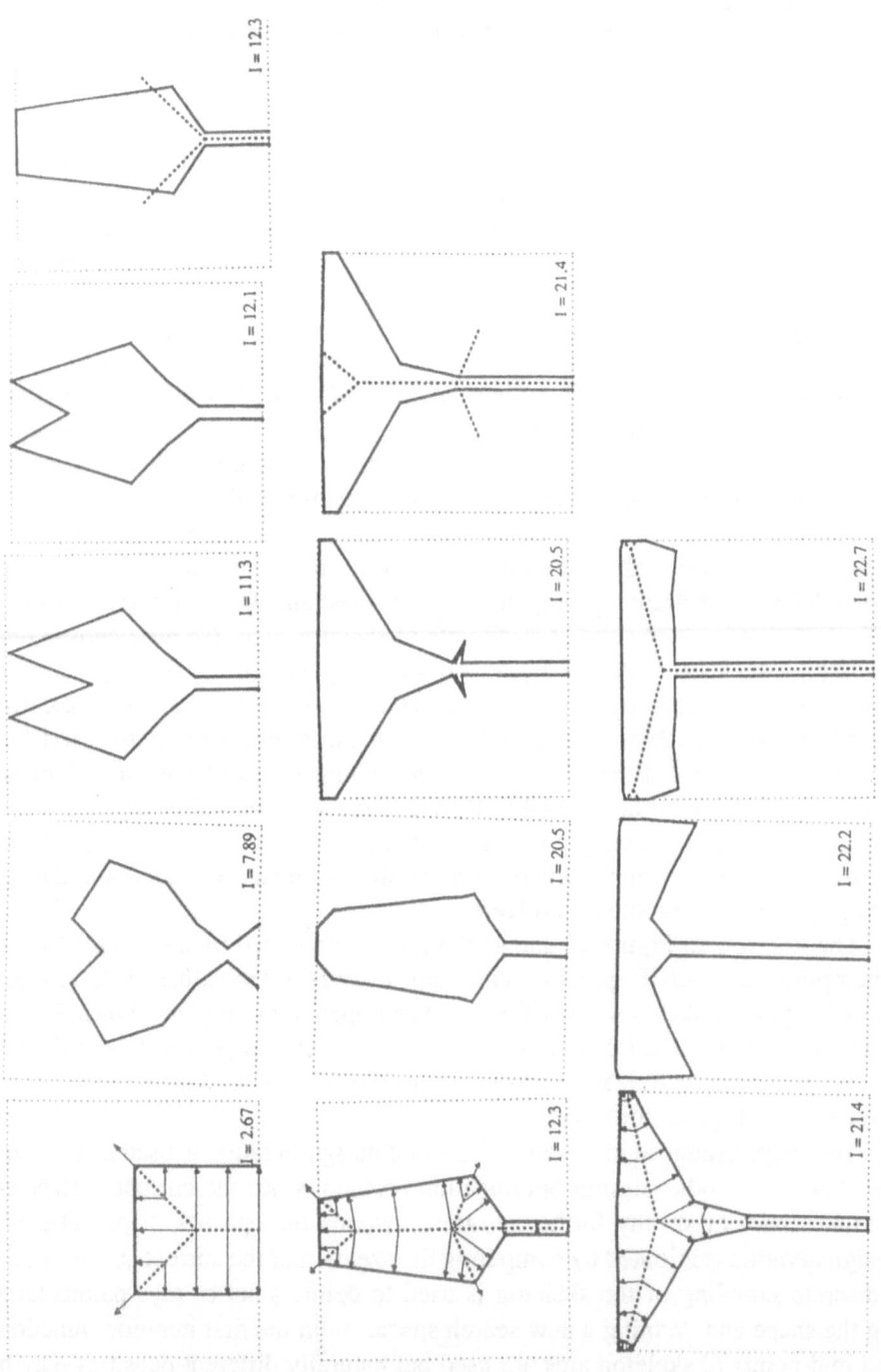

Figure 13.   Design history of a 2D cross section.

trol of an SQP procedure. Notice how "protrusions" in the starting shape have collapsed onto the skeleton and automatically disappeared: skeleton rays reached their lower bounds (i.e., hit the skeleton) during optimization. A third design iteration shown in the third row of Figure 13 gets very close to the well-known optimal wide flange shape.

5.2. DESIGN OF A CANTILEVER SUBJECTED TO A POINT LOAD

The goal of this problem is to find a shape that requires minimum material (weight) to transfer a given load (10 units) to a given set of supports, as formulated by Vimawala (1994). The locations of the load and supports are fixed and the shape is constrained to fit in a rectangular domain (5 units x 10 units) as shown in Figure 14. Additionally, internal stresses cannot exceed specified thresholds.

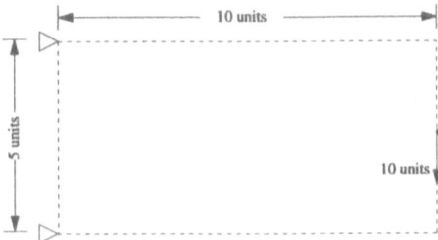

Figure 14. Locations of load and supports. Dotted rectangle indicates the limiting design domain.

Instead of using 2D plane stress finite elements to model the behavior of the continuous shape, a surrogate model—a model using truss elements—was used to approximate displacements and stresses. This approximation of the behavior was used to avoid the need for generating and adapting finite element meshes during the design processes and to reduce the overall computational cost. In this formulation, truss nodes on the model were defined as the tips and bases of the rays with truss members spanning between pairs of adjacent truss nodes. In order to model a homogeneous material, the stiffness $\frac{A_i E}{L_i}$ of all truss members was kept constant. The forces in the truss members thus depend only on their relative positions in the truss model. During each optimization, the tips of the rays move, changing the lengths and positions of the truss elements which are calculated at each iteration *inside* the optimization problem. Connectivity of the truss model, however, remains constant during each optimization. While this behavior is approximate and doesn't satisfy the elasticity equations, it is a useful model for shape generation during preliminary design.

The starting point of the design is the limiting rectangle. The truss model defined by the skeleton and the rays is shown in Figure 15(a). A numerical optimization problem with the ray lengths as variables is formulated and solved. Figure 15(b) shows the shape at the end of the first design iteration. As can been seen in the figure, design variables along two skeleton arcs on the right hand side reached their lower bounds simultaneously and were removed before starting the next design iteration, shown in Figure 16(a). Selected subsequent optimizations and reparameterizations are shown in Figures 16 to 19. After the second design iteration, the reduction in the area of the shape was small but changes in the shape which required new parameterizations justified the continuation of the design. The design continued until the tenth design iteration, when the shape only minimally changed. This shape was taken as the final design for the load transfer problem. The boundary of the truss model defined by the location of the ray tips represents the exterior outline of the 2D solid model as shown in Figure 20. This example also illustrates how behavioral requirements (e.g., maximal stress constraints) can be incorporated into skeleton-based design.

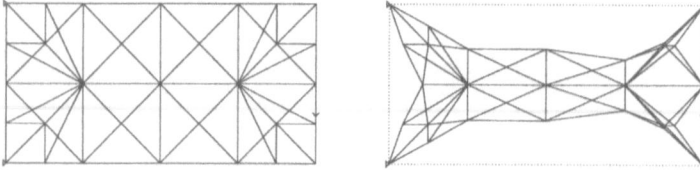

Figure 15. First design iteration: (a) initial shape (b) final shape.

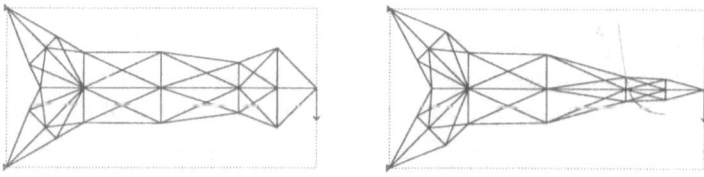

Figure 16. Second design iteration: (a) initial shape (b) final shape.

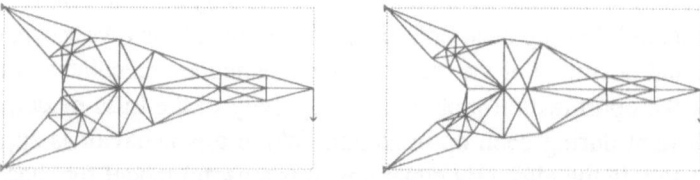

Figure 17. Fourth design iteration: (a) initial shape (b) final shape.

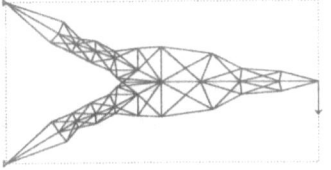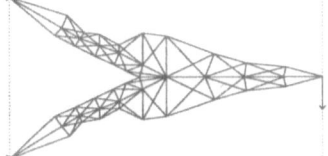

Figure 18. Seventh design iteration: (a) initial shape (b) final shape.

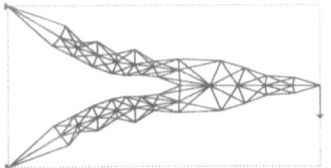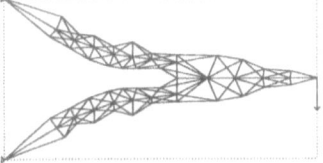

Figure 19. Tenth design iteration: (a) initial shape (b) final shape.

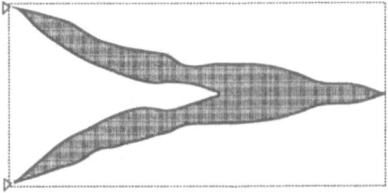

Figure 20. Final 2D solid shape.

6. Conclusions

We have described a procedure for the design of general shapes which overcomes several problems associated with current shape design procedures. The proposed method may be viewed as a "creative" design methodology because it does not preimpose a fixed parameterization on the design space nor does it preimpose a specific structure such as a Cartesian grid on the design domain. It does not, therefore, restrict the class of optimal shapes that can be obtained and always generates valid Euclidean shapes. The skeleton-based design process iterates on the construction of skeleton-based design spaces and formulation and solutions of numerical optimization search problems in these spaces. Through adaptive parameterization, large geometric changes as well as topological modifications may occur. Thus, the search space being explored is continually changing in response to trends in shapes changes and convergence may occur in a state-space completely or partially undefined at the outset of design. The methodology and examples from an implementation in two dimensions were presented. We are currently investigating three dimensional extensions and the coupling of this methodology with skeleton-based finite element mesh generation techniques (Stal, 1995) to allow a more general class of design requirements to be considered, such as constraints on stresses, displacements, and temperatures.

References

Brown, D. C. and Chandrasekaran, B.: 1985, Expert systems for a class of mechanical design activity, *in* J. S. Gero (ed.), *Knowledge Engineering in Computer-Aided Design*, North-Holland, Amsterdam, pp. 259–282.

Blum, H.: 1973, Biological shape and visual science (Part I). *Journal of Theoretical Biology*, **38**, 205–287.

Carlson, S. E.: 1994, Comparison of three non-derivative optimization methods with a genetic algorithm for component selection, *Journal of Engineering Design*, **5**(4).

Cagan, J. and Mitchell, W. J.: 1993, Optimally directed shape generation by shape annealing, *Environment and Planning B*, **20**, 5–12.

Coyne, R. D., Rosenman, M. A., Radford, A. D. and Gero, J. S.: 1987, Innovation and creativity in knowledge-based CAD, *in* J. S. Gero (ed.), *Expert Systems in Computer-Aided Design*, North-Holland, Amsterdam, pp. 435–465.

Chapman, C. D., Saitou, K. and Jakiela, M. J.: 1994, Genetic algorithms as an approach to configuration and topology design, *ASME Journal of Mechanical Design*, **116**(4).

Faugeras, O.: 1993, *Three-Dimensional Computer Vision: A Geometric Viewpoint*, MIT Press.

Gero, J. S.: 1990, Design prototypes: A knowledge representation schema for design, *AI Magazine*, **11**(4), 26–36.

Gero, J. S.: 1994, Exploration as a basis of creative engineering design, *in* K. Khozeimeh (ed.), *Computing in Civil Engineering*, Vol. 2, ASCE.

Goldberg, D.: 1989, *Genetic Algorithms in Search, Optimization, and Machine Learning*. Addison-Wesley, Reading, Massachusetts.

Maher, M. L.: 1994a, Creative design using a genetic algorithm, *in* K. Khozeimeh (ed.), *Computing in Civil Engineering*, Vol. 2, pp. 2014–2021. First ASCE Computing Congress (held in conjuction with A/E/C Systems '94).

Maher, M. L.: 1994b, Representation of case memory for structural design, *in* K. Khozeimeh (ed.), *Computing in Civil Engineering*, Vol. 2, pp. 2030–2037. First ASCE Computing Congress (held in conjuction with A/E/C Systems '94).

Papalambros, P. and Chirehdast, M.: 1990, An integrated environment for structural configuration design, *Journal of Engineering Design*, **1**(1), 73–96.

Schmidt, L. C. and Cagan, J.: 1994, Recursive annealing: A computational model for machine design, *Research in Engineering Design*.

Stal, D.: 1995, Three-dimensional Skeleton Generation from Geometric Triangulation. Master's Thesis, University of Washington, Department of Civil Engineering.

Turkiyyah, G. M., Storti, D. W., Ganter, M., Chen, H. and Vimawala, M. S.: 1996, An accelerated triangulation scheme for computing skeletons of free-form solid models, *CAD*, (to appear).

Vimawala, M. S. and Turkiyyah, G. M.: 1994, Computational Procedures for Shape Design. Techical Report, SGEM-94-5, Department of Civil Engineering, University of Washington.

Zhao, F., Louis, S. J. and Lenart, M.: 1994, Evolutionary methods for synthesis of truss topology, *in* K. Khozeimeh (ed.), *Computing in Civil Engineering*, Vol. 2, pp. 1816–1823.

first author electronic addresses

Altmeyer, J., altmeyer@informatik.uni-k.de
Bakhtari, S., shirin@bsr-consulting.de
Ball, N. R., nrb@eng.cam.ac.uk
Biljic, T., taner@ie.utoronto.ca
Chakrabarti, A., ac123@eng.cam.ac.uk
Coulon, C.-H., coulon@gmd.de
de Grassi, M., luca@anvax2.cineca.it
Dong, A., adong@jerry.me.berkeley.edu
Feijó, B., bruno@icad.puc-rio.br
Fruchter, R., fruchter@cive.stanford.edu
Fujii, H., haru@ori.shimz.co.jp
Gage, P. J., p-gage@adfa.oz.au
Gelsey, A., gelsey@cs.rutgers.edu
Goel, A., goel@pravda.cc.gatech.edu
Grabska, E., uigrabsk@cyf-kr.edu.pl
Grecu, D. L., dgrecu@cs.wpi.edu
Guan, X., x_guan@cad.strathclyde.ac.uk
Hower, W., walter@cs.ucc.ie
Koza, J. R., koza@cs.stanford.edu

Kundu, S., sourav@control.prec.metro-u.ac.jp
Marling, C., R., marling@alpha.ces.cwru.edu
Olivier, P., plo@aber.ac.uk
Peña-Mora, F., feniosky@iesl.mit.edu
Poon, J., josiah@arch.su.edu.au
Reddy, S. Y., sudha@pal.rockwell.com
Rogers, J. L., j.l.rogers@larc.nasa.gov
Rosenman, M. A., mike@arch.su.edu.au
Schmidt, L. C., lschmidt@eng.umd.edu
Schnier, T., thorsten@arch.su.edu.au
Schwabacher, M., schwabac@cs.rutgers.edu
Smith, I., smith@lia.di.epfl.ch
Smithers, T., ccpsmsmt@si.ehu.es
Stacey, M., m.k.stacey@open.ac.uk
Stal, D. M., dstal@ce.washington.edu
Stumptner, M., mst@vexpert.dbai.tuwien.ac.at
Sushkov, V. V., sushkov@cs.utwente.nl
Varma, A., anil@ton.berkeley.edu
Voβ, A., angi.voss@gmd.de

author index